# THE ROUTLEDGE COMPANION TO GLOBAL TELEVISION

Featuring scholarly perspectives from around the globe and drawing on a legacy of television studies, but with an eye toward the future, this authoritative collection examines both the thoroughly global nature of television and the multiple and varied experiences that constitute television in the twenty-first century.

Companion chapters include original essays by some of the leading scholars of television studies as well as emerging voices engaging television on six continents, offering readers a truly global range of perspectives. The volume features multidisciplinary analyses that offer models and guides for the study of global television, with approaches focused on the theories, audiences, content, culture, and institutions of television. A wide array of examples and case studies engage the transforming practices, technologies, systems, and texts constituting television around the world today, providing readers with a contemporary and multi-faceted perspective.

In this volume, editor Shawn Shimpach has brought together an essential guide to understanding television in the world today, how it works and what it means—perfect for students, scholars, and anyone else interested in television, global media studies, and beyond.

**Shawn Shimpach** is Associate Professor in the Department of Communication at the University of Massachusetts Amherst and Director of the Massachusetts Multicultural Film Festival. His research interests include the cultural history of film, television, and media; the social and institutional constructions of the media audience; genre theory and screen genres; and screen industries. He is author of the book *Television in Transition: The Life and Afterlife of the Narrative Action Hero.*

# THE ROUTLEDGE COMPANION TO GLOBAL TELEVISION

*Edited by Shawn Shimpach*

Routledge
Taylor & Francis Group
New York London

First published 2020
by Routledge
605 Third Avenue, New York, NY 10017

and by Routledge
2 Park Square, Milton Park, Abingdon, Oxon, OX14 4RN

First issued in paperback 2021

*Routledge is an imprint of the Taylor & Francis Group, an informa business*



Publisher's Note
The publisher has gone to great lengths to ensure the quality of this reprint but points out that some imperfections in the original copies may be apparent.

*Library of Congress Cataloging-in-Publication Data*
Names: Shimpach, Shawn, editor.
Title: The Routledge companion to global television / edited by Shawn Shimpach.
Description: New York, NY : Routledge, 2020. | Series: Routledge media and cultural studies companions | Includes bibliographical references and index.
Identifiers: LCCN 2019018452 (print) | LCCN 2019980257 (ebook) | Subjects: LCSH: Television broadcasting. | Television programs–Social aspects. | Foreign television programs–Social aspects.
Classification: LCC PN1992.5 .R68 2020 (print) | LCC PN1992.5 (ebook) | DDC 791.45–dc23
LC record available at https://lccn.loc.gov/2019018452
LC ebook record available at https://lccn.loc.gov/2019980257

ISBN 13: 978-1-03-208578-4 (pbk)
ISBN 13: 978-1-138-72434-1 (hbk)

Typeset in Bembo
by Swales & Willis, Exeter, Devon, UK

# CONTENTS

# FIGURES

# TABLES

# NOTES ON CONTRIBUTORS

**Ece Algan** is Senior Lecturer at the Institute of Media and Creative Industries at Loughborough University London. Her scholarly interests range widely, from global communication, media and politics, and local and community media to social movements, modernity, media ethnography, and audiences. She has conducted longitudinal ethnographic fieldwork research on local radio, emerging communication technologies, and youth for over a decade in Southeast Turkey. She is the co-editor of *Television in Turkey: Local Productions, Transnational Aspirations.*

**Ruoyun Bai** is Associate Professor of Media Studies at the Department of Arts, Culture and Media, University of Toronto (Scarborough). Her research interests include Chinese television culture, television as digital media, media control in the digital era, and scandal studies. She published a monograph, *Staging Corruption: Chinese Television and Politics*, in 2014, and co-edited two books, *TV Drama in China* (2008) and *Chinese Television in the Twenty-First Century* (2015).

**Jerome Bourdon** is Professor at the Department of Communications at Tel Aviv University and Associate Professor at INA-SUP School of Broadcasting in Paris. He researches the global history of television, the relations between media and memory (especially through the use of life-stories), Western representations of the Israeli–Palestinian conflict across media and genres, and the history of mediated presence, from correspondence to the Internet.

**Ergin Bulut** received his PhD from the Institute of Communications Research at the University of Illinois, Urbana-Champaign. He researches in the area of political economy of media and cultural production, video game studies, media and politics, and critical theory. His work has been published in *Media, Culture & Society, Triple C, International Journal of Communication, Critical Studies in Media Communication, Television and New Media*, and *Communication and Critical/Cultural Studies.* His book, *A Precarious Game: The Illusion of Dream Jobs in the Video Game Industry*, is forthcoming. He is an Assistant Professor at Koç Univeristy's Media and Visual Arts Department. Bulut is currently a visitor researcher at the Institute for Advanced Study, Princeton and a faculty fellow at Center for Advanced Research in Global Communication at Annenberg School for Communication, UPenn.

**Jean K. Chalaby** is Professor of International Communication at City University London and former Head of Sociology. He is the author of *The Invention of Journalism* (1998), *The de Gaulle Presidency and the Media* (2002), *Transnational Television in Europe* (2009), and *The Format Age:*

*Television's Entertainment Revolution* (2015). He is the editor of *Transnational Television Worldwide* (2005) and has published extensively in leading journals on a wide range of media-related topics

**Frederic Chaume** is a Professor of Audiovisual Translation at Universitat Jaume I, where he teaches audiovisual translation theory, dubbing, and subtitling; and Honorary Professor at University College London. He is author of the books *Doblatge i subtitulació per a la TV* (2003), *Cine y Traducción* (2004), *Audiovisual Translation: Dubbing* (2012), and co-author of *Teories Contemporànies de la Traducció* (2010). He has also co-edited several books and special journal issues (*Perspectives, Prosopopeya*) and is the director of the TRAMA book series (Publicacions de la Universitat Jaume I), the first collection of monographs on audiovisual translation.

**Aymar Jean Christian** is an Associate Professor of Communication Studies at Northwestern University and a Fellow at the Peabody Media Center. His first book, *Open TV: Innovation Beyond Hollywood and the Rise of Web Television* (2018), argues the web brought innovation to television by opening development to independent producers. His work has been published in numerous academic journals, including *The International Journal of Communication, Cinema Journal, Continuum,* and *Transformative Works and Cultures.*

**David Craig** is a Clinical Associate Professor at USC Annenberg School where he teaches multiple courses in media industries, management, and practice. With Stuart Cunningham, he has conducted research and published around the rise of the Social Media Entertainment industry and Creator Culture. He is a Peabody Fellow in the Peabody Media Lab, serves on the faculty at Shanghai Jiao Tong University, and is an Emmy-award nominated Hollywood film, television, and stage producer.

**Stuart Cunningham** is Distinguished Professor of Media and Communications, Queensland University of Technology. His most recent books are *Hidden Innovation: Policy, Industry and the Creative Sector* (2014), the co-authored *Social Media Entertainment: The new intersection of Hollywood and Silicon Valley* (2019), *Key Concepts in Creative Industries* (2013), *Screen Distribution and the New King Kongs of the Online World* (2013), *Media Economics* (2015), and the co-edited *Digital Disruption: Cinema Moves Online* (2012) and *The Media and Communications in Australia* (2014).

**Ranjana Das** is Senior Lecturer in Media and Communication at the Department of Sociology in the University of Surrey, Guildford. She has been co-author/editor of *The Future of Audiences* (2018) and *Provocative Screens* (2017). She has been funded by the Arts and Humanities Research Council, UK, the British Academy, and the Wellcome Trust, and has directed the CEDAR network on the future of audiences (2015–2018). Her current work focuses on parenthood, intimacies, and everyday life in platform societies.

**Alexander Dhoest** is Professor of Communication Studies at the University of Antwerp and Chair of the Media, Policy and Culture research group. His research focuses on the multiple connections between media, in particular television, and issues of social identity, including national, sexual, and ethnic identity. He has published widely on these issues in edited books and journals such as *European Journal of Cultural Studies* and *Critical Studies in Television.*

**Ayanna Dozier** is a PhD candidate in Art History & Communication Studies with a graduate option in Women & Gender Studies at McGill University. Her dissertation examines labor and critical fabulations in Black women's experimental short films in North America and the United Kingdom. Selected publications include: *Cléo, Feminist Media Studies, Another Gaze,* and *Liquid*

*Blackness*. Dozier is also a 2018–2019 Critical Studies Fellow in the Whitney Independent Study Program and a Joan Tisch Teaching Fellow at the Whitney Museum.

**Lyombe Eko** is a Professor of Journalism and Electronic Media at the College of Media and Communication, and a Humanities Fellow at Texas Tech University. He is a journalist and documentary producer who has worked in Cameroon and Kenya. He teaches courses in media law and ethics, media theory, contemporary issues in information technology, and international communication. His research interests are comparative media law and policy, international communication, and visual communication. He earned his PhD in Journalism from Southern Illinois University at Carbondale. He has published four books, numerous book chapters, law review journal articles, and peer-reviewed articles on comparative and international communication.

**Martin Fredriksson** is Associate Professor in the Department of Culture Studies (Tema Q), Linköping University. He has worked extensively with issues concerning the theory and history of piracy, commons, property rights, and the history of copyright. He has been visiting fellow at MIT, Western Sydney University, and Amsterdam University.

**Jorge A. González** is Professor and full-time Researcher at the Interdisciplinary Research Center on Science & Humanities (CEIICH) at the National Autonomous University of Mexico (UNAM). Since 2016 he has coordinated a Latin American Research Network on the Symbolic Dimension of Food Systems, and in 2018 he began coordinating the Communication for Development Latin American Strategy Research Team for UNICEF.

**Esther Hamburger** is Professor of Audiovisual History, Theory, and Criticism at the Department of Cinema, Radio and Television at the School of Communication and Arts of University of São Paulo. With a PhD in Anthropology from the University of Chicago, she has been a Visiting Scholar at Harvard University's David Rockefeller Center for Latin American Studies. She has been a Visiting Professor at the Federal University of Rio de Janeiro, and at University of Michigan, LACS. She has published in a wide range of book collections including *The Brazil Reader: History, Culture, Politics*; *A Companion to Latin American Cinema*; *The Routledge Companion to Media and Gender*; *Television Audiences across the World*; *El cinema de lo real*; and *Miradas cruzadas: sociedad, política y cultura*. She represents her university in the UNESCO Media Literacy and Intercultural Dialogue University Network. She has contributed to journals such as *Galaxia*, *Significação*, *Lua Nova*, *Novos Estudos*, *Television and the New Media*, and *Framework*. For 15 years she wrote regularly for *Folha de S. Paulo*. Her book *O Brasil Antenado: a sociedade da novela*, was nominated for the Jabuti Prize.

**John Hartley** was John Curtin Distinguished Professor at Curtin University, Western Australia. The author of many publications on cultural, media, communication, and journalism studies, his most recent research focuses on cultural science, the creative economy, creative citizenship, and open knowledge.

**Timothy Havens** is Professor and Chair of Communication Studies at the University of Iowa. His research interests include television studies; media globalization; race, ethnicity, and media; and critical analysis of media industries. He is the author of *Black Television Travels: African American Media Around the Globe* (2013) and *Global Television Marketplace* (2006); the co-author with Amanda D. Lotz of *Understanding Media Industries* (2011, 2016); and co-editor with Aniko Imre and Katalin Lustyik of *Popular Television in Eastern Europe Before and Since Socialism* (2012). He is a former Fulbright Scholar to Hungary.

**Annette Hill** is a Professor of Media and Communication at Lund University, Sweden, and Visiting Professor at King's College, London. Her research focuses on audiences and popular culture, with interests in media engagement, everyday life, genres, production studies, and cultures of viewing. She is the author of eight books, and many articles and book chapters in journals and edited collections, which address varieties of engagement with reality television, news and documentary, television drama, entertainment formats, live events and sports entertainment, film violence, and media ethics, including *Reality TV* (2015) and *Media Experiences* (2018).

**Darnell Hunt** is Dean of Social Sciences at UCLA and Professor of Sociology and African American Studies. He is lead author of the *Hollywood Diversity Report*, an annual series that examines the relationship between diversity and profit in the entertainment industry. The fifth annual report was released by UCLA in February 2018. Dr. Hunt has written and edited several books, including *Screening the Los Angeles "Riots": Race, Seeing, and Resistance* and *Channeling Blackness: Studies on Television and Race in America.*

**Nurçin İleri** completed her doctoral dissertation on the night, streets, and crime in the late Ottoman Istanbul at the History Department of Binghamton University. She continues to work on urban history, history of technology and science, and archives and cultural heritage studies in the late Ottoman Empire and Turkey. She is currently a Post-Doctoral Scholar at the LARHRA—Rhônes-Alpes Laboratory of Historical Research of the University of Grenoble-Alpes.

**Aniko Imre** is Professor of Cinematic Arts in the Division of Cinema and Media Studies of the University of Southern California. Her most recent book is *TV Socialism* (2016).

**Douglas Kellner** is George Kneller Chair in the Philosophy of Education at UCLA and is author of many books on social theory, politics, history, and culture. His most recent books are *American Nightmare: Donald Trump, Media Spectacle, and Authoritarian Populism* (2016) and *The American Horror Show: Election 2016 and the Ascendency of Donald J. Trump* (2017).

**Joe F. Khalil** is Associate Professor of Global Media and Communication at Northwestern University in Qatar. His publications include *Arab Television Industries* (with M. Kraidy); *Culture, Time and Publics in the Arab World* (with T. Sabry); *Arab Satellite Entertainment Television and Public Diplomacy*; and numerous scholarly articles. Prior to joining academia, he had extensive professional experience as a director, executive producer, and media consultant.

**Shanti Kumar** is Associate Professor in the Department of Radio-Television-Film at the University of Texas, Austin where he is also affiliated with the Department of Asian Studies, the Center for Asian American Studies, and the South Asia Institute. He is the author of *Gandhi Meets Primetime: Globalization and Nationalism in Indian Television* (2005), and the co-editor of *Global Communication: New Agendas in Communication, Television at Large in South Asia* (2012), and *Planet TV: A Global Television Reader* (2003).

**María Trinidad García Leiva** is a Professor and member of the Tecmerin Group at Universidad Carlos III de Madrid, where she co-leads the Audiovisual Diversity Task Force (diversidadaudiovisual.org). Her research interests include political economics in the field of communication and culture, with an emphasis on Ibero-American audiovisual industries, communication policies, and cultural cooperation.

**Ramon Lobato** is Senior Research Fellow in the School of Media and Communication, RMIT University, Melbourne. His research interests include video markets, piracy, and streaming. Ramon is the author of *Shadow Economies of Cinema* (2012), *The Informal Media Economy* (2015, with Julian Thomas), and *Netflix Nations* (2018).

**Nomusa Makhubu** is an art historian and an artist at the Michaelis School of Fine Art, University of Cape Town. Makhubu's research focuses on art interventionism, popular culture, and social engagement in African visual art. She was a fellow of the American Council of Learned Societies and an African Studies Association (ASA) Presidential Fellow in 2016. In 2017, Makhubu was a Mandela-Mellon Fellow at the Hutchins Center for African and African American Studies, Harvard University.

**Purnima Mankekar** trained as a cultural anthropologist and studies media and publics/public cultures with a focus on the politics of affect. She is currently completing a book titled *Future Tense: Affective Labor and Disjunctive Temporalities* (co-authored with Akhil Gupta.) She is the author of *Screening Culture, Viewing Politics* (1999) and *Unsettling India: Affect, Temporality, Transnationality* (2015). Her co-edited books include *Caste and Outcast* (2002) and *Media, Erotics, and Transnational Asia* (2013).

**Guillermo Mastrini** is Full Professor at the University of Quilmes and the University of Buenos Aires, and researcher at the National Scientific and Technical Research Council (CONICET) in Argentina. He holds a PhD in Information Sciences from Universidad Complutense de Madrid. He has specialized in the study of public communication policies and cultural industry economics.

**Marion McCutcheon** is a Research Associate with the Queensland University of Technology and an Honorary Research Fellow at the University of Wollongong's C3P Research Centre for Creative Critical Practice. A communications economist, she has worked in the Australian federal government in telecommunications and broadcasting policy advisory and research roles. Her current research interests include evaluating the economic and social benefits derived from cultural and creative goods and services—including crime drama series—and the role of the creative industries in economic systems.

**Divya McMillin** is author of *International Media Studies* (2007) and *Mediated Identities: Youth, Agency, and Globalization* (2009), and co-editor of *Place, Power, Media* (2018). Her research on media globalization has appeared in top-tiered journals and key anthologies. She is Professor of Global Media Studies, Executive Director of the Global Innovation and Design Lab and the Institute for Global Engagement at the University of Washington Tacoma. She received her PhD from Indiana University Bloomington in 1998.

**Cécile Méadel** is Professor at the Interdisciplinary Research Center for Media Analysis (CARISM) at Paris2 University. She researches the genesis and uses of communication technologies, from early broadcasting to Internet. On the quantification of audiences, she published the book *Quantifier le public: Histoire des mesures d'audience de la radio-télévision* (2010). Her most recent book in English is *Governance, Regulations and Power on the Internet* (2012, edited with E. Brousseau and M. Marzouki).

**Toby Miller** is Stuart Hall Professor of Cultural Studies, Universidad Autónoma de México–Cuajimalpa; Research Professor of the Graduate Division, University of California, Riverside, and

Professor in the Institute for Media and Creative Industries, Loughborough University London. The author and editor of over forty books, his work has been translated into Spanish, Chinese, Portuguese, Japanese, Turkish, German, Italian, Farsi, and Swedish. *How Green is Your Cell Phone?* (with Richard Maxwell) and *The Persistence of Violence* are in press. He is President of the Cultural Studies Association (US).

**Esther Milne** is Associate Professor of Media and Communication at Swinburne University, Melbourne. She is the author of *Letters, Postcards, Email: Technologies of Presence* (2010). With co-authors Aneta Podkalicka and Jenny Kennedy, she has recently published the first monograph study of the hit TV program *Grand Designs* entitled *Grand Designs: Consumer Markets and Home-making* (2018). Esther is currently working on a book entitled *Email and the Everyday: Stories of Disclosure, Trust and Digital Labour.*

**Lothar Mikos** is Professor of Television Studies in the Department of Media Studies at the Filmuniversity Babelsberg in Potsdam. He founded the Television Studies Section of the European Communication Research and Education Association (ECREA) in which he served as Chair from 2010 until 2016. His main areas of work are the global TV economy, television series worldwide, digital distribution, transmedia storytelling, (de-)convergence culture, popular television genres and formats, and qualitative audience studies. He has published widely on television, film, and popular culture.

**Asha Nadkarni** is Associate Professor of English at the University of Massachusetts Amherst, where she teaches courses in postcolonial literature and theory, Asian American literature and culture, and transnational feminism. She is the author of *Eugenic Feminism: Reproductive Nationalism in the United States and India* (2014).

**Jonathan Corpus Ong** is Associate Professor in Global Digital Media at the Department of Communication in the University of Massachusetts, Amherst. He is the author of *The Poverty of Television* (2015) and co-editor of *Taking the Square* (2016). He is the Co-Editor-in-Chief of the journal *Television & New Media.* His current project "Architects of Networked Disinformation" takes a production studies approach in mapping out industries of political trolling.

**Aneta Podkalicka** is a Lecturer and Researcher in the School of Media, Film and Journalism at Monash University, Melbourne. She is a co-author of Using Media for Social Innovation with Ellie Rennie (2018) and *Grand Designs: Consumer Markets and Home-making* with Esther Milne and Jenny Kennedy (2018); and a co-editor of *Leisure Cultures and the Making of Modern Ski Resorts* with Philipp Strobl (2018).

**Anna Potter** is Associate Professor and Deputy Head of School (Research) in the School of Creative Industries at the University of the Sunshine Coast. She is a researcher focusing on children's screen production cultures, distribution networks, and media and communication policy. She is the author of *Creativity, Culture, and Commerce: Producing Australian Children's Television with Public Value* (2015) and *Creativity Rebooted: Producing Children's Television In the On-Demand Age* (2020).

**Ousmane K. Power-Greene** is Associate Professor of History and Africana Studies at Clark University in Worcester, MA. He is the author of *Against Wind and Tide: The African American Struggle Against the Colonization Movement* (2014).

**Ana-Christina Ramón** is the Director of Research and Civic Engagement for the Division of Social Sciences at UCLA. Dr. Ramón is a social psychologist, who has worked on social justice issues related to equity and access in various arenas for over a decade. She is the co-Principal Investigator of the Hollywood Advancement Project and manages its graduate research team. She is also the co-author (with Dr. Darnell Hunt) of the annual *Hollywood Diversity Report* series.

**David Rowe**, FAHA, FASSA, is Emeritus Professor of Cultural Research, Institute for Culture and Society, Western Sydney University; Honorary Professor, Faculty of Humanities and Social Sciences, University of Bath; and Research Associate, Centre for International Studies and Diplomacy, SOAS University of London. His latest book is *Making Culture: Commercialisation, Transnationalism, and the State of 'Nationing' in Contemporary Australia* (co-edited, 2018). His work has been translated into Chinese, French, Turkish, Spanish, Italian, Korean, and Arabic.

**Andy Ruddock** lectures in Communications and Media Studies at Monash University. He has authored four books on media audiences, youth and media, and media theory. Andy is currently writing a book on cultivation theory in the digital age, and has published more than 40 chapters and articles on topics including political celebrity, school shootings, sport, alcohol, drugs, reality TV, social media in military relations, and media education.

**Shawn Shimpach** is Associate Professor in the Department of Communication at the University of Massachusetts, Amherst and Director of the Massachusetts Multicultural Film Festival. His research interests include the cultural history of film, television, and media; the social and institutional constructions of the media audience; genre theory and screen genres; and screen industries. He is author of the book *Television in Transition: The Life and Afterlife of the Narrative Action Hero.*

**Pawan Singh** is a New Generation Network Fellow in Contemporary Histories at Deakin University and the Australia India Institute in Melbourne. He has a PhD in Communication from the University of California San Diego. His research concerns questions of identity, representation and social justice in relation to film, television, the Internet, and popular culture within a transnational framework. His current project, which is funded by the Toyota Foundation and the Digital Identity Research Initiative of the Indian School of Business, examines contestations around biometric identity, privacy, mass surveillance, and recognition in the Indian context. His teaching includes courses in media studies including gender and sexuality in media, media stereotypes, how to read a film, bodies in the law, and gender and biomedicine.

**Jeanette Steemers** is Professor of Culture, Media and Creative Industries at the Department of Culture, Media and Creative Industries, King's College London. Her work has been funded by the British Academy, the Leverhulme Trust, and the Arts and Humanities Research Council (AHRC). She is currently working on an AHRC project that seeks to facilitate Arab–European dialogue on children's screen content in an era of forced migration.

**Ruth Teer-Tomaselli** is a Full Professor at the University of KwaZulu-Natal, Durban. She has held a UNESCO Chair in Communications for Southern Africa since 2002. She is a past Vice-President of the International Association for Media and Communications Research (IAMCR) and has served as a board member of the South African Broadcasting Corporation. Teer-Tomaselli has written widely on the history, regulation, and content of television in South Africa and across Africa.

**Susan Turnbull** is Senior Professor of Communication and Media at the University of Wollongong, Discipline Leader for the Creative Industries, and co-director of the Research Centre for Critical Creative Practice (C3P). Her publications include *The TV Crime Drama* (2014) and, with Stuart Cunningham, *The Media and Communications in Australia* (2014). She is Chief Investigator on the Australian Research Council grant: *Border Crossings: The Transnational Career of the TV Crime Drama* (DP 160102510) and a member of the ARC Linkage team for the project *Migration, Cultural Diversity and Television: Reflecting Modern Australia* (LP 150100202). She is currently developing a research project on the value of the web series to the screen industries.

**Graeme Turner** is Emeritus Professor in Cultural Studies in the Institute for Advanced Studies in the Humanities at the University of Queensland, Brisbane. He has published many works on the media and cultural studies, and most of his recent projects have focused on the analysis of television in the post-broadcast era: Books from this work include *Re-Inventing the Media* (2016) and (with Anna Cristina Pertierra) *Locating Television: Zones of Consumption* (2013).

**Ying Zhu** is Professor of Media Culture at the City University of New York with an appointment at the Film Academy of the Hong Kong Baptist University. She has published eight books, including *Two Billion Eyes: The Story of China Central Television*, *Television in Post-Reform China: Serial Drama, Confucian Leadership and the Global Television Market*, and *Chinese Cinema During the Era of Reform: The Ingenuity of the System*. She is a recipient of a National Endowment for the Humanities Fellowship, an American Council of Learned Societies Fellowship, and a Fulbright Senior Research Fellowship

# ACKNOWLEDGMENTS

Sincere thanks to the contributing authors—nothing like they warn you about, everything that could be hoped for in intellectual, professional collaborators—for support, eagerness to participate, a persistence of vision, brilliance, and ultimately, considerable patience. If this book in any way represents the power and the potential of taking television seriously and points to a future of study that is globally inclusive, it is because of the unsurpassed work of these leading scholars of television studies. Thank you as well to the team at Routledge for sharing in this vision and including global television in the ongoing *Companion* series. I am grateful especially for the invaluable support of Erica Wetter as well as the assistance provided by Mia Moran and Emma Sherriff.

Thank you for guidance and advice, once upon a time and still, Toby Miller, William Boddy, Anna McCarthy, Dana Polan, Vinicius Navarro, and David Irving. Thank you to the early readers of several chapters, Felicitas Baruch, Audrey Black, Mary Dickman, Joseph Eichner, Eean Grimshaw, Gichuhi Kamau, Kylie Lanthorn, Colby Miyose, Erica Tortolani, Stephen Warren, and Soo Oak Yoo, your comments and feedback made this a better book. Peer reviewers offered crucial feedback at a critical stage in the process, thank you for your suggestions and support. Thank you, as well, to Amelie Hastie for a generative conversation that greatly influenced the presentation of this book's division into sections. I am also indebted to my colleagues in the University of Massachusetts Amherst Department of Communication for early discussions, helpful advice and encouragement, particularly Mari Castañeda, Claudio Moreira, Marty Norden, Erica Scharrer, and Emily West.

The Department of Communication and Department Chair, Professor (now Associate Dean) Mari Castañeda, were instrumental in providing much needed and timely assistance in securing the completion of some translation work for this volume. Initial translation of one chapter was provided by the University of Massachusetts Amherst Translation Center.

To my first and fiercest reader, Elizabeth Horn, I thank you, along with Lily and Charlotte, always and forever, for your unwavering support, patience, good humor, and love.

# INTRODUCTION

## Global Television

*Shawn Shimpach*

Reports of the death of television have been greatly exaggerated.

Or have they?

Today digital technologies allow cable and satellite systems to carry hundreds of channels of television programming, from national broadcast networks to boutique outlets to local access. At the same time, digital technologies also allow for video on demand and for various means of signal storage and portable redistribution. Anyone with access to a "smartphone"—from school children to grandparents, from dissidents to aspiring cooks—can potentially shoot, edit, and distribute their own, homemade programming. Viewing today, meanwhile, may involve "binging" on an internet company's streaming programs with a tablet computer; finding a live broadcast of an Indian cricket match with a laptop in South America; tuning into a cable news channel screening live social media footage illicitly shot on protestors' mobile phones; or joining millions of viewers to watch and comment upon a Swedish college student's YouTube "channel," devoted to live and recorded videogame play accompanied by witty commentary and advice. The ownership of television companies and institutions around the world has been increasingly deregulated, privatized, and consolidated even while international treaties construing narrowly defined and strictly enforced definitions of intellectual property have proliferated.

As these initial examples suggest, technological change and global difference are, together, animating a moment full of potentials both promising and dire. All of this change may be very exciting.

But is it television?

### The End of Television?

Such questions are no longer strictly—ahem—academic. Evidence seems to mount each day suggesting that television may be soon a thing of the past. For a sense of this, consider that in early July of 2017, the British press was filled with stories about an episode of the long-running British Broadcasting Corporation (BBC) children's program, *Blue Peter*, that had attracted absolutely no viewers, zero. The immediate effect of these reports was the widespread questioning and, in turn, the publicly funded BBC's staunch defense of its commitment to a program so few—and apparently, on occasion, no one—wished to view (Heritage 2017; Stolworthy 2017). The broader effect was the renewed fervor of a favorite journalistic trope returned to regularly since the start of the twenty-first century in North America and Europe: predicting the inevitable demise of television. The oft repeated story—couched in nostalgic, gleeful, or resigned tones—tells of a generation coming of age in a digital environment of social media, internet streaming, and mobile interactivity that invariably culminates in their maturation and television's expiration (Mills 2017). From story to story this all happens either sooner or later than you might expect, as television suffers "death by a thousand clicks" (Smith 2017).

Certainly, the *Blue Peter* story would seem to portend the start of television's end. Regularly programmed by the BBC since 1958, *Blue Peter* well represents a "legacy" of "linear" programming. A magazine-style program with multiple segments and presenters each episode, covering topics from sports and crafts to pet care and omelette flipping, it has educated, informed, and entertained generations of British subjects as a function of the BBC's public service mandate. Yet in recent years the BBC's mission of public service has undergone multiple waves of scrutiny in the context of liberal market economics, competition, and changing definitions of the public it is meant to serve. Here it was in the digital age, having to defend production and distribution of one of its longest running programs, meant for children. The defense it proffered did not particularly bode well for the idea of television's ongoing vitality.

The BBC suggested that television is changing so rapidly that the markers of access and success television has relied on for decades can no longer account for the way people watch or the way programmers think about individual episodes (this, as defense for a program that has aired for nigh 60 years). A single television episode now exists over multiple broadcasts, digital streaming options, time-shifting, and, in any case, the BBC insisted, within a demographically focused, niche channel. Indeed in 2012 the BBC had moved *Blue Peter*, which at one time was viewed by 8 million people each episode, entirely off of its flagship BBC One channel to its digital children's channel, CBBC. Within this context, it suggested, the notorious episode of *Blue Peter* should be understood as the victim of an audience measurement sampling methodology—employed by the Broadcaster's Audience Research Board (BARB), with measurement "boxes" in 5,100 households nationwide—that was no longer adequate to the task of determining actual viewership. The television audience is now simply too fragmented, it claimed. In fact, the UK industry trade journal *Marketing Week* has suggested that "30% of all broadcast programmes register a 0.0 rating" in part due to the limits of BARB's audience sampling, which is not able to "represent smaller populations with statistical significance" (Chahal 2015). In any case, the BBC continued, the infamous episode was a repeated episode and had in fact been viewed by nearly 300,000 people in total if all the multiple broadcasts were added together along with views on the BBC's internet "over-the-top" (OTT) streaming delivery service, iPlayer (Heritage 2017). *Blue Peter* (and most other programs') viewership was down in an era of many screens and much competition, but it was indeed being viewed by considerable numbers. It was simply the case that people were no longer viewing it where and how they once did, together … at the same time … on television.

All of which is meant to suggest that it is unfair to continue to judge the BBC in terms of television. A zero-audience rating does not seem ideal, but we are urged to consider it in the context of the BBC's 2006 charter review assigning it the task of "helping to deliver to the public the benefit of emerging communications technologies and services" (quoted in Grainge 2017: 143). In a way, then, the fact that more people saw the episode on iPlayer than on legacy television was a sign that the BBC was fulfilling its mission, bringing its audience into a digital future of connected viewing through "TV-digital hybrid" technologies and programming (Grainge and Johnson 2018). Radio, television, and internet are now all part of the mission, indeed are increasingly conflated, as the BBC shapes itself into a digital public service broadcaster that "unites technology and editorial to meet a clearly defined audience need" (quoted in Grainge and Johnson 2018: 26). As the alibi for what was literally an unmeasurably small television audience, the iPlayer would seem to confirm, from one of broadcasting's stalwart institutions no less, television's utter diminishment, if not yet ultimate demise.

These questions of viewership too small to be measured through statistical sampling and audience need defined by charter, however, linger as markers of a national broadcasting legacy the BBC has not (yet) dispelled. They are not, for example, the kinds of questions that haunt Netflix, the US-based company that began as a mail-order DVD rental store and now offers subscription video on demand (SVOD) over the internet. Because it is digitally distributed to specific subscribers, the habits of individual viewers can be (and are) carefully collected—individually and directly, rather than through statistical sampling. Netflix watches you watching it. The "big data" that are collected inform programming algorithms. More than any other single attribute, this marks Netflix as a digital media company with an historically distinct relationship to its viewers. No one's viewing, however statistically insignificant, need be overlooked.

But neither are data merely used as direct numerical measurements of success. Instead, massive accumulations of detailed viewing data have resulted in new metrics by which a program's audience is measured, analyzed, valued, and anticipated. For Beijing-based iQiyi, this means the recent removal of its view count indicator, replaced with a "heat index" claimed to utilize artificial intelligence to consider multiple measures of viewer behavior, including interaction and feedback, quality, market performance, and shares on social media (Shijia 2018). For Netflix this means espousing disinterest in demographic categories that have shaped broadcasting, advertising, politics, and social identity for over a century. Rather than categories of age, gender or race, Netflix claims to favor the construction of what it calls "taste clusters." Its computers associate genre and textual qualities with affect, tone, and feelings. These are then matched with each individual viewer's habits (what you click on, what you watch and how much, what you watch next, etc.) to form more than 2,000 defined taste clusters and place viewers within "taste communities." How these textual qualities have been identified and measured, which taste community you may be part of, and how well all this actually works, Netflix will not reveal.

Netflix has, however, teased the idea of globally dispersed, demographically mixed "clusters" of linked individuals. This was a central part of the narrative imaginary in its science fiction program *Sense8* (2015–2018). That show's ethnically and sexually diverse characters, comprised of a multinational ensemble cast, propelled a story of individuals in Nairobi, Seoul, San Francisco, Mumbai, London, Berlin, Mexico City, and Chicago who discover they are psychically linked "sensates," strangers who, without prior agreement, now form a cluster, with forever interconnected lives and shared origins. Clusters supersede demographic considerations and cut across global divides. Created by Lana and Lilly Wachowski and filmed on location in eight different countries, *Sense8* seemed designed to produce a compelling imaginary of trans-demographic, global interconnection in sync with Netflix's own aspirations for marquee, big-budget global expansion through fantastical conceits (biological engineering and psychic connection within the program; seamless internet streaming and digital surveillance for subscribing viewers). The real limits of these fantastical conceits were exposed, however, as Netflix renegotiated ownership of the program, then cancelled the exorbitantly expensive show, and finally produced an additional, concluding episode after facing vocal fan backlash (Keegan 2018). All of which seems rather like a different kind of cluster, to be sure.

Still, this metrics revolution is not even an imaginable option for television broadcasters. Netflix utilizes its digital surveillance affordances to claim to provide consumers with the technology industry's twenty-first century libertarian dream/nightmare of individual "access, personalization and choice" (quoted in Lynch 2018; see also Zinoman 2018). Netflix claims to continually refine its formulae so that it will grow in proficiency as it seeks to offer tailored, if not bespoke, programming to each subscriber, rather than figuring audience "needs" in the abstract aggregate, or as a public, as broadcast media has done for decades.

Increasingly it claims to do this everywhere in the world, as its global imaginary of interlinked taste clusters combined with immense capital has facilitated a distinctly corporate model of globalization and remarkable worldwide expansion. Before 2010 Netflix was an entirely US-based company; by 2018 it was available in over 190 countries. In 2018 its international streaming revenues exceeded its US revenues for the first time as 73 million of its 130 million subscribers resided outside the US[1] (Brennan 2018). In most countries on the planet it is possible to subscribe to this transnational technology company's streaming service and view programming choices both local (and, increasingly, locally commissioned and offered in local languages) as well as international. While it is possible to do all this on your television screen, so long as it is connected to the internet, it is also just as possible and often more convenient to do so on your computer or mobile phone.

These examples illustrate the ways in which television's transformation into online, non-linear programming and internet companies' rapid expansions are typically presented with a sheen of inevitability, as a window into the very (and very near) future where television is no longer a factor. As broadcast audiences appear to fade, the BBC invests in digital streaming technologies, reframing its public service mandate. As high-speed internet access expands transnationally, Netflix emerges as a global media company. If there is anyone left in the living room with a boxy screen in the corner, adjusting the antenna for better reception, no one appears to be very much interested.

All this has not gone unnoticed by scholars of television studies. Drawing from a range of disciplinary precursors, from the social sciences, cultural studies, humanities, the arts, and studio production, television studies has remained a seemingly perpetually nascent academic discipline and an untidily productive amalgam of approaches that has never maintained a singular, consistent, nor stable definition of its object of study (Brunsdon 1998). Is television a physical object or a medium? An institution or an infrastructure? Is it defined by who makes it and when, or where and how it is viewed? Is it textual—and if so is that text an episode, a program, a series, a channel, or a flow? Is it a window into institutional or cultural power? Is it best understood through Marxism, feminism, critical race theory, or postcolonial perspectives? In answer to each of these questions, television studies has consistently answered: "well, yes, sometimes." Television *is* institutional *and* technological *and* social *and* textual. Indeed this irreducibility is what animates so much innovative analysis. Such a response certainly seems sensible and laudable, but it has also likely slowed television's perceived acceptance as a defined, legitimate object for scholarly investigations (Hilmes 2005), rendering both the object of study and the undertaking of that study contentious, contingent, and without specific place in the academies of higher learning. The increasing porousness of divides between television and digital online media has, therefore, been met less with surprise than with acceptance of the ongoing transmogrification of a forever unstable object. Perhaps "television" is not even what is really interesting after all?

Digital media has, of course, generated considerable excitement for the ways it is imagined (and promoted) to literalize and make visible empowering practices theorized more abstractly for decades by television scholars: choice, control, access, creativity, communication, interactivity, and personal expression are all democratically available, even encouraged, on new, flat, high-definition screens. No more accusations of passive reception with this media! Although at times still susceptible to some techniques of analysis devised for earlier iterations of television, this new media seems, from this perspective, clearly to exceed the limits of a medium delivered by one-way broadcast signals to a box in the corner. Even scholarship alarmed by the ways in which surveillance, convergence, commodification, and privatization come packaged along with these new potentials seems to recognize a clear break with the older form of television. Collections of scholarship hinting, therefore, at the eventual demise of television have emerged under titles such as *Television After TV: Essays on a Medium in Transition* (Spigel and Olsson 2004); *Television Studies After TV: Understanding Television in the Post-Broadcast Era* (Turner and Tay 2009); and a 2011 special issue of *The Annals of the American Academy of Political and Social Science* organized around the theme of "The End of Television? Its Impact on the World (So Far)." These collections engaged many of television studies' most influential scholars with the question of television's recent transformations and what they might mean for television and television studies' future (or lack thereof).

To be sure, digital media requires reconsideration of some of the foundational terms by which television has been defined. This new media typically fails to provide the "liveness," "flow," and "aggregated publics" as they were conceived in relation to the very essence of television as a distinct medium (Uricchio 2009). Even as North American and European programming proliferated in the first decades of the new century, garnering pronouncements of a new golden age of television, where an (over) abundance of quality-crafted, narratively complex, middle-brow programs have come to be called "peak TV," it is still technological change that has galvanized the moment; moving the field, generating debate, exciting new theories, and signaling the future. The programming is interesting and the subject of multiple analyses and interpretations, yes, but look at how we can access it now! Increasingly, the intellectual excitement and fierce engagements are looking toward the interactive, the non-linear, the makers and sharers, social media, and user-generated content. As William Uricchio notes, the multiple potentials of online media offer much more enticement for thinking beyond an apparent status quo than the diminishing possibilities of the established practices of the television industry: "if anything, the television industry has stuffed itself into an unnecessarily small conceptual space, and YouTube is providing a set of radical alternatives" (Uricchio 2009: 372). It is not for nothing that he makes this observation in a book that is not about television, but is instead called *The YouTube Reader.*

Acknowledging these developments, prolific television studies scholars Jonathan Gray and Amanda Lotz recently decided to make the call. Somewhere between 2012 and 2019, between the first and second editions of their *Short Introductions: Television Studies* textbook, they determined that television studies (if not yet television itself) had passed away. Whereas they end their first edition noting "we remain convinced of the remarkable centrality of television" (Gray and Lotz 2012: 143), their second edition omits this line in order to [SPOILERS] "conclude a book introducing television studies by arguing that it is now best subsumed within media studies." Essentially, given the rapid pace of convergence within and between formerly distinct media companies, combined with programming and technology convergences so that the same screen can be used to interact on social media, make beats and edit your podcast (and publish both online), view vlogs and feature films, and binge episodes—old and new—of television programming, it no longer makes sense to study each of these as if they are separate and distinct activities or self-contained texts. Ultimately, *Television Studies* concludes that "the heyday of television studies as a discrete, separate exploit is now past" (Gray and Lotz 2019: 144). Television can no longer be productively distinguished from other media. Death by a thousand clicks. After all, even Mark Twain did eventually die.

⋆ ⋆ ⋆

And yet, looked at another way, the mounting evidence could perhaps suggest that television nonetheless persists. As does its study. Upon closer inspection, in fact, most of the scholars within the initially alarmingly titled collections are analyzing and anticipating change, sometimes perhaps beyond current recognition, as a catalyst to think through and engage fervently in the study of television. Even in *The YouTube Reader*, Uricchio suggests that user-generated online videos do not, in fact, "pose a threat to the concept of 'seeing at a distance' that has long characterized television so much as to the institutional logics that have held it in a vice grip over the past few decades" (2009: 372). Introducing the volume she co-edited, *Television After TV*, Lynn Spigel observes that "television studies per se continues to be a varied and vibrant field of inquiry." She takes the occasion to warn against techno-futurist fetishization and a naive reverence for "anything that works differently from the old stuff," noting a scholarly tendency, remarkably aligned with the marketing of new media technologies, to believe that new media, "and especially if it works in ways mysterious to most people—is instantly considered for its radical utopian potential" (Spigel 2004: 11).

Graeme Turner and Jinna Tay, introducing *Television Studies After TV*, meanwhile argue that television indeed remains a vital, global medium even while technological change makes it "absolutely clear that we can no longer talk about 'TV' as if it were a singular entity" (Turner and Tay 2009: 3). Instead they point to the increasingly evident need to consider television in its specific histories and contingencies of place, use, and meaning. Such is a reminder that television's history is constituted by nearly continual change—it has always been an unstable object of study. Television originated as an idea for two-way, face-to-face communication. It emerged as radio with pictures. From a single channel, or maybe two, it eventually expanded to hundreds of channels. At different times in different places, television was a monochrome moving image and then it was color, it was state run or a public service and then, commercial. It was broadcast and then cable, analog and then digital, a box in a room and then many flat screens in many places. It was a constant, grainy, ever-present flow and then it was on demand or a high-definition, surround-sound, spectacle. Given such a history of absorption and transformation, it may well be that television will resist being subsumed within media and the optimistic promises and dire threats of new media will have been subsumed within television instead.

At the same time, it is growing clearer that the institutional vise grip is not so easily loosened. Every interactive "click" is also an opportunity for surveillance with the collection of big data harnessed to sell you more and better; every "shared" task and user-generated video is a means of off-loading labor costs onto consumers. Much of online, digital programming, meanwhile, from YouTube vlogging channels to streaming video companies to television channel OTT services, is segmented into discrete episodes available either to binge or tune in next time. Many of these are produced and accessed in familiar 30- or 60-minute frames. New media companies Netflix, Amazon, and Apple have all made headlines for demonstrating their commitment to the future of their endeavors by recruiting Hollywood film and television stars, writers, directors, showrunners, and producers with lucrative contracts to make online content that ends up looking like, getting promoted as, and winning critics and industry awards for being, television.

This is not simply an aside in the context of media studies, as recent industry reports indicate that 15% of global internet traffic is dedicated to Netflix streaming alone, while in the Americas, during peak evening hours, Netflix usage can account for 40% of all internet traffic. Worldwide, video is 58% of internet bandwidth traffic, with web browsing second at 17%, gaming at 8%, and social media 5%. In Europe, the Middle East, and Africa, YouTube accounts for the single largest use of internet bandwidth, followed by Netflix, while in the Asia-Pacific region, web-based video streaming uses the most internet bandwidth followed by Facebook video (Sandvine 2018; Spangler 2018). New media, the digital online, is taken up by an enormous amount of audiovisual streaming that looks very much like film and television.

There are other, even more direct signs that TV maintains a distinct cachet, even in the digital, online media world. In China, for example, Shanghai-based peer-to-peer streaming video company PPLive changed its name to PPTV in 2010 to better reflect its media streaming business, which has increasingly focused on distributing international and locally produced television programming. This change in name occurred at virtually the same time that state-run CCTV launched its online channels, collectively called CNTV. The Lagos-based "Nollywood" streaming service, sometimes referred to as the "Netflix of Africa," was launched in 2011 as iROKOtv. In

2016 Netflix's CEO declared to a trade conference that with Netflix we are "witnessing the birth of a global TV network" (quoted in Jarvey 2019). In London, the BBC's website describes BBC Three as "the first TV channel in the world to switch online in February 2016." Is TV disappearing into new media, or is it the other way around?

Much of the excitement over "spreadable" and "transmedia" storytelling in the first decade of the current century has diminished considerably as media companies are "focusing instead on using new media in ways that align with traditional programming, marketing and licensing practices" (Grainge and Johnson 2018: 25). Thus, viewers within the European Union increased time spent watching YouTube on television screens by 45% in one year, marking television as the company's "fastest growing device these days," while leading one executive to insist that "traditional broadcasters have always been part of YouTube" and that "the really sophisticated [broadcasters] understand that it's a way to reach an audience above and beyond just television" (Hern 2018). Just television, however, remains a consistently prevalent and utilized medium, with thousands of broadcast television stations operating around the world and thousands more delivered by cable and satellite. Media research and consulting firms, meanwhile, continue to remind global media buyers that "Traditional television—comprising broadcast and multichannel advertising and pay TV subscriptions—is still the largest component of the traditional media market" and that despite online competition, "traditional television has done relatively well because it still has the largest reach of any medium. Its audience has shown very little erosion compared with other non-digital media" (McKinsey & Company 2016: 22–23). Such advice is derived from examining television use at a global scale.

Nearly 60% of the world's population does not use the internet. Internet access and usage remain grossly uneven around the world. While more than 94% of individuals in the "developed world" use the internet, this figure drops to just over 41% in the "developing world" and is less than 18% of the population in the "least developed countries." Even for the digital generation, while 80% of the world's youth (ages 15–24) are online, nearly 9 out of 10 of those who are not live in Africa or Asia and the Pacific (ICT Facts and Figures 2017). And while the growth of internet access continues to expand at a remarkable rate, as Paula Chakravartty and Srirupa Roy have noted, "most of the world's population relies almost exclusively on broadcast media, especially television, for both entertainment and information." This leads them to conclude that "however contrary to popular perception, television is in fact the most significant 'new media' of the last two decades" (Chakravartty and Roy 2013: 350). Not only does television persist, but such evidence demonstrates how crucial it is to consider broader, or multiple, global perspectives before arriving at assumptions or asserting claims about television, its use, and its vitality.

## Global as Method and Subject

The "global" of global television in this *Companion*'s title therefore speaks to both the subject matter and the methods by which "television" may be understood. It is an expression of scope, to be sure, beginning with the recognition that "global television" is not singular, but multiple, diverse, and contingent. Just as "television" is a collection of industries, a set of constantly changing technologies, and many means of making, viewing, and interpreting; so, too, is "global" plural, referring to capitalism, political movements and governance, cultural identities and diasporas, social practices and concerns, technologies and uses, as well as an accumulation of locally specific traditions, beliefs, and experiences. Constituting television in global terms, therefore, involves attending to it all over the world without privileging a version of it in any one place in the world.

The challenge is clear from the ways that television remains a crucial site for observing, analyzing, addressing and redressing what Arjun Appadurai once identified as—and in many ways remains—a "central problem of today's global interactions": a tension between cultural homogenization and cultural heterogenization. As global communication expands, culture on one hand is observed to become increasingly westernized, Americanized, and/or commoditized, with television a central catalyst of these transformations. At least as rapid as this homogenization, however, culture and media "tend to become indigenized in one or other way" even as these dynamics operate on grossly uneven terrain (Appadurai 1990: 5–6). Television figures centrally in understanding these dynamics because it can be readily observed to be imbricated, complicit, and actively engaged in both processes of homogenization and heterogenization on virtually a daily basis.

Television in any particular instance is produced by and through the place in which it is encountered, the specific practices associated with it in that place, the meanings and uses and regulations and technologies that make it television (Pertierra and Turner 2013). Even as it is defined by its location, however, it is also part of a network of industrial, economic, political, technological, even aesthetic practices that link it, always, to other places. Global television is, therefore, conceived of in multiple ways. In some cases this implies comparisons, across and between national, linguistic, cultural, and socio-economic contexts. In some cases it implies travel and circulation and uptake with varying regards for different kinds of borders. In some cases it means, cumulatively, perspectives from many different locations (geographically, culturally, intellectually) around the world (Parks and Kumar 2003). Throughout, however, as Karin Wilkins notes, "it is not the medium that we deem global, but the human organization engaged in creating, sharing, seeking, and interpreting [it] that has the potential to be contributing to and conditioned by a global context" (Wilkins 2015: 284). Global television as a subject is constituted through the consideration of these multiple iterations and taking seriously their local specificities as well as their comparative similarities and differences in order to understand what connects and what separates television around the world.

Such an approach has the power to decenter narratives about television that emanate from only a single, historically privileged place, and to diminish universal claims about television based on the experiences of only one culture. Yet such an accumulative pluralist diversity does not effectively account for the reasons differences exist in such unequal ways around the world, nor does it interrupt the logics by which inequities have been produced and sustained. For this, "global" television also requires a global method that is both intentional and respectful (Wilkins 2015). Wilkins argues that global practice means that

> critical research agendas build on foundations that highlight power dynamics, manifest through structural economic conditions, political processes, social practices, and cultural rituals. When considering global contexts, it is not simply the spatial territory within which one is situated that matters, but one's access to a variety of forms of capital.
>
> *(Wilkins 2015: 285)*

Access to these various forms of capital (e.g. economic, cultural) remains inequitable, as "transnational linkages" between media spheres have been produced through a global history of colonialism, racism, and exploitation (Shome 2016).

More than simply recognizing and celebrating cultural difference, a global method may disrupt narratives and theories about the use, purpose, and meaning of television that have developed from a focus situated in specific (Western, Global-Northern) locations of geo-political and economic power, that present themselves as universal, and that allow "those in power to assert their own interests as conventional wisdom" (Wilkins 2015: 284). A global perspective contrasts "stagist" developmental presumptions of technological adoption and use, where modernity appears "first in Europe and then elsewhere" (Chakrabarty 2000) with the lived, multiple modalities and temporalities at play in everyday transnational media use and access. In places where a history of exploitation now manifests as radical economic disparity, it is not simply the case that development lags behind other parts of the world. The internet does exist, after all, as do mobile telephony and digital media, right alongside broadcast television and radio. It is a question of access and use, but also of priorities—which include, but are not exhausted by, economic priorities alone. The ongoing development of television around the world may, therefore, be recognized as a mesh of overlapping layers as well as distinctive pathways.

Hence, the Ghanaian online program *An African City* may look like a concession to neoliberal consumer choice and American-style "production values," yet it may also knowingly utilize these conceits to challenge ingrained assumptions that inform the expectations of "poverty porn," the potentials of African media, and the diversity of African women (Chávez and Cordes 2018: 203). The ethnically diverse cast comprising both the comically local and internationally cosmopolitan primary characters of the US broadcast situation-comedy *The Good Place* might seem like a ploy to aggregate various audience demographics. Yet within the global context of a staid television genre (sitcom), here consistently undermining genre expectations and its own situational premise as part of an ongoing exploration of ethics and philosophy, the characters also begin to register as a shorthand for implicating humanity as a global community in contractualist-derived duties and considerations of "what we owe to each other." Institutionally it is not necessarily surprising to observe Arab entrepreneurs setting up satellite television channels in Europe: from a European perspective they are the product of liberalization and privatization policies, yet from an Arab perspective they respond to regulatory, political, and economic restrictions on freedom of speech (Khalil and Kraidy 2009). Television's success, meanwhile, may be measured in terms of the number of viewers it gathers, or for the kinds of viewers and kinds of data about their viewing it produces, or for its ability to provide shared moments and experiences among close associates and between kin (Adejunmobi 2015: 125). Understanding global difference, therefore, involves not only an analysis of power and access to capital, but also of differential meanings, modes, times, and values that may be compared, but are not so readily hierarchized.

## The End of Televisions

A vital example of the intersection of technological change and global difference can be illustrated through an aspect of global television studies that remains woefully under-attended: the disposal of old screens. An effect of the acceleration of technological change has been the rapidly growing production of electronics waste (or e-waste) in the twenty-first century. In 2016 alone, global e-waste production was measured at 44.7 million metric tons, or an equivalent of nearly 13½ lb (6.1 kg) per inhabitant on earth. As one report helpfully illustrates, this is close to the equivalent of 4,500 Eiffel Towers of e-waste produced each year. The amount has been rising throughout the century and is expected to reach 52.2 million metric tons (or 15 lb (6.8 kg) per person) by 2021 (Baldé et al. 2017: 4). In 2016, only 20% was recycled through appropriate channels (Baldé et al. 2017: 2).

The production of this waste is not evenly distributed globally. One recent report noted that "the richest country in the world in 2016" generated nearly 33 times more e-waste than the poorest country (Baldé et al. 2017: 41). Nor is it valued and managed in the same ways in different places in the world. As a recent review of global e-waste practices explains, "Unlike developed countries, where products once discarded are considered 'waste' having no intrinsic value, 'waste' is considered valuable in developing countries" (Borthakur and Govind 2017: 109). In India, for example, obsolete electronics may still be valued as commodities and find new owners. Cambodia, indeed, emerged on the global electronics market in the early twenty-first century as neither a producer nor a final resting place for e-waste, but as a marketplace for new owners of this "waste." Growing local demand has been met by the importing of used electronics, most especially of televisions, from China, Finland, France, Hong Kong, Japan, Malaysia, Korea, Singapore, Thailand, and the US. These are resold locally or dismantled for parts. Yet, while government regulations have allowed for used electronic imports, they do not allow for imports intended for final disposal or recycling. Therefore, Cambodia has no recycling facility. At the end of their re-use, most e-waste is still considered too valuable to be disposed of and is instead more typically collected or purchased for illegal export and recycling in yet other countries (UNEP 2007: 81). There is considerable financial incentive from the perspective of individual livelihood; in a country where the average income was US$317 per year, second-hand television retailers earned about US$600 per year, while e-waste collection site workers made around US$360 - per year (UNEP 2007: 47).

The human and environmental costs of final disposal, meanwhile, are not evenly distributed either, with the disposal of used screens registering as a local, interregional, and global problem in different ways. Current e-waste regulations may allow for the export of e-waste from waste-producing wealthy nations to developing countries, so long as they are labeled a "donation" or for "recycling." In many of the places where this e-waste arrives, an interdependent network of official and informal recycling attends to these electronics. Informal recycling, in particular, can be dangerous for both human health and the environment, particularly because, "occupational safety and environmental protection are not prioritized." Instead the priority is on rapid processing, with poor children and women with few other income opportunities particularly vulnerable as "marginalized populations bear a disproportionate amount of the negative effects of improper e-waste practices" (Perkins et al. 2014: 291). As one report succinctly notes, "currently, e-waste scavenging provides a source of livelihood, albeit a risky one, to large numbers of people in developing countries" (Perkins et al. 2014: 290).

Risky indeed, because electronics are filled with toxic substances and those working to recycle them are put in direct contact with contaminants. A 2007 U.N. report offered one example of the specific risks to the environment and to the people working with this waste. In Guiyu (China), air pollution near e-waste recycling areas contained "the highest levels of dioxins ever recorded" exposing adults in the area to inhalation of dioxins at an amount "15–56 times higher than the World Health Organization recommended maximum" (UNEP 2007: 48–49). Richard Maxwell and Toby Miller, in what remains to date one of the few media studies accounts of e-waste as integral to an understanding of global television, list "risks to bones, brains, stomachs, lungs, and other vital organs, in addition to birth defects and disrupted biological development in children" as among the risks faced by individuals from the toxins contained in e-waste (Maxwell and Miller 2012: 3–4). Beyond the embrace of market logics and the flawed politics of finding revolutionary potential in every new media form, such a list demonstrates in tangible detail the literal human and environmental unsustainability of a still growing global fetishization of constantly new media.

This includes television in multiple ways, as in addition to the e-waste of used television sets and devices, as Maxwell and Miller note, new, flat-screen televisions can require more than three times the electricity of older, CRT sets, resulting in a dramatic rise in carbon emissions (Maxwell and Miller 2012: 81). Worse, such trends, rather than abating, may in fact be accelerating. An industry trade journal noted that in the European Union "in the first six months of 2018, only 33% of TV sets sold obtained the energy efficient class A+, showing a negative trend" that continues to increase the annual power consumption associated with TV sets as Ultra HD large screens continue to gain popularity (Krieger 2018).

For global television to fully account for technological change and global difference, the material impact on livelihoods, human bodies, and the environment is a crucial part of the conversation. It is an urgent example of the ways media technologies and global inequities in capital and political power intersect with human lives, and how this simply cannot be sustained, one way or another. The inequities produced by a legacy of exploitation that allows for waste to be exported to countries where the economics of individual livelihoods welcome the opportunity and disregard the risks, reveal ongoing, unequal access to capital, power, and resources on the global scale, even as they intimate the quite literal unsustainability of these inequities. The ways in which e-waste is valued and managed in different locations, meanwhile, suggests the variable temporalities, needs, and uses that can be revealed through a serious consideration of global media.

## Global Television Studies "Selfie"

This *Companion* is not an introduction to global television or global television studies, nor is it a completely comprehensive account of everything that is happening, everywhere in the world. The chapters in the pages that follow instead offer a snapshot of global television two decades into the twenty-first century and a "selfie" of global television studies right now. Like a selfie, they capture a moment in time that is revealing for what it portrays, but also for how it is framed, and the very fact of its existence and particular circulation. Together, these chapters mostly take television to be an established fact of daily life. They each seek to explain something significant about television, most often in the context of technological change and global difference. They provide us with histories and contexts for understanding television in its multiple specificities. Throughout this volume we encounter evidence of an industry as well as a field of scholarship adjusting to transforming conditions, with parallels as well as important distinctions between the parameters, metrics, terminologies, and implications at play. As a whole, the chapters that follow offer a transnational examination of, and engagement with, the lively, varied, and transforming practices, technologies, systems, and texts that constitute what television means today. Drawing on the legacy of television studies, but with an eye toward the future, this collection emphasizes both the thoroughly global nature of television and the multiple and varied experiences that constitute television in the twenty-first century. The hope is that this collection of research may offer something of a model or guide toward the future of television studies that is globally inclusive and will be of value to our own "taste cluster" of scholars, students, critics, viewers, and community members around the world.

The book is divided into five sections. For the reader ambitious enough to read this book straight through, cover to cover, it will become clear that the divisions between these sections are, ultimately, inevitably somewhat arbitrary. The concerns of each overlap with others. The organization of these sections, however, is intended to suggest the ways that a changing, transnational television actually matters to people (rather than, for example, reproducing the industry logic of producer–distributor–audience). Contributors to all sections include some of the leading scholars of television studies in the world, as well as new and emerging voices interested in the role of television in the world today. A few scholars offer perspectives from training and traditions not normally included in the formation of television studies. This collection of original essays is also designed to include perspectives and experiences from around the globe, covering six continents (apologies to Antarctica), but without limiting any one discussion necessarily by organization, design, category, or priority, to a single national or regionally defined scope. The global perspective on television and the multiple voices of television studies here emerge organically as these scholars have brought their own interests, knowledge, experiences, and contexts to bear on topics they find intriguing, urgent, and compelling about television at this moment.

Television today remains a crucial forum for democratic debate, a site for encountering and expressing new identities, and a place for making sense of cultural narratives. This *Companion* offers a guide to understanding what these look like, how they work, and what it means.

However television continues to transform and be (re)defined, there remains an enormous amount of programming to view. As of this writing, even *Blue Peter* is still in production, churning out new episodes—and someone, somewhere, in some way, is certainly watching.

## Note

1 See Netflix website: https://help.netflix.com/en/node/14164

## References

Adejunmobi, M. (2015) "African Film's Televisual Turn" *Cinema Journal* 54:2(Winter); 120–125.

Appadurai, A. (1990) "Disjuncture and Difference in the Global Cultural Economy" *Public Culture* 2:2; 1–24.

Baldé, C.P., Forti, V., Gray, V., Kuehr, R., and Stegmann, P. (2017) *The Global E-waste Monitor – 2017.* Bonn/Geneva/Vienna: United Nations University (UNU), International Telecommunication Union (ITU) & International Solid Waste Association (ISWA). Accessed https://collections.unu.edu/eserv/UNU:6341/Global-E-waste_Monitor_2017__electronic_single_pages_.pdf.

Borthakur, A and Govind, M. (2017) "Review: Emerging Trends in Consumers' E-waste Disposal Behaviour and Awareness: A Worldwide Overview with Special Focus on India" *Resources, Conservation and Recycling* 117; 101–113.

Brennan, L. (2018) "How Netflix Expanded to 190 Countries in 7 Years" *Harvard Business Review* (12 October). Accessed: https://hbr.org/2018/10/how-netflix-expanded-to-190-countries-in-7-years

Brunsdon, C. (1998) "What is the 'Television' of Television Studies?" C. Geraghty and D. Lusted, eds. *The Television Studies Book.* London: Arnold; 95–113.

Chahal, M. (2015) "What the Future of TV Looks Like for Marketers" *Marketing Week*, 15 July. Accessed: www.marketingweek.com/2015/07/15/what-the-future-of-tv-looks-like-for-marketers/

Chakrabarty, D. (2000) *Provincializing Europe: Postcolonial Thought and Historical Difference.* Princeton: Princeton University Press.

Chakravartty, P. and Roy, S. (2013) "Media Pluralism Redux: Towards New Frameworks of Comparative Media Studies 'Beyond the West'" *Political Communication* 30; 349–370.

Chávez, C. A. and Cordes, A. (2018) "Selling Subversion: *An African City* and the Promise of Online Television" *Television & New Media* 19:3; 191–207.

Grainge, P. (2017) "'Moment and Opportunities': Interstitials and the Promotional Imagination of BBC iPlayer" *Critical Studies in Television* 12:2; 139–155.

Grainge, P. and Johnson, C. (2018) "From Catch-Up TV to Online TV: Digital Broadcasting and the Case of BBC iPlayer" *Screen*, 59:1Spring 2018; 21–40.

Gray, J. and Lotz, A. (2012) *Television Studies.* Cambridge: Polity Press.

Gray, J. and Lotz, A. (2019) *Television Studies.* second edition. Cambridge: Polity Press.

Heritage, S. (2017) "Blue Peter Has Hit the Rocks. It's Time for the BBC to Sink It." *The Guardian*, 6 July; Accessed: www.theguardian.com/tv-and-radio/2017/jul/06/blue-peter-has-hit-the-rocks-its-time-for-the-bbc-to-sink-it

Hern, A. (2018) "YouTube: TV is Our Fastest-Growing Market" *The Guardian*, 14 September; Accessed: www.theguardian.com/technology/2018/sep/14/youtube-tv-is-our-fastest-growing-market

Hilmes, M. (2005) "The Bad Object: Television in the American Academy" *Cinema Journal* 45:1 (Fall); 111–115.

ICT Facts and Figures (2017) Accessed: (www.itu.int/en/ITU-D/Statistics/Documents/facts/ICTFactsFigures2017.pdf)

Jarvey, N. (2019) "Saudi Arabia Isn't the Only Trouble Spot for Netflix" *The Hollywood Reporter*, 9 January; 22.

Keegan, C. M. (2018) *Lana and Lilly Wachowski.* Urbana, IL: University of Illinois Press.

Khalil, J. and Kraidy, M. (2009) *Arab Television Industries.* New York: Palgrave Macmillan.

Krieger, J. (2018) "High-End TV Sets Drive Global Market" *Broadband TV News*, 29 August Accessed: www.broadbandtvnews.com/2018/08/29/high-end-tv-sets-drive-global-market/

Lynch, J. (2018) "Netflix Thrives By Programming to 'Taste Communities,' Not Demographics" *Adweek*, 29 July Accessed: www.adweek.com/tv-video/netflix-thrives-by-programming-to-taste-communities-not-demographics/

Maxwell, R. and Miller, T. (2012) *Greening the Media.* Oxford: Oxford University Press.

McKinsey & Company (2016) "Global Media Report 2016: Global Industry Overview" Accessed: www.mckinsey.com/~/media/McKinsey/Industries/Media%20and%20Entertainment/Our%20Insights/Global%20Media%20Report%202016/GMO%20Report_2016_Industry%20overview_v3.ashx

Mills, C. (2017) "Young Americans Will Be the Death of Cable TV" *New York Post*, 21 September Accessed: https://nypost.com/2017/09/21/young-americans-will-be-the-death-of-cable-tv/

Parks, L. and Kumar, S. (2003) "Introduction", L. Parks and S. Kumar, eds. *Planet TV: A Global Television Reader*. New York: NYU Press pp. 1-17.

Perkins, D. N., Brune Drisse, M.-N., Nxele, T., and Sly, P. D. (2014) "E-Waste: A Global Hazard" *Annals of Global Health*, July–August, 80:4; 286–295.

Pertierra, A. C. and Turner, G. (2013) *Locating Television: Zones of Consumption*. London: Routledge.

Sandvine (2018) "Global Internet Phenomena Report: Netflix is Approximately 15 per cent of Worldwide Downstream Traffic" 28 September Accessed: www.sandvine.com/blog/global-internet-phenomena-report-netflix-is-15-of-worldwide-downstream-traffic

Shijia, O. (2018) "iQiyi Scraps View Counts to Improve Popularity Gauge" *ChinaDaily.com*, 19 October Accessed: www.chinadaily.com.cn/a/201810/19/WS5bc9625ea310eff30328365e.html

Shome, R. 2016 "When postcolonial studies meets media studies" *Critical Studies in Media Communication* 33:3 245–263.

Smith, J. C. (2017) "Predicted the Death of Broadcast TV? Then Have I got News for You" *New Statesman*, 20 March Accessed: www.newstatesman.com/culture/tv-radio/2017/03/predicted-death-broadcast-tv-then-have-i-got-news-you

Spangler, T. (2018) "Netflix Eats Up 15% of All Internet Downstream Traffic Worldwide" *Variety*, 2 October Accessed: https://variety.com/2018/digital/news/netflix-15-percent-internet-bandwidth-worldwide-study-1202963207/

Spigel, L. (2004) "Introduction" L. Spigel and J. Olsson, Eds. *Television After TV: Essays on a Medium on Transition*. Durham: Duke University Press; 1–34.

Stolworthy, J. (2017) "Blue Peter in Danger as Recent Broadcast Fails to Record A Single Viewer" *Independent*, 5 July Accessed: www.independent.co.uk/arts-entertainment/tv/news/blue-peter-bbc-cbbc-episode-in-danger-episode-no-viewer-a7824111.html

Turner, G. and Tay, J. (2009) "Introduction" G. Turner and J. Tay eds. *Television Studies After TV: Understanding Television in the Post-Broadcast Era*. London: Routledge; 1–8.

United Nations Environment Program (UNEP), Division of Technology, Industry, and Economics, International Environmental Technology Center (2007) E-waste volume I: inventory assessment manual. Accessed: http://wedocs.unep.org/bitstream/handle/20.500.11822/7857/EWasteManual_Vol1.pdf?sequence=3&isAllowed=y

Uricchio, W. (2009) "The Future of a Medium Once Known as Television" P. Snickars and P. Vondeau, eds. *The YouTube Reader*. Stockholm: National Library of Sweden; 360–375.

Wilkins, K. G. (2015) "Articulating Global in Communication Research as Subject and as Practice" *Communication Research and Practice* 1:3; 283–290.

Zinoman, J. (2018) "The Netflix Executives Who Bent Comedy to Their Will" *The New York Times*, 9 September Accessed: www.nytimes.com/2018/09/09/arts/television/netflix-comedy-strategy-exclusive.html

# PART I

# Objects and Ideas

This section explores approaches to defining television as an object of study and to delineating current ways of thinking about television in a global context. This is not always a straightforward undertaking in an era marked by rapid transformation, where the definition of television changes seemingly before our eyes and is marked by stark differences at various places around the world. Chapters here offer and examine current critical theories of global television, its meaning, and its implications. Tellingly, television in the digital age is rather consistently defined here beyond or in excess of any single object. For John Hartley, television is best described not in terms of what it is, but instead in terms of what it is used for. His answer to "what is television?" leads to a consideration of technologies for the production of the modern subject. Purnima Mankekar agrees that we experience television in excess of specific material objects and, indeed, may best understand it in terms of the intersubjective relations and affect it produces and facilitates. Jorge A. González is similarly less interested in the material object of television per se and much more so in the technogenesis, or co-evolution, of human societies and screen technologies. The stakes, for him, are clear, as these technologies are produced through a global economics of inequity that positions Mexico with limited access to its own ecologies of symbolic meaning and exchange. Stuart Cunningham and David Craig define contemporary global television in terms of practice as much as object by tracing the use and circulation of moving image media produced and experienced on social media at key places in the world right now. These analyses are grounded in the history of television as an object and an idea. Indeed, history is critical to our understanding of global television. Timothy Havens, therefore, explores the historical tensions between disparate broadcast systems around the world, each of which has reflected a different ideological relationship between television and the state. Lothar Mikos examines a history of international cooperation and arrangements that demonstrates that the production, programming, and audience for television have been transnational (if not global) from the beginning. Toby Miller concludes the section by diagnosing television as an object and an object of study in order to identify important areas for action through which we may work toward addressing what television might ideally become and how we should study it. Throughout this section, theories, histories, and definitions of television intermingle, producing a fascinating, nuanced, and compelling portrait of the objects and ideas that animate television scholars today and inform the meanings of television in the sections and chapters that follow.

# 1
# WHAT IS TELEVISION?
## A Guide for Knowing Subjects

*John Hartley*

### Nothing to See Here?

When there's a commotion on the street, people stop to watch. A crowd gathers. Bystanders get involved. But then the security authorities show up. "Nothing to see here." "Move along." Same applies to TV. It gathered quite a big crowd; people stopped to watch; got involved. But now, we're told by the relevant authorities (Buonanno, 2016; Katz and Scannell, 2009), it's (nearly) over. Nothing to see here. Much ado about nothing. Move along: to "platform capitalism," if you please (Srnieck, 2017), where "the stack" (Bratton, 2016) will supply all your gawping needs. But, even as we're shoved further into the technological sublime, let's ask whether hanging around to watch is *itself* the thing to see. More formally: is there one thing that justifies television as an integrated object of study? The question is inevitably about knowledge (how we know) as much as it is about phenomena (what we know). TV's characteristics are multifarious, as are the disciplinary or scientific methods used to study it, as well as the critical positions from which it has been assessed. TV resists a singular ontology; and to watch it is to think with it – it's a site of everyday, practical epistemology, logic, metaphysics and aesthetics. If television does affect knowledge, do philosophers, guardians of ontology and epistemology, welcome it? Or do they react like security guards, protecting their stretch of the long road to knowledge?

Jason Mittell has reduced the object of study to six "functions" for television: it is "a commercial industry, a democratic institution, a textual form, a site of cultural representation, a part of everyday life, and a technological medium" (2009: 2). Concentrating on what TV is *for* is a good move, but can we go further? This chapter pursues the idea that what unifies and justifies television as an object of study is its *cultural function*, which is, I argue, *the making of the modern subject*. This is not to dismiss but to explain these industrial, institutional, formal, representational, quotidian, and technological functions at macro-system level, where TV has been used:

- to translate modern life into a shared, imagined, meaningful universe (Lotman, 1990)
    - for a given we-community or "deme" (Hartley and Potts, 2014),
        - by creating (making; representing; distributing; modelling; motivating; adjusting)
            - the modern subject (an unprecedentedly *knowing* subject).

That meaningful modern universe is both *parochial*, expressive of "our" group identity, especially in contrast to "they" – adversaries (internal and external) – and, at the same time, *universal*, capable of facing any uncertainty or problem: from daily surprises (news, sport), through abiding challenges and conflicts (drama, comedy), all the way up to myth and law-affirming events (ritual, festival). Meanings are made and marketed, shared among unprecedentedly large and heterogeneous populations, such that the personal act of enjoying fictions is also a group action, *constituting* the group's identity at macro-scale, as another of humanity's determining "fictions," in Yuval Noah Harari's sense (2015). Television is not an object in nature but is *constituted in discourse*, realized in social arrangements, and *reconstituted over time*. The TV audience is also a discursive "fiction" (Hartley, 1992: 101–18). The sense of co-subjectivity among millions who share a program or live event is imagined (Dayan and Katz, 1992). You can't meet the audience. But imaginary doesn't mean illusory. Harari argues that "fictions" are humanity's distinctive mode: by means of language and stories (Dor, 2015), *Homo sapiens* creates entities (systems and groups) that don't exist in nature – e.g. the law, religion, firms, money, rights, nations – and uses these to ensure cooperation in the growth of knowledge. The television audience is the largest such fictional group ever imagined, often thought of as coterminous with the species as a whole. Stories about it abound, but you'll never meet it face-to-face; and television is the most complete story-telling machine ever invented.

Meanings are carried to all quarters by what I've called the "twin energies" of modernity: *freedom* (intellectual emancipation, both individual and social) and *comfort* (material wellbeing, both household and market) (Hartley, 1996). These aspects of modernity have evolved and embedded themselves since the Renaissance and Enlightenment, taking the fruits – bitter and sweet – of Western expansion to the wider world: industrial capitalism, imperial and colonial power; struggles with persistent problems of identity, equality, mobility, and opportunity; never able to extend the mantle of modern freedom and comfort to all human agents or systems, even as the idea of it was globalized along with trade.

Television has globalized ahead of both freedom and comfort for most. Nevertheless, it's the fiction that binds: television has the knack of seeming to address and to include all of us (however construed); and when it doesn't, as inevitably will be the case, audience members may aspire to join the club, or they may protest at being neglected (or both). For all its fictional make-believe and uneven representational accuracy, we want television to be *true* about who *we* are. What makes television compelling as an object of study is not any particular textual or industrial forms (which come and go), intriguing as these continue to be, but the one thing that makes them all dangerous, risky and attractive: television's *popularity* (its population-wide reach). Its most important invention is the audience, this being the locus of the entire system's cultural functionality.

Television as a cultural form was established among the winners of World War 2. Now its audience is planetary, just at the moment when it is dawning on people everywhere, if not on their public representatives, that the planet itself is being shaped, possibly catastrophically, by our species' methods of valuing, achieving, exporting and enjoying modernity's fruits. What happens when *everyone, everywhere* aspires to modern freedom and comfort? What are the limits and alternatives to modernity? And how can we (audience, citizen, public, consumer, user, maker, species) decide what *collective action* is needed, and how do we know? Here, television lines up alongside other planetary communication systems (science, fiction, publishing, the press, social media) as both source of and solution to the problem: the very medium that brought us together in pursuit of the modern is – we must hope – the means by which we can learn its limits.

Popular broadcast television has been followed in turn by computational (digital) and networked (internet) forms, including apps, platforms and social media where, once again, the most important invention is not so much the technology, mind-boggling though that can seem, but *the user*, the metaphoric point of intelligibility and anxiety, around which technology, institution, and cultural function cohere. The user is routinely theorized as just another *Homo economicus*, that "self-contained globule of desire" (Veblen, 1919: 73) beloved of equilibrium economics, although "the user" (unlike "the audience") is endowed with agency or expertise to produce as well as to consume. But this abstract figure (like the audience) is still only significant in the aggregate. Users attract the usual aura of risk, danger and attractiveness when they go viral. Popularity (the "fiction" of group scale) drives the system, not individual behavior or even technological change.

Whether user-based media count as "television," or as something new, depends on whether the current decentralized, user-led but corporately globalized system of digital-mobile-internet connectivity is accepted as *part of modernity*, or as evidence of its demise. Television's successes and failures will tell us much about the prospects for modern life in the global era.

## "Round Up the Usual Subjects"?

One way of tackling a multivalent knowledge terrain would be to "round up the usual suspects" (*Casablanca*, 1942), not to constitute an *object* but a *field* of study. An iconic example of a literature review that constituted a new epistemological field is Alfred Kroeber and Clyde Kluckhohn's anthropological survey of 164 different concepts of "culture" (1952). Since I'm searching for television's *cultural function*, this seems a good place to start. Their effort to pin down what culture might mean in nature (as it were; i.e. not just in theory) was conducted in the post-World War 2 geostrategic context, just as television took off, where the newly dominant world power – the *pax Americana* – pursued global goals that were mutually incommensurable:

- egalitarian ideals (freedom for all);
- industrial ambitions (world markets);
- imperial responsibilities (world order).

US foreign policy operated across these contradictory objectives partly by trying to systematize knowledge of the cultures they encountered (a task of national intelligence that is still discernible in the CIA's annual *World Factbook*),[1] partly by funding new transnational institutions (the UN, Peace Corps, CIA), and partly by fictionalizing the tensions involved and projecting them into imaginary worlds for this planet's entertainment market, all the way from warmongering *Casablanca* to a prolonged reflection on the possibilities – philosophical rather than imperial – of non-lethal encounters with diverse others, as explored in *Star Trek* (Pearson and Davies, 2002: para. 10).[2]

Determining the "nature of culture" had been a perennial but parochial question for cultural critics for a century or more. It was made more urgent and more abstract by the times, because culture was not simply custom or national heritage but "live:" culture had played its part in totalitarian politics; culture was a *constitutional* issue for newly liberated, oppressed and self-determining groups; culture needed to be handled by policymakers, governors and occupation forces employed by the US Government, rendering it at once a universal experience and a source of local controversy. How should America behave in Europe, the Middle and Far East, and toward the so-called second world (Communist bloc) and third world (decolonizing and developing countries)? This question was political and strategic, of course, but it was also a question of culture: how far to rely upon, to respect or seek to change, the many conflicting cultures emerging from total war; and how far to continue to rely upon the informal entertainment market for strategic purposes, as had occurred systematically, and effectively, in "the Hollywood War" (Viotte, 2013).

Deciding what counts as culture could no longer be determined by magisterial readings, as had been the trend in European high literary and art criticism (e.g. Arnold, Ruskin, Burckhardt, Pater), whose purpose was to train discriminating judgement, especially among governing classes (Quiller-Couch, 1920), and whose values were thought to apply everywhere, at least by advocates. In a postwar, democratizing, pluralist, mixed economy, culture's definition needed to be derived not from Empire but from *usage*. The available tool for that was anthropology, itself rooted in the experience of modern imperial expansion.

Kroeber and Kluckhohn synthesized multifarious usages into one rule, thereby *codifying* culture, turning it from parochial history to universal scientific present tense. They were able to produce a serviceable definition, not quite an algorithm but nevertheless an abstract formula, of "culture systems," based on their function for "human groups":

> Culture consists of patterns, explicit and implicit, of and for behavior acquired and transmitted by symbols, constituting the distinctive achievements of human groups, including their embodiment in artifacts; the essential core of culture consists of traditional (i.e. historically derived and selected) ideas and especially their attached values; culture systems may, on the one hand, be considered as products of action, on the other, as conditional elements of future action.
>
> *(Kroeber and Kluckhohn, 1952: 181)*

Hmm, interesting: that might even work for television! TV didn't count as culture when this definition was published, but it was entering its heyday as the most popular pastime in the history of the world; part of everyday lived culture. It was regarded by cultural critics as a threat to judgement and taste, but TV too displays "patterns … behaviour … symbols … achievements … groups … artifacts … ideas … values." Like "culture systems," the *television system* is produced by past actions and that system determines future actions (i.e. causation is systemic, evolutionary, path-dependent and dynamic).

Kroeber and Kluckhohn's generalized abstraction of culture as a macro-system allowed critics to get off their high horses and look at contemporary life. Here's where cultural studies came in. French theorist Henri Lefebvre (2014) was among the first to perform the trick of linking the supposed trivia of modern "everyday life" with philosophical thought and questions about power, dominance and transgression, such that "the everyday" is a *causal* force in making the modern subject:

> The quotidian is what is humble and solid, what is taken for granted and that of which all the parts follow each other in such a regular, unvarying succession that those concerned have no call to question their sequence; thus it is undated and (apparently) insignificant; though it occupies and preoccupies it is practically untellable, and it is the ethics underlying routine and the aesthetics of familiar settings.
>
> *(Lefebvre, 2000: 24)*

The centerpiece of those familiar settings was the TV console: supplier of tellable stories; teacher of ethics, aesthetics and sequence. These were television's salad days of optimistic expansion, when *I Love Lucy* and *Sergeant Bilko* ruled the roosting suburbs, and popularity meant "population-wide." By 1955, a majority of households in the USA boasted at least one TV set; by 1960 it was 90 percent (Spigel, 1992: 1; 32). During that period, TV-sets were the hardware and TV-shows the software for promoting everyday suburban living, centered on *producing* "modernity, technology and the comfortable life" (Spigel, 1992: 18–26; 30ff.). There was little thought in the USA about how all this might play *beyond* Peoria, but play it did, across the Americas and

Europe, where the Americanization of everyday life was putting the skids under the only remaining non-Anglophone world empire, at least according to French critics from Lefebvre and the Situationists to the cinematic New Wave. "The *colonization* of everyday life" referred to the continuation of the structures of imperialism within domestic household routines: consumerism and technology promoted cleanliness, efficiency, interior isolation and American capitalism at the micro-scale (Ross, 1994; Olson, 2009). Among the responses to that process were the philosophical, critical and artistic ideas that led to the politicization of the everyday and thence postmodernism.

## Paedocracy

In this effort to "colonize the everyday," the most important agents of adoption were children. Lynn Spigel quotes *Advertising and Selling*, an industry journal, which reported as early as 1948 that "children not only exert a tremendous amount of influence in the selection and purchase of television receivers but that they are, in fact, television's most enthusiastic audience" (qtd in Spigel, 1992: 59–60).

When doubts arose concerning that enthusiasm, as they soon did, television studies did not yet exist to guide public debate (Gray and Lotz, 2011). In fact, the perceived problem was not *on* television; it was the juvenile "subject" lolling on the couch *watching* it. In the USA, it fell to existing social sciences to diagnose the upstart medium's invasion and colonization of children's leisure time (Wartella and Mazzarella, 1990), and its feared pathological effect on their worryingly malleable subjectivity. The pathologizing tone was set by psychiatrist Fredric Wertham's influential denunciation of comic-books, *Seduction of the Innocent*, which had a chapter on "Homicide at Home: Television and the Child" (1954: 353–83). Wertham found "bad things" even in innocent places: "crime is crime and violence is violence even in the patriotic setting of a Western locale or in the science-fiction setting of interplanetary space" (1954: 370). In short, the trouble with television was not television but culture. The child delivered to the screen was already "corrupted":

> The greatest obstacle to the future of good television for children is comic books and the comic-book culture in which we force children to live. If you want television to give uncorrupted programs to children you must first be able to offer it audiences of uncorrupted children.
>
> *(Wertham, 1954: 383)*

Hilde Himmelweit was the first career specialist in TV Studies in the UK, following her landmark study of *Television and the Child* (1958). Her *disciplinary* focus wasn't on television. She installed the child as the point of intelligibility for television studies, continuing the American trend (personified in such figures as George Gerbner, Sandra Ball-Rokeach, Ellen Wartella, Charles Osgood and Percy Tannenbaum), rather than the French one. Himmelweit enjoyed considerable academic celebrity and policy influence (Caplan, 1996), creating a path-dependency for television studies that was not rivalled until French Theory finally broke through in the 1970s and 80s. She was rather dismissive about changes in television "itself"; they were "superficial" compared with "children's lives": "The role of television in children's lives, the manner of children's reactions and the underlying principles that determine them remain constant in the face of the superficial changes in television itself" (Himmelweit, Oppenheim, and Vince, 1958: xiii). The book's confident subtitle, *An Empirical Study of the Effect of Television on the Young*, is contradicted in the telling: despite "the effect of television," childhood remains "constant," playing a "role" in children's daily lives, but by no means changing them. Himmelweit's *method* was "empirical," chiefly consisting of 4,000 interviews with children, parents and teachers in five

English cities. The researchers found that television was already the number-one leisure activity among children, who watched mostly adult shows, where they were exposed to Wertham's worst fears: sex, violence and bad language; although the children reported that they used TV mostly as a "time-filler" (Himmelweit, 1958: 15).

The interviews probed concerns about TV's impact on children's anxieties, outlook and education, finding that watching TV had a "solid and consistent" impact on their lives (Himmelweit, 1958: 245). The researchers were given access to children's IQ tests (a practice that's no longer acceptable: Livingstone and Bovill, 1999, Ch. 1: 4, fn5) and found that the "higher" a child's "intelligence" the less TV they watched, leading to generations of rationing for middle-class or aspirational-parented kids. Himmelweit recommended that parents should protect children from the "risks" of television; they should supervise program choices; and guide children toward "quality" programming (although the children didn't mention this), in order to learn how to distinguish "make-believe" from reality (as if semiosis is not real). Thus, right at the outset, television was deemed important – was constituted as an object of study – not because of its commercial, democratic, textual, representational, quotidian, or technological properties (Mittell, 2009), or its colonization of everyday life (Lefebvre, 2014), but because of its presumed role in forming the *modern subject – at risk.*

Forty years later, Sonia Livingstone, also at the London School of Economics, undertook a follow-up study, just at the moment when television was giving way – at least in terms of public disquiet – to "new media." Her research team "tried to put to one side the considerable public anxieties surrounding our key terms – 'children', 'youth', 'new' and 'media'". Instead, they "set out to contextualise children and young people's meanings and uses of new media in relation to their 'environment' or 'life-world'" (Livingstone and Bovill, 1999: Ch. 1: 2). Perhaps not surprisingly, they came to the same conclusions as did Himmelweit:

> The case for change should not be overstated. Each decade may see dramatic technological change, but in many respects children's lives are as they were ten or even forty years ago. Children grow up, watch television, ride their bikes, argue with their parents, study hard or become disaffected with school, just as they always did.
>
> *(Livingstone and Bovill, 1999: Ch. 1: 3)*

Where "larger changes" are discerned, "these are often only indirectly connected with new media technologies." They are left as open questions:

> Even larger changes are also at work, as globalising economic, political and technological developments challenge the autonomy of the nation state. How are we to link all these changes? Does lack of freedom to play outside influence time spent watching television, or does use of global media impact on consumerist values, or does children's new-found expertise with computers affect family authority?
>
> *(1999: Ch. 1: 3)*

Like Himmelweit, Livingstone is drawn not to "television" or even "media," but to "larger changes" that are seen as global and national, within which the modern subject faces universal pressures: freedom to play, consumer values, family authority. These questions about freedom and comfort impelled the direction of Livingstone's later research, focused around the large-scale *EU Kids Online* project.[3] Thus, "the child" – an abstract, universal fiction created by "public anxieties" – continues to drive and direct the research agenda in television studies, even while the real action seems to be going on at global scale, challenging the "autonomy" of nations and the "authority" of families, while "everyday life" itself is increasingly hard to separate from "media":

> New forms of engagement between user and medium may contribute to the gradual shift from a clear distinction between mass communication and face-to-face communication to a more diversified, participatory, active notion of mediated communication in everyday life.
>
> *(1999: Ch. 12: 11)*

Once semiosis and reality are recognized as one and the same, as French theory had realized from the start, then there are consequences, both for *modernity*, which was skeptical about that, and for *everyday life*, which becomes the site of struggle, not only between global colonization and local authenticity, but also between "order" and "transgression." Amanda Third and Philippa Collin return to Lefebvre's "everyday" as a "naturalised" but "totalising" sphere, consisting of "pressures and repressions at all levels, at all times and in every sphere of experience including sexual and emotional experience, private and family life, childhood, adolescence and maturity" (Lefebvre, 2000: 145). In other words, it's the site of "order." But, they note, it is also a bit wild:

> Lefebvre ... theorises the everyday as simultaneously a mechanism of dominant order—a tool of repression—and a site for the transgression and deconstruction of that order. For Lefebvre, the everyday is both "he point of delicate balance and that where imbalance threatens" (2000: 32).
>
> *(Third and Collin, 2016: 51)*

"Dominant order ... repression ... transgression ... deconstruction ... imbalance"; undermining dominance and unbalancing order even as it applies them to the general population – it all sounds a bit *postmodern*. So here's where television and philosophy intersect once again: television is not only an agent of a modern order that "evades critique," it is also an agent of deconstruction and transgression. Just as the act of watching television has fragmented across devices and platforms, and the identity of the viewer diversified, so the recognition of what modernity comprises and how it works has expanded to include its own erstwhile "other" – postmodernity.

What's new about television in the unpredictably unfolding twenty-first century is that its popularity is both greater than ever (across more countries, modes of consumption, and aggregate numbers of "eyeballs") and more fragmented and dispersed than ever. Even in its most successful national market, the USA, broadcasters no longer seek to gather populations into one simultaneous "imagined community" (Anderson, 2006), as *Dallas* or *M*A*S*H* once could, with *Dallas*'s famous "53 share" of TV sets (a majority of US households) for the 1980 "Who shot JR?" episode, or *M*A*S*H*'s finale in 1983, which scored nearly 106 million viewers. Now it's not drama but a special or fatal event – an assassination, terrorist attack, natural disaster, celebrity court-case or, most likely, a sporting final – that can bring impressive audiences together in the same time and place. Post-broadcast forms have internationalized these audiences, but such ratings only mask increasingly polarized demographic contours (Pew Research, 2014). TV now serves both to separate and to interconnect many disconnected knowledge-making "demic" groups; who know one another by semiotic and rhetorical means, not necessarily by national or ethnic boundaries.

## Television and Philosophy

Thus, postmodern uncertainty is not so much a departure from modernity as a way of integrating "how we know" with "what we know." The modern subject nowadays is not only the "universal" figure of Enlightenment-Industrial-Imperial freedom and comfort, but also *reflexive* (able to self-interrogate) and *positioned* (conscious of identity; partisan): thus, not universal at all. There's a strong relationship between general populations and the persuasion industries (both political and commercial), but it's not one of cause and effect. They can't make you do or think things; but they're in the business of making *you*. The idea that an industry can *make* something (subjectivity, or modern agency at population scale) that it does not *manufacture* (programs, etc.) may seem unempirical and postmodern from a positivist perspective. The role of language, media, textuality and discourse in shaping subjectivity cannot be observed directly: empirically, it seems to be an effect without a cause. However, we should recall that this is how television entered formal knowledge in the first place, through the figure most often thought to be damaged by watching television: the child. TV continues to be held responsible for "corrupting" children and is held to blame for social ills and evils, whether or not individual perpetrators have been watching.

Scholarship has been part of the problem; using the child as a handy metaphor for discounting popular media and culture as "childish," without inquiring too deeply into what enables children to construct and perform an identity (e.g. gender, ethnicity, etc.), relationships (trust; translation), meaningfulness and thence knowledge in modern times, using the discursive resources at hand (Butler, 1990). Philosophy was slow to take notice, because it denied that children "possess the capacities of *philosophical* subjects, that they have the attributes of 'knowing' subjects independent of their formation and training as social beings" (Hirst, 1979: 67). If you take modernity not only as Habermas's "incomplete project" for training these "social beings" (d'Entrèves and Benhabib, 1997: 39–55), i.e. not only as the outworking of Enlightenment communicative rationality, and not only as modern-*ism* in the arts, literature, architecture (etc.), but also as the modern-*ization* of the economy (*vide* Deirdre McCloskey, 2006, 2010, 2016), politics (Robert Darnton, 1995) and society (Immanuel Wallerstein, 2004; Henri Lefebvre, 2014), then the creation of modern subjectivity and its distribution across global populations becomes a most interesting phenomenon, even while avant-garde thought was criticizing the very notion of modernity in the name of postmodern doubt and skepticism about progress. We're used to the idea that modernity is a result of explosive revolutions (Industrial, French, etc.), but its distribution has relied too on banalities: markets, schools, watching telly. Children were "knowing subjects" throughout this part of the process. Some of them, on screen and off, scripted or spontaneous, *performed* "the capacities of *philosophical* subjects," as characters, comics and commentators in ads, shows and news, articulating modernity's happy smiling hope and bearing its despairing failures, depending on the genre.

Television systems wanted as many people to watch as possible. TV was good at ignoring demographic and partisan differences by appealing to "childlike" attributes in its address to audiences young and old – it was a "paedocratic" discursive regime (Hartley, 1992). But scholarship was skeptical of exactly that skill. Thus, the first problem to face early television scholars was that of the worthiness of the object of study. Professional philosophy was no more interested in television than it was in children (Hall, 1996; Hirst, 1979). When it did take an interest, the results were frequently disappointingly prejudicial (Botton, 2014; Bourdieu, 2001). Television shows were quite interested in philosophers, e.g. *Men of Ideas* (1978) or *The Great Philosophers* (1987), both by Bryan Magee for the BBC. Philosophers – especially French ones – were not shy about appearing on television (Chaplin, 2007). But TV was routinely dismissed as an "unworthy object" (Caughie, 1990: 55).

However, philosophy has been defined as the "systematic and critical study of fundamental questions that arise both in everyday life and through the practice of other disciplines" (Brown University, n.d.). If so, then it shares common cause with television, since both pursue questions that arise in everyday life and in other disciplines (Slade, 2002). Of course, there's room for debate about what might count as "systematic and critical" study. Do chat shows or journalistic interviews, dramatic scenarios and conflict, comedic critique or take-down, sporting contests and display, constitute such study? They might: critique may take vernacular forms. That's what TV

Studies has been working through since the 1970s. In short, traditional disciplinary knowledge has drawn too tight a boundary around its object and its methods, excluding both children and television, even as contemporary media were accumulating capabilities that might challenge its authority, and considering topics worthy of its most "systematic and critical" attention. With the emergence of interactive and multi-device digital media, this tendency could only accelerate, as viewers could binge-watch TV shows, with Google, Wikipedia, IMDb and social media at hand to settle arguments in a pleasingly Socratic manner.

## What Does it Mean to Be *Humans*?

This raises the question of how TV is used by contemporary viewers, makers and institutions, to consider the prospects and probabilities facing the future of the post/modern subject; a future that, all the experts agree, is uncertain. Where television in its expansive, aspirational phase was about *creating and teaching* the modern subject, television now is more reflexive, mindful of post-human, post-apocalyptic possibilities. Some of its best shows, in an un-systematized but accelerating wave of worry, are thinking through *what will become of the modern subject.* Perhaps TV always has pursued that question, not only through science and philosophy shows (gadgets, wonders, dangers, wacky costumes), but also in the guise of speculative entertainment, guided by *Star Trek, Buffy the Vampire Slayer, Game of Thrones*, and *Westworld* in the USA; or *Quatermass, Doctor Who*, and *Humans* in the UK.

As a global medium, television has been well-placed to ask the general question: *What does it mean to be human in an era of power?* – a question shared by cultural studies. Speculative TV drama has proven to be an efficient – that is, powerful, gripping, mind-blowing – machine for exploring problems of the modern (global) and postmodern (knowing) subject.

In the Cold War, enmity was imagined between adversarial cultures: e.g. the "Kohms" (Communists) *versus* "Yangs" (Yankees) of *Star Trek*'s "Omega Glory" episode, aired during the Tet Offensive in Vietnam, March 1968. Now, when real enmity seems to have turned inward, tribal, and intercommunal (i.e. post-imperial, for those who read Gibbon), bespeaking internal collapse in the midst of environmental catastrophe and mass extinctions, the *fictional* battle is pitched at species scale: humans *versus* the planet or other species; humanity *versus* the undead, aliens, robots or artificial intelligence. Honestly, who needs philosophy when you've got *Westworld* (HBO) and *Humans* (Channel 4 UK) to think with? That transatlantic pairing can serve to illustrate not only the importance of the questions raised in television entertainment, but also the strong contrasts to be found among their worldviews. Table 1.1 charts what I take to be profound differences between them.

How the relations between human and non-human agents are imagined invokes strong differences between US and European cultural preoccupations at system level. Both shows address an overall problem; an unresolved social anxiety and running sore of history, in which the boundaries of the human are not settled. In *Westworld* it is slavery; in *Humans* it is class. If a robot "host" (*Westworld*) can be owned, killed, sexually abused, and its consciousness (speech, character, actions) be tuned to the whim and pleasure of the owner, then what we are watching is *slavery* – AI as source of consolidated wealth and power; the other as ant. If, on the other hand, a "synth" (*Humans*) can, despite irreducible otherness, become part of a family, join with other synths in solidarity, struggle for rights, and feel love for a human child, then here is *class* – AI as the productive source of dynamic change, knowledge and meaning; "the other" as holder of rights and obligations.

The two series differ markedly in their narrative arcs, with characters driven by different motivations (individualism vs. mutual obligation); relationships based on sex (transmission) or love (translation); resolution achieved by death (winning) or compromise (cohabiting). *Westworld* (in its first season) adopts the narrative point of view of the owner/producer (how do I exploit; and will that destroy me?); *Humans* that of the user (how do I learn; and will that

*Table 1.1* Westworld vs. Humans

| | *Westworld (HBO, 2016)** | *Humans (C4/AMC, 2015–2016)** |
|---|---|---|
| *National culture* | USA | UK/Europe* |
| *… defining tension* | **Slavery**, power | **Class**, meaning |
| *Robot characters* | "Hosts" | "Synths" |
| | Dolores (Evan Rachel Wood) | Anita/Mia (Gemma Chan) |
| | Maeve (Thandie Newton) | Niska (Emily Berrington) |
| *… motivation of* | Individualism | Mutuality |
| *… self-realization of* | Making a decision/choice | Joining a group |
| *… relationship* | Sex, death | Love, accommodation |
| *Children? (human) †* | Little Boy (host) | Three Hawkins children |
| *Outcome (dénouement)* | Catastrophe (winning) | Compromise (cohabiting) |
| *Perspective on robots* | Power, owner ("they") | Culture, user ("we") |
| *Kagan ID ‡* | "Mars" (hegemony, force, US) | "Venus" (law, institutions, EU) |
| *Model of communication* | Transmission (Claude Shannon) | Translation (Yuri Lotman) |

* Both shows were preparing further seasons at time of writing: *Westworld* S2 and *Humans* S3 were released in 2018. The UK series *Humans* is adapted from a Swedish original by Lars Lundström (2012) called *Äkta Människor* (*Real Humans*), where the synths are called *hubots*.

† In Westworld S1 there are no *human* children, but one of the hosts is Little Boy (Oliver Bell), a copy of Anthony Hopkins' character Robert Ford (http://westworld.wikia.com/wiki/Little_Boy). In *Humans*, the children belong to the central Hawkins family: Sophie (Pixie Davies) is the youngest (aged 8–9); she welcomes the synths and is especially fond of Mia, who she named Anita. Older brother Toby (Theo Stevenson) is attracted to and protective of Anita; eldest daughter Mattie (Lucy Carless) is strongly motivated to discover more about the synths using her computer hacking skills; she becomes a major link between them and the human world. A child synth, Sam (Billy Jenkins), is introduced late in S2, to become more significant in S3.

‡ Kagan ID refers to Robert Kagan (2003), who famously argued that "Americans are from Mars; Europeans are from Venus"; see also *Policy Review*, 113 (2002), and 172 (2012).

complete me?). In the context of public disquiet about automation, the unsettling question for both shows – and you should watch both, preferably in the same timeframe – is this: what if the post-human other is not a technology but a class? Of course, the answer is already prepared: it's called *The Handmaid's Tale* (Hulu). Here we see humans reduced to a function, chattels of a power-elite, isolated and speechless, but nevertheless struggling to make common cause with strangers in a bid for emancipation. Offred (Elisabeth Moss) knows what it feels like to *become* the other – slave or class or robot. And, despite television's own institutional enslavement to the "dominant order", it's pretty clear whose side we're on, as a culture.

Speculative fiction augments reality, as post-broadcast television nurtures the post-human imagination. Today's corporate and cultural failures precipitate tomorrow's dystopia, where the AI classes may prove more humane than we are; and the destruction of humanity (as we know it) may prove to be our only hope. Thus, the question "What is television?" remains pertinent long after TV ceased to operate along strictly broadcast, mass-society lines, and well into the era of postmodern mediatization, with simultaneous globalization and fragmentation of platforms, languages, audiences and senses of collective self. Television as a world system continues to construct and reflect on modern subjectivity. Most compellingly, as we hang around to watch, it is making visible – and therefore open to question – how dangerous that subject is to the world and to other systems. It allows us to crowd-source our deliberations about what should become of the Western-oriented "we" of modernity.

There are many who have enjoyed the fruits of freedom and comfort but who fear for what their children may inherit at planetary scale. The prospect of the unfurling Anthropocene era, when we (*Homo sapiens*) become our own worst enemies, globalizing conflict, climate change and environmental catastrophe, and imposing cruelties on "they" classes from migrants to handmaids to robots, presents a doubtful future for our own children. Television has routinely pursued this unquiet thought through its most visible discursive elaboration, not philosophy directly (Wark, 2015), but the pop-culture streets and sets where philosophy and children habitually meet. What becomes of our children there? Do they succeed, or succumb?

## Notes

1. The CIA's own history of the *World Factbook* includes this justification:

> The need for more comprehensive basic intelligence in the postwar world was well expressed in 1946 by George S. Pettee, a noted author on national security. He wrote in *The Future of American Secret Intelligence* (Infantry Journal Press, 1946, page 46) that world leadership in peace requires even more elaborate intelligence than in war. 'The conduct of peace involves all countries, all human activities—not just the enemy and his war production.'
>
> *See: www.cia.gov/library/publications/the-world-factbook/docs/history.html*

2. For a flavour of the racial politics of *Star Trek* over its 50 years, see this interview with Nichelle Nichols (Lieutenant Uhura) given to *The Guardian* in her 80s: www.theguardian.com/tv-and-radio/2016/oct/18/star-trek-nichelle-nichols-martin-luther-king-trekker; and this review in the same outlet (by Sam Thielman) of a 2017–2018 TV/streamed revival of the franchise in *Star Trek: Discovery*:

> I would never accuse a contemporary television show of not being self-aware enough to elide the various unflattering cultural comparisons that a wildly otherized violent alien race is more or less guaranteed to call up. … *Discovery*, I guess in the name of a more grave and serious show about alien monsters and time travel, ostentatiously walks a darker path, and on that path are a lot of our worst tendencies.
>
> *See also: https://en.wikipedia.org/wiki/Star_Trek:_Discovery*

3. Some of the many publications arising from the EU Kids Online Project and associated research can be accessed here: www.lse.ac.uk/media@lse/research/EUKidsOnline/EU%20Kids%20Online%20reports.aspx; and here: www.lse.ac.uk/media@lse/research/EUKidsOnline/Home.aspx.

## References

Anderson, B. (2006) *Imagined Communities: Reflections on the Origin and Spread of Nationalism*, Revised edition. London: Verso.

Botton, A. de (2014) *The News: A User's Guide*. London: Pantheon.

Bourdieu, P. (2001) Television. *European Review*, 9:3, 245–256.

Bratton, B. (2016) *The Stack: On Software and Sovereignty*. Cambridge, MA: MIT Press.

Brown University (n.d.) Philosophy: What and Why? Accessible at: www.brown.edu/academics/philosophy/undergraduate/philosophy-what-and-why.

Buonanno, M. (ed.) (2016) Not Yet the End of Television. *Special Issue of Media and Communication*, 4:3. Accessible at: www.cogitatiopress.com/mediaandcommunication/article/download/661/661.

Butler, J. (1990) *Gender Trouble: Feminism and the Subversion of Identity*. New York: Routledge.

Caplan, S. (1996) Hilde Himmelweit (1918–1989). In N. Signorielli (ed.), *Women in Communication: A Biographical Sourcebook*. Westport, CT: Greenwood Press, 216–219.

Caughie, J. (1990) Playing at Being American: Games and Tactics. In P. Mellencamp (ed.), *Logics of Television: Essays in Cultural Criticism*, Bloomington, IN: Indiana University Press; London: BFI, 44–58.

Chaplin, T. (2007) *Turning on the Mind: French Philosophers on Television*. Chicago, IL: University of Chicago Press.

Darnton, R. (1995) *The Forbidden Bestsellers of Prerevolutionary France*. New York: W. W. Norton and Co.

d'Entrèves, M.P. and Benhabib, S. (eds.) (1997) *Habermas and the Unfinished Project of Modernity: Critical Essays on the Philosophical Discourse of Modernity*. Cambridge, MA: MIT Press.

Dayan, D. and Katz, E. (1992) *Media Events: The Live Broadcasting of History*. Cambridge, MA: Harvard University Press.

Dor, D. (2015) *The Instruction of Imagination: Language as a Social Communication Technology*. Oxford: Oxford University Press.

Gray, J. and Lotz, A. (2011) *Television Studies*. Cambridge: Polity Press.

Hall, S. (1996) Who Needs 'Identity'?, In S. Hall and P. Du Gay (eds.), *Questions of Cultural Identity*. London: Sage, 1–17.

Harari, Y.N. (2015) *Sapiens*. London: Harvill Secker.

Hartley, J. (1992) *Tele-ology: Studies in Television*. London: Routledge.

Hartley, J. (1996) *Popular Reality: Journalism, Modernity, Popular Culture*. London: Arnold/Bloomsbury.

Hartley, J. and Potts, J. (2014) *Cultural Science: A Natural History of Stories, Demes, Knowledge and Innovation*. London: Bloomsbury.

Himmelweit, H., Oppenheim, A. and Vince, P. (1958) *Television and the Child: An Empirical Study of the Effect of Television on the Young*. Oxford: Oxford University Press.

Hirst, P. (1979) *On Law and Ideology*. Basingstoke, UK: Macmillan.

Kagan, R. (2003) *Of Paradise and Power: America and Europe in the New World Order*. New York: Alfred A. Knopf.

Katz, E. and Scannell, P. (eds.) (2009) The End of Television? Its Impact on the World (So Far). *Annals of the American Academy of Political and Social Science*, 625. Accessible at: https://uk.sagepub.com/en-gb/eur/the-end-of-television/book234723#contents.

Kroeber, A.L. and Kluckhohn, C. (1952) *Culture: A Critical Review of Concepts and Definitions*. Cambridge, MA: Peabody Museum of Archaeology and Ethnology Papers, 47 (1).

Lefebvre, H. (2000) *Everyday Life in the Modern World*. London: Bloomsbury (first published in French, 1968).

Lefebvre, H. (2014) *Critique of Everyday Life: The Three-Volume Text* (first published in French, Vol I: 1947, Vol II: 1961, Vol III: 1981). London: Verso.

Livingstone, S. and Bovill, M. (1999) *Young people, new media*. Report of the research project *Children Young People and the Changing Media Environment*. London: LSE Department of Media and Communications: http://eprints.lse.ac.uk/21177/.

Lotman, Y. (1990) *Universe of the Mind: A Semiotic Theory of Culture*. London: IB Tauris.

McCloskey, D. (2006, 2010, 2016) *The Bourgeois Trilogy*, 3 vols. Chicago: Chicago University Press.

Mittell, J. (2009) *Television and American Culture*. New York: Oxford University Press.

Olson, L. (2009) *Modernism and the Ordinary*. Oxford: Oxford University Press.

Pearson, R. and Davies, M.M. (2002) A Brave New World a Week: *Star Trek*, Cult Television, Master Narratives and Postmodernism. In P. Le Guern (ed.), *Les cultes médiatiques: Culture fan et œuvres cultes*. Rennes (France): Presses universitaires de Rennes, 263–280. Accessible at: http://books.openedition.org/pur/24181.

Pew Research. (2014) *Political Polarization in the American Public: How Increasing Ideological* Uniformity *and* Partisan *Antipathy Affect* Politics, *Compromise and Everyday Life*. Accessible at: www.people-press.org/2014/06/12/political-polarization-in-the-american-public/.

Quiller-Couch, A. (1920) *On the Art of Reading: Lectures Delivered in the University of Cambridge, 1916–1917*. Cambridge: Cambridge University Press.

Ross, K. (1994) *Fast Cars, Clean Bodies: Decolonization and the Reordering of French Culture*. Cambridge, MA: MIT Press.

Slade, C. (2002) *The Real Thing: Doing Philosophy with Media*. New York: Peter Lang.

Spigel, L. (1992) *Make Room for TV: Television and the Family Ideal in Postwar America*. Chicago, IL: Chicago University Press.

Srnieck, N. (2017) *Platform Capitalism*. Cambridge: Polity Press.

Third, A. and Collin, P. (2016) Rethinking (Children's and Young People's) Citizenship through Dialogues on Digital Practice. In A. McCosker, S. Vivienne and A. Johns (eds.), *Negotiating Digital Citizenship: Control, Contest and Culture*. London: Rowman & Littlefield International, 41–59.

Veblen, T. (1919) *The Place of Science in Modern Civilisation, and other Essays.* New York: Huebsch. Accessible at: https://archive.org/stream/placeofsciencein00vebl.

Viotte, M. (2013) *La Guerre d'Hollywood 1939–1945: Propagande, patriotisme et cinéma.* Paris: Editions La Martinière. DVD. Accessible at: www.worldcat.org/title/hollywood-war-1939-1945/oclc/914462121.

Wallerstein, I. (2004) *World Systems Theory: An Introduction.* Durham: Duke University Press.

Wark, MCK. (2015) The Hacker: Interview by Richard Marshall. *3:AMMagazine.* Accessible at: www.3ammagazine.com/3am/the-hacker/.

Wartella, E. and Mazzarella, S. (1990) A Historical Comparison of Children's Use of Leisure Time. In R. Butsch (ed.), *For Fun and Profit: The Transformation of Leisure into Consumption.* Philadelphia, PA: Temple University Press, 183–185.

Wertham, F. (1954) *Seduction of the Innocent.* New York: Rinehart. Accessible at: www.scribd.com/doc/30827576/Seduction-of-the-Innocent-1954-2nd-Printing.

# 2
# WHAT WAS TELEVISION?
## The Global and the Local

*Timothy Havens*

For much of television's history, the medium was caught between two opposing geographical tendencies. On the one hand, television was inherently global: television distribution exceeded national boundaries; the processes and organization of production were borrowed from other locales; and the content itself featured a combination of domestic and imported programming. On the other hand, forces of localization ran through television as well: it was highly regulated and run by governmental or quasi-governmental agencies in most of the world, allowing issues of local cultural policy and preservation to be at least as important as profitability. In addition, television was overwhelmingly a domestic medium, owing at least in part to the small size of the screen and its poor image quality, which made television best suited to small groups of viewers, such as families. The domesticity of television seemed to require content that was largely familiar and quotidian, and consequently deeply rooted in local cultural references, practices, and dialects.

Television's dual geographies explain a good deal of the social debates that global television has raised over the past six decades, including debates about cultural integrity, cultural imperialism, and hybridity, which we will explore in the following pages. While television's global-local tendencies remain important for understanding television today, perhaps the greatest legacy of television's early years lies in the fact that it continues to shape our thinking about the social issues surrounding television globalization today, even as television itself has found ways to move beyond the global-local dichotomy. This chapter will end with some considerations of how television today has resolved this dichotomy.

### Early Globalization

While experiments in television broadcasting trace back to the 19th century (Edgerton, 2007), regular broadcasts only began to emerge in the late 1940s and early 1950 in much of the world. Once introduced, however, television broadcasting spread quickly. By the end of the 1950s, virtually every country of moderate wealth and size operated a broadcaster (Havens, 2006). Commercial stations in Mexico began broadcasting in 1950. The Italian public broadcaster, RAI, began transmitting in 1954, and in Thailand, the government began television broadcasting in 1952.

The explosion in television broadcasters in the 1950s led to a great deal of sharing of technology, practices, and programming between more and less established television markets. Executives and professionals from nations planning to establish television often attended workshops on technical standards, equipment operation, and program production in more established markets. Of course, Hollywood and London were the most popular centers for these workshops, but organizations such as the European Broadcasting Union, serving Western European broadcasters; the International Radio and Television Organization, serving Eastern European broadcasters; and the Asia-Pacific Broadcasting Union also developed at the time to assist member states with technical and programming issues.

The practical effect of all of this information exchange about television was that television broadcasting developed some general similarities worldwide. Four broad models of broadcasting emerged around the world: (1) the US commercial model featured regularly scheduled series with multiple episodes, an emphasis on popular culture, and the appearance of "commercials," or paid on-screen advertisements; (2) the BBC public service model featured mostly teleplays, documentaries, and other one-time programs, an emphasis on social education and cultural uplift, and shorter programming days than their commercial counterparts; (3) the Soviet model, which functioned much like the public service model, except that television was also used for overt propaganda and Communist Party promotion; and (4) the developing nation model, most clearly embodied by the Indian broadcaster Doordarshan, which focused primarily on news and educational programming related to agricultural and economic development.[1]

Not only were the management and technical requirements for each of these systems different, the programming tended to differ substantially as well. The commercial Hollywood system and the Soviet state system were the furthest apart, with the BBC public service model and the Doordarshan development model more similar, as Table 2.1 demonstrates.

It is important to note that these systems are ideal types, and that different broadcasters around the world often exhibited a mixture of multiple systems. By the time the BBC expanded to television broadcasting in 1946, it was already broadcasting several popular radio soap operas, which it brought over to television. Additionally, the commercial US system has, since the early years, been required by law to operate in the "public interest, convenience, and necessity." Finally, as Mihelj and Huxtable (2018) show, socialist broadcasters shared a number of features with public service broadcasters in the West.

*Table 2.1* Broadcast Systems

| | *Commercial* | *Public Service* | *Development* | *Soviet* |
|---|---|---|---|---|
| Main Purpose | Commercial profit | Citizenship and cultural uplift | Economic growth and education | Cultural uplift and party promotion |
| Control | Private, regulated corporations | Government | Government | Government |
| Predominant Programming Type | Popular entertainment, stripped daily/weekly | Educational and cultural, single episode and miniseries | Educational and agricultural | Educational and cultural, party propaganda, single episode and miniseries |
| Funding | Commercial Advertising | Tax (license fee) | Subsidy | Subsidy |

For our purposes, categorizing broadcasters into different types helps us understand the forces that drove early globalization and the social debates that arose in those years, many of which are still with us today. Television programs and schedules tended to develop in similar ways in nations that employed similar systems, and the exchange of programming and ideas within systems was common from the earliest years of broadcasting. That is, similar systems encouraged the globalization of television content, production practices, and corporate function.

Beyond these regional exchanges, certain kinds of television content seemed to exceed the boundaries of the nation and the region and find their way onto program schedules in virtually every corner of the globe. Chief among these were Hollywood films and popular US dramatic series. Television systems in most of the world, including even those operating under the Soviet system during the cold war, often bought rights to major US films and series and programmed them during prime spots. The same went for non-US program imports as well, though they tended to be less numerous than the US imports.

In part, the reliance on imported television programs stemmed from the costs of program production and the vast number of broadcasting hours that needed to be filled. Even for systems that only broadcast a few hours a day and a few days of the week – which was the case for many television systems in the 1950s – the costs of programming 20 or more hours of television per week were overwhelming. However, importing programming is almost always cheaper than self-producing, since one of the oddities of television as a cultural commodity is the fact that the majority of production funding goes into producing a "first copy" of a program, while additional copies are relatively cheap. Because of a vibrant post-war economy and highly commercialized film and radio industries, the United States quickly became a major producer of numerous hours of television that, because they could be sold cheaply abroad, became appealing to international broadcasters, often as "time-fillers." Together, these economic features of television programming encouraged the global circulation of content; that is, once a television series had been produced, it makes sense for both exporters and importers to allow it to circulate as widely as possible.

The 1977 US miniseries *Roots* provides a good example of how and why a television series embedded in a specific national setting could reach an international audience. *Roots* told the story of American chattel slavery from the perspective of a young man captured in Western African and sold into slavery in the United States. Subsequent episodes followed his life and the life of his descendants until the legal end of slavery after the Civil War. *Roots* ran for a total of 12 hours and was broadcast by ABC every night in prime-time for eight straight evenings. It became the highest rated US television series until 1983. Over the next four years, the series appeared in more than 100 foreign markets, where it often became a critical and popular success as well, eventually earning more money from international syndication than it did from domestic ad revenues.

Several elements of *Roots* helped the miniseries succeed abroad. First, the limited number of episodes, as opposed to most American series which ran episode after episode until they were cancelled, meant that the miniseries fit the scheduling practices of public broadcasters and socialist broadcasters much better than did other American fare. Second, the producer broke the series into multiple different lengths, including 12 one-hour episodes and six two-hour episodes, giving foreign broadcasters flexibility in how to schedule the miniseries.

As a story about slavery in the United States, however, *Roots* was perceived by many broadcasters outside of the United States as potentially foreign to their viewers. Of course, dubbing and subtitling helped viewers in non-English markets understand the series. In addition, broadcasters in some markets introduced the miniseries with a panel discussion about slavery, which sometimes touched on issues of violence against racial and ethnic minorities in the domestic market. Beyond these efforts, the historical facts of African slavery were well known enough to make the miniseries at least marginally intelligible to many.

To summarize, then: several elements of *Roots* made the miniseries ill-suited for non-American viewers, including the language of production, certain culturally specific elements of the story, and the way it was scheduled. On the other hand, translation into local languages, the limited run of the series, editing of episode lengths, and the practices of local broadcasters to explain some of the content for viewers all helped overcome the locally specific elements of the series.

As nation after nation introduced broadcast television and sought out cheap content to help fill hours, US television initially experienced a surge of popularity worldwide. However, as domestic services matured, bolstered by import quotas and local funding sources, local television content soon dominated the screen in most of the world, a trend that continued through the 1990s. One measure of the amount of global trade in programming during these early years is the amount of revenue that US producers generated from international outlets. During the 1950s, this percentage rose from 15 percent to nearly 40 percent by decade's end. In the early 1960s, however, this percentage began to decline, as broadcasters abroad ramped up their domestic production, and came to hover around 25 percent until the 1990s, when a new explosion in cable and satellite channels abroad ushered in a second era of television globalization.

## National Identity, Cultural Imperialism, and Television

Nation states in the early decades of television often intensified their localizing tendencies with regulations that favored domestic series and penalized imports. Because it was often funded by national governments, television broadcasting was thought to be central to the development of national identities and polities from the very start. A number of concerns and theories about global television that continue to animate discussions about global television today formed at the time. At the center of these concerns were (a) the forces that compel television programming to travel the globe and (b) the impact of imported programming on local viewers and citizens.

The earliest concerns about television globalization related primarily to news, and were framed as part of broader concerns about Western news media. At the time, Western news agencies, in particular Agence France-Presse, Associated Press, and Reuters, provided much of the world's news, and a number of newly independent nations, recently admitted to the United Nation's Educational, Scientific, and Cultural Organization (UNESCO), objected to the way that news about non-Western nations was reported and framed.

UNESCO created what came to be known as the "MacBride Commission," which issued a report on international communication in 1980, criticizing, among other things, control over global news reporting by Western nations and the imbalance in international flows of television entertainment programming. Both of these concerns pointed to a deeper worry about the cultural autonomy of small nations and cultures in a global television world (Carlsson, 2003).

Concerns about the unequal flows of television programs across borders reflected deeper worries about the influence of TV on viewers. Articulated by a number of researchers under what came to be called the "Cultural Imperialism Thesis," these worries mainly focused on efforts by the US and other Western nations to create cultural change among citizens around the world, changing them into willing consumers of Western products and political ideologies. Supporters of this position needed only to point to the vast availability of US television content and its popularity among viewers as evidence that these programs were having a destructive impact on local cultural autonomy. Worldwide, concerns that global television would erode cultural diversity and stifle domestic production in smaller nations ushered in a number of national regulations worldwide that limited the amount of imported television, offered subsidies to domestic producers, and sought to leverage funding from other small nations to co-produce content that could be domestically and internationally popular.

As researchers began examining the impact of US and Western television content on foreign viewers, the received wisdom about why people watch US programming and what its impact on them was began to change. Scholars in the US and the UK had long argued for an "active audience" approach to understanding viewers' relationships with television content, focusing on what people do with television, rather than what television does to viewers (Tomlinson, 2001). This same attitude began to find its way into studies of international audiences as well.

The *cultural imperialism thesis*, which focused on the way that US TV changes viewers abroad, began to be replaced with *hybridity theory*, which suggested that people use imported television to help understand, navigate, and enjoy an increasingly interconnected, technologized, and capitalist world. Drawn from biology, hybridity theory argues that imported cultures do not *replace* indigenous ones, but rather get integrated into indigenous cultures, altering but not replacing the original culture. Moreover, hybridity theory posits that all cultures are, in fact, the product of centuries of hybridization, and that present-day cultural exchanges are nothing new (Kraidy, 2005).

While hybridity theory accounts well for how audiences abroad make use of imported television programming, it has been criticized for being insufficiently detailed about the precise ways that cultures interact and what the consequences of those interactions might be. The real danger of hybridity theory lies in its tendency to catalog the endless ways that cultures around the world interact with imported Western media culture, without critiquing the systems and structures that lead to unbalanced cultural exchanges in the modern world. Some critics have pointed out that the main critique of the Cultural Imperialism Thesis was not that individual *programs* would create cultural change, but rather the exportation of an entire media *system* would. Returning to Table 1.1, these critics worry that the commercial Hollywood system has replaced every other way of organizing mass media in human societies (Schiller, 1991). With a handful of notable exceptions, this concern about the social impact of the media system writ large has mainly disappeared from academic and policy discussion of global television, even as commercial systems similar to those in the US have spread. At the same time, public service media outlets have fought, with some important successes, to reinvent themselves in the digital television era. The BBC, for instance, continues to produce high-end HD content and run the most popular streaming service in the UK, which it also makes available to viewers in the US and around the world.

Hybridity theory leads to a bit of a dead-end, only capable of claiming that the globalization of culture leads to cultural intermixing. But, some observers have put a finer point on the kinds of cultural intermingling that are occurring. Serra Tinic (2005), in her study of Canadian television, shows how particular, distinctive locales disappeared from globally focused television in the 1990s, in favor of generalized spaces such as home interiors that are thought to be universally familiar to viewers everywhere.

Perhaps the most agreed-upon consequence of early television globalization was the creation and spread of populist culture in nations around the world. Indeed, Tracey and Redal (1995) suggest quite convincingly that what we see in television trade is not the imposition of one nation's (or region's) culture on another, but rather the rise of populist cultures everywhere. Populism, in this context, refers to cultural practices that arise from and reflect the sensibilities of ordinary people, rather than society's economic or intellectual elites. Cultural expression in many nations was restricted to elite producers and consumers for much of the 18th and 19th centuries, in venues such as art museums, the theatre, and opera houses. For decades, elites around the world used mass media like television to transmit these elite cultural forms to the masses. Commercial television, by contrast, is produced and consumed by the masses, and its global spread is a triumph for the cultural tastes of ordinary people.

What theories of hybridity and populism bring to the fore is the fact that national cultures are not isolated enclaves. Instead, cultural differences *within* nations are sometimes more pronounced than differences *among* nations. A number of institutional and regulatory arrangements have, since the beginnings of television, sought to accentuate cross-national cultural similarities, even as these efforts were eclipsed by efforts to produce nationally isolated television culture for most of television's history.

## Navigating Cultural Difference

One of the lessons of television globalization, for both scholars and industry executives, is that the contours of cultural difference and similarity do not always run parallel with the contours of profit. Shared cultural attributes, such as language, race, gender, and sexuality, can help draw together audiences across the boundaries of national difference, while these same attributes can drive audiences apart when they are not shared. Unfortunately, for commercial media creators at least, the shared attributes of audiences are not always the most profitable ones. For instance, it might be *economically* appealing to try to target female viewers across the Middle East, but difference in gender cultures, religion, and language might make such efforts rather fruitless.

The television production and distribution industries developed a number of strategies in the 1960s that were designed to overcome differences of culture and language, while maximizing the savings and added production funding that television globalization affords. These strategies, which continue today, are co-production and re-formatting.

### *Co-Production*

Co-production refers to any arrangement where multiple corporations work together to create a product, but within global television and film, the term is generally restricted to productions that are produced by an international crew with funding sourced from more than one nation, which are intended to be viewed in more than one national market. In many parts of the world, national governments have official co-production treaties with other nations that specify the makeup of the cast, crew, and financing, and which make available government funding for eligible productions.

In Europe and Asia, co-production became popular in the 1980s and 1990s, as more and more commercial cable, satellite, and broadcasting channels began to launch. Co-production offered a way for cash-strapped producers to pool resources and create content that could compete with the high production standard of imported US and UK programming. Many of the initial efforts at co-production met with lukewarm popularity and critical acclaim. William Fisher (1990) sardonically offered a "recipe" for global television production in *Sight and Sound*:

> Take a story that crosses the borders of two or more Member States in the European Community ... Add a writer and director from one of these same Member States, then gently solicit investment from their domestic television networks ... YIELD: Two hours-plus of a deal-driven fictional screen entertainment, devoid of distinctive contour or flavor.

As Fisher's comments suggest, because television at the time was so tied to local cultural experiences and languages, the efforts of co-producers to engineer multinational popularity for the purposes of increasing funding were doomed to fail. With time, however, producers came to know better the similarities and difference between national television cultures, and developed strategies for navigating those differences. The Canadian-Irish co-production *The Tudors* (2007–2010) is a successful and prominent example of a series created under a co-production treaty with a multinational cast and crew, which was designed from the outset to play in Ireland, Canada, the UK, and the US (Roxborough, 2008). The series utilized a historical figure well known across English-speaking territories, King Henry VIII of England, highlighting the sensational elements of his story with elaborately recreated settings, costumes, and characters. The high production values, the careful attention to historical details, and the graphic depictions of sex and violence all combined to help the series stand out in the markets where it aired.

Most recently, streaming services like Netflix and Amazon have seen co-production as a strategy for breaking into foreign markets with popular local series, while leveraging the appeal that foreign stories and locales have for a segment of their subscribers everywhere. The Japanese series *Hibana: Spark* and the French series *Marseilles* are two prominent examples of this trend. Both series were designed to increase Netflix subscribers in Japan and France, respectively, and have achieved critical acclaim and buzz in the domestic markets, while also drawing viewers in the US and elsewhere.

Co-production, then, arises from the same philosophy of cultural protectionism that the NWICO effort did. Much of the financial appeal of co-production is rooted in the subsidies that national governments around the world provide to local television industries. But producers had difficulties realizing the potential of government funding until they were able to develop strategies of producing content that worked with, rather than against, the grain of cultural similarities across borders.

## Formatting

As a practice of the international television industries, formatting dates back to the beginnings of television, when broadcasters just getting started tended to borrow concepts and even scripts for television series from broadcasters in other countries. When Israeli TV began airing its own fictional series in the early 1980s, it borrowed sitcom plots and characters from established US and UK series (Shahaf, 2007). Even in the US, which had an active television production sector by the late 1950s, CBS bought scripts from the BBC to develop its 1972 series *Sanford and Son*.

In the late 1980s and early 1990s, however, formatting became a more formalized business practice in the global television industry. Formats are essentially remakes of television series in different national markets with local talent, as opposed to imports, which are finished programs aired in a different country, often translated and edited for local language and cultural preferences. In today's television markets, format producers specify every possible detail of the shows they make in order to help producers abroad recreate them as faithfully as possible, as well as to protect their intellectual property from unauthorized copying by producers abroad.

Formats are particularly well-suited to addressing the tendency for television to circulate widely for economic reasons, even as cultural elements restrict such circulation. Often, formats are based on popular series in one territory, which are then reproduced in other territories with local writers and talent. *Sesame Street*, the popular children's series, has been reproduced in over 120 different territories, beginning with the Brazilian format, *Vila Sésamo* in 1972. More recently, producers have begun creating series designed from the outset to appeal to viewers across multiple markets. The Endemol competition series *The Voice*, for instance, was designed from the beginning to be an international format that could compete with Fremantle's various *Idol* franchises.

Before formats could become popular ways to manage the localizing/globalizing tendencies of television, they needed to have some degree of legal protection. Legal protection keeps local producers from merely "stealing" formats from foreign broadcasters and reproducing them without paying the original creator. In the early years of television, unauthorized copying was a common practice around the world. Fairly quickly, though, broadcasters began developing informal compensation agreements among themselves in order to protect their intellectual property. As early as the late 1970s, groups of international television executives began to gather to discuss how best to protect their programming from unauthorized copying. In the late 1990s and the early years of the 21st century, the problem of protecting format copyrights became more urgent, as more and more formats entered the international markets. In response, the Format Recognition and Protection Agency was formed to protect formats and mediate disputes. Today, most television series in developed markets come with a production "bible" that both explains how to create local formats of the program and protects the program from unauthorized copying.

Co-productions and formats try to leverage production capital to work with the grain of cultural similarities in an effort to overcome the global-local tendencies of television programming. Newer delivery and display technologies, however, may eventually render obsolete the tension between television's localizing cultural tendencies and its globalizing economic ones.

## Television Becomes Transnational?

As we have seen, global television has slowly moved from the periphery to the center of television cultures in many parts of the world. Early regulations, industry structures, and technologies of transmission and reception lent themselves to nationally based television systems and cultures. However, the introduction of new transmission technologies – first cable and satellite, then, more recently, internet streaming – as well as large-screen HD display technologies, working in tandem with regulatory and industrial changes, may soon usher in what we might characterize as a truly transnational era of television. This era would be characterized by transnational media conglomerates that own channels and production companies in multiple countries, as well as television content that is simultaneously local and global.

Part of what made television such a locally specific medium was the size of the screen and its placement in the center of the domestic living room. These attributes meant that television became more of an aural medium than a local one, rooted not just in national languages, but local dialects as well. And its presence as the centerpiece of the home meant that the stories it told and its program schedule were rooted in local domestic customs, rhythms, and politics. What happens when television becomes a high-definition, individualized, asynchronous cultural technology? Streaming viewers are typically not distracted but focused, and are consequently willing to encounter uncertainty and foreignness in the content they consume. As a result, highly localized content can still find a substantial international audience: cultural specificity no longer works against widespread distribution. Streaming producers in Germany, for instance, employ both teams of local writers and teams of international writers to create content that will look genuine to domestic and international viewers, but still be decipherable for international audiences.

What we often see in contemporary global television is something more than simple hybridity. Rather the globalizing and localizing tendencies of television have slowly been harmonized, and no longer pull as strongly in different directions. There is, in fact, nothing natural or inevitable about people preferring culture that is familiar over culture that is foreign. And global television industries have begun to discover which segments of the world's viewers have such interests and to figure out ways to bring them foreign content that can turn a profit at the same time.

## Note

1. This model is adapted from the Four Theories of the Press, found in Siebert, et al. (1963). For further reading on comparative media systems, see Hallin and Mancini (2012).

## References

Carlsson, U. (2003). The rise and fall of NWICO. *Nordicom Review*, *24*(2), 31–67.

Edgerton, G. (2007). *The Columbia history of American television*. New York: Columbia University Press.

Fisher, W. (1990). Let them eat europudding+ European film coproductions on the rise. *Sight and Sound*, *59*(4), 224–227.

Hallin, Daniel C. & Mancini, Paolo, eds. (2012). *Comparing Media Systems Beyond the Western World*. Cambridge: Cambridge University Press.

Havens, T. (2006). *Global television marketplace*. London: British Film Institute.

Kraidy, M. (2005). *Hybridity, or the cultural logic of globalization*. Philadelphia, PA: Temple University Press.

Mihelj, S., and Huxtable, S. (2018). *From media systems to media cultures: Understanding socialist television*. Cambridge: Cambridge University Press.

Roxborough, S. (2008, November 20). "Global focus: Int'l co-production treaties," *Hollywood Reporter*, online: www.hollywoodreporter.com/news/global-focus-intl-production-treaties-123342 (accessed August 27, 2018).

Schiller, H. I. (1991). Not yet the post-imperialist era. *Critical Studies in Media Communication*, *8*(1), 13–28.

Shahaf, S. (2007). Welcome to the Sitcom school: A globalized outlook for the study of television history. *Westminster Papers in Communication & Culture*, *4*(4) 103–123.

Siebert, F., Peterson, T., & Schramm, W. (1963). *Four theories of the press: The authoritarian, libertarian, social responsibility, and soviet communist concepts of what the press should be and do*. Chanpaign, IL: University of Illinois Press.

Tinic, S. (2005). *On location: Canada's television industry in a global market*. Toronto: University of Toronto Press.

Tracey, M. and Redal, W. W. (1995) "The New Parochialism: The Triumph of the Populist in the Flow of International Television." *Canadian Journal of Communication*, 20:3 (March) ISSN 1499–6642. Available at: www.cjc-online.ca/index.php/journal/article/view/883. Date accessed: 28 aug. 2019. doi:https://doi.org/10.22230/cjc.1995v20n3a883

Tomlinson, J. (2001). *Cultural imperialism: A critical introduction*. London: A&C Black.

# 3
# OBJECTLESS TELEVISION[1]

*Purnima Mankekar*

For over a decade, transnational media across the world have reported, and evoked, a sense of publics on edge. For instance, some observers have attributed the rise of authoritarian populism leading to the election of Trump in the United States and the hegemonic triumph of the Modi regime in India to the chronic sense of crisis, unease, and anxiety created by transnational media. It is beyond the scope of this chapter to analyze the cultural and historical specificities of authoritarian populism.[2] However, it is worth noting that the intensity and speed with which affects of anxiety travel translocally and transnationally have been unprecedented owing, in no small measure, to the multiplication and density of media platforms.

In many parts of India, television is now watched on multiple screens and operates in conjunction with other communication technologies like the mobile phone and via platforms such as YouTube.[3] My contention in this chapter is that, even as it continues to constitute political landscapes, television has now become objectless: it cannot be located in any one object (namely, the TV screen) but, instead, is distributed through and across social formations via multiple nodes. In fact, I would argue that its affective potency has increased because of its objectless status. If television was the "mass" medium of the twentieth century, how has its dispersal through and across social formations changed how we respond to it and theorize it? My objective in this chapter is to outline a theoretical framework for examining the extent to which objectless television, in its interface with digital media, could contribute to the generation and circulation of affects of anxiety.

To borrow from Freud, anxiety signals a response to a traumatic situation or an impending trauma (1990 (1959):57).[4] Anxiety, much like television in the current conjuncture, also tends to be objectless. Unlike fear, which congeals around specific objects, anxiety has no object and therein lies its affective force. Like all affect, it is named as such retroactively. It has no single source or, indeed, any source; it is ambient, yet is experienced intimately and viscerally. The complex temporality of anxiety is evident in how it may *become* fear when provided with an object and, here, media play a crucial role. Anxiety is spectral rather than spectacular. A generalized sense of anxiety may provide the bases for the generation of fear sparked, in turn, by historical and mediatized "events" such as terrorist attacks.[5] In such contexts, anxiety as an affective formation and fear as a structure of feeling become mutually implicated.

Anxiety thus refers to that unnameable and objectless affective regime that becomes retroactively named as such; stated simply, anxiety is affective because it is generative of distinct structures of feeling, including fear, hate, or rage.[6] It is worthwhile to take a momentary detour to delineate what I mean by affect. I build on, but also diverge from, Raymond Williams' notion of the structure of feeling (1978); additionally, I draw on Deleuze (1997) and Massumi (2002) to argue that affect is distinct from feeling (the domain of individual subjectivity) and emotion (the domain of the linguistic) (see also Spinoza 1985). The following dimensions of affect are particularly pertinent to my argument about anxiety as an affective formation. Affect is neither located in an individual subject nor does it refer to the psyche or subjective feelings. Rather, affect is generative of subject formation: in this sense, then, affect produces subjects rather than the other way around. At the same time, affect "sticks" to and circulates between bodies and objects (Ahmed 2004). Affect is transitive; it gains potency through its circulation. Moreover, affect problematizes the purported split between private and the public. While felt viscerally and intimately, it is also ambient: it thus transects private feelings and public sentiments (Berlant 2008; Stewart 2007). Affect blurs the line between thought versus emotion, and mind versus body: it entails both forms of apprehension and bodily forms of feeling. I focus on affective regimes to foreground their material and institutional aspects: affects are neither free-floating nor unmoored from the sociohistorical conjunctures of which they are a part. Yet, affect is irreducible to ideology; it exists at the cusp of narrativization and lies beyond (or, perhaps more accurately, prior to) semantic articulation. Finally, undermining dichotomous conceptions of compliance versus resistance, affect is generative of action: it is the capacity to act and be acted upon. In what follows, I will outline how conceptions of affect enable us to theorize some of the ways in which objectless television might have become a node in the generation and circulation of anxiety.

While digital media have by no means replaced television they now feed off each other to produce reconfigured affective landscapes.[7] My chapter is intended to be evocative rather than explanatory: although it draws on data from India, it is conceptual rather than empirical. My objective is not to present an Indian case study to which we can "apply" current theories about television and thereby resurrect the tired binary of ("First World" = universalist) theory versus ("Third World" = particularistic) empirical data; rather, my goal is to theorize *from* the empirical material that follows in order to sketch the contours of an analytical framework that enables us to examine the implications of objectless television.

## It's All in the Feed: Objectless Television and Digital Affect

Deleuze argues that, after World War II, the disciplinary societies analyzed by Foucault ceded to societies of control. Eighteenth- and nineteenth-century societies, he insists, "initiate the organization of vast spaces of enclosure" (Deleuze 1992:3). Subject formation occurs in the passage from one enclosure to the next: the family, school, prison, and so on. He claims that these institutions, which used to be distinct analogical spaces, have been replaced by societies of control. Where disciplinary societies were "molds, distinct castings," control societies are marked by the continuous modulation of affect (Deleuze 1992:4). Deleuze's conception of the continuous modulation of affect across social formations seems an appropriate analytic for understanding objectless television. But instead of assuming that they are "finished" (Deleuze 1992:4), I prefer to explore how institutions (for example, the state and the market) might intersect. As I will outline below, theorizing objectless television entails looking carefully at the renegotiation of the boundary between the state and the market. Rather than conceptualize it as an ideological state apparatus, I theorize objectless television in terms of the generation of affective regimes such as those of anxiety.

To clarify: my concern is not with the "impact" or reception of objectless television because that would imply that television is separate from that which it is ostensibly impacting. Nor do I conceive of television as a "player" in political transformations. Instead, as I will briefly outline below, television is embedded in *and* constitutive of the social and the political. This will be followed by my analysis of a report on the entertainment industry released by the international consulting firm KPMG in collaboration with FICCI (Federation of Indian Chambers of Commerce and Industry). KPMG/FICCI produce annual reports that present comprehensive quantitative and qualitative data on media industries and their audiences in India. Available online, these reports are useful because of the information they provide to media planners as well as for the insights they offer media studies scholars into developments in the Indian media space and in the consumption habits of audiences surveyed. Rather than take its assertions at face value, I engage this report as an ethnographic artifact for what we may learn about the historical and institutional processes that have rendered television objectless and, equally importantly, for how it *imagines* it audiences. My goals in doing so are threefold: to point to the role of the state in preparing the ground for objectless television; to draw out the complex relationship between the state and entrepreneurial (private) capital; and to foreground the intertwining of the discourse of choice with that of the increasing "penetration" of mobile media into the everyday lives of viewers.

## Between the State and the Market

In India, as in many other parts of the world, television and politics have long been mutually imbricated. In the 1980s, satellites enabled state-run television or Doordarshan to become a national network: in so doing it produced a "national" audience with profound implications for how national belonging was reconstituted (Mankekar 1999). Aligning with the introduction of the liberalization of the Indian economy, Doordarshan introduced sponsored programs underwritten by the manufacturers of consumer goods. By displaying commodities never before seen on television, Doordarshan came to occupy a negotiated territory between the state and the market and became central to the constitution of middle-class citizen-consumers (Mankekar 1999; Rajagopal 2001). In November 1984, the televised spectacle of the body of the assassinated Prime Minister Indira Gandhi lying in state was followed by carefully orchestrated pogroms against Sikhs in different locations across the nation. The televisation of Hindu epics, together with other affectively charged discourses of belonging, community, and identity, enabled the consolidation of a strident Hindu nationalist state.

The discourse of choice, specifically the ability to choose one's sources of news and entertainment, became increasingly salient in the years to come. In the early-1990s, video news magazines played a crucial role in galvanizing upper castes against the government's proposal to expand reservations (or quotas) for lower-caste communities, thus demonstrating the power of televisual news media that were independent of the state. The advent of cable further weakened the monopoly of state-run Doordarshan by offering viewers more choices about what they could watch. These changes accelerated with the launch of transnational television networks. Rather than replace Indian programming with imported television shows, however, these transnational networks re-energized the domestic production of content and the "Indianization" of imported formats such as MTV. The proliferation of regional channels further undermined the notion of a single (and singular) national audience.

It should not surprise us that the discourse of choice that threaded through the foregoing developments articulated closely with the liberalization of the Indian economy. The Indian state assumed a crucial role in the proliferation of new media technology by formulating policies that encouraged private investors to invest in the telecommunications sector. As was true with some other efforts by the state to liberalize the economy, the rapid expansion of the telecommunications sector, and the growth of internet-enabled communications technologies were all undergirded by the valorization of neoliberal discourses of choice[8]: with television no longer the monopoly of the state, the privatization of sources of information and entertainment was facilitated. This entrenched the notion that viewers were free to choose what they wanted to watch and, eventually, how, where, and when they wanted to watch television. As the television screen began to be supplemented by other screens, this discourse of choice supported and was accelerated by the adoption of multiple screens. As I will argue later in this chapter, this very dispersal across multiple screens, platforms, and media did not simply offer viewers opportunities to "talk back" to content producers but also provided media planners and marketers with detailed data about their consumers: put another way, choice and surveillance became co-implicated.

## *Proliferating Screens*

The discourse of choice was further reinforced with the introduction of mobile telephony which changed not only what viewers could watch but, more profoundly, *how* they could engage with television. The growth of chat shows, reality shows, and game shows expanded opportunities for audience participation. These possibilities were engendered by the symbiotic relationship that was to develop between television and mobile phones. For instance, the game show *Kaun Banega Karorepati* (an "Indianized" version of *Who Wants to be a Millionaire*, 2000–2017) was particularly significant for how it invited audience members to engage with the show via their mobile phones.[9] This transformed how television industry professionals imagined and mobilized their audience(s) (Punathambekar 2010:246).

Soon viewers were able to choose *where* they could view television. The incorporation of new media technologies took an unprecedented turn when viewers began to watch television on multiple devices. Unlike Doordarshan or cable television, this meant that, for an increasing number of viewers, the locus of television was no longer the home or the family (cf. Mankekar 1999). Where, earlier, working-class and lower-middle-class viewers had tended to watch television within the home, usually with members of the family or community, they were now able to do so on their mobile phones or in cybercafés on platforms such as YouTube. Even as television remained a crucial source of information and entertainment, its reach began to exceed it as any single object. Put another way, through its dispersal across multiple platforms television became objectless.

With the shift in emphasis from state-funded development to privatization that occurred with the liberalization of the Indian economy, entrepreneurial capital assumed an increasingly significant role in the expansion of mobile telephony. In 2004, the Indian state facilitated the introduction of high-speed 4G services, the rapid expansion of internet and broadband access, decreasing data charges, and the proliferation of internet-enabled mobile phones. The year 2016 was a milestone in India's mobile internet history. This year was marked by the entry of a new company, Reliance Jio. Reliance Jio offered free data service plans to popularize its usage among consumers and attracted 100 million subscribers in less than six months (KPMG 2017:36). The mobile phone emerged as the most powerful medium of communication in India. As pointed out by Girish Menon, Director of Media and Entertainment, KPMG (India), by 2016 the mobile phone had transitioned "from being just an additional medium" and had made its way to the center of the media and entertainment space (KPMG 2017:1).

## *Penetration by Choice*

As television became objectless, the discourse of choice became closely intertwined with the proliferation or, in the words of media planners and producers, the "penetration" of television in the everyday lives and intimate spaces of viewers.[10] As viewers began to increasingly engage television content through their mobile phones, the adoption of internet-enabled mobile phones rose to 300 million users in 2016 and was expected to exceed 700 million by 2021 (KPMG 2017:27).[11] Manufacturers expanded the screens of many mobile phones which resulted in these becoming the screens of choice for media consumption. In fact, as asserted by Ajit Mohan, Chief Executive Officer of Hotstar, for many viewers the mobile screen became the primary screen: "The mobile is already playing the role of a primary screen for millions of users in cities today, and over time will become the second and third screen in households across the country" (KPMG 2017:42). The growing importance of mobile phones as screens heightened the role of social media feeds in viewers' engagement with television. The ubiquity of mobile phones also generated a rapid expansion of viewers' access to video streaming. As a result of the 4G uptake and the boost given to mobile telephony by the launch of Reliance Jio, video streaming apps saw an increase of 336 per cent (KPMG 2017:36). Consequently, consumers began to access content across a variety of platforms, with online video consumed via social feeds, User Generated Content platforms such as YouTube, and subscriptions of various free and paid Video on Demand platforms.[12]

Yet again, media producers and planners articulated these changes in terms of intertwined discourses of choice and penetration. In contrast to what media producers describe as the era of single-screen TV sets in homes, the proliferation of multiple screens enabled viewers to spend more time online and watch television on multiple platforms, and this transformed how content was generated. As one content producer pointed out,

> We are moving from a time where most content experiences were via analog media, to where all content experiences, besides live, are digital. Narratives no longer are beholden to length, genre or size of audience … the canvas for rich content experiences far surpasses what a TV-only screen could achieve.
>
> *(Sameer Pitalwalla, CEO and Co-Founder, Culture Machine Media Pvt. Ltd., quoted in KPMG 2017:29)*

In addition, as I suggest below, television programming no longer relied on assumptions of linear temporality. Content producers and advertisers claimed that, with the projected expansion of infrastructure, lower data prices, and increasing access, mobile media would be poised to penetrate into the daily lives of viewers and consumers in rural India as well. What were the implications of objectless television for emergent socialities? Social networking sites on the internet (such as, Facebook, Myspace, Orkut, websites such as bigadda.com and nukkad.com, and phone-based short messaging services [SMS]) articulate with existing forms of mediation to constitute new forms of sociality (Kumar 2010).[13] Likewise, mobile media technologies have engendered mobile publics around television through blogs, texting, and online fan communities (Punathambekar 2010:252).

My concern here is with how affective regimes *bridge* the domains of the public and the private. Furthermore, while ephemeral mobile publics are tethered to the time and space of the television event (Punathambekar 2010), objectless television reconfigures spatiality and temporality. Objectless television exceeds the enclosures of the family and the home: viewers engage with it on multiple screens, with their families or with their friends, with online communities of fans, and by themselves. Watching objectless television on mobile phones, for example, ostensibly enables privacy even as it blurs the distinction between private and public modes of engagement. Additionally, rather than being tied to the television event, digital media document, archive, and make content readily accessible and shareable. Objectless television produces streaming affects that ricochet off each other in feedback loops that are iterative, generative, and multiscalar.

## Streaming Affects

"Paralleling TV's temporal effects, anxiety is television's affect" (Patricia Mellencamp 1990:243).

Time and temporality have long preoccupied scholars of television. In a pioneering article, Mary Ann Doane argued that "the major category of television is time. Time is television's basis, its principle of structuration, as well as its persistence referent" (1990:222). Other scholars pointed to the alignment of television's schedules with capitalist formations of temporality.[14] Thus, for instance, network television was said to have been in synchrony with the work day and the work week (Browne 1987, in Doane 1990:232), and feminists analyzed how the "rhythms" of soap operas articulated and reinforced the rhythms of housework (Modleski 1979). Yet others theorized the ontology of "liveness" (Feuer 1983) and the temporality of crisis and catastrophe (Doane 1990).

Building on these scholars' theorizations of the relationship between television and time, I will next sketch how the dispersal of television, enabled by the proliferation of digital technology, reshapes formations of temporality and spatiality: we can now watch television programs all the time and all over the place on multiple, often portable, devices. For instance, the adoption of the mobile phone to watch television content has spatial and temporal implications. As noted above, watching television on one's mobile phone allows for personal space and privacy in public settings, but also demonstrates the penetration of objectless television into spaces of intimacy. Accessing content through mobile phones potentially allows round-the-clock usage with, for example, viewers watching content while commuting to work or while multi-tasking. Furthermore, as the KPMG report gleefully observes, viewers are no longer restricted by schedules centering around prime time because they can watch television later into the night than traditional TV and, more significantly, can take their devices into their beds (KPMG 2017:39). Objectless television thus spreads across the body politic because of its penetration into newer, often intimate, spaces in the everyday lives of its viewers.

Theorizing objectless television also involves tracing how interfaces between multiple platforms create feedback loops for the circulation of affect between viewers, viewers and producers, and between viewers and their devices. Social networking websites activate live streams enabling users to share content while also providing feedback via online chatrooms and fan communities. An increasing number of viewers engage content on two screens simultaneously, watching content on their primary device and expressing their views and interacting with other viewers on a second device. For example, as of 2017, 86 percent of Facebook users in India interacted with a second screen while watching programs, and users of Twitter were 66 percent more likely than average to interact with the online content of a show (KPMG 2017:47). The potentially endless reiteration of affects that stream across objectless television and social media have multiple implications. For one, viewers may now write (and, therefore, talk) back to media producers through the comments they circulate on social media platforms. These media also provide sites for the compilation of data on viewing habits and preferences. As stated in the report, "The concept of ratings is being disrupted continuously … Social media platforms are becoming evaluation engines where users give deeper insights on the content they consume or the advertisements they see" (2017:47, 48). At the same time, social media platforms enable media planners, marketing companies and producers to archive and mine data on the preferences of viewers/consumers.

### *Anxiety and the Temporality of Waiting*

Objectless television spans multiple and heterogeneous nodes of affect; anxiety is merely one significant such affective regime. Anxiety, Freud cautions, is not a simple matter (1990 (1959):60). It is, in the first instance,

> something that is felt. We call it an affective state ... As a feeling, anxiety has a very marked character of unpleasure. But that is not the whole of its quality. Not every unpleasure can be called anxiety, for there are other feelings, such as tension, pain or mourning, which have the character of unpleasure.
>
> *(ibid.)*

While anxiety produces distinct physical sensations, for now my concern is not with individual sensations but with anxiety as a socially and technologically produced affective regime that transects mind, body, and sociality. Freud's description of anxiety resembles television because it centers on indefiniteness and lack of object: "On a metaphoric level, indefiniteness and lack of object define television – electronic, erasable circuitry with no visible, material base until transmitted" (Mellencamp 1990:247–248).

The affinity between anxiety and television is intensified when television itself turns objectless. I am concerned with the temporality of *waiting* for catastrophe or crisis to occur. In many ways, anxiety is the opposite of crisis. But if crisis causes disruption, anxiety generates a sense of doom that is continuous and seemingly infinite precisely because it is objectless. Crisis represents a condensation of temporality.[15] In contrast, anxiety is about time hurtling forward while we wait, but it is a waiting that is rendered so banal that anxiety is itself normalized. Crisis is about the punctuation of the duration of waiting; it is precipitated by an event that, theoretically, may lead to resolution. Anxiety entails the *extension* of duration; it promises no resolution.

The continuous deferral of resolution entailed in the temporality of anxiety is freshly created through streaming affects and waiting for crises to occur. Freud insists that we need to trace anxiety "back" to danger (1990 (1959):101). Extending his conceptualization of the danger-situation, we might say that with streaming affects the apprehension of the danger-situation is iterative and recursive; it folds back upon itself to generate a looking back on past trauma and an expectation of the trauma to come. It is in this situation, then, that the "signal" of anxiety is given; anxiety is indexical of the scrambling of temporalities.

Objectless television entangles anxiety with banality. Ambient television (for instance, in waiting rooms, at airports and in airplanes, and in restaurants) foregrounds the interplay of banality and anxiety by making "potentially difficult or anxiety-provoking experiences seem like a routine, everyday reality, as mundane as the act of waiting" (McCarthy 2001:200). Nonstop coverage, video replay, and streaming affects on objectless television intensifies the twinning of anxiety and banality. Furthermore, processes of decontextualization associated with television are amplified by objectless television. Steven Shaviro's argument about post-cinematic affect holds well for objectless television: "The control society is characterized by perpetual modulations, dispersed and 'flexible' modes of authority, ubiquitous networks, and the relentless branding and marketing of even the most 'inner aspects' of subjective experience" (2010:6). The immediacy of affect, its reiteration, and its undulation across social formations is what makes objectless television particularly potent.

The KPMG report states that the media and entertainment industry uses inertial, motion, and image sensors to understand consumer preferences (2017:29). Objectless television represents a new form of ambient media that now "watch" us because they have penetrated our everyday lives. The Internet of Things (IoT) is beginning to compile and archive our browsing practices and our responses to what we watch.[16] These data are available for mining, ostensibly, for "customer service" and predictive analytics, but the data also enable marketers and content producers to solicit, guide, and refract the behavior of viewers/consumers. Theorizing objectless television necessitates tracking of how surveillance and "choice" have become ever more intertwined.

## Conclusion

I will close by recapitulating the key components of a theoretical framework that may enable us to analyze objectless television. First and foremost, studying objectless television entails taking affect seriously. This would involve broadening an analytical framework that focuses exclusively on "images" and "representation" to interrogate how objectless television generates feedback loops that circulate between subjects and across subjects and objects. To reiterate: anxiety is only *one* of many affective regimes generated by objectless television;[17] moreover, the forms and modalities of these regimes are heterogeneous across cultural and historical context. Inherently objectless, anxiety may escalate into fear when it finds an object. Objectless anxiety is spectral until it finds its objects: in the United States, these could be the bodies of brown men who could (potentially) be terrorists or people of color who no longer remain in their allotted place in a white supremacist racial order or, in India, in the bodies of Muslims who are lynched for eating beef or for allegedly engaging in "love jihad."[18] Anxiety is thus a useful optic because it is objectless. Like other affective regimes, it is diffuse, yet potent; it has no source, yet is sticky as it circulates across subjects and between subjects and objects; it involves an active waiting and is generative of agency; it is indexical of potentiality because it spurs individuals and collectivities to act in specific ways.

Crucially, theorizing objectless television necessitates foregrounding its relationship to institutions like the state and the market. The continuous modulation of affect is produced not by the dissolution of institutions but occurs *across* them in complex spaces of collusion, convergence, and contradiction. The conjunction of state policies with entrepreneurial capital enables the dispersal of television across multiple screens such that viewers may engage it at any time and in any place. Objectless television emerges from and is constitutive of specific political-economic and historical conjunctures: for instance, neoliberal conceptions of choice are inextricably entangled with processes by which objectless television penetrates the intimate spaces and everyday lives of viewers so as to disperse – undulate – across social formations.

Indeed, through its ubiquity in the everyday and the intimate, objectless television renders anxiety banal. Furthermore, objectless television produces streaming affects that acquire potency precisely through their movement across bodies, objects, and spaces. Objectless television reconstitutes spatiality by blurring the lines between the private and the public and the ambient and the visceral. The generation and circulation of anxiety as an affective regime is intensified because the temporality of waiting engendered by objectless television – watching for and anticipating the next event, the impending crisis, the looming catastrophe – reconfigures the duration, extension, and protraction of the temporality of (potentially) infinite deferral.

A theorization of objectless television would thus engage the configurations of affect, information, bodies, technology, and sensation produced by these new spatialities and temporalities. For instance, could we locate objectless television in a larger sensorial and affective landscape that spans billboards and shopping malls to digital media? Let me be clear: I am not suggesting that we ignore the formal and textual aspects of objectless television – we would need to analyze texts closely to understand how these media do their work. However, rather than bounce back and forth between text and audience looking for an object, we might interrogate whether the televisual text remains stable as it travels across multiple platforms. I would assert that the interplay between text and context – their inseparability and inextricability – is precisely what constitutes textuality.

I resist the temptation to conceptualize the cultural formations shaped by objectless television as homogeneous and totalizing and of viewers as divested of agency. Focusing on affect – the capacity to act and be acted upon – enables us to skirt the sterile trap of framing agency in terms of resistance versus compliance. Feedback loops, while ubiquitous with objectless television, are not seamless; they might be sidestepped, blocked, or doomed to failure. And if the temporality of television produces anxiety as one of its affective regimes, viewers are not passive recipients of what they watch and, indeed, may well be agitated and roused to action by it (cf. Mellencamp 1990; Spinoza 1985). The affect of agitation is amplified with objectless television when affects stream at us and through us; the anxiety generated by objectless television signals not merely a danger-situation but, equally significantly, indexes potentiality, lines of flight, and immanent rupture.

## Notes

1. My title reprises that of John Caughie's influential article, *Television Criticism: "A Discourse in Search of an Object"* (Caughie 1984 25 [4–5]: 109–120) but takes the argument in a different direction.

   My thanks to Tulika Verma for her research assistance for this chapter and to Akhil Gupta for his comments.
2. I would add the following caveats. One, these anxious publics are not, in any way, unique to the current conjuncture: the historical moments when anxious publics have articulated with political turmoil are too numerous to list here. Second, I would assert that, for many racial and sexual minorities, precarity and the anxious structures of feeling surrounding it have long been an everyday fact of life.
3. For an insightful analysis of the cultural impact of mobile phones in India, see Doron and Jeffrey (2013).
4. Freud conceives of anxiety as transformed libido; he argues that the first experience of anxiety occurs at birth with separation from the mother; anxiety is a reaction to loss and can only be felt by the ego (1990 (1959):71). While I do not aim for a psychoanalytic explanation of anxiety, I draw on Freud to interpret anxiety as "affective signal" of a perceived "danger situation" (1990 (1959):57).
5. On the affective production of "September 11, 2001," see Mankekar (2015).
6. As I have argued in Mankekar (2014, 2015), affect is related to structures of feeling but is also distinct from it. See also Massumi (2002) on the difference between affect and emotion.
7. For an early analysis of the relationship between television and digital media, compare with Spigel and Olsson (2004); see also Henry Jenkins 2006 on "convergence" media, and Kumar 2010 and Punathambekar 2010 on digitally remediated publics. My argument in this chapter owes a great deal to these scholars' pathbreaking work.
8. Elsewhere I have problematized the assumption that discourses of neoliberalism have become dominant in India (Mankekar 2015). For important discussions of the complexity of neoliberal governmentality in contemporary India, see Gupta and Sharma (2006) and Sharma (2008).
9. By 2011, viewers could send their answers to questions on the *Kaun Banega Karorepati* via SMS: the tagline for the show was "Anybody sitting at home can win the jackpot" ("*ghar baithe jeeto jackpot*").
10. I would like to flag the obvious phallic connotations of the term penetration; however, given the constraints of space, this will have to be the subject of another discussion. For an early critique of phallic discourses of capitalist penetration, see J. K. Gibson-Graham (2006).
11. It goes without saying that this was particularly true for those segments of the audience who had access to media. Initially, these consisted of urban middle-class viewers but, over time and with the lowering cost of data usage, this expanded to semi-urban and rural users of mobile phones. As of 2017, internet-enabled mobile phones were selling for half the price that they did in China (about INR 9,000). Age was also a significant factor: early adopters of mobile phones were within the age group 15–34 and constituted 70–75 per cent of users (KPMG 2017:35).
12. It was estimated that there were 299.7 million social media users in India in 2016 and this number was expected to reach 477.4 million by 2019. As of 2016, India had the second-largest number of users of Facebook with 166 million active subscribers, and had the highest number of users of WhatsApp with over 160 million active subscribers.
13. See, also, Ravi Sundaram's powerful theorization of cyberpublics (2000).
14. For instance, Mellencamp points out, "US network television is a disciplinary time machine … The hours, days, and television seasons are seriated, scheduled, and traded in ten-second increments modeled on the modern work week – day time, prime time, late night, or weekend" (Mellencamp 1990:240).
15. "Catastrophic time stands still," states Doane (1990:231).
16. The Internet of Things (IoT) refers to the ability to connect multiple devices such that consumer electronics are increasingly linked to the internet to facilitate the collection and exchange of data about consumers.
17. On the convergence of television and digital media in the production of hate, see Mankekar and Carlan 2019.
18. "Love jihad" refers to Hindu nationalist campaigns against Muslim men who, allegedly, seduce Hindu women and, eventually, convert them to Islam.

## References


Ahmed, Sara. 2004. *The Cultural Politics of Emotion*. London: Routledge.

Berlant, Lauren. 2008. *The Female Complaint*. Durham: Duke University Press.

Brunsdon, Charlotte. 1990. Television: Aesthetics and Audiences. In *Logics of Television*, ed. Patricia Mellencamp, pp. 59–72. Bloomington: Indiana University Press.

Caughie, John. 1984. Television Criticism: 'A Discourse in Search of an Object.' *Screen* 25 (4–5): 109–120.

Deleuze, Gilles. 1992. Postscript on the Societies of Control. *October* 59: 3–7.

Deleuze, Gilles. 1997. *Essays Critical and Clinical*. Minneapolis: University of Minnesota Press.

Doane, Mary Ann. 1990. Information, Crisis, Catastrophe. In *Logics of Television*, ed. Patricia Mellencamp, pp. 222–239. Bloomington: Indiana University Press.

Doron, Assa and Craig Jeffrey. 2013. *The Great Indian Phonebook: How the Cheap Cell Phone Changes Business, Politics, and Daily Life*. Cambridge: Harvard University Press.

Feuer, Jane. 1983. The Concept of Live Television: Ontology as Ideology. In *Regarding Television: Critical Approaches*, ed. E. Ann Kaplan, pp. 12–22. American Film Institute Monograph Series.

Freud, Sigmund. 1990 (1959). *Inhibitions, Symptoms, and Anxiety*. London: W. W. Norton.

Gibson-Graham, J. K. 2006. *The End of Capitalism (As We Knew It): A Feminist Critique of Political Economy*. Minneapolis: University of Minnesota Press.

Gupta, Akhil and Aradhana Sharma. 2006. Globalization and Postcolonial States. *Current Anthropology* 47(2): 277–307.

Jenkins, Henry. 2006. *Convergence Culture: Where Old and New Media Collide*. New York: NYU Press.

KMPG (India)-FICCI. 2017. *Media for the Masses: The Promise Unfolds*. https://assets.kpmg.com/content/dam/kpmg/in/pdf/2017/04/FICCI-Frames-2017.pdf. Accessed February 1, 2018.

Kumar, Shanti. 2010. Globalization, Media Privatization, and the Redefinition of the Public in Indian Television. *Bioscope* 1(1): 21–25.

Mankekar, Purnima. 1999. *Screening Culture, Viewing Politics*. Durham: Duke University Press.

Mankekar, Purnima. 2014. Television and Embodiment: A Speculative Essay. In *Television at Large in South Asia*, eds. Aswin Punathambekar and Shanti Kumar, pp. 185–200. Delhi: Routledge.

Mankekar, Purnima. 2015. *Unsettling India: Affect, Temporality, Transnationality*. Durham: Duke University Press.

Mankekar, Purnima and Hannah Carlan. 2019. The Remediation of Nationalism: Viscerality, Virality, and Digital Affect. In *Global Digital Cultures: Perspectives from South Asia*, eds. Aswin Punathambekar and Sriram Mohand, pp. 203–222. Ann Arbor: University of Michigan Press.

Massumi, Brian. 2002. *Parables for the Virtual*. Durham: Duke University Press.

McCarthy, Anna. 2001. *Ambient Television: Visual Culture and Public Space*. Durham: Duke University Press.

Mellencamp, Patricia. 1990. TV Time and Catastrophe, Or *Beyond the Pleasure Principle* of Television. In *Logics of Television*, ed. Patricia Mellencamp, pp. 240–266. Bloomington: Indiana University Press.

Modleski, Tania. 1979. The Search for Tomorrow in Today's Soap Operas: Notes on a Feminine Narrative Form. *Film Quarterly* 33(1): 12–21.

Punathambekar, Aswin. 2010. Reality TV and Participatory Culture in India. Popular Communication. *International Journal of Media and Culture* 8(4): 241–255.

Rajagopal, Arvind. 2001. *Politics after Television*. Cambridge: Cambridge University Press.

Sharma, Aradhana. 2008. *Logics of Empowerment: Development, Gender, and Governance in Neoliberal India*. Minneapolis: University of Minnesota Press.

Shaviro, Steven. 2010. *Post-Cinematic Affect*. Durham: Duke University Press.

Spigel, Lynn and Jan Olsson, eds. 2004. *Television after TV*. Durham: Duke University Press.

Spinoza, Benedictus de. 1985. *The Collected Works of Spinoza*. Princeton: Princeton University Press.

Stewart, Kathleen. 2007. *Ordinary Affects*. Durham: Duke University Press.

Sundaram, Ravi. 2000. Beyond the Nationalist Panopticon. http://cybercoolies.pbworks.com/f/Beyod%20the%20Nationalist%20panopoticon.pdf. Accessed January 2011.

Williams, Raymond. 1978. *Marxism and Literature*. London: Oxford University Press.

# 4
# GLOBAL SOCIAL MEDIA ENTERTAINMENT

*Stuart Cunningham and David Craig*

What counts as popular screen entertainment, especially for young people, is changing profoundly. What happens to television studies when "television" becomes for many the screen content delivered by global platforms such as YouTube, and increasingly Facebook, Instagram, Snapchat and Twitter and their Chinese counterparts Weibo, WeChat and Youku Toudo? Of course, it is not nearly as simple as broadcast television being eclipsed by platform-based content. Television *programs* continue to be watched in substantial numbers, just not necessarily on *TVs*. Leading television studies scholars Jonathan Gray and Amanda Lotz (2012) remind us that "Television is neither 'beating' nor 'losing' to new media in some cosmic clash of technology; rather, television is an intrinsic part of 'new' media" (2012, 3). At the level of content, this remains true, but the degree to which the lifeblood of television – advertising and brand alignments – is being diverted to the platforms, particularly Facebook and Google, and the degree to which the demographic profile of linear TV viewership is skewing dangerously old, makes the question we posed at the start of this paragraph increasingly and existentially pertinent,

Arguably one of the most innovative elements of this rapidly evolving world of popular screen entertainment is what we call "social media entertainment" (SME) (Cunningham and Craig 2019). We understand SME to be an emerging industry based on previously amateur creators professionalizing and monetizing their content by engaging media entrepreneurship across multiple social media platforms to aggregate global fan communities and incubate their own media brands. This "industry" is only a bit more than ten years old, starting soon after the acquisition by Google of YouTube in 2006, and concurrent with the launch of Twitter and their counterparts in China, Youku and Weibo. By 2017, it sees more than 3 million YouTube creators globally receiving some level of remuneration from their uploaded content and more than 5000 YouTube professionalizing-amateur channels with at least a million subscribers and over 376 million video views. "[T]he world has never before seen the likes of YouTube in terms of availability of non-infringing content" (Hetcher 2013, 45). And YouTube is but one of numerous platforms on which creators are leveraging their content and communities. In their study of "America's New Creator Economy" (Shapiro and Aneja 2018), the Re-Create Coalition counted nearly 15 million creators across nine platforms in the United States alone.

What is "global" SME? SME traverses the globe relatively frictionlessly because of the near-to -global span of the major digital platforms, and the fact that low-budget, professionalizing-amateur content is created primarily to be "spreadable" (Jenkins, Ford and Green 2013) which includes not being conventionally copyright-controlled. While a good deal of scholarly focus rests with Netflix, Amazon Prime, Hulu, and iTunes as challenges to established television (Lotz 2017), we argue that SME constitutes a more radical cultural and content difference to television.

Outside the US, a diverse and competitive global SME landscape has emerged rapidly. This chapter concentrates on the challenges SME poses for theories of media imperialism, cultural imperialism, and globalization and illustrates global diversity by profiling India (where SME is building out against the hegemony of Bollywood film and music, and offering an alternative to upscale demographics underserved by the great wasteland of television) and China (where SME enjoys greater economic sustainability and growth potential due to Chinese platforms eclipsing their Western counterparts in affordance innovation and e-commerce integrations).

## A New Take on Media Globalization

For the past two decades, the US has extended its seeming dominance of global entertainment to online as the corporate home of the largest digital television portals (HBO Go, Netflix, Amazon Prime) and social media platforms (YouTube, Facebook, Twitter) in the world. These portals and platforms have become familiar to and popular with populations around the planet with even greater ease and swiftness than previous epochs of Hollywood and the music industry's "cultural imperialism." By 2018, Netflix had about 100 million subscribers and could be viewed legally in every country on earth except China, Syria, and North Korea, while representing itself as "the birth of a new global Internet TV service." But these sheer numbers can be trumped by the social media platforms, such as YouTube, which had 1.5 billion users and operated in 76 languages in 2017 (Matney 2017). It is notable that 80% of YouTube traffic comes from outside the US, and 60% of creators' views come from outside their home country. In turn, YouTube is a distant second to Facebook with, by 2018, about 2.5 billion users, which represents more than half of the global online population.

But US companies don't have it all their way. This litany of warp-speed global reach needs to be framed as well against the great singularity of China, where state-based regulation has helped incubate, protect, censor, and build a parallel universe of digital and social media industries which in some respects are world-leading innovators. Parented by the Chinese BATs (Baidu, Alibaba, and Tencent), Chinese SME features an even more hyper-competitive landscape than in the West. The history and contours of this landscape are outlined by Michael Keane (2016) in his account of the rise of digital TV platforms (including iQiyi, Sohu, and LeTV) and by Elaine Zhao's (2016) survey of the proliferation of Chinese social media platforms (Youku Toudu, Weibo, and WeChat). And China is well ahead of the West in live-streaming mobile app development. All of these portals and platforms are competing for some 700 million Chinese mobile web users, roughly twice the entire population of the US.

Therefore, despite what might appear to be the fulfillment of Thomas Friedman's (2005) notion of seamless globalization captured in his book *The World Is Flat*, this new world of online entertainment is far from flat. What we must account for is *both* qualitatively greater frictionlessness in SME media globalization *and* the unevenness or lumpiness consequent on regulatory and political difference territory-by-territory and on the immense industrial and cultural diversity embodied in SME. Beyond the dimensions of platform scalability and state-based governance, other structural conditions inform the variegated manner in which SME operates around the world. One set of technological conditions, such as access to mobile phones, has fueled scale, while other technological conditions can inhibit them, for example, broadband speed. Further complications arise from local economic conditions, including the affordability of technology, growth of middle-class consumption-oriented economies, and the strength of national advertising markets. Other factors include the maturity, domination, cultural content, and representational diversity of national media industries.

Media globalization has been an enduring topic in film, media, and communication studies. Traditionally centered on questions of US "cultural imperialism" through widespread dissemination and popularity of its film and television output, debates of this long-established vintage have been staged, for example, around whether global television traffic is a "one-way street" (Nordenstreng and Varis 1974) or a "patchwork quilt" (Tracey 1988). Influenced by cultural studies' emphasis on viewer and audience agency, versions of "weak" rather than "strong" globalization have largely characterized recent discussion (Tomlinson 1999; Flew 2007; Straubhaar 2007) but also continue to contend with reassertions of "strong" cultural globalization (Boyd-Barrett 2015).

However, a reassessment of this debate is imperative in light of the global reach of YouTube and other major SME platforms, and the types of content they have spawned. On the one hand, it is possible to posit a new wave of media globalization based on the global availability and uptake of SME content which is relatively frictionless compared with national broadcasting and systems of film and DVD licensing by territory. It needs to be emphasized that platforms perform constant and widespread self-regulation in response to takedown notices concerning copyright, as well as to avoid controversy over content offending community standards by dealing with such blights as hate speech and revenge porn. It is estimated that Google alone handles 70–80 million takedown notices per month globally for copyright claims (Mills 2016). However, there is little or no imposed content regulation on the big social platforms – some of the world's largest information and communication companies – as their penetration proliferates globally.

On the other hand, the new professionalizing-amateur screen ecology embodies a huge step change in producer diversity, both in terms of amateur backgrounds and global locations. It is these differences that we stress between such content and platforms, and the legacy system of national broadcasting, film, and DVD release and licensing by windowing and territory. In the latter, established forms of media globalization enter territories with IP-controlled content, whereas platforms such as YouTube facilitate rather than control content, and exhibit much greater content, creator, and language and cultural diversity than traditional global media hegemons.

The implications of SME for contemporary accounts of media imperialism have hardly been broached. Mostly, the lead has been taken by exponents of *very* strong media imperialism theses. Critical political economists Christian Fuchs (2014) and Dal Yong Jin (2013), for example, insist on the strong continuity between earlier forms of media imperialism and today's version.

Fuchs' critique of social media views them entirely through the lens of their claims to democratize access to and participation in information exchange (rather than this chapter's focus on SME) while refusing any and all claims that social media may extend participation and offer outlets for alternative voices:

> The Internet and social media are today stratified, non-participatory spaces and an alternative, non-corporate Internet is needed. Large corporations colonise social media and dominate its attention economy. ... On corporate social media, the liberal freedom of association and assembly are suspended: big corporate and, to a lesser extent, political actors dominate and therefore centralise the formation of speech, association, assembly and opinion on social media.
>
> *(Fuchs 2014, section 5.1)*

For Jin, there is essential continuity across a century of imperialisms (Lenin's imperialism, cultural imperialism, information imperialism, and now platform imperialism). "The US, which had previously controlled non-Western countries with its military power, capital, and later cultural products" Jin argues, "now seems to dominate the world with platforms, benefitting from these platforms, mainly in terms of capital accumulation." He regards "the major role of intellectual property rights as the most significant form of capital accumulation in the digital age" (Jin 2013, 146).

A comparison of SME and the major professionally generated content (PGC) streaming services such as Netflix, Amazon Prime, Hulu, and iTunes highlights some critical distinctions. For Netflix, its aggressive global expansion requires it to negotiate with pre-existing rights holders in each new territory and often requires it to attempt to close down informal means of accessing its popular content such as VPN workarounds in such territories. As such, and together with its levels of investment in original intellectual property around which it builds a strong IP-based global brand, Netflix is rapidly resembling a Hollywood studio.

In contrast, SME content, once purged of infringing content, is largely born global and is created primarily to be "spreadable", in Jenkins, Ford and Green's (2013) felicitous term. Spreadability, for Jenkins and his co-authors, is contrasted with "stickiness" – aggregating content and attention in centralized places, typically through high-control IP regimes. The concept of spreadable media emphasizes the labor undertaken, and the value generated, by creators and audiences, and sets itself against techno-determinist notions of "virality." SME, in stark contrast to content industries in general (and Hollywood and broadcast television industries in particular), is not primarily based on IP control. Rather, YouTube elected to avoid the legally cumbersome traditional media model of owned or shared IP. YouTube also avoided paying fees for content as well as offering backend residual or profit participation. Rather, YouTube entered into "partnership agreements" with their content creators based on a split of advertising revenue from first dollar. Of course, as we argue elsewhere (Cunningham and Craig 2019), this has been no silver bullet for online creators, but it does serve to underline structural differences between main media and SME.

YouTube talks of being primarily a facilitator of creator and content in the many international markets in which it operates, and this goes to another fundamental meaning of spreadability. The key difference between traditional media operating multi-nationally and YouTube is that the former produces, owns or licenses content for distribution, exhibition or sale in multiple territories, while the latter seeks to avoid the conflation of YouTubers as the IP creators with YouTube as "platform" and "middleman" operating to facilitate linking of brands and advertisers with YouTube creators and MCNs. There are significant reasons for YouTube not taking an IP ownership position, which have to do with its continued status as a platform or online service provider rather than a content company. The US Digital Millennium Copyright Act 1998, in addition to criminalizing circumvention measures and heightening the penalties for copyright infringement on the Internet, created "safe harbor" provisions for online service providers (including ISPs) against copyright infringement liability, provided they responsively block access to alleged infringing material on receipt of infringement claims from a rights holder.

Online SME content is being distributed globally in ways that radically depart from time-honored principles and practices of territorial rights and traditional IP control. We are witnessing the rise of a nascent media industry that represents non-traditional media ownership. These are disruptive platforms which challenge our prior conceptions of media globalization, including national regulatory regimes. Furthermore, SME creators, including the native and entrepreneurial vlogger and live streamer, harness the affordances of platforms to engage in content innovation that bears little resemblance to traditional IP. Creator labor and practices operate in stark contrast to traditional media structures, whether media conglomerates, studios, networks, or producers.

Having said all that, however, the relatively frictionless globality seen in the operations of the major digital platforms is decidedly uneven. SME platforms are subject to significant "lumpiness" in business cultures and regulatory frameworks across the globe. YouTube reports that it is "localized" in multiple countries and numerous languages (i.e. it has a local presence, usually consisting of sales forces and government/public relations operatives). Of course, this is not a full index of the global reach – or limits – of YouTube. YouTube is accessed and used across much wider territory than localization data show, while a number of countries block, or restrict access to YouTube. Those that do so tend to also block access to Twitter and/or Facebook as well. North Korea (where Internet access itself is highly restricted) and China (with the exception of the Shanghai Free Trade Zone) block YouTube, Facebook and Twitter. Temporary blocking at a national level has always been an option used to deal with political and/or religious issues. Over the last few years, Pakistan has blocked YouTube on several occasions when it refused to remove an anti-Islamic video. Other nations, including Eritrea, Iran, Egypt, the Congo, Tajikistan, Syria, Iraq, Sudan, Bangladesh, Afghanistan, and Morocco, have also instituted temporary bans.[1]

In addition to the need to deal with global politico-cultural divergences of this order, localization is also a response to pre-existent entertainment content being subject to "location-based filtering that results from the unevenness of content-licensing deals across national jurisdictions" (Burgess 2013, 53). "Location-based filtering" is usually geo-blocking, which occurs when rights holders and/or content producers may not have the rights to show some content in different regions. One of the most prominent cases of geo-blocking is the long dispute between YouTube and GEMA, a performance rights organization in Germany, over payment of rights to performers of music that has resulted in German music videos or videos containing music being hard to access, or unavailable on YouTube in Germany. The seven-year stand-off was partially resolved in late 2016 when YouTube agreed to an undisclosed rate at which GEMA members would be remunerated for video streams.

Variegated media regulation and policy are helping to shape, restrict, censor, as well as assist, this emergent industry. Europe, the Middle East, and Africa (dubbed "EMEA" in the dialect of the platform globalizers) are much more regulated in terms of community standards, sponsorship and advertising than the United States. The relative free-for-all in branded content and sponsorship in the United States is by no means mirrored elsewhere. These are the same "glocalization" dynamics that have been scrutinized in multinational advertising debates for decades, with the difference being that the dollar-per-unit value is, at this stage of the monetization of digital content worldwide, much lower and thus the "education" of brands and advertisers needs to be that much more strategic.

As noted in introducing the chapter, we now provide brief vignettes of SME in India and China, the two largest countries by population, while the latter now has the largest economy in the world by purchasing power parity. These cases are indicative of global diversity and SME dynamism while suggesting that the strong media imperialism position makes little allowance for spatial variation globally and assumes that high-control copyright regimes are normatively the modus operandi of "platform capitalism" (Srnicek 2016).

## India

It is estimated that India will become the youngest country in the world by 2020 and already has the world's largest population of 10–24 year olds. According to a KPMG (2017) report, the number of Internet users in India had expanded to around 389 million in 2016. There has been a 62% increase in the broadband speed India has experienced from a very low and slow base to an average of 4mbps. India is experiencing a proliferation of Internet-enabled mobile phones. It has already exceeded 300 million and is expected to reach 700 million by 2021. 4G connections are predicted to grow at 38% cumulative annual growth rate. About 80% of the connections are expected to be on 3G or 4G by 2021 which is a 55% increase from 25% in 2016. Yet even as access speed is increasing, the cost of mobile telephony and streaming plans remains prohibitively high, despite the rapid growth of the Indian middle class, limiting access to online video mainly to urban dwellers and cosmopolitan youth.

The technological and economic limitations of the Indian mobile market may only temporarily inhibit the growth of Indian SME, whereas the lack of diversity and access to India's traditional film, television, and music industries may further accelerate it. Bollywood has been the dominant format for Indian film and music for decades, inhibiting new genres, stars, voices, and formats. Representing a limited form of vertical media integration, film music has dominated the Indian music industry, foreclosing alternative genres such as Indian rap, pop, or hip hop. Despite 850 channels vying for attention and the advertising rupee, Indian television remains focused on older audiences and traditional formats and content, including primetime soap operas and political news discourse. In numerous instances, our interviewees[2] derided hegemonic Indian popular discourse as "ABCD" – Astrology, Bollywood, Cricket and Devotional. Kumar (2016) sees Indian SME as "gradually chipping away" at the structural dominance of institutions such as Bollywood, television networks, and the celebrity culture that have remained the hegemonic forces within the sphere of cultural production in India.

As a consequence, millennials – at least those urbanized and educated – have turned off television, migrated online, and begun to create content outside these national and indeed global expressions of Indian-ness, grounded in hyper-local regionalism, millennial popular interests, and sharp satire. Indian SME has fostered a wave of new voices, genres, and formats, alternative to traditional Indian film, TV, and music. In many cases, this has seen the production online of scripted web series and satirical comedy which would be considered more mainstream in the West, but which have been radically underdeveloped in India. Having been rejected by Indian-based MTV networks, the creators and producers at The Viral Fever (TVF), for example, launched teen-oriented sitcoms such as *Permanent Roommates*, which Indian media outlets claimed was the second most watched scripted web series in the world at the time (Parthasarthi 2015). TVF, along with a raft of other native SME production companies, have emerged into this space launching scripted web series tailored to their youthful, urban, and middle-class viewers.

Operating outside of the Bollywood film and Indian television and music industries, talent harnessed social media platforms to promote performances by alternative musicians and stand-up comedians. As Indian SME rapidly evolved with venture capital investment, improvements in broadband and mobile access, and advertising support, entrepreneurial firms pivoted just as quickly into SME management and production companies. Only Music Louder (OML) started with music journalists and promoters using online digital zines and platforms to sell tickets to venues and festivals featuring music, comedy, and storytelling for the growing urban and youthful Indian middle class.

OML represents prominent comedy troupes like All India Bakchod (AIB) and East India Comedy (EIC). Operating in the less-regulated and -censored space of SME platforms, these troupes engage in social critique, for which they have both benefitted from recognition and suffered comparable backlash. AIB's notorious satirical video, *Rape: It's Your Fault* generated over 6 million views, while other social-issue topics tackled by the troupe include gay rights, sexism, recent efforts at demonetization by the Modi government, and even net neutrality. AIB has been resolutely opposed to the threat to net neutrality posed by Facebook's Free Basics program, even as Facebook's growth would seem to benefit the troupe which has 3.5 million likes on the platform.

Differentiating Indian SME intermediary firms reflects a rapidly maturing Indian SME industry. These firms are informed by the interests of their owners and the opportunities to cater to an underserved and diverse Indian market and, simultaneously, global cosmopolitan audiences rarely exposed to Indian culture outside of Bollywood film fare. Creator-centric Qyuki, owned in part by world-renowned Indian musician, A.R. Rahman, has secured multiple rounds of VC investment and YouTube partnerships to promote next-gen native online musicians such as India's first Bollywood A Cappella group, Raaga Trippin'. Culture Machine harnesses state-of-the-art digital technologies to create advertiser-friendly content with the push of a button across multiple platforms. Ping Network has partnered with YouTube and Tastemade to launch the Indian Food Network, fostering the careers of regional chefs, including rural housewives sharing their family recipes, for global food audiences. The South Indian dishes created by Preetha, the host of Dashan Curry, featuring recipes written in English and Hindi, have secured global audiences of millions.

The growth of and investment in Indian SME firms as well as empowered and diverse Indian creators suggest a highly dynamic content culture unbeholden to traditional notions of US cultural imperialism. However, like all small creative enterprises, Indian SME companies face challenges of how to grow. For several, it has been through buyouts. Mumbai-based YoBoHo, once an outsourced and under-paid service firm supplying Western animation, became the largest digital animation company in the world. In 2015, the firm was acquired by Broadband TV, a Canadian SME firm launched by Iranian-born Canadian Sharhzad Rafti, and, in turn, owned by RTL, the European media conglomerate. Culture Machine is now a wholly owned subsidiary of Singapore-based investment firm, The Aleph Group. Indian SME is already being globalized, come what may.

## China

As with India, the contours and evolution of China's digital and social media industries have been fueled by the growth in China's economy and middle class. Many recent innovations in global online commerce (e.g. QR codes, digital wallets), messaging, and live-streaming have been incubated and popularized in China. In many ways, China's singular and rapid development of its tech and online industries is due to the fact that:

> [I]t was able to fill a vacuum after the country essentially created much of its economy from scratch following the end of the Cultural Revolution … . Unlike in the United States, where banks and retailers already have strongholds on customers, China's state-run lenders are inefficient, and retailers never expanded broadly enough to serve a fast-growing middle class.
>
> *(Mozur 2016)*

For this reason, there is much more sustainable basis to monetization, as China can be very much characterized as an early-stage consumer culture. Indeed, drawing on some of Gabriel's (2006) work, we would characterize the zeal with which large populations have embraced the thorough commercialization of personalized social media as China's early-stage capitalist modernism. Outside the "first tier" cities on the east coast, there is significant fall-away of major branded bricks and mortar consumer outlets and thus e-commerce thrives. Competition among digital platforms is more intense than in the US as they are more unilaterally focused on mobile applications in a country which leads the world by a long way in mobile phone ownership and where the installed base of standard computers per capita is low.

China has notoriously built an alternative online ecosystem emerging from state-based intervention which includes not only banning YouTube, Facebook, Twitter, and Instagram but nurturing the growth of their own platforms (Keane 2016). The government's digital economic strategy incubated the massive tech giants, the so-called BATs, which, in turn, either birthed or adopted multiple highly competitive online TV platforms. As a consequence of this highly charged, iterative, competing landscape, a number of online TV platforms have since been aborted, including Ku6, while other platforms have pivoted aggressively toward a subscription based, PGC model (for example, iQiyi, LeTv, and Sohu). The latter platforms have done more than engage in a programming arms race over US and Chinese film and television content. Like Western SVOD platforms, China's iQiyi is distributing and producing ever more sophisticated fare, even partnering with Netflix to distribute content inside China (Brzeski 2017).

Zhao (2016) gives a sense of the rate of experiment and change in the short history of the online video space in China and its uneven professionalization. Initial enthusiasm for the possibilities of user-generated, amateur content saw a major correction when swathes of copyright-infringing content and platforms were closed down by the state in 2007–8. Platforms swung, pendulum-like, to PGC but many struggled with the cost of licensing increasingly expensive licit content. Zhao focuses on the current period of "resurgence and revalidation" of UGC. Online video production has rapidly come to offer alternatives to officially sanctioned institutions in cultural production, distribution and consumption. Many online video platforms, she argues, have their roots in amateur practices, facilitating flows of content unavailable in the official marketplace.

Thus, platforms such as Youku Tudou have moved away from the more expensive and competitive PGC portals to return to their original value proposition of user-generated content with social networking capabilities. Similar to the West, multiple social media platforms have launched to compete with or offer diverse affordances with Youku. China's gameplay platforms such as Duoyu and PandaTV have already outpaced Western equivalents like Twitch or YouTube Gaming, while accelerating China's booming e-game industry. The live-broadcasting affordances of these platforms super-charged the launch of over 100 live mobile applications in the past few years which, in turn, contributed to a swift backlash from China's censors (Custer 2016a). What the state incubates, they may also take away.

Despite rapid commercialization and expansion of the system, the state unrepentantly intervenes to censor content deemed "unhealthy" or "harmful" before production or force offending content to be removed from screens. Qing Zhou of Feidieshuo suggests this causes online entrepreneurs, as much as mainstream broadcasters, to be agile and creative in the way they commission and produce content, trying to "think ahead of" Chinese media regulation in its guise as both censor and protector – not only in the production of original online content, but in the repackaging of that content for traditional television audiences (Zhao 2016). One of our key Chinese informants, SME entrepreneur Heng Cai, forensically enumerates five ways the government seeks to manage online video: blocking the channel; licensing (to be a legal video distribution platform, one needs a permit called an "Internet broadcasting license"); censoring (which effectively means most video platforms self-censor); and subsidizing (the government hires "Internet commentators," who are paid by the propaganda department to write comments favoring the government). If one or more of these interventions fail, the final option is banning outright (Cai 2016).

Second-gen platforms like Tencent's WeChat, China's formidable messenger service, have reverse engineered platform development in the West. Facebook's purchase of WhatsApp messenger service for an exorbitant $19 billion may have represented a defensive effort by the West to block Chinese platform penetration outside their great firewall. In light of Chinese commitment to the soft power of media (Xi 2014), Zuckerberg barely thwarted the first salvo in the art of a social media war rapidly looming on the digital horizon (Lunden 2014).

Micro-blogging platform Weibo makes an interesting contrast with its US-based counterpart Twitter. Like Twitter, Weibo is a text-based micro-blogging service that has been threatened repeatedly with demise from various competitive threats, not least of which is WeChat (Custer 2016b). Unlike Twitter, Weibo integrated photos and a video player into the platform from the beginning, which helped nurture key opinion leaders (or KOLs), Weibo personalities from throughout the public sphere, including entrepreneurs, politicians, and celebrities, engaging millions of Chinese netizens around the globe. With the proliferation of smart phones with 4G speed fostering 700 million subscriptions, leaping over Youku, Weibo has become the premiere platform for short video content, helping foster the advertising-defined "influencer" (Wanghong 网红) economy (Zhao 2016).

The Chinese influencer economy affirms how these platforms have helped create the technological and commercial conditions upon which an alternative Chinese SME industry has emerged. In an interview with Di Zhou, Social Media Account and Program Operation Manager with Xinpianchang, Zhou stressed how Chinese SME features rapidly professionalizing-amateur Chinese content creators engaging in content innovation distinct from traditional Chinese film and television, like the aforementioned game players, food, fashion, and style vloggers, and a wave of social media celebrities. Bryan Shao (2016), Vice President of Cooperate Strategies and PGC operation at Youku, also stressed the popularity of online content steeped in China's long history, particularly those that differ from that offered through traditional TV. Once again, the Chinese state has engaged in tandem actions, nursing and disciplining these upstart micro-celebrities, like the Chinese vlog queen, Miss Papi. Wrist-slapped by censors for foul language, Papi issued a message of contrition to her 11 million followers, while simultaneously securing multi-millions in investment and brand integration (BBC 2016).

As in the West, the Chinese SME industry co-evolves alongside traditional Chinese media. In contrast to the US, but comparable to India, the industry exploits the lack of diversity in traditional film and TV to feature more professionally generated traditional content, albeit in more affordable formats than traditional scripted fare. One of the most popular shows on Youku is *The Luogic Show*, a history and social-issue themed talk show hosted by former CCTV producer Luo Zhen Yu. *Baozoudashijian* is an entertainment show hosted by Baozoumanhua, who remains anonymous, emerging onscreen solely in a *papier-mâché* mask, providing arguably a brilliant ploy to thwart state censorship.

As in the West, a new wave of intermediaries operating between platforms, creators, advertisers, and traditional media have emerged. Unlike the West, these firms also function as digital production companies generating original IP content across multiple platforms. As Qing Zhou of Feidieshou stated, "As advertising is our major source of revenue we seek to enlarge our influence. Our goal then is to put Feidieshou content across Youku, Weibo, iQiyi, Tencent, LeTV, Meipai, Miaopai, Baidu, Panda and Douyu" (Zhao 2016). Playing off the continued decline of youth audiences across CCTV, Feidieshuo has developed original animation designed for millennials, or rather 80s and 90s *balinghous* and *jiulinghuos*, featuring mature topics around relationships, sex, and social pressure missing on Chinese TV (Zhao 2016).

## Conclusion

This chapter has analyzed the global dynamics of SME, contrasting them to those of television, and illustrating them through the cases of India and China. Our analysis suggests the need for a revisionist account of strong versions of media imperialism. This is based on two arguments. First is the general observation that what distinguishes the concept of media imperialism from cultural imperialism often turns on the difference between influence in news and the infosphere on the one hand, and influence in entertainment formats on the other. Strong accounts of media imperialism such as those canvassed in this chapter typically focus on news and the infosphere whereas, in the realm of entertainment and popular culture, strong notions of domination are much harder to maintain. Second, SME content on the big social media platforms circle the globe largely independently of standard windowing and territorial licensing, and without standard IP control, suggesting a fundamentally different model of extreme spreadability *without* domination.

Having asserted the relative frictionlessness of SME globally based on platform "facilitation" rather than IP control, we have been careful to stress that it is mitigated by irreducible national cultural, market and regulatory specificity. Indian SME has been shaped nationally as an alternative to the long-standing hegemony of Bollywood, a hegemony largely exercised within Indian media and culture. Protected as well as censored by the state and fueled by China's transition to a consumption-based economy, the Chinese platforms and the SME industry based on them are well advanced and may look to go global themselves.

Globalizing SME is now beginning to attract some scholars' attention. For example, Yomna Elsayed (2016) and Mahommed El Marzouki (2017) have focused on the cultural politics and commercialization practices of Middle Eastern creators in Egypt and Morocco, respectively. Carlos Scolari and Damián Fraticelli (2017) examine the discursive practices of Spanish YouTubers and their influence on traditional Spanish media. These national case studies further highlight the multivariance of global SME on national, trans-cultural, and language axes.

Due to their vintage, debates in media and communication studies about media and cultural imperialism and globalization can develop a shop-worn quality. Paying attention to globalizing SME can help to rejuvenate the field as well as update television studies by focusing on what (young) people formerly known as (part of) the television audience are into.

## Notes

1. Data on these matters vary; however, Google posts traffic disruptions to their site: www.google.com/transparencyreport/traffic/disruptions/#expand=Y2015,Y2014, Y2013, Y2012.
2. As part of a wider study of global SME (Cunningham and Craig 2019), we interviewed numerous key informants including creators, SME companies, agents, and platform executives in India and China. Where interviewees are quoted, they are cited in the references.

## References

BBC (2016) "China Internet Star Papi Jiang Promises 'Corrections' After Reprimand." *BBC News*, 20 April. www.bbc.com/news/world-asia-china-36089057, accessed 16 June 2017.

Boyd-Barrett, O. (2015). *Media Imperialism*. London: Sage.

Brzeski, P. (2017) "Netflix Signs Licensing Deal with China's iQiyi." *The Hollywood Reporter*, 24 April. www.hollywoodreporter.com/news/netflix-signs-licensing-deal-chinas-iqiyi-997071, accessed 04 September 2017.

Burgess, J. (2013) YouTube and the Formalisation of Amateur Media. In *Amateur Media: Social, Cultural and Legal Perspectives*, edited by D. Hunter, R. Lobato, M. Richardson and J. Thomas, 53–58. London and New York: Routledge.

Cai, H. (2016) Co-Founder, Star Station TV, 31 May, interview with Stuart Cunningham and David Craig, China.

Cunningham, S. and Craig, D. (2019) *Social Media Entertainment: The New Intersection of Hollywood and Silicon Valley*. New York: New York University Press.

Custer, C. (2016a) "China's Live-Streaming Platforms are all Under Investigation for Sex, Violence, and Crime." *Tech in Asia*, 15 April. www.techinasia.com/chinas-livestreaming-platforms-investigation-sex-violence-crime, accessed 16 June 2016.

Custer, C. (2016b) "WeChat May be King, but Weibo Isn't as Dead as You Think." *Tech in Asia*, 28 June. www.techinasia.com/wechat-king-weibo-dead-data-shows, accessed 16 June 2016.

Elsayed, Y. (2016) Laughing Through Change: Subversive Humor in Online Videos of Arab Youth. *International Journal of Communication*, 10(2016): 5102–5122.
Flew, T. (2007) *Understanding Global Media*. Basingstoke: Palgrave Macmillan.
Friedman, T. (2005) *The World Is Flat: A Brief History of the Twenty-First Century*. New York: Farrar, Straus and Giroux.
Fuchs, C. (2014) *Social Media: A Critical Introduction*. London: Sage.
Gabriel, S. (2006) *Chinese Capitalism and the Modernist Vision*. London: Routledge.
Gray, J. and Lotz, A. (2012) *Television Studies*. Cambridge: Polity Press.
Hetcher, S. (2013) Amateur Creative Digital Content and Proportional Commerce. In *Amateur Media: Social, Cultural and Legal Perspectives*, edited by D. Hunter, R. Lobato, M. Richardson and J. Thomas, 35–52. London and New York: Routledge.
Jenkins, H, Ford, S. and Green, J. (2013) *Spreadable Media: Creating Value and Meaning in a Networked Culture*. New York: NYU Press.
Jin, D.Y. (2013) The Construction of Platform Imperialism in the Globalization Era. *tripleC*, 11(1): 145–172.
Keane, M. (2016) Disconnecting, Connecting, and Reconnecting: How Chinese Television Found Its Way Out of the Box. *International Journal of Communication*, 10(2016): 5426–5443.
KPMG (2017) Media for the Masses: The Promise Unfolds, Media and Entertainment, Industry Report. https://assets.kpmg.com/content/dam/kpmg/in/pdf/2017/04/FICCI-Frames-2017.pdf, accessed 21 August 2017.
Kumar, S. (2016) YouTube Nation: Precarity and Agency in India's Online Video Scene. *Interational Journal of Communication*, 10(2016): 5608–5625.
Lotz, A. (2017) *Portals: A Treatise on Internet-Distributed Television*. Ann Arbor, MI: Michigan Publishing, University of Michigan Library.
Lunden, I. (2014) "If WhatsApp Is Worth $19B, Then WeChat's Worth "At Least $60B" Says CLSA." *TechCrunch*, 11 March. https://techcrunch.com/2014/03/11/if-whatsapp-is-worth-19b-then-wechats-worth-at-least-60b-says-clsa/, accessed 16 June 2017.
Marzouki, M. (2017) "Commerce, Creativity and Youth Participation on YouTube Morocco." Paper presented at The 67th Annual Conference of the International Communication Association, Interventions: Communication Research and Practice, 25–29 May, San Diego, CA.
Matney (2017), https://techcrunch.com/2017/06/22/youtube-has-1-5-billion-logged-in-monthly-users-watching-a-ton-of-mobile-video/
Mills, C. (2016) "Google's Copyright Takedowns Have Grown by a Gazillion Per Cent in Ten Years." *Gizmodo*, 08 March. www.gizmodo.com.au/2016/03/googles-copyright-takedowns-have-grownby-a-trillion-percent-in-ten-years/, accessed 21 August 2017.
Mozur, P. (2016) "China, Not Silicon Valley, Is Cutting Edge in Mobile Tech." *The New York Times*, 02 August. www.nytimes.com/2016/08/03/technology/china-mobile-tech-innovation-silicon-valley.html, accessed 16 June 2017.
Nordenstreng, K. and Varis, T. (1974) "Television Traffic: A One-Way Street." *UNESCO Reports and Papers on Mass Communication No. 70*. Paris: UNESCO.
Parthasarthi, V. (2015) "Permanent Roommates: Going the Web Way." *The Hindu*, 18 June. www.thehindu.com/features/metroplus/permanent-roommates-going-the-web-way/article7329851.ece, accessed 16 June 2017.
Scolari, C. and Fraticelli, D. (2017) The Case of the Top Spanish YouTubers: Emerging Media Subjects and Discourse Practices in the New Media Ecology. *Convergence: The International Journal of Research into New Media Technologies*, doi: 10.1177/1354856517721807.
Shao, B. (2016) Vice President Cooperate Strategies and PGC Operation, Youku, 31 May, interview with David Craig, China.
Shapiro, R. and Aneja, S. (2018) "Unlocking the Gates: America's new creative economy." www.recreatecoalition.org/wp-content/uploads/2018/02/ReCreate-Creative-Economy-Study-Report.pdf, accessed 14 May 2018.
Srnicek, N. (2016) *Platform Capitalism*. Oxford: Polity Press.
Straubhaar, J. (2007) *World Television: from Global to Local*. Thousand Oaks, CA: Sage.
Tomlinson, J. (1999) *Globalization and Culture*. Chicago, IL: University of Chicago Press.
Tracey, M. (1988) Popular Culture and the Economics of Global Television. *Intermedia*, *16*(2): 19–25.
Xi, J. (2014) *The Governance of China*. Beijing: Foreign Languages Press.
Zhao, E. (2016) "Professionalization of Amateur Production in Online Screen Entertainment in China: Hopes, Frustrations and Uncertainties." *International Journal of Communication*, 10(2016): 5444–5462.

# 5

# SYMBOLIC ECOLOGIES

## Between Technologies, Screens and Society[1]

*Jorge A. González*

### Introduction

In these few pages I will consider, in provisional form, the specific relationship between a contemporary technological device—the screen—and its connection to the way we relate to the world through the meaning-making dimension of society. Instead of a set of fixed structures, I prefer to understand that dimension as a constant process for creating and re-creating any society. I have proposed the term *symbolic ecology* as a form to study the development of complex socio-historical relationships, comprising all levels of productive and reproductive life, simultaneously mediated by complex *signs* (information), complex forms of *association* (communication), and complex forms of *adaptation* (knowledge), by means of innovations capable of breaking dependencies and, so to speak, establishing protocols for action resulting in growth of capacities, at once symbolic, technological and organizational. Knowing means acting. Intelligent actions can help to solve a wide range of concrete and abstract problems.

I will therefore explore ways to formulate certain questions on the relative efficacy of screen technologies—understood as a vector rather than merely as a device—based on those three central and interconnected dimensions of human social life. Instead of simply reducing the current iteration of this domain to "everything that happens in cyberspace," I have already referred to the technological efficacy and development of these three dimensions as *cybercultur@* (González, 2015a, 2017).

### Technologies, Histories and Societies

The history of the relationships between communication and technology is the history of the development of humanity. Relations between technology and society are among certain constants in the historical development of humanity. Indeed, what distinguishes us from other social animals is not the use of tools, or the use of language, but, rather, that we remain the only species that not only manufactures tools, but also produces meta-tools, which is to say, tools that are specially designed *to manufacture other tools* (Cirese, 1984: 103–114). For example, a stone with certain characteristics may be *converted* into a tool when used by a bird or a monkey or as when we use it to open a coconut (which is more difficult to open by hand) or to more forcefully beat prey.[2] Yet it is only humans that would use a stone to shape and sharpen another stone into a spearhead for future hunting exhibitions. Only human beings produce these *special types* of (meta-) tools, which are essential for us to design and manufacture *other tools*.

So too with languages. Profuse ethological research has shown that several species (including dolphins, apes and elephants) develop and use relatively complex languages for coordinating their actions (de Waal, Freeman & Hall, 2005; Plotnik & Reiss, 2006; Reiss & Marino, 2001). However, the ability to use languages beyond merely naming things and events but also naming *other words and other languages*, it would so far appear, is unique to humans. Our species has developed a very complex syntax as an instrument for symbolically organizing and processing our experiences and actions in the world. The animal languages we know, even when they are very complex, are limited to the description of events in *the here and now*. It seems they cannot use language to refer to events not immediately present. Any abilities to coordinate their actions emerge when the necessity arises in the face of an event. If not required, language is not used. Thus, animals do not demonstrate evidence of being able to create possible, imagined, or alternative worlds. They cannot make art, religion, philosophy or science. Their language can only operate on objects within reach of their senses and immediate experience. Thus, their languages are only ever *presentational* (Del Río & Álvarez, 2011).

It is the development and use of meta-languages and meta-tools that have been unique to humans and essential to the development of our complex society. Only humans use (meta-) languages to represent not only the world's objects but also *other languages* and indeed *other complex symbolic realms*. Moreover, the historical development of these *second order* languages and tools has been a completely simultaneous process. These two elements have continuously intervened and intermixed with the result that they have given rise to all human cultures. It would be pointless to try to determine which of the two appeared first. The technologies developed to represent and to communicate our realities and our complex symbolic realms are part of the same technological developments we associate with the tools of civilization and, ultimately, of modernity. This continues to be true right into our current situation in which we are experiencing the blindingly rapid creation and diffusion of myriad contemporary technologies and platforms for networking.

### Childhood memories flash

I remember my grandmother's house in Colima (800 km west of Mexico City). When I was very little, at the beginning of the 1960s, I observed the coexistence of two technologies for illuminating the house: one—far from my experience up to that point in the capital of the country—was the oil lanterns that required an entire ritual of procedures to keep a room lit. The other—that was completely "natural" for me, but here mounted on the walls with cables that ended in a switch, a button, that simply by being pushed illuminated the entire space without smoke residuals or greater fuss, was the electric light bulb. Of course, those "buttons" helped change modes of relation in my grandmother's house. Time, and labor, and the very placement and intensity of light was changed. Indeed, the energy supply system was transformed (gradually, as homes and businesses adapted to the use of electricity, traveling oil salesmen began to disappear) and buildings were reconfigured and instruments that depended on the flow of that invisible energy that "provided light" made it possible to listen to the radio, hear music, quickly purée sauces, mix dough for pastries and make holes in the wall with greater speed and precision.

Indeed, at this stage in world capitalism, it has never been more appropriate to understand technological development as a *vector*. Which is to say, as a force with direction, with origin and implied destinations, and one that generates displacements of varying magnitudes, temporalities and intensities. Here I am not only referring to the area of material production, or social organization, but especially to representations of the world and life and how they are communicated and reconciled within symbolic realms. Increasingly, our daily life is surrounded and, at times, saturated by two devices that have left their marks on the past century and still affect the rhythm of the new millennium: buttons and screens.

Since the appearance of electric light and its progressive conquest of rooms and work spaces, the production of daily life in society has been completely modified, with ever growing deference to buttons and now screens.

## Screens: Light and Form *to Narrate* Something to Us

Screens have in fact been central to the organization of human society and the production of symbolic knowledge since long before electricity was tamed for domestic consumption. Some studies[3] indicate that even some cave paintings reproduce shadows that firelight likely distorted on the cave walls. In most cases, it seems, screens offered propitiatory or commemorative and, always, symbolic forms of representation. Since long before electric lighting we had already developed the practice of projecting shadows, which has origins in the East for ritual and liturgical purposes (in India) and was expanded for use in entertainment performances with plots and stereotypical characters (in China). At some point during the Early Middle Ages this practice found its way into Turkey and (with the expansion of the Ottoman Empire) from there it spread throughout all of Europe (Luengo & Alcalá, 1991).

A source of light from behind a white cloth, some objects and a simple plot. The effect of concentration and pleasure that viewing these productions generates, even today, is very moving. With screens, we have developed unique abilities to tell stories with characters and action that focuses the attention of the audience.

## Screens for Narrating Stories, History and Memories Linked to Feelings

With the technological developments of industrial modernity and the eventual domestication of electric energy, the ancestral technique of shadow theater was modified considerably. Within just a few short years following the first public screenings (1895), for example, people around the world began to consider it simply "normal" to attend "viewings" of moving-image films, mechanically reproducing images and movements across time and space (see Figure 5.1).

From exemplary scenes and sequences accompanied by live music, film exhibitions soon progressed to the narration of stories: some fantastic, some evidently realistic, others about love, still others of horror, action, violence. The universe of the imagination that we formerly constructed in our heads from the telling of stories and then the reading of books could now be seen and explored in all its complexity on the screen. Along with this addition to our imaginative realm, the composition of collective life and, particularly, our representations of it, also began to evolve.

In Mexico, we have few studies on the magnitude of such cultural and social transformations of this constantly accelerating interaction between the technology behind these enormous screens and the representational universes of love, evil, fear, humor and the presence of others (Gómez, 2005, 2007). Gradually, as a result of their very ability to graphically represent and mediate events and situations in motion, screens have become almost ubiquitous in our society: in the workplace, academia, the street, the market, public security, traveling, computers, communication between life partners, health … in short, mediation by the screen is now omnipresent. We have even fewer studies on the way small

*Figure 5.1* Sign for an exhibition of the Kinetophone (Colima City, 1913)
Source: Author's archives

screens, as additional devices of visual, if individual, gratification, accompany us in our daily lives and their scarcely documented cognitive and perceptual efficacy (Gorn, Chattopadhyay, Sengupta and Tripathi, 2004; Grödal, 1994; Hutchins & Alâc, 2004; Maass & González, 2005).

It is clearly observable that in many cases, even when present at an event, people prefer *watching* it on screens. Something almost magical and fascinating happens on screens, somehow more verifiable or personally engaging than the very events being depicted. Thus, people in the audience at soccer matches follow the action in front of them instead as it streams to their phones or is replayed on the jumbo-screen above the stadium. Similarly, when someone speaks at a large conference or performs at a music concert and behind the person there is a giant screen transmitting the presentation for the audience in the far back, or on television, or webcasting, the majority of the audience, wherever they are seated in that auditorium, despite being in front of the speaker, will watch the image on the screen. The ability of screens to capture *biological time—biotime*, understood as the amount of proportional time engaged to screens related to a daily set of social activities—is astonishing, and we must turn to an integrated combination of neurology, psychology and sociology of perception to understand it.

The attention screens command can be crucial when they mediate, for example, MRI diagnostics, the most detailed operations of microsurgery, video door phones, traffic monitors, or the remote landing of a probe on Mars. Indeed, from cellphones and digital cameras to security surveillance and drone piloting to scoreboards and billboards ... that is to say, all spheres of contemporary life, screen mediation is increasingly the very fabric of our everyday life in contemporary society and thus more important than ever to understand and come to terms with.

Even with the awareness of increasingly being seen, we are forever attached to the pleasure of seeing. Yet a body of research in the neurosciences is dedicated to surmising health risks and some indicates that activity mediated by screens can amplify underlying conditions and at times share some characteristics of addiction:

> Brain imaging research is showing that stimulating glowing screens are as dopaminergic (dopamine activating) to the brain's pleasure center as sex. And a growing mountain of clinical research correlates screen tech with disorders like ADHD, addiction, anxiety, depression, increased aggression and even psychosis. Most shocking of all, recent brain imaging studies conclusively show that excessive screen exposure can neurologically damage a young person's developing brain in the same way that cocaine addiction can.
>
> *(Kardaras, 2016: 4)*

At the least, such studies point to how much has yet to be learned about the role our bodies play in the mediation of culture through screens.

The culture industries themselves, in response to the shift toward the screen in entertainment, have adopted criteria seemingly more consistent with the qualities of the screen and the profitability of its products: for example, it is more important for an actress to *look good* on the screen than, say, to be an exceptionally talented singer. Screen appearance is more highly valued than any other "talent" or attribute. This aspect was one of various findings of our collective work on the relationship between *telenovelas* and Mexican society at the end of the 1980s (González, 2010). In this study my collaborators and I demonstrated more broadly that in a domestic society with a weak underlying civil society, in place of coffee houses and newspapers, television (in addition to religion) has become the principal means for metabolizing collective life.

Nothing of what I have written here so far is new. We can all agree, in general, on the overwhelming and mesmerizing presence of screens—small or enormous, functional or narcissistic, manipulative or pleasurable: the way social relations are made visible by these devices in social life has no comparison in our history. Certainly the consequences may be particularly serious: from (neurological, psychological and social) addictions generated by texting, extremely long sessions of on-line/off-line video gaming, constant and anxious checking of cellphones, and other such behaviors by children, young people and even adults all over the world, to the growing acceptance of a constant state of surveillance to the very means by which we relate to each other and organize our societies. Television is one of those screens and, for a time, was the most important in the establishment and definition of the population's cultural diet.[4] For new generations, video games and other forms of entertainment and social networking platforms (Whatsapp, Facebook, Twitter, Orkut, Linkedin, Meetup, Slack, Purple Squirrel, etc.), empowered by the Internet, all form part of the daily menu and social organization and acceptance focused through our screens As a matter of fact, those apps seems to be designed to become addictive to users.[5]

## By Wondering We Advance: Information, Communication and Knowledge (Again)

So should we be shivering with technophobia or celebrating technophilia? My position does not accept either of these two false options (González, 2017). Instead some important questions should be considered from the perspective I am proposing: How has this technological process affected our ability to relate experiences to codes, norms and the symbolic realm? In short, how has it affected our ability to think in a relational way, that means, our common *ecology of information*? How has this technological process, animated through the imposition of a neoliberal politics of "globalization," affected our ability to coordinate and organize our actions with others? In other words, how has this *screenification* of the world affected and how does it continue to affect our *ecology of communication*?

From the perspective I have developed and called cybercultur@ research and development, I argue that *all information and communication technology is simultaneously a technology of knowledge*. The implications of this formulation are so powerful that it is ignored or overlooked only to our great detriment. The details of concrete analysis, the complex relations historically established between information-communication and knowledge, demonstrate that devices empowered *by* screens and the Internet can function as technologies of knowledge, or else they will become *technologies of ignorance*. We use them or they use us. We can be empowered or else risk empowering devices of dislocation, deterritorialization, deregulation, disinformation, desensitization and disconnection.

This possibility is evident if we consider one of the most well-worn notions broadly used in contemporary studies: the field that defines itself as Information and Communication Technologies. "ICT" is now a commonplace, with the most important component of *knowledge* conspicuous by its absence. It takes for granted and teaches instrumentally, but not deeply nor critically. Instead, if we think about it, ICT is used often in place of more precise concepts for differentiating and integrating aspects of these technologies in this "globalized" world. It is a concept that blinds as much—or more—than it illuminates. It merges an undifferentiated assumption of practice and content in place of specific, concrete analysis of history, use, effect, impact and symbolic configuration.

Such is a symptom of a larger process coming to light amid all the glowing screens. The access one now has to information and very diverse symbolic forms (Thompson, 2005) within reach at all times on a tablet or any other small device, workable with one hand and connected to the Internet, is enormous. Portable screens, with their memory capacity, instant publication, manipulation of images, texts, sounds, textures and the ease they afford for uploading videos and images of the moment to the web, not only open a door to creativity but they are also linked to certain behaviors that indicate worrisome and numbing attitudes (without the ability to feel empathy) in the face of critical situations, such as accidents, people dying, muggings in the street, school bullying, sexual harassment and mob violence—in which one is merely encouraged to "share," "like" or "comment." The multiple forms of insults and disrespect that one can read in the comments sections on news pages, with alarming levels of insensitivity, with violent, obscene, intolerant content, are vast and increasing.

One can reasonably argue that this behavior is not directly caused by the devices and applications that enable it. That is true. However, given the speed of adaptation, increasingly indiscriminate access—limited only by the money it costs to buy a device (which is increasingly inexpensive)—and the aura of attraction surrounding new screen devices, the drive of information and communication enabled by these technologies without recourse to knowledge is nearly overwhelming. This may be by design. Ultimately information is not knowledge, and without the processing of information, knowledge is not generated. Processing is much more complicated and requires time. There is no such thing as "Fast Knowledge." The development of knowledge emerges slowly, as a whole, by conversing, collectively—not at all from within isolated solipsism—and it is only with knowledge, processed in this way, that we can *act to transform* the world. Knowledge is not merely a cognitive and rational process. It is simultaneously a process of *learning to act* in society, *learning how to manage oneself* with others in specific situations. It is a structure for action in the face of problems.

The so-called ICTs, shaped by a politics of forced "globalization," have been established as technologies of speed and data, ultimately producing ignorance rather than knowledge and, thus, operate as technologies of inaction. It is not that ICTs are ontologically "evil" by necessity but, once again, when we disconnect information and communication from knowledge, our ability to act is reduced, narrowed and depleted. This is partly due to lack of exercise, partly due to their addictive aspect (we cannot act if we cannot bear to break away from the screen) (Kardaras, 2016; Prospero et al., 2012), and, in part, as we demonstrated in Mexico in the 1980s, because in societies with an underdeveloped civil society—which is to say, with low citizen participation and awareness, citizens who are more accustomed to not hearing or making their own voices heard (Couldry, 2010), as well as with limited awareness of co-responsibility in the construction of social life and thus they are victimized, ignorant and reckless with regard to their own process of anthroponomic production, the production of their bodies and minds, (Bertaux, 1977)—a sense of the importance of the other in the construction of daily coexistence is normally precarious and superficially disposable.

This is the kind of "normal" social order in which the so-called "global media" operate and generate their own forms of profits, both monetary and symbolic. Only in cases where this social order is broken, suspended as in cases of catastrophes, during the first moments of awareness of this rupture, with confirmation that the suffering of the other could also be my own, are the conditions made evident that can generate a reason to reconfirm that society is an intelligent way to coordinate our actions and to help us live well (Lenkersdorf, 2008). The most progressive and subversive element in response to this "world" order (imposed and docilely accepted) of distributed ignorance—the most filled with memory, the present and possible worlds—is when the sentiment and clear perception emerges, spreads and activates people with the idea that "if others are well, then I am better" (González & Krohling-Peruzzo, 2018). Simultaneously the clear perception is generated that if I am well, this also means that others are (also) better. A proper win-win situation, of the type quite different from that so often evoked in the neoliberal rhetoric of world society: one winner, many losers.

Returning now to the development of cybercultur@, we will see that this involves at least *three* dimensions.

*1. The Culture of Information*

The first dimension, in which recklessness and lack of development is observed, occurs when enormous sectors of the population (as is taking place in Mexico) have a direct aversion to, and great difficulty in, developing original and relational thought, for instance, as with the resistance, even rejection of learning and using logics and math in elementary schools and the difficulty in understanding the practical and strategic importance of preserving and classifying social information (González, 2015a: 41). This is a formidable obstacle to being able to think and represent structures of relations and dynamics of processes, such as, courses of action in which structures are transformed through continuous functional mechanisms (García, 2000; Piaget, 1978). In other words, an inability to conceive transformations of temporal and relational structures that enable us to understand actions, interactions and phenomena. This has resulted, in part, from a disinvestment (both fiscal and moral) in education. In place of original and relational thinking, we perpetuate a simplistic, mutilated perception of the world. These kinds of mutilated visions have mutilating consequences in the real world.

*2. The Culture of Communication*

The second dimension resides in *our ability to coordinate actions* when, for different reasons (especially because of the poor culture of information), *we are not aware* of the fact that the social forms into which we organize ourselves (which tend to be vertical and authoritarian) are *inscribed in the product* of knowledge itself. We passionately suffer from the Procrustes complex, who cut or stretched his victims because he could not tolerate difference or a lack of fit, as it were (Volkoff, 1978). In contrast, I understand communication not only as the exchange of symbolic forms, but also, taking a concept from biology, based on all different forms of coordinating actions. The more social relations are horizontal, the more there is a culture of communication. So long as there is increasing cohabitation and efforts to reflexively horizontalize relations, there is less individual need to stand out from or be crushed by others. In fact, these non-hierarchical forms of organization promote different forms of collective intelligence. To protect individuals from predators, some species generate swarm intelligence by means of modifying their coordination of actions.

*3. The Culture of Knowledge*

The third dimension, the missing "k": the possibility for developing a new and expansive *culture of knowledge* is essentially inactive or simply cancelled out. This is the result of the grave discrepancies that have developed between the two other *cultures* (information and communication) and the inadequate tools for developing self-determined information and communication systems. So long as this acute problem is not even recognized, despite the fact that the construction of concepts is elementary for human beings, our abilities for action in the social world are directly depleted. Knowing, we can organize acting to change the order of things. Not knowing, we perpetuate the state of things as they are.

## To Know or to Be Known? That Is the Question

For a long time, Mexican society (and many other societies that have suffered under colonialism, adapting and adjusting to the social order of things) has developed an almost iron-hard vocation for serving as an *object of study*. This has entailed and been accompanied by the simultaneous repression and gradual dismantling of the material means and supports for producing studies in which Mexico figures as the *subject of knowledge*. Of course, Mexico is not exceptional in this respect. Many other countries of the so-called periphery in the world system have suffered similar "adaptations" in this invisible but efficient process of subalternity.

In other words, we cannot account (we haven't the means to know, only to be known from others outside) for what exactly is happening with this triplet of interdefinable cultural processes insofar as they cannot be separated or defined independently of the others (García, 2000: 68), yet they constitute our *symbolic ecologies* as they relate to global technologies. The genuine *directional force* of these technologies, as a vector that also possesses and generates, in the name of the entity that wields it, is enormous and crucial for the public recognition and *symbolic capital* produced through it (Bourdieu, 2001a: 152).

The consequences of this are of such a magnitude that, for example, in the case of a "technological training" program held by Mexico with hundreds of instructors from the National Program of Distance Learning, the already trained professors felt "inferior" to the computer and satellite decoders (González, 1999: 160). More recently, we verified elements of this same perception and conviction in the broad, national *non-usage* of the so-called digital community centers that formed part of the E-Mexico System from 2000 to 2006. These were centers armed with devices and installations designed and implemented for the "enforced inclusion" of the *digital poor*, the run-of-the-mill wretched, who have yet to "perceive" the benefits of "accessing" all that is now available thanks to the E-Government, that is, the *Information Society*

(González, 2015a: 316). It was a brutal failure, both economically and socially. No one ever appealed to those who were "technologically displaced." The entire project was undertaken in their name, but never directed toward nor inclusive of the actual people. Some of the collateral effects of this heteronomous interaction of the technological vector with our *symbolic ecologies* (information, communication and knowledge) is based on the idea that little by little this area of daily life which we cannot any longer even begin to understand, much less produce or control, will become still *broader* and more excruciating. We are becoming "processed by others," as Aníbal Ford (2000) put it so succinctly.

The collective ability to organize in order to use technologies as *generative platforms of knowledge*, meanwhile, is seen as utopian and practically impossible.

On the crest of this wave are the illusions of "transparency," the "user-friendliness" of devices, and "technology convergence." Screens are probably the most important interface encouraging these illusions. We have only to press a button, activate a command, or allow our iris to be read by a camera, to put our screens to use, but without our having to worry about what is actually happening inside that brilliant little *black box* (to our sight and knowledge), it is perhaps we who are being used by our screens. If we turn to the Internet, for example, we can see that this process has been transforming us into mere passive consumers of a technology that was "sold" to us as a special tool for accessing knowledge and information. Such "content" however, has been produced by others *for* and *about* us. But not *with* or *by* us.

It is a curious division of labor with geopolitical coordinates determined and marked by history, but also encouraged by an inability and incompetence for generating state policies on the production of scientific and technological knowledge. So-called "globalization" is not a natural phenomenon. "Globalization is here" is proclaimed as if it were the Spring or an approaching storm, however it is a deliberate policy of world action and effect that is presented to us (and we accept it) as the only possible solution to poverty (which is not "natural" either, of course, but rather is historically generated).

In Table 5.1 we can see how, in recent years, the position of Mexico has declined compared to other relatively similar countries since 1970. With a clarity that is rarely seen and confirmed, this observable change appears to be closely related to the scientific policies of those countries that were able to dedicate increasing percentages of their Gross National Product to the promotion of knowledge, in contrast to what took place in Mexico during that same period.

It is difficult to examine this comparison without establishing a directly proportionate relationship between investment in policies for the promotion of knowledge (research, higher education, technological development of innovations for change) and improvements in collective well-being.

Bourdieu had clearly warned us about this:

> Globalization is not a random, much less innocent event. […] it is convenient to refer to it as a "policy of globalization," or worldification, in which *a single form of defining and generating "well-being"* is imposed. The word globalization is thus a descriptive and prescriptive pseudo-concept that has come to take the place of the word "modernization," long used in American social sciences as a euphemistic way to impose a naively ethnocentric evolutionist model allowing the classification of different societies according to their distance from the economically more advanced society, that is, American society, erected as consequence and outcome of all human history …
>
> *(Bourdieu, 2001b: 105; emphasis added)*[6]

In Mexico's geopolitical condition, especially with respect to the United States and the rest of the world, we fulfill a secondary and subaltern role, obedient to the instructions of international

*Table 5.1* Empirical evidence of how not investing in knowledge production has relocated the GNP in Mexico, Brazil, Spain and Korea

(A) Investment in experimental research and development (R&D), and (B) Increase of per capita income (in American dollars per year)

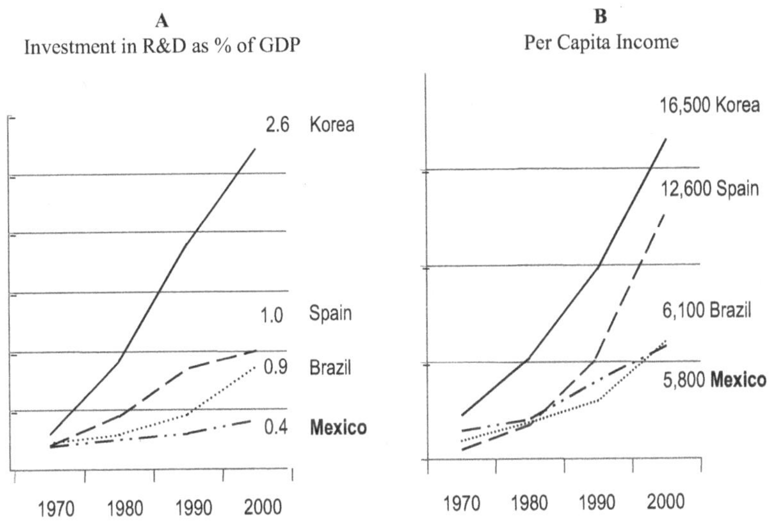


Source: González (2015b: 335)

organisms. We have thus postponed defining a policy that would render us strong and self-sufficient, both in terms of food production and the symbolic production of our very identity (Bertaux, 1977).

However, Mexico has become the world's *leading producer* of television sets, by following instructions from abroad, while remaining far behind other comparable countries in technological self-determination. Governments celebrate that:

> the northern border of Mexico has become the most important region for the production of televisions for the North American market, especially the cities of Tijuana, Mexicali (San Luis Rio Colorado), Ciudad Juárez and Torreón, which have developed extensive specialization in the production chain of consumer electronics, as demonstrated by the presence of thirteen large television production corporations (Zenith, Daewoo, LG Electronics, Hitachi, Goldstar, Matsushita, JVC, Mitsubishi, Thomson, Sony, Philips, Sanyo and Samsung), companies that generate more than twenty thousand jobs in assembly plants, in addition to constituting a powerful network of providers.
>
> *(Merchant, 2007: 18)*

This fills Mexicans with tremendous "pride" (affording them visibility) even though all of these brands are foreign-owned, assembling manufactured goods here largely as a means of exploiting inexpensive labor conditions. The origin of the technological vector remains Japanese or Korean or Chinese, North American or European. We are the target—the assembler and the displaced consumer—of this overwhelming force (accompanied, of course, by many other so-called "Global South" countries).

It seems that we fulfill, with unprecedented zeal, two roles in this tragicomedy of forced "globalization": (1) we provide cheap manual labor to assemble the inventions and brands of others who generate public policies (because we cannot trust in free-market competition) for the development of knowledge and (2) we are also the best conspicuous consumers of devices with screens, as reflected by the decade-long growth of the cellphone market in Mexico, where almost three-quarters of the population uses cellphones.

This cheap manual labor (in fact, *very* cheap labor) is excessive in Mexico (because the model of imposed economic precariousness impoverishes the labor force) and therefore we "export" it (perhaps I should say "we expel it") in the form of undocumented workers, which is one of two pillars supporting the Mexican economy (and a large part of the US economy as well):

> During the year the remittances that were sent by Mexicans to their families in our country reached a historic amount of 28.77 billion dollars, which is an increase of 6.6 percent with respect to 2016, when the amount was 26.99 billion dollars. The Bank of Mexico (Banxico) announced that its total operations for last year came to 93.42 million dollars, which is an increase of 2.04 percent over 2016. [...] the average remittance was approximately $308 [...] with the record amount reached by remittances in 2017, they were established as one of the main sources of foreign currency for Mexico, which is among the four major receptor economies of these types of money flows in the world.
>
> (Regeneración *newspaper, 2018)*

Independently of Information and Communication Technologies, incarnated as the convergent smartphone device (synthesizing what was previously sold separately, that is, Internet access and mobile phone in one device), the disappearance and negligence of the third inseparable cultural component is alarming: knowledge.

## Cibercultur@ and the Development of Knowledge

Just as this imbalance is apparent in the economy, the same takes place, *mutatis mutandis*, with the collective capacity to provide knowledgeable solutions to the practical and complex problems we face as a society. Despite the ubiquity and repetition of this refrain, it is *simply not true* to say that "having more and better computers generates better education."

If, instead, we do not cultivate and develop these three cultures of information, communication and knowledge—that I have termed "cybercultur@"—say, the increasing capacity of self-determination, powered by collective organization for generating adaptive and creative responses to our real problems, we will remain stuck, forever the cheap labor browsing the market as conspicuous and dependent consumers. It is quite a challenge to discover the arguments needed to show and explain how and at what cost we relate to the millions of omnipresent, glowing screens. Yet this work (endeavor) remains crucial and the perspective achieved from the margins of the globalized economy are at least as significant as those produced at its centers. Through research and development of emergent local knowledge communities, we are attempting to generate some leads and to develop more complex concepts for confronting the pseudo-theory on societies that have been globalized by "ICTs." These emergent communities are concerned with developing their own *information and communication systems* for generating *knowledgeable options* to confront their concrete and significant local problems. Gradually they will be able to connect with other communities that have undergone the same processes of exploitation, exclusion and ignorance: migration, poverty, unemployment, pollution, criminal and gender violence, hunger, malnutrition, environmental degradation, depletion of water supplies, the stripping of their lands by mining companies and so many others. A corrective will develop for what has been so far excluded from these processes. In the literature on the relationship between the poor (both digitally and non-digitally impoverished) and technologies of the "information society," I have been unable to find references that comprehend this crucial expropriated and apparently wayward dimension: knowledge.

The development of these emergent community's focuses on facilitating a horizontal dynamic that is very similar to Paulo Freire's findings of public education that allow them to collectively appropriate the technology of literacy to generate the ability to narrate and construct a self-determinant "we" (Lenkersdorf, 2008) and not one that is derived from outside impositions. Thus, a process of empowerment is begun through a new and fortified collective ability to re-narrate one's own past, to redefine one's present conditions, and to redesign one's future as well as other possible worlds. This can be accomplished through other forms of self-narration and becoming visible, emanating from the ability (previously disappeared and undeveloped) to generate and maintain systems of information, systems of communication/organization, and, finally, systems of knowledge, initially local and subsequently *situated*. When every emergent knowledge community becomes a hub in a network that connects them to other emergent communities, real alternatives can emerge. It is our turn as a country and as a society to reorganize ourselves and take charge of the cultivation and development of our symbolic ecologies in order to be able to relate to others in ways that are more just and less dependent. The tools are at hand. The challenge is to take advantage of these tools with the full force of the imagination that emerges from dialoguing with others.

## Notes

1 Some ideas and parts of this essay formed part of a previous version in Spanish. I am grateful to Shawn Shimpach's invitation and patience in anticipation of this project.

2 See the use of instruments by crows in New Caledonia: www.newscientist.com/article/2099246-crows-are-first-animals-spotted-using-tools-to-carry-objects/

3 See Gray (2010).

4 See the interesting book written by Jara and Garnica (2007) in which they report on the historical experience over the years in the quantitative and commercial study of television viewers in Mexico.

5 See the BBC News report "Social media apps are 'deliberately' addictive to users" www.bbc.com/news/technology-44640959

6 [Editor's Note]: This passage is a translation of a translation. The official English translation, by Loïc Wacquant, of this passage from Bourdieu reads as follows:

We must return here to the world "globalization." We have seen that, in a rigorous sense, it could refer to the unification of the global economic field or to the expansion of that field to the entire world. But it is also made to mean something quite different, in a surreptitious slide form the descriptive meaning of the concept, such as I just formulated, to a normative or, better yet, *performative* meaning. [...] The word "globalization" is, as we can see, a *pseudo-concept, at once descriptive and prescriptive,* that has supplanted the term "modernization," long ago used by American social scientists in a euphemistic manner to impose a naively ethnocentric evolutionary model according to which different societies of the world are classified in terms of their distance from the most economically advanced society, that is, U.S. society, instituted as the endpoint and end goal of all human history (Bourdieu, 2003: 84).

## References

Bertaux, Daniel (1977) *Destins personnels et structure de classes*, Paris, PUF.

Boden, Margaret (2004) *The creative mind. Myths and mechanisms*, London & New York, Routledge.

Bourdieu, Pierre (2001a) *Las estructuras sociales de la economía*, Buenos Aires, Manantial.

Bourdieu, Pierre (2001b) *Contrafuegos 2. Por un movimiento social europeo*, Barcelona, Anagrama.

Bourdieu, Pierre (2003) *Firing Back: Against the Tyranny of the Market 2*, London, Verso.

Cirese, Alberto (1984) *Segnicitá, fabrilitá, procreazione. Appunti etnoantropologici*, Roma, CISU.

Couldry, Nick (2010) *Why voice matters. Culture and Politics After Neoliberalism*, London, Sage.

Frans B. M. de Waal, Cassiopeia A. Freeman, Marietta Dindo, and Marisa J. Hall (2005) "The monkey in the mirror: Hardly a stranger," *PNAS*, vol. 102, no. 32, 11140–11147.

Del Río, Pablo & Álvarez, Amelia (2011) "La actividad como problema del desarrollo." *Cultura y Educación*, vol. 23, no. 4, Fundación Infancia y Aprendizaje, 601–619.

Ford, Aníbal (2000) "Procesados por otros. Diferencias infocomunicacionales y sociocultura contemporánea." In Carlos E. Valderrama (Ed.), *Comunicación–Educación. Coordenadas, abordajes, travesías*,Pp. 65-74 Bogotá, Siglo del Hombre.

García, Rolando (2000) El conocimiento en construcción. De las formulaciones de Jean Piaget a la teoría de los sistemas complejos, Barcelona, GEDISA.

Gómez, Héctor (2005) *Todas las mañanas del mundo*, Tesis de Doctorado, Universidad de Colima.

Gómez, Héctor (2007) "Estratos de Tiempo y Velocidad. La comunicación que hemos conocido, la comunicación posible," *Razón y Palabra*, vol. 57, junio-juliohttp://www.razonypalabra.org.mx/anteriores/n57/hgomez.html.

González, Jorge A. (1999) "Tecnología y percepción social: Evaluar la competencia tecnológica," *Estudios sobre las culturas contemporáneas*, vol. V, no. 9, 155–165, Universidad de Colima.

González, Jorge A. (2010) "Understanding Telenovela as a cultural front. A complex analysis of a complex reality." In Ilan Stavans (Ed.), *Telenovelas*, 68-78 Oxford-Denver-Sta Barbara, Greenwood.

González, Jorge A. (2015a) *Entre cultura(s) y cibercultur@(s)*, Mexico, UNAM. www.researchgate.net/publication/283720653_Entre_culturas_y_ciberculturs_Incursiones_y_otros_derroteros_no_lineales_1a_Edicion_electronica_2015.

González, Jorge A. (2015b) "Por una cultura de conocimiento," www.researchgate.net/publication/323218619_Por_una_cultura_de_conocimiento_2a_Edicion

González, Jorge A. (2017) "Technophilias and technophobias vis-à-vis research & development of cybercultur@," *Brazil: Media from the country of the future, Studies in Media & Communications*, vol. 13, 325–341. Emerald Publishing Ltd.

González, Jorge & Krohling-Peruzzo, Cicilia (2018) Agroecología y antroponomia en movimientos sociales de Brasil: "si los demás están bien, yo estoy mejor", *INTERDisciplina*, Vol 6, No 14, CEIICH-UNAM, pp. 205–235

Gerald J. Gorn, Amitava Chattopadhyay, Jaideep Sengupta, and Shashank Tripathi (2004 May) " Waiting for the Web: How Screen Color Affects Time Perception" *Journal of Marketing Research, 215* Vol. XLI, 215–225.

Goodman, Jarid & Packard, Mark (2016) "Memory systems and the addicted brain," *Frontiers in Psychiatry*, DOI: 10.3389/fpsyt.2016.00024.

Gray, Martin Paul (2010) *Cave art and the evolution of the human mind*, MA Thesis, Victoria University of Wellington, http://hdl.handle.net/10063/1640

Grödal, Torben (1994) *Cognition, emotion and visual fiction*, Copenhagen, University of Copenhagen.

Hutchins, Edwin and Morana Alâc (2004) "I see what you are saying: Action as cognition in fMRI brain mapping practice," *Journal of Cognition and Culture*, vol. V4, no. 3, 629–661.

Jara, Rubén & Garnica, Alejandro (2007) *¿Cómo la ves? La televisión mexicana y su público*, Mexico, AGB.

Jenkins, Henry (2005) "Welcome to convergence culture", Downloaded from *Receiver magazine*, www.receiver.vodafone.com. Vodafone Group.

Kardaras, Nicholas (2016) *Glow Kids: How Screen Addiction is Hijacking our kids*, St. Martin Press, NYCNew York City.

Lenkersdorf, Carlos (2008) *Aprender a escuchar. Enseñanzas maya-tojolabales*, Mexico, Plaza y Valdés.

Luengo, Rosa & Alcalá, Javier (1991) "El teatro de sombras chinescas," *Campo Abierto*, N/A no. 8, España : 176-192.

Maass, Margarita & González, Jorge A. (2005) "Technology, global flows and local memories: Media generations in 'global' Mexico." *Global Media & Communications*, Sage Volume 1: Number 2 August 2005.

Merchant, Marco Antonio (2007) *El comercio exterior manufacturero y los procesos de producción internacionalizados en México*, Madrid, ponencia en la IX Reunión de Economía Mundial.

Piaget, Jean (1978) La equilibración de las estructuras cognitivas Problema central del desarrollo.

Joshua M. Plotnik, Frans B. M. de Waal, and Diana Reiss (2006) "Self-recognition in an Asian elephant." *PNAS*, vol. 103, no. 45, 17053–17057.

Prospero, Oscar, et al. (2012) "El cerebro adicto", *Rev Med UV*, V. 2012, N. 1

Regeneración (2018) "1 de febrero," https://regeneracion.mx/en-2018-cifra-record-de-remesas-de-estados-unidos-a-mexico/

Reiss, Diana & Marino, Lori (2001) "Mirror self-recognition in the bottlenose dolphin: A case of cognitive convergence". *Proceedings of the National Association for the Advancement of Science*, vol. 98, no. 10, 5937–5942.

Santamarina, Francisco (1983) *Diccionario de mejicanismos*, Mexico, Editorial Porrúa.

Sayaz, Luis Fernando (2012) "Política científica para el siglo XXI. Prioridades para la agenda nacional en ciencia, tecnología e innovación," enGiovanna Valalenti (Ed.), *Ciencia, tecnología e innovación: Hacia una agenda de política pública*, FLACSO-México MEXICO CITY.

Taylor, A.H., Medina, Felipe S., Holzhaider Jennifer C., Hearne Lindsay J., Hunt Gavin R., Russell D. Gray, et al. (2010) "An investigation into the cognition behind spontaneous string pulling in new caledonian crows." *PLoS One*, vol. 5, no. 2, e9345. DOI: 10.1371/journal.pone.0009345

Thompson, John B. (2005) "The new visibility," *Theory Culture Society*, vol. 22, 31.

Volkoff, Vladimir (1978) *Elogio de la diferencia*, Barcelona, Tusquets.

# 6

# TRANSNATIONAL TELEVISION CULTURE

*Lothar Mikos*

## Introduction

In the second half of the 20th century, television became a mass medium and grew in importance to become the most widely used medium. Although television as an institution was regulated by individual nation-states, the creators of television programming always looked to other countries for orientation. According to Havens (2006, 16) "the international exchange of television programs goes back almost as far as television broadcasting itself." Programs were copied, programs were bought from other countries, and joint live broadcasts were organized. The first major international events in European television were the 1953 coronation of Queen Elizabeth II in Britain and the 1954 football World Cup in Switzerland.

The internationalization of television benefitted among other things from the fact that the medium was considered an extension of radio by the addition of pictures. Television was more or less fitted into the existing organizational structure of radio broadcasting. Radio had already established institutional structures in every country such as editorial offices, in which program forms such as news, talk shows, daily soaps and others had been developed, which could then be taken over by television. As Hilmes has noted,

> as the world's nations became network nations, the United States and Great Britain squared off as the dominant competing models for modern media-centred national cultures, influential across the globe. From Scandinavia to Japan and across the outposts of empire, nations modelled their broadcasting systems after Britain's BBC, sought its state ownership, regulation, quotas, tax structures, public funding and other methods to pull together the national culture and to resist de-nationalizing threats both internal and external.
>
> *(2013, 28)*

This look back at the history of broadcasting shows how closely national and international developments are interconnected.

An international broadcasting organization, the OIRT, had been founded in the Eastern bloc as early as 1946. The OIRT, also known as Intervision, was responsible for the international exchange of programs. Four years later an analogous body was formed in Western Europe as well: the European Broadcasting Union (EBU), based on the south coast of England (see Fickers 2012; Pollack & Woods 1959). Broadcasting agencies in 23 countries joined the EBU, whose purpose was to promote the exchange of news between these countries' broadcasting agencies. In 1954, the EBU founded the Eurovision network for the purpose of exchanging other programs besides news, especially in sports, and to facilitate joint programming. One of the network's first joint activities was the Eurovision Song Contest, which was first broadcast in 1956. Other attempts to develop joint European programs followed with the intention of promoting the European idea. After the dissolution of the Eastern bloc, the members of OIRT joined the EBU in 1993. These early examples of international cooperation for the purpose of programming exchange illustrate the international orientation and organization that characterized television from the beginning.

Indeed, beyond the programming exchanges organized through EBU and OIRT, programs were also imported bilaterally. The Chinese state television network CCTV, for example, began broadcasting Soviet programs in 1958 (Hong 1998, 46–7). Of course, the international relations between the world's television broadcasters were mostly confined within the political power blocs – the Eastern bloc on one hand and the industrialized Western countries on the other – and within geocultural regions such as the Arab countries and East Asia. The US was best able to extend its influence to Latin American countries, primarily because television there was organized, after the US model, as a private enterprise (cf. Hilmes 2013, 28). Therefore, television has been internationally oriented since its beginning.

The internationalization of television can be observed on a number of levels. According to Negrine and Papathanassopoulos (1991, 11), these include the organizational and regulatory level, the financial level, the content level and the level of reception and use. I would suggest concisely revising these notions to three main levels of transnational television: (1) production and distribution of television including TV technology, economy, media policy, and legislation, (2) television texts, and (3) television audiences.

Before going into more detail on the three levels, it is still necessary to clarify what is meant by transnational television culture. Transnationalism, however, is not necessarily synonymous with globalization. As Kearney asserts, "Transnationalism overlaps globalization but typically has a more limited purview. Whereas global processes are largely decentered from specific national territories and take place in a global space, transnational processes are anchored in and transcend one or more nation-states" (Kearney 2008, 273). Transnational television is anchored in the nation-state and national media legislation, and it is linked to the multidirectionality of flows and interactions. For instance, the European Union (EU) is a transnational space, and the media legislation of the EU is a transnational one. The Audiovisual Media Services Directive must be transposed into national law in each member state. So, there is a strong connection of European legislation and national legislation of the member states concerning media in general and television in particular. Also geolinguistic regions such as Latin America or Asia-Pacific can be called transnational, or the relations between the US and the UK in television drama are immanently transnational (cf. Weissmann 2012). Yet "transnationalization points out the salience of forces that cross borders, [even while] it also makes clear that borders persist" (Straubhaar 2007, 106). Therefore, local, national television has some regional, transnational and maybe global elements that interact with each other. As Chalaby notes, "The development of transnational television fully reflects this process of regionalization and its complex relationship with globalization" (Chalaby 2005, 5). Transnational television operates in a transnational arena, where the global, the local and the regional are both interconnected and transcended.

Culture, meanwhile, is not understood as a "whole way of life" (Williams 1968, 18), but as a signifying practice (cf. Hall 1997, 2016; Williams 1981), as a process of the production of meaning. Television culture is a practice of shared meanings and significant patterns of ideas, ideals and forms of communication regarding production and distribution, texts and audiences, and it is informed by global, transnational, regional and local structures and practices. As transnational television culture is crossing borders but anchored in national television cultures, it overcomes the distinction of the global and the local, and can be conceptualized as a social and cultural process in a transnational arena where agents, institutions and structures interact with one another.

## Transnational Production and Distribution Culture

As already mentioned, early television was integrated into the already-existing structures of radio broadcasting. In addition to the fact of nature that radio waves neither recognize nor yield to national borders, a transnational element was implemented from the beginning because "the development of shortwave radio technology created more opportunities than ever before for the transnational exchange of popular culture" (Hilmes 2013, 28). During the 1950s and 1960s television prevailed as a mass medium. In the 1950s, the countries that introduced television looked at existing programs for ideas for their own productions, and sometimes bought programs from other countries. The Soviet Union was the model for other countries in the Eastern bloc, while Western European and Latin American countries patterned their programming primarily after the United States, and to some extent after Great Britain. The Hollywood film studios were critical of television at first. As a result, companies that had sponsored radio serials, such as the soap and detergent producers Procter & Gamble and Colgate-Palmolive, were able to enter the new television market. The studios soon abandoned their aloof stance, however – especially once they discovered that money could also be made by producing, syndicating, and then exporting television programs to Latin America and Europe, since the television market overseas was booming (Dizard 1966, 163). There was a large demand to be served with programming exports. Television producers opened export offices in Europe, including the Television Film Export Committee, founded in 1959 by the export department of the Motion Picture Association of America (Havens 2006, 20), and most importantly the Television Program Export Association, founded in 1960 by the networks and independent syndicators (Dizard 1966, 159; Havens 2006, 20). "Supplementing these overseas sales offices, the major U.S. syndicators make extensive use of European trade fairs to display their products" (Dizard 1966, 159). Up to the 1980s, these trade fairs were a marketplace primarily for US programming. Accordingly, the proportion of US-produced shows in the other countries' television programming was high. In 1972, the proportion of US programs was 44% in Brazil, 48% in Australia, 62% in Jamaica, 28% in Hong Kong, 19% in South Korea, 41% in Lebanon, and 3% in France (Straubhaar 2007, 262). In the early 1970s, the US exported about seven times as many shows as France or Great Britain, and about 25 times as many as Germany (Nordenstreng & Varis 1974, 30 ff.). West German exports (and after 1990, exports from reunited Germany) were handled from 1965 to 1998 by the Trans-Tel company, a consortium of the two public networks ARD and ZDF and the Federal Republic of Germany. German programs were exported mainly to Africa, Asia and Latin America (ibid., 36). American television became a point of reference in Europe, although it was also met with criticism because European intellectuals were afraid of a "cultural levelling-down and threats to national culture" by American television formats and popular culture (Bondebjerg et al. 2008, 155). In Germany, for example, there was a public debate about the inferior quality of American crime series. Critics also argued that these series had nothing to do with German reality. Therefore, American TV shows were seen as a cultural invasion. On the other hand, the American influence on European television can also be seen as

> liberation and modernisation of old, European cultures that were not sufficiently able to integrate globalisation and new mainstream popular cultures such as those of young people. American entertainment genres such as quiz shows and talk shows were popular with large groups among the audiences of national television in Europe.
>
> *(ibid., 157)*

In the early years, producers, distributors and broadcasters were the most important players in international programming transfer (Moran & Keane 2006, 73). In the 1960s, the need was felt to bring these players together regularly at a fixed place. One of the first television markets was established in 1963 in Cannes, the trade fair Marché international des programmes de télévision (MIPTV) (Moran & Malbon 2006, 76–7), which was joined in the mid-1980s by a second fair, MIPCOM. In the 1960s and 1970s, MIPTV was dominated by US programming. Also in 1963, the National Association of Television Program Executives (NATPE) was founded in the US, and since then it has organized an annual programming market. The European television networks present their programming at events of their own. The largest of these is the BBC Showcase, which has been the annual export market for BBC Worldwide's programming since 1976 (ibid.). Havens (2006, 16 ff.) situates the first wave of globalization in the period from 1957 to 1972. This period saw an increase of more than 40% in the proportion of income from export sales in the total revenue of the US television industry, while at the same time European programming exchange in the Eurovision network increased from 261 hours in 1957 to 1138 hours in 1972. In the Eastern bloc's Intervision network, meanwhile, programming exchange increased from 170 hours in 1960 to 1124 hours in 1972 (Eugster 1983, 224 ff.). These figures alone show that television programming was being increasingly internationalized and transnational. Moreover, the figures do not include domestic productions based on ideas taken from other countries. Several popular Australian television programs in the 1960s, for example, were unlicensed imitations of BBC series formats (Havens 2006, 21). In Japan, American and British shows were frequently imitated and adapted to Japanese culture (Straubhaar 2007, 160). The same is true of many European countries' television networks, which imitated a number of US and British entertainment shows. Quiz show adaptations in particular were popular, as Bjork (2009) has shown for Sweden and Mikos (2016) for Germany. The European television networks had correspondents in the US whose duties included reporting on commercial television and its programming. In addition, programming directors from Europe travelled to the US and to the UK to find out about television shows and their production on the spot.

The formatting of television programs began in this period (Moran & Malbon 2006, 21 ff.), although this usage of the term "format" was not established until the 1990s. Latin American telenovelas as a specific narrative form evolved from Cuban adaptations of American radio soap operas, while television soaps in the US continued the original tradition from American radio. Other genres, adapted from radio or specific to television, developed, including game shows, quiz shows, talk shows, music shows, etc. (Straubhaar 2007, 157 ff.). Programs that followed the conventions of these genres were distributed around the world, both under license and in the form of unlicensed adaptations.

We can identify four varieties of format marketing in the global television market (Mikos & Perrotta 2013). (1) The buyer acquires the broadcast rights to a finished product – a "canned program" – and performs no modification except subtitling or dubbing as adjustment to the local television market. This is the usual case with television drama series and sitcoms in particular. (2) Another type of format marketing involves selling the rights to a series concept and a format outline. In this variant, no finished product is sold. Instead, each buyer produces the series, adapting it to local conditions within the limits of the agreed outline. An example of such a series adaptation is *Ugly Betty* (see McCabe & Akass 2013; Mikos & Perrotta 2013). (3) The third variant is the licensing of rights to quiz shows, game shows and reality shows. In the 21st century, such shows are developed as international brands (cf. Chalaby 2015; Moran 2009a; Moran & Malbon 2006). These shows have a logo used uniformly throughout the world, and their presentation, from dramaturgy to character constellations to design, follows uniform rules in all national adaptations. Only the contestants, the games and the quiz questions are adapted to local conditions. In the 1950s and 1960s, such shows were merely sold in the form of licenses to produce a local adaptation; no importance was attached to a uniform appearance and title. (4) The fourth variant is the sale of rights to broadcast sports events.

According to Havens (2006, 24 ff.), there was little change in the globalization of television between 1973 and 1985. The share of revenue from program exports in the US television industry stagnated, and even declined in some years. One important reason for this was that, by this time, production capacities had been established in many countries, so that the national television networks were able to draw more heavily on domestic productions. Similar developments occurred in Brazil and other Latin American countries, as well as in Hong Kong, Israel and South Korea. Domestic productions in those countries were primarily entertainment shows and soap operas or telenovelas (Straubhaar 2007, 161). The international programming market did not collapse, but it did stagnate.

This changed in the mid-1980s, however, as the deregulation of the television market and the introduction of privately owned networks created new growth, which was again boosted by digitalization in the mid-1990s. The second wave of globalization drastically changed the international television market (cf. Havens 2006, 26 ff.) and led to tremendous growth due to the proliferation of channels. More formats than ever were traded at the television fairs. Although the total revenues were rising as shows were bought and sold, there were also changes in the power structure of the industry. In the US television industry, international format sales no longer earned the share of revenue that they used to in the 1960s (ibid., 28). As the demand for programs has boomed, the number of domestically produced shows has also grown in many countries (cf. Straubhaar 2007, 162–3). As a result, the share of US productions in almost all national television markets has declined from 1972 to 2001: from 44% to 19% in Brazil, from 48% to 29% in Australia, from 62% to 53% in Jamaica, and from 19% to 6% in South Korea. Only in France has the market share of US productions grown from 3% to 18% (ibid., 262). Specific national television cultures which have developed in most countries are oriented toward transnational formats, but with local production, in part because licenses are still granted for a certain period and a certain geographical territory.

Although television was transnational from the beginning, there are three main developments that gave transnationality another boost, two technical and one legislative. In the 1980s the deregulation of television in Europe and the introduction of commercial television led to an increasing demand for content. The introduction of satellite broadcasting and its implementation as direct-to-home technology paved the way for a new era of transnational communication (cf. Brüggemann & Schulz-Forberg 2009, 697; Chalaby 2009, 21 ff.; Gripsrud 2007, 486 f.). Finally, digitalization simplified the production and distribution of television which resulted in a confusingly high number of television channels worldwide, distributed by cable, satellite or internet. In Europe, these developments were accompanied by legislation of the EU facilitating "the opening of the European market for transnational media ownership, production and consumption with a view to promoting the emergence of a European media market with European players who can compete on a global scale" (Brüggemann & Schulz-Forberg 2009, 697). So, a transnational European television culture came into being with its own standards and practices of signifying patterns of ideas and ideals. In a study on television buyers in four European countries, Kuipers (2011, 552) found that the buyers "formed part of a 'transnational class' [...] with considerable international experience." Their professional work is mainly informed by the international television market. For instance, they share a similar meaning of quality (cf. ibid., 553). Obviously, professionals in the European television market operate on the basis of shared meanings of quality. And this understanding of quality can be described as a transnational process of signification that makes it easier to import and export formats and to adapt formats to local television cultures.

During the decades from the 1990s onwards the media industry has faced dramatic changes. The notion of convergence describes the changes on the level of technology, economy, content and audience practices (Jenkins 2006; Mikos 2017). At the same time, an international television market has developed in which producers compete not only in national contexts but internationally for commissioning drama series. Since television series were only produced or commissioned by public broadcasters until the 1980s, private commercial broadcasters were added after the deregulation of broadcasting in Europe. In the 21st century, streaming platforms such as Amazon Prime and Netflix have been added as new players to refresh their offerings with TV series. In the US, this has pushed up the budgets of television series and, as a result, increased the production value of many series. In order for European television series to compete, they must also create increased production values – and they need more money for that. International co-productions are one way to generate a higher budget. Therefore, the first two decades of the 21st century saw an increasing number of international co-productions in television. There's a vivid scene of co-production of television drama series in the European television industry based on bi-national co-production treaties. The pooling of resources and gaining of higher budgets is necessary for the European television industry to be able to compete with US productions. International co-productions of television drama series

> were produced – in most cases – as a result of international treaties which allowed them to overcome previously restrictive barriers (high cost, limited market access) through financial incentives and a pooling of resources and markets. Their production and circulation, therefore, relied upon regulatory arrangements conducted by the state.
>
> *(Shimpach 2005, 341)*

A legislative and a cultural framework influences international co-productions, from the beginning on. The cultural contexts include the experiences of the producers with certain partners from other countries who are trusted because of a common production history (cf. Bondebjerg et al. 2017, 79 ff.). Confidence refers both to the skills as a producer and to the ability to communicate, because a complex co-production requires a lot of administrative work and a high degree of coordination, which requires communication – not only within the production team, but also with lawyers and representatives of banks, funding institutions and local authorities. The legislative framework is also important in the planning of the project, as the opportunities for funding and tax incentives have a significant impact on the budget. And, a high budget is more and more important in a highly competitive market in which audiences must be found and bound. International co-productions are part of a transnational television culture in which shared meanings are circulating.

## Transnational Television Texts

Television shows are exchanged and adapted in a number of countries outside their origin market. As already mentioned, the professionals in the transnational television industry have shared meanings of quality, and shared meanings of aesthetic and narrative standardization. Bielby and Harrington (2004, 94) have pointed out that the genre concept is a central mechanism of classification in cultural systems, "and as a culture industry, television relies heavily upon it to routinize the marketing and distribution of series." In other words, genres are presumed to be universal, standardized concepts that connect audience expectations with the routines of production and the conventions of representation in narrative and aesthetics. Early internationalization laid the foundation for the global television market, since television programs have always been exported or adapted internationally. Their international success depends not only on "series resonat[ing] locally abroad in culturally specific ways" (ibid., 83), but also on meeting the audience's genre expectations. The US and UK were always successful in exporting television shows. Some Latin American countries were successful in exporting telenovelas (cf. Thussu 2007). In the early 21st century, Scandinavian drama series were also internationally successful (Bondebjerg et al. 2017; Gamula & Mikos 2014; Redvall 2013). A number of these series and trends have been adapted in the US, illustrating Elke Weissmann's (2012, 192) point that, "US television will increasingly incorporate the transnational to sustain its dominance in the global market." International co-productions of TV drama series, especially, have to address transnational audiences by aesthetic and narrative standardization, genre conventions, and transnational actors and staff. The multiplicity of local adaptations of telenovelas and television drama series shows how these series become transnational brands (cf. McCabe & Akass 2013; Mikos & Perrotta 2013; Moran 2009b). What the development also shows

> is that (1) a global market is now regularly being considered as part of a program's initial run; and (2) the purpose of this initial run is, at least in part, to create "buzz" about the particular program. Buzz in this sense suggests an informal familiarity with aspects of the program that might build it into a "brand" that can be sold and leveraged in the future.
> *(Shimpach 2013, 201)*

In the global television market, series and shows are marketed worldwide. The formatting makes it easier to buy and sell new programs with a minimum of risk. The local adaptation of a television program is an attempt to minimize the risk involved in introducing a new show. Even such an adaptation is not guaranteed to succeed, however, since an advantage of the show in one country may turn out to be a drawback in another country. For example, the "Survivor" format was very successful in the USA, but did not really gain a foothold in Europe. On the other hand, the reality show "Big Brother" was more successful in Europe than in the USA. A format that is successful in one country may flop in another due to local and national differences in the viewers' culture, self-image and television habits. TV formats have a brand that is used uniformly throughout the world, and their presentations, from dramaturgy to character constellations to design, follow uniform rules across all adaptations (cf. Bielby & Harrington 2008; Mikos & Perrotta 2013; Moran 2009a, 2009b; Moran & Malbon 2006; Oren & Shahaf 2012). Only the contestants, the games, and the quiz questions adapt to local conditions. The uniform dramaturgical and aesthetic elements are specified in a format "bible." The shows are then produced to the same specifications by different production companies in the various local markets. Thus, these shows are simultaneously globally uniform and locally resonant. More and more television shows have become transnational texts addressing a transnational audience.

## Transnational Television Audiences

The international trade of television drama series and television formats, and their local adaptations and adjustment to local cultures through dubbing and subtitling, constitute a transnational audience. Social media platforms facilitate the communication of audiences about television shows. Fans of US TV drama series such as "Breaking Bad," "Game of Thrones," "House of Cards" or "Marvel's Agent of Shield" build a transnational community that is able to communicate via social media platforms. The results of an audience study on the "Topmodel" format in Germany and the US have shown that audiences cross geographical borders

> by creating a transculturality as mediatized practice through commonalities and similarities in the reception process. The viewers of the "Topmodel" format examined here can thus be called "trans-audiences" in this sense, because they are constituted by the similarities of their reception and appropriation, they constitute a viewership that can neither be clearly described as "global" nor as "local", but possesses an overarching status that is characterized by hybrid processes in media reception.
>
> *(Stehling 2015, 373)*

Transnational audiences follow "a 'non-resident' mode of media consumption" (Athique 2016, 11). Audiences are social bodies defined through media use. This becomes evident in fan cultures (cf. Hills 2002). Since television was transnational from the beginning, transnational audiences have formed around internationally traded television shows. As Athique (2016, 11) has noted: "In practice, any social body defined through media-use is likely to be shaped by transnational patterns of cultural consumption and association, since what most characterizes contemporary media is its multiple sources and its intertextuality." Global distribution platforms such as Amazon Prime Video and Netflix take the notion of transnational audience to another level by distributing television shows to their subscribers all over the world. They serve niche audiences of specific genres in several countries and thus promote transnationality.

## Conclusion

Transnational television culture is a social process with signifying patterns of production, distribution, legislation, aesthetic and narrative conventions, and media use. It consists of shared meanings of producers, distributers, creators and audiences. Due to national legislation television is transnational but "takes nationally specific forms" (Bignell 2013, 18). Even transnational legislation practices such as the Audiovisual Media Service Directive of the European Union are related to national legislation, also bi-national or multi-national treaties for co-productions are based on national laws. Transnational television is an arena in which the global, the local and the regional interact with each other. Transnational television culture exists within social relations of the nation-states, of transnational and global formations. "Media cannot be singled out from their historical and social contexts: media practices reflect and reproduce the transnational transformation of capitalism" (Artz 2015, 71). But:

> Media consumption differs from the consumption of other goods and services precisely because media texts are symbol systems that connect interior and exterior worlds, and as such they enable and constrain the production and circulation of meaning, and even our very imagination.
>
> *(Grindstaff 2015, 351)*

Transnational television culture reinforces shared meanings and shared cultural and social practices. It is a social and cultural process in a transnational arena where agents, institutions and structures interact with each other.

## References

Artz, Lee. 2015. *Global Entertainment Media. A Critical Introduction.* Malden and Oxford. Wiley Blackwell.

Athique, Adrian. 2016. *Transnational Audiences. Media Reception on a Global Scale.* Cambridge and Malden. Polity.

Bielby, Denise D. & Harrington, C. Lee. 2004. "Managing Culture Matters: Genre, Aesthetic Elements, and the International Market for Exported Television." In *Poetics*, 32, 73–98.

Bielby, Denise D. & Harrington, C. Lee. 2008. *Global TV. Exporting Television and Culture in the World Market.* New York and London. New York University Press.

Bignell, Jonathan. 2013. *An Introduction to Television Studies.* 3rd edition. London and New York. Routledge.

Bjork, Ulf Jonas. 2009. "'It's Better to Steal the Idea': Swedish Television Copies Programs from America, 1957–1969." In *Historical Journal of Film, Radio and Television,* 29, 2, 219–227.

Bondebjerg, Ib. et al. 2008. "American Television: Point of Reference or European Nightmare?" In *A European Television History.* Eds. by Jonathan Bignell & Andreas Fickers. 154–183. Malden et al. Wiley-Blackwell.

Bondebjerg, Ib. et al. 2017. *Transnational European Television Drama. Production, Genres, and Audiences.* Basingstoke. Palgrave Macmillan.

Brüggemann, Michael & Schulz-Forberg, Hagen. 2009. "Becoming Pan-European?: Transnational Media and the European Public Sphere."In *International Communication Gazette,* 71, 8, 693–712.

Chalaby, Jean K. 2005. "Towards an Understanding of Media Transnationalism." In *Transnational Television Worldwide. Towards a New Media Order.* Ed. by Jean K. Chalaby. 1–13. London and New York. I.B. Tauris.

Chalaby, Jean K. 2009. *Transnational Television in Europe. Reconfiguring Global Communications Networks.* London and New York. I.B. Tauris.

Chalaby, Jean K. 2015. *The Format Age. Television's Entertainment Revolution.* Cambridge and Malden. Polity.

Dizard, Wilson P. 1966. *Television. A World View.* Syracuse. Syracuse University Press.

Eugster, Ernest. 1983. *Television Programming across National Boundaries: The EBU and OIRT Experience.* Dedham. Artech House.

Fickers, Andreas. 2012. "The Birth of Eurovision: Transnational Television as a Challenge for Europe and Contemporary Media Historiography." In *Transnational Television History. A Comparative Approach.* Eds. by Andreas Fickers & Catherine Johnson. 13–32. London and New York. Routledge.

Gamula, Lea & Mikos, Lothar. 2014. *Nordic Noir. Skandinavische Fernsehserien und ihr internationaler Erfolg.* Konstanz. UVK.

Grindstaff, Laura. 2015. "From the Networks to New Media: Making Sense of Television Audiences." In *The SAGE Handbook of Television Studies.* Eds. by Manuel Alvarado, Milly Buonanno, Herman Gray & Toby Miller. 339–355. London et al. Sage.

Gripsrud, Jostein. 2007. "Television and the European Public Sphere." In *European Journal of Communication,* 22, 4, 479–492.

Hall, Stuart. 1997. "Introduction." In *Representation. Cultural Representations and Signifying Practices.* Ed. by Hall Stuart. 1–11. London et al. Sage.

Hall, Stuart. 2016. *Cultural Studies 1983. A Theoretical History.* Edited and with an introduction by Jennifer Daryl Slack and Lawrence Grossberg. Durham and London. Duke University Press.

Havens, Timothy. 2006. *Global Television Marketplace.* London. BFI.

Hills, Matt. 2002. *Fan Cultures.* London and New York. Routledge.

Hilmes, Michele. 2013. "The Whole World's Unlikely Heroine: Ugly Betty as Transnational Phenomenon." In *TV's "Betty" Goes Global. From Telenovela to International Brand.* Eds. by Janet McCabe & Kim Akass. 26–44. London and New York. I.B. Tauris.

Hong, Juanhao. 1998. *The Internationalization of Television in China: The Evolution of Ideology, Society, and Media since the Reform.* Westport. Praeger.

Jenkins, Henry. 2006. *Convergence Culture. Where Old and New Media Collide.* New York and London. New York University Press.

Kearney, Michael. 2008. "The Local and the Global: The Anthropology of Globalization and Transnationalism." In *The Transnational Studies Reader. Intersections & Innovations.* Eds. by Sanjeev Khagram & Peggy Levitt. 273–274. New York and London. Routledge.

Kuipers, Giselinde. 2011. "Cultural Globalization as the Emergence of a Transnational Cultural Field: Transnational Television and National Media Landscapes in Four European Countries." In *American Behavioral Scientist,* 55, 5, 541–557.

McCabe, Janet & Akass, Kim. (eds.). 2013. *TV's "Betty" Goes Global. From Telenovela to International Brand.* London and New York. I.B. Tauris.

Mikos, Lothar. 2016. "Germany as TV Show Import Market." In *German Television. Historical and Theoretical Perspectives.* Larson Powell and Robert Shandley, eds. 155–174. Oxford and New York. Berghahn Books.

Mikos, Lothar. 2017. "Transmedia Storytelling and Mega-Narration: Audiovisual Production in Converged Media Environments." In *Media Convergence and Deconvergence.* Eds. by Sergio Sparviero, Corinna Peil & Gabriele Balbi. 159–177. Basingstoke. Palgrave Macmillan.

Mikos, Lothar & Perrotta, Marta. 2013. "Global "Ugly Betty". International Format Trade and the Production of National Adaptations." In *The International Encyclopedia of Media Studies. Volume II: Media Production.* (ed. by Vickie Mayer). Ed. by Angharad N. Valdivia. 277–298. Malden. Wiley Blackwell.

Moran, Albert (ed.). 2009a. *TV Formats Worldwide. Localizing Global Programs.* Bristol and Chicago. Intellect.
Moran, Albert. 2009b. *New Flows in Global TV.* Bristol and Chicago. Intellect.
Moran, Albert & Keane, Michael. 2006. "Cultural Power in International TV Format Markets." In *Continuum: Journal of Media & Cultural Studies*, 20, 1, 71–86.
Moran, Albert & Malbon, Justin. 2006. *Understanding the Global TV Format.* Bristol and Portland. Intellect.
Negrine, Ralph & Papathanassopoulos, Stylianos. 1991. "The Internationalization of Television." In *European Journal of Communication*, 6, 1, 9–32.
Nordenstreng, Karle & Varis, Tapio. 1974. *Television Traffic – A One-Way Street? A Survey and Analysis of the International Flow of Television Programme Material.* Paris. UNESCO Press.
Oren, Tasha & Shahaf, Sharon (eds.). 2012. *Global Television Formats. Understanding Television across Borders.* New York and London. Routledge.
Pollack, Donald K. & Woods, David Lyndon. 1959. "A Study in International Communication: Eurovision." In *Journal of Broadcasting*, 3, 2, 101–118.
Redvall, Eva Novrup. 2013. *Writing and Producing Television Drama in Denmark. From "The Kingdom" to "The Killing".* Basingstoke. Palgrave Macmillan.
Shimpach, Shawn. 2005. "The Immortal Cosmopolitan. The International Co-Production and Global Circulation of Highlander: The Series." In *Cultural Studies*, 19, 3, 338–371.
Shimpach, Shawn. 2013. "'This Is What I Need, This Is What Will Travel': Television Programs in the Era of Transition." In *The International Encyclopedia of Media Studies. Volume II: Media Production.* (ed. by Vickie Mayer). Ed. by Angharad N. Valdivia. 192–214. Malden. Wiley Blackwell.
Stehling, Miriam. 2015. *Die Aneignung von Fernsehformaten im transkulturellen Vergleich. Eine Studie am Beispiel des Topmodel-Formats.* Wiesbaden. Springer VS.
Straubhaar, Joseph D. 2007. *World Television. From Global to Local.* Los Angeles et al. Sage.
Thussu, Daya Kishan. 2007. "Mapping Global Media Flow and Contra-Flow." In *Media on the Move. Global Flow and Contra-Flow.* Ed. by Daya Kishan Thussu. 11–32. London and New York. Routledge.
Weissmann, Elke. 2012. *Transnational Television Drama. Special Relations and Mutual Influence between the US and UK.* Basingstoke. Palgrave Macmillan.
Williams, Raymond. 1968. *Culture and Society 1780–1950.* Harmondsworth. Penguin.
Williams, Raymond. 1974. *Television. Technology and Cultural Form.* London. Fontana.
Williams, Raymond. 1981. *Culture.* London. Fontana.

# 7
# FUTURE PERFECT TV[1]—AND TV STUDIES

*Toby Miller*

## TV Studies

Television is blamed by pediatricians for making children weigh more than is fashionable in evidence-based public policy; criticized by some cultural critics for diverting the young, the middle-aged, the firm, and the infirm from attending to real matters of state or heart; valorized by corporations as a guaranteed source of revenue; and admired by neoliberal scholars for its populist impulses. In René Girard's words,

> [t]he conformism and the ethical agnosticism induced by media such as television could also produce forms of mimetic polarization at the mass level, making people more prone to be swayed by mimetic dynamics, inducing the much-feared populism in Western democracies.
>
> *(with Antonello and Cezar de Castro Rocha 2007: 249)*

Samir Amin referred disparagingly to an "average television-dulled citizen of the US" (2010), while for Giorgio Agamben, the "spectator who spends his evenings in front of the television set only gets, in exchange for his desubjectification, the frustrated mask of the couch potato, or his inclusion in the calculation of viewership ratings" (2009: 21).

A binary discourse in television studies continues to enchant participants from all backgrounds, who revel in arguments that go back centuries and are basically about how to govern societies and what to do about forms of life that invoke unruliness. Writing in the 1st century AD, Plutarch was already on the case, recounting this story about Solon: Having enjoyed what later became known as a tragedy, Solon asked the play's author, Thespis, whether he was not ashamed to tell such lies in front of so many people. When Thespis replied that there was no harm in speaking or enacting make-believe, Solon struck the ground angrily with his staff and exclaimed, "Yes, but if we allow ourselves to praise and honour make-believe like this, the next thing will be to find it creeping into our serious business" (Plutarch, 1976: 73).

Leaping forward to the 12th century, John of Salisbury warned of the negative impact of juggling, mime, and acting on "unoccupied minds ... pampered by the solace of some pleasure ... to their greater harm" (quoted in Zyvatkauskas, 2007). And when printed books began to proliferate in the early 18th century, critics feared a return to the "barbarism" of the post-Roman Empire; true erudition would be overwhelmed by popular texts, just as it had been by war (Chartier, 2004). By 1774, this was being imagined to occur at a psychological level. For example, when Goethe's *The Sorrows of Young Man Werther* appeared, its suiciding hero was deemed to have caused numerous mimetic deaths among readers, and the book was banned in many cities (Stack, 2003).

Ever since the invention of print and the spread of reading and writing, new gadgets and genres have brought with them marketing techniques and governmental interventions focused on high emotional reactions, accompanied by concerns about supposedly unprecedented and unholy new risks to rational conduct by audiences that in fact recur again and again: cheap novels from the 1900s; silent then sound film during the 1920s; radio in the 1930s; comic books of the 1940s and 50s; pop music and television from the 1950s and 60s; satanic rock as per the 1970s and 80s; video cassette recorders in the 1980s; and rap music, video games, and the internet since the 1990s. Scholarly studies into the supposed effects on audiences of these gadgets and genres have spun their web around everything from public policy to press coverage to moral panics to Edenic fantasies about an unalloyed relationship between citizens that can elude the gatekeeping power of state and capital. Young people in particular are simultaneously regarded as emotional dupes whose feelings will get the better of them through the high-tensile experiences of new technologies and stories, and/or as extraordinarily agile users of gadgets and genres, the new masters of the world (Miller, 2009).

TV is a relatively recent victim/criminal of this ongoing panicky discourse, because it is the entertainment *bodega* of the second half of the 20th century and since. The arguments remain the same. Only the names have changed to protect the ignorant. Attempts to attack and defend television follow a tiresomely familiar trajectory: It rots your brain/it expresses popular desire. It depoliticizes viewers/it reflects the material struggles of audiences. The dance between these poles appears unresolvable. Endless studies and claims are made on each side, with the refrains entirely predictable.

In addressing this, we're at least fortunate in television studies. The exclusively formal, stylistic, and ideological attention paid to texts in literary studies has never been deemed sufficient in our field—the immensely social nature of TV, like cinema, has largely militated against such reductionism. Endless moral panics about learning, lust, and lawlessness have meant that technologies, audiences, and regulations have been necessary components of our agenda. While those anxious origins may be rather dubious, their unintended consequences have left us arguably more agile than our colleagues in language departments; we look at infrastructure and technology as well as meaning, and hence can adapt to regulatory and technical changes and respond to issues of the public interest across a variety of topics.

That said, TV studies seems to be changing—becoming less social. Its new best friends are likely to be formalists, aesthetes, *littérateurs*, and ideologists rather than political economists, ratings mavens, policy wonks, psychologists, and environmentalists (Miller, 2010). Two things seem to have changed over the past 25 years. First, TV studies feels apolitical today in contrast with 1991, the last of the fabled International Television Studies Conferences. And second, bench science, AKA grant-getting, has set agendas, rather than social movements or ideological tendencies.

Because I lived most of the time between then and now in the US, but working outside the welfare-warfare nexus of Washington's grant system, I'm unused to this sort of thing. I'm familiar with big research projects, where people from around the world write things that fascinate and animate them for political reasons, and big, influential books are produced (Caldwell, 1995; Seiter, 1989). I am less *au fait* with the endless whirl of applying for grants, which has become almost an end in itself for people working in Australia and western Europe. Without these grants, many diligent, competent researchers would not make a living; and others, in more secure employment, would not have the time to study. But I see these people applying and succeeding—nice—then doing the type of work that their funding bodies want.

This leads to rather apolitical, administrative research into such topics as: how a small population can develop a TV industry; how to use new technologies to engage people; what is the map of co-productions; how much money "y" country makes from overseas drama sales; how executives of multinational corporations function; how YouTube stars emerge; how does Netflix work; and what are the difficulties faced by archivists?

All the above topics have important political aspects. But those elements are largely silent in publications produced about them. I surmise that this may be both because people diligently report on their sponsored research largely in the terms under which they applied for funding (i.e., minus politics) and because they structure their conclusions based on the need for further research (i.e., minus politics).

There are some important exceptions to this concern—mostly to do with the representation on- and off-screen of certain social identities. But apart from that, I read little in television studies about the basic political-economic exploitation of labor; the dominant structures of ownership and control; the regulatory deals that privilege oligopolies; the representational annihilation or distortion of minorities (think Polish and Colombian residents' absence from British TV drama); the exploitation of tax breaks by indolent local producers and eagle-eyed Hollywood ones in search of free money (loans with no interest or investment without equity); the social relations of technological innovation; and the environmental impact of making, watching, and disposing of television. These are matters that you might find covered in *Screen International*, *The Economist*, *Wired* magazine, *Hollywood Reporter*, *Television Business International*, or the *Financial Times*—none of them exactly claxons of socialism. The current scholarly context feels very different from the febrile commitment to social and cultural justice that dynamically animated TV studies in its origins and growth. Television studies began in its exciting politicized form not with tired media effects retreads but through dynamic efforts by socialist and feminist intellectuals to question cultural imperialism, social identities, and pleasure (Miller, 2010).

If we are to revive and develop that tradition, we need to view TV through twin theoretical prisms. On the one hand, it can be understood as a component of sovereignty, a cultural addition to ideas of patrimony and rights that sits alongside territory, language, history, and schooling. On the other hand, it is a culture industry, and subject to the rent-seeking practices, exclusionary representational protocols, and environmental destructiveness that characterize liaisons between state and capital. It is, therefore, potentially perennial and ephemeral, especially in an era when the ways it is made, watched, and archived keep multiplying. This becomes a significant task in an era that is supposedly seeing the television set being wheeled off to a retirement home for outmoded technologies.

## TV is Dead, Dying, Done

But wherever you are as you read this, television will not be far away. You can find it in the Seoul metro, the Mexico City bus system, an Australian cricket ground, your parents' home, a Cartagena hotel room, or a Hong Kong lecture theater. Screens abound, from your pocketbook to the wall, the ceiling, the corner, and the stadium. They bring images and sounds from elsewhere in order to tell journalistic, financial, historical, musical, dramatic, sexual, and athletic stories. That's television. And it is likely to continue, subject to change and disruption in both local and universal ways.

Nevertheless, folks persistently announce, predict, or incarnate the end of television. This generally takes the form of remarks such as "My children don't watch television," "Nobody I know does," or "Kids today aren't interested." From music to politics, television's day is supposedly over. This isn't only true of everyday talk, of course. We encounter it in scholarly, popular, and *bourgeois* media alike on an almost diurnal basis. A recent example, which extrapolates from one person's decisions to pronounce an entire cultural shift, is representative of a bizarre punditry that cannot stop itself despite almost two decades of saying the same, wrong, thing (Brockes, 2013). This is both immodest and empirically incorrect—but never let the facts get in the way of narcissism by spoiling a good rant.

My inner truck driver meets my inner Wittgenstein at such moments. Let's examine some numbers. Thanksgiving, a truly unholy holiday, saw US residents in the hundreds of millions watching the 2016 Macy's Day Parade, which commences at 9 am eastern—the highest rating since 2003. Millions more hung on to watch the dog show that followed NBC's official coverage (Lynch, 2013; Maglio, 2017).

Just weeks after that always awful day, the Consumer Electronics Show ushers in the new year. For all its cult of newness and touching dependency on built-in obsolescence, the Las Vegas convention acknowledges, yet again, that as far as advertisers are concerned, television remains the holy grail; it moves people, which means it moves product. Even digital specialists among marketers face the truth—people keep watching television, on a set, at home, with other people, based on the schedule constricted by networks. Ad agencies recognize that college students are promiscuous viewers, but there's nothing new in that. As always, as 30 years ago, once they graduate and get jobs, they subscribe to satellite or cable or blend those services with the internet and platforms that operate on demand. They don't *cut* cords; they *order* them (Poggi, 2014).

Worldwide, subscriptions to television via satellite and cable increased 8% to 800 million in 2012 (Friedman, 2013). Indian residents are more likely to own TV sets than have access to indoor plumbing, and politicians devote their advertising money to television ahead of all other options, drawing on its confessional qualities via close-ups and religion to appeal to voters in a highly personalized way. The number of Indian TV households grew by 11 million in 2012 (FICCI/KPMG, 2013). And in Australia, even a study that seems intent on demanding that multiple screens are the reality (almost bringing them into being through the will of desire) lies down in a post-orgasmic froth of exhaustion and admits that "all age groups continue to spend the majority of their screen time with the in-home TV set" (Nielsen Australia, 2013). People continue to want new TVs (although apparently not three-dimensional ones, despite corporate predictions) (C-Scott, 2017).

In Britain, the average viewer in 2013 watched "three hours and 52 minutes of linear television a day on a TV set" versus three minutes and 30 seconds on tablets, cell phones and laptops. Viewing of video on demand on devices other than televisions took up just 1.5% of overall televisual consumption—an increase of 1.2% on 2012, but still a minority pursuit (Cellan-Jones, 2014). In 2017, broadcast TV took up over three and a half hours per person per day—a very similar number (Ofcom, 2017). British advertisers and broadcasters spend £7.5 billion a year on making, buying, promoting, and distributing televisual content. Across age, class, and other forms of subjectification, the British watched more television in 2017 than in 2012. The post-adolescent young dedicate *more* time to watching TV: an increase of "1.4% year-on-year." The decline in viewing by "children aged 4–15 (3% 'year-on-year')" can be explained not by cybertarianism, but the political economy—by corporate, rather than consumer, choice: "children's TV has not been a priority for the largest commercial broadcasters" for many years (Broadcasters' Audience Research Board, 2017).

In Latin America, the now-venerable expression "the end of television" itself looks old, tired, and flawed: markets, cultures, politics, and policies alike find television more alive than ever, albeit in its usual state of technological, institutional, and textual flux. Advertising investment in TV continues to increase, governments still use television to promote general propaganda as well as daily agendas, football on screen remains wildly popular, and fiction programs, most notably *telenovelas*, dominate prime time, drawing large audiences aged between 25 and 60. While younger viewers watch television on a wider variety of screens and technologies, and do so at differing times, the discourse of TV remains an important referent in their audiovisual experiences. In addition, across age groups, divides persist between a minority with routine high-quality access to the digital world of technology and information and a majority without alternatives to the traditional audiovisual sphere, for whom cell phones, for instance, are at most devices for communicating with friends and family members (Orozco and Miller, 2017).

There simply isn't evidence that new technologies have displaced or are displacing the traditional cultural *bodega* of the last half-century. For example, owners of tablets such as iPads are the keenest consumers of television news worldwide. Their tablets are adjuncts, gadgetary partners, to the main source. Tablets stimulate people to increase their television watching. TV remains the dominant source of truth and object of consumption, in dual senses—the sets cost more, and we spend more time with them than other devices ("BBC World," 2013).

And consider US politics. TV rules the roost by a long, long distance when it comes to people seeking news during campaigns (Saad, 2013). In 2016, Donald Trump broke the mold by disobeying the norms of US electoral success. He didn't raise vast sums for television commercials from ordinary people or major donors. He used his Twitter account to scream at people and tendencies he didn't like, and was inarticulate, angry, ill-disciplined, and asinine in interviews and Presidential debates. So was this a case where we can say that television really didn't matter? Did a distinctly non-millennial figure draw a line through, and under, TV's political dominance? Not exactly.

Trump is unusual among politicians in being a longstanding prime-time television celebrity. His public face as a self-anointed hero of US capitalism derives as much from that as from his putative business acumen. He got lots of free TV coverage because his Twitter outbursts were like those of a spoilt child. Toys were ejected from the stroller and pacifiers spat out at a rate rarely seen in the post-War era. The white man went so far as to boast that "I don't even need commercials" (quoted in Dumenco, 2016a); but he spent almost US$25 million on them in the last week and a half of the campaign.

Trump's rants gifted a lifeline to news cyclists that was equivalent to the OJ Simpson trial or Bill Clinton's impeachment. It set ratings record after record for cable (Weprin, 2016). The press covered Hillary Clinton like the next president of the United States. The press covered Donald Trump like a future trivia question (and a ratings cash cow) (Ruttier and Poniewozik, 2016). Two of the three Presidential debates set ratings records. Over 84 million people watched the first—fewer than three million tweeted about it ("First Presidential," 2016). The General Election was the second-most watched TV event ever, well ahead of 2012 and just 0.1 million below the record set in 2008 (Bauder, 2016).

Why, then, do people persist in declaring an end to TV? There are several reasons—technological boosters fetishize newness; the proliferation of screen technologies and services suggests an entirely new world; journalists persist in pretending that, for example, Netflix, YouTube, and Amazon Prime are not producing television; regulators are too gutless to recognize Twitter and Facebook for what they are—communications companies—and hence worthy of content regulation; and as producers and networks increase popularity and power, they are all too happy to agree that their future is dire.

## Genre

TV has always been, and will always be, about genre. This is true in two ways. The original omnibus broadcast model of public broadcasters, state servants, and commercial leeches organized the day's regimen of viewing by generic hours—soap opera during the weekday, sports on weekend afternoons, and news in the early evening. With virtually worldwide deregulation since the 1980s, the proliferation of cable and satellite technology, and the subsequent emergence of what the Spanish call thematic channels—stations governed by a single genre—TV has moved into more specialized generic norms. This tendency has been derided as a threat to the better aspects of a common culture and the end of universally shared experiences. But it has also revealed the impact of interesting new pressures on television to democratize—which is to say, to move closer to perfection. I'll therefore look at the recent histories of TV sports and *novelas* to review some chaotic but discernible tendencies.

In Canada during the early 1990s, a beer company that owned The Sports Network (TSN) adopted "We Deliver the Male" as its cable motto (Sparks, 1992). And as late as 1998, a broadcast industry advertisement for ESPN in the US promised "More tackles, less tutus." A decade later, Fox Soccer Channel targeted men via commercials that emphasized regaining and sustaining hair and hard-ons, losing and hiding pimples and pounds, or becoming/adoring soldiers and sailors. But the station lost its hegemonic position in US TV football[2] to NBC once ratings for the English Premier League grew. And beyond that, numerous commercial and cultural changes have exerted tremendous pressure on the gender normativity of televised sport, weakening the seemingly rock-solid maleness at its core and suggesting wider changes to TV's future, as detailed below.

NBC also signed an extended multi-billion dollar contract to obtain exclusive US rights to live coverage of the Olympics. As is the case elsewhere, female spectators have long tuned to the Olympics, and now do so in ever larger numbers. Even back in the masculinist heyday of TSN, the 1992 Winter Games drew 57% of its US TV audience from women. Women's figure skating out-rated that year's men's baseball World Series and National Collegiate Athletic Association basketball championship game. And the women's technical skating program at the 1994 Winter Games drew the fourth-highest ratings of any program in US history. In the 1998 NBA play-offs, women preferred Game Seven of the Bulls-Pacers series to *Veronica's Closet* and *ER* (Miller, 2001).

Meanwhile, male spectatorship of TV sport in the US has been declining. The perennial success of network ratings, the NFL, saw 1998–1999 and 1999–2000 numbers for *Monday Night Football* at record lows—and a third of its audience was female. In 1999, more men aged 18–34 watched professional women's softball on ESPN2 than Arena football, the NHL, or MLS. The NFL suffered a 13% decrease in TV ratings in the five seasons from 1997 and Disney exiled *Monday Night Football* from ABC to ESPN in 2006 (Miller, 2010). The League increasingly relied for survival on women viewers, whose numbers doubled between 2007 and 2012 (McCarthy, 2012).

In 1995, more British women than men watched Wimbledon tennis on television, and the numbers were nearly equal for boxing. In 2009, the million people viewing the Women's FA Cup Final out-rated numerous men's cricket, rugby, and football fixtures, and the British Women's Open Golf drew larger audiences than the Ryder Cup. The majority of people watching women's sport on UK TV are men, and the 2017 Final of the Women's World Cup of cricket drew over 19.5 million Indian viewers (Chowdhury, 2017).

That all looks good in terms of opening TV sports up to the public. But the project of democratizing it is incomplete. For despite these trends, good old boys cling on to good old—and new—jobs. The hyper-masculine character of newsrooms and their biases remain issues. In the US, there are 48 male sportscasters for every female, and 94% of sports editors are men, as are 90% of their assistants (Schreiber, 2012). In 2008–2009, 9% of non-news Australian TV was dedicated to women's sport. The figure for men was 86% (Lumby *et al.*, 2010). But the tutus? Women comprise half of ESPN's US viewers, while TSN, which undertook to "deliver the male" a quarter of a century ago, now promises that "Sponsorship programs on TSN.ca can be tailored to your target audience."[3]

Similarly positive stories can be told about popular drama. In many parts of the world, soap opera has normalized extra-marital sexual pleasure, same-sex relationships, health issues, and progressive gender politics. It has been likened to "the 19th-century weekly sermon to a packed congregation," minus some of the moralism (Graham, 2000: 7). Latin American and Arabic *telenovelas* offer testing grounds for new subjectivities in an era of apparent choice but also widening inequality, often articulating women's constraints and freedoms, albeit in a sometimes explicitly anti-feminist, and never overtly feminist, form (Abu-Lughod, 2005; Acosta-Alzuru, 2003).

The *novela* is not only a site where social tensions are articulated and made manifest; it is also a "communicative bridge that links viewers across national, expanded regional, and global realms of transmission and reception, working to shape new cultural and intercultural communities" (Benamou, 2009: 152). *Novelas* provide audiences with the opportunity to invent histories, imagine selves, seek liberation, engage in personal reflection and reinterpretation, encourage innovative personal encounters, and experience new forms of communication. The symbiosis between audiences and these dramas can endure well beyond the moment of watching on a screen, gaining expression in private and public life, with families, neighbors, and co-workers (Martín-Barbero and Muñoz, 1992).

Prevailing social relations often determine practices of reception. Mexican women living in the US enjoy watching *telenovelas* exported from home, because they actualize affective and symbolic ties, but many men dislike the genre's sensationalized representation of their homeland. For other Latin American audiences, what matters is not geopolitics but a mixture of exoticism and repetition, a blend of new scenery with familiar sexual relationships. There are differences between urban and rural spectators, while indigent young female viewers' social imaginary may be expanded by the ideas of pleasure and autonomy that abound in the genre, as they compare their lives with *novela* representations (Brown, 2006, 2009; Concept Media, n.d.; Mayora Ronsini, 2014; Uribe, 2009).

I'm not arguing here that the magical machinery of supply and demand will deliver an Edenic world of TV gender relations via sports and soaps. I'm suggesting that the forceful feminist desire to participate in, govern, and consume genres is transformative, and that television indexes that shift. With a lot more advocacy, it can travel even further. But such advocacy needs to go beyond textuality.

## Labor

Mexico is the world's largest exporter of television sets, ahead of China, Thailand, Malaysia, and Việt Nam. The nation's young population, technological expertise, proximity to you-know-who, and taste for free-trade agreements make wages and other costs competitive in the New International Division of Cultural Labor.[4] TV manufacturing is undertaken in border *maquiladoras*, with components imported from San Diego, Germany, Japan, Taiwan, Malaysia, South Korea, and Thailand (Martínez-Siraitare, 2015). Major manufacturers there have included Sanyo, Sony, Samsung, JVC, Hitachi, and Panasonic. In 2001, twenty thousand people were employed, making over ten million televisions. Tijuana became "TV Capital of the World" and the *New York Times* headlined "A Boom Along the Border" (Malkin, 2004).

Yet this is not considered high-value work, most of which is undertaken elsewhere. It is dangerous, dull, and poorly remunerated.[5] *Maquiladora* wages consistently declined from 1993 to 2006, even as productivity increased. Two full-time workers in a Mexican plant receive just two-thirds of the pay needed to support a family of four, prior to medical and educational expenses, while employees are denied collective-bargaining rights and legal protection of privacy, health, and safety. Workers who make TVs can't afford to purchase them—flat-screen sets cost less in San Diego than in Tijuana, because tariff-free manufactures there are strictly for export (Hillyard, 2009).

Two decades ago, Human Rights Watch (1996) surveyed the numerous gendered assaults and discrimination women working in *maquiladoras* suffered. Matters haven't improved. The Centre for Reflection and Action on Labour Issues (2009) interviewed thousands of people in 2008 and 2009 across the electronics sector, disclosing systematic sexual harassment. And female employees are routinely classified as "temporary" so companies can elude regulations and deals that govern full-time labor. There is citizen activism on this score, notably Las Voces de la Maquila, the Colectivo Chilpancingo Pro Justicia Ambiental, the Environmental Health Coalition, and Greenpeace. In short, the future looks rosy for set production in Mexico, if not for those who do it, and continued advocacy is vital to protect and develop their rights.

## Environment

Just such advocacy is also crucial in one sphere where television really does have a dire influence on the world. That impact is missed by the dueling cavaliers and roundheads debating the pros and cons of media effects. The truly negative effect of television is not to do with violence or education, and it's not answerable through a rhetoric of celebration grounded in populist fandom. The obsession with consciousness that colors communication, television, media, and cultural studies alongside psychology, education, medicine, sociology, and assorted odds and sods doesn't really get to the nub of the matter. For the real damage done by TV is material.

In 2007, 207.5 million sets were sold around the globe, of which 56% were old-style, fat-screen TVs. The number for 2011 was 245.5 million, with a third being fat screens and the remainder the newer, leaner flat screens. Like electronics production more generally, TV relies on exorbitant water use and carcinogens. Most color televisions historically use cathode-ray tubes (CRTs), which send electrons from cesium cathodes into high-voltage electrodes that project onto phosphorescent screens and emit radiation to illuminate phosphors (Maxwell and Miller, 2012).

CRTs are made of zinc, copper, cesium, cadmium, silver, and lead. Major environmental problems occur both when they are made and when they are thrown away. Componentry seeps into underground water, leaving a legacy of heavy metals and toxic chemicals. This worsened with the 2009 transition to digital broadcasting in the US, when outdated analog sets, perhaps the hardest of all manufactures to recycle, were discarded.

Workers, activists, and scientists are aware of these issues. TV studies is not (Maxwell and Miller, 2012; Qiu, 2016). Again, activists lead the way for us to follow. For example, when the iPad was launched outside the US in 2010, protesters in Hong Kong responded by ritually burning photographs of iPhones. Similar demonstrations have occurred in India, Mexico, and other offshore assembly sites (Barboza, 2010; Students and Scholars Against Corporate Misbehaviour, 2010).

## The Future

Advocacy issues of the kind described above are frequently invisible to the conventions of capital. The annual corporate "TV of Tomorrow" conference finds numerous business leeches offering diagnoses and prognoses for the medium: spectators are no longer "passive observers" (when were they?); "consumers can find whatever they want to watch" (I rarely can); the future involves "a more on-demand experience" (complete with linguistic grotesqueries); and a "killer app" will appear "where the masses congregate" (i.e. Twitter or Facebook). But don't worry, some things remain constant: "advertisers can hear the cash registers ring" (Weisler, 2013) even as they hope to track down and discipline "cord-nevers" and protect "libraries that are being demonetized," all in search of "greener pastures" (Weisler, 2017). By the way, the price of these astounding insights, the cost of being there for yourself at these conferences to receive in person this remarkable analysis from executive stars—over a thousand dollars. I could have saved them the trouble. When my toddling daughter demands "Television!," she is referring to action on screen that she watches and listens to—TV today is about filmed motion and sound watched anywhere but cinemas and on anything, as she well knows. I hope you accept her position that we need to redefine TV to include all electronic devices bringing sound and imagery from afar, and my claim that we need more advocacy focused on questions such as labor and the environment. That said, TV studies will rightly also be concerned with meaning. So what should the future of television look like in textual terms?

Aimée Vega Montiel's examination of Mexican TV (2010) discloses that its address of the socially widespread gendered violence against women is always based on individual foibles and does not account for collective will and masculinity *en masse*. Of all the stories that could be told about decades and decades of conflict, the harsh treatment of women, grotesque rural inequality, and racial discrimination, only the pro-capitalist violence of the *narcos* has been glamorized on TV screens. And it has been dramatized in ways that play up both a rejection of traditional political institutions and a reassertion of traditional identities (Barrios *et al.*, 2018). For instance, there are few studies of gender and Colombian television. Most lack empirical evidence that might be used in the light of the social data adumbrated above to give us a snapshot of how TV relates to the demography and experience of its audience, though we know that women are generally represented in commercials as either sexual or domestic, and always dependent.

Related to that, it's noteworthy that longitudinal content analyses of news and current-affairs television in Colombia over the last half-century disclose that what began in the 1960s as programing dedicated to transcendental ideas of love of country, democracy, republicanism, morality, progress, and development were displaced by stories of violence, insurrection, insecurity, and a failed state (Narvaez Montoya and Romero Peña, 2017). Violence became a staple of TV by the 1980s, presaging the advent of *narco* fiction.

Because Colombia's struggle for women's equality has had a long and hard road to hoe, it is of the utmost importance that TV be aware of the constructive role it could play in destabilizing *machismo* and *Marianismo* and campaigning to stop violence against women. This is especially crucial in the recently "cool" genre of the *narconovela*, so lovingly lapped up at home and abroad, and so laden with misogyny at its very core. Popular knowledge of Escobar and his kind is stronger and deeper than familiarity with human rights, feminism, and democracy (Rincón, 2015).

The key lesson to draw from this contest is the need to construct new gender and racial relations (Hurtado and Sinha, 2016). A television system built on minimal difference and maximal repetition cannot play a part in changing that, unless it offers a multiplicity of perspectives on *narco* violence and commits to a feminist gaze among a variety of ways of seeing that put love, passion, self-determination, security, and care at their center.

Then there is the crucial issue of news and current affairs. It is easy and tempting to write off this genre, given, for example, the atrocious role played in 2016 by Colombian TV in defeating the peace referendum, Iraqi and Iranian television in broadcasting torture-induced murder confessions prior to trial, British TV in endowing the racists, xenophobes, and simpletons of Brexit with legitimacy, and US TV in feeding the white supremacy of the 2016 General Election. But then we think back to the moment in 1966 when the UK's Labour government convened meetings on homelessness, new charities emerged, and were given massive public support after the BBC had screened the docudrama *Cathy Come Home* (Ken Loach), despite hostility from the Corporation itself and the right of the *bourgeois* press (Hayward, 2006). Or the occasions in the 1960s and 70s when US networks covered the brutal way that African Americans were treated on a daily basis by authorities right across the country. That helped bring down the Nixon administration (Reed, 1999). And when we look today at the cavalierly positive way in which the US legal system is represented in televised drama, and juxtapose that with popular sources of news, TV remains both a problem and possibility for exposing the unethical conduct of prosecutors across the country (Davis, 2007).

## Conclusion

The various discourses on TV, be they popular or scholarly, don't speak to one another as they should. The consumerist appreciationism/pub talk of technology and television criticism is socially and environmentally irresponsible and reductive, while the serious tone of their social, psychological, and environmental equivalents lack articulation to communications technologies as objects of work and leisure.

Now is the moment when Dallas Smythe's (1981) fundamental insight into TV audiences—that they are constantly performing labor that is sold through their emotional engagement—should meet Heidegger's (1977) account of the role of foresters. They cut down trees that form the base material from which magazines, newspapers, and books are made. Then they buy them to read and relax, supposedly evidencing customer satisfaction and public opinion. TV studies must connect the work of making as well as watching television, without forgetting the realities of a simultaneously surrounding and submerging culture: "television is not conceived primarily on an artistic, purely aesthetic level, extraneous to present customs" (Adorno, 1954: 214). In Arjun Appadurai's words, we must follow "things themselves, for their meanings are inscribed in their forms, their uses, their trajectories" (1986: 5). That line of inquiry is in keeping with our foremost historian of media technologies, Roger Chartier (1994) and leading phenomenologist of the media, Horst Ruthrof (1997). The logic and life of the commodity sign is a constantly changing, attenuating, and accreting one. It urgently needs to be articulated to the science and ecology of the medium (Maxwell and Miller, 2012).

In short, TV should be understood in terms of its labor process, environmental impact, and textual pleasure. We need everyone involved in the medium to think about their fellow workers, whether they are governing, creating, consuming, teaching, or trashing television. Otherwise, the eco-crisis of labor and planetary exploitation unconsciously signified by the question of whether to "Junk the TV Set" will come to dwarf the question "What is Television?"

## Notes

1. I must depart from an expectation probably set up by my title. The future perfect tense describes activities that are already under way or will start and are finite. I'm not seeking to perform that kind of futurism in accordance with linguistic rules. Rather, I refer to television in the future that could be perfect, and how we might comprehend it.
2. I use the term "football" here to refer to the word used by 96% of the world's population to describe "association football," known to white Yanquis as "soccer."
3. www.tsn.ca/tsn-ca-help-section-tsn-ca-1.102539.
4. http://tacna.net/mexico-vs-china/.
5. www.youtube.com/watch?v=kPWLBuOp8GA.

## References

Abu-Lughod, L. (2005) *Dramas of Nationhood: The Politics of Television in Egypt*, Chicago: University of Chicago Press.

Acosta-Alzuru, C. (2003) ""I'm Not a Feminist...I Only Defend Women as Human Beings": The Production, Representation, and Consumption of Feminism in a *Telenovela*," *Critical Studies in Media Communication* 20, no. 3: 269–294.

Adorno, T. W. (1954) "How to Look at Television," *The Quarterly of Film Radio and Television* 8, no. 3: 213–235.

Agamben, G. (2009) *What is an Apparatus?* trans. D. Kishik and S. Pedatella, Stanford: Stanford University Press.

Amir, S. (2010) "A Misguided Critique of Aid," *African Review of Books* 6, no. 2: 7.

Appadurai, A. (1986) "Introduction: Commodities and the Politics of Value." In *The Social Life of Things: Commodities in Cultural Perspective*, ed. A. Appadurai, New York: Cambridge University Press, pp. 5–63.

Barboza, D. (2010, July 5) "Supply Chain for iPhone Highlights Costs in China," *New York Times* www.nytimes.com/2010/07/06/technology/06iphone.html?pagewanted=all.

Barrios, M. M., J. Arroyave, and T. Miller (2018) "Prime-Time *Narcos*: The Mafia and Gender in Colombian Television," *Feminist Media Studies* 10.1080/14680777.2018.1434223.

Bauder, D. (2016, November 10) "Election Night Narrowly Misses a TV Ratings Record," *Toronto Sun* http://torontosun.com/2016/11/10/election-night-narrowly-misses-a-tv-ratings-record/wcm/2d802757-433d-4c94-875e-322331bf2f4f.

"BBC World News and bbc.com Release World's Largest Global Study of News Consumption Habits across Multiple Devices." (2013, March 26) *BBC* www.bbc.co.uk/mediacentre/worldnews/news-consumption.html.

Benamou, C. L. (2009) "Televisual Melodrama in an Era of Transnational Migration: Exporting the Folkloric Nation, Harvesting the Melancholic-Sublime." In *Passion, Pathos and Entertainment: Latin American Melodrama*, ed. D. J. Sadlier, Chicago: University of Illinois Press, pp. 139–171.

Broadcasters' Audience Research Board (2017) *The Viewing Report: Our Annual Exploration of the UK's Viewing Habits* www.barb.co.uk/download/?file=/wp-content/uploads/2017/04/Barb-Viewing-Report-2017.pdf.

Brockes, E. (2013, October 27) "The Lost World of Watching TV on a TV," *Guardian* www.theguardian.com/commentisfree/emma-brockes-blog/2013/oct/27/lost-world-watching-tv.

Brown, L. (2006) "Invisible Maids: Slavery and Soap Operas in Northeast Brazil," *Agenda* 20, no. 70: 76–86.

Brown, L. (2009) "Pleasuring Body Parts: Women and Soap Operas in Brazil," *Critical Arts* 23, no. 1: 6–25.

Caldwell, J. (1995). *Televisuality: Style, Crisis, and Authority in American Television.* New Brunswick: Rutgers University Press.

Cellan-Jones, R. (2014, February 17) "TV's Changing? Not So Fast...," *BBC* www.bbc.co.uk/news/technology-26224625.

Centre for Reflection and Action on Labour Issues (2009) *Labour Rights in a Time of Crisis: Third Report on Working Conditions in the Mexican Electronics Industry* www.industriall-union.org/sites/default/files/migration/imf/RelatedFiles/09112509113366/CEREAL_REPORT-2009-ENGLISH.pdf.

Chartier, R. (1994) *The Order of Books: Readers, Authors, and Libraries in Europe between the Fourteenth and Eighteenth Centuries*, trans. L. G. Cochrane, Stanford: Stanford University Press.

Chartier, R. (2004) "Languages, Books and Reading from the Printed Word to the Digital Text," trans. T. L. Fagan, *Critical Inquiry* 31, no. 1: 133–152.

Chowdhury, S. (2017, August 6) "ICC Women's World Cup 2017 Final Was Most Watched Women's Sporting Event in India," *Sportskeeda* www.sportskeeda.com/cricket/icc-womens-world-cup-2017-final-most-watched-womens-sporting-event-india.

Concept Media, S. A. (n. d.) *Las Mil y Una Noches: Cuales son las claves de este éxito y cómo son sus televidentes* www.conceptmedia.com.ar/notas_mail/las_mil_y_una_noches.html.

C-Scott, M. (2017, February 7) "3D Television is Dead... So What Next?" *The Conversation* https://theconversation.com/3d-television-is-dead-so-what-next-72192.

Davis, A. J. (2007) *Arbitrary Justice: The Power of the American Prosecutor*, New York: Oxford University Press.

Dumenco, S. (2016a, October 28) "State-By-State Map: How the Clinton and Trump Camps Spent $595 Million on TV, Radio Ads," *AdAge* http://adage.com/article/campaign-trail/how-clinton-trump-camps-spent-595-million-TV-radio/306496/.

Dumenco, S. (2016b, November 4) "Surprise! Trump, Clinton and Their PAC Allies Just Booked Another $55.7M in Last-Minute TV, Radio Ads," *AdAge* http://adage.com/article/campaign-trail/trump-and-clinton-and-pac-allies-just-booked-tens-of-millions-in-last-minute-tv-radio-ads/306634/.

"Excesiva e inconstiutcional, multa de IFT a TV Azteca y TV Internacional: SCJN." (2017, October 25) *La Jornada* www.jornada.unam.mx/ultimas/2017/10/25/excesiva-e-inconstitucional-multa-de-ift-a-tv-azteca-y-television-internacional-scjn-5619.html.

FICCI/KPMG (2013) *The Power of a Billion: Realizing the Indian Dream* http://ficci.in/spdocument/20221/FICCI-KPMG-Report.pdf.

"First Presidential Debate of 2016 Draws 84 Million Viewers." (2016, September 27) *Nielsen* www.nielsen.com/us/en/insights/news/2016/first-presidential-debate-of-2016-draws-84-million-viewers.html?cid=socSprinklr-Nielsen.

Friedman, W. (2013, May 23) "Worldwide Pay TV on the Rise, Big Growth in Asia," *MediaPost* www.mediapost.com/publications/article/201062/worldwide-pay-tv-on-the-rise-big-growth-in-asia.html.

Girard, R. with P. Antonello and J. Cezar de Castro Rocha (2007) *Evolution and Conversion: Dialogues on the Origins of Culture.* London: Continuum.

Graham, D. (2000, December) "A Declaration of Independence," *Economic Affairs*: 7–12.

Gregory, P. R. (2013, March 12) "Why Doesn't The New York Times Expose Carlos Slim?" *RealClear Markets* www.realclearmarkets.com/articles/2013/03/12/for_155m_did_carlos_slim_buy_silence_from_the_times_100195.html.

Hayward, A. (2006, November 30) "Cathy Come Home," *Independent* www.independent.co.uk/news/media/cathy-come-home-422630.html.

Heidegger, M. (1977) *Basic Writings from Being and Time (1927) to the Task of Thinking (1964)*, ed. D. F. Krell, trans. Joan Stambaugh, J. Glenn Gray, David Farrell Krell, John Sallis, Frank A. Capuzzi, Albert Hofstadter, W. B. Barton, Jr., Vera Deutsch, William Lovitt, and Fred D. Wieck, New York: Harper & Row.

Hillyard, W. (2009, December 31) "Where is "Away?"," *Real Change* http://realchangenews.org/2009/12/31/where-away.

Human Rights Watch (1996) *Mexico: NO GUARANTEES: Sex Discrimination in Mexico's Maquiladora Sector* www.hrw.org/legacy/reports/1996/Mexi0896.htm.

Hurtado, A. and Sinha, M. (2016) *Beyond Machismo: Intersectional Latino Masculinities*, Austin: University of Texas Press.

Kestenbaum, D. (2012, November 29) "Why Mexico is the World's Biggest Exporter of Flat-Screen TVs," *Planet Money* www.npr.org/sections/money/2012/11/30/166180397/why-mexico-is-the-worlds-biggest-exporter-of-flat-screen-tvs.

Lumby, C., H. Caple, and K. Greenwood (2010) *Towards a Level Playing Field: Sport and Gender in Australian Media*, Sydney: Australian Sports Commission.

Lynch, J. (2013, November 25) "Americans' New Thanksgiving Tradition; Watching TV All Day Long," *Quartz* https://qz.com/150790/americans-new-thanksgiving-tradition-watching-tv-all-day-long/.

Maglio, T. (2017, November 24) "Macy's Thanksgiving Parade Ratings Deflate 7 Percent from 2016," *The Wrap* www.thewrap.com/macys-thanksgiving-day-parade-ratings-dog-show/

Malkin, E. (2004, August 26) "A Boom Along the Border," *New York Times* www.nytimes.com/2004/08/26/business/a-boom-along-the-border.html.

Martín-Barbero, J. and S. Muñoz eds. (1992) *Televisión y melodrama: Géneros y lecturas de la telenovela en Colombia*, Bogotá: Tercer Mundo Ediciones.

Martínez-Siraitare, H. (2015, May 29) "Television Manufacturing in Tijuana is King," *Tecma* www.tecma.com/television-manufacturing-in-tijuana/.

Maxwell, R. and T. Miller (2012) *Greening the Media*, New York: Oxford University Press.

Mayora Ronsini, V. (2014) ""Many People Are Just Dreamers": Telenovelas and the Ideology of Meritocracy," *Television & New Media* 15, no. 6: 551–561.

McCarthy, M. (2012, September 24) "Women's Top-Watched Show is Not What You Think," *AdAge* http://adage.com/article/cmo-strategy/nfl-continues-score-women-viewers/237374/.

Miller, T. (2001) *SportSex*, Philadelphia: Temple University Press.

Miller, T. (2009) "Media Effects and Cultural Studies: A Contentious Relationship." In *The Sage Handbook of Media Processes and Effects*, eds. R. L. Nabi and M. B. Oliver, Thousand Oaks: Sage Publications, pp. 131–143.

Miller, T. (2010) *Television Studies: The Basics*, New York: Routledge.

Narvaez Montoya, A. and A. C. Romero Peña (2017) "La nación en la televisión informativa Colombiana: 1960 y 2015. Del infractor inexistente al enemigo omnipresente," *Revista Folios* 46: 55–65.

Nielsen Australia (2013, June 27) *Australian Multi-Screen Report: Quarter 1 2013* www.nielsen.com/au/en/press-room/2013.html#sthash.FUC6ukjp.dpuf.

Ofcom (2017) *PSB Annual Research Report 2017* www.ofcom.org.uk/__data/assets/pdf_file/0019/103924/psb-annual-report-2017.pdf.

Orozco, G. and T. Miller (2017) "La Televisión más allá de sí misma en América Latina," *Comunicación y Sociedad* 30: 107–127.

Plutarch (1976) *The Rise and Fall of Athens: Nine Greek Lives by Plutarch*, trans. Ian Scott-Kilvert, Harmondsworth: Penguin.

Poggi, J. (2014, January 8) "Amid CES Gadgets, TV Still Works for Marketers," *AdAge* http://adage.com/article/consumer-electronics-show/amid-ces-gadgets-tv-works-marketers/290965/.

Qiu, J. L. (2016) *Goodbye iSlave: A Manifesto for Digital Abolition*, Urbana: University of Illinois Press.

Reed, A. Jr. (1999) *Stirrings in the Jug: Black Politics in the Post-Segregation Era*, Minneapolis: University of Minnesota Press.

Rincón, O. (2015) "Amamos a Pablo, odiamos a los políticos: Las repercusiones de *Escobar, el patrón del mal*," *Nueva Sociedad* 255: 94–105.

Ruthrof, H. (1997) *Semantics and the Body: Meaning from Frege to the Postmodern*, Toronto: University of Toronto Press.

Ruttier, J. and J. Poniewozik (2016, November 9) "After This Election Can the Media Recover?" *New York Times* www.nytimes.com/2016/11/09/arts/television/after-this-election-can-the-media-recover.html?mtrref=undefined&gwh=210CD4413BD0B0FFDFF3A2D53BBD5BEE&gwt=pay.
Saad, L. (2013, July 8) "TV Is Americans' Main Source of News," *Gallup* http://news.gallup.com/poll/163412/americans-main-source-news.aspx.
Schreiber, K. (2012, August 2) "Why Don't We Watch More Women's Sports?" *Greatist* https://greatist.com/fitness/why-dont-we-watch-more-womens-sports.
Seiter, E. (1989). *Remote Control: Television, Audiences and Cultural Power.* New York: Routledge.
Smythe, D. (1981) *Dependency Road: Communications, Capitalism, Consciousness, Canada,* Norwood: Ablex.
Sparks, R. (1992) "Delivering the Male": Sports, Canadian Television, and the Making of TSN," *Canadian Journal of Communication* 17, no. 3. www.cjc-online.ca/index.php/journal/article/view/678/584.
Stack, S. (2003) "Media Coverage as a Risk Factor in Suicide," *Journal of Epidemiology and Community Health* 57: 238–240.
Students and Scholars Against Corporate Misbehaviour. (2010) *Workers as Machines: Military Management in Foxconn* http://sacom.hk/archives/740.
"Traditional TV's Surprising Staying Power." (2017, February 9) *The Economist* https://media.economist.com/news/special-report/21716459-peak-tv-its-way-slowly-traditional-tvs-surprising-staying-power.
Uribe, A. B. (2009) *Mi México imaginado: Televisión, telenovelas y migrantes,* Colima: Editorial Porrúa/El Colegio de la Frontera Norte/Universidad de Colima.
Vega Montiel, A. (2010) "La responsibilidad de la television Mexicana en la erradicación de la violencia de género contra las mujeres y las niñas: Apuntes de una investigación diagnóstica," *Comunicación y Sociedad* 13: 43–68.
Weisler, C. (2013, January 7) "Experts Mull the TV of Tomorrow," *MediaPost* www.mediapost.com/publications/article/190486/experts-mull-the-tv-of-tomorrow.html#axzz2HKtOwsGf.
Weisler, C. (2017, December 8) "TV of Tomorrow Offers Exciting, Dystopian Future," *MediaPost* www.mediapost.com/publications/article/311336/tv-of-tomorrow-offers-exciting-dystopian-future.html.
Weprin, A. (2016, March 29) "2016 Election Coverage Keeps Shattering Cable News Ratings Records," *Politico* www.politico.com/blogs/on-media/2016/03/cable-tv-news-ratings-2016-election-221346.
Zyvatkauskas, C. (2007 February 3) "Theatre Critic," *Economist*: 18.

# PART II

# Audiences

This part addresses the ways television affects people by considering the viewer. It also examines the ways in which the viewer has been and is still studied, thus addressing the ways people affect television. This section, in other words, engages with contemporary global television audience studies. Size, dispersal, and typically private viewing circumstances have conspired to make the television audience notoriously difficult to study and understand. Neither scholars nor industry representatives can simply go have a look at the television audience. The industry has responded by producing elaborate means of measuring from a distance, through surveys, sampling and demographic categorizations in order to render audience behaviors less indeterminate. Scholarship follows some of the same procedures, while theorizing passive, susceptible, active, and creative behaviors for what people must be doing while the television is on. Technological change and new perspectives on global difference are propelling new considerations for understanding the audience. Indeed, as evidence that industry and scholarship alike seek to address these changes, we encounter in this part some emerging key terms that are shared by both, even while noting important distinctions in their usage. Shanti Kumar, for example, describes a growing interest among advertisers in influencing audience "affect" while suggesting that nevertheless it may offer a means of encouraging us to move past the long-standing debate over whether audiences are either passive receptacles of television or active agents and producers. Consideration of affect instead offers an approach to the television audience that examines how people feel and respond to it. Jonathan Corpus Ong and Ranjana Das, noting an emerging historical divide in audience studies, argue for a reconsideration of two keywords, not necessarily naturally paired: divergence and responsibility. With these they invite considered caution in addressing the social and ethical challenges posed by audience encounters with processes of datafication and disinformation. Jerome Bourdon and Cécile Méadel, meanwhile, look at the actual device, the peoplemeter, used around the world now to measure viewing and to produce—more for the need for agreement than for specific accuracy—a mutually acceptable measurement of the television audience. Anna Potter and Jeanette Steemers consider the transformations in programming aimed at a specific audience demographic, children, and how that group of viewers, and the engagements they have with televisual media, features at the center of television's public and privatized functions. Andy Ruddock theorizes the audience as itself an archive for future historians wondering what television meant to people in the early twenty-first century and suggests the evidence may not always corroborate official stories. Esther Milne and Aneta Podkalicka reconsider the neoliberal thesis diagnosing viewers of home improvement reality programming to suggest the idea of "engagement" might better get at the complicated—by no means overdetermined—enjoyments of viewing. Annette Hill focuses on reality food programming to further elaborate audience engagement—another of the terms shared by industry and scholarship—in order to find ways of researching that get at "both a pragmatic meaning of audience attention and a more subjective meaning of the relationships between people and media content." The chapters in this part all engage with the history of Television Studies to understand contemporary audiences, but each gravitates toward new approaches and new terms for understanding how people and television relate to each other today.

# 8
# THE AFFECTIVE AUDIENCE
## Beyond the Active vs. Passive Audience Theory Debate in Television Studies

*Shanti Kumar*

In *Slumdog Millionaire* (2008), the most celebrated of films about reality television, Jamal Malik, the hero, asks Latika, the heroine, why she and millions of other Indians want to always watch the TV show *Kaun Banega Crorepati*? (Indian version of *Who Wants to Be a Millionaire*?). Latika responds, in a seemingly fatalistic way, that everyone has to escape somewhere. Her response is particularly paradoxical since the reality of Latika's life in the film's narrative is that she is destined to be a prisoner of fate in one way or another. The surprising success of *Slumdog Millionaire* as a worldwide hit, and the rapid expansion of the "Millionaire" TV franchise as a global phenomenon offer great insights into the complex relationship between the television industry and its audiences. Much has been written about how the circulation of popular cultural texts like the *Millionaire* film and the TV franchise it represents has elicited many different responses in different contexts around the world (Bielby Denise and Harrington, 2012; Gehlawat, 2014; Tzanelli, 2016). But here I refer to Jamal's question about why people are so captured by TV, and Latika's response about escape as emblematic of the most common conversation viewers in India – and more generally viewers around the world – have about the relationship between television and its audiences: does television represent the social reality of its audiences, or does it provide them an escape from reality?

In this chapter, I revisit and revise the theoretical cul-de-sac on the relationship between television and its audiences that has vexed not only scholars in Television Studies, but also policy-makers and professionals in the television industry and fans and critics in popular culture. I provide an overview of the "passive audience" and "active audience" theories that have historically shaped the debate on television and its audiences; focusing in particular on the debates on television in India. I advocate a move away from the active/passive audience debate, and outline a theory of "affective audiences" by drawing non-representational theories of affect (Clough and Halley, 2007; Seigworth and Greg, 2009; Thrift, 2008).

According to Nigel Thrift (2008, 75), affect is often defined in relation to emotions and feelings such as "hatred, shame, envy, fear, disgust, anger, embarrassment, sorrow, grief, anguish, love, happiness, joy, hope, wonder." Scholars of emotion find affect to be a useful term in defining and describing feelings that theoretically cannot be classified as either activity or passivity (Karatzogianni and Kuntsman, 2012; Wetherell, 2014). For example, in the philosophical traditions of phenomenology, the term "activity" is generally used to describe something that is done consciously – like walking, sitting, whispering, shouting, etc. Passivity, on the other hand, is traditionally defined in terms of the absence of conscious action (Morris, 2016). But, when one feels an intense emotion like fear while watching a violent scene in a murder mystery or looking at a terrifying image in a horror movie and gets goosebumps on the skin, the goosebumps cannot be considered conscious action. Nor can they be considered passivity or inaction because the eruption of goosebumps is still a bodily "action" – even if it is non-conscious. Thus, affect, or affection, is often used as a term to describe a body's potential to act or to be acted upon in ways that are both prior to and beyond the conventional categories of (conscious) activity or passivity (Massumi, 1995).

Thrift finds that words like emotion and feeling are commonly used with reference to a person, or an individual body, and are thus not good translations of the term affect. Instead, he defines affect in more general terms of tendencies and lines of forces that can act upon or be acted upon relationally on a mass or a collective scale. Thrift (2008, 171) argues that affects such as anger, fear, joy and hope manifest themselves in everyday life on a mass scale in "the mundane emotional labor of the workplace, the frustrated shouts and gestures of road rage, the delighted laughter of children as they tour a theme park or the tears of a suspected felon undergoing police interrogation."

I find that Thrift's definition of affect holds rather well for describing the emotions or feelings engendered among the mass audience of mass media like television. For example, affect is manifest in the "mass hysteria" surrounding the lives and death of major movie and TV actors, or in the "fanatic" following that a rock band or a reality TV sensation can attract among a mass audience. Equally, the term affect can capture the thrills and chills that fans of a sports team can collectively feel during the telecast of a major sporting event, or the intense pangs of jealousy, hatred or love and admiration that mass audiences can feel toward a character in a soap opera as they collectively follow the narrative ups and downs on a daily or a weekly basis. In this chapter, I propose the category of "affective audiences" to move the debate on television and its audiences away from categories like active and passive. I also discuss how media producers and researchers in television, marketing and advertising industries are seeking newer avenues to target audiences in ways that go beyond representational strategies of conscious "looking" toward multisensory experiences of affect in non-representational terms.

## Active/Passive

Since the birth of broadcasting in the early twentieth century, countless academic studies, journalistic commentaries and policy reports have been published deriding mass media like radio and television broadcasting as escapist fantasy. The ideological suspicion that media critics harbored against broadcasting as an escapist medium was largely an outcome of the early debates in mass communications, where the emerging relationships between mass media and their audiences were framed in terms of linear cause-and-effect models. These models were then in vogue both in the "mass-culture" approach of Western European critical theory and in the "mass-society" approach in American communication studies (Lang and Lang, 2009).

In the mass-culture approach, popularized in the works of Frankfurt School scholars such as Theodore W. Adorno and Max Horkheimer (2007), the theory of passive audiences is based on a linear model of communication where media messages are used by their producers to inject the ideology of capitalism into the mass audiences in the same way that a hypodermic syringe can be used to inject a fluid into a body. Combining Marxist theories of ideology critique and Freudian theories of psychoanalysis, the Frankfurt School scholars argued that the masses are lulled into a false consciousness by the captains of the capitalist industries who use entertainment, advertising and marketing to induce the audiences to passively consume mass media messages. The role of the media scholar, the political activist and the policymaker alike, Adorno and Horkheimer argued, was to rid the passive audiences of their false consciousness by overturning the capitalist ideology at the base of the production process, and thus creating a more "enlightened" mass-mediated culture (Tarr, 2017).

In the mass-society approach, the passive character of the masses was clearly delineated in the linear communication models such as the "source-message-channel-receiver" model popularized by the works of Harold Lasswell (1936) and Paul Lazarsfeld (1933) among others, and in the "encoding-message-channel-decoding-receiver" model used by Shannon and Weaver (1971) at the Bell Telephone Laboratories. In the context of mass communications, where the dominant mode of interaction between senders and receivers of a message was not interpersonal, media scholars were transfixed on an urgent social need to comprehend the "effects" of mass media on their mass audiences (Lerner, 1958; McQuail and Windahl, 2013; Schramm, 1960). The framework of effects studies has undergone several significant transformations following criticism of its overreliance on linear models of transmission of knowledge from (Western) experts to (non-Western) masses. As linear models of one-way communication have been replaced by more complex multi-step models, and passive audience frameworks have given way to "uses and gratifications" theories, the influence of "effects" paradigm remains pervasive in communication studies, development studies and media industry research where sophisticated methods of statistical analysis and experimental psychology are still used to transmit messages to mass audiences, and to obtain feedback that can be relayed back to the source (Sparks, 2016; Tidd, 2010; Vishwanath and Barnett, 2011).

For example, in *The Diffusion of Innovation*, Everett Rogers (1962) sought to understand how an innovative message or a commodity spreads in mass society, and why some innovations are more readily accepted by a mass audience than others. Rogers thus argued that the main focus of the mass communications researcher was to intervene at the source of production to help the sender create innovative messages that would resonate better with mass audiences. A well-trained mass communication researcher, Rogers argued, could use social-scientific techniques, such as survey research and focus groups, not only to understand the effects of innovative media messages in mass society, but also to obtain feedback from the mass audiences and relay it back to the source, who can then adjust the message to better fit the needs of the audiences.

Rogers' influential model of diffusion of innovations has been used, revised and refined by many scholars working in Third World countries such as India, China, Brazil and Nigeria to aid the rapid modernization and development of audiences in underdeveloped areas such as villages and urban slums (Rogers, Ascroft and Röling, 1970; Singhal and Dearing, 2006). In India, for instance, mass media like newspapers, films, radio and television – and now digital media – are used both by the government and non-governmental organizations for the diffusion of scientific knowledge and technical innovations in various walks of everyday life such as agriculture, farming, health, housing, education and literacy (Kumar, 2009; Ramachandran, 1975).

For almost two decades, after its introduction in 1959, television in India was used by the government primarily as a medium for programming focused on developmental issues such as agriculture, literacy, national integration, education, health and family welfare. However, things began to change during the 1980s when the national network, Doordarshan, slowly began incorporating entertainment programming into its schedule. On July 7, 1984 Doordarshan began broadcasting *Hum Log* (We the People); a part-educational and part-entertainment television serial. As Arvind Singhal and Everett M. Rogers (1999, 75) write, "*Hum Log* was an attempt to blend Doordarshan's stated objectives of providing entertainment to its audience, while promoting, within the limits of a dominant patriarchal system, such educational issues as family planning, equal status for women, and family harmony."

The story of *Hum Log* revolves around the everyday activities of a north-Indian joint-family across three generations. The story of the *Hum Log* family was promoted by Doordarshan as the story of the "Indian" family, and the serial became India's first major commercial success on television. *Hum Log* paved the way for other "socially conscious" soaps such as *Buniyad* (The Foundation), and commercially sponsored sitcoms like *Yeh Jo Hai Zindagi* (That's Life). The telecast of *Ramayan* in 1987–1988 and *Mahabharat* in 1988–1989 marked the rise of a hybrid television genre that was part religious, part social, part dramatic and part soap-operatic. The television recasting of the Hindu epics *Ramayan* and *Mahabharat* that were already popularized on a mass scale in India through theater, literature, art and cinema, drew huge audiences across the nation and were viewed with ritual regularity in over 90% of all Indian television homes. Journalists and academic scholars alike reported with awe, amazement and even ridicule about how workers in factories or offices and families at home reorganized their daily and weekly routines around the Sunday morning telecasts of the religious epics. The general consensus among the critics was that mass media like radio and television – if not used appropriately – can turn audiences into passive "couch potatoes" – or to use Philip Lutgendorf's (1990, 163) cleverly Indianized phrase – "couch *pakoras*."

In a critical survey of the reception of religious television serials like *Ramayan* and *Mahabharat* in India in the late 1980s, Lutgendorf finds that, not unlike religion, television has long been assailed by media and cultural critics as an opiate for the masses. Lutgendorf argues that the elite criticism of religious television serials like *Ramayan* in India is reminiscent of how critics in the United States, such as Robert MacNeil and Neil Postman, likened television to the pleasure drug "soma" that is distributed by the government to the masses in Aldous Huxley's dystopian novel *Brave New World* (1932). In *Many Ramayanas*, Paula Richman (1991) recalls how noted historians such as Romila Thapar criticized the TV show's producers for reducing the rich diversity of religious epics into simplistic narratives and commodified objects for mass entertainment. Since Doordarshan is a government-run network, Thapar was concerned that a state-sponsored production of a religious epic would become the dominant-hegemonic narrative catering only to the desires of a middle-class consumerist society and displace other versions that reflected the concerns of a diversity of social groups in India. Although kings, village chiefs or other political elites had in the past acted as patrons for Ramayan performances, the television *Ramayan*, Thapar argued, was dangerous and unprecedented in its ability to authoritatively broadcast its hegemonic version to a mass audience on a national scale. While Thapar argued that the state-sponsorship of Ramayan leads to the homogenization of its narrative, Richman counters that the contributions in her book *Many Ramayanas* reveal how the popularity of the TV *Ramayan* rekindled the interest of audiences in the ancient epic and must thus be affirmed as the latest episode in an ongoing series of tellings and retellings of Ramayana.

The critiques of the passive audience theories and mass-society frameworks offered by Lutgendorf and Richman in their analyses of the responses to the television serial *Ramayan* in India are indicative of the theoretical turn in Television Studies toward "active audience" frameworks promoted by cultural studies scholars since the 1970s and 1980s. The most influential among active audience theories, is undoubtedly Stuart Hall's (1980) well-known Encoding/Decoding model. In this model, Hall develops a new paradigm for understanding the relationship between production and reception of television as a "text." In defining television as a "text," Hall posits that a television show or a message never represents a raw event or pure reality, but instead encodes the event as a representation of reality. This representation of reality is a "text" because it is mediated by the technological infrastructure, relations of production and the frameworks of knowledge of those who encode the text on the one hand, and those who decode it on the other.

One can only briefly summarize the richly theoretical analysis that Hall undertakes in this paradigm-shifting essay. Hall makes a series of very important points about how television functions as a system of unequal communication between the producers and audiences. Unlike previous theorizations in passive audience theories where communication was seen as a linear process from an active sender to a passive receiver, Hall argued that inequality of power relations between sender and receiver is inherently unstable because the meaning of a television show, an image or even a word cannot be fixed by the sender prior to the moment of reception.

Drawing on linguistic theories of polysemy – or the multiplicity of meanings in a sign – Hall suggests that meaning of a televisual text emerges in an active process of negotiation between the producers who create the message and audiences who use the message to make sense of experiences in their everyday lives. To be sure, the moment of encoding determines what meanings are put into a message by the sender/producer, but the moment of decoding is semi-autonomous because the message sent (encoded) by the sender/producer is not necessarily the message received (decoded) by the receiver/consumer who makes his or her own meaning. Based on the semi-determinant and semi-autonomous relationship between the moments of encoding and decoding, Hall develops three hypothetical reading positions that audiences could potentially take in making meaning of television texts. The first reading position, called dominant/hegemonic, emerges when audiences embrace the preferred meaning encoded by the producer of the text and accept the intended message. The second position, called negotiated, as the name suggests, is possible when the audiences accept some preferred meanings but reject others. The third position, called oppositional, occurs when audiences reject the preferred meaning or resist the intended message.

In the wake of Hall's encoding/decoding model, many scholars embraced the British cultural studies approach to examine the dynamic relationship between television and its audiences. In *Screening Culture, Viewing Politics*, Purnima Mankekar (1999) examines how Doordarshan, the state-sponsored television network in India, underwent rapid changes in relation to the emergence of a "national family" of upwardly mobile, middle-class consumers in the 1980s. In her analysis of Doordarshan's production of the Hindu epic *Mahabharat*, Mankekar focuses on the most powerful segment in the story – the public disrobing of Draupadi (the lead female character in the epic). Combining textual analysis of the episode with ethnographic observations, Mankekar demonstrates how the discourses of nationalism and gender overlap in divergent readings of the epic between the serial's (male) producers and its women viewers. Mankekar finds that the producers and the viewers saw Draupadi's character as an index of the position of women in Indian (Hindu) society, and in more symbolic terms as the female icon of the Indian nation. However, Mankekar's ethnography of female viewers also reveals that in addition to the hegemonic nationalist readings, many women viewed Draupadi's disrobing as a marker of women's vulnerability, and her rage as the rage of a woman wronged by a patriarchal society.

Using Hall's encoding/decoding model, Mankekar examines the inherently polysemic discourse of meaning making that is structured in dominance by the (male) producers at the moment of encoding but is also open to negotiated and oppositional readings in the semi-autonomous moment of decoding by a diversity of male, female, young and old, rich and poor, rural and urban receivers. Mankekar also describes how commercial programming on Doordarshan tried to identify with this powerful class of consumers by promoting and participating in the construction of a hegemonic image of the Indian Woman – *Nai Bharatiya Nari* – at the center of the national family. In television serials such as *Udaan* and *Rajani*, the New Indian Woman is represented as a modern, educated, middle-class consumer – and almost always Hindu. As the idealized representative of the New Indian Woman, she is capable of working both inside and outside the home, and is ultimately devoted to her family and her nation. Mankekar further interrogates the intertwined discourses of nationalism and gender in two serials on Doordarshan: *Param Veer Chakra* (1990) and *Tamas* (1988). While *Param Veer Chakra* was a fictionalized account of military heroes who have won India's highest medal of honor, *Tamas* was a dramatic serialization of the tragic tale of Partition in 1947, and communal violence between Hindus and Muslims in India and Pakistan. A common theme in these two serials on Doordarshan, Mankekar argues, is the state-sponsored anxiety to recast the violence of nationhood in terms of masculine ideals of patriotic duty and familial honor which are constantly mediated by hegemonic notions of the New Indian Woman and her idealized role in the nation and the family.

Due to the rapid globalization of media, and the arrival of transnational satellite and cable television channels such as STAR TV, BBC and MTV in 1991, several scholars began to analyze how the relationship between television and its audiences was fundamentally altered in India. For example, in an edited collection titled *Television Without Borders: Asia Speaks Out*, Anura Goonasekera and Paul S. N. Lee (1998) invited a group of scholars to examine how the arrival of transnational networks was influencing national culture and traditional values. While expressing concern about the potentially harmful effects of cultural imperialism from foreign channels on domestic audiences, the authors in the collection were also optimistic that the foreign programming could potentially deliver newer, more empowering content to viewers whose sole source of information on television until then had been under government control.

As the government's monopoly over audiences in national programming was actively contested by both foreign and domestic satellite television channels in India, several scholars advanced the "active audience" framework by conducting ethnographic studies to analyze the consumption practices of diverse audience groups. Others adopted literary and textual methods of reception studies to evaluate the diversity of readings that polysemic media texts could potentially engender in different cultural contexts. In *Transnational Television, Cultural Identity and Change: When STAR Came to India*, Melissa Butcher (2003) cautions against making conclusive generalizable statements about the effects of globalization on television audiences in India because different social groups, across generations, and even participants from the same focus group in her research offered different opinions of a program. Sharing insights from her focus group interviews Butcher writes:

> The afternoon soap opera *Swabhimaan* was considered "aspirational" by an upper income young person in Delhi, and "educational" by a young man who lived in a *basti*. Young women in rural Uttar Pradesh found it "frank" but also perceived role models in such programming. Participants who viewed the music channels also debated the identities of Channel [V] and MTV India, declaring preferences based on their perceptions and desire for more local "Indian" or more global programming.
>
> *(2003, 73)*

In support of her research findings that validate active audience theories, Butcher cites other scholars, such as Namita Unnikrishnan and Shailaja Bajpai (1996) who found that children's socio-economic backgrounds and education and cultural contexts had a major role to play in their responses to television programming. She also cites Robbin D. Crabtree and Sheena Malhotra (2000) who in their study of female audiences in Mumbai question the homogenizing power of Western media networks in India, and conclude that television viewing is an active and social process.

Countering theories of Western media imperialism in the changing media landscape in India, Kalyani Chadha and Anandam Kavoori (2012) argue that the hybrid and rhizomatic nature of Indian television resists a unitary definition. They discuss how the globalization of Indian television with the consolidation of transnational media corporations such as STAR TV, Disney, Time Warner and others, has been accompanied by the rapid growth of domestic companies such as NDTV, UTV and Network 18 as well as the establishment of regional-language networks like Sun TV and ETV. At the same time, foreign and domestic media companies have also worked to create joint ventures (such as Viacom and TV18's Viacom18 and NDTV and Time Warner's

Imagine TV). While alliances such as Viacom18 have been financially successful, others like Imagine TV, which was shut down in 2012, have failed. But the fact that such ventures are repeatedly attempted, Chadha and Kavoori argue, reveals the continually evolving rhizomatic hybridity of the Indian television marketplace.

In spite of the proliferation of a variety of transnational, national and regional-language media companies, Chadha and Kavoori find that essentially two formats have ruled television programming in India since the early 2000s: reality TV shows such as *Indian Idol* and *Fear Factor* and melodramatic soap operas popularly known as "*saas-bahu*" (mother-in-law-daughter-in-law) serials. So a viewer surfing through cable television channels during prime time can find any number of melodramatic serials or reality shows, but few other programming choices. Therefore, notwithstanding the variety of both foreign and domestic television channels and media companies, Chadha and Kavoori conclude that programming diversity remains rather limited in India.

In this contradictory terrain where a growing variety of channels provides a limited diversity of programming content, some scholars have cautioned against the use of active audience theories in Indian Television Studies. The main criticism of the active audience theories is that they grant too much agency to audiences in the process of meaning making, and often minimize the power that institutional structures in the media industries – such as transnational capitalism and patriarchal nationalism – have in framing the programming content as well as how audiences understand their everyday lives and their own cultural traditions. For example, surveying the representations of gender, sexuality and cultural diversity in advertisements from the 1990s to the 2010s, Maitrayee Chaudhuri (2014) contests the notion that the media landscape in India is constituted by a free flow of images and ideas in a hybrid, rhizome-like fashion. Instead, she argues that any analysis of gender, media and culture must take into account three constitutive frameworks: institutionalization of feminism; the rise of neoliberal economics; and the explosive growth of media and communications in popular and public culture. She further argues that this constitutive influence has been facilitated by new kinds of media research firms specializing in the production of knowledge about the emerging consumption patterns of Indian consumers to target their innermost desires and pleasures. Based on her analysis of the sophisticated consumer research undertaken by the media and advertising industries such as Ogilvy and Mather, Mudra, Erickson and TMRC, Chaudhuri contends that we should all be wary of researchers at these firms who claim that consumers are "self responsible subjects". Chaudhuri (2014, 149) writes, "These firms have been able to capture the complex mix of ideas, which form part of the lived reality of middle-class Indians. Chauvinism and feminism may thus both be deployed."

As an example, Chaudhuri describes the popular Axe Body Spray commercials that use sexually charged images to appeal to progressive feminist discourses of gender equality and sexual liberation along with regressive stereotypes of sexual objectification. For instance, in a commercial for Axe "dark temptation" body spray, a man covered in chocolate is licked, kissed and eaten by women who, as the ad's tag line suggests, find him – and by extension the body spray he wears – "irresistible as chocolate." Another example of the intersection of chauvinism and feminism in media culture that Chaudhuri cites is from an article on anti-ageing products titled "Like Daughters, like Mothers." In the article, the author suggests that consumer products such as anti-ageing capsules and creams, anti-cellulite oils, diet supplements, spas, rejuvenation centers and health foods are now available to women young and old alike. So mothers and daughters can now choose to share outfits, skincare products, fad diets, gym and yoga sessions, and pass off as sisters. By suggesting the confluence of progressive feminist ideals of solidarity in sisterhood and post-feminist ideas about self-care and individual liberty, researchers in consumer industries strive to shape women's desires for freedom of choice, as well as their perceptions about commodities and brands. Moreover, Chaudhuri is concerned that the "self responsible" consumer discourse promoted by the consumer research industry appears to resonate with active audience theorists in media studies who claim that audiences are "resistant actors" and not passive victims of global capitalism. In the process, Chaudhuri (2014, 146) claims, academia in conjunction with the corporate world of media leaves unchallenged "the industrialization of representation and communication."

Thus, the active vs. passive audience theory debate remains a dead-end in Television Studies as scholars vacillate between giving audiences too little credit or too much credit for their abilities to make sense of the many representations they encounter on television. In the next section, I revise the active/passive audience debate in Television Studies and argue that we need to move beyond this most vexing of theoretical dead ends. Contemporary transformations in Indian television, I argue, have made the active/passive debate less relevant, if not entirely irrelevant. Due to the globalization of media industries, the proliferation of transnational and domestic channels, hybridization of programming, and the diversification of audience engagement and experience with television, I argue that audiences can no longer be categorized as either active or passive but can only be understood in terms of the non-representational category of the affective.

## The Affective Audiences

In *Unsettling India*, Purnima Mankekar (2015) describes how the advertising-driven entertainment culture in India that began in the 1980s and rapidly accelerated due to globalization in the 1990s and 2000s, contributed to the emergence of new types of audiences who are no longer content to only "view" the programming that television provides them. Instead, Mankekar argues, audiences now begin to inhabit television programming and advertising in profoundly intimate ways to engage more affectively with the world of commodities and consumption that is so pervasive in their everyday lives. Drawing on the works of affect theorists such as Nigel Thrift and Patricia Clough, Mankekar coins the term "commodity affect" to describe a whole range of ineffable yet potent intensities that are engendered in the intimate interactions between commodities and consumers. Situating the relations of commodity affect in the conjunction between the desire to consume and erotic desire, Mankekar (2004, 418–419) claims that Indian television, advertising and marketing industries since the 1990s have exhibited an "unprecedented fascination with intimate relationships – particularly marital, pre-marital, and extramarital relationships – and contained new and varied representations of erotics (explicit as well as implicit)."

It is important to note that the notion of erotic in the context of affect should not be understood only in terms of images of sexuality or representations of sexual relations between people. Instead, the term denotes an assemblage of sexual, sensual and intimate relations, passions, fears and fantasies that could potentially act or be acted upon audiences' capacity to engage with other people, or images and things. There have always been many rich and diverse traditions of the erotic in art, architecture and literature in Indian culture that predate contemporary representations in advertisements, soap operas, sitcoms, talk shows, reality formats, Bollywood films, music shows based on films and MTV-style music videos.

Even though many of the older traditions of erotics are frequently invoked in the contemporary media environment, Mankekar argues that the newer representations of sensuality, intimacy and sexuality are often associated with Westernization and are therefore seen by nationalist critics and by some audience members as being antithetical to "Indian" or national culture. Reflecting on the diverse responses to the relationship between erotics and commodity affect she witnessed during her ethnographic observations, Mankekar reveals that some of her informants felt that gazing at commodities provided them a "window on the world" by introducing them to people and places far removed from their own lives. But beyond gazing at media representations, many of Mankekar's ethnographic subjects also connected affectively with commodities by yearning for them in ways that appeared to be laced with erotic desire. In some instances, Mankekar (2015, 115) writes,

> these pleasures were corporeal and synthetic, as in the haptic multisensorial pleasures that result when, for instance, one's eyes caress a silk sari displayed on television, or when one starts to salivate when looking at an advertisement for ice cream or a dish of *gulab jamuns*.

Mankekar's reporting of the erotic yearnings that her ethnographic subjects feel when the eyes caress an image of a dish of *gulab jamuns* or a silk sari on the screen, is reminiscent of Laura Marks' (2000) analysis of haptic pleasures in Shauna Beharry's experimental film *Seeing Is Believing* (1991). The example that Marks often uses in her writings on this film is a short sequence in which the eye of the camera repeatedly caresses over the surface of a photograph in which Beharry is wearing a silk sari that once belonged to her deceased mother. As the camera follows the folds of the sari, Beharry's voice on the soundtrack recalls the anger and confusion she felt when she could no longer recognize her mother in photographs after her death. But when she wore her mother's sari, Beharry says, a host of memories about her family's diasporic history flooded into her mind and she felt that she had climbed into her mother's skin. Marks (2000, 112) argues that Beharry's film attempts to "squeeze the touchability" out of representation by inviting the viewers to participate in a multisensory experience of an image that can occur affectively in the physical presence of an object in the image.

In "Video Haptics and Erotics," Marks (1998) outlines the multisensory nature of the haptic image, and describes how haptic visuality appeals to eroticism in very intimate, sensual and tactile ways. Drawing on an eclectic mix of scholars such as the art historian Alois Riegl, philosophers including Walter Benjamin, Frantz Fanon, Maurice Merleau-Ponty, Gilles Deleuze and Felix Guattari, and film scholars such as Miriam Hansen, trinh min-ha and Vivian Sobchack among others, Marks (1998, 332) defines the haptic as a combination of tactile, kinaesthetic and proprioceptive functions. She explains that in haptic visuality, "the eyes themselves function like organs of touch." In some cases, Marks argues, haptic visuality appeals to what escapes optic visuality altogether. But what escapes in the optical sense of representation is revealed in other ways in a non-representational sense. In *Touch: Sensuous Theory and Multisensory Media*, Marks (2002) insists on the materiality of the non-representational realm – described as the "virtual" in Deleuzian terms – by drawing attention to aspects of the image that escape our symbolic recognition. For example, while watching a movie, Marks (2002, xi) argues, "we may become entranced by a gesture, a lock of hair falling across an actor's face, the palpability of sunlight spilling through a break in the wall." In an information age saturated with imagery, Marks worries that mass media like film, video, television and digital technologies make audiences very good at mastering the optical codes of visual representation at the expense of other senses of visuality that we do not fully comprehend or have clearly defined categories for. To escape the mass-mediated regimes of representation, and the concomitant categories of activity and passivity that define our relationship to the visual image, Marks suggests that media producers and audiences alike must experiment with the multisensory realm of the haptic image.

## Conclusion

To conclude, I would like to return to the scene that I began with from the film *Slumdog Millionaire* where Jamal asks Latika why she and millions of other Indians like to watch television and Latika responds that everyone has to escape somewhere. In light of the above-mentioned distinction between optic and haptic senses, I would argue that Latika did get it right in some ways – television can and does provide escape from the dominant regimes of representational visuality into the non-representational realms of erotic sensuality. When evaluating the pleasures of watching *Kaun Banega Crorepati? (KBC)* for viewers like Latika, media critics have often dismissively characterized the fans' worship of reality television shows as a form of escapism that is characteristic of passive audiences.

When *KBC* first aired in India on the StarPlus Channel in 2000, it instantly became the biggest hit on Indian television with chartbusting audience ratings and fan following not seen since the telecast of *Ramayan* and *Mahabharat* on Doordarshan in the 1980s. While the host of the Indian version of the *Millionaire* show in *Slumdog Millionaire* was played by Bollywood superstar Anil Kapoor, the host of *KBC* on StarPlus was Amitabh Bachchan who is, without doubt, the greatest movie star of all time in Indian cinema. In the first two years of KBC, StarPlus received around 200,000 calls a day from potential contestants who were eager to be in the hot seat next to the "Big B" as Bachchan is popularly known in India ("Survival of the Fittest," 2001). The viewers at home were just as thrilled to tune in to *KBC* for an opportunity to vicariously share in the intimate experience of being right next to the great Bachchan himself. A viewer named Vandana Puri described the visceral sense of anticipation and excitement in her family as the appointment time with *KBC* approached in the evenings thus:

> All of us sit glued to the Star TV network. Hushed silence. Anyone who makes a wisecrack is snubbed with hostile glances. Each family member is ready with the answer before the guy on the hot seat can reply. *KBC* has indeed cast a magic spell on the Indian viewers.
>
> *(Puri, 2001)*

The "magic spell" cast by KBC on Puri's family and many other families in India can be best understood in terms of what Nigel Thrift (2010) has called the magical technologies of public intimacy. Thrift argues that media producers, advertisers and marketing executives have always been drawn to the "magical" allure of glamor, style and celebrity to make a commodity stand out from other similar commodities in a crowded marketplace. Describing how glamor works through and for commodities, Thrift (2010, 297) writes:

> For all its breathtaking qualities, glamour does not conjure up awe. It operates on a human scale, in the everyday, inviting just enough familiarity to engage the imagination, a glimpse of another life, utopia as tactile presence. ... Glamour is about that special excitement and attractiveness that characterizes some objects and people. Glamour is a form of secular magic, conjured up by the commercial sphere.

Elsewhere, I have written about how Bachchan – in spite of his superstardom and god-like celebrity status in India – had the uncanny ability on *KBC* to create extremely intimate personal connections at a human level with the contestants on the stage and with the audiences at home alike (Kumar, 2005). Here I invoke Thrift to describe how the glamorous allure of Bachchan's celebrity operated on a human scale as a magical technology of public intimacy and was instrumental in enabling *KBC* to stand apart from other similar commodities in the commercial sphere and emerge as the most successful reality show on Indian television. For example, in an online discussion group on *KBC* at mouthshut.com, a post by "dhrumil 83" proclaimed:

> KAUN BANEGA CROREPATI might have had been copied from "WHO WANTS TO BE A MILLIONAIRE". But to tell you the truth the copied version is better than the original one. SIMPLE ANSWER – It has AMITABH BACHAN [sic] in it. He is the one the greatest.

But critics insisted that there was nothing unique or "Indian" about *KBC* because the Indian version was similar to the many versions of *Who Wants to Be a Millionaire?* produced in more than 30 countries under a franchise agreement with the London-based Celador Productions. A posting on mouth shut.com by "Amrita" reads, "Before I start my review, let me educate the members here that *KBC* is an exact copy of American show *Who Wants to Be a Millionaire.*" Amrita's desire to "educate" other members of the online community about the fact that *KBC* is a copy of the more famous American version of the *Millionaire* format is reminiscent of the concerns expressed by passive audience theorists about the unenlightened passivity of "couch pakoras" discussed in the previous section. But where advocates of the passive audience theories go wrong, I argue, is by assuming that the magical allure of reality television shows like *KBC* is a form of escapism *from* reality for audiences. However, as numerous active audience studies show us, television cannot be thought of as an escapist medium that takes us away from the reality of representation if reality is always-already a mediated re-presentation – in the dual sense of representation that Hall describes. Instead, what I have argued in this chapter is that television does not function only in the representational realm of optic visuality, but instead takes viewers into non-representational spaces of haptic visuality by foregrounding the tactile presence of affect in the multisensual practices of everyday life.

Drawing on non-representational theories of erotics, haptics and the magical allure of affect, I have outlined a framework of "affective audiences" that I argue provides new avenues in Television Studies to move away from the active vs. passive audience debate. I conclude that television audiences can no longer be categorized as active or passive since audiences are also being targeted by the media industries in the affective registers of emotions, erotics and haptic sensuality. Therefore, the concept of "affective audience" is significant for Television Studies not only because it is a useful theoretical construct for the reconceptualization of audiences beyond activity/passivity, but also because affect is now a thriving framework in the media industries for targeting and monetizing consumers in a hyperglobalized commodity culture.

## References


Bielby Denise, D. and Lee C. Harrington. 2012. *Global TV: Exporting Television and Culture in the World Market.* New York: New York University Press.

Butcher, Melissa. 2003. *Transnational Television, Cultural Identity and Change: When STAR Came to India.* Thousand Oaks, CA: Sage Publications.

Chadha, Kalyani and Anandam Kavoori. 2012. "Mapping India's Television Landscape: Constitutive Dimensions and Emerging Issues." *South Asian History and Culture* 3 (4), 591–602.

Chaudhuri, Maitrayee. 2014. "Gender Media and Popular Culture in a Global India." In *Routledge Handbook of Gender in South Asia,* edited by Leela Fernandes, 145–159. New York: Routledge.

Clough, Patricia Ticineto and Jean Halley. 2007. *The Affective Turn: Theorizing the Social.* Durham, NC: Duke University Press.

Crabtree, Robbin D. and Sheena Malhotra. 2000. "A Case Study of Commercial Television in India: Assessing the Organizational Mechanisms of Cultural Imperialism." *Journal of Broadcasting & Electronic Media* 44 (3), 364–385.

Gehlawat, Ajay (editor). 2014. *The "Slumdog" Phenomenon: A Critical Anthology.* New York: Anthem.

Goonasekera, Anura and Paul S. N. Lee. 1998. *TV without Borders: Asia Speaks Out.* Manila: Asian Media Information and Communication Centre.

Hall, Stuart. 1980. "Encoding/decoding." In: *Culture, Media, Language,* edited by Stuart Hall, Dorothy Hobson, Andrew Love, and Paul Willis, 128–128. London: Hutchinson.

Horkheimer, Max and Theodor W. Adorno. 2007. "The Culture Industry: Enlightenment as Mass Deception." In: *Dialectic of Enlightenment,* 94–136. Stanford, CA: Stanford University Press.

Karatzogianni, Athina and Adi Kuntsman (editors). 2012. *Digital Cultures and the Politics of Emotion: Feelings, Affect and Technological Change.* New York: Palgrave Macmillan.

Kumar, Ramesh S. 2009. *Consumer Behaviour and Branding: Concepts, Readings And Cases-The Indian Context.* New Delhi: Dorling Kindersley.

Kumar, Shanti. 2005. "Innovation, Imitation, and Hybridity in Indian Television." In *Thinking Outside the Box: A Contemporary Television Genre Reader,* edited by Gary R. Edgerton and Brian Geoffrey Rose, 314–335. Lexington: University of Kentucky Press.

Lang, Kurt and Gladys Engel Lang. 2009. "Mass Society, Mass Culture, and Mass Communication: The Meaning of Mass." *International Journal of Communication* 3, 998–1024.

Lasswell, Harold D. 1936. *Politics: Who Gets What When and How.* New York: McGraw-Hill.

Lazarsfeld, Paul F. 1933. *On Social Research and Its Language.* Chicago, IL: University of Chicago Press.

Lerner, Daniel. 1958. *The Passing of Traditional Society: Modernizing the Middle East.* New York: Macmillan.

Lutgendorf, Philip. 1990. "Ramayan The Video." *The Drama Review* 34 (2), 127–176.

Mankekar, Purnima. 1999. *Screening Culture, Viewing Politics: An Ethnography of Television, Womanhood, and Nation in Postcolonial India.* Durham, NC: Duke University Press.

Mankekar, Purnima. 2004. "Dangerous Desires: Television and Erotics in Late Twentieth-Century India." *The Journal of Asian Studies* 63 (2), 403–443.

Mankekar, Purnima. 2015. *Unsettling India: Affect, Temporality, Transnationality.* Durham, NC: Duke University Press.

Marks, Laura U. 1998. "Video haptics and erotics." *Screen* 39 (4) Winter, 331–348.

Marks, Laura U. 2000. *Skin of the Film: Intercultural Cinema, Embodiment and the Senses.* Durham, NC: Duke University Press.

Marks, Laura U. 2002. *Touch: Sensuous Theory and Multisensory Media.* Minneapolis, MN: University of Minnesota Press.

Massumi, Brian. 1995. "The Autonomy of Affect." *Cultural Critique* 31 (Fall), 83–109.

McQuail, Denis and Sven Windahl. 2013. *Communication Models for the Study of Mass Communications.* New York: Routledge.

Morris, David. 2016. "Affect as Transcendental Condition of Activity versus Passivity, and of Natural Sciences." In: *Phenomenology and Science: Confrontations and Convergences*, edited by Jack Reynolds and Richard Seabold, 103–120. London: Palgrave Macmillan.

Puri, Vandana. 2001. "Sawal Hai – Chote Mian Ya Bade Mian?" *Meghadutam.* May 26, 2001. Accessed March 12, 2004. www.meghadutam.com/tdis.php

Ramachandran, R. 1975. *Spatial Diffusion of Innovations in Rural India: A Case Study of the Spread of Irrigation Pumps in Coimbatore Plateau.* Mysore: University of Mysore.

Richman, Paula. 1991. "Introduction: The Diversity of the *Ramayana* Tradition." In: *Many Ramayanas: The Diversity of a Narrative Tradition in South Asia*, edited by Paula Richman, 3–21. Berkeley, CA: University of California Press.

Rogers, Everett M. 1962. *The Diffusion of Innovations.* New York: Free Press.

Rogers, Everett M., Joseph R. Ascroft, and Niels G. Röling. 1970. *Diffusion of Innovations in Brazil, Nigeria, and India.* Ann Arbor, MI: University of Michigan Press.

Schramm, Wilbur (editor). 1960. *Mass Communications.* Urbana, IL: University of Illinois Press.

Seigworth, Gregory J. and Melissa Greg. 2009. *The Affect Theory Reader.* Durham, NC: Duke University Press.

Shannon, Claude E. and Warren Weaver. 1971. *The Mathematical Theory of Communication.* Urbana, IL: University of Illinois Press.

Singhal, Arvind and James W. Dearing (editors). 2006. *Communication of Innovations: A Journey with Ev Rogers.* Thousand Oaks, CA: Sage Publications.

Singhal, Arvind and Everett M. Rogers. 1999. "The *Hum Log* Story in India." In *Entertainment-Education: A Communication Strategy for Social Change*, 73–104. Mahwah, NJ: Lawrence Erlbaum Associates.

Sparks, Glen G. 2016. *Media Effects Research: A Basic Overview.* Boston, MA: Cengage Learning.

"Survival of the Fittest." 2001. *India Today*, October 10. Accessed March 12, 2004. www.india-today.com/webexclusive/dispatch/20011010/dhawan.html

Tarr, Zoltan. 2017. *The Frankfurt School: The Critical Theories of Max Horkheimer and Theodor W Adorn.* New York: Routledge.

Thrift, Nigel. 2008. *Non-Representational Theory.* New York: Routledge.

Thrift, Nigel. 2010. "Understanding the Material Practices of Glamor." In: *The Affect Theory Reader*, edited by Melissa Gregg and Gregory J. Seigworth, 290–308. Durham, NC: Duke University Press.

Tidd, Joseph (editor). 2010. *Gaining Momentum: Managing the Diffusion of Innovations.* London: Imperial College Press.

Tzanelli, Rodanthi. 2016. *Mobility, Modernity and the Slum: The Real and Virtual Journeys of* Slumdog Millionaire. New York: Routledge.

Unnikrishna, Namita and Shailaja Bajpai. 1996. *The Impact of Television Advertising on Children.* New Delhi: Sage Publications.

Vishwanath, Arun and George A. Barnett. 2011. *The Diffusion of Innovations: A Communication Science Perspective.* New York: Peter Lang.

Wetherell, Margaret. 2014. *Affect and Emotion: A New Social Science Understanding.* Thousand Oaks, CA: Sage Publications.

# 9

# TWO CONCEPTS FROM TELEVISION AUDIENCE RESEARCH IN TIMES OF DATAFICATION AND DISINFORMATION

## Looking Back to Look Forward

*Jonathan Corpus Ong and Ranjana Das*

### Introduction

We write this chapter as communication researchers who came of age learning about the achievements of television audience studies and contemplating its continuing relevance at a point in time when social media had just arrived, and when the transformation of audiences in the age of the internet was at the heart of exciting conversations around us. But as we write this chapter, we are motivated anew to revisit and reconsider the conceptual repertories from television audience reception studies, for it is highlighted exemplarily, if shockingly, by the Cambridge Analytica controversy and the resultant Facebook hearings that, once again, "audiences are now being newly fought over as the pawns in the games of powerful others" (Livingstone 2019). These fights over audiences, their data, their privacy, their safety, among much else, are part and parcel of a new socio-technological moment we find ourselves in – a moment scholars are varyingly approaching as times of datafication, or dataism, theorized succinctly by van Dijk as a condition where "masses of people – naively or unwittingly – trust their personal information to corporate platforms" (2014, p. 197). As scholars around us are occupied with a spate of new challenges surrounding digital disinformation such as "fake news" (Marwick & Lewis 2017; Ong & Cabanes 2018), algorithmic injustices (Gillespie 2018; Noble 2018), and data- or techno-colonialisms (Couldry & Mejias 2019; Madianou 2018), we find similar debates replaying themselves, as we look back at television audience studies today. As Livingstone notes in her thoughtful account of datafication and mediatization:

> [I]t is important to remember John Hartley's (1987) critique of the concept of "the audience" as the invisible fiction invented by the industry to create docile subjects. The flip side of this implied quantifiable and commodified audience is an implied all-powerful media industry that will never succumb to the rule of law, the norms of civil

> society, or the public interest. Promoting these fictions may be in the media's own interests but it does not serve those of the academy or the public.
>
> *(Livingstone 2019)*

So, we pause today, to look back at the long and rich history of television studies in terms of its focus on audiences as socio-culturally situated subjects, and we begin to think of two key concepts, keywords if one will, which we suggest are worth retaining within our repertories, today. These might not quite be the most visible and most circulated keywords, but we suggest these hold critical value in contemporary times. We first consider *divergence* – the seemingly simple concept but largely fraught with critiques that arose out of television audience studies. Divergence, as we discuss below offers us a promise today in the age of big data to note with care, the value of contextual messiness. It also offers us, through its critiques, a reminder of the power of platforms, within and against which relatively powerless audiences must operate. Next, we consider *responsibility* – which underlines how audiences not only have agency or rights but, crucially, also moral obligations across their diverse activities of mediated participation. Responsibility asserts that audiences are not entities wholly exploited or powerless but through an ever-expanding array of technologized actions – searching and connecting, but also trolling or doxing – are fundamentally moral actors in a shared mediated public sphere. Of course, in considering *responsibility*, we remain conscious of the many responsibilities that platforms, with great power, bear, and to which they must be held accountable. But in this piece, we focus on *responsibility*, our second keyword, in terms of its invitation to us to engage with the question of justice, as we have to acknowledge that any discussion of moral norms, standards, and regulation needs to confront issues of historical inequality and structural oppression across lines of race, class, gender, sexuality, and ability.

We argue, therefore, that it is important to continue to hold onto the central tension that drives television audience studies research: on the one hand, its sensitivities to pluralism, diversity, and radical contextualization (e.g., Radway 1983), and on the other hand, its concern for norms, values, and moral economies (e.g., Silverstone 1994). We need to hold on to this tension as our field is caught in a pendulum swing back toward old assumptions of hypodermic needle media effects, as in some recent writings about how audiences are duped by fake news and filter bubbles, sharply criticized by Paula Chakravartty and Srirupa Roy (2017). Recent concepts of dataveillance (van Dijk 2014), data colonialism (Couldry & Mejias 2018), and technocolonialism (Madianou 2018) also characterize mediation as enforcing totalizing logics of domination, particularly toward vulnerable communities such as refugees who are dehumanized in the process of becoming data points. These accounts are convincing and alarming, but certainly not the last word.

The authors who write this paper came together as members of a generation of audience researchers who began their working lives at the birth of social media, who were trained to look back toward reception analysis with electronic media to make sense of analyzing new media audiences. This meant the coming together, perhaps, of the two pathways Curran once presented as oppositions in describing empirical reception studies – "a reversion to previous received wisdoms rather than a reconnaissance of the new" (Curran 1990, p. 135). In seeking to do both, today this generation finds itself once again at another point of socio-technological transformation where the core ambitions of audience research – to do research on the side of the audience (Ang 1996) – need restating, at the brink of the potentially transformative Internet of Things (IoT), mediating the life worlds and practices of audiences as individuals and communities, and becoming an increasingly realistic possibility. When we use the word *transformative* here, we combine hope and skepticism alike, evading, hopefully, the hype that seems to surround us. These ambitions of doing audience research that tailors itself to transforming communicative conditions, but nonetheless also continues to do research *on the side* of the audience, must note that critical questions about media regulation, surveillance, privacy, and essentially inequalities of power, are beginning to overlap across conversations on social media and on the IoT (Dencik 2017; Deuze et al. 2012; Dourish & Bell 2011; Mansell 2012; Noble 2018). In recognizing generational positions in this narrative thus, far from being determined, even softly, by technology (see Stalder 2006), one must listen carefully to Bolter and Grusin's work on remediation (1998), Livingstone's account of the mediation of everything (2009), and parallel conversations on media life (Deuze 2009). As we read the history of the IoT (Ashton 2009) we can see, for instance, glimpses of what went before it – for instance, ubiquitous or pervasive computing (Ark & Selker 1999). In selecting the keywords we do, from television audience studies, we are conscious that we write this chapter at the rise of rapid developments in connected gadgets, highly individualized digital experiences, the connection of "things," and, of course, the availability of previously unprecedented amounts of data to analyze. Like all new socio-technologically transformative moments, these developments now sit at the heart of often contrasting discourses, varyingly optimistic and pessimistic, like debates from the earliest days of the internet (see Volume 1, Issue 1 of the journal *New Media & Society* 1999).

Countering utopian narratives about the promises and potentials of technological advancements, come findings from audience researchers who remind us of what has essentially been the longstanding duality between materialism and sociality discussed within science and technology studies (cf. Woolgar 2002). Such research states that audiences and users "might resist implied user practices, renegotiate functions of interfaces and even force media companies to change some of their restrictive settings" (Dhaenens & Mollen 2017, p. 27). Once again, it seems necessary to re-investigate and re-iterate the societal, political, and even intellectual importance of audience agency, literacies, and interpretive work in the face of emerging technological conditions, perhaps even more so than ever before, just as it remains imperative on audience analysts to engage with the study of the very architectures and affordances of these emerging material spaces (Hutchby 2001) whose biases range from the implicit to the intentionally obscured.

For each concept below, we reflect on its value, first and foremost, in the socio-technological conditions of today, following it up with reflections on some of the critiques these concepts drew as well, for we find value in both.

## Divergence

One of the key achievements of television audience reception studies was the lesson that audiences diverge from authorial intention in making meaning, and that audiences, socio-culturally located, diverge from each other in their meaning-making work. This impetus in television audience studies, came from both sociological and cultural studies approaches, throughout the 1980s and 1990s, where interpretative work was contextualized within relations of structure and power (see Lotz 2000 for a review; see Ang 1985; Bobo 1995; Brunsdon 1997; Morley 1980; Press 1991; Press & Cole 1999; Radway 1984). Television audience studies was marked by the rise of genre-specific, often ethnographic studies of the interpretations of film and television texts in contexts, and the valuable pursuit of pleasure, resistance, critique, play, and identity (see Allen 1995; Ang & Hermes 1991; Bailey 2005; Brown 1990; Long 1986; Morley 1992, 1993). Radical contextualism offered by Radway (1988) 20 years ago held two possibilities. One was the promise of contextual richness which is still today being interestingly adopted by many audience ethnographers, leading to thick accounts of cultural reception in everyday life (e.g., Bird 2003). The other, as Ang put it, in different words, was a feeling of endlessness in this journey (Ang 1991). We suggest that these lessons around divergence, seemingly simple, are often forgotten in contemporary analyses of people's engagement with platforms, or in sweeping generalizations around big data derived "patterns" revealed about audience behavior and practices, and hence, a lesson worth returning to. Hear, for instance, the moral panics echoed in recent studies that overstate how social media's filter bubbles or microtargeted ads have duped voters in western democracies, for a rearticulation of the "hypodermic needle" perspective of media effects for a digital media age (for a review, see: Chakravartty & Srirupa 2017).

But equally, amid the importance of remembering the critical relevance of divergence, agency, and context, we are reminded of the many critiques, the very word divergence accumulated in the heyday of television reception studies: "The claim that audiences have the ability to create their own empowering responses to mass mediated texts loses little of its force when it is acknowledged that the polysemic freeplay of discourse has been overestimated" (Condit 1989, p. 108).

Voices from critical-cultural studies, for instance those writing in rhetorical and textual analyses traditions in general (Condit 1989), and those within political economy studies (Dahlgren 1998), brought the earliest of critiques, that divergence and polysemy had been over-celebrated, and over-glorified – the case for audiences' active agency having been taken too far forward, and real, lived issues of power being mis-read. Reading this critique in the context of newly emerging and intrusive interfaces, it seems particularly instructive to pay attention to the issues shaping and sometimes even restraining audience and user agency today, with the rise of what the CEDAR network has called "intrusive media" – interfaces, for instance the most ubiquitous social media platforms, which are designed to boost the contribution of user data and labor. As Dhaenens and Mollen (2018) note, these newer, intrusive forms of mediated communication are marked by four characteristics:

> *Exploitation* which is used as a generic term to assemble such phenomena as free labour and encompasses research that targets the economic interests of media companies; *formativity* which describes how specific conceptions, roles and types of agency become preconfigured for audiences and their engagement with media in their everyday life within the software interfaces and the algorithmic functioning of intrusive media; *pervasiveness*, which refers to the increasing ubiquity, embeddedness of and reliance on digital software-based media in people's everyday life, requiring them to display and adopt complex and differentiated ways of handling and managing their engagement with media, and last, *exclusion* which refers to the power imbalance between producers and providers of digital media platforms and their users and audiences.
>
> *(pp. 25–26)*

These textual features may not only work to anticipate agency into standardized preconfigurations, but they also then feed into wider, core structural issues of power, within which user and audience agency is imbricated. So, for instance, amid the huge participatory and creative potentials often discussed with regard to interfaces such as YouTube, what new forms of hidden labor arise (see Fuchs 2015)? How does creative agency get co-opted? These and other questions draw attention to the old concerns around over-celebrating agency and divergence, albeit in a different context: "The inferences derived from reception analysis as a whole have not pointed to new directions. In some cases, they have resulted in old pluralist dishes being reheated and presented as new cuisine" (Curran 1990, p. 151).

Curran's well-known rebuttal of the claims to apparent novelty by what he called the "new revisionism" (1990) that was active audience studies contained critiques that the endless line of empirical reception projects, presenting broadly the same kinds of findings on audience agency, interpretive work, critical decodings, and resistance, in the context of their everyday lives, was occasionally repackaging older, pluralist knowledge achievements, for instance, those emerging from within gratifications research. He goes further, to note that the so-called new revisionists' celebration of individual decodings had led to a case for the destruction of public service broadcasting across Europe and deregulation in general. The latter presents the case that the focus on what was then perhaps perceived, or even presented, as limitless polysemy (see also Condit's critique above, or Seaman's (1992) critique as well), led to the idea that, since decodings were endlessly diverse and versatile, cultural producers and media institutions ultimately get a pass when asked challenging questions of harm and offense (see Das & Graefer 2017).

Seaman (1992), in the critique of the supposed "pointless populism" of reception studies (with which many audience analysts will rightly disagree), makes a similar point that,

> the problem does not lie with audiences, but rather with a system of mass communication that systematically excludes certain forms of programming and imagery, in favor of a profoundly restrictive and highly interest-driven selection. The problem is not with audience interpreting practices, but what is available for interpretation.
>
> *(1992, p. 308)*

So Curran's warning early on, in 1990, that the "new revisionism" had led to a perhaps unintended push toward the dislocation of responsibility from the producing and regulating institutions behind texts, is worth bearing in mind today, in the age of Web 3.0, where the social-democratic roles and responsibilities of those behind emerging technologies need to be kept firmly in focus for those behind intrusive architectures to be held accountable. This is a straightforward reminder, echoed outside of pluralist traditions, within critical-cultural scholarship in rhetoric, for instance. As Condit notes,

> the audience's variability is a consequence of the fact that humans, in their inherent character as audiences, are inevitably situated in a communication system, of which they are a part, and hence have some influence within, but by which they are also influenced.
>
> *(1989, p. 120)*

Whether the so-called new revisionism had indeed simply led to a string of projects without advancing theory is now dubitable, for in the decades that have followed the publication of the piece, scholars have repeatedly returned to tell the story of the field, to make sense of its repertoires, and a considerable amount of reflexive stock-taking has taken place. This needs to continue as the logical next step of the audience agenda as we enter yet another phase of socio-technological transformations. So, with divergence, we seek to make a two-fold argument. First – that we retain it, and not lose sight of contextual diversity and messiness, amid the rise and rise of big data led methods. But equally, we remain mindful that we must not allow the existence of divergence to become a route through which the power of platforms goes unchallenged, and where harmful content goes unregulated, simply because people diverge in their meaning-making work.

## Responsibility

In productive tension with the concept of divergence, therefore, is our second keyword of responsibility. In the face of today's populist political currents, the rise of fake news, and the opaque operations of algorithmic formulae that entrench or even deepen socio-cultural divides, we should ask three questions: What normative standards, if any, do we hold audiences in evaluating the terms of their participation in a digitally mediated "space of appearance" (Silverstone 2007)? Additionally, what is the responsibility of platform owners toward their users and subscribers – however dispersed, divided, and divergent they might be across time and space, cultural background, and political affiliation? Finally, as audience researchers, many of whom are dedicated to the spirit of ethnographic research, how can we make sure that the "deep stories" of the people we meet aim toward building bridges and breaking down "empathy walls," to borrow the words of the feminist sociologist Arlie Hochschild (2016), so that our interventions truly work "in the best interests of audiences" (Ytre-Arne & Das 2018)?

Responsibility as a central and all-encompassing concern that implicates the variety of morally implicated agents interlinked across screens and interfaces was the late Roger Silverstone's challenging invitation in the early 2000s when he expanded the agenda of media ethics. It is worth revisiting his invocation for audience ethics:

> If audiences are active and if the notion of activity has any meaning at all, then they must be presumed to have to take responsibility for those actions. If audiences refuse to take that responsibility, then they are morally culpable. And we are all audiences now.
>
> *(Silverstone 2002, p. 777)*

Indeed, some audience researchers have directly engaged with Silverstone's invitation in recent scholarship around distant suffering and mediated humanitarianism. A group of scholars in this area of research has creatively challenged the methodological choice of textual and visual criticism of narratives and photographs of natural disasters by insisting on a project of mapping out the diverse modes of moral engagement, and conversely, denial strategies that audiences express in the face of mediated events of tragedy (see, e.g., the 2015 special issue "Audiences in the Face of Distant Suffering", edited by Stijn Joye and Johannes von Engelhardt in the *International Communication Gazette*). The aim to catalogue divergences often through thick description moves beyond "pointless populism" by keeping divergences in tension with normative criticism, acknowledging that media representations nevertheless serve as important "cultural resources" that may facilitate public actions of memorialization (Kyriakidou 2014) or witnessing (Ong 2014), just as social media architectures often constrain but do not preclude cosmopolitan action (Madianou 2013; Pantti 2015; Scott 2014).

In light of contemporary populist formations and the controversial contributions of new digital weapons from anonymous political trolls (Bradshaw & Howard 2017) to entrepreneurial teenage platform workers in Macedonia (Silverman & Alexander 2016) to unholy alliances of right-wing nationalists (Marwick & Lewis 2017), critical researchers need to address the insidious ways in which platform owners, tech designers, digital laborers, and media users deny their complicity to the current climate of information pollution and political polarization. Complicity, for Silverstone, is when media producers "fail to reflect on the limitations of their practice, and fail to communicate these both to their subjects and their audiences" and also when audiences "uncritically accept the media's representational claims, and insofar as their knowing acknowledgement of its limitations remains tacit" (Silverstone 2007, p. 21). In an era where it is normalized creative practice for online influencers to blur the lines between factual and promotional content when they sell their "digital estates" (Abidin 2017) to corporate brands, it is unsurprising how digital influencer culture becomes weaponizable in political campaigning especially as politicians aim for attention hacking and media manipulation (Marwick & Lewis 2017). In the project of Jonathan Corpus Ong and Jason Cabanes, the concept of the "disinformation interface" underscores the porous boundaries separating "paid troll workers" from the unpaid supporters and political fans who exuberantly perform their support for political figures and influencers. They found that disinformation production is not masterminded by evil villains but involves the dispersed promotional labor of ordinary people who are complicit in normalizing digital witchhunting and cyberbullying (Ong & Cabanes 2018). The question of complicity challenges us to consider how various agents express various moral justifications in their everyday engagement with the mediated world: *it's not me, it's somebody else who is responsible. It's always the other who is villainized as the troll or the purveyor of fake news.* In the case of Cambridge Analytica, the important set of questions scholars have asked is: did these shady consultants and data analytics gurus exploit the vulnerabilities of a fragile ecosystem, or did they simply use Facebook as it was truly meant to be used?

In the case of digital disinformation, the keyword responsibility invites a social analysis of the contemporary political moment that assigns moral agency and commitment to the public world to all participants in a mediated environment ("we are all audiences now"). It is encouraging that emerging policy directives underlining the need for collaborative interventions assign responsibility to various sectors from government to journalists to big tech (e.g., Phillips 2018; Wardle & Derakhshan 2017). It is crucial to articulate more clearly what responsibilities we assign to ordinary people, particularly in debates on media literacies, data justice, and ordinary ethics (e.g., Dencik 2017). We should also push back on tendencies toward "pointless populism," such as in studies that end up downplaying the social harms of fake news by citing that only a certain portion of the population actually spread them (Guess, Nyhan & Reifler 2018). We should also continue to explore with greater sensitivity the "deep stories" and calculated rationalities behind the formation of populist publics.

## Discussion

Our two keywords, divergence and responsibility, are not the most obvious partners. At first glance they appear to sharply divide between a celebration of difference or disruption and a conservative exhortation toward universal principles. We view this not as a contradiction but as a creative tension, a dialectic, that we need to recover from some of the most insightful work in the tradition of television audience studies to engage anew today. Taken together, these concepts invite caution in the social critique we produce when engaging with recent anxieties around media manipulation and data privacy. Such a critique requires, following the keyword divergence, an acknowledgment of the multiple positionalities in the account of audience experience (but not to the extent that we allow those holding massive power in platform societies to evade attempts to hold them accountable), and following the keyword responsibility, a value judgment about the relationships of power that manifest in the mediated interaction such that we can assign diverse degrees of moral culpability to all media participants. We found thinking about these keywords in this manner productive, as researchers dealing with transmedia environments amid datafication, looking back constantly to make sense of the present and the future. This is not because we concluded that, somehow, television audience studies *must* be re-positioned in the age of datafication in order to continue to be meaningful, for, far from it, television itself continues to be a powerful and fascinating medium itself undergoing a plethora of transformations. We found this task useful because it helped us draw a scholarly line, a strand of intellectual lineage, inheritance, or heritage, if one will, between datafication today, and the rise of television audience studies decades ago, for across these diversely mediated communicative conditions, some patterns still ring loud, clear, and true. These keywords direct our scholarly energy to drawing out the opaque risks and vulnerabilities baked in to new technological transformations and logics of datafication and accounting for audiences' constraints and variable literacies to exert influence, reject deceptive content, and demand recognition in polyphonic media environments. While media and communications scholars have much to be anxious about both top-down insidious operations as well as unintended digital harms posed by the unholy alliances among big tech firms, populist political players, and unscrupulous digital disruptors, audience studies should double down in its mission to reject the "impasse of disappointment" (Henderson 2013, p. 134) and recuperate the everyday as the site of diverse, messy, and generative possibilities.

## References


Abidin, C. (2017). Hook ups, break ups, make ups: Influencer couples as digital estates, Paper Presented at Digital Intimacies: Connections and disconnection, RMIT, Melbourne.

Allen, R. C. (Ed.) (1995). *To be continued–: Soap operas around the world.* London: Routledge Psychology Press.

Ang, I. (1991). *Desperately seeking the audience.* London and New York: Routledge.

Ang, I. (1996). Ethnography and radical contextualization in audience studies. In J. Hay, L. Grossberg & E. Wartelly (Eds.), *The audience and its landscape.* Boulder: Westview Press.

Ang, I. & Hermes, J. (1991). Gender and/in media consumption. In J. Curran & M. Gurevitch (Eds.), *Mass media and society* (pp. 307–328). London: Edward Arnold.

Ang, I. (1985). Watching Dallas: soap opera and the melodramatic imagination. trans. *Della Couling (New York: Methuen), 11.*

Ark, W. S. & Selker, T. (1999). A look at human interaction with pervasive computers. *IBM Systems Journal, 38*(4), 504–507.

Ashton, K. (2009). That 'internet of things' thing. *RFID Journal, 22*(7), 97–114.

Ashton, K. (2009). That 'Internet of Things' thing. *RFID Journal, 22*(7), 97–114.

Bailey, S. (2005). *Media audiences and identity: Self-construction in the fan experience.* New York: Palgrave Macmillan.

Bird, S. E. (2003). *The audience in everyday life: Living in a media world.* New York: Routledge.

Bobo, J. (1995). *Black women as cultural readers.* New York: Columbia University Press.

Bolter, J. & Grusin, R. (1998). *Remediation: Understanding new media.* Cambridge & London: The MIT Press.

Bradshaw, S. & Howard, P. (2017). Troops, trolls and troublemakers: A global inventory of organized social media manipulation. *Computational Propaganda Research Project.* Retrieved from http://comprop.oii.ox.ac.uk/wp-content/uploads/sites/89/2017/07/Troops-Trolls-and-Troublemakers.pdf

Brown, M. E. (Ed.) (1990). *Television and women's culture: The politics of the popular.* London: SAGE.

Brunsdon, C. (1997). *Screen tastes: Soap opera to satellite dishes.* London: Routledge.

Chakravartty, R. & Srirupa, R. (2017). Mediatized populism: Inter-Asian lineages. *International Journal of Communication, 11*, 4073–4092.

Couldry, N. & Mejias, U. A. (2019). Data colonialism: Rethinking big data's relation to the contemporary subject. *Television & New Media, 20*(4), 336–349.

Condit, C. (1989). The rhetorical limits of polysemy. *Critical Studies of Mass Communication, 6*, 103–122.

Couldry, N. & Mejias, U. (2018). Data colonialism: Rethinking big data's relation to the contemporary subject. *Television & New Media* 20(4): 336-349.

Curran, J. (1990). The new revisionism in mass communication research. *Europen Journal of Communication, 5* (2–3), 135–164.

Dahlgren, P. (1998). Elusive audiences. In R. Dickenson, R. Harindranath & O. Linné (Eds.), *Approaches to audiences* (pp. 298-319). London: Edward Arnold 298-319.

Das, R. & Graefer, A. (2017). Provocative screens. In Das, R and Graefer, A co-authored book *Provocative Screens* (pp. 113–126). Cham: Palgrave Macmillan.

Dencik, L. (2017). Surveillance realism and the politics of imagination: Is there no alternative? *Krisis: Journal for Contemporary Philosophy.* Retrieved from http://krisis.eu/surveillance-realism-and-the-politics-of-imagination-is-there-no-alternative/

Deuze, M. (2009). Media industries, work and life. *European Journal of Communication, 24*(4), 467–480.

Deuze, M., Blank, P. & Speers, L. (2012). A life lived in media. *Digital Humanities Quarterly, 6*(1), 1–15.

Dhaenens, F. & Mollen, A. (2018). Audiences' coping practices with intrusive interfaces: researching audiences in algorithmic, datafied, platform societies. In *The Future of Audiences* (pp. 43-60). Cham: Palgrave Macmillan.

Dourish, P. & Bell, G. (2011). *Divining a digital future: Mess and mythology in ubiquitous computing.* Cambridge & London: The MIT Press.

Fuchs, C. (2015). The digital labour theory of value and Karl Marx in the age of Facebook, YouTube, Twitter, and Weibo. In Fuchs, Christian, Fisher, Eran (Eds.), *Reconsidering value and labour in the digital age* (pp. 26–41). London: Palgrave Macmillan.

Gillespie, T. (2018). *Custodians of the Internet: Platforms, content moderation, and the hidden decisions that shape social media*. New Haven & London: Yale University Press.

Guess, A., Nyhan, B. & Reifler, J. (2018). Selective exposure to misinformation: Evidence from the consumption of fake news during the 2016 US presidential campaign. Retrieved from www.dartmouth.edu/~nyhan/fake-news-2016.pdf.

Henderson, L. (2013). *Love and money: Queer, class, and cultural production*. New York: New York University Press.

Hochschild, A. (2016). *Strangers in their own land: Anger and mourning on the American Right*. New York & London: The New Press.

Hutchby, I. (2001). Technologies, texts, and affordances. *Sociology*, *35*(2), 441–456.

Joye, S. & von Engelhardt, J. (2015). Audiences in the face of distant suffering: An introduction to the special issue. *The International Communication Gazette*, 77(7), 603–606.

Kyriakidou, M. (2014). Distant suffering in audience memory: The moral hierarchy of remembering. *International Journal of Communication*, *8*, 1474–1494.

Livingstone, S. (2009). On the mediation of everything: ICA presidential address 2008. *Journal of Communication*, *59*(1), 1–18.

Livingstone, S. (2019). Audiences in an age of datafication: critical questions for media research. *Television & New Media*, *20*(2), 170–183.

Lotz, A. (2000). Assessing qualitative television audience research: Incorporating feminist and anthropological theoretical innovation. *Communication Theory*, *10*(4), 447–467.

Madianou, M. (2013). Humanitarian campaigns in social media: Network architectures and polymedia events. *Journalism Studies*, *14*(2), 249–266.

Madianou, M. (2018). *Technocolonialism: Digital innovation and data in humanitarian practice, presented at Migration and Mobility in a Digital Age: Paradoxes of Connectivity and Belonging Conference, Heyman Center for the Humanities*. New York: Columbia University Press.

Mansell, R. (2012). *Imagining the Internet: Communication, innovation and governance*. Oxford: Oxford University Press.

Marwick, A. & Lewis, R. (2017). *Media manipulation and disinformation online*. Data & Society. Retrieved from https://datasociety.net/pubs/oh/DataAndSociety_MediaManipulationAndDisinformationOnline.pdf

Morley, D. (1980). *The nationwide audience: Structure and decoding*. London: British Film Institute.

Morley, D. (1992). *Television, audiences and cultural studies*. London & New York: Routledge.

Morley, D. (1993). Active audience theory: Pendulums and pitfalls. *Journal of Communication*, *43*(4), 13–19.

Noble, S. U. (2018). *Algorithms of oppression: How search engines reinforce racism*. New York: New York University Press.

Ong, J. & Cabanes, J. (2018). *Architects of networked disinformation: Behind the scenes of troll accounts and fake news production in the Philippines*. Newton Tech4Dev Network. Retrieved from http://newtontechfordev.com/wp-content/uploads/2018/02/ARCHITECTS-OF-NETWORKED-DISINFORMATION-FULL-REPORT.pdf

Ong, J. C. (2014). 'Witnessing' or 'mediating' distant duffering? Ethical questions across moments of text, production, and reception. *Television & New Media*, *15*(3), 179–196.

Pantti, M. (2015). Grassroots humanitarianism on Youtube: Ordinary fundraisers, unlikely donors and global solidarity. *International Communication Gazette*, 77(7), 622–636.

Phillips, W. (2018). *The oxygen of amplification: Better practices for reporting of far right extremists, antagonists, and manipulators*. Data & Society. Retrieved from https://datasociety.net/output/oxygen-of-amplification/

Press, A. L. (1991). *Women watching television*. Philadelphia: University of Pennsylvania Press.

Press, A. L. & Cole, E. (1999). *Speaking of abortion: Television and authority in the lives of women*. Chicago: Chicago University Press.

Radway, J. (1984). *Reading the romance: Women, patriarchy, and popular literature*. Chapel Hill: University Carolina Press.

Radway, J. A. (1983). Women read the romance: The interaction of text and context. *Feminist studies*, *9*(1), 53–78.

Radway, J. (1988). Reception study: Ethnography and the problems of dispersed audiences and nomadic subjects. *Cultural Studies*, *2*(3), 359–376.

Scott, M. (2014). The mediation of distant suffering: An empirical contribution beyond television news texts. *Media, Culture & Society*, *36*(1), 3–19.

Seaman, W. R. (1992). Active audience theory: Pointless populism. *Media, Culture & Society*, *14*, 301–311.

Silverman, C. & Alexander, L. (2016). How teens in the Balkans are duping Trump supporters with fake news. *Buzzfeed*, 14 November. Retrieved from www.buzzfeed.com/craigsilverman/how-macedonia-became-a-global-hub-for-pro-trump-misinfo?utm_term=.ujw9E6pjL#.dwjMv74Bg

Silverstone, R. (1994). *Television and everyday life*. London: Routledge.

Silverstone, R. (2002). Complicity and Collusion in the Mediation of Everyday Life. *New Literary History* 33 (4): 761–780

Silverstone, R. (2007). *Media and morality: On the rise of the mediapolis*. London: Polity.

Stalder, F. (2006). *Manuel Castells: The theory of the network society*. Cambridge, UK: Polity.

van Dijk, J. (2014). Datafication, dataism and dataveillance: Big data between scientific paradigm and ideology. *Surveillance & Society*, *12*(2), 197–208.

Wardle, C. & Derakhshan, H. (2017). Information disorder: Toward an interdisciplinary framework for research and policy making. Retrieved from https://rm.coe.int/information-disorder-toward-an-interdisciplinary-framework-for-researc/168076277c

Woolgar, S. (2002). *Virtual society? Technology, cyberbole, reality*. Oxford: Oxford University Press.

Ytre-Arne, B. & Das, R. (2018). An agenda in the interest of audiences: Facing the challenges of intrusive media technologies. *Television and New Media*.

# 10

# GLOBALIZING THE PEOPLEMETERED AUDIENCE

*Jerome Bourdon, and Cécile Méadel*

This chapter recounts and explains the global rise of quantitative, continuous, individual television audience measurement (TAM), based on the technology of the peoplemeter. The peoplemeter (PM) combines the passive technology of the audiometer (registering the channel on which a set is tuned) with a special remote control where each member of the home has to push an individual button when viewing. Pioneered in the 1970s (although not yet named), the PM was adopted in the Western world (Europe and North America) by the mid-80s, then in all major television markets across the Americas and Asia, and finally a few markets in the Middle East and Africa. PM-based audience measurement is now deployed in some 80 countries, worldwide (Eurodata TV 2016), and the device continues to be updated and improved but also challenged (see conclusion).

Media research at large has typically neglected to consider the making of quantitative audiences. The reasons are well known (Bourdon & Méadel 2014). Scholars of various theoretical ilks tend to share a suspicion *vis-à-vis* the commercial purposes of such data and their accuracy (and still do to a large extent), while commercial organizations may be reluctant to make their expensive data and inner workings available to scrutiny. In critical communication studies, Dallas Smythe's (1977) analysis of the measured audience as a "commodity" sold to advertisers remains influential. Academic work about ratings, as well as the academic use of ratings numbers, has therefore long been limited to the US (Webster et al. 2013, first edition 1991), with a few exceptions (e.g. Bjur 2014; Méadel 2010; Souchon 1981), while academic attacks on ratings have been numerous, including from illustrious quarters (e.g. Bourdieu 1998). However, especially in the disciplines of media economics and media management, there is a growing body of work on the role and making of ratings. Indeed, in a world where new forms of quantification are gaining much attention (big data), international aspects of audience measurement are beginning to receive new and more attention (Balnaves, O'Regan & Goldsmith 2011; Bourdon & Méadel 2014; Taneja 2013).

## What Kind of Truth?

In this academic landscape, our own view of TAM is agnostic. We will not discuss (as Ang did in 1991) the validity of the data regarding "real" viewing practices. For us, audience ratings provide a form of *procedural* truth, a concept borrowed from theorists of law (van de Kerchove 2000). As in courts, the type of reasoning involves a more or less convincing argumentation intended to obtain adhesion in a variety of ways (for the PM, globalization has been no small part of this argumentation). Of course, audience measurers (here we mean both the companies directly involved in the production of figures, and, within them, the individuals whose professional identity is intimately linked to the production of figures, those we call the ratings wizards below) also aspire to produce the *substantial* truth akin to that of the scientific laboratory, based on constraining and impersonal demonstration (resorting to both statistical and technological knowledge). Their work would be facilitated if people (media professionals, but also the media and media audiences) used those numbers with the trust given to scientific research, ignoring the complex machine that produces the numbers, and relying on numbers as a faithful reflection of audience behaviour. But this is not ultimately within their reach. One reason is that ratings are appropriated far beyond the specialized world that produces them; they therefore receive mostly sceptical media coverage while the cultural and artistic world is highly critical of ratings. Finally, including in the most "commercial" of regimes (the USA), the state is involved in the fabric and consumption of data (Bourdon & Méadel 2015). So, ratings are a form of highly visible, socially exposed data (in this sense they might be compared to unemployment figures).

Ratings are not a scientific measurement produced by trial and error in the scientist's laboratory, but a "market information regime", used by stakeholders to make sense of a specific marketplace (Anand & Peterson 2000). This information regime has also been globalized, which involves a form of standardization, aiming to render "the world equivalent across cultures, time, and geography" (Timmermans & Epstein 2010).

## The Rise of the Metered Public

The rise of television ratings was preceded by a long history of the progressive quantification of the public, starting with the US whose case is known best (Igo 2008). The most well-known technology of quantification of the public is the opinion poll based on random sample, promoted by Gallup before the Second World War and quickly exported to a number of Western countries. However, including in the US, statisticians had long been using techniques other than random sampling in order to represent the whole population (for example: the questionnaire-based survey, or the focus on knowledgeable spokespersons supposedly able to report on fellow-listeners/voters/consumers).

The ancestor of the PM is the radio audiometer, patented in the USA in 1929, a device that aimed to automatically record the time devoted to each wavelength as the radio was being used in people's homes. The engineer A.C. Nielsen, founder of the eponymous company in 1923, acquired the patent in 1936. Meanwhile, Nielsen's competitor, Arbitron, pioneered the audiometer for use in television viewing in the New York region in 1947. The measuring instrument gradually improved and (combined with the continued use of viewing diaries) was rechristened audimeter, soon becoming the favoured method for measuring national American TV audiences.

In the rest of the world, dominated by public (service) television with a few exceptions (mainly Latin America, with much early US penetration), quantitative data were not neglected. In the 1920s, quantitative surveys of radio audiences were already used in various Western countries (Huth 1937). The use of quantitative data became systematic after the Second World War, so much so that rapidly, in the public broadcasting services, some felt they had to "resist" a growing pressure to quantify audiences and rank programmes, especially by providing alternative kinds of data (Ang 1991; Bourdon 2011). For example, the first head of audience research at the BBC (1936–68), statistician Robert Silvey, was wary of what he called the "quantitative fallacy": i.e. assuming that the higher the rating, the better the programme (Schwarzkopf 2014). He included in the viewing diaries he used as the basis for audience research a scale of enjoyment from 1 to 5. The BBC was also the first European broadcaster to measure audiences continuously: a representative sample of the population had to be recruited for this. This "panel" has remained a key component of TAM. Its members (panellists) have to remain anonymous, they must be "trained" to learn how to operate the devices used (filling viewing diaries, pressing buttons, sometimes both), and are monitored at regular intervals.

After the USA, the richest television markets were the first to adopt the audimeter. In the UK, in 1959, advertisers and commercial television (ITV, the first commercial TV network in Europe) commissioned a survey based on a Nielsen adapted audimeter developed by the British company Attwood (again combined with diaries). This competed with Silvey's system until the 1980s (Schwarzkopf 2014). In Germany, in 1963, the public broadcasters ARD and ZDF commissioned a new audience measurement system (based on an adaptation of the Attwood audimeter combined with weekly interviews from a separate panel) from the company *Infratam*, which had been founded by three partners: *Infratest*, the leading public opinion researcher in Germany, Attwood (England), and Nielsen (USA) (Vollberg 2014). A similar system was used in the Netherlands, with an audimeter developed by a subsidiary of Philips for the German-based company GfK Intomart (since 2014, GfK). In southern Europe, the audimeter was introduced only in the early 1980s, together with the deregulation of broadcasting and the rise of the competition. The reason for this seems to lie in the characteristics of media models in those countries (Hallin & Mancini 2004). In southern Europe, public broadcasters evolved in a highly political environment, and depended heavily on an unstable state, that is to say, critically, on changing governments. Broadcasters in northern Europe operated with more independence, in a more stable environment, and were also more connected to international evolutions, including regarding audience measurement.

In 1975, German public service broadcasters commissioned the first version of what was not yet known as the PM. The "peoplemeter" was first discussed internationally in the mid-1980s, when it was introduced in the UK and in France, then in Italy, and, in the course of 10 years, in all the Western world (including Latin America) and most major television markets. Although commercial competitive television started in the US, Nielsen did not move that market to the PM until 1987, and then only after being challenged by four competitors (Buzzard 2012). Nielsen had no reason to replace its traditional method with a costly one that would produce new data that was not comparable with the previous, long series of figures (a crucial issue for all recurring statistical measures, which make any change difficult). Eventually, the entrance of competitors on the market, especially AGB, forced the company to mobilize its considerable resources and modernize its techniques.

Until the fast rise of the PM, the variety of national and international methods had been remarkable: viewing diaries, usually self-filled, with computerized treatment starting in the 1970s; faster and economical but less exhaustive instant phone surveys; computer assisted telephone interviewing or CATI, pioneered in the 1980s, still much used in commercial surveys; and audimeters, usually combined with viewing diaries. Putting together a PM panel is a very expansive operation, usually requiring the agreement (and the financing) of all major stakeholders. It went together with the deregulation of television, and the rise of commercial channels, including in the former Soviet bloc in the 1990s. Competition and deregulation prompted television markets to look for a faster, individual, unified technology for audience measurement. The PM, which had been developed some ten years earlier in Germany, then appeared as the "cutting edge", "state-of-the-art", and its quick expansion made this, to a certain extent, a self-fulfilling prophecy.

## Methodological Decisions: Toward a Global Definition of the Viewer?

Behind what Bogart (1966: 50) has called "the monolithic self-assurance with which the statistical uncertainties of survey data were transformed into beautiful, solid, clean-looking bar charts", myriad decisions had to be made. For example: Should viewers be told to press when they "watch" (still the German convention) or when they are "present in the room" (now the international standard)? But then, what is a room? Should visitors (and how many) have a special button on the remote? Should the panel be renewed, because of members' "burnout", or not renewed, because members have to be "trained"? As viewers may forget to press the button, all measurers provide for some control: regular phone checks, or, especially, when there is a behaviour considered as anomalous; but what should be the frequency of controls?

Moving to figures: what indicators should be used? Again, there has been an internationalization: the *audience (or market) share* (ratio of viewers for a channel/a content, to the number of people watching television) is now most widely quoted, before the *rating* (ratio of viewers to the population with a TV set in the home), while the *reach* (ratio of viewers watching a channel a given minimum time, to the population with a TV set in the home), long defended by the BBC (as a way to check if they were actually reaching the population they were supposed to serve), has lost much ground.

In order to gain a sense of geographical variety, we analysed the basic 2015 data for 60 major television markets (Eurodata TV 2016). Although the PM is by far the dominant technology, CATI was used in parts of the Middle East (Egypt, Jordan, Iraq) and Africa (Cameroon), diaries in Pakistan, face-to-face in Ivory Coast. The minimum age taken into consideration has been homogenized but there still are differences: 60% start from 4, the rest being dispersed between 10, 8, 7, 5 or … 0 years. The ratio of the number of panellists to the total viewing population varies. It depends on resources, obviously, but also on the stage of implementation of continuous audience measurement, which starts with panels, which tends to grow under the pressure from stakeholders for more precise results. A single individual can "speak" for a minimum of 1,500 viewers, and a maximum of 50,000 viewers. This kind of comparison is often used to question the reliability of figures. This is not our point. Small panels can effectively represent huge populations. However, the smaller the panel, the higher the margins of errors, and the lower the quality of the resolution for subpopulations: this is where the problem starts, especially when precision is needed for sub-audiences, regional or sectorial.

In short, even before we consider the relatively new question of multiscreen viewing and the technology used to identify the channel (see below), the "beautiful charts" mentioned by Bogart reflect many different ways of measuring, and very different qualities of measurement.

## Managing TAM: The Dominant Model and its Avatars

All over the TV-viewing world, the key aims for measurers, no less or even more than precision, are consensus and stability (Méadel 2015). Different stakeholders, with different interests, are involved in the operation (which, again, makes it very different from scientific work in the laboratory): advertisers, commercial and public broadcasters, and the state as regulator (or supervisor). Building consensus, and making it durable, is hard work. In order to reach stability and consensus, the Joint Industry Committee (JIC) has slowly emerged as a good institutional arrangement. Initiated in the UK for other media, it has become the dominant arrangement worldwide. The most famous JIC is the British BARB, the Broadcasters Audience Research Board. It was born in 1982 to unify competing systems operated by the BBC on the one hand, and the commercial ITV network, on the other.

The JIC is a small organization, usually a non-profit, where representatives of broadcasters (public and private) and the advertising industry (both advertisers and agencies) meet at regular intervals. It is a forum where various interests can work out compromises with each other. It does not manage measurement directly, but is supposed to ensure fair competition by publicly inviting bids for measurement operations ("putting them out to tender") at regular intervals (every few years). In addition, to better ensure reliability and avoid conflicts of interest, the "ideal" JIC (that is, the one that operates on the richest markets) entrusts the complex operations of measurement to different companies. The main company (mostly Nielsen, TNS, or GfK, depending on region) implements audience measurement; a second one should perform the so-called "establishment survey" which provides the demographic basis for recruiting the panel; a third one should match the "raw" data (usually minute by minute) with the programme schedules. Those technical aspects are important as they affect results. Presentation can inflate (or deflate) measurement of an individual programme, for example, by dividing shows into shorter programmes.

Although the JIC is dominant, two other situations can be found: the commercial provider sells data directly to interested parties, which obtains in the US, but also in peripheral markets directly penetrated by the multinationals of measurement (for example, Nielsen in the Philippines, Ipsos in some Middle-Eastern countries, TNS in Kazakhstan). In some countries, meanwhile, there is no JIC but, instead, a so-called "Media Owner Contract" between the measurer and major media owners (sometimes joined by advertisers). In France, Médiamétrie is a unique, monopoly measurer incorporating representatives of broadcasters and the advertising industry which, elsewhere, would meet in a JIC.

The functioning of the JIC is based on trust between stakeholders. Trust must be built and maintained, especially when competition is scarce, and the same measurer may find itself in a de facto monopoly position with no real competition. In addition, in some countries, the measurer takes care of matching data with the programming schedule, as well as measurement (e.g. Israel). To reinforce trust, all measurement systems emphasize control, supervision, and certification, in order to forestall complaints about errors or bias. Those mechanisms have been increasingly internationalized: for example, certification by ISO, or the code of ethics by professional marketing bodies, mainly ESOMAR.

## Competition and Concentration

In the 1990s and 2000s, the marketing industry became both more concentrated and globalized, which involved complex restructuring and rebranding. This affected audience measurement directly. AGB and Nielsen, former rivals in the US market, created the global entity AGB Nielsen Media Research in 2004. In Latin America, the Brazilian marketing giant IBOPE (created in 1942 and later strengthened by its alliance with TV giant Globo) became involved in audience measurement in most Latin American countries in the 1990s. The few remaining markets are controlled by AGB Nielsen. In 2014, the Kantar Group acquired the IBOPE media division which changed his name to Kantar IBOPE Media. Formed in 1965 in the UK, Taylor Nelson acquired the French marketing and polling company Sofres in 1993 and became TNS, which was acquired in 2008 by the multinational, UK-based firm, WPP; TNS then became part of WPP's Kantar group and was rebranded Kantar TNS in 2016.

In 2015 (Eurodata TV 2016), out of 60 television markets, Nielsen ranked first (20), followed by TNS Kantar (11), Germany-based GfK (with the telecontrol technology, first introduced in Germany in 1983) and the Paris market research firm Ipsos (5 each). Médiamétrie, meanwhile, reaches a handful of French-speaking TV markets in Africa. Those globalized players compete for TAM on national markets, often partnering with local companies. Sometimes, they choose to work together; for example, Kantar and Nielsen cooperated in India. In most markets, whether small or large, the same measurer tends to operate for long stretches of time as stability is one of the outcomes sought by interested parties in the first place.

This global industry also convenes in international conferences linked to various organizations (e.g. ASI, Advertising Seminars International, created in 1991, or the annual meetings of ICom, the "Global Forum for Marketing Data and Measurement", since 2004). An early forum was the GEAR, the Group of European Audience Researchers of public broadcasters, created in 1967. Among present organizations, EGTA, the European Group of Television advertising, created in 1974, now boasts more than 130 members in 38 countries. In this milieu, "gurus" or "wizards" (as they are often called) of TAM (mainly Western males with a background in engineering or statistics), are respected figures. For example, Peter Meneer, head of research at the BBC from 1979 to 1997, audited the Israeli system in 2002 (Bourdon & Ribke 2014); after a long career in research at SSR (the Swiss public service broadcaster), where he developed the telecontrol meter, the engineer Matthias Steimann became an independent entrepreneur in 2004 and was soon involved in the development of the Admap company in India (New Rating System 2004); Alberto Colussi pioneered a local audimeter in Italy in 1976, partnered with AGB in 1981 in order to successfully compete against Nielsen and Telecontrol in the bid invited by the new JIC, Auditel, and became chairman of the AGB Nielsen management board in 2005.

The very adoption of such terms as "guru" or "wizard" for this personnel, highly qualified in engineering and/or statistics, is an interesting symptom of the characteristics of the market of information regime studied here. On the one hand, it makes high claims at rationality and scientificity, on the other hand, as figures are highly public and appropriated in various social arenas, users of figures who have no time to consider the apparatus which validates the figures find themselves treating experts as some sort of magicians, a metaphor which reoccurs each time "mysterious" experts have to work in a highly public space: one may think of the operators of the first computer, the ENIAC, who were nicknamed "the priesthood" (Levy 1984: 5).

## Fault Lines and Conflicts

Indeed, the apparent stability of continued measurement and apparent ease with which the wizards pull out figures from their hats hides hard work, especially to overcome "built-in" fault lines between certain categories of actors. First, public TV stations often criticize the system when they experience a severe drop in audiences, due both to the change of measuring methods and the rising competition, for example, Turkey's TRT in 2008 (Major Ratings Fraud 2011). When the PM is introduced, these public operators might suggest that it does not serve ageing audiences uneasy with the PM technology (Bourdon & Ribke 2014); such a claim was also made in the US (Jenson Adams 1994). In the longer term, they also want to make sure that audience data accurately reflect the behaviour of the whole population which they are charged with serving, so they will push for full national representation, as opposed to commercial actors, especially advertisers, who care more for urban, middle-class audiences with purchasing power. This leads to claims of some viewers being preferred and others ignored, e.g. in Brazil or India (Chakrabarti 2014). In 2001, when the new Russian JIC first looked for a company to measure Russian audiences, public channels supported the candidate AGB-Italia (partnered with a local company), which offered more complete coverage, and commercial channels favoured TNS, more focused on their commercial targets – TNS won the bid (Davydov & Johansson 2014).

Second, there have been tensions between commercial channels. This depends, among other things, on the specific reach of each channel, bringing us back to the question of representation. New emerging commercial channels have an interest in publicly questioning the existing system. In 2002–3, the new commercial Israeli Channel 10 forced a renewal of the whole panel, after claiming that "old" panellists were not open to viewing a new channel. In India (Chakrabarti 2014), the private network Zee was the main critic of the existing system and a driving force in the establishment of the Indian JIC, the BARC (Broadcasters Audience Research Council) in 2010.

Third, the state may intervene as regulator, as direct manager, or as supervisor, especially in times of acute crises or major reorganizations (the British BARB was born from a recommendation of a parliamentary commission). In the US, dissatisfaction with the system was ripe in the late 1950s. A first investigation of the ratings services by the Federal Trade Commission blamed them for "exaggerating the size of their samples and arbitrarily modifying the figures" (Beville 1988: 243). The 1963 "Harris hearings" (named after the representative who initiated them) led to a major review of the system and the creation of the Broadcast Rating Council (now Media Rating Council, still quite active) to accredit the companies conducting measurements.

At worst, TV measurements can be challenged in court, especially for allegedly accepting bribes from producers or channels for tampering with results, or leaking names of panel members who can be, in turn, secretly encouraged to view certain channels or shows. (Rumours of) such leaks play a crucial part in attempts to destabilize the system. Since the 1980s, leaks have occurred in France, Greece, the Philippines, India, and Israel. In 2007, ABC-CBN, one of the Philippines' television networks subscribing to the Kantar TNS service sued the dominant measurer, AGB Nielsen Philippines (Desiderio 2017). In 2011, accusations of bribes were levelled at AGB Nielsen in Turkey (Major Rating Fraud 2011). In Israel the new commercial Channel 10 organized and publicized a leak through one of its current affairs programmes (Bourdon & Ribke 2014), and succeeded in having the whole panel renewed against the opinion of the majority of the members of the IARB (Israel Audience Research Board), the Israeli JIC.

Political contexts may directly affect TAM, although open state intervention (beyond the "natural" regulatory role) is rare. In 2014, Argentinian president Cristina Fernandez de Kirchner, claiming that the IBOPE peoplemeter (introduced in 1993) underestimated the audience of public channels, sponsored a new system, with the help of a network of universities (Obario 2014). This initiative floundered with the change of government in 2016. In authoritarian regimes, for example Russia, where the state pays close attention to the audience of its own media, local measurers and JIC obviously may be submitted to pressures (Davydov & Johansson 2014).

## A Natural Monopoly?

The global view adopted here allows us to put into perspective the thesis, well-argued by Napoli (2005), that TAM is a natural monopoly. Napoli combines J.S. Mills' classical justification of natural monopoly (very high costs of entry to the market) with a pragmatic justification – the need for a single currency for buying and selling audiences, necessary for the smooth running of the information regime. In addition, from a policy point of view, the state may have an interest in the production of statistics for regulating the market (Bourdon & Méadel 2015).

However, this may apply better to stable, well-regulated markets, than to countries with changing landscapes and weak regulation, where our fault lines may work against a durable monopoly, and several of the trade currencies (or audience measurements) may be available for long periods. Especially in large markets, clients may consider that different systems bring different types of information, for example in India where Admap (providing daily results) and TAM (weekly ones) coexisted for a number of years (Chakrabarti 2014; Taneja 2013). Different powerful broadcasters may find an interest in different measurements, as in the Philippines since 2009 (Desiderio 2017), but also in Australia in the past (Balnaves et al. 2011).

## Conclusion: The Resilient Peoplemeter

Is the reign of the system born in the mid-1980s in Europe coming to an end? TAM has been submitted to a number of challenges. An early challenge was "timeshifting", or delayed viewing, which started with the wide availability of the VCR in many markets in the 1980s. BARB pioneered the measurement of timeshifting within a certain time lag (e.g. within three days or seven days). "Timeshift" viewing now refers to VOD and recording on a hard drive, while "catch-up" is used for TV content viewed on internet platforms. In 2015, of our 60 markets, 30 measured only live TV, 12 timeshifting, and 18 catch-up and timeshifting (Eurodata TV 2016).

In addition, today the variety of viewing platforms and devices makes more difficult the automated identification of the signal/frequency/channel/content. With phone surveys and diaries, viewers were asked to identify the channel. With early (people)meters, channels were identified through analogical frequency. The vast majority of PM systems now used audiomatching (also known as fingerprinting): the channel is identified through a comparison between soundtracks and a database of sound samples. The more sophisticated systems have just moved to watermarking, which supposes the collaboration of broadcasters: a special mark (inaudible to viewers) is introduced in the sound channel, which allows the identification of more parameters (including the devices or platforms used, timeshift and catch-up, etc.). First used for the regular TV screen, this information is or will be progressively exploited for measuring computer, tablet, and smartphone viewing.

The problems of delay and channel identification are closely connected to the variety of screens used for viewing. While second TV sets could be equipped with PMs, other screens cannot resort to a specific remote control for viewers to press on. Various technologies are available: mainly, a small sensor set next to the computer or TV set in the home, or a portable peoplemeter (both coupled with watermarking). First experimented in the US by Arbitron in 2005 (for both radio and TV), the portable PM was once seen as the holy grail of exhaustive measurement. In 2015, it was used for national TAM only in Canada, and combined with the fixed PM in Norway. Compared to the PM, the PPM reintroduced an element of passivity (not having to press a button): which also meant that simple contact with the media is considered enough. If you are around the set (including radio), then you are considered to be watching/listening. But panel members must agree to carry a device (pager? watch?). Sensors or PPM also raise the problem of detection of the sound signal (from what volume? what about noisy environments?). Once again, the holy grail has turned out to be just another method, more precise in some respects, less in others.

It is the industry's consensus today that present systems cannot (yet) be based on a "single panel with single device" system and must still use "the reliable, panel-based TAM as a basis", but integrated into a "hybrid product: a single currency based on more than one measurement" (Egta 2016); this combination of various sources may be unsettling for data users who can be reminded of the time when viewing diaries and audimeter data had to be mixed. Today's systems often resort to several panels. Some combine panel data with census data: census (also known as Return Path Data) refers to the complete collection of consumption data (for example, through a cable distributor's set-top box) for a channel or a piece of content. Census, however, is device-centric and not viewer-centric (unlike the PM or the PPM).

For example, in France, Médiamétrie (Méadel 2017) has two panels, its TV panel and an internet panel with Nielsen (launched in 2002). In 2013, Médiamétrie also entered a partnership with Google. Google uses a meter taking the form of a proxy Wifi with a wireless router, to propose a four screen (laptop, desktop, tablet, smartphone) panel which is combined, for TV sets, with classic PM TV data transmitted by Médiamétrie. The so-called dovetail project of the BARB is similar.

Another challenge is the use of big data for television, especially social media conversation about TV programmes (social TV analytics) (Kosterich 2016). In the US at least, this is more and more used as complementary information to panel-based data. Then again, it shows us that the history of audience measurement is not linear. Social TV analytics brings us back to the question of measuring not simply exposure (as with the PM or the PPM), but some element of appreciation or attention. Today, for programmers, this means more channels can be "measured", and that so-called TV hits reflect more diversity, and maybe economic viability.

In a world of competing channels, competing devices, and overabundant data, the question of trust remains central. It seems that TAM organizations – that is, well-established commercial providers, mostly with the mediation of a JIC, using an improved PM – are still in a position to enjoy credibility and a sense of transparency among various stakeholders and will remain at the centre of national and globalized TV market information regimes for the foreseeable future. Increasingly, various private data providers will use big data to offer complementary figures. But it is doubtful that this kind of data can replace existing measurement, or even be integrated in a consensual manner into existing arrangements: "Big Data" is just a broad term for various ways of exploiting various and heterogeneous digital resources, which is not based on a recognized, universally accepted method of regular extraction of representative data for a given population.

For television scholars, this resilience of a form of representation of the public which has been much vilified or ignored, demands more critical attention as it is not only a question of market information: peoplemetered audiences play a crucial part in the sale of advertising space, but also in the decision of broadcasters about production and scheduling, and in the power games between broadcasters, advertiser and regulators; last but not least, audiences use ratings as mirrors of their own behaviour. Further research is needed on ratings, neither accepting them wholesale nor reducing them to a vile system of manipulation, but understanding both their continuous construction and numerous uses in our changing media landscapes.

## References


Anand, N., & Peterson, R. A. (2000) "When market information constitutes fields: Sense making of markets in the commercial music industry", *Organization Science*, 11(3): 270–284. doi:10.1287/orsc.11.3.270.12502

Ang, I. (1991) *Desperately seeking the audience*, London: Sage.

Balnaves, M. O'Regan, T. & Goldsmith, B. (2011) *Rating the audience: The business of media*, London: Bloomsbury.

Beville, H. M. (1988) *Audience ratings: Radio, television, cable*, Hillsdale, NJ: Lawrence Erlbaum.

Bjur, J. (2014) Thickening Behavioral Data. New uses of Ratings for Social Sciences, in J. Bourdon & C. Méadel (eds.), *Television audiences across the world*, Basingstoke, UK: Palgrave Macmillan. pp. 248–263.

Bogart, L. (1966) "Is it time to discard the audience concept?", *Journal of Marketing*, 30: 47–54.

Bourdieu, P. (1998) *On television*, New York, NY: New Press.

Bourdon, J. (2011) *Du service public à la téléréalité, une histoire des télévisions européennes*, Paris: INA.

Bourdon, J. & Méadel, C. (eds.) (2014) *Television audiences across the world. Deconstructing the ratings machine*, Basingstoke, UK: Palgrave Macmillan.

Bourdon, J. & Méadel, C. (2015) "Ratings as politics: Television audience measurement and the state, an international comparison", *International Journal of Communication*, 9: 2243–2262.

Bourdon, J. & Ribke, N. (2014) "Ratings, the state and globalization: the politics of television audience measurement in Israel", *Media, Culture and Society*, 38(2): 159–174.

Buzzard, K. (2012) *Tracking the audience: The ratings industry from analog to digital*, New York: Routledge.

Chakrabarti, S. (2014) Power Games: Audience Measurement as a Mediation between Actors in India, in J. Bourdon & C. Méadel (eds.), *Television audiences across the world*, Basingstoke, UK: Palgrave Macmillan. pp. 113–131.

Davydov, S. & Johansson, E. (2014) Market Requirements and Political Change: Russia between Two Worlds, in J. Bourdon & C. Méadel (eds.), *Television audiences across the world*, Basingstoke, UK: Palgrave Macmillan. pp. 179–197.

Desiderio, L. (2017) "ABS-CBN, GMA contest TV ratings leadership", *The Philippine Star*, February 4.

EGTA European group of television advertising (2016) *Advances in Hybrid Television Audience Measurement.* http://00_egta_insight_04_TV_measurement_edition_2.pdf
Eurodata TV Wordwide (2016) *One TV year*, Paris: Médiamétrie.
Hallin, D. & Mancini, P. (2004) *Comparing media models*, Cambridge: Cambridge University Press.
Huth, A. (1937) *La radiodiffusion, puissance mondiale*, Paris: Gallimard.
Igo, S. (2008) *The averaged American: Surveys, citizens, and the making of a mass public*, Cambridge, MA: Harvard University Press.
Jenson Adams, W. (1994) "Changes in ratings patterns for prime time before, during and after the introduction of the people meter data", *Journal of Media Economics*, 7(2): 15–28.
Kosterich, A. (2016) "Reconfiguring the "Hits": The new portrait of television program success in an era of big data", *International Journal on Media Management*, 18(1): 43–58. doi:10.1080/14241277.2016.1166431
Levy, S. (1984) *Hackers: Heroes of the computer revolution*, New York: Doubleday.
Major ratings fraud probe strikes popular Turkish TV series (2011) *Sunday's Zaman*, Istanbul, December 11.
Méadel, C. (2010) *Quantifier le public. Histoire des mesures d'audience à la radio et à la télévision*, Paris: Economica.
Méadel, C. (2015) "Moving to the peoplemetered audience. A sociotechnical approach", *European Journal of Communication*, 30–1: 36–49.
Méadel, C. (2017) Interview with J. Méric, Director of Measurement Systems Developments, Paris: Médiamétrie, September 6.
Napoli, P. M. (2005) "Audience measurement and media policy", *Communication Law and Policy*, 10(4): 349–382. doi:10.1207/s15326926clp1004_1
New TV rating system to challenge TAM monopoly (2004) *Times of India*, August 5. http://m.timesofindia.com/business/india-business/New-TV-rating-system-to-challenge-TAM-monopoly/amp_articleshow/803398.cms
Obario, M. (2014) "El Gobierno lanzó su propio sistema de medición de audiencias audiovisuals", *La Nacion* (Buenos Aires), October 10.
Schwarzkopf, S. (2014) The Politics of Enjoyment: Competing Audience Measurement Systems in Britain, in J. Bourdon & C. Méadel (eds.), *Television audiences across the world*, Basingstoke, UK: Palgrave Macmillan. pp. 33–52.
Smythe, D. (1977) "Communications: Blindspot of western marxism", *Canadian Journal of Political and Social Theory*, 1(3): 1–27.
Souchon, M. (ed.) (1981) *Three weeks of television in the world*, Paris: UNESCO. http://unesdoc.unesco.org/images/0006/000659/065907eb.pdf.
Taneja, H. (2013) "Audience measurement and media fragmentation: Revisiting the monopoly question", *Journal of Media Economics*, 26(4): 203–219. doi:10.1080/08997764.2013.842919
Timmermans, J. & Epstein, S. (2010) "A world of standards but not a standard world. Towards a sociology of standards and standardization", *Annual Review of Sociology*, 36: 69–89. doi:10.1146/annurev.soc.012809.102629
van de Kerchove, M. (2000) "La vérité judicaire", *Déviance et Société*, 24(1): 95–101.
Vollberg, S. (2014) Germany: Pioneering the Peoplemeter, in J. Bourdon & C. Méadel (eds.), *Television audiences across the world*, Basingstoke, UK: Palgrave Macmillan. pp. 102–110.
Webster, J. G., Phalen, P. F. & Lichty, L. W. (2013) *Ratings analysis: Audience measurement and analytics*, New York: Routledge.

## Further reading

Kelly, J. P. (2019). Television by the numbers: the challenge of audience measurement in the age of Big Data. *Convergence: The International Journal of Research intoNew Media Technologies*, 25(1): 113–132.
Nelson, J. L., & Webster, J. G. (2016) "Audience currencies in the age of big data", *International Journal on Media Management*, 18(1): 9–24. doi:10.1080/14241277.2016.1166430
Taneja, H., & Mamoria, U. (2012) "Measuring media use across platforms: Evolving audience information systems", *International Journal on Media Management*, 14(2): 121–140. doi:10.1080/14241277.2011.648468
Webster, J. G. & Taneja, H. (2015) "Media exposure measurement", Oxford Bibliographies in Communication. doi:10.1093/obo/9780199756841-0164J
Webster, J. G. (2016) "Why study measures of exposure? From exposure to attention", *Communication Methods and Measures*, 10(2–3): 179–180. doi:10.1080/19312458.2016.1150974

# 11

# TRANSFORMING MARKETS FOR CHILDREN'S TELEVISION INDUSTRIES

*Anna Potter and Jeanette Steemers*

## Introduction

Children's television has long been subject to powerful globalizing forces, with largely US-owned channels such as Disney, Nickelodeon/Viacom and Cartoon Network/Turner distributing screen content, and its associated merchandise, to millions of children worldwide (Potter and Steemers 2017). The advent of digital transmission, and the rapid proliferation in dedicated children's channels that followed, further opened up national markets in the early 2000s. The introduction and increasing uptake of internet television services in the late 2000s, including Netflix, Hulu, Amazon Prime and YouTube, saw existing processes of media globalization intensify. Despite the popularity of globally distributed screen content, from Disney's *Frozen* to YouTube's *Little Baby Bum*, locally produced, age-specific television that situates children in their own cultural context, reflecting their daily lives back to them, is still considered important by policy makers, children's media advocacy groups (see Steemers and Awan 2016) and international organizations such as Unicef (2011). Almost all children watch television, regardless of their personal, cultural or social circumstances, and many countries make efforts to support domestic screen production, because of its perceived contribution to national cultural representation. This is particularly the case in those countries with a tradition of public service broadcasting (PSB) in Western Europe, and also in Australia and Canada.

Notwithstanding the importance of local screen content to young audiences, transformations in television's distribution and consumption have disrupted longstanding funding models and policy instruments designed to support domestic screen production. Children continue to watch animation series and TV drama, but increasingly they are accessing these online rather than as part of a broadcast schedule. Their viewing range has also expanded to include vloggers (YouTube personalities), how-to videos, watching other children and young adults play games such as Minecraft, and the unpacking of toys by adults or other children (Craig and Cunningham 2017). These shifts in viewing are particularly evident for older children, an audience that has been neglected by linear broadcasters. According to a 2017 report by UK regulator, the Office of Communications (Ofcom), 21 percent of five to fifteen year olds view YouTube unboxing videos, and a further 39 percent like to watch others playing games (Ofcom 2017, 85). The most popular content for older children aged 12 to 15 includes music videos (26 percent), followed by funny videos or pranks (23 percent) and their mobile phone is the device they would miss the most, compared to children aged 5 to 11 who would miss the TV set the most (Ofcom 2017, 2).

In this emerging viewing environment, legacy media, including public service broadcasters, now compete mostly for children's attention with Over the Top (OTT) providers such as YouTube and Subscription Video on Demand (SVOD) services such as Netflix, Hulu and Amazon Prime.. The US-owned SVODs have large production budgets and consider children important drivers of subscriptions. YouTube offers an abundant supply of 'free' content, including through its own children's app, launched in 2015 in some territories, much of which is highly commercialized and beyond the regulatory reach of national agencies.

The arrival of SVODs and online providers represents one further step in the development of screen content for children. During the late 2000s the increasingly platform-agnostic child audience fragmented across new and legacy providers. So too did advertising revenues and production budgets. Yet the policy and industrial implications of market fragmentation and the threat to local content are often overlooked, obscured it seems by the apparent over-supply of globally appealing content offered by predominantly commercial providers, sourcing content from North America and Asia (Westcott 2011).

This chapter begins by tracing children's television's transition from a scarce resource, primarily funded (with the exception of the US) by public service and state-owned broadcasters to an increasingly complex, commercialized and global enterprise. Focusing on the developed economies of North America, Western Europe and Australia, it then explains how technology has transformed children's viewing habits in these countries, and the range of screen content and viewing platforms to which they have access, challenging longstanding policy and production norms. New means of content distribution, particularly the rise of internet television services offered by a range of providers, are then analyzed, and their impact on children's screen production and consumption discussed. This chapter concludes with some thoughts about what the future might hold for children's screen content 'after TV' and where research is lacking.

## Children's Television from Yesteryear

Although in many countries children are considered 'a special audience with distinctive characteristics and needs' (Buckingham 2005, 468), prior to digitization and the internet, supplies of children's television and opportunities for them to view were scarce. In countries with mixed systems of government-funded and private broadcasting, including the UK, Germany, France, Australia and Canada, PSBs were usually charged with responsibility for providing children with age-specific content, although their ability to provide 'quality' programming varied. For the British Broadcasting Corporation (BBC), in the years following the Second World War, television was very much seen as a public good. Children were young citizens for whom a wide range of appropriately uplifting content should be provided that mirrored provision for adults as public service broadcasting 'in miniature' (Buckingham et al. 1999, 17). As newer, free-to-air commercial providers such as ITV emerged in the UK in the 1950s, these channels were also given public service obligations relating to children, because of their access to what was then scarce channel spectrum (Steemers 2010). Without this kind of regulatory intervention, the cost of domestically produced children's drama, compared with cheaper (generally US) imports, would have been unattractive to most commercial networks, because it was not attractive to advertisers.

In the US, however, a country without publicly funded media, television has always been highly commercialized and children tend to be seen primarily as consumers of advertising and program merchandise, rather than as citizens in the making (Pecora 1998). Although the not-for-profit Children's Television Workshop (later Sesame Workshop) enjoyed great success with the preschool educational series *Sesame Street* (1969–) which originally aired first until 2015 on the

US Public Broadcasting Service (PBS), PBS's fragile funding structure means it has never been able to compete with commercial providers in the US. Animation associated with toy brands and other merchandise remained the dominant genre available to US children, content which also had a competitive advantage in markets outside the US because it had covered its costs domestically. By the mid-1990s, some PSB children's shows like the UK's *Teletubbies* and Australia's *Bananas in Pyjamas* had become lucrative vehicles for international merchandising, but generally the largely European PSB model eschewed the overt commercialism of US companies.

Children's television, like screen media for adults, was transformed in the 1990s, when digitization and the end of spectrum scarcity led to a rapid proliferation in television channels. A worldwide trend toward the deregulation and privatization of media industries and the liberalization of the regulatory regimes in which they operated helped accelerate the pace of globalization (Cunningham and Jacka 1996). A series of mergers in media industries rendered the Disney Corporation the second largest media company in the world by 1995; this period also saw the emergence and expansion of US-based subscription channels including The Disney Channel, Nickelodeon and Cartoon Network. These children's services from US corporations moved rapidly into new international markets opened up by deregulation and advances in technology (Pecora and Lustyik 2011). They also benefited from the programming libraries of their parent companies, economies of scale in production and distribution, and established strategies of content monetization, including the repurposing of content for different platforms (Potter 2011).

Recognizing the importance of retaining the child audience at a time of fragmentation and transformation, PSBs also took advantage of the end of spectrum scarcity to launch their own children's channels. German broadcasters ARD and ZDF introduced KinderKanal (KiKa) in 1997. The BBC introduced CBBC and preschool channel CBeebies to UK audiences in 2002, while Australia's ABC launched the country's first free-to-air children's channel in 2009. In the mid-2000s, television and screen industries faced further disruption with the establishment of online television streaming services such as Netflix, Hulu and YouTube, all of which distributed children's content. Platform-agnostic child audiences quickly became accustomed to viewing screen content on demand and on multiple devices, including mobile phones and tablets. As market power became increasingly concentrated in the US, domestic screen industries in Europe and Australia faced further disruption and difficulties in funding local children's content.

## Transformations in Consumption of Children's Television

As early adapters of technology, children have been quick to embrace streamed television services on laptops, tablets and mobile phones. However, when considering transformations in the consumption of children's television, a key factor is what exactly is meant by 'the child audience'? This is important because industry and society conceptualizations of childhood are constructed (Messenger Davies 2010, 7) and do not necessarily align. Yet industry conceptualizations of what constitutes the child audience are important in determining to what extent resources (such as time, space and financial investment) are allocated to children's content.

In earlier times, when the distinction between childhood and adulthood seemed more certain, the BBC Children's Department, for example, used to cater for children and young people up to 16, as did many other broadcasters. Broadcasters with limited children's slots catered for both younger and older children, and scheduling was determined by their availability to view. Preschoolers might enjoy programming in the morning or at lunchtime, and older children enjoyed programming for their age group after school or on Saturday mornings. Children had to satisfy themselves, however, with content that was decided and rationed by adults and confined to particular time slots.

The United Nations defines a child as anyone under 18 (UN 1989), yet most linear broadcasters, both public service and commercial providers, rarely cater for children over 12. Within broadcasting, conscious decisions have been made to reduce provision for older children, who are thought to be difficult to reach. Mainstream broadcasters do target older children as part of the wider and more profitable aggregated family audience for shows such as the children's talent show format *Little Big Shots* (2016–), which originated on US network NBC, or the BBC science fiction series, *Doctor Who* (1963–). But in general, drama and factual programming for older children have increasingly become the preserve of a few PSBs (D'Arma and Steemers 2012; Ofcom 2007, 2018) with many commercially funded generalist broadcasters, including those in Australia, Denmark, New Zealand, the US, the UK and Ireland retreating almost entirely from commissioning content for this demographic because it makes no commercial sense to cater for teenagers. Where regulators have sought to persuade broadcasters to cater for an older child audience, success has been limited. For example, UK commercially funded PSB, Channel 4, is supposed to serve older children, but has continually not met its statutory requirements in spite of annual rebukes by the regulator Ofcom (Ofcom 2016).

In recent years, the most lucrative section of the children's market has been preschool children aged five and under. This was not always the case, but the rapid proliferation of dedicated preschool channels made this age group commercially attractive. Preschool children and their parents have been enthusiastic consumers of merchandise related to top shows such as *Dora the Explorer, Paw Patrol, Teletubbies* and *Peppa Pig.* These revenues proved crucially important to the funding strategies of a limited number of shows, given advertising around preschool shows is often restricted. Yet there have been fewer of these global hits in recent years as younger children, like their older counterparts, shift their interest to other activities and embrace watching screen content on new platforms. As preschool viewing also fragments across many children's broadcast channels and thousands of YouTube offerings, it has become more difficult to sustain new shows and build the larger scale of audiences necessary to support global campaigns in licensed merchandise. With fewer global hits, the market for licensed merchandise has become dominated by a small raft of transnational companies – Disney, Turner, Lego – with deeper resources to sustain development, access their own distribution outlets and create global brands.

Children and young people still watch television shows, but their viewing habits are changing with a shift away from the TV set. In the UK, for example, viewing by children aged four to 15 of broadcast TV on a TV set fell seven hours a week to 10 hours between 2011 and 2017 (Ofcom 2018, 9). The amount of time spent by UK children aged five to 15 watching TV on a TV was surpassed in 2016 by the time children spend online at 15 hours a week (Ofcom 2016, 5). Children's increasing engagement with tablets and smartphones has not decreased overall viewing, which is shifting to online platforms. Nonetheless, these changes in children's consumption of screen content are having an impact on the underlying economics, resource allocation and sustainability of professionally produced content for children.

Contemporary children's viewing habits present challenges for traditional channels which feel impelled to respond or risk losing current child audiences, as well as future audiences accustomed to viewing screen content online, on demand and on multiple devices. For traditional channel providers, decisions need to be made about new strategies for the online consumption of their offerings that reflect how children are accessing and engaging with content. This means looking beyond linear programming and traditional distribution methods. For example, in 2017 the BBC took the decision to reallocate over 25 percent of its children's budget by 2019/20 to online content. By creating the tools that will allow it to reach child audiences via streaming and apps, with 'new forms of content and interactivity' (BBC 2017a) the BBC hopes children will engage with content in a more personalized way, as they do on social media. For television channels, this strategy represents a costly undertaking, but without responding they risk losing large parts of their audience to the short-form videos that attract millions of children on YouTube.

## Transformations in Distribution

As the internet became an increasingly popular, accessible and cost-effective means of distributing and consuming children's content in the mid-2000s, linear broadcasters and pay-TV providers have quickly established catch up, on-demand services. At the same time, the internet facilitated the swift emergence and global expansion of SVOD and OTT services, which compete aggressively with national free-to-air and pay-TV systems. As the new streaming services enthusiastically pursued the child audience, many free-to-air advertising-funded networks continued to reduce their investment in domestic children's content, because of what they perceived as disruption to their business models caused by the transition to digital regimes, and the concomitant loss of advertising revenues. Determined to maintain its market dominance in the face of these new streaming services, in 2014, Disney purchased YouTube's Maker Studios and is due to launch its own SVOD service in 2019.

By 2016 Netflix was available in 190 countries and had more than 75 million subscribers around the world, although notably China, the world's largest market, alongside North Korea, Syria and the Crimea, remain beyond its reach. Recognizing the importance of children for attracting families to their services, and determined to build brand loyalty at every stage of their subscriber life cycle, these streaming television services have made a sustained global play for the child audience. In 2017 Netflix promised to commit $8 billion to original production, including children's content (Whittingham 2017). That same year, at Mip Junior, the international market for children's content held annually in Cannes, France, Netflix announced more investment into children's content, building on 37 original series it had produced to date (Littleton 2017). But Netflix rarely elaborates on the size of its child audience, and its investments are aimed at global rather than local audiences.

Nevertheless, Netflix's evolution from a mail order DVD rental business in 1997 to an online streaming service 20 years later has provided some new funding for children's screen content. Recognizing its own inexperience in children's content production, it has embraced co-production arrangements with established PSB rivals, including the BBC, Germany's ZDF and the ABC in Australia, in order to secure international rights, and PSB production expertise. The SVOD's first British children's originals were live action dramas and included a remake of *The Worst Witch*, a co-production with the BBC and ZDF in 2016. Netflix has also commissioned animation, such as the Australian CGI-animation series *Bottersnikes and Gumbles* (2015) and a remake of *Watership Down* (2017), both produced with BBC involvement. For Netflix, co-producing content with trusted broadcasters such as the BBC has two advantages. First, it establishes a connection with the BBC brand for quality and excellence in programming. It also allows the SVOD access to development, production and editorial expertise developed over decades by the PSB with established producers.

Another SVOD, Amazon Prime, available in 200 countries in 2017, has replicated Netflix's strategy of investing in original content to attract subscribers to its service, including content made for children. Although Amazon's exact subscriber base is not known, the company uses its Prime Entertainment offerings to attract audiences to its main retail site, and children's TV is one of its biggest growth areas (Landau 2016). In mid-2016 Amazon Prime launched its first three original children's programs, made by its Amazon Studios division; in 2017, its original commissions slate included the animated series *Ronja, the Robber's Daughter*, produced by prestigious Japanese animation house Studio Ghibli. In order to increase its educational offerings, Amazon Prime also has exclusive rights to a range of PBS programs from the US. In addition to their investments in original content, SVODs such as Amazon Prime, Netflix and Hulu also provide additional licensing opportunities for content that has already aired on other outlets, and are willing to pay competitive licence fees to ensure access. The refusal of these SVODs to release viewing data can be frustrating for producers, however, for whom ratings are a key indicator of the success of their content.

In contrast to Netflix and other SVODs, YouTube, which is wholly owned by Google, invests little in original content, relying instead on amateur and professional content creators for much of its children's offerings. Children all over the world have embraced YouTube as an alternative source of content; in 2016, 35 of the 100 popular YouTube channels were those aimed at young audiences. Between them these channels attracted 8.6bn views in 2016 (Dredge 2016). User-generated content targeting children includes toy unboxing channels, nursery rhyme channels, vlogging, how-to instructional videos, and gaming channels (Minecraft), in addition to animation, mini-films, songs and other short-form content. Successful YouTube channels and their stars can make considerable amounts of money from the advertising revenues associated with their offerings (controversial Swedish vlogger PewdiePie made an estimated $7m in 2016). For many traditional content producers of animation and drama, however, YouTube is a means of reaching audiences without necessarily generating actual income from their content.

When YouTube was first introduced in 2005 it was branded as a site for the posting and sharing of user-generated (or amateur-produced) content. However, its need to generate advertising revenues has seen the OTT service increase the range and breadth of its professionally produced content (Tryon 2015). Clearly aware of the importance of children to its advertising revenues, in 2015, YouTube introduced a children's mobile app, YouTube Kids, funded by paid advertising and largely unregulated by national bodies. Marketed as a 'walled garden' for children, it attracted public opprobrium for its overt commercialism. Beyond the children's app, YouTube also offers branded channels from toy manufacturers such as Fisher Price and Lego, and fast food provider McDonalds. Growing concerns around YouTube content include unofficial, inappropriate and disturbing representations of popular characters like Peppa Pig or Elsa from *Frozen* (BBC 2017b; Bridle 2017), as well as the lack of transparency about how children's data are collected and used for targeted advertising on the main YouTube platform, which is more popular with children than YouTube Kids (Wired 2018).

## Transformations in Production

Changes in consumption and distribution have inevitably impacted the production ecology of children's television, at both the national and transnational levels. This has been most keenly felt at the level of funding. The funding of children's broadcast television has always been precarious because of the limited revenue possibilities from advertising to a small target audience with limited purchasing power. This is compounded by restrictions in many countries about what can be advertised (with bans on certain products such as junk food or, indeed, complete advertising bans in some countries) and when. The international availability of low-cost animation series produced over many years for North American and Asian markets has further diminished the incentive to invest in local productions. The availability of vast amounts of content, which can be acquired below production cost and then cheaply dubbed and repeated year after year for successive generations of children, has made it particularly difficult to sustain production in smaller or less affluent markets. As a market failure 'genre', therefore, children's television production has often been supported in many countries by positive regulatory interventions, which help sustain domestically produced content (Steemers 2017), situating children in their own cultural context, and contributing to their sense of identity and citizenship. Empirical evidence about the cultural and educational value of domestically produced children's television has been hard to gather, difficulties compounded by the complex way in which children now consume services on different platforms and devices, generating data that is proprietary and not open to scrutiny. This, in turn, makes it difficult to understand the role and value of public service media and domestically produced content for children (Livingstone and Local 2017, 75). Even if there is only limited evidence that watching domestic content is more beneficial for children than watching imports, what matters in policy terms is the extent to which policy makers believe this to be the case and whether they are prepared to intervene at a national level (Steemers 2017).

First and foremost, of these support measures has been public service broadcasting, whose remits usually contain a commitment to serve children. However, while some PSBs in Scandinavia, Belgium, the Netherlands, the UK and Germany are recognized for their commitment to children, others (in Australia, New Zealand and Canada) have struggled to justify public support for children both within their organizations and from governments (Potter 2011; Zanker 2017). In other countries – particularly in eastern and southern Europe – PSBs have not always catered to the needs of children because of poor funding and limited political support (D'Arma and Labio 2017; Lustyik 2013). Other initiatives to support production for children have included transmission and production quotas (Australia, Canada and France), levies on commercial broadcaster revenues (France, Canada), expenditure obligations (France, Canada), independent contestable funds (UK, Denmark, Ireland and New Zealand) and a host of tax incentives designed to attract international investment (Steemers and Awan 2016).

Production in its traditional form for linear channels has been handicapped by the small number of players that actually commission original first-run content for children. Transnational commercial providers such as Disney, Nickelodeon and Turner commission very little from local producers outside the US, relying instead on their tried and tested franchises which support other commercial activities and ancillary products (D'Arma and Steemers 2012). The emergence of SVODs such as Netflix and Amazon does not alter this situation, because they too are primarily commissioning for a global audience. Their input into commissioning, even if it is higher budget, does not plug the commissioning hole left by decreased commissions from both transnational providers and national broadcasters (either commercial or public service). The decline in funding for locally produced content and the resulting decline in commissions has led to a decline in the number of European, North American and Australian production companies that produce for children.

Production of children's content can be characterized by three tiers of producers, with the production enterprises of large corporations such as Disney, Turner and Nickelodeon at the top. They have produced primarily for the US market, but have outsourced development and production to other suppliers in North America and, increasingly, overseas (Europe and Asia) to reduce costs and to benefit from tax incentives. Production by national broadcasters (either public service or commercial) is in decline and these too have outsourced largely to independent producers, but with reduced budgets for children's content. For example, in 2017 PSB RTE in Ireland closed down its in-house production capabilities; and the BBC since its new Charter in 2016, has opened up all its children's commissions to competition from independent producers. A mid-level tier of producer-distributors has survived, supplying content to US transnationals, national broadcasters (both private and public) and increasingly to the SVODs. However, online distribution for more high-end productions on platforms such as YouTube does not generate sufficient revenues to cover production costs. For these producers, a careful calculation needs to be made between the extent to which they develop their own properties (and hold onto valuable IP to grow their businesses) and the extent to which they work for hire. Yet, the recent history of medium-sized children's production companies is littered with companies that grew quickly on the basis of one or two large hits, usually in the preschool realm, before crashing under financial strain before they were absorbed by other larger players. Recent examples include the UK's

HIT Entertainment (*Bob the Builder, Thomas the Tank Engine*) which was acquired by US toy company Mattel in 2012; Entertainment Rights (*Postman Pat*) which went into administration in 2009, and whose catalogue was subsumed into Classic Media, before it was acquired by US-based Dreamworks in 2012, itself absorbed by NBC Universal in 2016; TV Loonland declared insolvent in Germany in 2009; and Chorion (*Octonauts, Noddy, The World of Peter Rabbit*) whose assets were sold off in 2011 to meet debts. In autumn 2017, the high risks associated with developing children's properties, combined with limited returns from broadcast sales and uncertain revenues from ancillary rights, has made production a precarious business. For a sustainable model of production, companies need to be able to balance production of their own IP with work for hire, while also securing subsidies from a range of different tax incentive schemes.

However, while traditional producers working within a broadcast funding model are struggling, YouTube has generated many alternative providers and channels such as Funtoys Collector, Little Baby Bum, and Russian animation channel Masha and the Bear. These are frequently the most watched channels on YouTube. Channels like Little Baby Bum, which upload vast amounts of low-cost animation every week, offer a business model that differs markedly from the broadcast model. Often focused on songs and nursery rhymes and aimed at a younger audience, Little Baby Bum's compilation productions, which work well on tablets and mobile phones, are made on small budgets and dubbed for multiple markets. Production and dubbing are done on a work-for-hire basis, with animation outsourced to Asia, and on a more accelerated time-scale than broadcast quality productions. The expectation of income generated from licensed merchandise does not form the core of revenues, which instead are generated by views and advertising, including for products which might not be permitted on broadcast television. For producers who have relied previously on broadcast exposure for their high-end content, exposure on YouTube is not an option, because it impedes future sales to broadcasters and SVODs, who want exclusivity.

The apparent success of these approaches suggests that internet-distributed television services such as YouTube and Netflix are going to continue to upend longstanding production funding and distribution arrangements for children's television. But despite their growing popularity, these services sit outside national regulatory frameworks, which raises concerns, especially among parents, about the type of content children can access on them. In contrast, parental trust in PSB brands like the BBC remains high. Appropriately funded PSBs are likely to become even more important to the future of domestically produced children's content in increasingly disrupted and transforming television production ecologies, particularly culturally specific content that reflects children's daily lives back to them.

## Conclusion

Two recent television studies books, Spigel and Olsson's *Television After TV* (2004) and Turner and Tay's *Television Studies After TV* (2009) both ask the questions 'what was television/studies, what is television/studies, and what will television/studies become in the future'? And yet none of the contributions in either book deals in any depth with one of television's most important audiences and interest groups: children. This is surprising given that children's content and the child audience remain significant policy objects in many countries. It is also surprising given that, to a great extent, children today will determine the norms and settings of production, distribution and consumption of television and television-like services in the future. As the post-broadcast landscape develops with online and on-demand services becoming established in the space formerly known as television, this chapter has shown how many different forms of television are addressing, representing, engaging with or ignoring children and adolescents in North America, Europe and Australia.

There is no doubt that children's television in these countries, like television for adults, is undergoing a period of rapid technological, regulatory and economic change, transforming the ways in which content is produced, distributed and experienced by child audiences. The turbulence is disturbing for the industry because it undermines funding models that have always been fragile, particularly for domestically produced content for small child audiences. Despite the perceived value of locally produced television by policy makers and children's advocacy groups that reflects children's cultural context back to them (Messenger Davies and Thornham 2007), the production and distribution of broadcast quality children's television is now an increasingly globalized and complex business involving myriad funders, subsidies, tax incentives and a host of ancillary rights.

The pre-eminence of US-owned channel brands such as Disney, Nickelodeon and Cartoon Network *is* being challenged by new market entrants. At one end of the spectrum these are high-end SVOD services such as Netflix and Amazon Prime who are investing in internationally attractive original children's content to attract subscribers. At the other end of the spectrum there are YouTube channels with millions of viewers for low-cost, high-volume content. Some professional producers are adapting to these new production ecologies, finding new ways to fund content and reach audiences. Very large transnational companies co-exist alongside niche providers of children's content, who appeal directly to child audiences via YouTube and social media.

These transformations in children's screen content have significant implications both for children's media industries and for the largely national policy and funding instruments intended to support locally produced content for child audiences. The new situation raises questions about whether it matters that domestic children's television industries are struggling, if children themselves are actively seeking out viewing possibilities that they want, engaging with content that offers them interactive possibilities and, in some but not all cases, also making their own content and stories.

The answers to these questions are not yet evident, but there is certainly scope for more and better research of industry developments that goes beyond the remit of this chapter. One avenue is the extent to which some industry players are beginning to work *with* rather than *for* children and young people to produce content that better reflects their lives and aspirations, rather than adult and public service perceptions of what they need. Another underexplored avenue is the role of children's screen industries in emerging markets, such as India and China, which are generating large amounts of computer-generated content for online distribution internationally. In this respect. this brief chapter remains somewhat unbalanced. As a starting point, it has set out to outline the challenges facing children's television industries in markets with public service broadcasting, starting with the short-lived baseline of traditional post-war linear television, before assessing the separate challenges in respect of consumption, distribution and production. Given the short history of children's television and the rapid pace of change, many questions remain unanswered, underpinning the increasing need to track these transformations, and their implications in a wider range of countries, whose children's screen industries merit further consideration.

## References

BBC. (2017b) "BBC announces biggest investment in children's content and services for a generation." Accessed November 2, 2017. www.bbc.co.uk/mediacentre/latestnews/2017/bbc-investment-childrens

BBC (British Broadcasting Corporation). (2017a) "The disturbing YouTube videos that are tricking children." 27 March. Accessed August 2, 2018. www.bbc.com/news/blogs-trending-39381889.

Bridle, J. (2017, November 6) "Something is wrong on the Internet." *Medium.* https://medium.com/@jamesbridle/something-is-wrong-on-the-internet-c39c471271d2

Buckingham, D. (2005) "A special audience? Children and television," in J. Wasko (ed.) *A companion to television*, Blackwell: London. pp. 468–486.

Buckingham, D. (2009) *The impact of the commercial world on children's wellbeing*. London: Department for Children, Schools and Families & Department for Culture, Media and Sport.

Buckingham, D., Davies, H., Jones, K. and Kelley, P. (1999) *Children's television in Britain*. London: British Film Institute.

Craig, D. and Cunningham, S. (2017) "Toy unboxing: living in a (n unregulated) material world," *Media International Australia* 163 (1), pp. 77–86.

Cunningham, S. and Jacka, E. (1996) *Australian television and international mediascapes*. Melbourne: Cambridge University Press.

D'Arma, A. and Labio, A. (2017) "Making a difference? Public service broadcasting, distinctiveness and children's provision in Italy and Spain," *International Journal of Television* 8 (2), pp. 183–199.

D'Arma, A. and Steemers, J. (2012) "Localisation strategies of US-owned children's television networks in five European markets," *Journal of Children and Media* 6 (2), pp. 147–163.

Dredge, S. (2016) "YouTube channels benefit as kids look beyond traditional TV." Accessed 2 November 2017 www.theguardian.com/media/2016/jul/11/youtube-channels-kids-tv-minecraft-gamers-vloggers

Landau, N. (2016) *TV outside the box: trailblazing in the digital revolution*. New York: Focal Press.

Littleton, C. (2017) "Netflix to boost international kids content as viewership spikes." *Variety* 15 October. http://variety.com/2017/tv/news/netflix-kids-guillermo-del-toro-trollhunters-andy-yeatman-1202590366/ Accessed 2 November 2017.

Livingstone, S. and Local, C. (2017) "Measurement matters: difficulties in defining and measuring children's television viewing in a changing media landscape," *Media International Australia* 161 (1), pp. 67–76.

Lustyik, K. (2013) "From a socialist endeavour to a commercial enterprise: children's television in east-central Europe," in A. Imre, T. Havens and K. Lustyik (eds.) Popular *television in Eastern Europe during and since socialism*, London: Routledge. pp. 105–122.

Messenger Davies, M. (2010) *Children, media and culture*. Maidenhead: Open University Press.

Messenger Davies, M. and Thornham, H. (2007) *Academic literature review: 9. The future of children's television programming*. London: Ofcom.

Ofcom. (2007) *The future of children's television programming*. London: Ofcom.

Ofcom. (2016) *Children and parents: media use and attitudes report*. London: Ofcom, UK. www.ofcom.org.uk/__data/assets/pdf_file/0034/93976/Children-Parents-Media-Use-Attitudes-Report-2016.pdf.

Ofcom. (2017) *Children and parents: media use and attitudes report*. London: Ofcom. www.ofcom.org.uk/__data/assets/pdf_file/0020/108182/children-parents-media-use-attitudes-2017.pdf.

Ofcom. (2018, 24 July) *Children's content review: update*. London: Ofcom.

Pecora, N. (1998) *The business of children's entertainment*. New York: Guilford Press.

Pecora, N. and Lustyik, K. (2011) "Media regulation and the international expansion of Nickelodeon," *Journal of Children and the Media* 5 (1), pp. 4–19.

Potter, A. (2011) "It's a small world after all: New media constellations and Disney's rising star – the global success of High School Musical," *International Journal of Cultural Studies* 15 (2), pp. 117–130.

Potter, A. and Steemers, J. (2017) "Children's television in transition: policies, platforms and production," *Media International Australia* 163 (1), pp. 6–12.

Spigel, L. and Olssen, J. (eds.) (2004). *Television after TV*. Durham and London: Duke University Press.

Steemers, J. (2010) *Creating preschool television*. Basingstoke: Palgrave Macmillan.

Steemers, J. (2017) "Public service broadcasting, children's television, and market failure: the case of the United Kingdom," *International Journal on Media Management* 19 (4), pp. 299–314.

Steemers, J. and Awan, F. (2016) *Policy solutions and international perspectives on the funding of public service media content for children*. London: CAMRI.

Tryon, C. (2015) "TV got better: Netflix's original programming strategies and binge viewing," *Media Industries Journal* 2 (2), pp. 104–116.

Turner, G. and Tay, J. (2009) *Television studies after TV*. Abingdon: Routledge.

Unicef. (2011, November) *Communicating with children*. www.unicef.org/cwc/files/CwC_Final_Nov-2011.pdf

United Nations. (1989) *The United Nations Convention on the Rights of the Child*. Article 1 - Available here https://www.unicef.org.uk/what-we-do/un-convention-child-rights/

Westcott, T. (2011) "An overview of the global animation industry," *Creative Industries Journal* 3 (3), pp. 253–259.

Whittingham. (2017) "Netflix plans $8bn content spend in 2018." www.c21media.net/netflix-plans-8bn-content-spend-in-2018/17October

Wired. (2018) "Groups allege YouTube is violating law that protects kids." *Wired*, 9 April. www.wired.com/story/groups-allege-youtube-is-violating-law-that-protects-kids/.

Zanker, R. (2017) "The future isn't coming; the future is here: The New Zealand Children's Screen Trust's engagement with media policy for children," *Media International Australia* 163 (1), pp. 55–66.

# 12
# UNDERSTANDING AUDIENCES
## Television Publics as "Cultural Indicators"

*Andy Ruddock*

### Introduction: The Contemporary Relevance of Television Audience Studies

When future historians investigate Australian thought about terrorism in the 21st century, many of the people that they meet will be television audiences. An archive of free-to-air news broadcasts, *TV Online*, hosted at the Royal Melbourne Institute of Technology, holds stories about public opinion on extremism, broadcast in pivotal years such as 2014. In that year, a gunman claiming an ISIS affiliation murdered two hostages during a siege in a Sydney coffee shop. Carried live on TV, the siege and its aftermath set the scene for Australians to react to domestic terror onscreen. Their words, on this and other terror themes at this time, are now matters of historical record, thanks to the RMIT archive. The resource is historically significant because in the early 21st century, Australians continued to get their news from free-to-air TV in large numbers. Ratings from Sunday 29 July 2018, for example, showed the following: the most popular TV show in the country, including those for subscription stations, was Channel 7's free-to-air nightly news, which drew 1.2 million viewers. Free-to-air news programming accounted for ten of the top 20 shows on that day. The lowest rating of these, 7's *Weekend Sunrise*, still drew more viewers than the most popular subscription program. When Fox Sport's screening of a live Rugby game drew 240k viewers, the largest audience for any pay channel on that day, it still failed to match the 253k scored by 7's breakfast news show. What all of this means is that news reports from *TV Online* record what Australians are most likely to have heard about terrorism in the early 21st century, and what they are most likely to have seen people like them say about terror – its causes, its relevance, and its policy implications – when queried by journalists.

This chapter uses these data to consider how television audiences are involved in writing political history. Its aim is to show that TV audience research creates cultural histories of media power. Using Gerbner's "cultural indicators" concept (1973), this chapter will argue that "audience" captures TV's role in creating, sharing and recording political consciousness. Whatever its future, television's capacity to attract publics as audiences will shape historical memory. I make the case by studying how ordinary Australians made news about terror in the years 2014–2016, using the RMIT archive as a case study source.

The argument has the following structure. First, I outline idealised scholarly accounts on the relationship between television and political participation. Second, I describe how scholars view participating audiences as integral to strategy and governance in TV industries. Third, I explain why these developments suggest the value of understanding audiences as "cultural indicators" (Gerbner, 1973) of changing media practices. Finally, I deploy archive data to show how this method of interpreting audiences exemplifies television's enduring political significance.

## Historical Perspectives: Audiences, Publics, Politics

### *Audience Power*

TV audience research studies how societies create, share and normalise specific ways of understanding the world. One important theme, here, has been how television invites its viewers into the process. So, let us begin by outlining how media researchers have explained relations between audience, publics and politics.

For some scholars, audience research illustrated how ordinary people used television to think and talk their way into history. Audience research in the late 1980s considered television's efficacy as a medium of public views and aspirations. The operating question was; how did society become more participatory when people acted as audiences (Allor, 1988; Ang, 1989; Bird, 1988; Grossberg, 1988)? John Fiske's essay "Meaningful Moments" (1988) made this point especially clearly. Fiske defined the purpose of TV audience studies as "understand-(ing) concrete, contextualised moments of semiosis as specific instances of more general cultural processes" (1988, 250). "Semiosis" meant the blending of television content and audience action into processes of social communication. TV was an especially powerful medium of social communication because it invited the "collapse" of the distinction between text and audience; the moment of "semiosis" where viewers turned content into meanings, conceiving television and stories as resources for everyday thinking. Scholarship began to concur that, at their best, therefore, television audiences were creators of culture (Carey, 1989; Hall, 1980); television was a social technology that let ordinary people register what were ultimately politicised sentiments publicly as audiences.

One influential example comes from Henry Jenkins' early work on "slash" literature (1992) which demonstrates the potential of such semiosis. Famously, Jenkins discovered that the TV series *Star Trek* inspired its fans to produce volumes of written fiction. A familiar theme in this oeuvre concerned a romantic relationship between male leads, Kirk and Spock. This fan practice had political repercussions. In a period when gay TV characters were either absent or pathologised, the creation and circulation of stories about healthy same-sex relationships were a significant contribution to public discourses about sexuality. Slash registered audiences who viewed sexuality in broader terms than mainstream popular culture allowed. Gay fans enjoyed acceptance and community denied to them elsewhere. Since then, academics have scrutinised slash as an exemplar of semiosis. The "repositioning texts outside the borders of heteronormativity" (Dhaenens et al., 2008, 335) has become a synecdoche for the power of audiences. In particular, the rise of the internet and social media has given fans of cult TV (Jones, 2004), science fiction (Scodari, 2003) and sport (Gong, 2017) the capacity to rewrite television content to tell different stories about gender and sexuality (Skains, 2010). This is tremendously important, given global concerns about sexuality and human rights.

"murtagh899" (2012) provided an example of this activity in her/his YouTube video *One Thing Slash Fandom*. The video remixes scenes from hit TV shows, such as *Merlin*, *Supernatural* and *Sherlock*, to represent romantic relationships between their male protagonists, accompanied by the song *One Thing* from British boyband *One Direction*. The producer describes her/his motivations in a way that's worth considering:

> I really did this just for fun. While having my sister convert the clips and gifs for me it turned out sorta like this and then I just fooled around with it, so I hope that people will like it anyway, even if it doesn't make much sense. lol. I guess the idea was for when if the characters ever met what would they act like around each other and if friendship or something more would happen.
>
> (murtagh899, 2012)

It's the playful tone here that captures Fiske's view on the political import of TV audiences. "murtagh899" suggests that "repositioning texts outside heteronormativity" is a routine pleasure, energised by the addition of social media to the imagination that has always been TV's core appeal. Stories about same-sex relations are a handy vehicle for having fun by playing around with TV using digital tools – which rather suggests that TV fans have succeeded in embracing same-sex relations in culture's language. There's something deeply political about the notion that one can use TV to write slash stories without any firm political intention in mind. The concept that this is an unremarkable thing to do has tremendous significance in a world pressing for marriage equality, for instance. This is one example of what audience-driven semiosis looks like, and also why it remains a relevant concept, even given the digital turn, the rise of social media, and the development of an array of new media habits.

## TV Power

But TV audience research also remains pertinent because the question of who benefits from popular creativity remains vexed. Global media industries count on viewers who act. Television drama thrives not just because of a new focus on quality for the industry, but because online forums allow more intense engagement with content that affords the chance to socialise with others (Simons, 2015). The rise of multi-screen practices, where people use TV with other devices, enhances the potential for semiosis, as people convert television narratives into social conversations, and even make content, through doing things like tweeting directly to live programming (Agirre et al., 2016; Gómez Aguilar et al., 2015; Kim et al., 2016). But this carries two different effects on public discourse. Certainly sharing, editing and rewriting TV narratives registers viewers' voices in history. But it also secures the power of the medium; TV creates a world where the quickest and easiest way to engage with politics is by becoming an audience. Despite a vastly changing media ecology, TV's "mustering" power remains the hub of debates on media and democracy (Livingstone, 2015).

Bolin (2012) argued that TV remains socially significant because viewers work for themselves, each other, and media industries at the same time. Succinctly, audiences count because they produce value for television and society. People like murtagh899 help TV with free labour. Which doesn't stop it from being a resource in the reproduction of non-heteronormative stories. We are speaking about two modes of cultural production, which have always coexisted (Bolin, 2012). The key, then, is not to define audiences once and for all. It is more useful to appreciate how different definitions of the phenomenon reflect television's multiple effects on the production and circulation of political meanings.

All of this is to say that there remains considerable contemporary political currency in scrutinising the idea that "audience" is itself a television story that affects cultural production and consumption. Late 1980s debates on the problems of defining and studying audiences have taken on a new political energy in the Big Data age. In the 1980s, Ang (1989) explained how sophisticated rating tools failed to capture the cultural significance of audience. But more recently, Bourdon and Cécile (2015) have argued that these insights have had little political impact, since "data become a social metrology that speaks for the television audience, seen as a synecdoche for the entire population" (2255). Looking at France, Israel and Germany, the authors note these states read television audiences as political publics, assuming close connections between entertainment and political tastes. Governments read ratings as cultural indicators of deep-seated political preferences. Audience metrics are real, insofar as they ground plausible stories connecting audiences, publics and political policy. Taken together, Bolin and Bourdon and Meadel's thoughts lend contemporary relevance to historical work on defining audiences. Writing plausible *stories* about television viewers – who they are, and what they want – remains a vital political skill.

At the same time, this offers new methodological possibilities in contemplating the relationship between TV, people and cultural history. Audiences are now content too. When Turner (2010) coined the phrase "the demotic turn", he referred to television's reliance on ordinary people as content providers, through new genres such as reality TV. His concept is vital to grasping the links between television, audiences and history, that I am outlining here. Turner described a situation where TV effortlessly swallows its viewers, well versed in the medium's central message. Public desires to be on screen respond to the conviction that celebrity offers a quick and easy way to self-fulfilment. Many suspect that lives performed for media are the only ones worth living. Thus, although we see ordinary people making content on our screens, these actions are scripted by stories, and the people we see are in this sense responding audiences, clamouring to be part of the "media centre" (Couldrey, 2003).

These developments lend TV audience research a new historical gravity. When future historians seek evidence of public sentiments during times of crisis, many will turn to digital TV archives. There, they will find audiences talking and talk about audiences. They will discover, in short, competing audience stories, closely aligned with the political issues of the period that they are researching.

In illustration, the chapter now turns to an analysis of an archive of Australian Television news. What follows explains how news programming about terror drew audiences into their production at a significant moment in Australian media history. The period 2014–2016 witnessed a series of heavily televised terror attacks both abroad and in Australia, that prompted considerable concern about media coverage. Such concern was expressed both in and through television itself. For example, when morning TV host Sonia Kruger voiced her desire for a Muslim immigration ban, due to personal fears about terror, commenters fretted that such interventions could stoke Islamophobia. Around this time, then, Australia faced a political challenge created by the concern that television had, among other things, created an audience for Islamophobia. Speaking on the ABC, journalist David Marr claimed that televised appearances by Kruger and other figures such as One Nation Senator Pauline Hansen misrepresented public levels of fear about terror (The reshuffle on Monday, 2016). We can say, then, that in this period, Australian political discourse was littered with myths about audiences; the question of what Australians really thought about terrorism was pursued as a matter of considerable political capital. Against this history, it is worth asking; when Australians wanted to know what their peers thought about terrorism, what were they told when they turned to TV for answers?

## Method

The method used here draws on the idea that the public's onscreen presence is a "cultural indicator"; an index of TV's changing relationship with publics and societies. "Cultural Indicators" is a phrase coined by George Gerbner (1970, 1973), who pioneered work on television's role in political socialisation during the broadcast era in the United States. Gerbner's method for researching the enormous political effects of small screens was to identify common symbols within the medium, then to compare the political views of "heavy" versus "light" viewers, to find the outcomes of exposure. For instance, analyses of prime time American TV during the 1970s showed that images of violence against women, people of colour, children and the elderly riddled entertainment. The more exposed audiences were to these tales, the more likely they were to believe that the world was a dangerous place where others could not be trusted. Gerbner argued that any research on TV's impact on audiences had to begin with a systematic understanding of the medium's message. Hence: "cultural indicators would be the periodic analysis of trends in the composition and structure of message systems cultivating conceptions of life relevant to socialization and public policy" (Gerbner, 1970, 69). While this definition targeted the broadcast era, it's even more apposite to an age where audiences are therefore also onscreen, a literal part of the programming message. The cultural indicators idea becomes even more evocative in light of the recent TV audience research mentioned earlier. Turner (2010) argued that the routine inclusion of ordinary people in television programming represented a major structural content shift. Talk about audiences, and the desire to harness their energy to media production, are also vital elements in understanding how TV works today. Hence the methodological conclusion; there is value in studying how audiences appear in the news on momentous political subjects, such as terrorism, as leading cultural indicators of public readings of those topics.

What, then, might our future historian actually find, when delving into Australian TV archives? Putting all this into action, the TV Online database was searched using the terms "terrorism" and "public opinion", with results limited to 2014–2016. This period saw landmark examples of domestic televised terrorist events in Australia. The search returned 25 stories. Naturally, this is a small sample, but it can be taken as indicative of what future historians are likely to find when they look at what Australians thought about terror and extremism. In that regard, it is significant that the analysis found diverse issues that nevertheless conformed to a significant story about television audiences.

The stories were loaded into the qualitative data analysis software package Nvivo. They were transcribed and then coded according to three broad categories: What did politicians say about public opinion; what did TV journalists say about public opinion; and, when they spoke, what did members of the public have to say on the topic?

These questions were developed to place the sample within a history of thought on violence, audiences and politics. When asking if television risked stoking fear among audiences, with disastrous political consequences to follow, Marr, in fact, raised a question that has been central to understanding connections between the medium and political consciousness. In the 1970s and 1980s, George Gerbner, Larry Gross, Michael Morgan and Nancy Signorielli pioneered the argument that television made its viewers afraid by bombarding them with violent tales of a "mean world" (Gerbner, 1994, 1995, 1998; Gerbner et al., 1980). So, studying television and fear addresses core processes in how the medium also cultivates political publics through the constitution of audiences. It also, in this regard, provides an opportunity to seek moments of semiosis; if the medium includes the public as producers of content, what can we tell about their reaction to cultivation processes from their presence in television's story on terror?

## Results

Results of the archive search are listed in Table 12.1.

The first thing to note is that state-subsidised broadcasters (ABC and SBS) outnumber commercial news operations (7, 9 and 10); 17 stories come from the former, nine from the latter.

On the whole, the stories depicted Australians who feared fear more than terrorism. Although a desire for tougher action against radicalism was noted (The majority of voters, 2016) far more attention was paid to the risks of "toughness". Stories including comments from Federal, State and local government officials referenced the desire to avoid Islamophobic responses (Backlash concerns; Member's of Abbott's government; Mosque dispute; Terrorism experts). When it came to television commentary on the public mind, one report did note how plans for a mosque in

*Table 12.1* Australian free-to-air reports on terrorism and public opinion, 2014–2016

| *Story* | *Source* |
|---|---|
| Turnbull-Shorten bipartisanship on inclusiveness | SBS |
| The majority of voters say the Government should do more to combat radicalization | SBS |
| Fight over plans for a multi-level mosque on a small Sydney block | 7 News |
| Thousands of people across Australia have traveled to Western Sydney for the annual Mawlid celebration | SBS |
| … claims tonight the Grand Mufti has sent a poor message to young Muslims | SBS |
| Tony Abbott's Government are urging him to do more to help refugees fleeing conflict | ABC |
| Sydney siege happened just months after anti-terror laws had been introduced in Australia | ABC |
| Death threats have been made against the Magistrate who granted bail to the man who carried out Sydney Siege | ABC |
| Siege shrine: Field of flowers: Just four days after Man Haron Monis took over Sydney's Lindt Cafe | ABC |
| City shrine: In the wake of the tragédy, Martin Place has transformed into a makeshift memorial | 9 Network |
| United in grief: These are pictures just in from Martin Place | 9 Network |
| Radical Islamists or peaceful group? Hizb ut-Tahrir could be banned | ABC |
| … teenage terror suspect Numan Haider has been buried | 10 Network |
| … pleas for calm are being echoed around the country | SBS |
| Backlash concern: Police and political leaders are calling for calm | ABC |
| The Prime Minister has revealed Federal Parliament has been talked about as a potential terrorist target | 10 Network |
| Not since the Cronulla riots has the mood been quite so volatile on our streets | 10 Network |
| Family day: Coming together | ABC |
| … our involvement in the fight against Islamic State militants could increase the terror threat | 10 Network |
| Terrorism experts are warning the high risk of an attack on Australian soil could last for years | SBS |
| Among the immediate effects of the raised terrorism alert level is tighter security at airports | 10 Network |
| They're united in condemnation of Islamic State's latest atrocity | SBS |
| Australians traveling to Syria or Iraq might soon be forced to prove what they were doing there | ABC |
| Mosque dispute: Council clash | ABC |
| Attorney General George Brandis has rejected the former Labor government's assessment | SBS |
| Although the Federal Government denies it, our involvement in the fight against Islamic State militants could increase the terror threat here at home | 10 Network |

regional Victoria had sparked fears about the imposition of Sharia law (Mosque dispute). But the reference paled into insignificance against many other stories noting concerns about infringement of civil liberties (Sydney siege, 2016; The Majority of Voters, 2016), Muslim community condemnations of extremism (Although the Federal Government; Family Day; United in Grief) and simple grief about the Martin Place incident (City Shrine, 2014; Death Threats, 2014).

If anything, then, stories about the public mood stressed stoicism. Significantly stories *from* the public, the audience who were making sense of extremism on air, were different. When people spoke in reports, they expressed a true diversity of views. There was little evidence that Kruger had struck a chord; people proclaimed more concerns about racism, civil liberties, the need to support one another, the downplaying of the terror threat, and pride in public response to terror. *However*, fear and sorrow were the things most likely to be expressed. Sixty-five people spoke across the 25 reports, and the most common thing they did was emote; ten expressing fear, nine grief.

This finding gets to the nub of the crucial yet ambiguous role of television audiences in writing political history. Fear took many forms; people worried about travel (Among the immediate effects), the presence of Mosques (Fight over plans), and the safety of women wearing hijabs (Family Coming Together). They also "worried" about things that had nothing to do with terror. Some residents speaking on the controversy over a new mosque in a Sydney suburb stated that their concern was about parking (Fight over plans). Fear, then, assumed many hues. And other emotions were displayed. For one person, public commemorations of the Lindt siege inspired civic pride (City shrine), while others expressed the optimistic view that Australia was simply too wonderful a place to have to worry about terror (Terrorism experts warning).

So, when our future historian comes to this data set, she will find that television audiences felt all kinds of things about terror, but mostly, they felt. And in this, she will sense the operating tensions in audience studies. She might also grasp why television audiences became players in controversies over media and democracy. The idea that television news limits audiences' political agency has been expressed robustly in sophisticated empirical work. Lewis, Wahl-Jorgensen and Inthorn's review of thousands of TV broadcasts found viewers in the news usually expressing feelings, not judgments (2005). Television tended to portray the public as emotionally overwhelmed audiences.

Lewis, Wahl-Jorgensen and Inthorn's work lends an insightful perspective that illuminates the significance of this small case study, reflecting back on semiosis and political consciousness. It enables a description of how the inevitable ambiguity of "audience" is precisely the thing that makes it such a useful tool in considering the matters at stake in mediatised politics. On the one hand, the notion that audiences have always been a somewhat inconvenient truth for media industries is retained by their vocal presence in these news stories, urging other viewers to understand that the risks of terror include the rise of bigotry, and indeed the creation of unnecessary panic. For example, when SBS reporter Helen Isbister reported "a so-called lone wolf attack like in the Boston bombings is a major fear", a woman inconveniently responded on air that "A terror threat in Australia just doesn't seem real to me". Here was a classic moment where the audience subverted power, in a Fiskean sense. In a story that led with "Terrorism experts are warning the high risk of an attack on Australian soil could last for years", the first thing that viewers saw were people like them saying they were unperturbed; suggesting that the report didn't fully reflect what was on the public mind at that time. So, when SBS reported that research showed Australians were worried about Islamic extremism (Turnbull-Shorten bipartisanship), audiences were there to say this wasn't all they were worried about, and other things bothered them more. Future historians will not see a locked down opinion, thanks to viewers present onscreen who reacted in different ways to the message that their nation faced a new terror age.

That said, their presence speaks to television's power to marshal social life. Many of the expressions of emotions came thanks to the public mourning of the Sydney Siege, in this time period. If there could be no guarantee of what the public would say on camera, their presence as grieving audiences, responding to televised tragedy, was unsurprising. The idea that television's main effect has been to transform audiences into performing extras for media spectacles (Abercrombie and Longhurst, 1999) has been underlined by the ubiquity of televised events (Couldrey, 2003), and the familiar refrain of celebrity culture (Turner, 2010). Succinctly, TV's political power has everything to do with making the public appear, no matter what they say. The important thing to consider is not what television makes people think – in many ways, that is quite beside the point of its relationship with its viewers. Instead, the essence of TV power lies in its ongoing ability to seduce viewers into conflating the categories of audience, performer and historical subject; the easiest way to say "I was there" in history is to respond to TV. On TV. Television has solved an enduring problem in cultural history by turning historical subjects into public audience/performers. Writing of the Industrial Revolution, E. P. Thompson (1963) famously announced a thirst for a new kind of history, focused on the ordinary experience of social change. Through the capacity to produce and preserve stories about people reacting to news events, such as terror attacks, terror fears and Islamophobia, television sates this thirst with easily accessible stories on how publics feel.

## Critical Issues/Topics: Television, Audiences and History

Whatever television's fate, the lessons of television *audience* research will endure, thanks to its novel take on historically significant questions about how ordinary people become involved in politics via culture. There are many interesting questions about TV audiences: Who are they? What do they do? Why do they matter? To whom do they matter? Consequently, "audience" has become a story about the social function of media. Different audience stories paint alternative portraits of media power in action. In this sense, the noun audience has become a cultural indicator; a sensitising topic used to address challenges regarding media, justice, democracy and equality. Battles between competing definitions of the audience have been a feature of media studies for some time (e.g. Allor, 1988). This is because all of our hopes and fears about media are articulated in relation to ordinary people. Despite the forces of convergence and digital media, the constitution of publics as TV audiences remains a topic of tremendous political gravity.

In all, television audience research has established the link between media and cultural history, and is, as such, an insightful method for scrutinising politics in mediatised societies. The audience still explains the nature of political subjectivity and media power. Approaching audiences as "cultural indicators" reconciles vexed questions of where power sits between audience and industry. And, perhaps most crucially, TV preserves viewers as enduring historical subjects – people whose ideas will remain as a record of what it was like to experience historically significant moments. Whatever future historians make of the political past, they will be guided by TV-mustered audiences. In that sense, Gerbner's central point; that television matters most as an authoritative story teller par excellence, remains compelling and contemporary.

This chapter has made this argument by noting how the combined forces of "performing audiences" and digital archiving preserve television's status as a central, and enduring voice. "Television" and "Audience" are both "threatened species", in some respects, thanks to multimedia competition and forms of interaction where "user" appears a more suitable noun. Set against this, recent Australian experiences offer insights on why television and its audiences remain forces to be reckoned with. Perhaps free-to-air TV's capacity to draw audiences ain't what it used to be, but when it comes to news, it remains considerable. Similarly, since TV news routinely relies on publics for content, audiences' capacity to subvert narratives by failing to "act on cue" – by saying things like they aren't worried at all by terror – is a familiar theme in public discourse. But this doesn't change the fact that these "disruptions" are only possible because of how television orchestrates public responses in coherent, preservable forms. In fact, the creation and proliferation of TV archives may mean that the medium's political legacy as an organiser of political thought is taking on subtle yet compelling new forms.

## References

5 City Metro Network Overnight Ratings. 2018. *Decider TV.* Retrieved from https://decidertv.com/ratings/

Although the Federal Government denies it, our involvement in the fight against Islamic State militants could increase the terror threat here at home. 2014, September 14. *Ten Network News.* [TV] Retrieved from https://search-informit-com-au.ezproxy.lib.monash.edu.au/search;action=doSearch.

Abercrombie, N. and Longhurst, B. 1999. *Audiences.* London: Sage.

Agirre, I. A., Arrizabalaga, A. P., and Espilla, A. Z. 2016. "Active audience? Interaction of young people with television and online video content." *Communication & Society* 29 (3):133–147. doi:10.15581/003.29.3.133-147.

Allor, M. 1988. "Relocating the site of the audience." *Critical Studies in Mass Communication* 5 (3):217–233.

Ang. I. 1989. *Desperately Seeking the Audience.* London: Routledge.

Bird, S. E. 1992. "Travels in nowhere land: Ethnography and the "impossible" audience." *Critical Studies in Mass Communication 9*(3): 250–260.

Bolin, G. 2012. "The labour of media use." *Information, Communication & Society* 15 (6):796–814. doi:10.1080/1369118X.2012.677052.

Bourdon, J, and Cécile, M. 2015. "Ratings as politics. Television audience measurement and the state: An international comparison." *International Journal of Communication (19328036)* 9:2243–2262.

Carey, J. 1989. *Communication as Culture.* Boston, MA: Unwin Hyman.

Couldrey, N. 2003. *Media Rituals: A Critical Approach.* London: Routledge.

Dhaenens, F., Van Bauwel, S., and Biltereyst, D. 2008. "Slashing the fiction of queer theory: Slash fiction, queer reading, and transgressing the boundaries of screen studies, representations, and audiences." *Journal of Communication Inquiry* 32 (4):335–347.

Fiske, J. 1988. "Critical response: Meaningful moments." *Critical Studies in Mass Communication* 5 (3):246–251.

Gallardo-Camacho, J., Lavín, E., and Fernández-García, P. 2016. "Sports television programmes and their relationship with the social audience on Twitter in Spain." *Revista Latina de Comunicación Social* (71):272–286. doi:10.4185/RLCS-2016-1095en.

Gerbner, G. 1994. The· Politics of Media Violence: Some Reflections. In C. Hamelink & O. Linne (eds.), *Mass Communication Research: On Problems and Policies* (pp. 133–146). Norwood, NJ: Ablex.

Gerbner, G. 1995. The cultural frontier: Repression~ violence~ and the liberating alternative. In P. Lee (ed.), *The Democratization of Communication* (pp. 153–172). Cardiff: University of Wales Press.

Gerbner, G. 1998. "Cultivation analysis: An overview." *Mass Communication and Society 1* (3-4):175–195.

Gerbner, G., Gross, L., Morgan, M., and Signorelli, N. 1980. "The mainstreaming of America: Violence profile #11." *Journal of Communication 30*:10–29.

Gerbner, G. 1970. "Cultural indicators: The case of violence in television drama." *Annals ofthe American Academy of Political and Social Science* 388:69–81.

Gerbner, G. 1973. Cultural Indicators: The Third Voice. In G. Gerbner, L. Gross and W. Melody (eds.), *Communications Technology and Social Policy* (pp. 555–573). New York: John Wiley.

Gómez Aguilar, M., Paniagua Rojano, F. J., and Farias Batlle, P. 2015. "The behaviour of the television audience on social networks. An approach to its profile and the most talked-about programmes." *Revista Latina de Comunicación Social* (70):539–551. doi:10.4185/RLCS-2015-1058en.

Gong, Y. 2017. "Media reflexivity and taste: Chinese slash fans' queering of European football." *Communication, Culture & Critique* 10 (1):166–183. doi:10.1111/cccr.12140.

Grossberg, L. 1988. "Wandering audiences, nomadic critics." *Cultural Studies, 2*(3): 377–391.

Gwenllian-Jones, S. and Pearson, R. E. (Eds.). 2004. *Cult Television.* Minneapolis, MN: University of Minnesota Press.

Hall, S. 1980. Encoding/decoding. In S. Hall (ed.), *Culture, Media, Language: Working Papers in Cultural Studies* (pp. 129–138). London: Hutchinson.

Jenkins, H. 1992. *Textual Poachers.* New York: Routledge.

Kim, K., Cheong, Y., and Kim, H. 2016. "The influences of sports viewing conditions on enjoyment from watching televised sports: An analysis of the FIFA World Cup audiences in theater vs. home." *Journal of Broadcasting & Electronic Media* 60 (3):389–409. doi:10.1080/08838151.2016.1203320.

Livingstone, S. 2015. "Active Audiences? The debate progresses but is far from resolved." *Communication Theory (10503293)* 25 (4):439–446.

Lewis, J., Inthorn, S., and Wahl-Jorgensen, K. 2005. *Citizens or Consumers?* London: Routledge.

Murtagh899 (Producer). 2012. One Thing Slash Fandom. Retrieved from https://www.youtube.com/watch?v=nNTSLf4ekMc

Scodari, C. 2003. "Resistance Re-Examined: Gender, Fan Practices, and Science Fiction Television." *Popular Communication: The International Journal of Media and Culture 1*(2): 111–130.

Simons, N. 2015. "TV drama as a social experience: An empirical investigation of the social dimensions of watching TV drama in the age of non-linear television." *Communications: The European Journal of Communication Research* 40 (2):219–236. doi:10.1515/commun-2015-0005.

Skains, R. L. 2010. "The shifting author-reader dynamic." *Convergence: The Journal of Research into New Media Technologies* 16 (1):95–111.

Thompson, E. P. 1963. *The Making of the English Working Class*. London: Gollancz.

Turner, G. 2010. *Ordinary People and the Media*. London: Sage. Videography


## New Videos


Although the Federal Government denies it, our involvement in the fight against Islamic State militants could increase the terror threat here at home. (2014, September 14). *Ten Network News*. [TV] Retrieved from https://search-informit-com-au.ezproxy.lib.monash.edu.au/search;action=doSearch

Among the immediate effects of the raised terrorism alert level is tighter security at airports, at government buildings and also at the football as hundreds of thousands of people prepare to attend finals across the country. [online]. Ten 5pm News (TEN NETWORK); Time: 17:00; Broadcast Date: Friday, 12 September 2014; Duration: 1 min., 58 sec. Availability:≤http://search.informit.com.au.ezproxy.lib.monash.edu.au/documentSummary;dn=TSM201409120111;res=TVNEWS≥ [cited 28 August 2017].

Backlash concern: Police and political leaders are calling for calm amid concerns last night's events could lead to a backlash against Victoria's Muslim community. [online]. ABC News Victoria (ABC1 Melbourne); Time: 19:00; Broadcast Date: Wednesday, 24 September 2014; Duration: 2 min., 13 sec. Availability: ≤http://search.informit.com.au.ezproxy.lib.monash.edu.au/documentSummary;dn=TSM201409240188;res=TVNEWS≥ [cited 28 August 2017].

City shrine: In the wake of the tragedy, Martin Place has transformed into a makeshift memorial, Lizzie Pearl is there. [online]. Nine News (NINE NETWORK); Time: 18:00; Broadcast Date: Wednesday, 17 December 2014; Duration: 4 min., 34 sec. Availability:≤http://search.informit.com.au.ezproxy.lib.monash.edu.au/documentSummary;dn=TSM201412170015;res=TVNEWS≥ [cited 28 August 2017].

Family day: Coming together: The Government's plans to send Australian troops to the Middle East and news of a third beheading by Islamic State overnight have raised concerns about the implications here in Australia. [online]. ABC News NSW (ABC1 Sydney); Time: 19:00; Broadcast Date: Sunday, 14 September 2014; Duration: 2 min., 6 sec. Availability:≤http://search.informit.com.au.ezproxy.lib.monash.edu.au/documentSummary;dn=TSM201409140139;res=TVNEWS≥ [cited 28 August 2017].

NSW votes: The Sydney siege happened just months after anti-terror laws had been introduced in Australia. [online]. ABC News NSW (ABC1 Sydney); Time: 19:00; Broadcast Date: Friday, 20 March 2015; Duration: 2 min., 3 sec. Availability:≤http://search.informit.com.au.ezproxy.lib.monash.edu.au/document Summary;dn=TSM201503200103;res=TVNEWS≥ [cited 28 August 2017].

Sydney siege: Death threats: Its understood death threats have been made against the magistrate who granted bail to the man who would go on to carry out the Lindt Cafe siege. [online]. ABC News Victoria (ABC1 Melbourne); Time: 19:00; Broadcast Date: Friday, 19 December 2014; Duration: 2 min., 9 sec. Availability:≤http://search.informit.com.au.ezproxy.lib.monash.edu.au/documentSummary;dn=TSM201412190201;res=TVNEWS≥ [cited 28 August 2017].

Siege shrine: Field of flowers: Just four days after Man Haron Monis took over Sydney's Lindt Cafe, more questions are being raised about the nation's gun laws. [online]. ABC News SA (ABC1 Adelaide); Time: 19:00; Broadcast Date: Thursday, 18 December 2014; Duration: 2 min., 42 sec. Availability:≤http://search.informit.com.au.ezproxy.lib.monash.edu.au/documentSummary;dn=TSM201412180089;res=TVNEWS≥ [cited 28 August 2017].

The fight over plans for a multi-level mosque on a small Sydney block is fast becoming a national talking point. [online]. Seven News (SEVEN NETWORK); Time: 18:00; Broadcast Date: Sunday, 10 January 2016;

Duration: 1 min., 55 sec. Availability:≤http://search.informit.com.au.ezproxy.lib.monash.edu.au/documentSummary;dn=TSM201601100097;res=TVNEWS≥ [cited 28 August 2017].

The reshuffle on Monday was overshadowed by Monday morning TV in particular the reaction of host Sonia Kruger to the deadly attacks in Nice. [online]. Insiders (ABC1); Time: 09:00; Broadcast Date: Sunday, 24 July 2016; Duration: 8 min., 7 sec. Availability:≤http://search.informit.com.au.ezproxy.lib.monash.edu.au/documentSummary;dn=TSM201607240025;res=TVNEWS≥ [cited 18 August 2017].

Terrorism experts are warning the high risk of an attack on Australian soil could last for years, with the cost of increased security filters down to the public. [online]. World News Australia (SBS TELEVISION); Time: 18:30; Broadcast Date: Saturday, 13 September 2014; Duration: 2 min., 43 sec. Availability:≤http://search.informit.com.au.ezproxy.lib.monash.edu.au/documentSummary;dn=TSM201409130093;res=TVNEWS≥ [cited 28 August 2017].

There are claims tonight the Grand Mufti has sent a poor message to young Muslims with his comments on the Paris attacks. [online]. World News Australia (SBS TELEVISION); Time: 18:30; Broadcast Date: Tuesday, 17 November 2015; Duration: 2 min., 22 sec. Availability:≤http://search.informit.com.au.ezproxy.lib.monash.edu.au/documentSummary;dn=TSM201511170194;res=TVNEWS≥ [cited 28 August 2017].

Turnbull-Shorten bipartisanship on inclusiveness comes as research from one of the country's top universities shows Australians are worried about Islamic extremism. [online]. World News Australia (SBS TELEVISION); Time: 18:30; Broadcast Date: Monday, 10 October 2016; Duration: 2 min., 25 sec. Availability:≤http://search.informit.com.au.ezproxy.lib.monash.edu.au/documentSummary;dn=TSM201610100019;res=TVNEWS≥ [cited 18 August 2017].

United in grief: These are pictures just in from Martin Place where there's been a huge outpouring of grief hundreds of people leaving flowers for the victims of the deadly siege. [online]. Nine News (NINE NETWORK); Time: 18:00; Broadcast Date: Tuesday, 16 December 2014; Duration: 2 min., 33 sec. Availability:≤http://search.informit.com.au.ezproxy.lib.monash.edu.au/documentSummary;dn=TSM201412160038;res=TVNEWS≥ [cited 28 August 2017].

Vote 2016: The majority of voters say the Government should do more to combat the radicalization of young people. [online]. World News Australia (SBS TELEVISION); Time: 18:30; Broadcast Date: Tuesday, 21 June 2016; Duration: 1 min., 52 sec. Availability:≤http://search.informit.com.au.ezproxy.lib.monash.edu.au/documentSummary;dn=TSM201606210212;res=TVNEWS≥ [cited 28 August 2017].

# 13

# *GRAND DESIGNS* AND *THE BLOCK*

## Audience Engagement and Modes of Consumption through Lifestyle Reality TV in Australia

*Esther Milne and Aneta Podkalicka*

### Introduction

This chapter explores the material and symbolic significance of the British hit television program *Grand Designs* in creating viewing publics and new markets. We discuss what the program means to audiences and more broadly for emerging forms of audience research. As a point of contrast and comparison we explore the popular Australian competition renovation show, *The Block*. Examining how people make sense of these programs could lend itself to traditional textual analyses asking questions about, for example, the way architects are represented or gender is framed. However, our focus is more on the material purposes to which these programs, or media more generally, are put and how this is achieved. Talking to Australian renovators and building practitioners we discovered that these programs occupy multiple cultural registers and do not remain solely at the level of representation. We need, therefore, a wider critical lens that has the capacity to recognise the ways in which *Grand Designs* and *The Block* function as cultural objects circulating beyond the screen.

If the discursive or representational focus has substantially shaped the literature on TV audiences, it has been particularly noticeable within lifestyle or reality TV scholarship, with neoliberalist interpretations of these media texts leading the field. This matters because the media landscape is increasingly dominated by the reality TV format. One 2016 advertising study reveals that reality TV accounts for half of the top 50 programs on free-to-air television in Australia with live sport a close second. Of these, cooking shows are the most popular, just ahead of renovation programs where *The Block* features as the highest rating reality TV show in Australia (Hickman 2016). With this in mind, we are keen to trace alternative analytical trajectories and possibilities, arguing that the governmentality frame can foreclose productive analysis. Once a viewing position or entertainment genre is deemed as reproducing the norms of market logics, individualism or entrepreneurship, all that's left to do is submit to its inexorable force and stubbornly predictable outcome. Indeed, one could argue that the very criticism commentators make about the seeming inevitability of neoliberalism – it presents itself as the only model for social relations – is actually often at work in the analysis of lifestyle and reality formats themselves. So complete and fully formed is the critical blueprint that it's very difficult to peer round its edges to appreciate any other vision. In this chapter we take our cues from Beverly Skeggs and Helen

Wood's empirically grounded study in which they show how to 'recognise the role of textuality in the social and sociality in the text' (Skeggs and Wood 2012, 215).

The chapter begins with a brief survey of the literature, followed by an outline of our methods and sources of data. We then introduce our two media sites, the highly successful British program *Grand Designs* and Australia's longest running renovation program, *The Block*. We select these programs because they emerged as key points of reference in our research on media and home renovations in Australia. Moreover, these have been explicitly juxtaposed as contrasting examples by audiences when we asked them about their viewing patterns. The remainder of the chapter focusses on the rich and complex role played by these media texts in the *material* everyday lives of their audiences.

## Existing Literature

While the definitions of lifestyle media, reality TV and property or real estate TV often overlap, certain programs will seem easier to locate under a specific heading. For example, *Grand Designs* is not usually understood as a reality TV format in the scholarly literature but nonetheless is presented as such by a TV-based content provider, such as the subscription channel 'Lifestyle' in Australia. Boundless, the production company of *Grand Designs*, is known as the creator of reality TV programs such as *My Kitchen Rules* and *The Apprentice* but it also produces programs such as *Great British Railway Journeys* which, like *Grand Designs*, is more factual and documentary in approach. As we discuss below, part of the success of *Grand Designs* is how the program manages to side-step some of the more obviously tabloid elements of the reality format while retaining its widespread popularity and accessibility. *The Block*, however, sits firmly within the 'makeover' or 'gamedoc' sub-category premised on competition.

Scholars have been debating the definitions and scope of reality TV and lifestyle media formats for decades. In *A Companion to Reality Television*, editor Laurie Ouellette observes that trying to define the genre conclusively 'may not be the best way to pinpoint and address what is most salient about the reality phenomenon' since 'reality television as a whole revels in generic hybridity and borrows extensively from other televisual forms' (Ouellette 2014, 5). Key to its success is often its expansive reach as Mark Andrejevic explains: 'As it has grown and differentiated beyond blockbuster formats such as *Survivor* and *Big Brother*, reality television has revealed its omnivorous character: its ability to enfold a broad range of social life and redouble it for entertainment and profit' (Andrejevic 2014, 50). In order to grasp its role in the contemporary media landscape a sense of its history is necessary. For Ouellette current configurations of the reality TV genre can be traced to the early days of TV itself and beyond:

> While the term 'reality television' is relatively new, the history of factual and unscripted entertainment featuring ordinary people dates to the origins of the medium (and, before it, radio and cinema) … Attention to the use of ordinary people in early experimental and social realist cinema, the work of guerrilla video artists, participatory media advocates and alternative television makers of the 1970s, and the social reform tradition of documentary provide further context for tracing historical continuities and breaks between the past and the reality television phenomenon today.
>
> *(2014, 5)*

If the inclusion of 'ordinary people' is clearly a defining feature of the genre, how to characterise their actions and meaning to audiences has been harder to capture. This is often because there are competing theoretical frames brought to bear on the signifying practices of the format. Dominating the critical imagination during the last decade has been a focus on ideas of governance:

how lifestyle and reality TV are inflected by the dynamics of neoliberalism and 'governing at a distance'. In their highly influential work *Better Living through Reality TV: Television and Post-Welfare Citizenship*, Ouellette and Hay (2008) establish a compelling argument that reality TV emphasises principles of privatisation that directs responsibility to individuals, rather than the state, to fix wide ranging social, economic and health issues. Participants on screen model for viewers self-care as they fix their bodies, finances, diets and houses. Stripped clean of the social and financial inequalities wrought by race, gender and class, the genre focuses on individual obligation; it is a moral not political failing that cast members must try to solve. As the authors explain, at a 'time when privatization, personal responsibility, and consumer choice are promoted as the best way to govern liberal capitalist democracies, reality TV shows us how to conduct and "empower" ourselves as enterprising citizens' (Ouellette and Hay 2008, 2).

Ouellette and Hay's thesis has had a profound and lasting effect on lifestyle media and reality TV scholarship. Generating fruitful lines of enquiry, scholars have explored a wide range of governance discourse discussing its national, cultural, racial and gendered instantiations across our screens. In 'Moral entrepreneurship and individual responsibility in *Jamie's Ministry of Food*', for example, Joanne Hollows and Steve Jones (2010) explore how celebrity chef Jamie Oliver promotes individual responsibility for good nutrition and how this particular strand of UK neoliberalism contrasts with US models. For Catherine Squires there is an unequivocal, indeed 'toxic', link between the so-called 'post racial' moment of the 2008 victory of President Obama and neoliberal notions of individual choice performed in the complex worlds of reality TV. As she explains, 'postracial discourses dovetail neatly with neoliberal discourses because of their shared investment in individual-level analysis and freedoms'. Efforts to 'ameliorate injustice ... is always suspect, particularly remedies that require intervention in the marketplace, a realm imagined to be divinely ordered by individual-level acts of choice' (2014, 266).

Similarly, Hunter Hargraves advances an intriguing argument about the tanning obsessions of the docusoap *Geordie Shore*. What she calls 'Tan TV' echoes the changing configurations of whiteness set in the context of a 'neoliberal state intent on dismantling public welfare programs and, at least rhetorically, on moving "beyond" race' (2014, 284). Body size and image is, of course, a significant preoccupation of lifestyle media and its critics. For the authors of '*The Biggest Loser*: The Discursive Constitution of Fatness' neoliberalism is seen as particularly eviscerating in its cruel and disingenuous attacks on the body. As they explain, *The Biggest Loser* (*TBL*) works as a 'powerful public pedagogy' that 'provides the platforms for a government of the self, a component in the neo-liberal reinvention of "welfare" that promotes choice, personal accountability and self-empowerment' while 'at the same time, masking social forces that position people into the dejected borderlands of consumer capitalism'. In its rhetorics and visual language, the program 'classifies the obese, overweight and physically unfit as personal moral failures, immoral and irresponsible citizens, socially, morally and economically pathologized outsiders' (Silk, Francombe and Bachelor 2009, 369).

While neoliberalism and governmentality have proved rich analytical resources for investigating the meanings of reality programming and makeover TV, they are not without their drawbacks. For one thing, critics sometimes seem reluctant to acknowledge the sheer breadth and diversity of the genre(s). In his book, *Why Voice Matters: Culture and Politics after Neoliberalism*, Nick Couldry (2010) advances an elegant argument about how the contemporary neoliberalist workplace, with its exhortations to its employees that they demonstrate 'authenticity' and 'passion', maps onto reality TV formats. And the argument is convincing until it turns to specificities. It is certainly persuasive that the values of workplace performance, familiar to us with phrases such as 'team conformity' and 'being positive', structure the formats of reality TV. As he explains, contestants 'must be positive, banishing any thought of contradiction, in the same way that employees must be "passionate" to ward off reflections about the contradictions which their performance entails' (Couldry 2010, 78).

However, as any fan of reality TV will tell you, the screen is often set ablaze by participants 'pushing back', resisting, arguing and deceiving the unseen authorial hand of the show runner. Pieces to camera might explain or justify a certain cast uprising or muffled audio will suddenly reveal people's complaints or refusals. While some may maintain this is the result of deliberate casting or scripted on-screen performance, we argue there is more nuance at play when reading those programs. In fact, a pivotal element of the format is the exposé often made by the participants following their appearance on the show. Here we might see the particular constraints or demands of a program laid bare by contestants and enthusiastically critiqued by audiences as they swap stories on social media (White 2014). Such multidirectional patterns of engagement trouble the notion of unmitigated control that reality TV purportedly exerts over its subjects and audiences. Therefore, you could counter with the argument that reality TV is where the negotiations between power figures play out. People rehearse their work environments and all those micro contests, on the sets and screens of reality TV. And this gets to a central problem within the neoliberal critique.

As Brenda Weber argues, although 'discourses of American citizenship underpinned by a logos of neoliberalism' populate the genre, 'the makeover is too vast, complicated, and internally contradictory to be summarized exclusively though the theoretical apparatus provided by neoliberalism' (2009, 51). Likewise, for Skeggs and Wood, the 'almost wholesale seduction of media analysts by ideas about governmentality' is problematic because it draws attention to the individual as 'an insular and singular unit rather than the social as a process in formation' (2012, 4). As Skeggs and Wood have done, many researchers have turned their attention to conducting empirical studies of audiences. Are they so easily duped and persuaded by the state as the neoliberal critique of reality TV suggests? The answer is 'no' according to Kate Holland, Warwick Blood and Samantha Thomas who write that governmentality perspectives 'tend to gloss over the question of how people position themselves and assert their values in relation to makeover TV, whether, for example, they take up or resist the biopedagogies and conceptions of self presented to them' (2015, 18). Their study involved interviews with 142 people 'classified "obese" or "morbidly obese" on the basis of their self-reported height and weight' to gauge their reactions to *TBL*. Disputing a governmentality perspective does not mean unthinkingly embracing a particular reality TV show. The authors found biting criticism for *TBL* in their audience sample illustrated by terms such as 'demeaning, humiliating, exploitative, disgusting, cruel, gimmicky, vile, lowest common denominator, commercialized, and degrading' (21). Furthermore, Holland, Blood and Thomas conclude that:

> obese people's views of *TBL* do not support the idea that it constitutes a form of governing at a distance by encouraging people to incorporate the technologies of the self shown on the program into their own lives, thereby translating the goals of authorities (or experts) into their own goals.
>
> *(28)*

For others, the key reason for the dominance of the governmentality frame is the very provenance and scope of the scholarship on reality TV, grounded in the US media system (Lunt 2014). To internationalise the media studies field would mean asking whether, as Miller and Kraidy put it, 'neoliberalism [is] a valid universal framework for reality television, or does it reflect North Atlantic premises that all nation-states share similar political-economic arrangements?' Our interpretation seeks to extend 'the theoretical repertoire' (182) of reality television studies by examining how two popular and contrasting series – *Grand Designs* and *The Block* – figure into their audiences' everyday life and consumer cultures in the Australian context.

## Methods and Sources of Data

This chapter draws on a mix of TV ratings and interview research conducted with home renovators, both self-professed fans and fervent critics of the show, together with production and industry personnel working in the built environment in Australia. Some of this work was carried out in conjunction with the Australian Cooperative Research Centre for Low Carbon Living (CRC LCL). Our empirical data comprise a national survey (n = 156 responses) that was conducted in 2015, and focus groups with home renovators (5) which ran in Melbourne during 2015. Through the survey and focus groups we asked people about their media use in getting ideas and tracking progress of their renovations. We also ran focus groups (3) and conducted interviews with industry practitioners and sustainability experts (3) to explore the role of media in their building, design and environmental communication practices. Television ratings were supplied by the Australian TV audience measurement company, OzTAM Pty Limited and sourced through the Annual Reports of the Australian Broadcasting Corporation (ABC).

## Real Estate TV Formats: *Grand Designs* and *The Block* in Australia

In this section we briefly explain the tropes and format of these two programs. While their formal differences can be marked, for audiences they represent comparable sites to shape renovation practice. We also map audience engagement by outlining the ratings enjoyed by the two programs.

*Grand Designs* is a British program, produced by Boundless Productions, a subsidiary of Fremantle Media, that has been on air since 1999, when it was first broadcast on Channel 4. It now screens internationally in over 130 countries including Germany, Australia, New Zealand, Taiwan, Poland, Norway and South Africa. In the UK, *Grand Designs* continues to be broadcast on advertising-supported public service broadcaster Channel 4, while in Australia it screens on the national public service broadcaster ABC. In some countries such as Norway or the US, *Grand Designs* is available on Netflix. *The Block* is a distinctly Australian series, tapping into Australia's so-called 'obsession' with home renovations (Allon 2008). It was created in 2003 and is produced by Watercress Productions (seasons 4–11) and Cavalier Productions (seasons 12–present). *The Block* has won numerous awards and the format has been sold internationally to the UK, Israel, New Zealand and Belgium.

*Grand Designs* and *The Block* represent distinctive narrative approaches. *Grand Designs* depicts projects that can involve building a dream house from scratch, renovating and heritage conservation while *The Block* focuses on large-scale renovations of the whole 'block' (hence the name) by pairs of contestants rather than individual owner-occupiers. *Grand Designs* follows participants on their journey of making a family home, including those with low budgets and green aspirations, unlike *The Block* that has an unapologetically commercial agenda with product placement routinely displayed in characters' shopping and renovating practices. *The Block* is a 'gamedoc' with each season focussing on a building site located in either Melbourne or Sydney. Contestants compete to achieve the highest bid at auction by renovating the often uninhabitable property and participating in various challenges along the way. Extra narrative tension is generated because the contestants live on site while renovating, a situation that increases the stress both to participants and to their builders. Also adding to the drama is the extremely brief time frame which constrains the renovators and their trades. Each season works toward the climactic moment when the properties are thrown open to the public. Interested viewers get to inspect the renovations as they might do when making an offer. The final episode sees the properties 'go under the hammer' as contestants compete to win the show by securing the highest price at auction. These auction episodes traditionally attract audiences in the millions as we explain below.

*Grand Designs* reaches its narrative apex within each episode, with a house reveal accompanied by lyrical commentary of its educated and well-spoken presenter, Kevin McCloud – described as 'a kind of David Attenborough of the building site' (Aitkenhead 2008) – who in the majority of cases, praises the building effort of the program protagonists. *The Block*'s dedication to the Australian vernacular is pronounced in the way the show represents the 'low-brow' social interactions between the contestants and tradespeople working alongside them or when some characters subject themselves, aided by an expert, to getting up to speed with the urban elite 'classy' interior design aesthetic and lingo. The judges' scoring for each consecutive project (the kitchen or bedroom, for example) in the lead-up to the auction, can be as laudatory as unforgiving of the amateur-renovators' efforts.

The different stylistic and aesthetic features of the two shows reflect and reinforce those divergent preoccupations and subjectivities. *Grand Designs* offers effectively composed wide shots of the landscape and intimate close-ups of participants, offering careful reflections on the process intertwined with none too rare poetic monologues of the smartly dressed presenter Kevin McCloud. Where builders are sometimes seen on screen in *Grand Designs* these appearances are relatively infrequent. By contrast, on *The Block* building practitioners or 'tradies' are a defining characteristic of the program's narrative structure. So much so, that Keith Schleiger, the building site foreman, has become a highly recognised member of the cast. His taciturn demeanour and forthright opinions about the properties play an indispensable role in creating suspense. Will he suddenly close down a particular renovation project? Has the necessary Occupational Health and Safety paperwork been filled in correctly? In addition, a substantial amount of screen time is devoted to the process by which the contestants first engage tradies and then liaise with them over the course of the build. Successful tradies who manage to deliver cost effective and timely renovations have been rehired in multiple series; their names, like David Franklin, the popular landscaper, joining the permanent team of on-air personnel. It goes without saying that such high profile and regular screen time probably helps these tradies in their own commercial businesses.

What the two programs do have in common is the shared interest in representing the human drama of building or renovating, although this plays out with significant contrasts to the ways in which the process itself, taste, expertise and voice are represented. *Grand Designs* can be argued to be aligned with popular, well-attuned public pedagogy as informative and educational, not least through its association with Channel 4 as a publically owned, hybrid public service broadcaster (Lunt 2014; Podkalicka, Milne and Kennedy 2018). *The Block*, on the other hand, is positioned as pandering to the Australian common denominator and commercial values, ostensibly advertising consumer products to the viewer who is relentlessly directed to online shopping websites associated with the program. Kavka and Weber (2017, 4) have demonstrated how the term 'redneck' can be manipulated 'for the purposes of entertainment [as] an obvious act of commodification' within reality TV. *The Block* can be seen as tactically 'bogan' at times, which in the Australian everyday talk is a derogatory notion denoting the lack of refinement of low socioeconomic classes. Beyond the conspicuous differences, however, both programs have had consistently high audience figures, prompting the critical comparison.

## Ratings

During its 20 years on screen in Britain, *Grand Designs* regularly attracts viewer figures in the millions. In Australia, *Grand Designs*, together with its franchises and spinoffs, are available on free-to-air and subscription TV. In general, *Grand Designs* is aired on the national, free-to-air platform the ABC, although at various times it has also been available through subscription TV. When shown on the ABC, *Grand Designs* averaged audiences of 913,000 in 2015 and 619,000 in

2016. These figures will often increase sharply with certain seasons proving particularly successful over the years. In 2013, for example, Series 12 which included the 100th episode of *Grand Designs* averaged just over 1 million viewers. That episode itself, featuring the iconic 'Derelict Water Tower' project, drew 1,076,000 audience members, the highest rating of the series. Also popular with audiences are the specialised versions and spin offs such as *Grand Designs Revisited* which returns to the sites of well-known episodes to find out how the properties and renovators have fared since the initial project was undertaken. *Grand Designs Revisited* peaked at 1.5 million viewers in 2015 and 1.3 million in 2016. Another ratings success is *Grand Designs: House of the Year* where Kevin McCloud visits the houses nominated for the prestigious Royal Institute of British Architects' Award. In 2016, *Grand Designs: House of the Year* averaged audience numbers of 737,000 in Australia. Figures such as these may mean, in some instances, that the programs have won their timeslots over the commercial channels, a notable achievement for the ABC. The local franchise, *Grand Designs Australia*, is available through the subscription network Foxtel on its Lifestyle Channel, with 31,000 viewers in 2015 and 29,000 in 2016. These relatively low audience numbers are actually on a par with other property-based programs aired on subscription TV. For example *Selling Houses Australia* attracted on average 32,000 weekly viewers in 2015 and *Location Location Location Australia* garnered 25,000 for the same time period.

Where some audience figures for *Grand Designs* might seem relatively moderate, *The Block* has regularly beaten all competition. It screens on Channel Nine, one of three commercial, free-to-air national networks in Australia. In 2010 *The Block* drew an average of 1.2 million nightly viewers peaking at nearly 1.8 million people for the auction finale. Over the next six years, *The Block* has maintained or increased this level of interest, with the 2011 final attracting nearly 3.4 million viewers, the 2012 auction episode watched by over 3 million viewers and the 2016 final viewed by nearly 2.3 million audience members. To give some context, it regularly beats its closest free-to-air commercial TV rival, *House Rules*, which reached 2,070,000 for its 2014 final compared to *The Block*'s final for that year of 2,687,000 viewers. Success in audience figures often translates to material impact with the program credited for driving Australia's 'renovation obsession'. One study estimates there is a $251 billion boost to quarterly renovation expenditure when a season of *The Block* is screened (Aggelli and Melles 2015). *The Block* enjoys a vibrant social media footprint generating high traffic across multiple platforms including Twitter, Facebook, Instagram and Pinterest. Significantly, many of these conversations are instigated by the users rather than emanating through official advertising communications. Finally, in terms of demographics it's worth noting that spanning the period of 2010 to 2016, *The Block* is most popular with the 35–49 age demographic which contrasts with *Grand Designs* which is dominated by the 50–64 bracket.

## The Significance of *Grand Designs* and *The Block*

In our own qualitative research, *The Block* and *Grand Designs* came up as a particularly important source of entertainment and popular conversation about home renovations and home-making. The shows were often juxtaposed in their on-screen depiction of the building or renovation process, with *The Block* criticised for its unrealistically short timelines that it takes the participants to turn a pile of mess into a fashionable property. This fictional representation with the focus on competition and the 'trashiness' is what entertains audiences: 'I get to see people having fights on TV, I love that', observed one home renovator in a focus group. The research revealed that Australian audiences rely on the program for some styling ideas, when for example a couple of home renovators confessed to changing their original design plans after watching the finals of *The Block* (see Hulse et al. 2015; Podkalicka et al. 2016). They specifically opted for a 'double shower' as, by their own admission, they had 'the space' so 'why not?' But the format is frustrating for building practitioners and designers who complain about their clients' unrealistic expectations, modelled on the show. As one tradesperson put it: 'They might subconsciously think a bathroom renovation takes three days, so they're like ... "What are you talking about? You can do it in 24-hours." That's just not possible. Yeah, there's some bad things coming out of that.'

The negative impacts of the show were also raised in relation to the absence of on-screen depictions and conversations about environmental sustainability, about why the installation of energy-efficient materials or appliances should matter, and in connection with thoughtless waste. One committed green interviewee lamented: 'I think that in fact there's some pretty abhorrent things that they do, you know, putting in and ripping something out and putting it back in again'.

*Grand Designs* has not avoided public criticisms either. Over the years, the program has been decried for its dominant focus on large-scale, expensive 'grand' projects that remain outside of the realm of the average homeowner (Aitkenhead 2008). However, our empirical research revealed that *Grand Designs* was considered 'more factual' and informative than *The Block*. It was valued for the generous time devoted to documenting the houses under development, with on-screen representations of the building challenges and tried solutions – and not being fixated on following the 'couples bickering about yellow versus red', as one of home renovators explained. The perceived differences between *Grand Designs* and *The Block* were so stark for some research participants that they insisted on not categorising the former as reality TV (Podkalicka, Milne and Kennedy 2018)

## Discussion

One of the arguments informing the neoliberal interpretation is the very role that lifestyle TV plays in encouraging consumption or consumerism. Since these programs rely on sponsorship deals, direct product placement and advertising are rife. But again, to grant totalising force to the neoliberal – consumption-stimulating – frame would be to disregard the empirical nuances that define the ways in which audiences use the program. In the case of *Grand Designs* (UK) in Australia, for example, one interviewed designer told us:

> I do like *Grand Designs sometimes* because they do have some fun stuff on that. I'm more of a head in the cloud, I haven't got pay TV so I don't watch the other programs. But I do find I'm shouting at the television and telling them "That's a lot of crap!" And there are people that take Kevin to task on the internet and say, well, you talked about this building having enough solar panels to power itself and then people do the numbers and say "Kevin!" (emphasis ours).

The key take-away from this observation might indeed be the caveat of 'sometimes'. It encompasses the recognition of the active audience reading and situated and often reflexive uses of the program in everyday life. Popular shows such as *Grand Designs* and *The Block* affect conversations and the ways consumption practices are shaped but not uniformly and by no means holistically. As we have argued, the neoliberal framework has dominated much of the scholarship on reality and lifestyle TV. While it has usefully emphasised the mobilisation of private responsibility and enterprise to solve social problems, what often accompanies these approaches is a barely concealed contempt for the genre. Even within those analyses which seek to contest the ascendancy of neoliberalist interpretations, the associated condemnation of the genre seems difficult to shake. Returning to our discussion of the literature, we are intrigued by the almost palpable internal conflict at work in Miller and Kraidy's book *Global Media Studies* when it turns its attention to consider reality TV. Here the authors argue trenchantly for a re-evaluation of the critical lens through which it has traditionally been understood since the local, economic, political and social conditions of its production across global territories demand greater nuance. As they explain:

> [D]enouncing or celebrating a genre such as reality TV for being trivial, popular, or neoliberal misses the point. Closer inspection is required of the infrastructures of production and meaning that both bind it together and differentiate it as a concept.
>
> *(2016, 193)*

Such calls for greater critical sophistication are dampened somewhat by blithe pronouncements such as the 'origins of reality TV lie in the propaganda ministry of the Nazi party of the 1930s' (175). It is a fascinating piece not so much for the conclusions it reaches but for the dialectic practised within its pages. Back and forth it goes; sometimes highlighting the governmentality argument in which reality TV is 'suffused with neoliberal deregulatory nostrums: individual responsibility, avarice, possessive individualism, hyper-competitiveness, and commodification' that is 'played out in the domestic sphere rather than the public world' (176). While at other times the authors maintain that the format allows for complex national debate across global sites: as 'pan-Arab, pan-African, and Indian contexts demonstrate, ideological contests animated by reality television in many parts of the world embody progressive social and political values – gender equality, citizenship, and political pluralism' (182).

That reality TV is contradictory and multi-faceted is not news. It's hardly surprising that cultural theorists (ourselves included) might equivocate in their determinations. For readers, audiences and analysts alike reality TV presents itself as a conundrum: on the one hand 'voice', on the other hand 'Donald Trump'. So our contribution here is to insist more firmly that this ambiguity be taken seriously as one of the interpretative prisms through which to view these formats. Our respondents were often shy to 'admit' their emotional investment in either *The Block* or *Grand Designs* but once emboldened by others, the discussion became lively and funny. In one memorable focus group two different renovators were planning to apply to be contestants on *The Block*. Other focus group participants were quite puzzled by this choice: why subject yourself to such scrutiny during an already demanding renovation? However, the budding contestants were adamant it could work to their advantage. Now they had learned the 'rules' of both the program and renovation, they could and almost *should* harness this knowledge. Based on the strength of their renovation, one participant had been strongly encouraged by friends to apply. But her partner had significant reservations that she could not allay:

> A lot of our friends were like, you need to apply, you need to apply. And we were going to do an application, but I think it was still too fresh for my partner. He was more stressed out by the process than what I was … when *The Block* started he was like, 'You can't watch this show because it's too real for me' … And I was like, 'Come on. We've done this.'

In recounting this interaction about her partner, this renovator quipped 'He's like, "No, brings back bad memories." I was like, "You weren't even here."' For the other renovator, appearing on the show could actually be liberating, a chance to prove their mastery of skills and perhaps even gender. As she put it 'we are actually going to apply … we love doing things. We don't need any men in our lives. It's just like, we are the men and women in the household'.

Our research, then, chimes with Skeggs and Wood (2012) who write insightfully of the important conceptual work enacted by affect and ambiguity. As they explain:

> [T]here is a contradiction between what the televised field of exchange demands (heightened performance for sensation and attracting viewers for profit) and the demands of governmentality (inciting self-regulation) which opens up a gap into which many of our viewer's reactions enter … Our middle class respondents did not believe that advice was directed to them, and our working class respondents often became outraged by the advice on offer. They did sometimes like 'tips' but only if these fitted into their aesthetic, social and domestic lives.
>
> *(221)*

For Skeggs and Wood, this ambiguity reveals the bluntness of governmentality instruments for understanding reality and lifestyle formats. Instead, what their research suggests is that not only do viewers resist and vacillate themselves between, for example, feeling sympathy for a TV participant and rushing to judge their fashion sense, the genre itself exposes the mechanics of what they call the 'normative performative' which makes 'almost every practice subject to performance evaluation'. Simultaneously, audiences occupy the position of 'the judge' but also become conscious of being themselves judged (222).

Such ambiguity clearly animates the genre and, as discussed, was often at work in the ways in which Australian renovators used these programs in both symbolic and material ways. The insights from our research reinforce the value of empirical qualitative methodologies, to supplement the traditional textual analysis of television representations and media industry studies. Our interviews and focus group discussions have demonstrated how the shows provide a lexicon for home renovators to use in their 'real' interactions with practitioners and also offer ideas for materials and building methods to be used in everyday renovation projects. More than anything the research confirmed the active interpretations on the part of audiences, which in the context of plural and diversified media landscape extends opportunities for scrutiny, participation and dialogue.

## Conclusions

Drawing on the empirical research we have shown that it is important to consider both *Grand Designs* and *The Block* beyond their televisual aspects of representation, storytelling and aesthetics. The insights about the symbolic *and* material uses made of these programs by audience members would be impossible to glean through textual analysis only. Inspired by the reference to a particular sustainability product on screen, for example, renovators might insist to a builder this be tracked down and tried. While neoliberal arguments could easily be applied to the responses from our audience members, by showing how the programs function as engines of individualisation, competitiveness and consumerism, for example, we find this too bald an analytical method for uncovering more nuanced audience engagement. With its potential for a 'cookie cutter' style, the neoliberal critique of reality TV seems unable to account for either the truculence of its viewers or the diversity of its genre. Indeed, so total is its reach as an explanatory force that some media scholars have been compelled to ask, in rhetorical tones, 'is reality TV neoliberal?' (Redden 2018). What gets lost in enquiries dominated by the neoliberal reading is the specificity of the encounter. While we do live in neoliberal times, the pertinent question for social and media critique needs to also include what follows – by what means, and why? In response, our study has attempted to contribute interpretations that are not wholly subservient to the inexorable force of the market with reality TV as its chief uncompromising commercial and governing agent. It is also self-evident that the content delivery, reach, business models and audiences of mainstream, terrestrial TV are rapidly evolving. Since, as discussed, reality TV programming is fast becoming the major output of free-to-air TV itself, it makes as little sense to talk of a singular analytical approach to this format as it would to TV in general. One intervention is to focus on local cross-cultural analyses (Kavka and Weber 2017). Our call is to add to this by resisting the uncritical acceptance of the meanings or market desires of the genre, without paying due attention to the diversity, complexity and indeed nuance with which audiences make sense and negotiate the influence of reality TV in their everyday lives.

## References

Aggelli, Aggeliki and Gavin Melles 2015. "Greening The Block: Sustainability in Lifestyle Mainstream TV". In *Unmaking Waste 2015 Conference Proceedings*. Adelaide.

Aitkenhead, Decca. 2008. "Man about the House." *The Guardian*. 11 October. www.theguardian.com/artanddesign/2008/oct/11/design-architecture.

Allon, Fiona. 2008. *Renovation Nation: Our Obsession with Home*. Sydney: University of New South Wales Press.

Andrejevic, Mark. 2014. "When Everyone Has their Own Reality Show." In *A Companion to Reality TV*, edited by Laurie Ouellette, 40–56. West Sussex, UK: John Wiley & Sons.

Couldry, Nick. 2010. *Why Voice Matters: Culture and Politics after Neoliberalism*. London: Sage.

Hargraves, Hunter. 2014. "Tan TV Reality Television's Postracial Delusion." In *A Companion to Reality TV*, edited by Laurie Ouellette, 283–305. West Sussex, UK: John Wiley & Sons.

Hickman, Arvind. 2016. "AdNews Analysis: The top 50 TV Programs of 2016." 29 November. www.adnews.com.au/news/adnews-analysis-the-top-50-tv-programs-of-2016#Ap1hkHQAePpV3IFp.99.

Holland, Kate, Warwick Blood and Samantha Thomas. 2015. "Viewing the Biggest Loser: Modes of Reception and Reflexivity among Obese Pople". *Social Semiotics*, 25: 1, 16–32.

Hollows, Joanne and Steve Jones. 2010. "'At Least He's Doing Something': Moral Entrepreneurship and Individual Responsibility in Jamie's Ministry of Food". *European Journal of Cultural Studies*, 13: 3, 307–322.

Hulse, Kath, Aneta Podkalicka, Esther Milne, Tomi Winfree and Gavin Melles. 2015. "'I'd Just Google It': Media and Home Renovation Practices in Australia." www.lowcarbonlivingcrc.com.au/sites/all/files/publications_file_attachments/rp3021_project_report2016.pdf.

Kavka, Misha and Brenda Weber. 2017. "Introduction: Transnational Gender Cultures and Reality TV". *European Journal of Cultural Studies*, 20: 1, 3–9.

Lunt, Peter. 2014. "Reality Television, Public Service, and Public Life A Critical Theory Perspective." In *A Companion to Reality TV*, edited by Laurie Ouellette, 501–515. West Sussex, UK: John Wiley & Sons.

Miller, Toby and Marwan M. Kraidy. 2016. *Global Media Studies*. Chicester: Polity Press.

Ouellette, Laurie, ed. 2014. *A Companion to Reality TV*. West Sussex, UK: John Wiley & Sons.

Ouellette, Laurie and James Hay. 2008. *Better Living through Reality TV: Television and Post-Welfare Citizenship*. Malden, MA: Wiley-Blackwell.

Podkalicka, Aneta, Esther Milne, Kath Hulse, Tomi Winfree and Gavin Melles. 2016. *Hashtag Sustainability? Home Renovators' Media World*. Melbourne. www.lowcarbonlivingcrc.com.au/resources/crc-publications/crclcl-project-reports/rp3021-project-report-2-hashtag-sustainability.

Podkalicka, Aneta, Esther Milne and Jenny Kennedy. 2018. *Grand Designs: Consumer Markets and Home-Making*. London: Palgrave Macmillian.

Redden, Guy. 2018. "Is Reality TV Neoliberal?" *Television & New Media*, 19: 5, 399–414.

Silk, Michael, Jessica Francombe and Faye Bachelor. 2009. "The Biggest Loser: The Discursive Constitution of Fatness". *Interactions: Studies in Communication and Culture*, 1: 3, 369–389.

Skeggs, Beverley and Helen Wood. 2012. *Reacting to Reality Television: Performance, Audience and Value*. London: Routledge.

Squires, Catherine R. 2014. "The Conundrum of Race and Reality Television." In *A Companion to Reality TV*, edited by Laurie Ouellette, 264–282. West Sussex, UK: John Wiley & Sons.

Weber, Brenda R. 2009. *Makeover TV Selfhood, Citizenship, and Celebrity*. Durham: Duke University Press: 51.

White, Mimi. 2014. "House Hunters, Real Estate Television, and Everyday Cosmopolitanism." In *A Companion to Reality TV*, edited by Laurie Ouellette, 386–401. West Sussex, UK: John Wiley & Sons.

# 14
# ENGAGING WITH REALITY TELEVISION

*Annette Hill*

## Introduction

The argument of this chapter is that audience engagement with reality television tells us something about the multiple ways people connect or disconnect with television in the context of their everyday lives. Drawing on qualitative audience research of the global talent format *MasterChef*, the findings show audiences cognitively and affectively engage with this reality television series within the place and time of their local reception contexts. For reality television, there is not one type of audience, but rather people switch between different modes of engagement, performing the roles of viewers and users, fans, critics and participants; and there is not a general picture of a global audience but rather local and transnational audiences. Discourses of audience engagement in industry and academic contexts try to label people as one thing or another. But audiences are often performing several roles at the same time, highlighting the complex dynamics of engagement at work in entertainment television.

## What Is Engagement?

Engagement refers to the various ways we encounter and experience media within society and culture. Engagement is a broad term and has often been used in common sense ways to signal audience attention in television and social media. The synonyms of participation, or experience, are also commonly used when speaking of audience engagement, and this makes the meaning of engagement a slippery concept that somehow addresses subjective states of being. To be engaged with the media means more than being taken up with, diverted by, or reactive to a cultural artefact. This chapter argues for a more concrete meaning of engagement that captures the power of the concept as constituting in itself forms of agency.

Often the concept of engagement signals a common sense notion of people's interest, attention and involvement in media. A typical way of using the term engagement might relate to social media engagement, the number of tweets and re-tweets around a reality TV show, for example. When the term engagement is used in relation to screen culture and entertainment it usually means audience involvement as a dimension of human experience, that is to say the subjective realities that shape our relationships with media content, artefacts and events.

Within political discourses of engagement, there has been a focus on rational and subjective dimensions of being political and participating in civic cultures. In *Media and Political Engagement* (2009) Peter Dahlgren charts the affective turn in political communication and media and cultural studies (see Gregg and Seigworth 2010; Dahlgren 2013). He argues that both rational and subjective dimensions are crucial to models of civic subjectivity and reflections on normative visions of democracy. Current research on political engagement addresses passionate politics, or social media engagement and political activism. Zizzi Papacharissi (2010) explores what she calls the private sphere of democracy in the digital age, where blog and Twitterspheres contain affective networked publics. This body of work constitutes a more politically orientated perspective on engagement in relation to media and democracy, and although this is not the focus of this chapter on television audiences it is a useful theoretical foundation on engagement as political agency.

Within the media industries, including public service and commercial media, engagement is a word that is shorthand for audience and consumer attention, used as an industry metric in conjunction with daily, weekly and monthly ratings. In this sector, audience information systems, from ratings data to social media analytics, attempt to measure and secure engagement for audiences and consumers of cross-media content (Napoli 2010). There is also recent policy engagement by advocacy (children's media groups, citizens groups) and industry players (broadcasters, industry associations, producers), who have become intensely involved in debates about securing a better financial basis for media content. This engagement in policy runs parallel to cultural engagement: engagement by audiences, fans, users and so forth, whose choices about what they engage or disengage with influence the ways in which media policies and strategies around content are being formulated by both policy-makers and industry (see Hill and Steemers 2017). Such an area intersects with consumption, an aspect of engagement that relates to commercial logics, most often in the form of advertising.

Within audience studies, there are quantitative and qualitative research in different types of interaction, engagement and disengagement with multiplatform brands, social media, sports and other live events. This kind of multi-method and multi-site research draws on a tradition of audience studies within popular culture that looks at processes and practices, that is to say the dynamic and often contradictory ways people encounter and experience cultural artefacts (see Hermes (2005) and Hill (2015), among others). John Corner has argued in his book *Theorising Media* (2011) that there is an increasing interest in subjectivity, and notions of engagement, interaction and participation in media industry and academic research. As television and related social media work across different distribution flows and media devices it's important to consider the multiple modes of engagement of audiences for contemporary screen culture. He notes how different kinds of engagement and involvement occur within a changing 'economy of attention' within which commercial activities, media productions and the politics of media industries have to operate. In this economy, visibility is clearly a pressing requirement, although one that does not by itself guarantee engagement (Corner 2017; Hill and Steemers 2017). This suggests researchers need to critically reflect on the relationships between media, such as television and social media, and other factors, such as everyday life, that help shape the structures and processes of engagement.

One of the things to bear in mind when using the term engagement are the many synonyms that are often referred to by researchers in relation to this term: attention, interaction, participation, affect and emotion, subjectivities and identities, or experience, all make use of conceptual developments from other fields in shaping the meaning of engagement for different media contexts. It is worth briefly addressing some of these related concepts to engagement to understand recurring themes within the meaning of this term (see Dahlgren and Hill 2018).

If we take participation, this can be used to mean something similar to engagement. There is participating in the media, for example as a contestant in a reality TV series, a caller for a radio show, a speaker in a podcast, and participating through the media, which means that cultural artefacts are a resource for thinking and feeling about something or someone, and transforming reactions or interactions into participatory practices. For Nico Carpentier (2011) engagement is something different from, although important to, participation within democratic societies; this is a useful distinction when we want to see how political engagement is a pre-stage to participation in politics. But, for popular culture, engagement is a subjective state that is in and of itself important to understand and research.

By subjective state, we can understand this in relation to other synonyms often used when talking about engagement, such as affect, emotion and experience. Affect is the collective side of emotions, involving infrastructures within industries, society and culture which shape shared media experiences. One way of understanding affect is to look at the earlier work of the cultural scholar Raymond Williams. According to Sharma and Tygstrup (2015: 2) affectivity is 'what tinges or colours the way in which we take part in the environments we find ourselves placed into' (2015: 14). For Williams (1978), structures of feeling give expression to prevailing cultural currents and moods of a given historical moment; they are implicit and inchoate, yet can still impact on people's cultural horizons. Affect connects with emotion in that it is the dynamics at work in the expression of emotions and feelings. These subjective states help to shape our lived experiences (Frosh 2011). We should bear in mind when researching engagement that this is enmeshed within our media experiences. As John Corner (2017, 5) explains:

> It is the experiences, both shaping and shaped, which variously precede, inform and then follow media engagements that are often the real matter at issue. Research into media engagement is often, if only partly, an inquiry into the realm of the experiential and its contemporary cultural resources, with all the challenges that implies.

More specifically in relation to audience engagement there is an urgent need to find useful ways of researching engagement that capture both a pragmatic meaning of audience attention and a more subjective meaning of the relationships between people and media content. For the purposes of this chapter, a dynamics of engagement signals the different interests, feelings and involvements – from love, hate to indifference – which shape the subjective dimensions of media audiences. If we extend our meaning of engagement to different settings that include both material infrastructures, such as industrial settings, and the symbolic interactions and subjective experiences of everyday life, we can widen our horizon of understanding of how and why people engage and disengage with cultural artefacts such as reality television.

## The Dynamics of Audience Engagement

The dynamics of engagement addresses the cognitive and affective modes of engagement of audiences and users, participants and producers, highlighting the different positions and intensities of engagement in various contexts. This is a sense of engagement as multiform, where engagement is based on core elements but experienced in diverse ways, such as positive and negative engagement, or disengagement. Positive engagement typically might include emotional identification with a character in a television series, for example, inviting sympathy and empathy, voting for the underdog, or sending encouraging tweets. Negative engagement might involve emotional dis-identification with a character in a television series, closing down sympathy, or trash talking on Twitter. These cognitive and emotional modes of engagement often work in tandem, and professionals and performers are fully aware of how to craft both positive and negative emotions even in the same character, thus inviting intense feelings from audiences. For example, hate watching a scripted reality series such as *Keeping up with the Kardashians* (E!) is one example of negatively engaging with television as an entertaining experience.

There is little research on how disengagement happens and why it is a routine feature of our media experience. For example, audiences may disengage from a TV series because it is too emotional, or they have become bored with a storyline. Disengagement can be sudden, a brusque disconnect with a series, or it can happen gradually over time. Disengagement is also a means of interpreting how audiences opt out on a regular basis, sometimes due to the simple fact there isn't enough time in the day and they need to make room for other content, but also due to disaffection and even anger with an entertainment brand.

Media engagement is highly dependent on context. There is the context of time, including fleeting engagement with a live event, or embedded engagement with a brand over a sustained period. For example, a reality talent show is often based on live events, but can also be a recurring feature of a television season, and as such a brand can be based on the immediacy of live, short-form engagement, and at the same time be based on audience expectations that the live experience is embedded in routines, year on year (see Hill 2017). Sporting events, such as football seasons and tennis tournaments, all rely on mixing different kinds of fleeting engagement and embedded engagement with institutional, seasonal and everyday constructions of time. How people make time for television content is perhaps one of the most pressing issues for engagement, for example storing up time to make space for television drama as binge viewing, compared with integrating live television into everyday life and family routines.

Similarly, the context of place and space is also significant to media engagement. The places of media engagement indicate how audiences find media content, including media distribution and broadcast, commercial and streaming platforms, as well as illegal distribution and piracy sites. Although audiences often access media through mobile devices they are engaging with content in a specific place, for example, a region or country, or in the home, or on the train. The spaces of media engagement include digital spaces across television and social media, and the spaces of everyday life. In terms of popular culture, how audiences access content and then embed it in their everyday routines is a major aspect of understanding engagement, especially given the myriad spaces people can pick and choose from for mobile viewing.

## Engaging with Reality Entertainment

Reality television is a container for a range of diverse programmes, series, formats and events in which elements of documentary, talent shows, gameshows, talkshows, soap operas, melodramas and sports, mix together to produce sub-genres. We can broadly define reality television into two distinct spaces that draw on various sub-variants of other genres across fact, drama and entertainment (see Hill 2015). There is the 'world' space of television programmes set in hospitals, airports, or hotels, which are often described as 'fly on the wall', 'docusoap', or 'reality soap' to signal the mix of observational style documentary and soap opera elements within this style of reality television. The other kind of reality content includes the 'entertainment' space of programmes set in specially designed studios, houses or locations. Many examples of the 'entertainment' space of reality television can be found in competitive reality formats such as *Big Brother* (Endemol Shine) and talent shows such as *Strictly Come Dancing* (BBC). Such 'format' programming represents a business model for the protection of a creative idea and the selling of this idea in a form that can be replicated in different regions. This type of reality television is usually described as 'competitive reality', 'lifestyle' and 'factual entertainment' to signal the mix of entertainment (talkshows, lifestyle and competition) within these series. The inter-generic space of these series and formats, set in created for television locations, usually contain participants as contestants who are both performing as themselves and competing in a reality contest. Often these formats contain celebrities and professionals, such as chefs, food critics, singers or music producers, and there can be celebrity versions of the formats that work alongside amateur versions.

The case of global talent format *MasterChef* (Endemol Shine, created by Franc Roddam in 1990) highlights the dynamics of engagement for varieties of national and transregional audiences, from at home audiences to social media users, and audiences as participants in the competition itself. The premise of *MasterChef* is a cooking talent show where amateurs audition and attend bootcamp before being whittled down to a select group who participate in a tough cooking competition judged by professional chefs. One person is crowned the *MasterChef* winner. The format had been around for a decade but a reboot in Australia changed the nature of the brand from a gameshow to a talent show. According to the official website (Endemol Shine 2017), the format is: 'produced in over 50 territories worldwide and broadcast in over 200 territories, *MasterChef* is watched globally by over 250 million viewers and has transformed over 100 amateur chefs into professionals.' There is the original amateur version of the format, a celebrity version, a professional version, a spin off that includes children, and an all stars series including previous winners. There is also a *MasterChef* online universe that contains different information and entertainment elements, from recipes and how to videos, to social media promotions, cookbooks, travel and lifestyle and other consumer products.

According to official website, 'the success of *MasterChef* is down to its highly adaptable format that taps into a global appetite for watching everyday people fulfil their dreams of achieving something extraordinary' (ibid.). This gives the global format various engagement points with transnational audiences; it is crafted to contain strong emotional engagement with ordinary people competing in the talent show, a typical feature of the reality talent show, and it is crafted to engage local audiences in their food cultures, a typical feature of scalable reality formats that are made to measure for large and small territories (Hill and Steemers 2011). Summed up by an executive producer, 'It gives you that country's relationship with food. People watching it should feel like it is their show' (Warwick 2013).

In researching *MasterChef*, the case study used a pragmatic and situated approach to the study of the production of the format and audience engagement with it across three countries: the UK, Denmark and Sweden (2013–2016), in collaboration with Endemol Shine; the case study was part of the project *Media Experiences*, funded by the Wallenberg Foundation (see http://mediaexperiences.blogg.lu.se). The qualitative empirical data in this chapter refer to British and Danish audiences. The data included 16 production interviews with executive and creative producers; production observations of the format during the filming and editing process; 27 interviews with contestants in the auditions and semi-finals; and 30 audience interviews with individuals and small groups (1–3 persons in interviews of 40–60 minutes), aged 18–55+, containing a mix of males and females, from a range of professions, such as student, waitress, soldier, accountant, teacher and pizza baker. There was also a ratings and social media analysis and website analysis of the official and unofficial *MasterChef* online universe. The fieldwork was designed by Annette Hill, and conducted by Julie Donovan, Tina Askanius and Koko Kondo.

According to Tasha Oren (2013), *MasterChef* signals the shift in food television from cooking competition, or gastro porn, to formats gone wild, what she playfully describes as culinary S&M. By this she is referring to the stress tests, the emotional highs and lows, and the critique of food within competitive reality. She argues that competitive food formats have shifted emphasis from teaching audiences how to cook to being critical of food culture. The food format 'encourages audiences' investment in their own expertise as critics, diners, foodies and even wanna-be professional chefs. FoodTV, in turn, feeds back into a web-powered, gastro-culture and critique-economy where appraisal outranks delight' (2013: 33). While Oren's comments may apply to a critical analysis of the format, and its representations in America (think Gordon Ramsey), local variations of the format signal the significance of regional contexts to engagement. Audience research for the amateur version of the format suggests something different is taking place; not a hyper critical mode of engagement with food criticism, but an emotional mode of engagement with the value of cooking from scratch in local food cultures.

In terms of the dynamics of engagement, we can see a primarily positive engagement. One viewer commented on their emotional engagement with a contestant who went on to win: 'She really did cook with her heart and soul, you really felt like she was bringing something of herself and her history, putting it all into it' (35-year-old British female project manager). Another viewer spoke of character engagement: 'I had tears in my eyes when I watched, it, we were so rooting for him' (46-year-old British female public health consultant). There was a sense of the British version as a pro social brand, something that represented the BBC's ethos of entertainment for the public good. People described it as an invitation to celebrate food, not necessarily critique it:

> *MasterChef* is an invitation to dinner, really. When you invite people to dinner and you make something really nice for them, you make the effort, the look on their faces just says it all … Feeding people is something very important.
>
> *(46-year-old British female public health consultant)*

Such a comment highlights the sociability of audience engagement with *MasterChef*.

There were examples of negative engagement with the format. Viewers criticised the dramatic use of editing: 'Is he going to get his pot to boil in time, is he, is he? There is always going to be some fake drama in there' (34-year-old New Zealand female art director). Regular viewers criticised the series for being formulaic, signalling boredom with the repetitive nature of the format: 'The people feel real. The format maybe feels a bit contrived' (35-year-old British female project manager). Others critiqued a reliance on emotions in the storytelling for this kind of talent show: 'in the recent series this one woman was crying all the time and I grew really tired of that. In the end I came close to stop watching' (32-year-old Danish female unemployed). People were wary of being manipulated by the producers, drawing attention to the affective structures of talent formats and wary of what they called 'emotional porn', something they saw as a negative feature of reality television in general. They positively engaged with *MasterChef* when it avoided being overly critical or dramatic, comparing it to other talent formats where there was perceived to be 'this negativity that counts as entertainment, and to me that is a bit sad' (42-year -old Danish female actress).

By far the most significant finding from the audience research was the embedded engagement of the series in both countries, where people routinely made a social ritual of watching, eating and cooking when *MasterChef* was on live television. At a time when audiences can choose to watch content anytime, anywhere (with licence fees, subscription, or get arounds for geo-blocking), this was an unexpected finding that signalled the power of live television for social ritual viewing. In Denmark, for example, the scheduling of the series on the commercial channel TV3 during weeknights matched the rhythms of people unwinding after a busy day. The typical term used to describe *MasterChef Danmark* was cosy, or hygge TV. The metaphor of actually eating *MasterChef* was common in viewers' discussions, for example snacking on the show, a mid-week treat, something nice and delicious. As this young viewer explained: 'I get home, I take some food and I go to my bed with the iPad and then I watch. I do that almost every day' (10-year-old Danish schoolgirl). *MasterChef* was a marker point in everyday family routines, signalling a quiet moment when parents and children could unwind alone or together: 'It is all about timing … we've finished eating and the kids are running around like wild dogs, we can put this on and sit down. It fits into the rhythm of the family' (47-year-old Danish male engineer).

The ten-year-old schoolgirl who watched the series after school highlights the sense of *MasterChef* mentioned earlier by another viewer as an 'invitation to dinner.' This sociability was a form of embedded engagement with younger viewers as well as their parents: 'I have a boy who is completely crazy about it. He'll be seven soon and he loves watching it with me. So that just makes it even nicer, the two of us together on the couch' (32-year-old Danish female shop assistant). Students described cooking a meal in dormitories and watching the series together as a social ritual. Working mothers described watching *MasterChef* with the family, and sometimes saving the show as a treat to watch alone; one mother watched it on catch-up digital services while in the bath. Other parents reflected on the format's emphasis on cooking good food from scratch, something they felt was a significant value for food culture that they had not necessarily learned growing up in Britain or Denmark. For example: 'My seven-year-old daughter is inspired by *MasterChef* to cook which is quite good because I couldn't inspire her to cook' (46-year-old British female public health consultant). Or, as another person reflected:

> it is in our culture to share good meals … I hope it will bring back a focus on food as a unifying force that will get us together around the table so we can talk about our day, our thoughts and feelings.
>
> *(32-year-old Danish female shop assistant)*

When it came to social media, viewers often shared pictures of food they had cooked after watching the series, or screen shots of standout dishes made by amateurs, spreading the brand as food inspiration. Instagram was a favourite, with the tag *MasterChef* most often used when uploading a home cooked dish. Few viewers spoke about using social media for gossip about amateur participants, critique of food culture, following celebrities, or promoting their favourite contestants, but rather circulating user-generated content about good food. If we consider the entertainment space of *MasterChef*, British presenter Gregg Wallace is perhaps best associated with the cheeky fan mashup video Buttery Biscuit Base; this video by Swede Mason was uploaded in 2011 with the note 'editing doesn't get tougher than this' – the video has been seen by nearly 10 million viewers to date (www.youtube.com/watch?v=IfeyUGZt8nk). This sweetly funny homage to the show signals the primarily positive and rather endearing relationship of viewers with the series: 'If you leave it 15 years and you ask "what's *MasterChef*?" the answer will be "Buttery Biscuit Base." That says everything about it' (34-year-old British male charity worker).

We should note that viewers were not uncritical of the connections between the format and media, society and culture. Audiences were aware of the hypercritical social media engagement related to *MasterChef* judges, participants, or scandals, but they differentiated between the entertainment and real-world spaces of the format. For example, the judge Gregg Wallace within the series in the UK invited a primarily positive mode of engagement within the entertainment space of the talent show; the celebrity Gregg Wallace invited more negative modes of engagement in the real world. In 2013 Wallace was involved in a minor scandal. A member of the public contacted him on Twitter: '"Hi Greg," ran the message. "I am cycling just over 180 miles in 2 days for Macmillan Cancer Support. Any chance of a retweet?" The presenter's reply was simply: "Gregg?" "No worries mate," retorted the charity bike rider. "It's only people with cancer. You worry about your extra G."' (Hyde 2016). The next year when Wallace appeared in the reality talent show *Strictly Come Dancing* he was the very first celebrity to be voted out by the public. This is an example of how audiences and users switch between positive and negative engagement within the context of the series and celebrity promotion. The performance of Wallace as a television presenter in the entertainment space of the series invites a positive emotional engagement with his personality and relationship with the contestants, a relationship carefully crafted by producers and editors; but the performance of Wallace within the real-world spaces of social media, where Wallace is his own producer, invites negative engagement with his celebrity brand and lack of awareness of civic participation.

Overall, the intended message of the brand as a made-to-measure format is that 'people watching it should feel like it is their show' (Warwick 2013), and this is predominantly what we found in our research – an emotional investment in the amateur series as part of the social ritual of food cultures in Britain and Denmark. People made it their show, feeling a sense of invitation to engage with the local versions of the series within their family routines, reflections on food and practices of cooking and eating. We should be wary of seeing audiences as naive, manipulated by the television industry to follow and like the show because that is what they are told to do; rather we should see the viewing practices of audiences as they differentiate between the value of the series in the entertainment space of reality television, and what their engagement with the series suggests about why this talent show matters to their lives and their experiences of local food cultures.

## Engagement Matters

Engaging with the media, in the context of society and culture, is a significant psychological investment in something or someone that matters in that moment and over a longer period of time. A dynamics of engagement signals 'a larger range of psychological orientations to the world and to the artefacts within it' (Corner 2017). This is why engagement matters; it tells us about the connections across reason and rationality, affect and emotion, and why people connect or disconnect with popular culture.

The case study of global reality talent format *MasterChef* illustrates how the notion of audience engagement needs to be expanded to include the multiplicity of entry points to engagement with reality entertainment, where varieties of audiences, participants and internet users, all work at engaging with content across traditional and digital platforms. The case study illustrates the dynamic character of engagement, underscoring the cognitive, affective and emotional engagement of audiences with reality entertainment. The contexts of engagement are paramount in the ways audiences engage with this kind of entertainment in the place of reception in particular countries, and the everyday practices of audiences and what they value about the series in relation to food, cooking and lifestyle. The significance of time is also paramount in the scheduling of the series and the time audiences make to engage with this kind of cooking show. All of these modes of engagement and contexts to engagement illustrate how the institution of television can shape engagement as embedded in everyday routines, something hard won but easily lost in the new landscape of television content across broadcast and digital spaces.

### *References*

Carpentier, Nico. (2011) *Media and Participation: A Site of Ideological-Democratic Struggle*, Bristol: Intellect.

Corner, John. (2011) *Theorising Media*, Manchester: Manchester University Press.

Corner, John. (2017) 'Afterward: Reflection on media engagement' *Media Industries*, Volume 4, No.1, 1–6.

Dahlgren, Peter. (2009) *Media and Political Engagement*, Cambridge: Cambridge University Press.

Dahlgren, Peter. (2013) *The Political Web: Participation, Media, and Alternative Democracy*, Basingstoke, UK: Palgrave Macmillan.

Frosh, Stephen. (2011) *Feelings*, Abingdon: Routledge.

Gregg, Melissa and Seigworth, Gregory J. (eds.). (2010) *The Affect Theory Reader*, Durham NC: Duke University Press.

Hermes, Joke. (2005) *Re-reading Popular Culture*, London: Blackwells.

Hill, Annette. (2015) *Reality TV*, London: Routledge.

Hill, Annette. (2017) 'Reality TV Engagement: Reality TV Producers and Audiences for Talent Format Got to Dance' in *Media Industries*, Volume 4, No. 1, 1–17.

Hill, Annette and Steemers, Jeanette. (2011) Big Formats. Small Nations: Does Size Matter? in *Small Among Giants*, eds Lowe, Gregory Ferrell, and Nissen, Christian, Gothenberg: Nordicom, 203–18.

Hill, Annette and Steemers, Jeanette. (2017) 'Introduction to Media Engagement' in special section of *Media Industries*, Volume 4, No. 1, 1–5.

Hill, Annette, Steemers, Jeanette, Roscoe, Jane, Donovan, Julie and Wood, Douglas. (2017) 'A Dialogue Across Industry and Academia' in *Media Industries*, Volume 4, No. 1, 1–12.

http://www.endemolshinedistribution.com/masterchef-formats/

Hyde, Marina. (2016) 'Gregg Wallace Falls Out of the Spotlight and Into the Deep Fat Fryer' in the *Guardian*, accessed online 29 October 2017: www.theguardian.com/lifeandstyle/lostinshowbiz/2016/sep/01/gregg-wallace-falls-out-of-the-spotlight-and-into-the-deep-fat-fryer.

Napoli, Phillip. (2010) *Audience Evolution*, New York City: Columbia University Press.

Oren, Tasha. (2013) 'On the Line: Format, Cooking and Competition as Television Values' in *Critical Studies in Television*, Volume 8, No. 2 (Summer 2013): 20–35.

Papacharissi, Zizzi. (2010) *A Private Sphere: Democracy in a Digital Age*, Cambridge: Polity Press.

Sharma, Devika and Tygstrup, Fredrik (eds.). (2015) *Structures of Feeling: Affectivity and the Study of Culture*, Munich and Boston: de Gruyter.

Warwick, Paula. (2013) Interview with Julie Donovan when Head of International Productions at Shine, London November 8th 2013.

Williams, Raymond. (1978) 'Structures of Feeling', *Marxism and Literature*, Oxford: Oxford University Press, 128–135.

## Further Reading

Corner, John. (2011) *Theorising Media*, Manchester: Manchester University Press.

Hill, Annette. (2015) *Reality TV*, London: Routledge.

Hill, Annette. (2019) *Media Experiences: Engaging with Drama and Reality Entertainment*, London: Routledge.

# PART III

# Information, Programs, and Spectacle

Part III offers a range of varied approaches to the content of television, the actual reason we watch, and why we care about television in the first place. This part, therefore, considers multiple aspects of global television programming. Here, authors analyze television texts and flows, ideologies and representations, formats and spectacles. The section is organized to consider some of the most significant transnational genres of television first, with chapters on the resilience and adaptability of the telenovela (Esther Hamburger), the travels and re-interpretations of Nordic-noir crime dramas (Susan Turnbull and Marion McCutcheon), and the global popularity and infrastructures of distribution of televised sports (David Rowe). This part next offers analyses of specific programs and offers multiple means of considering a television program at very close range within a global context. The situation comedy *Outsourced* is scrutinized for how it works through, symbolically and narratively, the realities and contradictions of globalized economies and displaced labor practices as they are experienced (Asha Nadkarni); the landmark mini-series and its more recent remake, both called *Roots*, are tracked as forms of popular historiography, the impact of which shaped multiple national discussions and identities around the origins and maintenance of oppressive, immoral, racist social and economic structures (Ousame K. Power-Greene); the long-form music video, Beyoncé's *Lemonade*, is considered in the context of a history of alternative forms of access to televisual production, providing a crucial space for reworking and navigating complex, multiple, gendered, global representations of Blackness (Ayanna Dozier); and the historical drama *Payitaht Abdulhamid* is analyzed for its instrumental mobilization in the Turkish government's dual desire to both establish cultural hegemony and consolidate its populist style of government (Ergin Bulut and Nurçin İleri). The part concludes with two forms of spectacle associated with very different iterations of media stardom. Priyanka Chopra's global popular ascendency is examined as a rare example of seemingly transcendent multi-media, globally mobile, international stardom (Pawan Singh), while the ongoing media spectacle that has produced Donald Trump's reactionary populism is situated and lambasted as a reality horror show (Douglas Kellner). By considering television programming at these different scales and through these different approaches, this part engages in the multiple ways television matters. It is always viewed locally and yet experienced as part of a global context. Its genres and programs engage politics in terms of governance, in terms of the relations between and among peoples, and in terms of the formation and representation of subjectivities and identities. What is foregrounded and what is marginalized give us clues as to how and what is made to matter and become meaningful.

# 15
# TRANSNATIONAL MEDIATION, TELENOVELA AND SERIES

*Esther Hamburger*

Diversification of the audiovisual landscape has led to a decrease in telenovelas' ratings. The increasing number of digital platforms for distribution of audiovisual contents has turned cinema, television, and the internet into hot sites of transnational competition. Around the world, structures of production change, yet the economics of audiovisual content making seems to be growing. While new platforms open new venues for local indie production (Piñon 2016, 2017), previous network structures remain in place. Diversification happens through post network production of local series, documentaries, and other contents for subscription television[1] or for digital platforms such as Netflix and/or YouTube, Brazil being the second country in the world, after the United States, in number of accesses to both platforms. Even though Brazilians are eager to access post network contents, neither cable television nor the internet are yet available through the entire national territory. In 2014 subscription television came short of reaching 30% of Brazilian households, but due to the current economic crisis, the number of subscribers has decreased to 25%.[2] Internet connection continues to grow, reaching in 2018 around 70% of households. Cellular phones are responsible for most of them; although internet connection through television sets and tablets has increased, these numbers are still small.[3] New platforms, legislation, and state grants have opened up a diverse virtual space, which is not yet consolidated. Opening up for independent production has stimulated a strategic de-concentration in an industry dominated by few networks which both produce and broadcast.

Nevertheless, free and accessible throughout national territory, network television remains remarkably popular. Within network television, the telenovela remains the most profitable television product, and the primary product through which networks compete. Diversification has occurred within the field of the telenovela itself, as international networks that used to import have started to produce their own titles. Diversification also occurs with imported religious-related titles. Turkish soaps, for example, problematize Muslim tensions with Western modern gender relations. Even though these series present low production values, are shot in faraway places, and are dubbed into Portuguese, they have captivated segments of viewers. Meanwhile, the Brazilian Record network has produced and exported the biblical *Ten Commandments* as a telenovela.[4]

The resilience of the telenovela has to do with form. These series can be thought of as proto-interactive, or as digital *feuilleton* in the sense that they are written shortly before being produced and aired. French Brazilian literary scholar Marlyse Meyer (1996) has emphasized the feedback system that goes on with this bicentennial genre which was inaugurated in the *fait divers* sessions of early French newspapers. Contemporary digital *feuilletons* present multi-plot and multi-genre dramatic structures, in melodramatic mode (Gledhill 1987; Williams 2012; Xavier 1996) rather than classic linear plots. Despite a common melodrama structure, the genre also somehow blurs the lines of fiction and documentary, especially when stories are set in contemporary times, and in well-known places. Moreover, during the second half of the 20th century, when television appeared as a modernizing and integrating national force, viewers learned about different ways to experience transforming social relations from telenovelas.

In the late 20th century, Latin American and Brazilian telenovelas have received attention in international scholarly literature. Coming from a range of disciplines in the Humanities including Communications, Film Studies, History, Anthropology, Political Science, Literature, and Cultural Studies, scholars in the United Kingdom (Rowe and Schelling 1991), the Netherlands (Vink 1988), France (Mattelart 1990), Finland (Varis 1988), Australia (Sinclair 1999), and the United States (Straubhaar 1982; Herold 1988; Kotak 1990; McAnany 1992; McAnany and Potter 1993; Skidmore 1993; Allen 1995; Baldwin 1995; Lopez 1995), among others, have written about the genre. This body of work debates the potential of Latin American television for disrupting the North–South unidirectional flow of contents and of aesthetic forms; its relations with popular and/or non-hegemonic cultures. The specificities of the family structured television industries; the relations between television contents with social and/or political events are also debated. The telenovela has been treated as a discrete specific Global South sort of phenomenon that ranges from counter-hegemonic content that might be thought of in conjunction to literature, and politics (Vink 1988; Mattelart 1990; Rowe and Schelling 1991), but it has also been thought of in association with the lowbrow end (Allen 1995; Baldwin 1995) of what Andreas Huyssen called the great divide (Huyssen 1986).

In *To Be Continued: Soap Operas around the World*, Robert Allen (1995) associated telenovelas to soap-operas, since both series are broadcast daily and associated with melodrama. Allen defines soap-operas as *open* in the sense that each title runs for years, without closure. Stories unfold as does daily life. In the same volume, Ana Lopez and Jesus Martin-Barbero agree with the melodramatic base of telenovelas, nonetheless they stress that telenovelas have a clear *closure*, lasting an average of six months. Other differences between soap-operas and telenovelas relate to broadcast time slots and to audiences. Modleski (1986) associates soap-operas – which are broadcast during the day – with mimicking the repetitive and fragmented structure of domestic labor. Characters age as their viewers do, in ways that suggest that diegetic and extra-diegetic temporalities run in parallel. According to Modleski, these series' audiences were 90% female. Broadcast on prime time, telenovelas have reached audiences at least 40% male. Although male viewers don't like to admit they watch programs they identify with women, they have demonstrated good knowledge of both ongoing and past titles (Hamburger 2003; Aidar et al. 2013).

On the continent, a wide range of also diverse research has been dedicated to the subject (Kehl 1986; Leal 1986; Manhaes Prado 1987; Prado 1987; Barbero 1988; Leal and Oliveira 1988; Faria 1989; Barbero and Munoz 1992; Hamburger 1999; Martin-Barbero 1993, 1995; Faria and Potter 2002; Lopes 2002, 2017; Almeida 2003, 2012; Hamburger 2008). From pioneering works in the social sciences, this literature has marked the connections between commercial television and consumerism in a highly unequal society. The literature has grown, many studies of reception stress the different possible readings of the same telenovelas, the daily life routine with these popular television programs, the melodramatic base of the genre. Specialized researchers have developed case studies of specific titles, they have also mapped specific authors, topics, gender and transgender across time. Comparative quantitative systematic data on Latin American telenovela production has been provided by Obitel.

In the early years of the 21st century, despite this accumulated specialized literature, emergent television studies, with rare exceptions, has looked mostly at the renewed Hollywood series. Benavides' account of *Xica da Silva*, *Bette la fea* and Mexican narco-dramas provide welcome suggestive exceptions here (Benavides 2008). The above-mentioned works by Juan Piñon can also be considered as exceptions as they provide suggestive information on recent connections between indies and transnational production companies across Latin America. Other studies emphasize the work of directors who made a difference on film, on network television series, and telenovelas (Carter 2018). These authors move in the direction of mapping connections between contemporary Latin American and global production and circulation of audiovisual contents, including telenovelas, a direction this chapter aims to reinforce.

I am interested in grasping the dynamics of the relations of a field that includes the tensions between network and post network television (Lotz 2007a, 2007b) at the fringes of the Western world (DaMatta 1990). Perhaps the melodramatic base telenovelas share with series such as HBO's *The Wire* (Williams 2012) or with Netflix's *3%* allows for a common ground on which to mark finer distinctions. *Brazil Avenue* is an example of a telenovela that has disrupted the limits producers and marketers imagine for these titles. In so doing it reached Northern prime time slots and highbrow press. It also reached across class, across gender, and across generation audiences. Its consumerist glamorous world speaks of a hegemonic configuration which had recently included large segments of the population who had previously not been classified as "C" in the social scale that television ratings are, in turn, based on.[5]

Like in the telenovela system, the process of making contemporary series incorporates feedback when producing future seasons. Also, like telenovela series, it can allude to ongoing political processes. Credit sequences situate narratives in well-known national landscapes. Daily (but Sunday) prime time telenovela broadcast favors intense involvement with soaps. Binge-watching also favors intensity, despite the annual gap between seasons. When a telenovela does well it reverberates for a couple of years in different countries. In telenovela production, there is no time to think or to elaborate much. Diegetic time does not coincide with extra-diegetic time, although one or another intersection between the two might occur. An extreme case of expanded time happened in the telenovela, *O Grito* (*The shout* written by Jorge Andrade 1975–76). Globo Network aired this story for six months in the late 10 pm slot, but within the diegesis, the whole story happened during one single night (Anzuategui 2013). Time is not uniform throughout the story, i.e. stories might move fast in certain parts, to expand in other parts.

Working in telenovela means being under pressure for the duration of the story. Writers, directors, cast, prefer to work in series, because they have more time to create. Not having that time demands improvisation, but also recurrent use of repetition, reiteration, formulae. While series allow for writing and re-writing, each episode having its own specific director and scriptwriter, writing a telenovela demands producing one episode a day with the same teams of directors and scriptwriters.

Although telenovelas' ratings have decreased, in 2012 one title, *Brazil Avenue*, managed to re-enact the national reverberations Brazilian telenovelas used to provoke during their golden age, roughly from the early 1970s to 1980s and early 1990s. After reaching high ratings at home, *Brazil Avenue* reverberated as an export for another three years. As multiclass, multigender, multiage, transnational phenomena, cases such as this suggest that telenovelas still constitute shared repertoires, i.e. repertoires through which viewers problematize their differences and similarities. In a segmented contemporary world, this genre is still able to bridge gaps.

After 30 years with the most democratic constitution in Brazilian history, concentrated and diversified television content production, and broadcast models battle each other as the country goes through economic and political crisis. On the film front, indie production includes directors and cast from poor segments of the population. These productions confront race and class discrimination from new perspectives. Netflix produced its first series outside of the United States in Brazil. From many choices, in a relatively structured field, the North American VOD platform chose a project by recently graduated young professionals who wrote about their own experience of applying to a very competitive college program, one in which only 3% of the candidates are approved. Even though this reference is not explained in the series, which has a more abstract environment, its premise hit an anxiety that is common to their generation on a global scale. The series narrative includes black and LGBT characters, departing from the mostly white telenovela realm. Netflix has since produced other indie series, and as I write it is about to produce a series by YouTube's successful funk channel Kondzilla. Both cases suggest the ways in which digital platforms challenge network television's concentrated structures.

## Mediating the Nation

References to national issues have been present since early on in literary adaptations, and/or in telenovela stories based in historical events. This national appeal turned denser after 1968, when Tupi network introduced some of the conventions of the new cinemas to telenovela production, where they acquired different meanings. The 1950 pioneer network (named after the main native Brazilian language) regained audiences with a breakthrough contemporary story, shot on location in São Paulo. The use of colloquial dialogue stressed relations of contiguity between the diegetic realm of the narrative, and extra-diegetic audience's world. Under harsh military censorship subsequent titles produced and aired by Globo Network (inaugurated in 1965, and by 1970 the leader in national and international audience preferences, a vocation its global appealing name suggests) made allusion to national performance at international sports competitions such as Formula 1 in 1969 *Bride's Veil*, or soccer during the 1970 World Cup in *Brother's Courage*. During the 1970s and 1980s, allegoric references to the paradigmatic figure of the "coronel" present in the literature as the patriarch who mediates between local northeastern municipalities and the workings of the national state (Nunes Leal 1977 [1949]) permeate telenovelas. Sheriffs, mayors, their wives, their lovers, and priests were also recurrent characters. Censorship has not hindered stories from favoring the expansion of possibilities in the realm of sensuality and gender relations. Divorce appeared in telenovelas before it was legal (Fadul 1993; Fadul et al. 1996). Along the years, female protagonists gained terrain. Very popular stories have helped to sell fashion, as well as commercial products advertised within the narrative and/or during commercial breaks. Spin offs include books that are eventually cited in the stories, soundtrack albums, and so on and so forth.

Toward the end of the 20th century, references to the Brazilian nation became explicit. Especially elaborated electronic credit sequences mobilized national colors and songs. These openings framed intimate melodramatic narratives in space and time. In the 9 pm slot, most titles are set in contemporary times, and in Brazilian territory. Well set in time and space, stories revolve around complicated romantic involvements and identity questions, including LGBT issues. Characters have identity issues, so does the nation.

During the 1970s and 1980s, a main opposition between "tradition" and "modernity" encompasses plots and sub-plots defined according to dramatic oppositions set in terms of gender, generation, class, and place (Hamburger 2008, 2011). A de-politicized sense of "modernity," related to the display of technology devices, fashion, and liberal family and gender relations, can perhaps be associated to what Miriam Hansen called vernacular modernism, in reference to industrial North American film (Hansen 1999). If the notion is not clear for the case of early Hollywood, it is less so in the South American, late 20th-century television case. Nonetheless, the possible parallels are tempting. Like in the studio system, networks controlled both production and broadcast; they also kept their main cast under contract (Schatz 1991). Moreover, stories were set in contemporary urban settings which were presented as sites of "modern" life. Modern here is understood in relation to technological constant update, especially with respect to means of transportation and to means of communication. Building what Stefan Zweig phrased as "the country of the future" also refers to liberal anti patriarch stories. Brazilian titles, unlike other Latin American examples such as Venezuelan (Piñon 2017) or Mexican (Baldwin 1995), have expressed shifting gender and transgender relations, as well as increasingly liberal ways of portraying sexuality (Almeida 1993, 2000, 2003; Hamburger 1999, 2013). Specialized literature has associated liberal gender relations and the small number of children per telenovela family with the sharp, unpredicted, and unplanned decrease in birth rates (Faria 1989; McAnany 1992; McAnany and Potter 1993; McAnany and Pastina 1994; Kotak 2004 [1990]; Kottak 2016).

Telenovela has glamorized life, stimulated consumerism, and reproduced race discrimination (Araújo 2000). The contrast between the space to display gender and the inexistence of a space to problematize race reveals the configuration of Brazilian virtual space. As a series made for women but watched by national audiences, telenovela admitted the expression of provoking gender roles probably because at the time these issues were classified as belonging to intimacy, to the private domain of the home. Race, in its turn, has been considered as a political issue since independence. The myth of racial democracy has justified non-recognition of race discrimination. Race has remained a taboo, and telenovelas have expressed a white version of a country that is more than 50% black (Araújo 2000). Contemporary series such as *3%*, and films such as *White Out, Black In* (2014) bring up race and class discrimination (Hamburger 2017). Unfortunately, they are still not accessible to the millions of viewers who watch network television.

## *Avenida Brasil* or *Brazil Avenue*: A Nation for Transnational Viewership

From late March to mid-October 2012, an average of 46 million viewers watched *Brazil Avenue* from Monday to Saturday, at 9 pm. Produced by Globo Network, this telenovela reached an average of 39% national rating or 64% share.[6] *Forbes* magazine estimated the series cost $45 million and earned $1 billion in the Brazilian national market alone (i.e. partial profit estimated to value 20 times the investment[7]).

In the first two decades of the 21st century, a period that marks diversification of the global (Lotz 2007a) as well as the Brazilian television scenes (Hamburger 2005, 1999), other telenovelas reached better ratings. Nonetheless, *Brazil Avenue* has been a financial and export phenomenon. *Brazil Avenue* is the Globo Network's most exported telenovela so far.[8] The title was licensed to 132 countries and dubbed into 19 languages.[9]

Set in contemporary times in the city of Rio de Janeiro, *Brazil Avenue* updates some of the familiar conventions of the genre. Rather than a sort of modernizing discourse titles of the 1970s and 1980s displayed, this title speaks from an already modern nation, where after 30 years of improving (but yet far from ideal) conditions, large parts of the population have had access to some education and healthcare. The story focused on the inclusion of this new segment of the population in the consumer market. Moving up the social ladder meant moving the story. It also meant having access, and commenting on – not necessarily praising, indeed often mocking – the consumption, eating, and living styles of the wealthy.

*Brazil Avenue* tells the story of the visceral, back-and-forth clash between two obsessive women who, despite each other, managed to move up the social ladder: Carminha, the wicked stepmother seemingly taken straight out of Grimms' Fairy Tales, who benefited from her husband's death and left her stepdaughter in a dumping ground; and the now grown-up stepdaughter, Nina/Rita, who seeks revenge and justice. As is familiar in this genre, viewers are aware of the villain's true nature and know about their evil actions, even if many of the other characters do not, at first. These two women antagonists come from different generations, represent different trajectories in class and gender mobility, but each act to move the story. The male heroes, meanwhile, are soccer players, father and stepson, but relegated to naive, good, and narratively passive innocence.

Visually, the story is illustrated by carefully chosen costume colors and constructed mise-en-scène. Rather than the conventional main opposition between two social groups – a millionaire and a low-middle class character – this telenovela brings three social groups, each situated in a main setting, each corresponding to a different social class: the hip south side of the city where the upper-class dwellings are located; the fictional suburb of Divino; and the literal dumping ground where abandoned kids have been taken care of by an energetic female character.[10]

The abandoned stepdaughter, Nina/Rita (Débora Falabella), seeks vengeance by coming back from Argentina, where she was raised by a wealthy man who adopted her, to infiltrate her stepmother's mansion, disguised as a cook. After gaining the family's confidence, the outsider manages to dominate her enemy. Her empowerment is marked by a wicked tour, in which, in her black uniform, she confronts her employer in the maids' room. The already popular villain Carminha (Adriana Esteves), as usual wearing white, in an inversion of the most obvious bad-black v. white-good color meaning association. The effect is heightened by the use of dramatic low light, full of shadows, creating a dynamic within the sequence that deserves mention on its own. The sequence is a rare 1 minute and 40 seconds chunk, rather long by telenovela standards. But the subject matter is provocative. After all, it rubs salt into the wound of an institution that helps structure Brazilian society: the mistress–maid relationship, which has been transformed in many ways, due to new legislation, and to new demands by young women who have now more years of schooling and other expectations in life than to become a nanny.

In tune with this new energy coming from a more demanding workforce, two years before the economic, and soon thereafter political, crises interrupted three decades of poverty diminishment in one of the most unequal countries in the world, this telenovela promoted hard-working characters from Divino, fictional popular west neighborhood, who were outsiders in the glamorous coastal east side of Rio de Janeiro, a marvelous city. In line with the already mentioned reduction of poverty resulting from inflation control, health, education, and income redistribution policies, *Brazil Avenue* compares, at times humorously, the eating habits, housing, and leisure activities of these characters from "emergent" segments, with those of the wealthy residents in the south area of the city.

*Brazil Avenue* became a topic of discussion during lunch and dinner times, at street fairs, in taxis, in business meetings, in the main daily newspapers, on the phone, and on Facebook. The increase in audience ratings temporarily interrupted the recent trend of a slow but steady decline, which has not yet threatened the leadership of Brazilian telenovelas in Latin American countries.[11] *Brazil Avenue*, as well as other albeit less talked about titles, confirms viewers' interest in telenovelas, a format that has been present on broadcast TV channels since the inauguration of TV in Brazil in 1950. Telenovelas have been aired daily since 1963, and since 1970 – and until 2016, when newscasts transmitted more realistic daily instalments of national drama, and in so doing, have for a while defeated these series – have constantly been among the most watched programs in Brazil, as well as the ones that provide the largest profits. *Brazil Avenue* consolidated a new generation of telenovela professionals. Author João Emanuel Carneiro, helped by a group of four young writers, and the directors Amora Mautner and José Luiz Villamarim, updated the genre to the era of interactive narratives and digital media.

Silvio de Abreu, one of the masters of telenovela scriptwriting, once said that telenovela is about "rooting".[12] His definition recognizes the proto-interactive dimension of the genre. Spectators do not support one character only. There is no single protagonist. The multi-plot structure allows for variable and flexible identification. In *Brazil Avenue*, the intense sequence of events provides abundant raw material for the intense performance of the cast. The situations and dialogue are provocative, frequently having an effect within the narrative in relation to specific characters and often also containing extra-diegetic references, as if including viewers almost directly in the conversation, often in a humorous and politically incorrect ways. Some may root for the unscrupulous villain, others for the sincere and funny stud who divides his love among three women. Or for the hoochie social climber, who ends up betraying herself by falling in love with her escort gay husband. By offering a range of unconventional characters, the telenovela has fed and has been fed by controversy. Examples of extra-diegetic references include debating soccer two years before the 2014 World Cup in Brazil; or presenting characters who recently ascended to "C" class. In so doing, the story was apprehended as referring to, at the time, ongoing social political processes.

In *Brazil Avenue* the sports-based metaphor of rooting is literal. The main character, Tufão (Typhoon), played by Murilo Benicio, is a good man, devoted to his family and faithful to his origins; but he is cuckolded by his wife Carminha (Adriana Esteves), the great villain of the story, an emerging and prejudiced person from the poor neighborhoods who mistreats employees and bites the hands that feed her. Obsessed with revenge, Nina (Débora Falabella) turns out to be Carminha and Tufão's son's sweetheart. Viewers root for either one of these dominant female protagonists.

Nina is a modern cosmopolitan who rides a motorbike, speaks foreign languages, and is a gourmand. Early in the story, before her machinations absorbed all of her time and energy, she would recommend readings to her patrons. She mentioned, for instance, *Primo Basilio*. The Portuguese novel by Eça de Queiroz, a highbrow classic of 19th-century realism, anticipates the topic she wants to deal with in relation to her stepmother. In the novel's plot the maid of the house discovers the secret adulterous love affair of her mistress, and blackmails her. This and other literary references work within the diegesis as clues provided by the maid/stepdaughter in her attempt to open the eyes of the characters to the wiles of the shrew she intends to unmask. References also address viewers. These hints suggest possible sub-texts. Understanding these sub-layers of meaning might not be essential to the understanding of the narrative, but they enrich viewers' experience. Among other self-referential resources, literary quotations indicate the writer's erudition and cultural background, as well as his previous experiences as a writer of film scripts (including 1998's *Central do Brazil* (Central Station), by Walter Moreira Salles). These quotations challenge viewers who share the writer's literary repertoire to participate in a sort of game where the task is to decipher the possible connections between the extra-diegetic references and the internal world of the narrative, and the relationship between these levels and the unfolding of the plot and sub-plots.

Carminha excels in being excessive. She is excessive in being evil. She is excessive in her shortsighted ambition, restricted to cash and consumption. She is excessive in her pettiness and bad taste. The excellent performance of actress Adriana Esteves, recognized by audience and critics, reverberates in and outside the diegesis through the self-reflexive lines of her worst enemy, Nina. For example, Nina, amid her vengeful attack, orders the stepmother to obey her command to pretend that nothing has happened. The character recognizes the power of her antagonist (and, by implication, the talent of Adriana Esteves) when she returns the comments collected by the narrative and stresses: "You have to keep your performance impeccable as always." Other lines like this draw attention to the artificiality of the television spectacle. Tufão says that his family "could be part of a TV show." The character also observes that excessive drama "seems like a telenovela."

The interplay of sub-texts and text multiplies the layers of possible readings, increasing the telenovela's ability to mobilize viewers from diverse segments of the public. By making explicit diegetic references to extra-diegetic objects (such as to the novel *Primo Basilio*), the telenovela associates its bicentennial melodramatic *feuilleton* structure to a classic novel that can be bought in form of book or series (on DVD). Prime time broadcast, in its turn, stimulates reverberations beyond the usual niche restricted to places and times imagined as feminine and domestic. Examples of this expanded reverberation include front page articles in main daily newspapers such as *O Estado de S. Paulo*, or space in foreign prestigious newspapers such as *El Pais*.

International broadcast of North American series once stimulated a whole wave of reception studies (Ang 1988; Liebes and Katz 1990). At the time Brazil was considered an exception for having native programs that were more successful than foreign series (Varis 1988). Even though this Global South, Portuguese-speaking country developed an export television industry, during the early 21st century, diversification of platforms, programs, producers, and broadcasters has caused a decrease in telenovela ratings, and certain acceptance of foreign series. Some of the foreign series that have found viewers in Brazil are broadcast on network television; others on subscription cable channels; others on Netflix. Most of these series are North American – e.g., *Twin Peaks I and II, Six feet Under, The Sopranos, Friends, Girls, House of Cards*, and so on. *Brazil Avenue* brought Brazilian production – albeit temporarily – back to the "talk of the nation." As a sort of *Le bourgeois gentilhomme* of the 21st century, this telenovela temporarily reiterated the distinguished potential of the local television industry, the already mentioned literature, especially foreign works, have argued for.

## Transnational Travel: Welcoming Sensual, Young, Diverse, Soccer and Dancing Global South

"*Oi, Oi, Oi, I want to see everybody dancing; nobody can stay still; it is crazy*," sings Portuguese Lucenzo in a 2010 clip shot in Portugal. The old urban colonial-like ruined streets, as well as the mixed black, mulatto, and white young dancing crowd, includes bodies of mixed skin tones, a combination of architecture and people that could be in Salvador's Pelourinho (the ancient place for slave torture where Michael Jackson shot his 1996 clip: *They Don't Care about Us*). It could also be Havana, with its frozen in time atmosphere. Or perhaps somewhere in Africa, where at least part of the musical conventions come from. The *reaggeton kuduro* rhythm mixes Caribbean and African rhythms in ways that make it similar to what in Brazil is known as *funknejo*, a mixture of *funk* and *sertanejo*, itself a musical genre associated with the dry northeastern Brazilian heartlands of the "Sertão."

In 1976 when the first telenovela was exported to Portugal, Brazil reversed the direction of colonial media fluxes. The first export title was *Gabriela*, adapted by Walter George Durst from Jorge Amado's novel. Globo sold the production to RTP, Portuguese Public Television, where its liberal values made history in the immediate post-Salazar era. Roughly 40 years later, *Brazil Avenue*'s 179 episodes[13] achieved record ratings not only in Portugal, but in other countries such as Argentina, Spain, and France, in some cases on prime time with a credit sequence inspired in the performance by the Portuguese singer.

*Brazil Avenue* has generated a transnational thrill expressed in more than 400,000 tweets. *Nina y Carmina*, the main antagonists in the plot, became trending topics worldwide. In March 2015, three years after the initial Brazilian release, the title was still on the air in 27 countries, 17 in Latin America. In Mexico, it was aired in two channels simultaneously (TV AZTECA and MundoFOX). In France, it was aired at least three times on prime time. The telenovela has been downloaded from international sites thousands of times, has received hundreds of comments, and was featured in foreign publications such as *Forbes, New Yorker, The Guardian* and *El País*.

With *Brazil Avenue*, Brazil reaffirmed the position, rare for a Global South country, of a worldwide content exporter, including to the Northern hemisphere. Since the 1970s and

1980s telenovelas have traveled in the counter-flux. As an against-the-grain culture industry production, Brazilian and other Latin American telenovelas take part in this coming-from-the-South flux of television contents. According to Geraldo Casé, who in 1992 was in charge of Globo Network's international department, until that moment exports did not represent actual profit. Globo made programs for the national market and that is where their main revenue came from.[14] More than 30 years later the situation seems to have changed. In May 2014, the Brazilian government, still under President Dilma Rousseff, recognized for the first time the economic potential of the national audiovisual industry by awarding Globo Network, in conjunction with the Brazilian Foreign Commerce Association, the *Foreign Commerce of Service Highlight* prize.[15]

Even if native networks did not plan to become transnational exporters, transnational circulation seems to be intrinsic to telenovela, especially within Latin America. In Argentina, when the final episode of *Brazil Avenue* was aired, shortly after the end of the 2014 World Cup in Brazil, the telenovela beat the local program, which had the highest audience until then. In this southern, Spanish-speaking neighbor *Brazil Avenue* was entitled to a public screening of the last episode, with the live presence of part of the cast. This combination of digital screening and live performance echoed a similar 1965 ceremony in Rio de Janeiro's Maracanã stadium organized to include viewers in a live celebration of the last episode of the first television version of Cuban 1948 legendary radionovela by Felix Caignet, *Right to Be Born.*[16]

Telenovela transnational circulation involves a time gap between national and international broadcast that persists even in digital times of instantaneous communication. While the *feuilleton* feedback process happens during the original production-broadcast system, when international broadcast happens this process is already over. Versions of the already broadcast series vary. The duration of each episode varies. The number of episodes also varies from place to place. Some networks exhibit telenovelas in daily time slots. Others follow a weekly prime time convention. The question here is: if the strength of the genre is increasingly based in this digital media feedback process, how to explain the following international reverberation?

Perhaps the repercussion a series achieves during its national original broadcast stands as a reference for the foreign reverberation of that title. Export titles undergo specific changes in relation to each specific buyer. Changes involve re-editing in order to adjust to time slots, different local censorship policies, different imagined tastes and taboos. This process of change, to my knowledge, has not yet been studied. One can speculate about possible answers to that question on the basis of the knowledge already accumulated about the workings of a flexible genre, one that combines formal conventions that are 200 years old, with references to contemporary times and places. Only a few attempts to place this popular television genre in relation to other film and television genres have been made.

Telenovelas share some of the cinematic features recent television studies have found in United States series (Butler 2010). Telenovela includes *inter-textual* references to other telenovelas, as well as *intermedial references* to music, books, cinema, and so on. Certain characters draw attention to the artificiality of this *feuilleton* universe through terrific *non-naturalistic performances* and *reflexive* lines of dialogue. Like Northern series, recent telenovelas also contain *cinematic* decoupage, lighting, costumes, or sharp and precise dialogue, *elements* that help to thicken the plot. Perhaps the contemporary US series share, at least to some extent, what Martin-Barbero has called *permeability* between the industrial production and daily life (Martin-Barbero 1995), a quality other scholars also have called attention to (Lopez 1995), one that links the genre to the 19th-century *feuilleton* (Ortiz 1987; Ortiz et al. 1988; Meyer 1996) that, in its 20th-century electronic version, entailed a whole industrial feedback process which I have mapped for the case of Brazil in order to show the clash of native theories, and the gaps that result from this distorted relational process (Hamburger 1999).

Series can also be thought of as *opened*. The number of seasons is not predetermined. Furthermore, continuation or not has to do with feedback. Future stories, as well as characters' destinies, relate, at least in part, to feedback from viewers, critics, and ratings. Telenovela scripts differ though. Rather than the classic linear structure, a telenovela script is better represented in the figure of a network full of connections. Parallel plots and sub-plots make up for comic and tragic moments. Parallel plots sometimes grow in importance as the relationship between actor or actress, character, and viewers click. Characters who move the narrative are more numerous and their prominence shifts in unexpected ways, as characters might disappear and reappear, new characters might appear, and so on. Multi-genre, and multi-plot telenovelas account for intricate, nonlinear structures. Parallel plots offer diversified gender models, as if to promote a range of possibilities for viewership identification and rooting. The range of positive male characters in *Brazil Avenue* for instance, include, in Divino, a gay soccer player, a stereotyped Italianized macho, and an understanding, good provider, retired soccer star protagonist, who is a cuckold.

## Contemporary Digital Television

*House of Cards*' fifth season Twitter teaser line in Portuguese was *It's hard to compete (Tá difícil competir*[17]), in reference to the local political crisis in Brazil, which favored newscasts over series. Netflix's pioneer original production about the North American political system is popular in Latin America.[18] Even though it is a North American company, the largest VOD platform in the world is stronger among audiences out of the United States than among national viewers. Despite the language difference, Brazil shares with the United Kingdom the position of largest Netflix market outside of the US.[19] No wonder the VOD platform has produced its first outside of the US series in Brazil. From all possible players in a longstanding consistent audiovisual scene, Netflix has commissioned *3%*, an independent production created by Pedro Aguillera and co-directed by J.H. Crema, Daina Gianechini, and Dani Libardi, a group of recent audiovisual graduates from the University of São Paulo who started to work on the project while still in school. The first season was headed by experienced Uruguayan-Brazilian director and *City of God* cinematographer César Charlone. The second season was headed by young film and television director Philippe Barcinski. While I write, the third season is in production employing a whole team of recent graduates.

Netflix does not release numbers, which probably means ratings are a lot lower than network television ratings. Nonetheless, like telenovelas, *3%* has reached transnational Northern, as well as Southern, audiences. Although the made-for-VOD series benefits from the native familiarity with television series such as telenovelas, it defines different aesthetic conventions. Unlike what became the rule in Brazilian telenovela, *3%* manages to be contemporary without defining clear time or space references. By speaking metaphorically, and in the abstract, about the creative team's experience, *3%* speaks metaphorically about the many different violent and arbitrary application processes young men and women go through in different parts of the world, and which, at least to some extent, define their future. The series focuses on an abstract radical *gincana*-like selection process that defines the ones that will cross the border to *the other side*, a sort of utopian land of freedom, welfare, in a dream-like tropical, healthy environment, but not happiness. Inhabitants of the other side include characters pertaining to various social groups including social class, color, gender, and ethnicity. Stylized sets mark the mainland as a sort of contemporary metropolitan region with a sort of global visual look with ruined buildings where both blacks and whites live. Nationality is not even referred to, but inequality and poverty are confronted. Somewhere in between precise definitions of time and space – such as the ones that connect the narrative of Brazilian, Colombian, and other telenovelas to series like *House of Cards* or *Breaking Bad* – and fantastic realms with no reference to extra-diegetic places or times – such as *Game of*

*Thrones – 3%* seems to refer to virtual transnational audiences from within. A sort of abstract critical metaphor for a disenchanted, impoverished, post cold war, and perhaps post capitalist world, the series achieved some transnational repercussions.[20]

According to British HIS consulting, Brazil is the sixth largest online television market in the world.[21] In 2017, for the first time, the investment in internet publicity was larger than the investment in television ads. Brazil is also YouTube's second market after the United States.[22] The popularity of these worldwide web platforms in this Southern country at the "fringes of the Western world" (DaMatta 1991) reinforces longstanding anthropophagic[23] appropriation of narratives and technologies. Early master of film and television conventions connect with the promise of democratization of knowledge and education that is currently renewed by the internet. Local commercial melodramatic series have dominated the Brazilian television industry. For the last 50 years they have continuously updated the display of the most recent electronic devices, motorbikes, cars, planes, helicopters, telephones, and computers. They have proposed constantly renewed fashion design for furniture, hair styles, and clothes. They have renewed expression of liberal gender roles, including LGBT. Over the years they have marked up their narrative universe as being the Brazilian nation. More recently they seem to have expanded beyond national and language frontiers, to include a sort of Ibero-Afro-Brazilian transnational landscape, a sensual dancing southern Latin portion of the Global South. Beyond this consumerist, mostly white, series, contemporary indie productions express diverse aesthetic elaboration and format. Nonetheless, the price of subscription cable television and broadband internet connection currently restricts viewership, despite the higher popularity of Brazilian series when compared to foreign contents.[24] As a result the 21st century starts with the threat of digital divide, a divide that would reinstate the 19th-century divide posed by the emergence of what Walter Benjamin called the technical media, and by the mass culture they produce. Andreas Huyssen has defined response to the threat of the divide as a strategic goal of the avant-garde. Huyssen has also noted, however, that the 20th-century culture industries ultimately absorbed the aesthetic forms with which the avant-garde confronted them. Telenovelas' use of montage and self-referentiality seems to confirm the diagnosis. In order to face this erosion of the divide, rather than trying to rebuild structures according to previously defined distance-×-contiguity relations between viewers and artists, Huyssen envisages critical thinking that might look at processes, a strategy contemporary artists also embrace in their attempt to move beyond preconceived high modernism notions of art and industry (Huyssen 1997). The question about how to bridge divides is crucial in times when populism might once more be isolating artists and intellectuals from crowds.

Television and post-television content tend not to be taken into account when analyzing current challenges the world faces against social inequality, religious and ethnic discrimination, violence, and populist authoritarianism. Nonetheless, the advent of social media as channels of communication that seem to escape the control of institutional powers has brought the media to general attention. The point is that content matters. And the understanding the different meanings of that content may have for different people in different places in the world presents a problem that qualitative and quantitative research might help to solve. The challenge for democracies here and elsewhere is to keep repertoires shared by citizenship, taking into account diversity as well as dissent.

## Notes

1 Subscription television is known in Brazil as *TV a cabo* or cable television. Network television is free and aired by satellite.

2 https://exame.abril.com.br/economia/numero-de-clientes-de-tv-por-assinatura-recua-4-em-12-meses/ (access July 18, 2018)

3 www.abta.org.br/dados_do_setor.asp (access July 18, 2018). Cellular access to the Internet might not favor television content.

4 https://zww.theguardian.com/world/2015/may/29/brazil-biblical-soap-opera-ten-commandments (access November 16, 2017)

5 [Editor's Note]: This is in reference to a prevalent market classification of social classes used in Brazil, where "C" indicates relative middle income and typically education past high school (Aidar et al, 2013).

6 http://folhadatv.blogspot.com.br/2013/04/veja-audiencia-detalhada-de-avenida.html. among other sites that replicate the same information. (access November 22, 2017)

7 www.forbes.com/sites/andersonantunes/2012/10/19/brazilian-telenovela-makes-billions-by-mirroring-its-viewers-lives/#37070b8d44c0 (access November 16, 2017) Story estimates total earning based on the price of 30-second ads. There is no official information on television network's earnings.

8 http://memoriaglobo.globo.com/programas/entretenimento/novelas/avenida-brasil/curiosidades.htm (access November 16, 2017)

9 https://brasil.elpais.com/brasil/2014/05/19/cultura/1400512519_321606.html. (access November 16, 2017)

10 Dumping grounds are places for garbage discharge. Poor people live nearby and work as collectors. For film accounts of these places see *Mitt hen är Copacabana* (Arne Sucksdorff, 1965); *A Margem* (Ozualdo Candeias, 1967); *Boca do Lixo* (Eduardo Coutinho, 1992); *Estamira* (Marcos Prado, 2005), Oscar nominated *Extraordinary garbage* (Lucy Walker; João Jardim, 2010), and *Entre Vales* (Philippe Barcinski, 2012), among others. In contrast with this relative abundant film presence, only this telenovela has presented a dumping ground, already in times when these former sites of mere abjection were perhaps re-valued as places for recycling.

11 For comparative data on telenovelas' ratings in Latin America see Lopes, M. I. V., and Guillermo Orozco (coordinators) (2017) *One Decade of Television Fiction in Ibero America. Ten Years of Obitel (2007–2016).* Porto Alegre: Globo and Editora Sulina.

12 Abreu, S. (1992). Personal communication. E. Hamburger.

13 http://memoriaglobo.globo.com/programas/entretenimento/novelas/avenida-brasil/curiosidades.htm (access November 16, 2017)

14 Personal communication August 1992.

15 http://natelinha.uol.com.br/noticias/2014/05/29/governo-premia-rede-globo-por-exportacoes-de-novelas-entenda-75544.php.
www.sescsp.org.br/online/artigo/9042_O+BRASIL+NAS+TELINHAS+DO+MUNDO.
(access November 16, 2017)

16 The original Cuban *feuilleton El derecho de nacer* written by Félix Caignet has been adapted and re-adapted many times around Latin America, first on radio and later on television. In Brazil, after being aired on the radio, the story was adapted to television shortly after the introduction of the video tape, from December 1964 to August 1965, by TUPI network in São Paulo, and by TV RIO in Rio de Janeiro. Another regional version aired in southern Curitiba, capital of Paraná state, in 1966. National versions followed in 1978 and 2001.

17 https://twitter.com/houseofcards/status/864992970994368512 (access November 16, 2017)

18 www.americasquarterly.org/content/news-quiz-did-it-happen-brazil-or-house-cards (access November 16, 2017)

19 www.bloomberg.com/news/articles/2017-07-17/netflix-scorches-subscriber-forecasts-with-overseas-top ping-u-s (access November 16 2017)

20 www.newyorker.com/culture/cultural-comment/a-brazilian-thriller-that-exposes-the-sinister-side-of-merit ocracy (access November 16 2017)

21 www1.folha.uol.com.br/ilustrada/2017/02/1858978-brasil-e-o-sexto-mercado-de-tv-pela-internet.shtml (access November 16 2017)

22 www.bloomberg.com/view/articles/2017-10-25/youtube-rules-brazil-s-brave-new-media-world (access November 16 2017)

23 In the 1920 Brazilian modernists metaphorically appropriated the indigenous ritual to suggest the ways in which they swallow, digest, and transform foreign contents. For a contemporary study of this concept see Stam, R. (2019). *Transnational cinema, global media: towards a transdisciplinary commons*. London, Routledge.

24 http://noticiasdatv.uol.com.br/noticia/series/por-que-game-thrones-esta-fora-da-lista-das-25-series-mais-vistas-da-tv–13599?cpid=txt (access November 16 2017)

## References


Abreu, S. (1992). Personal communication. E. Hamburger.

Aidar, T., H. Almeida, and E. Hamburger (2013). *Imagining TV Audiences in Brazil: Measuring Television Audiences Globally: Deconstructing the Ratings Machine.* C. Méadel, B. Jerôme. Hampshire, UK and New York, Palgrave Macmillan.

Allen, R., Ed. (1995). *To be Continued: Soap-Operas Around the World.*

Almeida, H. B. (2003). *Muitas mais cousas: telenovela, consumo e gênero.* São Paulo, Anpocs/Edusc.

Almeida, H. B. (2012). "Trocando em miúdos – gênero e sexualidade na TV a partir de Malu Mulher." *Revista Brasileira de Ciências Sociais* **27**(79): 125–137.

Almeida, H. B. D. (2000). "Ethnography in Montes Claros, Gender Constructions in Relation to TV content, Cebrap/Unicamp."

Ang, I. (1988). "(Not) Coming to Terms with Dallas." *Global Television,* Eds. C. Schneider and B. Wallis. New York, Wedge Press: 69–77.

Anzuategui, S. (2013). *O Grito de Jorge Andrade: A Experiência de um Autor na Telenovela Brasileira dos Anos 1970.* São Paulo, Ateliê Editorial.

Araújo, J. Z. (2000). *A negação do Brasil.* São Paulo, Senac.

Baldwin, K. (1995). "Montezuma's Revenge: Reading Los Ricos También Lloran in Russia." *To be Continued, Soap-Opera Around the World.* Eds. R. Allen. London and New York, Routledge: 285–300.

Barbero, M. J. (1988). "Matrices culturales de la telenovela." *Estudios sobre Culturas Contemporaneas* **2**(4–5): 137–162.

Barbero, M. J. and S. Munoz (1992). *Television y melodrama: Generos y lecturas de la telenovela en Colombia.* Bogota, Tercer Mundo Editores.

Benavides, O. H. (2008). *Drugs, Thugs and Divas.* Austin, University of Texas Press.

Butler, J. (2010). *Television Style.* New York and London, Routledge.

Carter, E. (2018). *Reimagining Brazilian Television.* Pittsburgh, University of Pittsburgh Press.

DaMatta, R. (1991). *Carnivals, Rogues and Heros: An Interpretation of the Brazilian Dilema.* Notre Dame, University of Notre Dame.

Fadul, A. (1993). *Serial Fiction in TV: The Latin American Telenovelas.* São Paulo, Núcleo de Pesquisa de Telenovelas, ECA-USP.

Fadul, A., E. McAnany and O. T. Morales (1996). "As temáticas sócio-demográficas nas telenovelas brasileiras (1980–1995)." 20a. International Conference in Social Communication Research. Sidney, Australia.

Faria, V. E. (1989). "Políticas de governo e regulação da fecundidade: conseqüências não antecipadas e efeitos perversos." *Ciências Sociais Hoje* **5**: 52–103.

Faria, V. E. and J. E. Potter (2002). "Televisão, telenovelas e queda de fecundidade no Nordeste." *Novos Estudos Cebrap* **62**: 21–39.

Gledhill, C., Ed. (1987). *Home Is where the Heart Is – Studies in Melodrama and the Woman's Film.* London, British Film Institute.

Hamburger, E. (1993). *Telenovelas in Brazil and the "Feminization" of the Gaze: Preliminary Analysis.* Series The Social Impact of Television on Reproductive Behavior in Brazil. Working Paper n.05. São Paulo, CEBRAP.

Hamburger, E. (1999). "Politics and Intimacy in Brazilian Telenovelas." PhD, University of Chicago.

Hamburger, E. (2003). "Politics and representation: television in a São Paulo favela." *Framework* **44**(1): 104–115.

Hamburger, E. (2005a). *O Brasil antenado, a sociedade da novela.* Rio de Janeiro, Jorge Zahar Editor.

Hamburger, E. (2008). "Performance, Television and Film: Bus 174 as a perverse case of appropriation of the means of constructing spectacular audiovisual form."*Observatorio (OBS*) Journal* **7**: 1–11.

Hamburger, E. (2011). "Telenovelas e Interpretações do Brasil." *Lua Nova* **82**: 61–86.

Hamburger, E. (2013) "Telenovelas, Gender and Genre." *The Routledge Companion to Media and Gender,* Eds. L. S. C. Carter, L. McLaughlin. London, Routledge: 150–165.

Hamburger, E. (2014). "Beto Rockfeller, a motocicleta e o engov." *Significação, revista de cultura audiovisual* **41** (41): 1–23.

Hamburger, E. (2017). "Brazilian Film and Television in Times of Intermedia Diversification." *Companion to Contemporary Latin American Cinema,* Eds. S. Hart, R. Johnson and M. Delgado, London, Wiley: 375–391.

Hansen, M. (1999). "The Mass Production of the Senses: Classical Cinema as Vernacular Modernism." *Modernism/Modernity* **6**(2): 59–77.

Herold, C. M. (1988). "The brazilianization of Brazilian TV: A Critical Review." *Studies in Latin American Popular Culture* **7**: 45–57.

Huyssen, A. (1986). *After the Great Divide: Modernism, Mass Culture, Post-Modernism*. Bloomington, Indiana University Press.

Huyssen, A. (1997). *Memórias do Modernismo*. Rio de Janeiro, UFRJ.

Kehl, M. R. (1986). "Eu vi um Brasil na TV." *Um país no ar: história da TV brasileira em três canais*, Eds. M. R. Kehl, A. H. D. Costa and I. F. Simões. São Paulo, Brasiliense/FUNARTE: 167–323.

Kotak. (1990). *Prime Time Society*. Ann Arbor, University of Michigan Press.

Kotak. (2004 [1990]). *Prime Time Society*. Ann Arbor, University of Michigan Press.

Kottak, C. (2016 [2009] [1990]). *Prime Time Society: An Anthropological Analysis of Television and Culture*. London, Routledge.

Leal, O. F. (1986). *A leitura social da novela das oito*. Petrópolis, Vozes.

Leal, O. F. and S. Oliveira (1988). "Summer Sun." *Theory, Culture and Society* **5**: 81–99.

Liebes, T. and E. Katz (1990). *The Export of Meaning. Cross-Cultural Readings of Dallas*. Oxford, Oxford University Press.

Lopes, M. I. V. and G. Orozco (coordinators) (2017). *One Decade of Television Fiction in Ibero America. Ten Years of Obitel (2007–2016)*. Porto Alegre, Globo and Editora Sulina.

Lopes, M. I. V., Sílvia Helena Simões Borelli E Vera da Rocha Resende (2002). *Vivendo com a telenovela: mediações, recepção, teleficcionalidade*. São Paulo, Summus.

Lopez, A. (1995). "Our Welcome Guests: Telenovelas in Latin America." *To be Continued: Soap Operas Around the World*, Eds. R. Allen. London and New York, Routledge: 256–275.

Lotz, A. (2007a). If It's Not TV,What Is It? *The Case of U.S. Subscription Television. in Cable Visions:Television-Beyond Broadcasting*, Eds. C. C. Sarah Banet-Weiser, and A. Freitas. New York, New York University Press: 85–102.

Lotz, A. D. (2007b). *The Television Will be Revolutionized*. New York, New York University Press.

Manhaes Prado, R. (1987). "Mulher de Novela, Mulher de Verdade." Doctoral, Federal University of Rio de Janeiro.

Martin-Barbero, J. (1993). *Communication, Culture and Hegemony: From the Media to Mediations*. London, SAGE.

Martin-Barbero, J. (1995). "Memory and form in the Latin American Soap Opera." *To be Continued, Soap-Operas Around the World*, Eds. R. Allen. London and New York, Routledge: 276–285.

Mattelart, A. e.M. (1990). *The Carnival of Images: Brazilian Television Fiction*. New York, Bergin & Garvey.

McAnany, E. (1992). "The Telenovela and Social Change: A Brazilian Communication/Demography Challenge." Conference on the Telenovela and Social Change in Brazil. The University of Texas at Austin/Population Research Center.

McAnany, E. G. and A. L. Pastina (1994). "Telenovela Audiences: A Review and Methodological Critique of Latin America Research." Communication Research **21**(6): 828–849.

McAnany, E. G. and J. E. Potter (1993). "Telenovelas and Reproductive Behavior in Latin America." Annual Meeting of the Population Association of America. Cincinnati, Ohio.

Meyer, M. (1996). *Folhetim. Uma história*. São Paulo, Companhia das Letras.

Modleski, T., Ed. (1986). *Studies in Entertainment: Critical Approaches to Mass Culture*. Bloomington, University of Indiana Press.

Nunes Leal, V. (1977 [1949]). *Coronelismo: The Municipality and Representative Government in Brazil*. Cambridge, Cambridge University Press.

Ortiz, R. (1987). *A moderna tradição brasileira*. São Paulo, Brasiliense.

Ortiz, R., S. Borelli and J. M. Ramos (1988). *Telenovela: história e produção*. São Paulo, Brasiliense.

Piñon, J. (2016). "Complex Labor Relations in Latin American Television Industries." *Precarious Creativity: Global Media, Local Labor*, Eds. S. Kevin. Oakland, University of California Press: 132–145.

Piñon, J. (2017). "Jane the Virgin." ReVista Harvard Review of Latin America **Fall**.https://revista.drclas.harvard.edu/book/jane-virgin

Prado, R. M. (1987). "Mulher de Novela e Mulher de Verdade – Estudo sobre Cidade Pequena." Mulher e Telenovela Master of Science Dissertation, Museu Nacional/Universidade Federal do Rio de Janeiro.

Rowe, W. and V. Schelling (1991). *Memory and Modernity – Popular Culture in Latin America*. London, Verso.

Schatz, T. (1991). *O Gênio do Sistema*. São Paulo, Companhia das Letras.

Sinclair, J. (1999). *Latin American Television*. Oxford, Oxford University Press.

Skidmore, T., Ed. (1993). *Television, Politics, and the Transition to Democracy in Latin America*. Washington, DC, Woodrow Wilson International Center for Scholars.

Stam, R. (2019). *World Literature, Transnational Cinema, and Global Media: Towards a Transartistic Commons*. Londonand New York, Routledge.

Straubhaar, J. (1982). "The Development of the Telenovela as the Pre-eminent Form of Popular Culture in Brazil." *Studies in Latin American Popular Culture* **1**: 138–150.

Varis, T. (1988). Trends in International Television Flow. *Global Television*, Eds. C. Schneider and B. Wallis. New York, Wedge Press: 95–107.
Vink, N. (1988). *The Telenovela and Emancipation: A Study on Television and Social Change in Brazil.* Amsterdam, Royal Tropical Institute.
Williams, L. (2012). "Mega-Melodrama! Vertical and Horizontal Suspensions of the "Classical"." *Modern Drama* **55**(4): 523–543.
Xavier, I. (1996). "From the Religious Moral Sense to the Post-Freudian Common Sense: Images of National History in Brazilian Telefiction." 5th Conference on Latin American Popular Culture. Tulane University, New Orleans.

# 16

# OUTBACK NOIR AND MEGASHIFTS IN THE GLOBAL TV CRIME LANDSCAPE

*Susan Turnbull and Marion McCutcheon*

'What are you missing if you're not watching international television?' journalist Mike Hale (2018) demanded of his readers in the *New York Times* in March of 2018. 'A significant slice of the peak-TV pie', he replied to himself – since according to Hale's reckoning there were 108 new seasons of non-American scripted shows available on 'traditional TV or streaming' in the US at that time. Hale's select list of what to choose from this cornucopia of scopic delights includes four crime dramas: the fourth season of the Danish/Swedish co-production *Bron/Broen* (2011–18); the second season of the Swedish 'hybrid' police procedural *Jordskott* (2015–17); *OCTB*, a web series about the Organised Crime and Triad Bureau of the Hong Kong Police available on Netflix, and *The Mechanism* (2018–) from Brazil, directed by Jose Padilha, who also directed the first two episodes of the popular Netflix drug-kingpin series, *Narcos*: *The Mechanism*, which Hale notes, somewhat condescendingly, 'aspires to some of the quotidian procedural force of *The Wire*'.[1]

As Hale's review reveals, and as many have already pointed out – including, for example Lotz (2017), Bodey (2017) and the Australian television industry executives interviewed in Given, Brealey and Gray (2015) – what to watch on television, or your computer, iPad or mobile phone, has become infinitely more complicated since the early days of broadcast television when choice was limited not just by time and place, but also the country in which people lived and even the region, especially on geographically dispersed continents such as Australia. In many ways, audiences are the real beneficiaries of this new streaming landscape, especially in the US given its alleged history of being almost completely impervious to foreign imports, especially subtitled ones, on free-to-air network TV, with the noble exception of PBS where one could at least find select British imports.

Australian audiences, on the other hand, have continued to look both ways when it comes to the American and British shows that have always been prominent in the Australian TV landscape, where the need to ensure the continuing existence of nationally produced scripted drama has hitherto been maintained through the implementation of a quota system. The new streaming platforms, however, have made such attempts to protect the local screen industry even more urgent and challenging. Getting international players such as Netflix and Amazon to invest in 'local' drama content would appear to be the next big hurdle for smaller national television industries such as Australia's, in a market where the increasing popularity of streamed services appears to be further entrenching the dominance of the United States (Aguiar & Waldfogel, 2018; Cunningham & Jacka, 1995; Make It Australian, 2018).

While the reinvigorated global traffic in television content made for those who can afford to access it might mean much more choice from content made in countries other than the US and the UK, what it might mean for the producers, distributors and the creative personnel who make television is still uncertain (Bakøy, Puijk & Spicer, 2017). This is especially true in countries such as Australia, where the television industry and its audience are relatively small, where free-to-air television is hanging on to its increasingly fragmented audience by a reality TV thread, and where public service television, wherever it has managed to survive, is reimaging itself as a digital content provider and relocating online (Turner, 2016). There are megashifts occurring in the global TV landscape, especially when it comes to the television crime drama as it adapts to new influences and makes its way into new markets, shifts that this chapter will illustrate from the perspective of what once might be perceived as the periphery.

As disruption brings mounting uncertainty in local markets, production companies are increasingly reliant on overseas sales (Groves, 2018a). In Australia, as in other smaller TV markets, the commissioning of scripted drama, especially crime drama, is now reliant on the promise of reaching international audiences, providing producers with opportunities to develop content that is competitive in the international marketplace and thereby increasing production costs: in 2017–18 for example, the average cost of Australian-produced adult drama series increased by 13.5 per cent in real terms, to A$1.567 million per hour (Screen Australia, 2018). Examples of the wave of Australian drama series that have found success overseas include *The Code* (2014–16), *Miss Fisher's Murder Mysteries* (2012–), *Jack Irish* (2012–), and *Mystery Road* (2018–), as well as *The Kettering Incident* (2016) and *Secret City* (2016–18). While the first four of these were produced by and for the Australian Broadcasting Corporation (ABC), Australia's major public broadcaster, *Kettering* and *Secret City* were both produced by Foxtel, majority owned by former Australian Rupert Murdoch's News Corp, and intended to sit alongside other 'quality drama' offerings on its Showcase channel (Penny Win, personal communication, 2 November 2017). While in the past co-productions intended for an overseas market tended to be somewhat frowned upon in Australia because of the creative compromises they might necessitate, the situation appears to have changed. Indeed, overseas success is now considered a hallmark of quality, a cause for celebration, and a reason for Australian audiences to watch, as in the case of *The Kettering Incident* which won a much-touted Series Mania Award at a festival in France before its air date, this award becoming a key feature of the show's extensive promotion in Australia, as in Bastow (2016) and BBC Worldwide (2016).

As a genre, the TV crime drama has always crossed borders, whether this be the vigorous trade between the UK and the US in the first decades of the television industry established on the foundations of an already established trade in film (Weissmann, 2012), or the more recent global reach of the *CSI* franchise, touted as the most popular TV series in the world at the turn of the millennium (Turnbull, 2007); or even the international success of the British, arguably 'cosy', *Midsomer Murders*, an unlikely hit in Denmark (Batty, 2013). What has changed more recently, however, is both the point of origin and the direction of the flow in TV crime drama especially in Anglophone markets. Where once innovations in the crime genre might have been imagined to have travelled from centre to centre (US to UK and vice versa), or from centres to the margins (US and UK to Australia or Canada), it would appear that the margins, and not just the English-speaking ones, are now perceived as exciting and influential centres of content production (Creeber, 2015; Jensen & Waade, 2013).

In seeking to identify how this change in the direction of the traffic has emerged, this chapter will explore how a recent wave of Scandinavian crime dramas have had a significant influence on the production of TV crime drama in Australia. While the influence of Nordic or Scandi noir, as it is somewhat loosely described, has already been identified in British and

American crime dramas, such as *Broadchurch* (2013–17) and *Happy Valley* (2014–) in the UK, or the American remakes of both *The Killing* (2011–14) and *The Bridge* (2013–14), we will explore how and in what ways the global success of Scandinavian crime drama has played a key role in the conception and development of three recent Australian crime dramas: *Secret City* (2016–18), *The Kettering Incident* (2016) and *Mystery Road* (2018). To understand the genesis of these series and their position in both the domestic and international markets for television content, we have spoken with the producers and commissioners of these series and surveyed the trade press, news media, and online articles and discussion boards. We used the coding package NVivio to identify themes and topics in our transcripts and collated articles, which allowed us to focus on our area of interest here: the ways in which these three series in particular have embraced what are perceived to be some of the key features of Nordic Noir in terms of mood, the use of landscape, themes and production strategies in an Australian context in the hope of finding new audiences overseas.

It might be noted that in Australia, as in many other countries, the emergence of a taste for Scandinavian crime fiction can be traced back to the series of internationally successful crime novels written by left-wing journalists Maj Sjöwall and Per Wahlöö, who were themselves influenced by the police procedurals of American author and former policeman, Ed McBain. This last point suggests that attempting to identify any point of origin in the rhizomatic roots of crime fiction or television crime drama is always a complex process (Turnbull, 2014). Projected as a ten-book series with the overarching title, *The Story of a Crime*, Sjöwall's and Wahlöö's crime novels following the career of policeman, Martin Beck, were intended to draw attention to the failures of the post-war social democratic ideal in Sweden: a critique that was subsequently taken up in the work of more recent internationally successful Swedish crime writers such as Henning Mankell and Steig Larsson whose books have also been adapted for the screen. The focus on social issues is hardly surprising since crime fiction routinely draws attention to the fault lines in the society within which it is set.

After winning an American Edgar Award for best crime fiction in 1971, one of Sjöwall and Wahlöö's books, *The Laughing Policeman*, was adapted into a film starring Walter Mathau and relocated to San Francisco. This adaptation illustrates an important proposition: while crime fiction has routinely crossed national borders, historically there has been a perception that it needs to be relocated and reimagined on screen in order to succeed, as was indeed the case for the TV series and films based on the work of Mankell and Larsson and the more recent American and British adaptations of *Bron/Broen/The Bridge*. Meanwhile, back in Sweden, all ten of Sjöwall and Wahlöö's Martin Beck series were adapted into films as well as a TV series, with the seventh season of *Beck* (1997–) appearing as recently as 2018 and available on BBC4 in the UK and in Australia on the second public broadcaster, SBS, with its remit to represent a multicultural society. As Hansen and Waade suggest, the figure of Martin Beck is 'not only a pivotal game changer' in the history of Scandinavian crime fiction, but a close analysis of the numerous adaptations that have ensued 'tells a representative story of the development of Nordic Noir as a brand and production strategy' (Hansen & Waade, 2017, p. 127). This is a story that must inevitably include the ways in which series such as *Beck* and the Emmy Award-winning *Rejseholdet/Unit One* (2000–04) reflected the aesthetic of a number of successful American shows in the 1990s, in particular *NYPD Blue* (1993–2005) and *The X-Files* (1993–) (Hansen & Waade, 2017, p. 132). The relationship between American crime fiction, crime drama and the emergence of what is now identified as Nordic Noir is complicated. While Scandinavian dramas have long incorporated American influences, they are now themselves influencing television crime drama production internationally.

As an illustration of how a taste for crime from Scandinavia is being accommodated in Australia, consider the slate of offerings made available on the free-to-air streaming platform SBS OnDemand under the heading Nordic Noir (Table 16.1).

*Table 16.1* SBS OnDemand: Nordic Noir (10 October 2018)

| *Title (as listed)* | *Language* | *Original network* | *Dates* | *Format (seasons × episodes)* | *Notes* |
|---|---|---|---|---|---|
| *Angleby* | Swedish | SVT1 | 2015 | 1 × 12 | |
| *Below the Surface* | Danish/ English | Kanal 5 | 2017 | 1 × 8 | |
| *Blue Eyes* | Swedish | SVT | 2014 | 1 × 10 | |
| *Dicte* | Danish | TV2 | 2012–16 | 3 × 10 | |
| *Eyewitness* | Norwegian | NRK | 2014 | 1 × 6 | Remake USA network 2016 |
| *Farang* | Swedish | TV4 | 2017 | 1 × 8 | |
| *Greyzone* | Danish/Swedish | TV2/C.More | 2018 | 1 × 10 | |
| *Jordskott* | Swedish | SVT | 2015–17 | 1 × 10; 2 × 8 | |
| *Lilyhammer* | Norwegian/US | NRK1/Netflix | 2012–14 | 3 × 8 | |
| *Mammon* | Norwegian | NRK1 | 2014–16 | 1 × 6; 2 × 8 | |
| *Midnight Sun* | Swedish/ French | SVT/Canal Plus | 2016 | 1 × 8 | |
| *Modus* | Swedish | TV4 | 2015– | 2 × 8 | Based on a character created by Norwegian crime writer, Ann Holt |
| *Monster* | Norwegian | NRK | 2017 | 1 × 7 | |
| *The Bridge* | Danish/Swedish | SVT1/DR1 | 2011–18 | 3 x 10; 1 x 8 | |
| *The Lawyer* | Swedish/Danish | SFStudio/Viaplay/ TV3/StudioCanal | 2018 | 1 × 10 | |
| *The Legacy* | Danish | DR1 | 2014–17 | 1 × 10; 2 × 7; 3 × 9 | Not a crime drama |
| *The Team* | English/Danish/ Dutch/German | 11 different production companies listed in IMDB | 2015– | 2 × 8 | |
| *Trapped* | Icelandic | RUV | 2015– | 2 × 10 | |
| *Valkyrien* | Norwegian | NRK1 | 2017 | | |

What's interesting about this list of 19 titles is the sheer diversity of the dramas that are designated Nordic Noir. Indeed, the Danish series *The Legacy* (2014–) is not even a crime drama, with Eva Novrup Redvall describing it as 'a character-driven family series with a complicated inheritance discord as the motor' (Redvall, 2013, p. 116). This somewhat arbitrary aggregation thus points to the sheer vagueness of Nordic Noir as a term, despite consistent efforts to identify a set of shared characteristics (Agger, 2012; Bergman, 2014; Creeber, 2015; Hill & Turnbull,

2017; Jensen & Waade, 2013). As Hansen and Waade usefully point out, the term Nordic Noir now has little analytical value at all, having become a global brand instead (Hansen & Waade, 2017, p. 6). However, what the deployment of the term Nordic Noir reveals in this context is that the programmers at SBS OnDemand are well aware that there is an audience for Scandinavian drama in Australia. Furthermore, this is an audience that precedes the emergence of the digital streaming platform, given that SBS – the nation's second public broadcaster with a remit to reflect multicultural Australia – first started showing Danish crime series such as *Unit One/Rejseholdet* (2000–04) and *The Eagle/Ornen* (2004–06) as part of its regular broadcast schedule in the early noughties.

Tracing the recent history of Danish-language television on SBS, Jensen and McCutcheon (2017) reveal that this grew from 20 hours a year in 2001–02 to a peak of 157 hours in 2009–10, with the political drama *Borgen* (2010–13) and the family melodrama *The Legacy* attracting significantly higher ratings with female audiences. As to why audiences were attracted to these Danish shows, according to Jensen and McCutcheon's focus group research, the series were perceived as more 'authentic', 'poignant' and 'distinctive' compared to both American and British imports. Even though these Danish dramas might be 'culturally removed' or distant, they were perceived to be 'emotionally proximate', the experiences of the Danish characters mirroring the emotional lives of the respondents in Australia despite their very different geographical location. On the other hand, the series were also considered attractive because of their 'otherness', including the Nordic weather and landscapes. Respondents also considered the Danish series reflected a much more progressive cultural perspective on issues such as gender equality, with Jensen and McCutcheon identifying this as an 'aspiration to modernity', borrowing a term from Koichi Iwabuchi.

While Jensen and McCutcheon's research reveals just how distinctive these Danish series were thought to be in 2016, looking back at the first season of *Unit One* it is evident that it too owes a debt to a number of American crime series of the time. With its pacey titles sequence, industrial office setting, a moody blue palette, episodic structure and ensemble cast, *Unit One* might be described as *NYPD Blue* meets *CSI* in Copenhagen. This is hardly surprising, since Scandinavian crime dramas are themselves in conversation with the wider world of TV crime. In her exemplary study of the production context and history of the drama series produced by the Danish public broadcaster DR, subtitled 'from *The Kingdom* to *The Killing*', Redvall recounts how DR producer Sven Clausen was sent on a research trip to Sweden, England and the US in the 1990s to study the working methods operational within these different production cultures. Clausen, and writer Stig Throsboe, were apparently inspired by a talk given by Steven Bochco, the writer and producer of the series *NYPD Blue*. After a visit to Fox Studios in Los Angeles, Bochco's production strategies then became the template for managing their own next production, *Unit One/Rejseholdet* (DR 2000–04) (Redvall, 2013, p. 67).

As Redvall explains, it was as a result of this trip, and others, that the 15 dogmas that underpinned the DR approach to making television emerged. As outlined by the Head of Drama at DR, Ingolf Gabold, and according to Redvall's translation, the first of these was the concept of 'one vision'. This establishes 'the author' as the 'one with the vision to drive the fiction onwards when it comes to a particular project', a role that has sometimes been compared with that of the showrunner in the American context. The second insists on the importance of a 'dual narrative'; that is, as well as a 'good story', any DR drama must also contain a plot with ethical/social themes reflecting DR's role as a public service broadcaster. The third suggests that DR productions will not be made entirely 'in house', and that there should be a 'cross over' between writers and directors drawn from the DR fiction department and those drawn from the industry more broadly, with the fourth dogma affirming this 'crossover' in terms of the production staff as well. While the fifth dogma deals with the lines of communication between the producers, directors and the head writer, the sixth and seventh dogmas affirm the freedom of the producer to make the appropriate creative choices in choosing personnel. The remaining eight dogmas also deal with communication issues, not forgetting a commitment to spend 2 per cent of the overall DR budget on 'developing new drama' (Redvall, 2013, p. 69).

As many commentators have pointed out, the 'breakthrough series' in terms of the success of DR dramas in the UK on Channel 4 was *Forbrydelsen/The Killing* (2007–12) (Redvall, 2013, p. 159). This was followed by *Borgen* (2010–13) and *Bron/Broen/The Bridge*, series that have also had a profound impact on the recent evolution of television crime drama in Australia even though *Borgen* is best described as a political drama. In seeking to illustrate the impact that the DR series broadcast by SBS have had in the Australian context, we will focus on three recent drama series: *Secret City* (2016–) and *The Kettering Incident* (2016) both produced for the subscription service Foxtel, and *Mystery Road* (2017) produced for the ABC. As will be argued, these three shows not only echo the DR series in terms of their production ethos and style, but also generically and thematically, as in the case of the hybrid crime/supernatural series *The Kettering Incident* and the Swedish series *Jordskott* (2015–). What this analysis serves to reveal are the ways in which producers of the television crime drama in Australia negotiate the expectations that a crime drama should reflect national concerns and interests, while also reflecting global trends in order to secure their transnational currency.

At Foxtel, a subscription platform with an obligation to spend 10 per cent of its drama expenditure budget on Australian drama, head of drama Penny Win began working toward aligning their production slate with audience expectations and international trends in 2012, examining local and international markets and assessing what should motivate their work over a five-year time horizon (Mathieson, 2017). In an interview with the authors in November 2017 that also included Foxtel's director of content, Ross Crowley, and director of business affairs movies and content, Tony Pollitt, Win described how in building from a minimal production slate, Foxtel 'wanted aspirationally to be HBO'. According to Win, this entailed establishing a relationship of trust with the production community in Australia in the hope that this would attract 'great ideas' for future drama productions. The kinds of shows they wanted, Win added, would be 'noise makers' and 'thought breakers': shows that would attract the attention of the media, media critics, social media and get people talking about them at barbecues (rather than the watercooler, given this is Australia). While achieving high ratings with a mass audience was never the main goal, attracting a 'smart' and 'intelligent' audience prepared to pay high subscription rates for premium content was clearly a major part of the game plan. As Win explained: 'We want to be critically successful. The production values have to be incredibly high. Everything we commission sits on the Showcase Channel where we have the-best-of-the-best HBO' (Penny Win, personal communication, 2 November 2017).

In outlining their global strategy, Crowley suggested that because Australia is the fourth-largest Anglophone market in the world, there is an opportunity to feed content into 'the Anglosphere of global products'. Furthermore, Crowley added, because Australia absorbs 'great British drama as much as we absorb great American drama', there is the potential to generate 'a hybrid that works in both spaces'. While this ambition has long been the goal of many Australian content producers in film as well as television before the emergence of Foxtel, Australian success overseas has hitherto been limited to the cult series, *Prisoner Cell Block H* (1979–86) and a swag of Australian soap operas. When it comes to the films and TV that win awards in the Anglophone world (the Emmys and the Baftas), Australian content is conspicuously absent.

In attempting to describe the audience to which these hybrid shows might appeal, Crowley struggled to find the right term. Dismissing the concept of 'tribe', he arrived on the term 'niche' in an attempt to identify the global audience for 'noir':

> Arguably one of the conversations is there is a – and I don't know if tribe is the right thing – it's a niche. If you are making gothic, you are making it for a global gothic audience. If you make noir, you're making it for the global noir audience, regardless of where you are born.
>
> *(Ross Crowley, personal communication, 2 November 2017)*

In other words, as far as Crowley was concerned, there is a global niche audience for Nordic Noir which Foxtel is well placed to serve.

Asked about what to make to fill this niche, Win was quite explicit in rejecting any suggestion that Foxtel might aspire to make what she called 'scandi noir'. As she explained:

> There was that time four years ago when everybody that came in was pitching scandi noir and toward the end I was getting a bit irritated saying we're not Scandinavia. We can't make scandi noir, that's Australian noir if you're going to make it. They were borrowing all the tropes and structure of *The Killing* and all those ones, but it didn't work [...] The reason scandi noir was really successful is because they didn't try to be anything other than what they were. And so it travelled really well, so everyone thought of we can replicate it but you can't – you've got to be yourself for it to be successful internationally.
>
> *(Penny Win, personal communication, 2 November 2017)*

In light of these comments, Win's account of the origins of Foxtel's drama series *Secret City* is informative and somewhat contradictory as it reveals how as a 'creative champion' she steered this production through in ways that enabled it to incorporate a number of 'scandi elements' while producing a drama that was also quintessentially Australian. *Secret City* is a political crime drama and a political thriller that references Australia's precarious relationship with China and the US at the same time as it incorporates elements that clearly reference DR series such as *Borgen* and *The Bridge*.

According to Win, the inspiration for this series came from a conversation she had with producer Penny Chapman from Matchbox Pictures during which Win expressed a desire to see a 'really great political thriller' on Australian television. Chapman then recalled *The Marmalade Files*, a book set in Canberra written by two well-known Australian journalists, Steve Lewis and Chris Uhlmann, who had already approached Chapman with the project. Win subsequently read the book and 'loved it', but also had a very clear idea of how it needed to be adapted:

> [I]t was a satire and of course Harriet was Harry, a guy. I wanted it to be a thriller, not a drama. It was a lot of development, it meant a rewrite of the script to get it to be a thriller ... At that time my great love was *Borgen* so it was very much sitting in that space.
>
> *(Penny Win, personal communication, 2 November 2017)*

Having changed both the gender of the central protagonist (now played by the cool blonde Anna Torv) and the tone of the book (dropping much of the satire) to accord more effectively with her inspiration, *Borgen*, Win was also intent on achieving a particular 'look'. As she explained, Win was keen for Canberra, Australia's capital city, to feature as a 'character' and be made to look as it had never looked before. This vision was shared with director Emma Freeman and her director of photography, Mark Wereham, and eventually achieved through the use of long camera movements, drone shots, carefully framed Canberra architecture, wintry weather and modern interior design that managed to appear both modern and very Danish since director Emma Freeman, Win observed, was very influenced by Danish architecture. Also important was the framing of the bridge over Lake Burley Griffin seen during the dramatic opening scene as a young man runs toward the camera before jumping into the water in order to avoid his pursuers. Shot a night, making dramatic use of the long lines of the bridge, the scene clearly echoes the use of the Oresund bridge during the opening sequence of *Bron/Boren/The Bridge.*

Now available on Netflix, the first season of *Secret City* was initially seen by very few Australians on the Foxtel Showcase channel with its limited subscriber base. When pressed as to why there was so little advance publicity and 'noise' around the launch of *Secret City* in 2016, it became apparent that this was an uncomfortable topic for our Foxtel interviewees, leading to a suspicion that there may have been 'real' political tensions at play in Canberra at the time of its release relating to ongoing diplomatic issues with the US and China. While China is Australia's foremost trading partner and America an important military ally, when it comes to the rivalries between the two super powers as played out in naval operations in the South China Sea, Australia can be caught somewhat in the middle. Despite *Secret City*'s lack of visibility, Foxtel were clearly happy enough with the original to commission a second season, *Secret City: Under the Eagle*, to be released in 2019 (Foxtel, 2018).

By way of contrast, Foxtel's other major crime drama series, *The Kettering Incident*, was launched with considerable fanfare and extensive media coverage. Frequently mentioned in the pre-publicity was the fact that *Kettering* had won a Special Jury Prize at the Series Mania Festival in France in 2016, thus being one of the very few Australian drama series to win an international television award – ever. Such international recognition may well have played a role in the global circulation of *Kettering* which sold exclusively to Amazon Prime, and in 2017 was Australia's top international sales-earning adult drama series as reported by Screen Australia (BBC Worldwide, 2016; George & Tansley, 2018). Despite this international success, Foxtel declined to pick up a second season of the show for reasons that were unclear. Once again, this proved to be a sensitive topic in our interview with Foxtel.

According to Win, *The Kettering Incident* owed its existence to writer and producer Vicki Madden, who had had a long career in the Australian television industry before moving to the UK where she worked with Lynda La Plante on her series *Trial and Retribution* (1997–), and as story producer on the long-running British crime drama *The Bill* from 2003 to 2004, *The Clinic* (2008) and the *Doctor Who* spin-off *K9* (2009–10). Upon return to her native Tasmania, Madden came up with the concept of *The Kettering Incident*, a drama series with both crime and supernatural elements that Win described as a cross between the American cult series *The X-Files* with 'a little bit of *Twin Peaks*' and the British crime drama series, *Broadchurch*, in terms of its structure. Nordic Noir was also an influence:

> Vicki got that because she'd worked overseas and she's seen it. She knew what she was doing. She's totally influenced by the scandis and she's been over there a number of times and met them all but she knew it had to be Tasmanian. And that's why it was successful, it didn't try to be scandi, but it definitely was noir, making it true. It's being true to yourself and not trying to be anything else is what travels really well.
>
> *(Penny Win, personal communication, 2 November 2017)*

The series that Win doesn't mention, nor does Madden make any mention of this in any of the interviews we have read to date, is the Swedish series *Jordskott* (2015–17) which also mixes elements of the crime drama with the supernatural in a forest setting. Furthermore, while both *Jorskott* and *Kettering* reference *Twin Peaks* in both overt and covert ways, there are many other notable parallels between the Australian and the Swedish series.

For a start, both series are premised on the mystery of a missing (white) girl in the past and the investigation to find out what happened to her in the present. In the case of *Jordskott*, this investigation is led by the girl's mother, police detective Eva Thörnblad (Moa Gammel) who returns to her home town of Silverhöjd when a boy is declared missing under similar circumstances. Eva discovers a town in crisis as the local logging company, formerly owned by her father and in which she has shares, threatens to destroy the forest. In *Kettering*, Dr Anna Macey (Elizabeth Dubecki) returns home to Kettering after a number of years overseas to discover what happened to her best friend who disappeared 15 years ago. She too discovers a town in crisis as another girl goes missing amid a turf war that is escalating between the local logging company and the environmentalists. While the most obvious similarity is that of the setting – the Swedish forests surrounding the town of Silverhöjd look remarkably similar to the Tasmanian forests surrounding Kettering and both series include a storyline that references the environmental threat posed by logging – there are also similarities in the casting and portrayal of the two central protagonists, Eva and Anna. Both are slim, blonde and troubled, clad in shades of grey and blue that reflect the muted noirish colour palette of the series as a whole. There are, however, significant differences, with *Jordskott* delving into the mystical world of sylvan folklore and managing rather more story arcs involving the minor characters, while *Kettering* focuses more intently on Anna's personal journey while veering into the realm of science fiction.

What the similarities between these two series reveal in their deployment of a lost white girl and blonde female protagonist – recent crime series tropes that are explored in more depth by Klinger (2018) – is just how intertextual and hybrid the realm of the television crime drama now is, especially in countries such as Denmark and Australia where the creative personnel may travel overseas to seek inspiration, as in the case of the DR executives cited above, or watch TV series from elsewhere, as is revealed by Win's comments about *Borgen*. However, although the global dominance of the British and American crime shows is now being challenged, we cannot overlook the influence of what we might describe as American 'ur' shows such as *Twin Peaks* and *The X-Files* that have clearly influenced a whole generation of creative personnel including, as Redvall notes, *The Killing* creator Søren Sveistrup himself (Redvall, 2013, p. 24). Given that Denmark and now other countries in Scandinavia are producing shows that are finding a global niche audience, one that the would-be producers and creators themselves clearly watch, it is salutary to discover how these multiple intertextual references are woven together in different contexts to produce series that may combine these familiar elements in ways that are also completely original as in the case of Australian 'outback noir' series, *Mystery Road*, our last case study.

The inspiration for *Mystery Road* (ABC, 2018), the television series, came from the eponymous feature film from 2013 and its sequel, *Goldstone* (2016), both crime thrillers deliberately drawing on the Western genre and written and directed by indigenous film maker, Ivan Sen. Both films were widely acclaimed – *Goldstone* opened the 2016 Sydney Film Festival – and both were connected by an epic story arc in the nature of a saga, with Aaron Pedersen as indigenous detective Jay Swan who traverses black and white rural Australia while single-handedly dealing with the most violent of offenders in spectacular shoot-outs in monumental landscapes that Western director John Ford would have loved.

*Mystery Road* was produced by Bunya Productions – a partnership between Ivan Sen, David Jowsey and Greer Simpkin – for the ABC. The public broadcaster's reasons for commissioning *Mystery Road* were multi-fold – while its charter obliges it to 'broadcast programs that contribute to a sense of national identity and inform and entertain and reflect the cultural diversity of the Australian community', its ever-tightening budgetary constraints mean that each dollar it invests in content creation needs to yield recognisable benefits. The ABC's commitment to *Mystery Road* came off the back of increasing investments in indigenous drama programming, including commissioning the ground-breaking socio-realist *Redfern Now* (2012–13) and the sci-fi Aboriginal superhero series, *Cleverman* (2016–17). Such content is a core business for the ABC: promoting and contributing to culturally significant issues and activities is fundamental to its content strategy and remit, with the ABC maintaining a series of reconciliation action plans, put in place under a national framework to build relationships with Aboriginal and Torres Strait Island peoples, communities and organisations and the broader Australian community. The ABC's latest plan 'commits the Corporation to a long-term course of action to improve opportunities for Indigenous Australians in the areas of employment, content, community links, and cultural understanding and respect' (ABC, 2018).

Ensuring that *Mystery Road* would find an international audience was a goal for the production team from the outset, with the distributor responsible for global sales, All3Media International, involved in the project from early in development (Fry, 2018), and the series showing at the Series Mania festival where *Kettering* had had such success. With an international audience in their sights, as well as a domestic audience already engaging with indigenous stories, during development the Western-style crime thriller of the original movies moved closer to Nordic Noir, recognisably following at least some of the dogmas adopted by the Danish broadcaster DR. One vision, for example, was achieved by renowned Aboriginal director Rachel Perkins, who, along with script producer Michaeley O'Brien, maintained the connection with Sen's films by mining the themes and the slow-burn plotting of his work, while ensuring a double narrative by weaving in new characters and storylines about an indigenous community that delivers on the aims of the ABC's reconciliation action plan (Goodall, 2018). In commissioning the series from Bunya Productions, the ABC was, in effect, enabling them to produce their first television series, which was, according to Jowsey, 'a massive undertaking for the company' (Bizzaca, 2018). Bunya, in turn, hired indigenous crew wherever possible and employed locals as interns, 'giving back a little' (Bizzaca, 2018).

It was an approach that is yielding dividends. The series was hugely popular, both on the main ABC broadcast channel and on its iView streaming and catch-up service, attracting an audience of 800,000 at launch – an impressive feat given it was up against an interview with the protagonists of a political sex scandal on a competing commercial channel – and only weeks into its run was ranked as the most-watched non-children's series on the ABC's catch-up service iView (Groves, 2018b), even though audiences appeared to be watching it in traditional delayed-gratification serial style, rather than binging it in one big hit (Quinn, 2018). The ABC has committed to a second series, and at the time of writing *Mystery Road* is leading the nominations in the small-screen categories in the Australian Academy of Film and Television Awards, with director Rachel Perkins paying tribute to the ABC:

> You can make a high-end drama, you can shoot it in the far reaches of Kununurra and it's got something to say about our country … The ABC getting behind that, developing it, that's why our national broadcasters are so important.
>
> *(Maddox, 2018)*

While *Mystery Road* takes inspiration from recent Nordic Noir, its location in a rural indigenous community in the hot and dry north-west Kimberley region of Australia, ensures that it maintains a clear connection with Sen's films and its own unique voice. In addition to its socially aware storyline, elements that connect it to its Nordic Noir influences include much stronger roles for women and the use of landscape within the storytelling. The introduction of a female detective, Emma James, played by Judy Davies in her first-ever turn as a policewoman, provides an opportunity for the series to explore internal community dynamics in a different way from that of the Sen movies – who could understand the motivations of the local people best, the coconut cop (black on the outside, but white inside) or the red-headed, white rancher's daughter who has grown up knowing these people more intimately? – providing Jay Swan with a testy colleague who challenges him in situations where reaching for a gun or walking away are simply not options.

Although the first *Mystery Road* film begins with the death of a young girl, the TV series twists the 'lost white girl' in significant ways. This time it is two young jackeroos who are lost, one black one white, and it is through an investigation of their relationship to each other, to the community and to the outside world that the hidden truths and lies that simmer beneath the surface will be brought to light.

The look and feel of the series led to it being identified in the press as outback – rather than Nordic – noir, thus differentiating it from other Australian noir productions. *Guardian* press reviewer Luke Buckmaster qualified this slightly by describing *Mystery Road* as 'outback neo-noir', thus signalling its reinvention (Buckmaster, 2018). Distributor All3Media was clearly keen to capitalise on the potential sales value of the term, with the Vice President of Acquisitions, David Swetman, extolling the series for its 'sumptuous, cinematic outback noir character' while proposing that it is 'a ground breaking, genre-defining drama' (Fry, 2018). When it comes to identifying just what 'outback noir' might be, it is significant to note that unlike the dominant soft blues, greys and mossy greens of *Secret City* and *Kettering*, *Mystery Road* features a colour palette of rusty red soil, dusty green spinifex and brilliant blue skies. Writing in the Fairfax Press, TV reviewer Deni Enker enthused about that landscape thus:

> [A] vast night sky sparkling with stars, expanses of baked earth, bulbous boab trees, imposing mountain ranges and outcrops of red rock ... the kind of epic setting beloved by cinema, especially westerns, and rarely seen on TV: ancient, monumental, remote, spectacular.
>
> *(Enker, 2018)*

TV blogger David Knox was also enthusiastic about the series, suggesting that while director Rachel Perkins brought a 'slow-burn Scandinavian style to the piece', the use of rock and roll as well as bluegrass on the soundtrack reminded us that *Mystery Road* was also uniquely Australian (Knox, 2018). Given that neither rock and roll nor bluegrass originated in Australia, Knox's comment points to the fact that quite diverse cultural influences and tropes can be combined in ways that are themselves entirely original. In other words, 'outback noir' may be simply an attempt to brand the series in ways that will sell it to that much desired global niche audience.

From Ed McBain to *Beck*, *Twin Peaks* to *The Killing*, *Borgen* and *Bron/Broen/The Bridge* to *Secret City*, *Jordskott* to *The Kettering Incident* and all of the above to the uniquely Australian genre hybrid that is *Mystery Road*, the contemporary crime drama series continues to draw on the televisual past even as it seeks to reinvent itself for the present: a present in which a TV series can now find an international audience by addressing issues and concerns that are local, national and global. As is evident in many of the comments from the producers and reviewers cited here, one of the outcomes of the new streaming era is the emergence of a global niche audience eager to consume content that is immediately recognisable in terms of its generic inheritance, while also completely original in terms of its location and/or take on the genre. This is an era in which the value of content from 'small countries' such as Denmark and Sweden and those who have traditionally been on the periphery of the global trade in TV, such as Australia, is at last being recognised and valued. Whether that recognition will translate into the production of more locally produced content, however, depends on whether broadcasters maintain their will to both commission and acquire new series from small country producers. With public service television services, such as DR in Denmark and the ABC in Australia, under continual threat from neo-liberal governments, and the new streaming providers such as Netflix or Amazon under little obligation to produce local content unless quotas or taxes are imposed, the future for the television crime drama continues to be somewhat uncertain. However, given the current taste for 'international television' in America, and encouraged by a recent report from Screen Australia – the Australian federal government's key funding body for screen production – there is hope. According to Graeme Mason, Chief Executive Officer of Screen Australia, in 2018 *Mystery Road* was the most watched drama series on the streaming platform ABC iView 'in history' while also securing broadcast deals in the US and the UK. 'Quite simply,' Mason concludes, 'Australian audiences want local drama, and the global market is hungry for our stories' (Screen Australia, 2018). We can only hope this is true and that a market-driven future will ensure the ongoing circulation of 'international television' for a global niche audience that wants to watch quality crime dramas that continue to reinvigorate the genre.

## Note

1. The two other series are a Belgian soccer series called *The Score* and a British school and family dramady called *Ackley Bridge*.

## References


ABC. (2018). Corporate responsibility. Retrieved from https://about.abc.net.au/how-the-abc-is-run/what-guides-us/corporate-responsibility/.

Agger, G. (2012). Nordic noir on television: *The Killing I-III. Cinéma & Cie, 12* (19), 39–50.

Aguiar, L., & Waldfogel, J. (2018). Netflix: global hegemon or facilitator of frictionless digital trade? *Journal of Cultural Economics*, 42 (3), Published online.

Bakøy, E., Puijk, R., & Spicer, A. (Eds.). (2017). *Building Successful and Sustainable Film and Television Businesses: A Cross-National Perspective*. Bristol, Chicago: Intellect.

Bastow, C. (2016, 10 June). The kettering incident review: Tasmanian gothic thriller par excellence. *The Guardian*. Retrieved from www.theguardian.com/tv-and-radio/2016/jun/10/the-kettering-incident-review-tasmanian-gothic-thriller-par-excellence.

Batty, D. (2013). Midsomer murders keeps danes glued to TV. *The Guardian*. Retrieved from www.theguardian.com/media/2013/may/02/midsomer-murders-denmark-hit.

BBC Worldwide. (2016). *BBC Worldwide North America Announces Amazon Prime Video as the Exclusive U.S. Premium Subscription Streaming Home of the New Australian Drama, The Kettering Incident* [Press release]. Retrieved from www.bbcwpressroom.com/sales-and-co-productions/press/bbc-worldwide-north-america-announces-amazon-prime-video-exclusive-u-s-premium-subscription-streaming-home-new-australian-drama-kettering-incident/.

Bergman, K. (2014). *Swedish Crime Fiction: The Making of Nordic Noir*. Milan: Mimesis/Degenere.

Bizzaca, C. (2018). 'Bunya: Sweet Country, Mystery Road & More', *Screen Australia Screen News*, 17 January, accessed from https://screenaustralia.gov.au/sa/screen-news/2018/01-17-bunya-sweet-country-mystery-road-and-more on 20 Auugust 2018.

Bodey, M. (2017, 18–24 February). The death of broadcast television. *The Saturday Paper*. Retrieved from www.thesaturdaypaper.com.au/news/media/2017/02/18/the-death-broadcast-television/14873364004246.

Buckmaster, L. (2018, 13 August). Aaron Pedersen: Is the Mystery Road star one of the greatest actors of his generation? *The Guardian*.

Creeber, G. (2015). Killing us softly: Investigating the aesthetics, philosophy, and influence of Nordic noir television. *Journal of Popular Television, 3* (1), 21–35.

Cunningham, S., & Jacka, L. (1995). 'Australian Television in World Markets', in *New Patterns in Global Television: Peripheral Vision*, J. Sinclair, E. Jacka, & S. Cunningham (Eds.), New York: Oxford University Press,.192–219.

Enker, D. (2018, 24 May). The Judy factor: Davis returns to Australian TV on ABC's Mystery Road. *The Sydney Morning Herald.* Retrieved from www.smh.com.au/entertainment/tv-and-radio/the-judy-factor-davis-returns-to-australian-tv-on-abcs-mystery-road-20180516-h1059s.html.

Foxtel. (2018). FOX SHOWCASE - Foxtel's new premium on demand drama destination [Press release]. Retrieved from https://community.foxtel.com.au/t5/Foxtel-Press-Releases/FOX-SHOWCASE-Foxtel-s-new-premium-On-Demand-drama-destination/m-p/253553.

Fry, A. (2018). Hot picks: Mystery Road. *Broadcast.* Retrieved from www.broadcastnow.co.uk/international/mystery-road/5127895.article.

George, S., & Tansley, R. (2018). *International TV Sales Snapshot for 2017.* Retrieved from www.screenaustralia.gov.au/sa/screen-news/2018/06-18-international-tv-sales-snapshot-for-2017.

Given, J., Brealey, M., & Gray, C. (2015). *Television 2025: Rethinking Small-screen Media in Australia.* Hawthorn, Victoria. Retrieved from http://hdl.handle.net/1959.3/400201.

Goodall, J. (2018). Out there. *Inside Story.* Retrieved from https://insidestory.org.au/out-there/.

Groves, D. (2018a). 'The kettering incident' was Australia's top-selling drama overseas last year. *IF Magazine.* Retrieved from www.if.com.au/the-kettering-incident-was-australias-top-selling-drama-overseas-last-yeaar/.

Groves, D. (2018b). 'Mysery road' sets a new high for ABC's iView. *IF Magazine.* Retrieved from www.if.com.au/mystery-road-sets-a-new-high-for-abcs-iview/.

Hale, M. (2018). Critic's Notebook: TV Shows to Watch From Hong Kong, Brazil, Sweden and More. *The New York Times.* Retrieved from https://nytimes.com/2018/03/15/arts/television/international-tv-hong-kong-brazil-sweden.html.

Hill, A., & Turnbull, S. (2017). Nordic noir. *Oxford Research Encyclopedia of Criminology.* Retrieved from doi:10.1093/acrefore/9780190264079.013.294

Jensen, P. M., & McCutcheon, M. (2017). *Ratings Behaving Madly: Danish TV Drama's Fortuitous Success in Australia.* Paper presented at the ANZCA 2017, The University of Sydney.

Jensen, P. M., & Waade, A. M. (2013). Nordic noir challenging the language of advantage: Setting, light, and language as production values in Danish television series. *Journal of Popular Culture, 1* (2), 259–265.

Klinger, B. (2018). Gateway bodies: Serial form, genre, and white femininity in imported crime TV. *Television & New Media, 19* (6), 515–534.

Knox, D. (2018). 'Mystery Road', *TV Tonight*, 1 June, accessed https://tvtonight.com.au/2018/06/mystery-road.html on 1 August 2019.

Lotz, A. D. (2017). Linking industrial and creative change in 21st century US television. *Media International Australia, 164*,.10-20.

Maddox, G. (2018, 30 October). Ladies in black, Mystery Road top Australian academy award nominations. *The Sydney Morning Herald.* Retrieved from www.smh.com.au/entertainment/movies/ladies-in-black-mystery-road-top-australian-academy-award-nominations-20181028-p50cho.html.

Make It Australian. (2018). Open Letter.

Mathieson, C. (2017, 6 July). Foxtel has a dramatic plan to make HBO-quality shows. *Sydney Morning Herald.* Retrieved from www.smh.com.au/entertainment/tv-and-radio/foxtel-has-a-dramatic-plan-to-make-hboquality-shows-20170629-gx1nl5.html.

Quinn, K. (2018, 14 June). A formidable package put to exceptional use. *The Sydney Morning Herald.* Retrieved from www.smh.com.au/entertainment/tv-and-radio/mystery-road-is-the-supergroup-of-indigenous-television-20180611-h118ig.html.

Redvall, E. N. (2013). *Writing and Producing Television Drama in Denmark: From The Kingdom to The Killing.*

Screen Australia. (2018). *Drama Report 2017-18.* Retrieved from file:///C:/Users/mario/Dropbox/Border%20Crossings/Sources%20general/Screen%20Australia%202018%20Drama-report-2017-2018.pdf.

Turnbull, S. (2007). 'The Hook and the Look: *CSI* and the Aesthetics of the Television Crime Series', in *Reading CSI: Crime Television Under the Microscope*, M. Allen (Ed.), London: I.B. Tauris, 15–32.

Turnbull, S. (2014). *The TV Crime Drama.* Edinburgh: Edinburgh University Press.

Turner, G. (2016). Surviving the post-broadcast era: The international context for Australia's ABC. *Media International Australia, 158* (1), 17–25.

Hansen, K. T. & Waade, A. M.(2017). *Locating Nordic Noir: From Beck to The Bridge.* Basingstoke: Palgrave Macmillan.

Weissmann, E. (2012). *Transnational Television Drama: Special Relations and Mutual Influence between the US and UK.* Hampshire and New York: Palgrave Macmillan.

# 17

# GLOBAL SPORT TELEVISION

## Seamless Flows and Sticking Points

*David Rowe*

### Introduction: The Possibility of Global TV Sport Flow

Imagining sport without television would be analogous to considering popular music without the communication technologies that convey it to distant homes, entertainment venues and elevators. Indeed, many screen sport viewers may enjoy consuming sport at a distance through sophisticated audio-visual media more than being restricted to their line-of-sight inside a stadium (Levy, 2017). It is for this reason that Lawrence Wenner (1998) coined the neologism 'MediaSport' to capture this enduring, high-intensity institutional interpenetration. Like music, sport has long been performed by expert practitioners and appreciated by spectators who were physically co-present, but the logistics of real-time, place-based attendance limited the capacity to experience them (Rowe, 1995). The development of recording technology ameliorated this problem for music. Regarding sport, the arrival of its most important medium – live television – was crucial because it provided the means by which its fleeting, pulsating moments can be experienced in ways that afford, across space and time, a sense of sharing the stadium experience with those who were there. Even if time zones and other impediments open a temporal gap between the action moment and its delayed reception, the recorded event can still be experienced as if it were live – indeed, this is the technique deployed by US-based television networks such as NBC, which held back many live Olympic events until convenient for the east coast US prime-time market while presenting them 'as live' (Chappell, 2017). NBC now broadcasts most events live. Away from the live or as live moment, sport also offers a vast, global televisual library of footage that can be manipulated and molded in many ways, including through highlight clips, documentaries, film drama, chat and quiz shows (Rowe, 2004). Thus, sport on television is not restricted to the reproduction of live sport encounters. But all such programming is to varying degrees subordinated to the 'having been there' experience of live audio-visual mediation on which it relies for the immediate interest to stimulate other modes of sports discourse and treatment.

Sport, then, can be regarded as perfect material for global television which, having overcome the restrictions imposed by analogue broadcast technology, is now characterized by a flexible and evolving mix of analogue and digital technologies. It is easily packaged and re-packaged, flowing like lava into the proliferating spaces created by digital networks, commanding massive viewerships or servicing the smallest market niches. It can command huge broadcasting rights and substantial other media rights fees, be 'bundled' cheaply along with many minor interest offerings, or 'pirated' to satisfy the voracious appetite of a televisual apparatus that is dependent on high-volume content. While in its most prized form, the live broadcast of a contest, the sport text combines specific forms of commentary, editing and presentation that are directed toward identified audiences and so customized for them (such as for a national team competing in the FIFA World Cup or the Olympic Games), its reliance on action conducted according to an agreed set of (principally global) rules and procedures means that it can be decoded easily without too many contextual cues. Thus, sport television is simultaneously local and global, national and transnational in the case of major, non-indigenous sports. In this chapter I demonstrate the ways in which sport television both flows around the world and 'lodges' in particular national environments according to how different television fields are organized in terms of private and public ownership, legal regulation, technologies, tastes and so on. This uneasy, unstable co-existence of the global and the local mirrors that between 'television as we knew it' and what it is becoming (Turner and Tay, 2009). The once unchallenged dominance of a 'box in the corner' providing sport content via nationally regulated airwaves and cable is now only one (although still the predominant single delivery technology) among many screen sport options. Thus, it is misleading to talk of global television and its consumption as if they were governed by a universal logic or following a specified pathway. More accurately, the television sport landscape is rough and only partially charted, while, to extend the physical geographical metaphor, having its own microclimates, tranquil zones and volcanic eruptions. In mapping these variations, it is important to examine sport television's origins in national broadcast systems before analyzing the ways in which it can at least partially detach itself from them.

## Sport Television Contexts

Pierre Bourdieu (1984) has influentially theorized the division of cultural forms into a range of fields in which agents contest possession and control over various forms of capital (especially economic, social and cultural). In the case of the television and sport fields (once largely separate, now, as noted above, indissolubly linked) these were largely national in nature. As nation-states developed, nationally based cultural industries helped both to constitute and service them. It is not necessary here to rehearse accounts of the ways in which national sports developed and national television systems rendered the nation to itself culturally through sport broadcasting (see Whannel, 1992). Instead, it can be observed briefly that the sport–television nexus first took root in a national context that became, in some cases, international by means of sporting competitions between nations. These international contests facilitated a sense of national identity and purpose in competing with other nations, but also required international cooperation in the organization of sport events and their televisualization. Especially for multi-sports events such as the Olympic and Commonwealth Games or major single-sport contests such as the FIFA World Cup and Grand Slam tennis, international protocols needed to be developed for shooting, editing, videotape format, satellite communication and so on. These sport events, though dauntingly large, were intermittent and so constituted only a relatively small proportion of televised sport programming. The continuing expansion of the 'media sports cultural complex' (Rowe, 2004) occurred not in the first instance because such signature sport events could claim to be global, but for the prior reason that sport became an integral element of both free-to-air and subscription television within nations.

In most countries, public service broadcasters (that is, those funded by taxation and other revenues raised from citizens rather than by the paying customers of private corporations) took the lead in sport (certainly in most of Europe and Asia) while in others it was the commercial sector (notably in the United States) (Scherer and Rowe, 2014). Whatever the initial public or private complexion of national sport television systems, technological enhancements, including the progressive replacement of black-and-white by color television and the widening and flattening of screens, made the live mediated sport experience increasingly vivid and 'life-like'. Its increased value as an audience lure created a booming market in sport broadcast rights in which exclusive access to premium live sports such as association football (soccer), gridiron, rugby union, basketball, tennis and baseball was sold for escalating sums of money to broadcasters (Evens et al., 2013). National public service broadcasters, therefore, increasingly gave way to, in the first instance, commercial free-to-air and subscription broadcasters in mainly national markets. The arrival of multi-channeling allowed this sporting material to occupy whole television schedules, a development that was rapidly advanced by the coming of digital television (Hutchins and Rowe, 2012). This easy availability of copious quantities of televised sport produced, in several mature markets, saturation (Miller et al., 2003, 2011). As free-to-air and subscription television became increasingly dependent on sport and the costs of acquiring rights escalated, media corporations sought to connect their rights across territories around the world in the process of becoming international and even global. This development laid the foundation of what could be called global sports television – the circulation of sports texts through a range of pathways inserted into the television schedules and channel spaces made possible by the development of televisual technologies (Rowe, 2011).

However, there is a vast difference between making particular sports available in mediated form around the world and actually finding significant audiences for them. For example, Australian rules football, which is the dominant code of football in that country and is only played professionally there, is circulated around the world for the benefit of ex-patriates, small fan groups in other countries, and international broadcast partners looking to fill sport television slots. Its governing body, the Australian Football League (AFL), can on this basis claim that it is, in the technical sense, a global sport. This discourse of the global is emphasized in promotional statements regarding the AFL's international online streaming service: 'Global footy fans outside Australia can watch every match of the 2017 Toyota AFL Premiership season LIVE and on demand with the Watch AFL Global Pass' (AFL, 2017a). In assembling a media network across the world, this quintessentially Australian sport acquires a global televisual sheen. The AFL has international broadcast partnerships with BT Sport (UK), Econet (South Africa, but with a Pan-African reach), FOX Sports (Australia, with a US parent company), Global Eagle Entertainment (US, for the international airline industry), GZTV (China), Orbit Showtime Network (Middle East and North Africa), Sky Television and TVNZ (New Zealand), TSN (Canada), Viasat (Russia) and Australia Plus (Australia owned, covering Asia and the Pacific). Yet, even in this ultra-contemporary commercial context, the historical traces of national public service television are revealed in the last case.

Australia Plus, 'a unique multi-platform media service delivered online, on mobile, on social media and through an extensive international television service', is a 'targeted subscription television service reaching more than 140 million homes across 39 countries in Asia and the Pacific with an established and unique digital and social media network that engages millions of people a month in some of the world's largest markets' (AFL, 2017b). This service was re-launched in 2014 under the aegis of the national public broadcaster, the Australian Broadcasting Corporation (ABC), as the latest incarnation of the regional service Australia Television International and, later, the Australia Network, which had received some funding from Australia's Department of Foreign Affairs and Trade, been run for a time by the commercial Seven Network, and had carried content (including sport) from public and commercial free-to-air and subscription broadcasters. This interweaving of the national, the regional and the global in public service and commercially provided sport television illuminates the key principle that what can be called global sports television remains shaped by its particularistic histories.

This point is no less valid in the contrasting cases of the major national leagues of sports that are played and are popular with spectators in many countries, such as the National Basketball Association (NBA) and the English Premier League (EPL), which have expensively purchased broadcast rights and are extensively marketed in regions where they are developing player pools, fan bases, and broader audiences, especially in the Asia-Pacific region (Falcous and Maguire, 2005; Rowe and Gilmour, 2010). The globalization of sport television is not an even process through which there is a consistent and predictable sport diffusion and take up of copyrighted content. A range of contingent factors shapes the relative success of sports in different places. A conspicuous case of this variability can be seen in the three key US sports of gridiron (American football), baseball and basketball. All have internationalization strategies in terms of participation and viewership, including playing major games overseas in order to market themselves (Koba, 2013). All can be accessed on television in many countries and have been promoted to new viewers in them, although often only via subscription to a bundle of sport and non-sports channels. Gridiron has probably been the least successful in this regard, perhaps because its rules are for many arcane, it requires substantial protective equipment (including metal headwear) in order to play it, and is associated (despite its very substantial proportion of African American footballers) with a strong image of White America (McGregor, 2016). It has been successful in attracting the attention of global television viewers, though, in one conspicuous instance – the US's National Football League's (NFL's) Super Bowl – in which the sport contest, barely watched or understood by a substantial section of its diverse audience, is only part of a spectacle involving celebrities, high-profile halftime musical interludes by acts of the status of Justin Timberlake, Madonna, The Rolling Stones, Beyoncé, Prince and Paul McCartney (Real and Wenner, 2017), and its expensive signature advertisements. The success of the Super Bowl as a television spectacle, it should be observed here, indicates an important dimension of global sport television – as in the case of the lavish, dramatized Opening and Closing Ceremonies of mega sport events such as the Olympics (Tomlinson and Young, 2006), sport per se may be incidental for many viewers being exposed to program content that is classified as being of a sporting nature.

Baseball is also coded as a quintessential American sport (with its attendant association with White nostalgia) and, while its World Series playoff (spread across up to seven games) does not attract a global television audience of over 160 million like the Super Bowl, its international outreach has been more extensive in countries such as Cuba, the Dominican Republic, Korea and Japan (Gmelch and Nathan, 2017). Apart from its relative sporting simplicity when compared to gridiron, its greater success in international participation and media spectatorship can be partially explained by the much higher proportion of foreign players in the US's Major League Baseball (MLB) (about 25 percent) than in the NFL (less than 5 percent) (Berry, 2013; Gaines, 2011). This manifestation in sport of the 'New International Division of Cultural Labour' (Miller et al., 2001) does not only bring international baseballers to the US domestic game, but also television viewers from the same countries who wish to follow their compatriots. In saturated domestic sport and media markets like the US, this supply of generally less expensive imported athletic labor from countries such as the Dominican Republic and Venezuela can be accompanied by the export of sport television to affluent nations that also provide baseballers to MLB such as Canada, Japan, South Korea and Australia.

Basketball and its premier professional league, the NBA, has a stronger global reach, taking advantage of the ease of participation (requiring no more than a small court area and a basket), a shorter timespan per game, and astute marketing that has used the image of African American 'cool' in a manner akin to that of the popular music industry (Boyd, 2008). The NBA has, in addition, been highly strategic in its approach to social media (Kilgore, 2017). It also has a substantial number of international players (over 100) who also bring many television viewers with them, most influentially in the case of the now-retired Chinese basketballer Yao Ming (Farred, 2006). A particularly striking current example of such inter/transnational mobility is Giannis Antetokounmpo, whose parents left Nigeria for Greece as undocumented immigrants in the early 1990s before moving to the US in support of his basketball career (Stein, 2017). While it has a prominent place on traditional television and its broadcast rights are valuable properties (its 2014 10-year domestic rights contracts alone were worth over US$26 billion), the NBA, like other leagues such as the NFL and EPL, has experienced some fluctuations in television viewing. However, it is compensating for changing viewership patterns and practices through the sale of global digital rights. As the NBA commissioner, Adam Silver, has stated:

> We will continue to see [US]$2bn spent in the NBA. From a relative standpoint, the value of premium sports content is becoming that much more valuable compared to other types of entertainment.
>
> It's the closest thing to a sure thing ... compared to the amount of investment it takes to create a hit show ...
>
> Digital media companies are by definition global and there is an opportunity to do deals with them in 200 countries around the world where basketball is followed.
>
> (quoted in Ahmed, 2017)

In this regard, basketball's prominence in television and in the broader digital sphere (and, indeed, on the 'field of play' itself) outside the country in which the game was invented resembles that of a sport that has been resisted by some in that country as 'un-American' – association football (soccer) (Schwartz, 2016). This so-called 'world game', in contrast to the more nationally introspective sports of American football and baseball (with the latter's primary domestic competition ironically called, as noted, the World Series), symbolizes a more cosmopolitan outlook. The main men's domestic league, Major League Soccer (MLS), is not as powerful as these sports, but has grown significantly since reincarnating the national domestic competition in 1993 (Belson, 2017), while the National Women's Soccer League (NWSL), in contrast to the men's, has established itself as the world's best domestic league and the global hub of the women's game (U.S. Soccer, 2017). MLS has also used global sport celebrities such as Englishman David Beckham and a range of international players, the most currently conspicuous of whom is the Swedish Zlatan Ibrahimović, to enhance its appeal to potential participants, stadium attendees and television viewers, while the NWSL has also attracted many leading overseas players, such as the Brazilian Marta, as part of the global flow of players and watchers in association football.

Soccer's broad popularity as sport television is demonstrated by its comprehensively larger global media audiences for its major events than for the three main US sports, which are also out-rated by the most important international cricket contests (Pennsylvania Sports, 2017). Indeed, there is substantial penetration of other televised association football leagues into the US market, with the EPL especially popular in the United States among college students, ex-patriates and football fans with a range of backgrounds (notably Hispanic), who have been responsive to its clever and systematic global marketing of a combination of English working-class authenticity and internationally mobile footballer glamor (Andrews, 2009). Selling the EPL's international rights in lucrative markets such as the USA and the economically dynamic Asia-Pacific region, and cultivating a following in Africa and the Middle East, has been greatly assisted by the longstanding profile of leading club brands such as Manchester United, Arsenal, Chelsea and Manchester City, all of whom have received large infusions of international capital. Other major European leagues, such as La Liga (Spain), Serie A (Italy) and the Bundesliga (Germany), have also garnered substantial US and other viewerships that have symbiotically developed alongside long-form international club competitions such as the UEFA Champions and Europa Leagues. It can be concluded from this discussion that the possibility of global sports television can only be made material through a combination of established sport interest, potential to expand that interest, sophisticated marketing and, of course, available technology. The last is especially important because of the rapid diffusion of sport screen options other than broadcast television.

## Different Screens, New Tensions

It is often forgotten that, despite its omnipresence in affluent societies, much of the world's population still does not have a television set in their homes. Before it will be possible for many of them to do so, some of those who own television sets in the West will have discarded them in favor of other screen technologies, while many in Africa are likely to bypass broadcast television altogether by using the internet (via mobile broadband) and mobile telephony (Wallis, 2016) to watch sport. In the first instance, mobile technologies have been mainly used in Africa for 'low-tech' person-to-person communication. However, it can be expected that, just as has occurred in the Global North, the expansion of the smartphone enables viewing of 'sportainment' texts, including live contests. The growth of networked sport media, rather than of television per se, is intimately linked with the flourishing of global mega sports events. Thus, for example, the 2008 Beijing Olympics, 2010 South Africa FIFA World Cup, 2014 Brazil FIFA World Cup, 2016 Rio Olympics and 2018 PyeongChang Winter Olympics, have all seen advances in the use of a complex mix of screen technologies that have placed sport at the center of communication and representation (Hutchins and Rowe, 2017a; Matheny, 2018).

Traditional television has been the bridgehead to new sporting spaces, but it now takes its place among the many other technologies that mediate sport. Amid this complex mix of technologies, delivery systems and forms of revenue, the intellectual property apparatus surrounding screen sport becomes a particularly intense battleground (Blackshaw et al., 2009). In the free-to-air television environment in which sport first emerged, copyright was of variable importance. A national government would normally allocate spectrum to a public broadcaster or sell it to a commercial television network. Public broadcasters, at least before the emergence of public/private hybrids (such as Australia's Special Broadcasting Service (SBS)), were motivated by symbolic nation building rather than profit (Whannel, 1992). Providing major sports on television for free across the country was a dimension of cultural citizenship (Scherer and Rowe, 2014) that nurtured a sense of national cohesion that became even more conspicuous during international mega sport events. As universal access was consistent with the philosophy of, say, the International Olympic Committee (IOC), free-to-air broadcasters, such as the European Broadcasting Union (EBU) consortium, tended to be favored as they offered the easiest access for the greatest number of viewers. Although public service broadcasters did not like their copyright being infringed by 'pirates', defense of it was not their *raison d'être* because they were not dependent on a financial return on their investment. The same could not be said, though, of commercial free-to-air broadcasters. For example, NBC first broadcast the Olympics in 1964, and has spent billions of dollars in maintaining its position as the world's most important Olympic broadcaster (although, like all others, it receives its primary broadcast 'feed' from the IOC-owned host broadcaster Olympic Broadcasting Services (OBS)). NBC's major economic interest in an event such as the Olympics is to maximize the size of its audiences – especially their affluent consumer components – for exposure to the product and service advertising that underwrites its sport broadcasts. The IOC, the recipient of broadcast rights revenue, has a similarly deep interest in protecting its intellectual property, especially given that it is the 'liveness' of sport that is crucial for assembling vast audiences. The digitization and diversification of media technologies, it should be noted, has led the IOC to be less wedded to broadcast television as the most reliable means of reaching a mass audience across the world.

The intricacy of intellectual property arrangements in global media sport runs from the very top in sport contests such as association football's UEFA Champions League (as might be expected) even to the lower divisions of the English Football League (EFL). There are 72 league clubs in the EFL, many of whom only have modest crowds of a few thousand at live events and would not, therefore, merit a significant rights contract with a broadcaster. However, advances in media technology have enabled services to be offered such as 'iFollow [which is] a digital live streaming and content platform that will revolutionise the way football fans overseas can follow their EFL teams' (EFL, 2017). For the modest price of GB£110 (US$144), iFollow provides 'overseas fans with the opportunity to live stream up to 46 live league games featuring their club within the regular EFL season' (EFL, 2017). However, the intricate fabric of international rights agreements, and their sale of even minor games across continents, means that some will not be available in certain territories. For example, in October 2017 a League 1 match (that is, in the third tier of professional football in England) could not be watched live by the iFollow subscribers to Plymouth Argyle and Fleetwood Town, as announced on the former club's website:

> OVERSEAS supporters will not be able to watch our game this Saturday against Fleetwood via iFollow Argyle as per normal.
>
> This is because ITN productions will be on hand at Home Park to film the game as a major outside broadcast, to be screened in dozens of territories worldwide.
>
> As per the agreement with broadcasters in overseas territories, and as noted on the iFollow site, games being broadcast in this manner will not be screened on iFollow. Subscribers will be able to follow the game via audio commentary.
>
> We do not have an exhaustive list of where the game is being screened, but we encourage you, in Europe on [*sic*] particular, to check local listings to find the game, which will kick off at 3pm, UK time.
>
> In the Americas, the game will be screened on channels such as Sportsmax, ESPN and Sky Mexico; In Africa, on Kwese Sports and Fox Sports; In the Middle East and North Africa on beIN Sports. #
>
> Domestic iFollow subscribers are not affected.
>
> *(McNichol, 2017)*

So, this game, which attracted a live crowd of less than 9,000 people, was being made available audio-visually to subscribers to other television services in some places in the world. However, if they were domestic iFollow subscribers, different arrangements pertained. 'Domestic' as defined here is the UK and Ireland, despite the fact that the former includes Scotland and Northern Ireland, with no EFL teams, as well as Wales, which does, while the latter is actually a different nation-state. These viewers, for GB£45 (US$59) a season, do not receive live games with visuals, only having 'access to live audio commentary, as well as extended highlights and exclusive behind the scenes videos in HD, across desktop, mobile, tablet and app' (EFL, 2017). Yet, other EFL teams, from the larger (such as Aston Villa, Hull City, and Middlesbrough, all of whom have recently been in the EPL) to the very small (Accrington Stanley and Forest Green Rovers), have opted out of iFollow and have their own live streaming arrangements for international and domestic fans (although live audio commentary is universally available).

This relatively minor case has been detailed because it exposes the careful patrolling of intellectual property rights as they relate to the sports media even at a comparatively low level of sport competition. At higher levels, copyright conflicts are intense, such as in cases where sport fans upload live sport action from sports stadia using live video streaming apps like Periscope or, as in one recent case, deploying Facebook Live to re-broadcast without charge a pay-per-view boxing match (Hutchins and Rowe, 2017b). Despite concerns over widespread piracy, the emergence in sports of competing 'over the top' (OTT) services suggests that competition for global sports media rights will intensify, as is illustrated by the movement of the Perform Group into this domain:

> The long-awaited OTT move on premium sports rights has not come from the big US internet players but from a British company – albeit one owned by a Russian-born American billionaire, Leonard Blavatnik. Perform Group's low-cost premium service is initially being rolled out in Germany, Austria and Switzerland in Europe, and in Japan. It represents a considerable gamble that existing pay-television customers in those markets will be prepared to pay an additional cost – the monthly fee has yet to be made public – on top of a hefty pay-TV subscription. To back the gamble, Perform has invested heavily in exclusive premium rights, including Spain's La Liga, Italy's Serie A and France's Ligue 1, plus top US sports, such as basketball's NBA and American Football's NFL. Perform's most eyecatching deal came in December, when it snatched the rights to England's Premier League from incumbent rights-holder, pay-television platform Sky Deutschland, paying an increase of almost 200 per cent on what Sky had been paying.
>
> *(Dunne, 2016)*

Live screen sport, then, remains a highly prized commodity, and is becoming increasingly integrated with networked social media like Facebook and Twitter (Hutchins and Rowe, 2012). It is also possible that broadcast rights might be acquired directly by them just as it has been by telecommunications companies such as BT and Optus. Large technology companies like Google and Apple, or content streaming operations like Netflix, have the undoubted capacity to advance the globalization of sport texts, but in the context of very different business models based, respectively, on advertising and subscription. Whether this provision would still constitute sport television may be of principal interest to taxonomists and industry media sport industry players, but it is still of major concern to those interested in the rights of cultural citizenship that pertain to mediated sport. This domain of cultural citizenship is usually focused on the rights pertaining to key manifestations of national culture but, as noted above, the global circulation of mediated sports texts is now common, extending the reach of those who produce and distribute them. The costs to citizens around the world may variously be pay- replacing free-to-air sports television, or more intrusive advertising accompanied by intensified surveillance of viewer consumption practices. In both such cases, and in related areas of culture and economics, cultural citizenship is at issue in contexts ranging from the minutely local to the macroscopic global.

## Conclusion: Scrambling for the Sport Screen

This chapter has considered the possibility of the emergence of a global media sports cultural complex. It has registered a definite shift toward a free-flowing global sport television that uses, but is no longer dependent on, the satellite broadcasts that first heralded its arrival. The national television systems that once governed sport television through a relatively orderly series of arrangements with national governments, media corporations and sports organizations have been disturbed, though not entirely supplanted, by the transnational connectivity of media corporations such as 21st Century Fox, ESPN and beIn and the high-intensity global promotion and marketing campaigns by sports rights holders from FIFA and the IOC to the EPL and NBA (Sage, 2015). However, this new international TV order has itself been disturbed by the cross-border and -platform flows provided by highly mobile digital media technologies controlled by non-broadcast companies (such as Google and Apple) and passed through social networks (such as Twitter, Facebook and its subsidiary, Instagram) populated by 'co-producer' citizens. In this context, media regulation by nation-states recedes somewhat in the face of privately owned and acquired intellectual property control, especially in the form of live mediated sport texts. For this reason, questions of fair use and piracy of such texts become paramount. Thus, it is misleading to assume that the capacity for such screened sport material to flow freely necessarily enhances access to it by all viewers and users – its monetization in most cases demands its restriction to those willing and able to pay for it, irrespective of the transition 'from broadcast rationing to digital plenitude' (Hutchins and Rowe, 2009).

It is at this point, though, that the nation-state retains its relevance where it can exercise its capacity to intervene in media sport markets in the name of national cultural citizenship. Many national governments, although to highly variable degrees, still mobilize 'anti-siphoning' laws and regulations that prevent some key mediated sports events from being taken out of non-paying environments and made available only for a fee (Scherer and Rowe, 2014). They – and sometimes via supranational blocs like the European Union – are also able to use competition laws to prevent, or at least mitigate, undue concentration of media sport ownership and control. Citizen-fans, as noted, can exercise their own judgments in various ways, by rejecting mediated sports that are artificially forced upon them as part of a process of global expansion out of congested or exhausted domestic markets, by infringing copyright where it is felt that pricing is exploitative or access unreasonably restrictive, and sometimes by recasting and redistributing them in creative expressions of mediated fandom (Hutchins and Rowe, 2012). This friction produced by commercial rivalry, technological innovation, government supervision and sport fan assertiveness is manifest within, between and beyond nations.

Global sport television, irrespective of how it is organized and delivered, occupies a dynamic, untidy and conflictual media space. Where it asserts its globality it is of the complex kind described by John Urry (2003: 103) in which 'there are pockets of order (or ordering) within a sea of disorder'. It is the product, like sport itself, of jockeying for position both within and outside the rules, of deploying strategies and tactics designed to produce success for competing teams, of struggling to control the dynamics of the 'field' (Bourdieu, 1984), and of striving for outcomes that are subject to the uncertainties of a game that, while seemingly predictable, is still capable of surprise.

## References

Ahmed, M. (2017) 'NBA Looks to Score Big with Global Digital Rights Revenues', *Financial Times*, 12 January, www.ft.com/content/ad84f056-d81b-11e6-944b-e7eb37a6aa8e?mhq5j=e6

Andrews, D.L. (2009) 'Of American Slouch and English Fervour', *Flow*, 5 December, www.flowjournal.org/2009/12/of-american-slouch-and-english-fervourdavid-l-andrews-university-of-maryland/

*Australian Football League* (AFL). (2017a) 'Global', www.afl.com.au/global

*Australian Football League* (AFL). (2017b) 'International Broadcast Partners – Television', www.afl.com.au/tv-radio/international-broadcast-partners

Belson, K. (2017) 'As Appetite for Soccer in U.S. Grows, So Does M.L.S', *The New York Times*, 8 August, www.nytimes.com/2017/08/08/sports/soccer/mls-expansion-nashville-cincinnati.html

Berry, A. (2013) 'More than One Quarter of Players Born Outside US', *MLB.com*, 1 April, http://m.mlb.com/news/article/43619160/more-than-one-quarter-of-mlb-players-born-outside-us/

Blackshaw, I., Cornelius, S. and Siekmann, R. (2009) *TV Rights and Sport: Legal Aspects*. The Hague: T.M.C. Asser Press.

Bourdieu, P. (1984) *Distinction: A Social Critique of the Judgement of Taste*. Cambridge, MA: Harvard University Press.

Boyd, T. (2008) *Young, Black, Rich, and Famous: The Rise of the NBA, the Hip Hop Invasion, and the Transformation of American Culture*. Lincoln, NE: Bison Books/University of Nebraska Press.

Chappell, B. (2017) 'The Olympics' TV Time-Delay Is Going Away, NBC Says', *NPR*, 29 March, www.npr.org/sections/thetwo-way/2017/03/29/521913449/the-olympics-tv-time-delay-is-going-away-nbc-says

Dunne, F. (2016) 'OTT's European Power Struggle and What it Means for Sports-Media Rights', *SportBusiness International*, 1 June, www.sportbusiness.com/sportbusiness-international/ott%E2%80%99s-european-power-struggle-and-what-it-means-sports-media-rights

English Football League (EFL). (2017) *iFollow*, www.efl.com/iFollow/

Evens, T., Iosifidis, P. and Smith, P. (2013) *The Political Economy of Television Sports Rights*. Basingstoke: Palgrave Macmillan.

Falcous, M. and Maguire, J. (2005) 'Making it Local? NBA Expansion and the English Basketball Subculture', in M. Silk, D. Andrews and C. Cole (eds) *Corporate Nationalisms: Sport, Cultural Identity and Transnational Marketing*. Oxford: Berg, 13–34.

Farred, G. (2006) *Phantom Calls: Race and the Globalization of the NBA*. Chicago: Prickly Paradigm Press.

Gaines, C. (2011) 'Sports Chart of the Day: The International Origins of NFL Players', *Business Insider Australia*, 18 November, www.businessinsider.com.au/chart-international-origins-of-nfl-players-2011-11?r=US&IR=T

Gmelch, G. and Nathan, D.A. (eds) (2017) *Baseball Beyond Our Borders: An International Pastime*. Lincoln, NE: University of Nebraska Press.

Hutchins, B. and Rowe, D. (2009) 'From Broadcast Rationing to Digital Plenitude: The Changing Dynamics of the Media Sport Content Economy', *Television & New Media*, 10(4): 354–370.

Hutchins, B. and Rowe, D. (2012) *Sport Beyond Television: The Internet, Digital Media and the Rise of Networked Media Sport*. New York and Abingdon, UK: Routledge.

Hutchins, B. and Rowe, D. (2017a) 'The Tyranny of Perpetual Innovation: Global Mobile Media, Digital Communications and Television', in N. Schulenkorf and S. Frawley (eds) *Critical Issues in Global Sport Management*. London: Routledge, 238–251.

Hutchins, B. and Rowe, D. (2017b) 'Prize Fight Over Live-Streamed Sport Will go on Long after the Final Bell Sounds', *The Conversation*, 8 February, https://theconversation.com/prize-fight-over-live-streamed-sport-will-go-on-long-after-the-final-bell-sounds-72494

Kilgore, A. (2017) 'How the NBA's Embrace of Social Media Might Help it Someday Surpass the NFL', *The Washington Post*, 8 June, www.washingtonpost.com/news/sports/wp/2017/06/08/how-the-nbas-embrace-of-social-media-might-help-it-someday-surpass-the-nfl/?noredirect=on&utm_term=.f4c4c867e853

Koba, M. (2013) 'US Pro Teams Give "Away Game" a Whole New Meaning', *CNBC Sports Business*, 9 October, www.cnbc.com/2013/10/09/us-sports-overseas-pro-teams-give-away-game-new-meaning.html

Levy, D. (2017) 'Is Watching Sports on TV Actually Better than Being at the Game?', *Bleacher Report*, 10 January, http://bleacherreport.com/articles/923563-is-watching-sports-on-tv-actually-better-than-being-at-the-game

Matheny, S. (2018) 'The Winter Olympics Come to Next Gen TV Broadcasting', *PILOT™*, 22 February, https://nabpilot.org/the-winter-olympics-come-to-next-gen-tv-broadcasting/

McGregor, A. (2016) 'Peyton Manning: The NFL's Great White Hope', *Sport in American History*, 4 February, https://ussporthistory.com/2016/02/04/peyton-manning-the-nfls-great-white-hope/

McNichol, R. (2017) *Plymouth Argyle Football Club* 'Fleetwood Game – iFollow Details', 7 October, www.pafc.co.uk/news/2017/october/fleetwood-game—ifollow-details/

Miller, T., Lawrence, G., McKay, J. and Rowe, D. (2001) *Globalization and Sport: Playing the World*. London: Sage.

Miller, T., Rowe, D. and Lawrence, G. (2011) 'The New International Division of Cultural Labour and Sport', in J. Maguire and M. Falcous (eds) *Sport and Migration: Borders, Boundaries and Crossings*. London: Routledge, 217–229.

Miller, T., Rowe, D., Lawrence, G., and McKay, J. (2003) 'The Over Production of US Sport and the New International Division of Cultural Labour', *International Review for the Sociology of Sport*, 38(4): 427–440.

*Pennsylvania Sports* (2017) 'Super Bowl Myth Exposed: These are Actually the Most-Watched Sporting Events', 5 February, www.pennlive.com/sports/index.ssf/2017/02/super_bowl_myth_exposed_these.html

Real, M.R. and Wenner, L.A. (2017) 'Super Bowl: Mythic Spectacle Revisited', in L.A. Wenner and A. Billings (eds) *Sport, Media and Mega-Events*. London and New York: Routledge, 199–217.

Rowe, D. (1995) *Popular Cultures: Rock Music, Sport and the Politics of Pleasure*. London: Sage.
Rowe, D. (2004) *Sport, Culture and the Media: The Unruly Trinity* (Second edition). Maidenhead, UK and New York: Open University Press/McGraw-Hill Education.
Rowe, D. (2011) *Global Media Sport: Flows, Forms and Futures*. London and New York: *Bloomsbury* Academic.
Rowe, D. and Gilmour, C. (2010) 'Sport, Media and Consumption in Asia: A Merchandized Milieu', *American Behavioral Scientist*, 53(10): 1530–1548.
Sage, G.H. (2015) *Globalizing Sport: How Organizations, Corporations, Media, and Politics are Changing Sports*. London and New York: Routledge.
Scherer, J. and Rowe, D. (eds) (2014) *Sport, Public Broadcasting, and Cultural Citizenship: Signal Lost?* New York: Routledge.
Schwartz, D.A. (2016) 'Soccer as Un-American Activity: A Thematic Analysis of Online US Media and Flopping in the World's Game', *Soccer & Society*, www.tandfonline.com/doi/ref/10.1080/14660970.2016.1199425?scroll=top
Stein, M. (2017) 'The Unspeakable Greatness of Giannis Antetokounmpo', *The New York Times*, 3 November, www.nytimes.com/2017/11/03/sports/basketball/giannis-milwaukee-bucks.html
Tomlinson, A. and Young, C. (eds) (2006) *National Identity and Global Sports Events: Culture, Politics, and Spectacle in the Olympics and the Football World Cup*. New York: State University of New York Press.
Turner, G. and Tay, J. (eds) (2009) *Television Studies After TV: Understanding Television in the Post-Broadcast Era*. London and New York: Routledge.
U.S. Soccer. (2017) 'World's Best Feature in NWSL and Tournament of Nations', 24 July, www.ussoccer.com/stories/2017/07/24/21/23/20170724-news-wnt-worlds-best-feature-in-nwsl-and-ton2017
Urry, J. (2003) *Global Complexity*. Cambridge: Polity.
Wallis, W. (2016) 'Smart Africa: Smartphones Pave Way for Huge Opportunities', *Financial Times*, 27 January, www.ft.com/content/aba818a6-c392-11e5-808f-8231cd71622e?mhq5j=e6
Wenner, L. (ed) (1998) *MediaSport*. London: Routledge.
Whannel, G. (1992) *Fields in Vision: Television Sport and Cultural Transformation*. London: Routledge.

# 18

# NEOLIBERAL MULTICULTURALISM, *OUTSOURCED*

*Asha Nadkarni*

When the U.S. network NBC added the comedy series *Outsourced* to their fall 2010 primetime line-up, they were taking quite a gamble. For one, the series was the first to feature a cast composed primarily of actors of South Asian descent. But even more striking was the comedic situation upon which the show is premised: business process outsourcing from the United States to India. Accordingly, *Outsourced* opens with U.S.-based company, Mid-American Novelties, moving its telemarketing call center and white middle manager Todd Dempsey from Kansas City to India. Unsurprisingly, cultural misunderstandings and various hijinks ensue as Todd and his Indian co-workers adjust to each other and their new work environment. *Outsourced* thus presents viewers with several questions, as succinctly posed by an article announcing the series: "will [*Outsourced*] help humanize Indian workers for U.S. audiences, or reinforce stereotypes? Does the arrival of the show mean that Americans have accepted offshore outsourcing as routine? And how realistic will it be?" (Thibodeau 2010). The cancelation of the show after just one season would seem to have answered these questions in the negative.

Nonetheless, I turn to this short-lived series to think through the work it was doing in its particular cultural moment of the late 2000s, with special attention to its representation of the racial politics surrounding outsourcing. Specifically, I want to examine how *Outsourced* participates in what Jodi Melamed calls neoliberal multiculturalism, considering the ways it brandishes a "multicultural formalism" wherein difference becomes abstracted to "[represent] neoliberal economic policies as multicultural rights and global capitalism itself as antiracist in spirit" (Melamed 2011, 148). Indeed, a sitcom about outsourcing aimed at a mainstream U.S. audience would seem to be a perfect example of neoliberal multiculturalism. However, both *Outsourced* and neoliberal multiculturalism itself need to be situated within longer histories of economic development in South Asia and of U.S.–Indian neo-imperial relations. Thus in turning to *Outsourced* I suggest that while on the one hand the show is a familiar "fish out of water" tale of cultural clash, on the other hand it is easily assimilable within the framework of a U.S. multiculturalism gone global. In taking up this particular cultural object to think through the shifting conversations around outsourcing, moreover, I suggest that media representations such as the series become sites for working through, symbolically and narratively, the realities and contradictions of globalized economies as they are experienced. As a review in the *Hollywood Reporter* puts it, ideally

*Outsourced* presents viewers with "a chance to grin in the face of modern economic realities while having fun with the U.S.-India culture crash" (Pedersen 2010). In this sense, the failure of the show can be viewed as a failure to domesticate the experiences of outsourcing properly for a U.S. audience.

Although it was based on a successful 2006 independent film of the same name, the television version of *Outsourced* is different in both style and substance. In each, our protagonist, Todd Dempsey, travels to India with the call center he manages, and movie and situation comedy alike use this premise as their central drama. But the similarities stop there. The film is a romantic comedy in which Todd (played by Josh Hamilton) goes on a journey of self-discovery, thus resurrecting the time worn narrative of "finding oneself" in India. While the film certainly plays on the same cultural stereotypes as the sitcom (jokes about gastrointestinal distress, accents, and cultural practices abound), they are played much less broadly and function more to show Todd's initial discomfort with India than to comment on Indian cultural mores as comically different. Indeed, the trajectory of the movie follows Todd's acceptance of India and Indian culture as he comes to know himself and fall in love with his co-worker, Asha. In contrast, the sitcom depends primarily on the comedy of culture clash, and as such even though Todd's relationship to India deepens and changes over the course of the series the comedic frisson of the show relies on Todd 's continued alienation with India even as he forges bonds with his co-workers. Finally, there is the crucial difference that the film was shot on location in India and utilized a largely Indian cast, while the sitcom was shot on a Hollywood sound stage with primarily American and British actors of South Asian descent. The movie thus operates in a realist idiom that the sitcom eschews. Because it was made to appeal to a much wider and mainstream audience than an indie film, I focus on the television version of *Outsourced* to interrogate the work it is doing (or failing to do) to manage the contemporaneous crisis of offshore outsourcing.

## Techno-Orientalism, White U.S. Masculinity, and U.S.-Indian Imperial Circuits

Business process outsourcing, or the practice of contracting out certain formerly internal operations to other companies, is one of the more visible and remarked upon aspects of globalization. In particular, the offshoring of operations to foreign sites where wages and facilities are cheaper has primarily been talked about as a threat to U.S. workers. Accordingly, a 2004 special issue of *Wired* magazine on outsourcing describes the Indian labor force as "an artificial intelligence, the superbrain that never arrived *in silico*. No wonder [U.S.] workers tremble" (Anderson 2004). This rendering of Indian workers as terrifying machines trades in long-standing discourses of "techno-Orientalism" – what David S. Roh, Betsy Huang, and Greta A. Niu define as "the phenomenon of imagining Asia and Asians in hypo- or hypertechnological terms in cultural productions and political discourse" (Roh, Huang and Niu 2015, 2). Figuring Asians as examples of "an artificial intelligence" makes them both threatening and containable insofar as it denies their humanity; in this sense *Outsourced* both mobilizes and works against techno-Orientalist stereotypes by strategically granting humanity to some of the Indian workers it portrays (Todd's team at Mid America Novelties) and denying it to others (the machine-like "A-Team" employees working for companies such as Apple that I discuss in the next section).

But even as techno-Orientalism seeks to contain the threat posed by outsourcing, it cannot entirely do so. Importantly, in both *Outsourced* and in the *Wired* special issue, outsourcing is viewed as posing a specific menace to white masculinity. The *Wired* feature story begins precisely with this affront: "meet the pissed-off programmer … . He's the guy – and yeah, he's usually a guy – launching Web sites like yourjobisgoingtoindia.com and nojobsforindia.com" (Pink 2004). Here, as in *Outsourced*, outsourcing is dramatized through the dilemma of the white male protagonist who has been displaced. But whereas the *Wired* story imagines it as a pitched battle between the "pissed-off programmer" and the Indian workers who have replaced him, *Outsourced* views it as rife with comic potential. Thus, while film and television show rely on old stereotypes of India, they also animate new ones related to the mimetic qualities of Indian labor. *Outsourced* thereby uses comedy to minimize the affective labor necessarily performed by Indian call center workers, dramatizing outsourcing as both a crisis of, and ultimately a salve to, white U.S. masculinity alone. In other words, the dilemma of outsourcing is only a dilemma for our white male protagonist; the very real tolls such work takes on the Indian workers are effaced.

By focalizing the experience of outsourcing through a white, middle-class U.S. man, moreover, *Outsourced* obscures how the contemporary migration of jobs is linked to earlier labor migrations of people, which are in turn embedded in legacies of labor extraction from the South Asian subcontinent. In contrast to this, I argue that the contemporary moment of outsourcing speaks to longer imperial connections between the U.S. and India. As both an ally of and competitor with British colonialism in India, the United States positioned itself as the modernizing agent in India in the years leading up to Indian independence, and then continued to operate as a modernizing force in the region through the guise of development. Development (both in the form of direct aid and of the transfer of technical knowledge) served a number of linked goals for the U.S.: it was intended to promote U.S. values and institutions, open up new markets for U.S. capital, and serve as part of a Cold War containment strategy. In South Asia, this meant that development aid to the region was supplemented by a military alliance with Pakistan to counter the threat of Indian non-alignment (Latham 2010, 69–70). Additionally, the United States was involved in the Afghan–Pakistan region through supporting (via provisions of money and weapons) proxies fighting the Soviet Union, such as the Taliban and Al-Qaeda.

At the same time, modernization – a doctrine largely conceived in and advanced by the United States, even if it is not synonymous with Americanization – was crucial to both the Indian and Pakistani postcolonial states. In this sense developmentalism's progressivist creed at once rearticulated many of the commonplaces of colonialism, and became a key articulation of postcolonial nationalisms (Nadkarni 2014, 136–139). Indeed, the direction of state-led development planning in India is an important part of the story of outsourcing. In the wake of independence and in a drive to make India a global force in science and technology, Jawaharlal Nehru instituted much of the existing infrastructure for technology through the expansion of technical institutes. This had the unintended result of creating the labor pool for the so-called "brain drain" of educated Indians to the UK and North America, as countries in the global North recruited a skilled labor pool trained in India. Additionally, this state-sponsored emphasis on science and technology also created the infrastructure for the contemporary phenomenon of the outsourcing of information technology (IT) labor to India. With the liberalization of the Indian economy beginning in 1991, the groundwork for outsourcing to India was complete (Prashad 2001; Reddy 2015).

Although modernization theory fell out of favor by the 1970s (for a variety of reasons, not least of which was that it didn't work as predicted), the rise of neoliberalism in the late 1970s/early 1980s replaced the global vision of modernization theory with one that was remarkably similar. The key difference between modernization theory and neoliberalism is the role of the state – instead of the state manipulating development, all that was necessary was to let capitalism run its course, thus "replacing the liberal, visible hand of the modernizer with the classical, invisible hand of the market" (Latham 2010, 158). Nonetheless, even if the neoliberal stance toward the state considers it most useful for getting out of the way, there are various state structures, such as tax exemptions, that support the outsourcing industry in India. Therefore, one of the many contradictions of the current moment is that neoliberal appraisals of the state as effective only through its absence "[fail] to consider that the current situation [of outsourcing] is the direct result of at least two generations of state-sponsored investment in scientific and technical education" (Sharma and Gupta 2006, 4).

The cultural narrative of imperiled U.S. jobs lost to cheap overseas labor conceals this longer and more complicated history. Accordingly, *Outsourced* presents outsourcing primarily as a crisis for Todd and the other displaced U.S. workers. The pilot episode opens with Todd walking into his Kansas City office after completing management training, only to find it emptied of all personnel. His boss, Jerry, explains, "Costs were through the roof. You know, pensions, health care. We needed to do a little right sizing … . Yesterday we outsourced the entire call center to India." The violence of this move is underscored by a brick flying through the window, with a note attached to it that says "you bastard." Jerry laughs, adds the brick to a sizeable pile, and cheerfully remarks, "That's for me." Despite the humor of the moment (the note, the visual gag of the pile of bricks by Jerry's desk), the very real anger and helplessness felt by workers whose jobs have been outsourced is clearly referenced, as is Todd's own precarious position as a worker. When the boss informs Todd that keeping his job is contingent on his running the call center in India, Todd attempts to fight back by saying perhaps he will open a rival novelty company. The boss flippantly replies, "Don't you owe like forty grand in student loans?" In this moment it is clear that the once commonplace markers of the so-called "American dream" (health insurance, a pension, an affordable education) are what are really being outsourced. Todd's helplessness in the face of this is reinforced as the scene cuts to show him in India, being driven through the streets in an auto rickshaw.

## Manufacturing U.S. Affect

The products sold by Mid America Novelties deliberately emphasize the irony of outsourcing its call center to India: in addition to joke items such as fake vomit, the company also sells patriotic products, such as red, white, and blue condoms and a coffee mug proclaiming "America is #1." Given the nature of these objects, it becomes doubly important for the Indian call center workers to learn about U.S. culture and mask their accents. After all, even if these products are in fact *produced* in China (a fact that Todd references in episode 107), the people buying them want their patriotic purchases uncomplicated by such realities of global capital. The pilot underscores this when call center worker Manmeet takes his first call and does not disguise the fact that he's in India. As he reads back the order number he clarifies by stating: "K as in Krishna, P as in Punjabi, R as in Ramayana." The irate caller's response, before he hangs up, is: "Where am I calling? Is this India? Am I calling freakin' India to get a mug that says 'America is #1?'" Manmeet responds to say, "No, we're in Detroit. City of motors and Black people."

I pause on this moment – one that is played for laughs but which also acknowledges a real disquiet about outsourcing (a disquiet parsed in racialized ways that reflect a transition from an industrial to a service economy) – to think more deeply about the nature of call center labor. Manmeet and his co-workers are not simply selling novelties, they are also selling their performance of U.S. affect (even if in this case that performance is unsuccessful). This false production of self, however, is part of the very process of self-improvement that call center labor supposedly offers its workers, as outsourcing work becomes an opportunity for financial, cultural, and personal self-improvement. Indeed, in *Outsourced* all three of these come together in the Assistant Manager Rajiv's plotline, as the salary he will earn when he takes over Todd's job will allow him to marry: the final episode of the series features his wedding. Thus call center work is represented in *Outsourced* as being about personal choice and self-improvement: in a word – development.

As Purnima Mankekar argues in *Unsettling India* (2005), the affective labor of producing either a "neutral" or explicitly "American" affect is an act of impersonation that becomes an aspirational act of self-making, or personation. Theorizing call center work through Michael Hardt's notion of affective labor (i.e. "labor that produces or manipulates affects"), Mankekar demonstrates how the call center workers' "self-constitution and self-regulation as laboring subjects [reveal] the convergence of cognitive labor, embodiment, and affect" (Mankekar 2015, 202–203). In particular, by tracking how such workers are forced to conform to U.S. time zones, ways of speaking, and cultural norms, Mankekar argues that "the normative imperatives of neoliberalism … produce the laboring subject as entrepreneurial subject" (Mankekar 2015, 220). In other words, the self is one of the products being produced through outsourcing labor. But arguably it is only being produced through an effacement of self (as signaled by the fact that call center workers adopt American names and norms), and it is a production that can never be entirely complete.

I now turn to two moments from the sitcom to think about the production of U.S. affect. The first shows Todd's call center workers performing their knowledge of different U.S. cultural productions, because they are not, as Todd puts it, "really getting the cultural stuff:" Manmeet does a monologue from *Glengarry Glen Ross*, Madhuri does something inaudible, and Gupta performs The Pussycat Doll's hit song "Don't Cha." The second of these moments takes place in the cafeteria and is a performance of a different sort by one of the so-called "A-Team" workers (call center workers employed by companies such as "Intel and Apple," who wear suits, have been to the United States, and have studied different regional American accents). Charlie – another white, U.S.-born, male call center manager who befriends Todd – calls over one of the A-Team members and asks him what grits are. The A-Team worker replies in a spot-on Southern accent: "Well, grits is just ground up corn. I'll tell you what, my mom used to make the best grits. I'd be sitting on my front porch with my hound dog Freedom sipping on some sweet … " Charlie cuts him off and shoos him away, to which he replies (again in a Southern accent) "Y'all take care."

The comedy of these scenes, where the characters attempt to make themselves over in an "American" image, comes from impersonations that are either imperfect, or too perfect. In the case of Gupta, his buffoonery is funny precisely because of the disconnect between his performance of "Don't Cha" (complete with sexy dance moves) and his physical appearance. As he enthusiastically dances, sings, and rolls his hips, his co-workers avert their eyes and make protesting noises. His performance also plays on U.S. stereotypes of South Asian men as emasculated and asexual – thus the irony of him singing "don't cha wish your girlfriend was hot like me." On the opposite end of the spectrum we have the "A-Team" worker, whose uncanny ability to mimic not just an American accent, but a Southern accent (in which he displays his knowledge of grits and talks about his hound dog "Freedom") is funny because it's both impressive and "creepy." In fact, it is so creepy that Charlie likens the "A-Team" workers to "pod people," thus recoding the threat of communism analogized in the 1956 film, *Invasion of the Body Snatchers*, into a neoliberal nightmare.

By translating these forms of affective labor into humor, *Outsourced* erases the real hardships of call center work and the impersonations it requires. It likewise fabricates an essential condition of call center work by showing it as taking place during regular daytime working hours, even though in reality this is work done at night. Indeed, as Shehzad Nadeem and others have argued, one of the reasons that call center work takes such immense physical and emotional tolls is because workers must adjust their schedules around U.S. working hours (Nadeem 2011). Such erasures allow the affective labor the call center workers perform to be played simply for laughs – laughs that are almost always on the Indian workers and, by extension, on the actors of South Asian descent who play them. *Outsourced* was notable for employing (both in front of the camera and behind it) an unprecedented number of people of South Asian descent. Unlike the film version of *Outsourced*, which was shot in India and featured Indian actors, most of the actors in the series were born in either North America or Great Britain, and thus were performing an Indian accent, or what Shilpa Dave calls "brown voice" (Dave 2013). Paradoxically, the very U.-S. accents and affects that call center workers are striving to perform come naturally to the actors, the majority of whom were born in North America. There is thus a multiplicity of cultural performances at work here, as the essentially performative nature of call center work is at once being underscored and effaced by the performances of the actors themselves.

## Stereotyping and Multicultural U.S. Citizenship

Arguably the broad nature of the situation comedy as a form both relies upon and excuses the use of stereotyping. Actor Rizwan Manji, who plays Rajiv, describes how he felt "uncomfortable" with the character he played on the dramatic series *24* because he "felt like it was perpetuating a stereotype." In contrast he argues that "[*Outsourced*] is a sitcom where it's supposed to be broad and it's supposed to be funny" (Frederick 2010). In this formulation because humor depends on stereotypes the use of stereotypes cannot be offensive. A similar logic was deployed in online conversations about that show, as Mary Grace Antony demonstrates in her analysis of *Outsourced.* Citing a discussion between viewers "Hari K" and "Anon" she writes:

> Although Hari K voices concerns that uninformed audience members may mistake the stereotypes on *Outsourced* for factual and authentic representations of a foreign culture, his fears are dismissed by Anon who notes several exaggerated stock characters that clearly do not accurately represent Americans.
>
> *(Antony 2013, 16)*

But while stereotypes may and do pertain to both the Indian and the American characters, what is different is the ability of U.S. viewers to read them as such. For instance, when watching the show *The Office* (a U.S. remake of a British workplace situation comedy), the audience understands simplifications as simplifications because they are presumably familiar with U.S. work culture. Even though *Outsourced* was originally aired directly after *The Office* and accordingly could be viewed as just another show about a dysfunctional workplace, the essential foreignness of *Outsourced*'s setting makes such simplifications much harder to parse. Therefore, despite the fact that the actors and writers on the show repeatedly attest to the fact that they feel they are "making history" by being a part of "the first show about our community in America," the issue of stereotyping persists (SAJA HQ 2011).

But even putting the issue of broad cultural stereotyping to one side, *Outsourced* uncritically mobilizes familiar gender stereotypes to figure the relationship between the United States and India. This is consistent with representations of outsourcing in the medium in which it has most often been portrayed: documentary film. Documentaries, as Aimee Rowe, Sheena Malhotra, and Kimberlee Perez attest in *Answer the Call,* make different claims to verisimilitude than the broad comedic strokes of a situation comedy in order to "generate a truth effect that shapes and manages U.S. anxieties over globalization" (37). But even so, documentaries and sitcom alike attempt to "manage Western anxieties over outsourcing by recirculating familiar relations of race, gender, and heterosexuality" (33). Looking at documentaries such as the 2004 *The Other Side of Outsourcing* and the 2006 *1-800-INDIA*, *Answer the Call* argues that these works safely frame outsourcing in developmentalist terms in which India is always lagging behind. One of the ways they do this is by using a white male narrator to construct India in general, and the Indian female call center worker in particular, as the object of a male gaze that sutures the disparate temporalities each represents through heterosexuality. Thus the Indian female call center worker as a stand-in for Indian progress contains the threat of a modernizing India by reinstantiating the temporal lag between the U.S. and India (she is always in the process of developing but not yet developed), thus modeling India's properly subordinate status to a dominant U.S. masculinity.

In *Outsourced*, this narrative of feminist development is illustrated by call center worker Madhuri's story. In the pilot, Madhuri is portrayed as being so soft-spoken and shy that she is virtually inaudible. By the end of the episode, however, Madhuri literally finds her voice as she successfully sells fake dog poop and vomit. The scene suggests that the products are beside the point – the triumph is that Madhuri is finally speaking and is successfully upselling by convincing the caller to buy the vomit. But arguably the fact that she is selling fake excrement and vomit is very much to the point, as these novelties are symbols of bodily abjection that show the instability and porosity of bodily boundaries. It is thus significant that the moment of supposed feminist empowerment is also a moment of abjection – showing at once Madhuri's capacity to develop and for that development to be undone.

That this moment is husbanded by Todd is similarly to the point in that, as the authors of *Answer the Call* argue, "Indian women in call centers stand in for progress, or a movement toward development, while white men signify a temporal point of arrival, or a telos of progress" (*Answer the Call* 34). This movement toward development is dramatized in a later episode of the series, when Todd discovers Madhuri's beautiful singing voice and encourages her to participate in a singing competition (which she wins by singing The Bangles' 1988 hit "Eternal Flame"). The implication of the episode is that she could go on to the next level of the competition and become a star. But when encouraged by Todd she demurs, saying "I am already living my dream. I have a great job. I make enough money to provide for my entire family. I am happy with the way things are." Here "tradition" becomes a barrier to development as Madhuri is folded back into a narrative of "traditional values" that cause her to place her family's wellbeing over her individual achievements, thus betraying the neoliberal individualism advocated by Todd.

But even as Todd is both model and midwife of some form of developed subjectivity, he is not simply the white male master of the universe; in fact, as a comedy that derives its humor from "clashing cultures" would dictate, Todd is constantly getting it wrong. The show's pilot shows Todd making a series of gaffes; mispronouncing his co-worker's names or making inappropriate jokes about them (saying to Manmeet, "Man meat? It must be hard to chat on the internet with a name like 'Man meat'"), mocking the cow in the courtyard, making loaded comments about people's head coverings. Even so, his very ability to wrestle with difference and his good intentions in doing so (his cultural insensitivity is portrayed as ignorance, not malice) show him to be the proper multicultural subject who is actually more open to "difference" than his Indian co-workers – and in fact this openness functions as yet another sign of his modernity in contrast to his Indian counterpart's stubborn adherence to certain "traditional" practices. It likewise contrasts his attitude to the disgruntled caller who hangs up on Manmeet in the pilot episode.

Returning again to Melamed's idea of neoliberal multiculturalism, as a "multicultural U.S. citizen" Todd stands in "as an emblem for the most universal and legitimate form of global citizenship" (153). That this mode of global citizenship must be learned through consuming various kinds of American products – be they pop culture, or the "kitsch" sold by Middle American Novelties – furthermore attests to the fact that turning workers in the Global South into global citizens must be achieved through market means. Thus, the odd way, as Melamed describes it, that "being able to buy anything from anywhere" is figured as a form of freedom that can redress the legacies of colonialism and continued global structures of economic equality. For instance, when Todd's Indian co-workers are confounded by the products in the catalogue because they don't understand, as Manmeet puts it, "why . . . Americans need these things," Todd pulls out "jingle jugs," a disembodied pair of breasts mounted on a plaque that jiggle to the sound of sleigh bells. Todd's response to Manmeet's query is that

> We don't … there is no purpose. In America you can do whatever you want. You could be the president. Or a scientist. Or you could even invent novelties like … [he pulls out "jingle jugs"]. This is exactly what I'm talking about. See, maybe no one needs this but in America no one can stop you from making it. This is the definition of freedom. This is jingle jugs.

This series of strange equivalences in which being able to become president is the same as inventing a senseless and sexist novelty flattens out the idea of freedom into a neoliberal market logic in which to be free is to be able to buy and sell anything regardless of its value.

Finally, not only does the show thematically explore multicultural neoliberalism, it is indeed a product of it, as the very presence of this show in the NBC line-up attests to a multicultural U.S. audience open to consuming difference. Thematically, it flattens out how the different subject positions occupied by the characters are inserted into larger structures of global capital and power – even as it is ostensibly about precisely those forces. At the same time, as a product the show itself fits very neatly into a framework of a U.S. multiculturalism gone global. Hence the NBC slogan at the end of the trailer promoting the show before it debuted: "more colorful." Ostensibly this slogan, which NBC introduced in 2009, references the feathers of the corporate mascot peacock (introduced in 1956 to highlight early color television broadcasting), but it also seems to reference a specific idea of U.S. multiculturalism that has arguably been promoted by NBC at least since they began airing *The Cosby Show* in 1983.

## Conclusion

Even if we can fit *Outsourced* into a narrative about multicultural progress and acceptance, the very proliferation of South Asian bodies on the screen signals that other most predominant stereotype related to South Asia – that of South Asians as terrorists (and, of course, outsourcing and terrorism are also linked through a rhetoric that asserts both outsourcing and terrorism as prime threats to U.S. futures). Contemplating the differences between representations of outsourcing and terrorism, Mrinalini Chakravarty suggests that cultural productions

> that visualize terror and its aftereffects usually conform to the generic codes of gothic horror and tragedy. The effects of cultural mistranslations, malaise, and disaffections in [cultural productions] dealing with flexible and mobile work cultures more commonly find expression through romantic comedy and comedies of error.
>
> *(2014, 212)*

Nonetheless, the distinction between these genres is not entirely straightforward, as outsourcing and terrorism present separate but linked threats, both in their relationship to global capital and to South Asian bodies. The difference is that while terrorism is spectacular, outsourced labor is most often spectral – what Jean Comaroff and John Comaroff theorize as "zombie or 'ghost labor'" (quoted in Chakravorty 2014, 211). Thus both comedy and anger come out in the moments where outsourced labor is made visible through accents and inappropriate cultural cues and references. It is this friction (i.e. those moments when the national and spatial coordinates of global capitalism are revealed) that creates *Outsourced*'s particular mix of comedy and unease stemming from well-worn tropes of cultural clash.

In the pilot the threatening aspects of outsourcing are at the fore. The first instance is in the aforementioned scene in which former workers of the now closed Kansas City call center show their discontent by throwing bricks through the window at their former manager. The second moment is the seemingly unnerving presence of a Sikh call center worker, whose unsmiling demeanor and turban unsettle Todd at crucial moments. But this unease is also apparent in the frisson between the "America" being sold, who is selling it, and to whom it's being sold. *Outsourced* consistently mocks the consumers of Mid America Novelties' products, portraying them as white and lower class: as Todd states in the film, his job is to "sell kitsch to rednecks." It likewise mocks the efforts of the call center workers to sell the products, and mocks the Indian workers who see the products as potentially worth purchasing. *Outsourced* thus positions Todd, despite his moments of cultural insensitivity, between the "redneck" consumers of the products and the Indian employees who sell them as the global multicultural citizen to be emulated. Insofar as the viewer is asked to identify with Todd, his pedagogic purpose is to model the appropriate attitude toward outsourcing and toward "other" cultures in general.

By way of closing, I turn to a scene from episode 113, "Training Day," that encapsulates the tension between U.S. multiculturalism and U.S. racism. The cold open of the episode shows us the call center in crisis: phones ringing off the hook, employees apologizing to customers in hushed tones. Gradually the crisis is revealed to be that there has been a mistake at the factory and an order of ghost costumes has been sent out with pointed, rather than rounded, hoods, thus making them resemble Ku Klux Klan robes. As the call center workers assure the callers that they will be issued a full refund, Rajiv, the manager of the center, appears in one of the robes, saying that the costume "is still perfectly scary." Todd's reply is, "in a very different way than we intended." The crisis is resolved when Gupta announces that an "exalted wizard" has said that he will buy them – but he wants to know if they are flame retardant. Rajiv checks and answers, "not in the slightest," but Todd makes the split decision to sell them to him in any case.

This joke works in part because throughout the series *Outsourced* borrows on stereotypes about the "redneck" purchasers of Mid America Novelties' products – the stereotype here is that they are racist (and certainly as much is implied in the fact that the first caller we are introduced to hangs up on Manmeet once he realizes he has called India). Although most of the people on the lines are complaining about the costumes, the fact that an exalted wizard happens to have a Mid America Novelties catalogue lying around his house is very telling as to the assumptions the show is making about the consumers of the products, the so-called "rednecks." Once again the solution is presented through the multicultural U.S. citizen as represented by Todd – sell Klan members flammable robes because it serves them right. Todd's decision thus speaks to Todd's difference from the people who buy his products, his status as antiracist, his ability to make the sale, and neoliberal capitalism's ability to serve as the vehicle of antiracism. At the same time, the visual gag of Rajiv being the one wearing the robes at once references a history of racial terror and displaces that history onto South Asian bodies in a way that links the white nationalist terrorism of the Ku Klux Klan to forms of terrorism associated with brown people.

In that sense, looking back at *Outsourced* reveals the twinned failures of development and multiculturalism to contain the threat of modernizing South Asia on the one hand, and of brown bodies in the U.S. on the other. Notably, it was the success of the 2008 film *Slumdog Millionaire* (itself a celebration of neoliberal individualism in the Global South) that allowed the show to be greenlit. Producers also felt that *Outsourced*'s prospects were improved by the increased visibility of South Asians on network television, particularly Mindy Kaling of *The Mindy Project* and Aziz Ansari of *Parks and Recreation* (SAJA HQ 2011). But perhaps in the final analysis the issue is not that the South Asianness of *Outsourced* got in the way of its viewership, but rather that the economic situation it was premised on was too bleak. Two years into the recession U.S. audiences might have felt less sanguine about outsourcing than in 2006 when the film was released; it is worth noting that outsourcing was one of the issues on which Donald Trump campaigned in 2016, and it is not much of a leap to think that within the imaginary of the show the callers on the end of the line might be wearing "Make America Great Again" hats, and that "Mid America Novelties" might be selling them. In this light the failure of the show can be read both as a failure of the neoliberal multiculturalism it promotes, and as anger at being the butt of an economic joke that "Mid-America" found to be anything but funny.

## References

Anderson, Chris. 2004. "The Indian Machine," *Wired*, February 1, 2004. www.wired.com/2004/02/india/.

Antony, Mary Grace. 2013. "'Thank You for Calling:' Accents and Authenticity on NBC's *Outsourced*," *Journal of Intercultural Communication Research* 42, https://10.1080/17475759.2013.799084.

Chakravorty, Mrinalini. 2014. *In Stereotype: South Asia in the Global Literary Imaginary*, New York: Columbia UP.

Dave, Shilpa. 2013. *Indian Accents: Brown Voice and Racial Performance in American Television and Film*, Bloomington, IL: University of Illinois Press.

Frederick, Brittany. 2010. *TVOM*. October 26, 2010. www.tvovermind.com/rizwan-manji-outsourced-exclusive-interview/.

Latham, Micheal. 2010. *The Right Kind of Revolution: Modernization, Development, and U.S. Foreign Policy from the Cold War to the Present*, Ithaca, NY: Cornell University Press.

Mankekar, Purnima. 2015. *Unsettling India: Affect, Temporality, Transnationality*, Durham, NC: Duke University Press.

Melamed, Jodi. 2011. *Represent or Destroy: Rationalizing Violence in the New Racial Capitalism*, Minneapolis: University of Minnesota Press.

Nadeem, Shehzad. 2011. *Dead Ringers: How Outsourcing is Changing the Way Indian Understand Themselves*, Princeton, NJ: Princeton University Press.

Nadkarni, Asha. 2014. *Eugenic Feminism: Reproductive Nationalism in the United States and India*, Minneapolis: University of Minnesota Press.

Pedersen, Erik. 2010. "Outsourced—TV Review," *The Hollywood Reporter*, October 14, 2010. https://www.hollywoodreporter.com/review/outsourced-tv-review–30057

Pink, Daniel. 2004. "The New Face of the Silicon Age," *Wired*, February 1, 2004. www.wired.com/2004/02/india/.

Prashad, Vijay. 2001. *The Karma of Brown Folk*, Minneapolis: University of Minnesota Press.

Reddy, Sujani. 2015. *Nursing & Empire: Gendered Labor and Migration from India to the United States*, Chapel Hill, NC: University of North Carolina Press.

Roh, David S., Betsy Huang, and Greta A. Niu. 2015. *Techno-Orientalism: Imagining Asia in Speculative Fiction, History, and Media*, New Brunswick, NJ: Rutgers University Press.

SAJA HQ. 2011. "TV: NBC's 'Outsourced' – Meet the Stars and Executive Producer" [podcast], *BlogTalkRadio*. www.blogtalkradio.com/saja/2011/03/17/tv-nbcs-outsourced.

Sharma, Aradhana and Akhil Gupta. 2006. *The Anthropology of the State: A Reader*, New York: Blackwell.

Thibodeau, Patrick. 2010. "Is America Ready for NBC's *Outsourced?*" *Computerworld*, July 29, 2010. www.computerworld.com/article/2519710/it-outsourcing/is-america-ready-for-nbc-s-outsourced-.html.

# 19

# ROOTS

## Here and There, Then and Now

*Ousmane K. Power-Greene*

In January 1977, a televised adaptation of Alex Haley's *Roots: The Saga of an American Family* aired on the American Broadcast Channel (ABC) over eight consecutive nights, breaking all records at the time for viewing audience. Not only was it the longest made-for-TV film (over 9 hours), but it ushered in the concept of a "miniseries." Beyond this revolution in form, *Roots* brought the history of slavery into the living rooms of millions of American viewers. Over the next two years, *Roots* would also be viewed globally in countries such as the United Kingdom and Ireland, France, West Germany, Italy, and Denmark (Haley 1976; Ball and Jackson 2016).

First published by Doubleday in 1976, Alex Haley's book, *Roots: The Saga of an American Family*, begins with the story of Kunta Kinte, a western African adolescent kidnapped and enslaved in the United States during the eighteenth century, and continues to follow his descendants all the way into the twentieth century. Haley's book purported to trace his own genealogy, blending fictionalized accounts of slavery with his own family's history. One reason the book had such a big impact on readers was because Kunta Kinte struggled against his captors rather than accepting his status as a "slave," and this challenged a common belief among Americans that enslaved Africans were docile and compliant. For example, once Kunta Kinte is bought, his slave owner gives him a new name, "Toby," which Kunte Kinte refuses to acknowledge. His resistance to the name Toby is emblematic of his refusal to accept his status as a slave. This sense of agency challenged the stereotype that associated African American men with meekness and self-sacrifice, as exemplified in Harriet Beecher Stowe's 1852 novel, *Uncle Tom's Cabin*. Kunta Kinte, thus, became as influential a cultural signifier as Stowe's character Uncle Tom, but for portraying the opposite traits. In this sense, the book revised the image of African Americans and used Haley's own intergenerational search for identity to frame the 200-year struggle against racial oppression.

Since the 1970s the name "Kunta Kinte" circulated in popular culture through a variety of media forms, especially Hip Hop. For example, in rap artist Ice Cube's iconic battle rap, "No Vaseline," the names "Kunta Kinte" and "Toby" are employed in service of Cube's criticism of his former bandmate, M.C. Ren. Ice Cube raps:

> So don't believe what Ren say
> Cause he's going out like Kunta Kinte
> But I have a whip for ya, Toby
> Use to be my homie now you act like you don't know me.

In this case, Cube uses the appellation "Kunta Kinte" and "Toby" for his former bandmate M.C. Ren to characterize him as a "sell-out" for hiring a white manager and abandoning the resistant ideology that had defined NWA (Cube 1991).

Other rap artists have drawn inspiration from Kunta Kinte's resistance in more direct ways, as well. Kendrick Lamar's 2015 hit song, "King Kunta," was also inspired by the character of the *Roots* miniseries. In an interview, Lamar explained that the song was about

> the story of struggle and standing up for what you believe in. No matter how many barriers you gotta break down, no matter how many escape routes you gotta run to tell the truth. That's what I think we all can relate to.

Characters in *Roots*, as well as the circumstances surrounding these characters, provide artists and audiences with a metalanguage for talking about racism and black identity in the United States and all around the world. *Roots* the TV miniseries narrated the history of slavery, altering the global discourses around race, slavery, and U.S. history more than any other TV show or film (Moor, Bensman and Van Dyke 2006: 215–216).

Despite popular attention to the way *Roots* shaped American audiences, scholars have only recently written about *Roots*' impact as a global television phenomenon that informed how audiences all over the world thought about the history of slavery and race relations. In the United Kingdom and Ireland, for example, *Roots* hit the black British community like a tsunami, showing the way TV can shape how uncomfortable topics and issues enter public discourse. In this case, *Roots* offered entire nations the terms, names, and concepts related to a specific history that had been neglected. This impact was particularly acute among people of African descent living in the black Atlantic (Gilroy 1993). Although the Civil Rights and Black Power movements in the United States and Britain compelled many African descendants to embrace their own roots, raising many people's sense of racial consciousness, the TV miniseries *Roots* led to a broad search for identity within mainstream American and British life.

This chapter explores this very issue: How did Alex Haley's miniseries *Roots* shape a sense of African consciousness beyond the United States? In what ways did *Roots* influence how people of African descent living in the United Kingdom and Northern Ireland respond to the history of Atlantic world slavery? Why did *Roots* make such a large impression on this specific British audience, and what does it say about the way in which the experience of enslavement for African Americans became the dominant version of such an experience in the popular imagination? Finally, this chapter considers the limitations of the 2016 remake of *Roots* to shape global audiences given the transformation of TV and media in the Digital Age.

## Tracing Haley's *Roots*

Having gained wide acclaim for collaborating and writing the *Autobiography of Malcolm X* in 1965, Alex Haley began researching his own family history in order to write a book that traced the black American experience from capture in West Africa until the present day. Critics, by and large, lauded *Roots* after it was published, and, as one reviewer argued, Haley wrote "his entire story from the Black perspective which is sorely needed to correct the distortions and fill the void left by the omissions of 'objective' white historians, the winners in the war of human degradation – slavery" (Arnez 1977: 367–372). Haley tapped into the new "bottom-up" approach to the study of African American history, from slavery to freedom, which decentered the interpretations of white elites and politicians in favor of accounts by the enslaved. Haley drew from the work of Carter G. Woodson, W.E.B. Du Bois, and Sterling Stuckey to provide his book with details about folktales, burial practices, and aspects of African culture retained by black Americans that was outside the gaze of white slaveholders in the United States.

Beyond the walls of the academy, the black consciousness movement led by activist-intellectuals, such as Angela Davis, primed audiences for *Roots*. In London, for example, Stokely Carmichael's 1967 visit after the release of his book *Black Power* inspired a Black Power movement that culminated in the 1970 Black Power march down Lancaster Road, West London (Brown 2013; Wild 2015).

*Roots* was more than a mere corrective, however. Some believed the book had the power to change American society by thrusting racism as a historical construction into the consciousness of the nation, as Harriet Beecher Stowe's *Uncle Tom's Cabin* did a century before. Yet, Haley's task was quite different than Stowe's in that he sought to use his book to reorient the American imagination away from the subservient Uncle Tom trope, and toward black characters who demonstrated the sort of agency found in Richard Wright's (1938) pathbreaking collection of short stories, *Uncle Tom's Children*. Wright's stories served as a response to Stowe's meek, subservient central character, Uncle Tom. Haley shared Wright's desire to craft a story with black resistance and agency as a central component. From the start, the protagonist of Alex Haley's *Roots*, Kunta Kinte, struggles against his captors, which served for audiences as an important example of his agency in the face of the effort to "break" Kunta Kinte, mentally. One reviewer noted the important psychological consequences of slavery that Haley sought to address, writing "Notwithstanding the horrors of the stealing, the beatings, the burnings, the castrating, the murders, the psychological warfare was the most grievous of all crimes wrought upon a people" (Sundquist 1986: 45–85).

Despite its praise, the book also garnered criticism and outrage from scholars and writers. Most notably, novelist Margaret Walker and folklorist and novelist Harold Courlander claimed Haley plagiarized parts of their two novels. While Courlander won his suit, identifying passages that resembled ones he'd written in his 1967 novel, *The African*, Walker lost hers because the judge did not see specific instances in Walker's novel that Haley copied. Other critics argued that the novel misconstrued West Africa's contact with Europeans, and thus, Kunta Kinte's ignorance about, for example, guns or ships, was not plausible (Taylor 1977: 203–204).

When the miniseries came out, scholars called attention to the improbability of Kunta Kinte being kidnapped by roaming slave catchers (Thornton 1992). While this had been a popular belief at that time, historians pointed out how rare it was for Europeans to trounce through West Africa searching for Africans to enslave. In fact, African rulers dictated the terms of the trade in flesh well into the nineteenth century. Scholars picked apart other historical inaccuracies in the miniseries, but what some found most vexing was the white ship captain, absent from the novel, written into the script because producers imagined there must have been at least some sympathetic whites who wrestled with the inhumanity of the slave trade (McFadden 1977).

Rather than blame Haley's novel or the miniseries for the humiliating and brutal treatment of the black characters, audiences in the United States and around the world responded with frustration, indignation, and anger that this history had not been taught to them. Much of these strong feelings came from black people's sense that the education system intentionally withheld the truth about slavery in order to preserve the myth that the United States and Europe had been a beacon of "civilization" and Christian morality over the past 400 years.

This response pushed politicians in the U.S. to facilitate a national re-education about this history, inspiring an industry in and of itself. An estimated 250 colleges and universities had courses that examined aspects of the miniseries, driving a broader exploration into slavery and American history in communities across the nation. Thirty cities established "Roots weeks" in order to promote a city-wide dialogue about the miniseries. These diverse audiences ensured that the miniseries and the book cut deep into the American public mind in ways that inspired people to learn more about the history of slavery with the hope that this would shape white racial attitudes, tackle racism as a structure, and open the path forward from America's racist past (Van Deburg 1984).

## Making of Television History through Black History

When president of ABC Entertainment, Fred Silverman, and executives at ABC agreed to green-light *Roots* as a miniseries, they were well aware that they could make broadcast television history. In order to give the miniseries the best possible chances at financial success, the executives cast well-known actors, such as John Amos from the *Mary Tyler Moore Show* and *Good Times*, Edward Asher from the show *Lou Grant* who played a white slave ship captain who struggled over his role as a slave trader, football star O. J. Simpson, Robert Reed from *The Brady Bunch*, and well-known poet Maya Angelou (Wolper and Troupe 1978).

At the Emmy Awards, *Roots* dominated the evening, collecting eight awards of the 40 nominations it received. Those associated with the film also won a Peabody Award, Humanities Award, and Directors Guild Award, among others. Over the next few years, the rights were sold to screen the film globally, making *Roots* a world-wide sensation. Between 1977 and 1980, *Roots* aired in over 50 nations, from Nigeria to West Germany. Indeed, *Roots* shaped global audiences' imaginations in ways on a par with how it influenced American viewers, even laying the groundwork for audiences in West Germany to discuss the Holocaust after the airing of the 1978 miniseries, *The Holocaust* (Dreisbach 2009). Forty years later, no miniseries has yet had as much of an impact on global television audiences as *Roots* (Moor, Bensman and Van Dyke 2006).

Perhaps its greatest achievement was the way it shaped how people learned the history of slavery and the slave trade, two topics that American and British people knew little about. *Roots* did this in three principal ways. First, the miniseries pushed African American history into the public imagination and fundamentally changed the way in which all people regarded American history and the history of slavery in the Atlantic world. Second, it thrust into the fore the brutality of slavery, simultaneously upsetting and compelling audiences. Even with two pathbreaking histories of slavery, Kenneth Stampp, *The Peculiar Institution: Slavery in the Ante-Bellum South* (1956) and Eugene Genovese, *Roll, Jordan, Roll: The World the Slaves Made* (1976) that overturned U.B. Phillip's argument in *American Negro Slavery* (1918) that slaveholders were benevolent, most viewers were unaware how instrumental violence was to maintaining the institution. Third, it compelled African people in the diaspora to confront this brutality, and this led some to rethink their own identity as it relates to the history of their nation. Thus, the audience reception of *Roots* is as important as the artistic production of the miniseries, the literary merit, or historical accuracy of the book. *Roots*, then, offers the student of global television an important vantage point from which to consider the impact of television and film on society.

## Global Audience Response to Roots: Afro-Caribbeans in the United Kingdom

Although *Roots* the miniseries had a major impact on people of African descent in the U.S., it may have had a greater impact on black Britons, many of whom had Caribbean backgrounds. Such a comparison is worthwhile for several reasons (Ball and Jackson 2016). First, Afro-Caribbeans share with African Americans the experience of living in society as a minority. For this reason, Afro-Caribbean audiences reacted in ways that illustrate, perhaps most clearly, how truly significant the miniseries was for how it changed the way they thought about their own heritage, as well as the way they thought about white British colleagues and peers. Second, many black Americans and black British share a history of exploitation and enslavement by the British Empire. Like white Americans, *Roots* compelled white British elites, intellectuals, and politicians to make these sorts of connections a part of a national dialogue. Third, the United States Civil Rights and Black Power movements influenced how Afro-Caribbeans, as well as Irish Catholics in Northern Ireland, regarded the importance of this history for contemporary race and ethnic relations. Although scholars have for decades examined the impact of *Roots* in the United States, scholars of Global Media Studies have only recently begun to explore the way in which American cultural imperialism actually disrupted reified social relations that emerged within European nations after World War II. These postwar societies had ignored histories that called attention to the foundational role the Transatlantic Slave Trade had on contemporary racial inequities. As media scholars Robert M. Entman and Andrew Rojecki argue, programs like *Roots* were "quite clearly not just entertainment but also consciously produced political communications whose creators anticipated reactions from elites and activists as well as advertisers and audiences" (Entman and Rojecki 2001).

For black Britons coming of age in the 1970s, who had much less exposure to the history of slavery than black Americans, *Roots* was a revelation. Theology Professor Robert Beckford recalled, "I was completely shocked the first time I watched *Roots*," because growing up in the midlands this history was new to him (Kwei-Armah 2007). However, his parents believed viewing *Roots* introduced a difficult story that they believed had to be told. Beckford claimed that *Roots* remained "compulsive viewing in my house and it became more important than going to church." Analyzing the importance of watching *Roots* in black communities and within familial environments offers the student of global television an opportunity to consider the impact of media beyond the sort of critical analysis that drives much of Global Media scholarship. The very act of sitting together and viewing the film, provided Beckford's parents with the chance to reckon with an uncomfortable truth about the black experience in Britain and the black Atlantic world. For some black Britons, the film served as a form of community healing as much as an opportunity to vent anger about white British people's role in the slave trade (Kwei-Armah 2007).

Recognizing this impact, the British Broadcasting Company (BBC) conducted a study of audience responses to *Roots* in the UK, which documented these impressions (Smart 2014). According to the study, some black British viewers were pleased by the honest portrayal of the brutality of slavery, but on the other hand they felt humiliated by the dehumanization on display that led white colleagues or co-workers to ask questions or make comments they found uncomfortable. Black people in the UK reported having tense, at times awkward, interactions and discussions with whites at work and school over what they learned about slavery through the miniseries. Discussing the history of slavery in the privacy of their own living rooms was one thing, but such conversations percolating beyond the private realm in places of employment and school was virtually unbearable for some. Reflecting back on this, black British audiences recalled that some of their peers denounced the miniseries for bringing up uncomfortable truths about slavery that caused them to stop watching the miniseries after the first few episodes.

On the other hand, some black British viewers felt a sense of personal pride for the resistance displayed by the black characters in *Roots* who were neither comic nor docile, which were common stereotypes that dominated British television and film industries. Lenny Henry recalled the way in which the children who attended his school brought to school a resistant spirit that came from watching *Roots*. Henry explained, "I remember going to school on the Monday and people somehow didn't mess with you that day because all the black kids had this look in their eyes that said you better back off." Henry recalled black students being outraged after learning through *Roots* the centrality of violence for maintaining slavery. Doreen Lawrence remembered that "going in work you looked at people completely different, you'd begin to have a mistrust of white people and that took a long time to go." Thus, the impact on adults mirrored that of children, leading some, such as Lawrence, to rethink race relations in ways that placed contemporary social interactions within a historic context (Kwei-Armah 2007).

*Roots* also led Caribbean immigrants in West London to rethink their connection with African people. As Doreen Lawrence points out, some black people had never considered themselves African. "Until Roots came out," she explained, "I would have never seen myself as a descendant of a slave. That was never part of my background that I learnt growing up in the Caribbean" (Kwei-Armah 2007). *Roots*, then, drove black British audiences to conceive of themselves as members of the African diaspora and as a people who shared a history of slavery with other people of African descent. Up until they watched *Roots*, they felt no sense of kinship with other Africans and no desire to search for their own "roots." But Haley's personal journey connected black British audiences with the history of slavery and compelled them to rethink their own family history in ways that offered an imagined kinship among Africans and people of African descent in the diaspora. As novelist Kwame Kwei-Armah reflected:

> Looking back, for me, the most important thing about *Roots* was not just that it pointed me in the direction of my heritage but it showed me the power of a story to change perceptions and lives. And that is probably why I am a writer today.
>
> *(Kwei-Armah 2007)*

## *Roots* Then and Now

Not only did the TV miniseries *Roots* extend Haley's authorial reach beyond U.S. borders, it has also transcended the era in which it was originally aired. In the mid-2010s, executives believed it was time to remake *Roots* for a new generation. The Wolper Organization and Marc Toberoff with A&E studios and executive Phillip Noyce hired three directors – Mario Van Peebles, Thomas Carter, Bruce Beresford – to make an updated version of *Roots* (Dunn 2016). This team also included actor LeVar Burton as one of the co-executive producers. Burton, who played a teenage Kunta Kinte in the original version, admitted publicly his skepticism about making an update. Yet, he ultimately decided it was worthwhile, having argued after the release of *12 Years a Slave* in 2013: "The bottom line for me is if one soul is moved irrevocably toward the side of humanity, then it's worth it" (Benson 2013). Despite Burton's involvement, the remake of *Roots* did not come close to shaping global discourses on race, slavery, and history as the miniseries had in 1977. Nevertheless, it did receive a positive critical reception, winning four Primetime Emmy Awards and executives at A&E Studios claimed the film was viewed across media platforms by over 14 million people.

The *Roots* remake was clearly shaped by how some viewers responded to the original version of the miniseries. For example, the directors of the 2016 remake chose not to film bare-chested black women in Africa as a way to provide viewers with an "authentic" or "realistic" portrayal of West African society. Even if slight, this alteration illustrates the way in which audience reception to the first film influenced the choices made by contemporary directors. Indeed, this choice suggests a "call and response" between writers and the audience whereby writers and the directors made decisions based on how people responded to the original screened version of the book. Here, the audience reception of the first screened version and updated scholarship about slavery shaped the new interpretation of *Roots*.

On the other hand, the 2016 remake had more on-screen violence than the original TV miniseries. Perhaps, because the film aired on cable television, the directors were far less restricted by FCC regulations related to on-screen violence. The 2016 remake includes heads impaled, blood splattering from bullets piercing skin, and, perhaps the most gruesome, the on-screen shot of white captors cutting off the foot of Kunta Kinte. But such revisions may have more to do with an effort on the part of the directors to place white violence into the center of the narrative, rather than black resistance. As historians have established over the past two decades, slavery was an institution where extreme acts of brutality served as a crucial lever for control. The original miniseries, of course, included acts of brutality. Yet, the on-screen violence in the 2016 version suggests that the writers and directors chose to emphasize this aspect of the institution of slavery over, for example, the confinement of one's liberty as expressed in the original version by the chains Kunta Kinte was constantly forced to wear (Dunn 2016).

Despite these specific points of comparison between audience responses to the 1977 version and the 2016 version, the impact of these two televisual interpretations of the novel must be considered through the lens of the emergence of cable television and contemporary Digital Age viewing practices. The reboot of *Roots* airing on the cable network A&E had to compete with many more television programs than were available to audiences in the 1970s. Also, those deemed "target" audiences for *Roots* may have missed episodes because of the NBA playoffs and other competing programs. But, even if as many people tuned in to the remake of *Roots* in 2016 as they had in 1977, common audience practices, such as "channel surfing" during commercials, had the potential to shape how impactful the remake could have been on audiences. One can imagine, likewise, that the viewing experience of older and younger people may have been shaped by practices, such as checking cell phones, status updates on Facebook, Twitter, or snapchatting away with friends. These examples serve to highlight the reason assessing the impact of these two versions of *Roots* is so difficult to measure (Lawson 2017).

In addition, black British audiences today have learned more about slavery and the slave trade than those in the 1970s because of changes in the British school system's curriculum (Lipsett 2008). Institutions, such as the International Slavery Museum in Liverpool, established in 2007, provide the British public with more opportunities to learn this history than were available in the 1970s. Even if teachers gloss over the history of slavery for young people, this history is much more of a facet of global media and popular culture now than back in the 1970s. For this reason alone, *Roots*' potential impact on audiences in 2016 may have been blunted.

Today, the miniseries itself is an outdated form, and has been replaced by streaming sites like Amazon with shows that offer similar opportunities for conversations about slavery, such as *Underground*. The remake of *Roots* competed with many more televisual experiences that document slavery and the Transatlantic Slave Trade. Even with the acclaim of feature films, such as *12 Years a Slave* (2013), as well as other widely popular films such as *Django Unchained* (2012), there still has yet to be a film, television show, or digital series that comes even close to reshaping global discourse about slavery and racism like the original *Roots* miniseries (Ball and Jackson 2016).

## *Roots*' Legacy

Henry Louis Gates, Jr.'s PBS series "Finding Your Roots," illustrates, more so than films and narrative shows, how ordinary and legitimate retracing one's "roots" has become a part of black and white people's personal development. Gate's television series has even featured prominent African Americans, such as Oprah Winfrey, tracing their own roots on a personal journey toward a feeling of completeness. This version of history – telling a story through an individual – continues to inspire audiences all over the world, and it even sparked an entire industry in Ghana around bringing African

Americans there to tour Cape Coast Castle and Elmina, two of the more popular sites where enslaved Africans left the continent for the Caribbean and the Americas (Polgreen 2005).

Ancestry.com, as well as other internet sites, cater to this interest in tracing one's DNA to Africa as well. Like heritage tourism, an entire industry exists that allows an individual to purchase a DNA kit to take the appropriate measures in order to trace his or her DNA. Scholars have raised all sorts of questions about the scientific validity of such tests, and whether or not DNA testing can ever provide a person with a specific location or knowledge of one's "ancestors'" culture (Koenig, Lee and Richardson 2008). But, even if such tests were accurate, the broader question Alex Haley's own genealogical journey raises is what does identifying one's own history mean for an individual in the Digital Age? This also begs the question whether or not history has the ability to provide people with what Afro-Caribbean bibliophile Arthur Schomburg called a "usable past," that can shape the way an individual thinks about him or herself in the world today (Locke 1925).

Clearly, what inspired Haley to write *Roots* continues to fascinate people in the twenty-first century. More so than the awards or its place in popular culture, *Roots* set black Americans and black Britons on a path toward a sense of connectivity with the past and a search for a way to impact their present like nothing else. The idea of fulfilling one's "destiny" drives people to connect with ancestors and their perceived heritage. This holds true today as it did when *Roots* came out in the 1970s. Perhaps the most important and enduring significance of *Roots* is the audience response; the way in which black British audiences discovered an aspect of themselves that provided a site for unity and shared experience that had not existed previously among Africans throughout the diaspora. By studying the *Roots* miniseries, one can see the impact of television to shape the way audiences think about how history can create discourse about the relevance of one's individual and national past in ways that other global media events cannot (Havens 2013).

## References


Arnez, N. (1977). From His story to Our Story: A Review of "Roots". *The Journal of Negro Education*, 46, No. 3, pp. 367–372.

Ball, E. L. and Jackson, K.C. ed., (2016). *Reconsidering Roots: Race, Politics, and Memory*. Athens: University of Georgia Press.

Benson, E. (2013). The *Roots* Rubber-Band Effect. *The New Yorker*. New York: New York.

Brown, M. (2013, December 27). Britain's Black Historians Power Movement is At Risk of Being Forgotten, Say Historians. *The Guardian*. Retrieved from www.theguardian.com/world/2013/dec/27/britain-black-power-movement-risk-forgotten-

Cube, I. (1991). *No Vaseline. On Death Certificate* [CD]. Los Angeles: Priority Records.

Dreisbach, T. (2009). Transatlantic Broadcasts: Holocaust in America and West Germany. *Penn History Review*, 16, No. 2, p. 7.

Dunn, S. (2016, May 29). Why the *Roots* Remake Is So Important. *The Atlantic*. Retrieved from www.theatlantic.com/entertainment/archive/2016/05/why-the-roots-remake-is-so-important/484661/

Entman, R. and Rojecki, A. (2001). *The Black Image in the White Mind: Media and Race in America*. Chicago: University of Chicago Press.

Genovese, E. (1976). *Roll, Jordan, Roll: The World the Slaves Made*. New York: Vintage.

Gilroy, P. (1993). *The Black Atlantic: Modernity and Double-Consciousness*. Cambridge: Harvard University Press.

Haley, A. (1976). *Roots: The Saga of The Saga of an American Family*, Alex Haley. Garden City: Doubleday Press.

Havens, T. (2013). *Black Television Travels: African American Media around the Globe*. New York: New York University Press.

Koenig, B., Lee, S. and Richardson, S. (Eds.) (2008). *Revisiting Race in a Genomic Age*. New Brunswick: Rutgers University Press.

Kwei-Armah, K. (2007, March 23). Going Back to My Roots. *BBC News Magazine*. Retrieved from http://news.bbc.co.uk/2/hi/uk_news/magazine/6480995.stm

Lawson, M. (2017). Roots Revival: How Does the New Kunta Kinte Compare to the Classic? *The Guardian*. Retrieved from www.theguardian.com/tv-and-radio/2017/feb/08/roots-revival-how-does-the-kunta-kinte-remake-stand-up

Lipsett, A. (2008, August 26). Slavery History Lessons to be Compulsory. *The Guardian*. Retrieved from www.theguardian.com/education/2008/aug/26/slavery.schools

McFadden, R. (1977, April 10). Some Points of 'Roots' Questioned; Haley Stands By Book as a Symbol. *New York Times Magazine*.

Moor, B., Bensman, M. and Van Dyke, J. (2006). *Prime-Time Television: A Concise History*. Westport, CT: Praeger Publishers. 215–216.

Newcomb, H. (Ed.) (1997) *Encyclopedia of Television*, volume 3, M-R. New York: Firzroy Dearborn. 1953–1955.

Polgreen, L. (2005, December 27). Ghana's Uneasy Embrace of Slavery's Diaspora. *The New York Times*. Retrieved from www.nytimes.com/2005/12/27/world/africa/ghanas-uneasy-embrace-of-slaverys-diaspora.html

Schomburg, A. (1925). "The Negro Digs Up His Past," in A. Locke (Ed.) *The New Negro*. New York: Simon and Schuster.

Smart, B. (2014). The BBC Television Audience Research Reports, 1957-79: Recorded Opinions and Invisible Expectations. *Historical Journal of Film, Radio and Television*, 34, No. 30 452-462.

Stamp, K. (1956). *The Peculiar Institution: Slavery in the Ante-Bellum South*. Berkeley: University of California Press.

Taylor, Q. (1977). Review of *Roots: The Saga of an American Family*, by Alex Haley. *Minnesota History*, 45, No. 5, pp. 203–204.

Thornton, J. (1992). *Africa and Africans in the making of the Atlantic world, 1400-1680*. Cambridge: Cambridge University Press.

Van Deburg, W. (1984). *Slavery and Race in American Popular Culture*. Madison: University of Wisconsin Press.

Wild, R. (2015). "Black Was the Colour of Our Fight": The Transnational Roots of British Black Power," in R. D. G. Kelley and S. Tuck (Eds.) *The Other Special Relationship Race, Rights, and Riots in Britain and the United States*. New York: Palgrave Macmillan 25-46.

Wolper, D. and Troupe, Q. (1978). *The Inside Story of TV's "Roots"*. New York: Warner Books.

Wright, R. (1938). *Uncle Tom's Children*. New York: Harpers.

Yarborough, R. (1986). *Strategies of Black Characterization in Uncle Tom's Cabin and the Early Black Afro-American Novel in Sundquist, E.* New Essays on Uncle Tom's Cabin. New York: Cambridge University Press.

# 20

# THE MUSIC VIDEO'S COUNTER-POETICS OF RHYTHM

## Black Cultural Production in *Lemonade*

*Ayanna Dozier*

### Introduction

Although it has been three years (and counting), I still think about the image of Beyoncé wearing cornrows, a gold ankh necklace, a fur stole, a two-piece grey bodysuit with matching ankle boots flicking her braids back as the sound of the tapping cymbal emerges before she turns to the camera and states, "Who the fuck do you think I is?" This arresting audiovisual moment of Black American vernacular dialogue (the "incorrect" is versus am) with visual "urban" aesthetics (cornrows) made me reconsider my understanding of Beyoncé's work, and it is a moment that I have repeatedly come back to in the years since my encounter. This study will examine Beyoncé's 2016 visual album *Lemonade* to see how Black cultural production, as a global diasporic practice, works through alternative (and often experimental) audiovisual images (like the music video) utilizing television's expansive digital properties to reconfigure our engagement and understanding of Blackness. Informed by Stuart Hall's seminal article "What is the Black in Black Popular Culture?", this chapter examines the global circulation of *Lemonade* to assert that the expansive technological changes of television work dialogically across the Black diaspora to alter the quotidian experience of Blackness. My use of quotidian here and throughout refers to what Sharon Holland writes in *The Erotic Life of Racism* (2012) as the unremarkable encounters of racism, the encounters that are diminished because they are not spectacular in nature and thus go, literally, un-remarked. Such quotidian encounters demonstrate that many people in their day-to-day encounters with Black individuals "don't create meaning as much as [they] reproduce it" through racist acts. Thus, television, and its growing digital expansive properties via phone and computer applications, is a medium ripe with the potential of transforming (and impacting) those daily (quotidian) encounters. As media studies scholar Herman Gray notes in *Watching Race: Television and the Struggle for Blackness* ([1995] 2004), television creates "spaces and practices where Black artists and cultural workers use the new digitally based technologies, old forms of representation, the articulation of markets, the brand, and the logo to construct different notions of Blackness and Black cultural practice" (xxiii). The combination of digitally based technologies and old forms of representations offer Black cultural producers possibilities for reconfiguring and manipulating these mediums to present alternative, technologically innovative explorations of Blackness (Wynter, 1992). Black cultural production defines the practice of making culture that collectively speaks to spaces, subjects, and experiences that are racialized as Black. This may include a globally dispersed Black diaspora of the people who have been or are descended from the kidnapped, forcibly removed, exiled, and migrated people from common geographical zones, such as Africa, the Caribbean, Australia, etc., stemming from similar encounters with European colonial invasion. Black culturally produced music videos eschew the question of global community and outreach, instead taking that as a given, considering the conditions of the Black diaspora. In this way, then, Beyoncé, as an international popular Black artist can deliver a specifically Black televisual film to a global audience in ways that other artists cannot. She is therefore in a critical position, as a renowned Black cultural producer, to, as Hall writes, create a dialogue of the shared histories and cultures that Black individuals have following the (ongoing) period of European colonialization, with *Lemonade*.

*Lemonade* is a visual music album that chronicles the reconciliation of a marriage following suspicions of infidelity, produced as a collaborative directorial effort between Beyoncé and Kahil Joseph (with added artistic contribution by director Melina Matsoukas and others). Told over eleven chapters that feature spoken word interludes ("Intuition," "Denial," "Anger," "Apathy," "Emptiness," "Accountability," "Reformation," "Forgiveness," "Resurrection," "Hope," and "Redemption"), the film accompanies Beyoncé's sixth solo studio album *Lemonade*. Aesthetically, *Lemonade* draws from the wealth of global Black diasporic culture (and not just Black American culture) to reflect upon the expansive and quotidian nature of Blackness as a site for continual possibility and reconfiguration. Temporally speaking, *Lemonade*'s use of Black diasporic culture—including the Igbo landing/flight, Nigerian culture, Yoruba mythology, police brutality in the United States, and much more—demonstrates how Blackness disappears into memory while remaining firmly in the present (Lepecki 2006, 127). This reflects what film theorist Kara Keeling (2007) notes as the "problem" with the visual representation of Blackness. The colonial memory-imagery renders Black images as "problems" available to public memory via perception (Keeling 2007, 43). This is to say, that Black images operate as troubling encounters whose occupation and bodies are always made "known" to viewers via public (colonial) perceptions of Blackness. *Lemonade* troubles this "memory-image" of Blackness by presenting images, movements, and content that work in dialogue with the myriad songs to destabilize the standard, seemingly known, images of Black bodies that are not available to public memory based on perception alone.

The distinction that the music video has within Black culture is worth threading out in further detail as it provides Beyoncé a historical legacy to which the aesthetic, narrative, and distribution innovation found in *Lemonade* can be traced back. I view this as a necessary point to address as it moves Beyoncé away from the singular, individual artist and places her in explicit archival history of a larger Black audiovisual collective. The formal aesthetic innovation found in *Lemonade* is pulled from Black expressive culture. Therefore, the formal gathering of expressive cultural practices across the Black diaspora help us view *Lemonade* as an example of what Sylvia Wynter writes as the "counter-poetics of rhythm" (Wynter 1992, 260).

Counter-poetics, an undertheorized term by Sylvia Wynter first appeared in a 1992 essay on Black film aesthetics and critique, "Rethinking 'Aesthetics': Notes Towards a Deciphering Practice." In that article, Wynter points to Black music videos by way of arguing that a more experimental approach to film's aesthetic is needed to break the current model that reproduces hegemonic meaning through its symbolic coding of whiteness as a positive "true" value and Black bodies as death or a "zero value" (1992, 252–253). The hybridity of the music video, then, in addition to pulling from the rich long history of rhythm and blues (from the Black diaspora) provides a unique space to see contemporary depictions of Black bodies directing, crafting, producing, and circulating a refusal of cinema's coding system of value through its counter-poetics of rhythm. In Wynter's article she examines how storytelling has constantly been (re)made to repeat the image of white bodies as the dominant, "accurate" body at the exclusion and erasure of indigenous, Black, and Brown bodies (Dozier 2017, 4).

Counter-poetics defines the work of individuals commenting, engaging, and challenging hegemonic structures of being in the world, that is questioning the argument and construction of whose body and history are coded with "positive" values that work in the service of reproducing colonial discourse and images. Such work is not meant to provide answers but rather is to question what we as individuals have "inherited from imperial Europe, the possibilities and limitations of purely Western science and knowledge systems, and how humanness can be recognized as connective and interhuman" (Mignolo 2015, 122). The rhythm in Wynter's statement refers to Black studies' relationship with rhythm, specifically how rhythm informs and makes up Black expressive culture, which in turn extends itself to our manner of speech, appearance, etc. Beyoncé's look for "Don't Hurt Yourself" recounted at the beginning coupled with her use of Black Vernacular English (also known as African American Vernacular English) is an aesthetic audiovisual interrogation of the audience's awareness of not only who Beyoncé is but the rich Black diasporic histories, traditions, and cultures that would beget her and her work.

*Lemonade* works as a significant Black cultural production to place itself in dialogue with digital technology and "old forms of representation" through its use of the digital visual album format. The music video provides a space for Black cultural producers and artists to manipulate audiovisual's formal and narrative aesthetics to experiment with the (re)presentation of Blackness and Black culture—where they can work through the historical significations of Black identity and culture utilizing contemporary technological advancements and modes of dissemination. This is not to suggest that film does not do this, but rather that television is at the vanguard of new practices in mass media dissemination to a global audience. As Elizabeth Jacka, John Sinclair, and Stuart Cunningham write, the invention of the satellite television enabled an increase in global media content that could appear in the home, this increase of global content becomes more pronounced by the ways in which bodies, cultures, identities, histories, can now be mediated to us via the privacy or confines of our homes and bedrooms (1996, 4).

The increase of global dissemination of Black American music videos while effective in reaching a global audience could be viewed as problematic. Noted Black feminist scholar Patricia Hill Collins asserts that some culturally specific music videos are then stripped of their histories when presented abroad (2005, 31). Hill Collins expresses her dissatisfaction with the music video genre through a critique of Black masculinity in hip-hop videos. She writes,

> Camera angles routinely are shot from a lower position than the rapper in question, giving the impression that he is looming over the viewer. In real life, being this close to young African American men who were singing about sex and violence and whose body language included fists, angry gestures, and occasional crotch-grabbing might be anxiety provoking for the typical rap and hip-hop consumer (most are suburban White adolescents). Yet viewing these behaviors safely packaged within a music video protects consumers from any possible contact with Black men who are actually in the videos. Just who are these videos for?
>
> *(2005, 31)*

While Hill Collins is correct to assert that this global dispersal can create conditions of white exploitation of Black bodies and culture, such an analysis erases Black consumers from the "perceived" global audience. It also denies the power of the music video in combining body movements, camera angles (formal aesthetics) with the sonic properties of the song itself to expressively portray an encounter that, in many ways for Black bodies across the diaspora, could only creatively perform via the music video. Much of Hill Collins' critique is, rightfully, centered on the repetition of images of Black culture that reproduce hegemonic discourses around the subordination of the Black body. In her focus, though, we lose the possibilities in which Black artists creatively use the musicality of the music video to aesthetically embed and work through Black expressive culture in a manner that questions those histories. In this way, I argue that *Lemonade* does not reproduce meanings but creates new meanings, narratives, and histories to emerge around Black women through the use of counter-poetics of rhythm found in the narrative and formal experimentations of the music video. Like my arresting moment with Beyoncé during the song "Don't Hurt Yourself," *Lemonade* confronts its audience head on by questioning what we *really* know about Black women in the diaspora and how much of what we know is based on the repetition of hegemonic discourse.

## Black Cultural Production, the Music Video and Its Counter-poetics

Through their hybrid audiovisual format, music videos have served as an opportunity for Black entertainers to experiment with their identities and cultural histories since the late 1970s. Black artists have long been invested in utilizing the music video to exercise space or freedom of movement as it was a genre that was, as media scholar Carol Vernallis writes, "*the* laboratory: while commercials and films … tended toward tightly controlled client-author supervision and careful storyboarding, a music video director or editor might try anything" (Vernallis 2013, 5). In its experimentation with the form, Black artists doubled that freedom and extended it to the portrayal of the Black body as well, this experimentation of the body was an affordance as music video's audience has always been rooted in the home consumer and not the theaters (although this is changing). As Black feminist scholar Aisha Durham writes, the music video offered potential for a "repeated performance that increasingly mirrors radio airplay due to media synergy. The formal and content aspects of the music video—fashioned as performance and seduction—is layered to encourage multiple viewings as well." This structure has contributed to the success and transformation of numerous Black artists including Beyoncé (Durham 2012, 38). In so doing, music videos, even amateur ones, have enabled the increased visual presence of Black bodies onscreen and has thus been used to challenge harmful stereotypes and affirm Black culture, such as *Lemonade*.

The MTV origins of the production, promotion, and circulation of the music video may appear to be a dated history or of a bygone era to a contemporary music video audience. Many scholars have noted that while the history of the music video predates the invention of the television set, it was MTV—which still stands for music television despite its recent shift toward scripted young adult content and reality TV—who ushered in the mainstream visibility of the music video in the early 1980s (Durham 2012, 37). Like mainstream radio stations, music videos that appeared on MTV dictated the success and visibility of an artist. Thus, like all forms of mainstream media engagement then, the struggle over the production of Blackness was fraught within the genre of the music video. For Black cultural producers, the unique hybridity of the music video via its ability to create immersive content that enabled the production of Black images with a song was nothing new (Duke Ellington, Bessie Smith, and Billie Holiday all created music film shorts in the 1930s [Durham 2012, 38]). What was new about the MTV platform was the increased visibility and distribution of such audiovisual material and, in that way, Black artists saw the music video of MTV as a way to wrestle control of the audiovisual image of Blackness by producing clips that complicated and transformed the limits and awareness of the Black body via film's aesthetics (some examples include; Grace Jones' "La Vie en Rose" [1977], Michael Jackson's "Thriller" [1982], and Prince's "When Doves Cry" [1984]). Initially, MTV suppressed its playing of Black artists but, after much public ridicule and pressure from other artists, MTV began incorporating videos by Black artists into its rotations, which would change the trajectory of the genre in the years to come (Tannenbaum and Marks 2011, 168).[1]

Nowadays, the production, promotion, and circulation of the music video is far more, what Vernallis would state as, "unruly" in her book *Unruly Media: YouTube, Music Video, and the Cinema* (2013). The platforms by which individuals consume the music video is a far cry from the centralized network and distribution of MTV, which at one point included the popular *Making the Video* program (a reality show/docu-series on the making of a music video). Instead, individuals can rely on multiple platforms to watch the latest music video. In this age of dissemination, the music video itself has begun to mirror the distribution platform it is on, such a mirroring has been argued by some scholars to be an effect of digital technology/media upon the creation of audiovisual content (Shaviro 2017; Vernallis 2013). In *Digital Music Videos* (2017) Steven Shaviro, for example, writes how the use of digital intermediate (DI) affects and shapes how directors capture color and texture in music videos because DI, a process that only became common after 2005, can control and transform color and texture on a pixel by pixel basis (2017, 13). Vernallis additionally writes that the quick editing shots and sensory overload that is featured in the video are a direct result of shooting digitally (you shoot more information in a quicker fashion), and that this information overload found in the music video is meant to, on one hand, destabilize the viewer by immersive means while additionally using the multiple shots to demonstrate that there is more to the world established in the music video (2013, 94).

Additionally, the distribution platform of YouTube (and other similar sites such as Dailymotion) affects how music videos are conceived. YouTube, as a hosting platform, presents an archive of the ordinary and spectacular. Such a hybrid gathering of information (in which anything can become viral) bleeds into the assemblage and creation of contemporary music videos, which begin to mirror that content. For example, Beyoncé's music video for "7/11" (2014) was inspired by the YouTube dance videos by Pinoy, Gabriel Valenciano. This brief outline of the effect digital technology has upon the contemporary music video reveals how it is affected by its circulation properties and, in turn, is becoming a genre with the explicit purpose of creating embodied responses, reactions, and choreographies in relation to what we have seen.

The circulation of the music video format allows for a shared public dissemination of Black representation in spaces where they are often prohibited or legislatively locked out. In the case of *Lemonade*, the film was initially broadcast on the network HBO (Home Box Office) in the United States on April 23, 2016 during a 72-hour period where access to their content was free (this also coincided with the season six premiere of *Game of Thrones*). It was then again broadcast on HBO's international services on television (but not via its digital on-demand platforms of HBO NOW or HBO GO) on June 18, 2016. While numerous consumers across the world experienced *Lemonade* before its international premiere, either via its appearance on Tidal (a global music and entertainment monthly streaming platform co-owned by Beyoncé and her husband Jay Z), album purchase, or illegal download, the second premiere emphasized what its initial tagline stated as a "world premiere event."

The medium of television and by extension new global digital platforms that feature serial content such as Netflix, premium television apps, Hulu, and YouTube, have the power to impact the quotidian nature of individuals' media consumption habits and thus potentially steer audiences' creative ambitions toward the creation of new meaning as opposed to the repetition of established meaning. It is no surprise then that this medium has historically been a site that Gray has described as the "struggle for Blackness," or the representation of Blackness. The digital shift in television platforms (as described above) means that the televisual representation of Blackness now shapes and informs our interpretation of Black existence on a global scale. *Lemonade* demonstrates an overt attempt to insert alternative (re)presentations of Black womanhood on television and thus, emphasize and utilize the power that televisual images "take on at the level of everyday life and common sense" (Gray [1995] 2004, 7). Beyoncé's focus on Black womanhood throughout *Lemonade* is informed by how Black women's bodies and labor bear the "gross insult and burden of spectacular exploitation in transatlantic culture," and yet are subsequently rendered by "hegemonic hermeneutics" as known, simplistic sites of inquiry (Brooks 2006, 7). *Lemonade* reveals that Beyoncé is "not just another Black woman with traceable African roots, but a woman who wants her audience to know that she understands and appreciates that her essence is drawn from an ancient and eternal pool of rich, cultural manifestations" (Okoroafor 2016).

The prestige of releasing an entire visual album onto the premium network channel HBO was a tactful move by Beyoncé. By premiering the film there (then making it available via her co-owned streaming service Tidal later that evening), Beyoncé was placing her film in dialogue with "high culture." When scholars, critics, producers, and actors discuss the "golden age of television" as a central locus where quality storytelling can be found, HBO is consistently listed as one of the channels that produces and distributes "high quality" work that contributes to that narrative. What is unique about Beyoncé's decision to screen *Lemonade* on HBO is that the channel serves merely as an (initial) hosting site, not unlike the use of hosting a video on YouTube (or, in recent years, Apple Music, another monthly music and entertainment streaming service). HBO does not own *Lemonade* and it was swiftly pulled from their on-demand services (such as HBO GO and HBO NOW) within a week of its release in the United States. This model eschewed the traditional or even contemporary digital model by which a platform distributes and circulates the music video by way of receiving profits from ad revenue. It reveals that Beyoncé was far more invested in casting a wide net of (prestige) visibility and media attention toward her film before directing that attention toward the digital circulation properties (television apps and Tidal) that enable free flow of content and dissemination to occur in a more pronounced manner. Vernallis notes that these distinctions of platform or network viewing ultimately blur in our digital era, creating a swirl by which information travels up and down, sideways and then up again and across platforms, truly immersing us for a moment as we can be consumed by the music video's ability to not only transfix us but transfix our bodies at any moment in time: on your phone, on your home screen, on your laptop, etc. (Vernallis 2013, 2).

The musicality further adds to the music video's ability to transfix our bodies in a specific moment based on music's ability to stop us in our tracks. The editing, body, and narrative aesthetics are all organized in relation to the rhythm of the beat. The musicality of the music video and its manipulation of formal and narrative aesthetics to match the sonic thumping of the beat has begun to bleed into our televisual and cinematic experiences as well, as entire scenes in a film or tele-series are organized by a sonic arrangement that starts with a tick or beat by way of highlighting to the audience that you need to pay attention to what you hear in this next sequence (two examples are the use of synth music in *Stranger Things* [2016–] and Led Zeppelin's "The Immigrant Song" in *Thor: Ragnarok* [2017]). Vernallis states that rhythm drives the aesthetics of the music video, which in turn often "avoids a ground because the sound wafts it along," meaning that establishing a sequence of events between cuts is not necessary because the beat dictates the next image, even if it is out of sequence (2013, 5).

An evocative example of the above-mentioned sonic "cueing up" in *Lemonade* can be heard in the song "Freedom." The song starts by matching the crackling thumping of the horns with restless images; Beyoncé trying to navigate the ruins of a plantation, a thunderstorm, and a newborn baby on a bed, before going silent to a black and white shot of Beyoncé on a stage in front of the mothers of victims of police brutality (or gun violence) in the United States (Sybrina Fulton, Lezley McSpadden, and Gwen Carr) and emerging Black actresses and singers from across the diaspora (Chloe x Halle, Zendaya, Amandla Stenberg, Lisa-Kaindé Diaz and Naomi Diaz). The silence cue from the cacophonic cluster of horns and images alerts the audience to listen carefully now. And we do, as we hear an emotive a cappella rendering of "Freedom" by Beyoncé before the horns are foregrounded once again during the chorus. The lyrics of this first verse state:

> Tryna rain, tryna rain on the thunder
> Tell the storm I'm new
> I'ma walk, I'ma march on the regular
> Painting white flags blue

Lord forgive me, I've been running
Running blind in truth
I'ma rain, I'ma rain on this bitter love
Tell the sweet I'm new
I'm telling these tears, "Go and fall away, fall away"
May the last one burn into flames

The emphasis on Black Vernacular English (or slang) found in the usage of "I'ma" and "tryna" instead of "I'm a" (or "I am a") and "trying to" is an effective way in which Beyoncé weaves culture through all facets of the film. The distinct emphasis of cueing up to listen to her sing these words in this manner creates a dialogue for thinking critically about the ways in which Blackness is policed in the public, from image to speech.

The rhythm dictating the flow of the images is pertinent to the music video as it structures the frequent split images and sequences that simply appear within the video without context. These seemingly random images that appear and disappear suggest a larger world in the video that is populated by other bodies beyond our viewing experience. This immersion tactic has been skillfully used by Black artists, long before the digital, to suggest a larger world within the video, between cuts, populated by other Black artists via the cameo appearance. The cameo appearance by another Black artist in the music video would be a blink and you miss it moment, forcing the audience to watch it again and to imagine the relationship between these artists (that of the singer/rapper and the cameo artists) in the video. The relational use of a cameo appearance by another Black artist in a music video would also convey to the audience that this was an artist in conversation with other Black artists (Lil Kim's 1996 music video for "Crush on You" features a blink and you miss it cameo by the late Aaliyah). This type of cameo appearance appears in *Lemonade* several times over, from the twerking of Serena Williams in the song "Sorry," to the slew of Black actresses and singers that appear in the "Freedom" segment of the film.

If rhythm structures the arrangement of the images that appear, the specific use of that rhythm plays a crucial role in shaping how we receive the music video as a whole. Sound possesses the ability to temporally pull listeners' emotions into the fold as they may "wish to hold onto what has unfolded in the past, while simultaneously staying in the saddle of time and reaching for the future" (Vernallis 2013, 6). The music video's musicality immerses audiences into the visual temporal rhythms of the song itself. For *Lemonade*, a Black culturally produced video, the long-form music video pulls the audience into a web of diasporic hauntings, abuse, betrayal, reconciliation, and, ultimately, healing. The images do not always flow together, but they flow in service of the rhythmic structures of the film, the music, the poetry, and the voice (speech). The emphasis on rhythm's dictation of filmic structure and the fact that *Lemonade*'s rhythm is grounded in Black culture demonstrate how the film creates an immersive experience for creating new meaning around Black women's histories through the aesthetic use of a counter-poetics of rhythm.

## Rhythm and Culture in *Lemonade*

If the music video was a laboratory in the 1980s, its current Prometheus-like genre of rhythmic hybridity and cinematic fluidity is a productive way to convey the destabilizing function of spatial hauntings. By spatial hauntings, I am referring to the structures of feeling that preserve and remain in place after trauma. The music video in the context of Black culture is a prime example of how the counter-poetics of rhythm are used to create a doubling effect that simultaneously destabilizes us from normative poetics in the world and urges us to invent new truths around bodies, representation, and their history instead.

While there are numerous arresting scenes in Lemonade, one sequence effectively conveys the performance of spatial remembrance. During the song "Love Drought," Beyoncé and her dancers conduct a processional march on the Igbo (also spelled as Ibo by some) landing, the space where the mass suicide of captive Nigerian Igbo people occurred on the shores of St. Simons Island, Georgia in 1803. Just prior to this song, Beyoncé recites a poem by British-Somali poet Warsan Shire, whose poetry threads the interludes between songs together in *Lemonade*. Beyoncé speaks,

> He bathes me until I forget their names and faces.
> I ask him to look me in the eye when I come
> home. Why do you deny yourself heaven?
> Why do you consider yourself undeserving?
> Why are you afraid of love?
> You think it's not possible for someone like you,
> but you are the love of my life.

The last stanza is spoken over a shot of a crying Beyoncé in the empty Superdome in New Orleans, a place known for football games and concerts but also known as the refuge center following Hurricane Katrina where three people died during the storm. When the beat and chord progression is heard in the opening notes of the song, the film swiftly cuts to a series of feet walking through water. The editing of this march is in synch with the backing drum beat heard throughout the song. Even when Beyoncé begins to sing the lyrics, the camera does not break its gaze on watching these women march into the sea from afar. The beat in "Love Drought" guides the formal and narrative aesthetics for that moment. As the song continues, we encounter a series of disjointed shots of Beyoncé in a bed of flowers covered in gold dust, laying on a reclined chair, the marshes and Spanish Mosses of the island, and several shots of Beyoncé interlocking hands with her dancers with their arms outstretched as they face the ocean. By the end of the first chorus of "Love Drought" the beat takes a drastic left turn and we hear a heavy 808 drum, this sonic change is marked by a distinctive visual as well. Suddenly, we are swooped away from the lush colors of Beyoncé covered in gold dust on beach at sunset surround by red and pink flowers to an extreme close-up of Beyoncé's face with facial markings in Black and white. As the beat drums along the camera begins to shift and turns upside down around Beyoncé before returning to match her eye level. This shift is followed by a brief two second clip of women reaching out their hands toward Beyoncé (in the same B&W sequence) as if she were a golden calf of some sorts. The film swiftly moves us to the next chapter, "Forgiveness."

Beyoncé dives deep into the pool of Black culture to weave it into her film, as Okoroafor (2016) stated, *Lemonade* revealed that Beyoncé was "not just another Black woman with traceable African roots, but a woman who wants her audience to know that she understands and appreciates that her essence is drawn from an ancient and eternal pool of rich, cultural manifestations." One of the citations of Black culture that Beyoncé manifests in the film quite clearly is Julie Dash's film, *Daughters of the Dust* (1991). *Daughters of the Dust* holds significance in film and Black culture as it is the first feature film (let alone experimental film) by a Black woman to be distributed in theaters in the United States. The film is told from the perspective of the unborn child who recounts the last time her family were together before they migrated from Georgia to the North at the turn of the twentieth century. The family are descendants from the surviving members of the Nigerian Igbo tribe, those who were prevented from committing suicide in 1803. The movie further makes this connection by taking place on St. Simons Island in Georgia. *Daughters of the Dust* relies heavily on experimental aesthetics to convey shifting time and movement onscreen. Dash stated in an interview that her use of experimental aesthetics was in dialogue with the fact that, for her, traditional narrative and formal aesthetics (such as continuity editing, and the clear introduction of characters) would not be of service to a narrative on Black women who, often, do not see themselves accurately represented in those traditional aesthetics (Dash 1992). *Lemonade* cites key frames of *Daughters of the Dust* in dialogue with music to create a new interpretation behind that citation. The chapters "Reformation," "Resurrection," and "Hope" show the influence of Dash's film through the mise-en-scène of the plantation, women sitting in the Spanish moss trees, and the setting of St. Simons Island for "Love Drought." These aesthetic citations also serve the purpose of furthering our awareness of the original source and, additionally, place Beyoncé and her film in a shared dialogue with the aesthetic intentions and impact of Dash's film.

The late Toni Cade Bambara offers some sage words for the opening shot of *Daughters of the Dust* that I believe are necessary to repeat here for they immediately frame Dash's experimental aesthetics in dialogue with the creation of a new, alternative visual history to emerge around Blackness, specifically with regard to Black women. Bambara writes:

> Following the credits, a boat glides down a thick, green river. Standing near the front of the boat is a woman in a long white dress and a large veiled hat. The image is familiar from dominant cinema's colonialism-as-entertainment genre. But we notice that this woman stands hipshot, chin chocked, one arm akimbo. These ebonics signify that filmmaker Dash has appropriated the image from reactionary cinema for an emancipatory purpose. She intends to heal imperialized eyes.
>
> *(1992, xii)*

Bambara's words can be applicable to making sense of how Black women received *Lemonade*, in that the dissonant images, the citations, the music, the free-flow aesthetics spoke to many with an intention of creating a new narrative on our lives as opposed to the repetition of hegemonic images that structure our quotidian interactions. Bambara's words also capture my immediate response toward Beyoncé's arresting look, sound, and question directed at the audience of "Who the fuck do you think I is?" This specific arresting scene in *Lemonade* in dialogue with the cornrows and Black Vernacular English heard in that scene and throughout the film was a counter-poetic of rhythm meant to destabilize our colonial perception of Black women by embedding Black expressive culture into the music video's audiovisual aesthetics. The re-materialization of Black culture and memory into the film's aesthetics in *Lemonade* offers an opportunity to make sense of Black subordinate positions and its colonial histories and spaces in the world (Hall 1992). By placing herself on the plantation, the Igbo landing, and other spaces specific to Black culture, Beyoncé is tapping into the feelings left behind on the land to make amends or to transform those energies into something new in her film.

## Conclusion

Digital platforms have ushered in a new set of global television politics and options for Black individuals across the diaspora to shape and construct their (re)presentational bodies. Prior to this increased global circulation of television content, the daily audiovisual representation of Blackness circulated through the music video by mirroring radio play with the visual. The fact that music videos were *the* primary channels in which Black individuals could occupy television is not accidental as anti-Black racism informs what type of Blackness can be portrayed and consumed on the small screen (Durham 2012). Since the mid-1980s, music videos for Black individuals provided a crucial space to re-work and navigate the global (re)presentation of Blackness. Outside of cinema, music videos presented an opportunity to persuade and counter global anti-Black representations of Black bodies for Black Americans largely, but not exclusively.

Given this history, Beyoncé's visual album *Lemonade* works through this fraught contestation of visual Blackness to present an exploration of Black womanhood informed by the Black diaspora. Beyoncé's performance allows her to work through Black culture and spirituality, in addition to conveying the unliveable states of haunted feelings, like suspension but also trauma and pain. The energies that remain from these sites and feelings are transformed via *Lemonade* into a document of reconciliation for Black women in the diaspora. Performance and hybrid audiovisual forms for Black culture are pedagogical practices for navigating colonial rule, but it is a negotiation that needs to exist in the popular sphere while speaking to a global subordinate group. The counter-poetics of rhythm in *Lemonade* refer to the aesthetic embedding of Black expressive culture (music, voice, image, etc.) into the film in a manner that does not repeat hegemonic constructions of Black bodies but forces us to invent new histories instead by questioning what we know about Black histories in the world and how we came to know it. The colonial memory-imagery renders Black images as "problems" available to public memory via perception (Keeling 2007, 43). The musicality of the music video, then, is a unique position to aesthetically convey Black expressive culture in a manner that communicates such a counter-poetic of rhythm that momentarily disrupts the prior colonial perceptions of Blackness through its creation of new histories mimed from Black culture. This is not specific to the United States and Black cultural production works to tap into a type of intra-African dialogue in the hope of creating a discourse that can speak to Black individuals on a global scale. The music video and its contemporary digital iterations such as *Lemonade*, demonstrate Black culture's ability to adapt and pioneer innovative hybrid technological forms while using a medium that maximizes their visibility as Black artists to reach, more importantly, a global Black audience.

## Film References

*Daughters of the Dust*. 1991. Directed by Julie Dash. Produced by Julie Dash, Arthur Jafa, Lindsay Law, and Steven Jones. New York: Kino International. DVD.

*Lemonade*. 2016. Directed by Beyoncé Carter and Kahlil Joseph. Produced by Parkwood Entertainment. New York: Parkwood Entertainment. DVD.

*Thor: Ragnarok*. 2017. Directed by Taika Waititi. Produced by Marvel Entertainment. Burbank: Walt Disney. DVD.

## Note

1 In an interview with Carolyn Baker a former programmer at MTV she states: "I said, 'We've got to play James Brown'. And Bob said, 'The research says our audience thinks rock n' roll started with the Beatles'. I came through the civil rights movement. I was a member of SNCC. I believe in opening doors. The partial line at MTV was that we weren't playing Black music because of the 'research'. But the research was based on ignorance. I told Bob that to his face. We were young, we were cutting edge. We didn't have to be on the cutting edge of racism" (Tannenbaum and Marks 2011, 168).

## References

Brooks, Daphne. 2006. *Bodies in Dissent: Spectacular Performances of Race and Freedom, 1850–1910*. Durham: Duke University Press.

Collins, Patricia Hill. 2005. *Black Sexual Politics: African Americans, Gender, and the New Racism*. New York. London: Routledge.

Dash, Julie. 1992. *Daughters of the Dust: The Making of an African American Woman's Film*. New York: The New Press.

Dozier, Ayanna. 2017. "No Happy Returns: Aesthetics, Labor, and Affect in Julie Dash's Experimental Short Film, *Four Women* (1975)." *Feminist Media Studies*. DOI: 10. 1080/14680777.2017.1326561.

Durham, Aisha. 2012. "Check on it: Beyoncé, Southern Booty, and Black Femininities in Music Video." *Feminist Media Studies*, 12(1): 35–49.

Gray, Herman. [1995] 2004. *Watching Race: Television and the Struggle for 'Blackness'*. 2nd ed. Minneapolis: University of Minnesota Press.

Hall, Stuart. 1992. "What is the 'Black' in Black Popular Culture?" In *Dia Center for the Arts: Discussions in Contemporary Culture* (Number 8), edited by Gina Dent, 21–33. Seattle: Bay Press.

Holland, Sharon Patricia.2012. *The Erotic Life of Racism*. Durham: Duke University Press.

Keeling, Kara.2007. *The Witch's Flight: The Cinematic, the Black Femme, and the Image of Common Sense*. Durham: Duke University Press.

Lepecki, André. 2006. *Exhausting Dance: Performance and the Politics of Movement*. New York: Routledge.

Mignolo, Walter D. 2015. Sylvia Wynter: What Does it Mean to be Human? In: *Sylvia Wynter: On Being Human as Praxis*. edited by Katherine McKittrick, 106–123. Durham: Duke University Press.

Okoroafor, Cynthia. 2016. "The African Influences in Beyoncé's Visual Album, *Lemonade*, Explained." *Ventures*, April 26 2016 http://venturesafrica.com/features/the-african-influences-in-beyonces-lemonade-album-explained/.

Shaviro, Steven. 2017. *Digital Music Videos (Quick Takes: Movies & Popular Culture)*. New Brunswick, NJ: Rutgers University Press.

Tannenbaum, Rob and Craig Marks. 2011. *I Want My MTV: The Uncensored Story of the Music Video Revolution*. New York: Plume.

Vernallis, Carol. 2013. *Unruly Media: YouTube, Music Video and the New Digital Cinema*. Oxford: Oxford University Press.

Wynter, Sylvia. 1992. Re-thinking 'Aesthetics': Notes Towards a Deciphering Practice. In: *Ex-iles: Essays on Caribbean Cinema*, edited by Mbye B. Cham, 237–279. Trenton, New Jersey: Africa World Press, Inc.

# 21

# SCREENING RIGHT-WING POPULISM IN "NEW TURKEY"

## Neo-Ottomanism, Historical Dramas, and the Case of *Payitaht Abdulhamid*

*Ergin Bulut and Nurçin İleri*

On 9 February 2017, the Turkish president delivered a speech during the Culture and Tourism Ministry's Special Awards Night. Along the historical lines of Turkey's right-wing politics, President Erdogan lamented that the realms of education and culture were the weakest links in his party's more than a decade long uninterrupted tenure (Diken 2017). This disappointment regarding the failure to achieve cultural hegemony resurfaced not too long after the presidential referendum in April 2017. Now speaking at the 38th General Assembly of Ensar Foundation, Erdogan said:

> We have been in government for 14 years without interruption. Still, we have problems regarding social and cultural power. Undoubtedly, there have been promising developments. There is now more interest in *Imam-Hatip* (religious) high schools. Quran, Ottoman Turkish are now elective courses in all schools. Still, we have much to do in terms of cultivating the generations that we dream of and our nation needs and demands.
>
> *(DHA 2017)*

Given the impact of networked humor and satirical popular culture during the Gezi Uprising in the summer of 2013 (Yesilyurt 2016), one can understand the President's desire to establish cultural hegemony. Acutely aware of popular culture and media hegemony's vitality for political power in the highly polarized context of Turkey, the governing party patiently built its own partisan media (Bulut 2013; Yesil 2016). Yet, the existence of this news media empire has not necessarily resulted in a solid cultural hegemony, leading to stricter Internet regulations or reactions toward creative use of social media (Bulut 2016a; Koçer 2015; Tunç 2015).

To counter this lack, the governing Justice and Development Party (JDP/AKP) resorted to a few strategies such as founding its own troll army (Bulut and Yoruk 2017; Saka 2018) and production of historical dramas that are "real" as opposed to the "fictional" ones that are full of "inaccuracies" – especially the *Magnificent Century*[1] – and therefore disliked by the governing party's leadership (Batuman 2014). The President took a decidedly visible role in this cultural war involving TV dramas, paying visits to the sets of conservative productions such as *Filinta: Bir Osmanli Polisiyesi* (TRTHaber 2015) and *Dirilis*, even inviting the cast of *Dirilis* onto his presidential plane (Haber10 2017), and breaking fast with the stars of another, *Payitaht Abdulhamid* (Haberler.com 2017). This show in particular seems to have captured his political imagination. In April 2017, he appeared on the state channel TRT, the home channel of *Payitaht.* The anchorman asked a question about the show, and he responded:

TURKISH PRESIDENT: *Payitaht* is still behind *Dirilis*. Right? *Dirilis* is ahead.

INTERVIEWER: Yes, in terms of ratings. Do you follow the ratings as well?

TP: Partly. But my grandkids are passionate about this. Their primary preference is *Dirilis* and then comes *Payitaht.* And of course, all those political games in that show, and the insidious traps designed against Sultan Abdulhamid II. We watch them as a family and evaluate it on political terms and see that the same games are played on the contemporary world's political scene. The very same game that the West is playing against us. Only the time period and the actors are different.

This answer illuminates the convergence of historical and contemporary reasons behind our endeavor to critically analyze *Payitaht* in the current populist context. This debate has two dimensions: political economy and geography of soap opera productions (Bulut 2016b; Celik Rappas and Kayhan 2016) and Turkey's assertion to emerge as a cultural soft power in the region within the context of "neo-Ottomanism" (Carney 2014; Dağlıoğlu 2017; Kraidy and Al-Ghazzi 2013; Onaran 2017; Yörük and Vatikiotis 2013). Here, we are interested in the latter.

Neo-Ottomanism is broadly defined as "a widespread nostalgia for things Ottoman" (White 2002, 30) but is currently a tool of political mobilization and feelings. Indeed, since the 1980s, neo-Ottomanism has been a "powerful ethnic Turkish amplitude by positioning Turkey at the center of a new Imperial project to 'lead' the Muslim world" (Yavuz 1998, 23). In that sense, neo-Ottomanism differs from 19th-century Ottomanism, which aimed to foster a secular state-centric loyalty, including both Muslim and non-Muslim subjects of the Ottoman state. Turkey's ruling JDP heavily mobilized various aspects of its Ottoman heritage in different realms including education, marketing, architecture, and popular culture. Undoubtedly, media, particularly popular culture, have proven vital to establish this cultural hegemony (Arsan and Yıldırım 2014).

Then, it is no coincidence that TRT is producing a show on Sultan Abdulhamid II, the main character in *Payitaht.* Abdulhamid II is a divisive figure in Ottoman historiography. His acts during his reign have produced various interpretations of his personality as either "the Red Sultan" or "the Almighty Sultan" (Özbek 2002; Zürcher 2008). Until the 1890s, the Western media described Sultan Abdulhamid II as a reformist autocrat. However, following the escalation of problems with Armenians in East Anatolia after the 1890s and the suppression of Armenian rebellions with force, Abdulhamid's image turned into one of a bloody-minded Oriental despot (Hanioğlu 2016). Other historians went beyond this dichotomy by emphasizing Abdulhamid II's social policies on education, public health, and security and therefore highlighting his visionary character (Shaw 1989). Moreover, the survival of the empire depended on his capacity to pursue a balanced policy between international powers and his success in mobilizing the Muslims outside the Ottoman lands (Akyol 2017). The struggle over the historical iconography of Sultan Abdulhamid II to a great extent has been overcome[2] in Ottoman historiography, although ideological strife between secular and Islamic groups continues (Deringil 2009; Özbek 2004).

The current Turkish president is equally a debated figure due to his populism. In this regard, *Payitaht* brings the struggle over the meaning of the "past" and "what really happened" together with contemporary political struggles in Turkey, including its heavily censored media system (in which the current discourse regarding "post-truth" emerged well before Brexit and the election of Trump). In this context, as one watches *Payitaht*, it becomes hard to distinguish whether it's the state-sponsored news bulletin, a conspiratorial debate show, or an artefact of popular culture.

We are not interested in a debate between "real" and "fake" history. Our motivation to analyze this show stems from our attention to the instrumental mobilization of popular culture for the government's dual desire to both establish cultural hegemony and consolidate its populist style of government. We define populism as a thin-centered ideology that divides the society into two groups – the corrupt elite and the people – and approaches politics as "the general will of the people" (Mudde 2004). As a thin-centered ideology, populism exists in an impure form, mixed with other ideologies and for that very reason has many manifestations from left-wing to right-wing and in-between (Canovan 2005; Moffitt 2016; Müller 2016). In the Turkish context, populism as a thin-centered ideology specifically refers to the pragmatic articulation and implementation of diverse ideologies and political practices ranging from Islamism and nationalism to liberal multiculturalism. It's common to hear politicians express nationalist statements on one day and completely multicultural statements on another, mainly for populist reasons. Essential to populists in all their manifestations is their claim to discursively construct and bring into being "the people" around social demands. Then, for populism to exist, there has to be an articulation of people's unsatisfied political demands. A populist leader brings together "the underdogs" and constructs them as "the people" against "the elite" (Laclau 2005).

This political construction of the people relies heavily on dualisms and examples of such dualisms abound in the show. For instance, in Episode 12, Sultan Abdulhamid's right-arm, Tahsin Pasha, asks Abdulhamid when "the dirty games of these dishonorables (the Armenian groups that raided the Ottoman Bank) will end." Abdulhamid responds: "It will end when we give up our faith, only when we become what they are. That is to say, never." This dualism between "us vs. them" seems only natural given the show's emotional economy, which works to further consolidate these dualisms. Below, we examine them as they appear in Season 1.

## The Context

*Payitaht* is a drama about Sultan Abdulhamid II's reign in the historically turbulent context of the late 19th-century Ottoman Empire. The show narrates Abdulhamid II's struggle to unify the Empire's subjects and globally dispersed Muslims against internal (non-Muslim communities, traitor government officers) and external (Western powers, portrayed with a highly anti-Semitic emphasis) enemies. The Empire, as Abdulhamid II expresses in Episode 12, is in a survival mode:

ABDULHAMID: The existence of our state is endangered. gangs raided a bank. Streets in the capital city are not peaceful.

ABDULHAMID'S WIFE BIDAR: My Sultan, what do these rascals want?

ABDULHAMID: It's clear. They want to devastate our state. This state is built upon four foundations. First, religion of Islam. They attack our women's veils and put animal droppings in front of our mosques. They dare to put a crescent on Hagia Sophia. They massacre Muslim communities and attack our beloved Islam. Second is the government of Turkey. Our state has been ruling for six centuries and even nine centuries if you consider the Seljuq Dynasty. Now the British, together with the Russians, French and the Italians, have an eye on our government. The evil of the enemy never ends. But they forget that courageous men amongst Oghuz Turks are infinite. I am a sultan, who is also of Oghuz descent. Third is the capital of Islam. Our precious city, heralded by our Prophet, is burning. British vessels are approaching Istanbul with their cannons fully loaded. These three foundations are under attack. So, what is our fourth foundation?

PRINCE BURHANEDDIN: The Ottoman dynasty.

ABDULHAMID: Some members of our dynasty, meaning you, are here. When Islam was under attack, the Turks have given a hand and upheld it. When the civilization order of Turks was targeted, the capital of Istanbul helped with its urban civilization. When our dynasty was attacked, the believers of Islam helped with their revitalizing breath. Now, our only anchorage is the Ottoman dynasty.

A Russian ambassador's remarks in a meeting of foreign ambassadors during a strategic attack on the Ottoman Bank[3] confirms Abdulhamid's concerns in Episode 12. The Russian ambassador highlights his desire to wake up to an Istanbul without Abdulhamid. However, Abdulhamid is a resilient sultan, constantly disappointing these enemies. He is information-savvy and always one step ahead of the enemy, acting like a detective. Indeed, he is shown to understand the content of telegraphs simply by listening to the sound of the machine without even reading the document. He pre-empts the enemies' conspiracies. He is conscientious, compassionate, but also reckless if necessary. In fact, he does not refrain from slapping the British ambassador in front of all other foreign ambassadors in the very first episode of the series. In Episode 7, he calls himself a "subject whose debt to the Nation of Islam is infinite." In Episode 15, his brother Sultan Murat describes Abdulhamid as the convergence of the Empire's most important sultans. He has inherited Fatih Sultan Mehmet's intelligence, Yavuz Sultan Selim's courage, and Kanuni Sultan Suleyman's justice.

Abdulhamid's fight against the Western powers is narrated through characters whom the audience can easily figure out due to both historical and contemporary right-wing and xenophobic political discourses in the country regarding minorities. The audience can easily figure out the context for these enemy figures because one of the major narratives regarding the collapse of the Ottoman Empire is linked with how the West historically united in their efforts to terminate the Ottoman reign. To simplify a highly complicated history, these narratives need identifiable – and very likely non-Muslim – personas. As far as the show itself is concerned, they include the Zionist leader and Jewish journalist/intellectual Theodor Herzl; Sara Hedeya, an English-born Jewish spy; Jewish lawyer and politician, Emanuel Karasu; an Istanbul-born Greek arms dealer, Basil Zaharoff; Abdulhamid's brother-in-law, co-conspirator Mahmut Pasha and his son; and "West-lover," Prince Sabahaddin. These are only pawns, though, as it is soon revealed that they are manipulated by a secret hand: Mr. Parvus. The role of the secret hand is revealed in a literal fashion in the Episode 6. As Herzl declares the Greater Israel with Sara Hedeya, Abdulhamid II literally detects the hand on her shoulder (see Figure 21.1).

This image of the mysterious hand is telling because it signifies foreign powers, which many citizens believe to constantly work toward undermining the Ottoman Empire back then and Turkey in the contemporary moment. As far as populism is concerned, this "hand" provides a functional visual narrative that explains the roots of domestic problems through international conspiracies. Specifically, it signifies Jewish families and circles that, again, many in Turkey believe to be the force behind how the world is generally run.

Although one can possibly approach the show from a number of contested issues including media and information, economic development, and "truth vs. fiction," we primarily examine the dual construction of the Other as both the unruly enemy (internal and external) and Abdulhamid's object of paternal multiculturalism.

*Figure 21.1* Abdulhamid, using a magnifier to detect the hand on Sara's shoulder, as Mr. Herzl declares the Greater Israel in Episode 6

## Enemies United: Jews, Armenians, Ottoman Greeks, the West, and the Internal Traitors

In Turkey's historical imaginary, the West occupies an ambivalent space. On the one hand, Westernization is the ultimate destination. On the other hand, the Western powers are not to be trusted as they are believed to have ulterior motives to economically, culturally, and politically undermine the Ottoman Empire and Turkey. The show is no exception in its construction of the Western threats personified through, primarily, non-Muslims in and outside the Empire. As Abdulhamid tells Tahsin Pasha in Episode 6, the war is between "the faithful and the infidel," whose "puppets" include Mr. Herzl, Mr. Karasu, Sara Hedeya, Zaharoff, Mahmut Pasha and his son.

Economic affairs are central to this war. As Theodor Herzl highlights in Episode 15, "the wars of the modern era are economic" and Herzl's role, in a highly anti-Semitic caricatured manner, is to economically undermine the Ottoman Empire, overthrow Abdulhamid II and found the Greater Israel. He believes that this can be achieved by possessing economic power. In Episode 14, Herzl shares his plans with Sara to disenfranchise the Ottoman Empire, as well as buy Israel from the Ottoman sultan, whose country is in bad shape in economic terms. The entire Season 1 represents Herzl as a greedy, bloodthirsty Jewish intellectual who has no boundaries in reaching his goals. At one point Herzl explains:

> Jews will not be servant to anyone. They will found their own banks and states with their own money. Soon, bloody news will come from Jerusalem. And the sultan will desperately watch Palestine slowly slide away from his hands. The Ottoman sultan might be able to manage the Russians with the British and the British with the Germans. However, the Jews will unite all the other nations and fight against the Ottomans, Sara.

When necessary, Herzl will mobilize the Armenians in order to reach his goals. He promises to provide arms for the Armenians. In Episode 7, the Armenian leader Karekin Pastirmaciyan is perplexed and asks Herzl, "Why are you helping us?" Herzl replies, rather directly, "A Jew will do something only when it benefits him. This is just business."

Herzl is not alone in undermining the Empire. His team-mate Sara is a British Jew, who helps Herzl in his conspiracies against the Ottomans. Sara has a vital role as she receives her orders directly from Mr. Parvus. Not dissimilar from Mr. Herzl, Sara is also pragmatic and greedy: "All I care about is money. Small profits will not suffice. All or nothing." And there is Mr. Emanuel Karasu, an Ottoman Jewish lawyer who internally targets the Empire. Yet, Karasu's focus is not economy but culture. In Episode 12, as Prince Sabahaddin and he watch the raid against the Ottoman Bank, they dream about a new Ottoman state:

KARASU: We need to overthrow Abdulhamid through a silent revolution. Without any guns or blood.

SABAHADDIN: Mr. Herzl and my dad have ideas and they fight. You too have ideas but are just watching.

KARASU: You think so? I am designing, Sabahaddin.

SABAHADDIN: What?

KARASU: I am assembling the committee to overthrow Abdulhamid. I am planning that the youth will carry that committee into Abdulhamid's palace. Promise me Sabahaddin.

SABAHADDIN: Promise what?

KARASU: That you, as a member of the dynasty, will not forget me when we overthrow Abdulhamid and you are sworn in.

SABAHADDIN: Sure, but how?

KARASU: We will build new ideas. We will start from the schools and cultivate a youth that is fervent and enthusiastic. This will be a revolution without guns. We will have our men in every level of the state. And with only one sign, a brand new Ottoman, a new people, and even a new religion.

If Mr. Herzl's project is to found the Greater Israel, the British Empire's political goal is to establish her global dominance. As Figure 21.2 illustrates, in a meeting with a trusted solider Celal and Tahsin Pasha, Abdulhamid reveals the "new world order,"[4] which, in Abdulhamid's mind, refers to the foundation of the Greater Israel on Ottoman territories.

ABDULHAMID: This is the new world order that the British call global monarchy. Their goal is to bring the Ottomans down and construct the new world order. And this is (pointing to Israel on the map) the castle of the new order. The only thing that makes this Jewish state valuable is that it is the castle of the global monarchy in the Middle East.

CELAL: Who is the shah of this new world order?

ABDULHAMID: The British Queen.

TAHSIN PASHA: And the vizier?

ABDULHAMID: That's even more important that the Queen. We'll find out.

CELAL: And the pawns?

ABDULHAMID: That's why I pretended that I was poisoned, in order to find the internal traitor among us.

Again, this map, shown in Figure 21.1, visualizes the global conspiracy against the Ottomans. The conspiracy, which still powerfully resonates with many Turkish citizens and is a productive ingredient in nationalist performances, involves the construction of the Greater Israel, as well as the acceleration of the Ottoman Empire's collapse desired by the Western powers, mainly Britain. Abdulhamid II believes the Ottomans are the target because they reject

*Figure 21.2* In Episode 4, Abdulhamid reveals the "new world order"

being a part of "the new world order." Although Abdulhamid resists the "new world order," Mahmut Pasha and Prince Sabahaddin strongly desire to join this new system by replacing Abdulhamid II with Sultan Murat, who, they believe, would easily accept the demands of the West. Throughout the first season, Mahmut Pasha and his son conspire against Abdulhamid but in a highly cartoonish manner. Mahmut Pasha is depicted as a greedy Pasha who talks a lot, eats a lot, and drinks wine. He makes secret deals with Herzl, Sara Hedeya, and Zaharoff, an Ottoman Greek arms dealer, who profits from wars as he sells arms to the British, the Russians, or simply whoever has money. Prince Sabahaddin's primary role is to trick Abdulhamid's son, Prince Abdulkadir. Both Mahmut Pasha and Sabahaddin are admirers of the West. They deeply despise Abdulhamid's Islamist regime and everyday religious practices, including prayer, fasting, and chants.

Finally, there are the Armenians, the empire's "loyal subjects," who end up rebelling against the Empire. Ordinary Muslims in the show believe that Armenians are the puppets of the Russians and the British. Episode 11 depicts this highly charged environment. Zaharoff wants Hiram, the hit man of all the powers against Sultan Abdulhamid II, to provoke Muslims against Armenians by attacking Muslim women's veil in public spaces. According to Zaharoff, this is the only way to provoke a Muslim attack against the Armenians and ensure the occupation of Istanbul by the British and the Russians.

Although Turkey had experienced a contested period of relative liberalization and media reform between 2002 and 2010, it was quickly replaced by authoritarianism (Somer 2016). The resurgence of such hostile representations against non-Muslims, however, can be best understood through Fethi Acikel's notion of "sacred subalternity" (1996). Our preference of this notion has to do with the Turkish political right's historical tendency to claim the status of both the "oppressed" and the "righteous victim." The claim to being oppressed produces the value of a subaltern and victimized subjecthood, which in turn paves the way for an arena to make further political claims (Baştürk 2014). Drawing on critical theory and psychoanalysis, Acikel defines "sacred subalternity" as a synthesis of material practices through which "masses that have been socially, culturally and imaginatively uprooted and thereby lost their material ground due to dispossession transform into a repressive-neurotic ideology that represents the masses' will to power" (Acikel 1996, 155). As Acikel suggests,

> the ideological success of "sacred subalternity" as a synthesis relies much on its ability to mobilize the negative energy of the masses within a larger political framework ("little America," "Big Turkey," "Turkish World from the Adriatic Sea to the Chinese Wall") and also to spread the faith in resolving social malaise with the help of a bigger goal.
>
> *(1996, 157)*

Through this notion, we are able to avoid seeing populism in contemporary Turkey as a deviation. Rather, this notion helps us to understand and historically embed the resurgence of xenophobic representations within institutional structures and legacies. Ultimately, the question of what is "true" and what is not "true" history is bypassed, reminding us once again how the popular form itself is a political construction, as is historical narrative.

## Paternal Multiculturalism

In *Payitaht*, "the Other" is not deployed only for political demonization. "The Other" also serves as an essential tool of Abdulhamid's, and by implication, neo-Ottomanism's paternal multiculturalism. We understand paternal multiculturalism as a key component of contemporary neo-Ottomanism and define it as a nostalgic gesture toward a glorified past that no longer is. Aiming to recuperate the Empire's protective role of the non-Muslims, *Payitaht*'s paternal multiculturalism seeks to recover the multicultural past of the Empire and instrumentalize it to the ends of cultural hegemony.

The period in which Abdulhamid reigned was particularly turbulent. The show both acknowledges this and also highlights the resilience of the Empire. In Episode 6, Abdulhamid orders his Pashas to care for certain troublesome geographies and assigns them to focus on Africa, Turkistan, Damascus, Anatolia, the Balkans, Albania, and Bosnia. One Pasha asks about Hejaz. Abdulhamid responds:

> I've got Hejaz. Its pain will be enough for me. Macca and Medina are on my shoulders. Be relieved. In these geographies, you'll serve not as Pashas but as subjects. You will help in accordance with the commands of God and the Prophet. Remember, we had discussed a dream? You will turn that dream into reality. You will turn Islamic geographies into gardens of peace.

This Islamism is not a hindrance to the Empire's compassionate approach toward non-Muslims, though. In fact, non-Muslim communities are vital to the multicultural Empire and the conflict with the non-Muslims has two main reasons: the Western powers' provocations and the non-Muslim minorities' failure to appreciate the Empire's affection toward them. In Episodes 11 and 12, Sultan Abdulhamid emphasizes how the Armenians were provoked and deceived by the British, French, Russians, and even the Italians. Yet, such provocations are baseless. As opposed to domestic and international claims regarding the oppression of non-Muslims' freedom of speech and religion, the audience learns that non-Muslim subjects are free to exercise their religion and have always been protected by the state. For instance, in Episode 5, Abdulhamid pays a visit to a student protest and asks the young protestors:

> Are you not free? In our reign, people are free to worship as they wish. You may go to a mosque. For Christians, the largest churches are free. Turkish, Kurdish, Bosnian, Albanian, Ottoman Greeks, Armenians, Circassian. What nation's peace has been disturbed in our state?

This paternal multiculturalism also sets the boundaries of who counts as a desirable or undesirable minority. Those who acknowledge the Empire's compassion for non-Muslims are the desirable ones. Theodor Herzl's father Jacob Herzl, for instance, is a desirable Jew, as he recognizes how Abdulhamid's "grandfather Sultan Beyazit also saved the lives of Jews exiled by the Christians" for which the Jews "are grateful." In Episode 7, Abdulhamid's officers Celal and Kemalettin pre-empt a conspiracy and save Jewish lives in Jerusalem. A Jewish man approaches a British journalist and asks him to report that an Ottoman emperor saved their lives and that the state of Israel will only bring death to Jewish communities.

Paternal multiculturalism is also didactic. Throughout the show, Abdulhamid teaches non-Muslims their history. In Episode 10, he receives information about non-Muslim minorities' plans to replace the crescent in Hagia Sophia with a cross. Infuriated, Abdulhamid goes to Hagia Sophia to defend it, since it belongs to "the state, to the nation, to all Muslims." In his encounter with the group aiming to replace the crescent with the cross, Abdulhamid speaks:

> Aren't you Yanni, the goldsmith in Grand Bazaar? Once, you came to us, crying. Your shop was robbed. You went bankrupt, saying that you desperately needed help. The state helped and you were relieved. Now, are you here to pay your debt with that worthless knife in your hand?

Then, Abdulhamid turns to Kavafis, another Ottoman Greek, and reminds him how he mobilized the talented doctors to help his pregnant wife. Then, he reminds another man in the group of the state's reconstruction of the walls of a local church. Then Abdulhamid's most spectacular words follow:

> Didn't our state rush to help you and take you under our wings no matter what nation you belong to? Didn't your ancestors, during the conquest of Constantinople, say that they wanted to see a Turkish turban rather than a drunk Latin? Do you think you can replace the crescent with the cross despite all this? Islam is Allah's order. There is justice in Allah's order. The Ottoman Empire is built on this justice. It doesn't harm its subjects. It is never unfair to its subjects; never meddles with one's religion; never despises anybody because of their religion; never has an eye on one's honor. On top of Hagia Sophia is the flag of this justice.

Abdulhamid ends his lesson with a serious reminder: "The flag is crescent. Any hands leaning towards that crescent will be broken."

The state's level of care for its subjects is omnipresent, almost panoptic, and sometimes seemingly without limits. Watching over his subjects from a balcony, Abdulhamid sees Onnik, a disabled Armenian child. Interested in carpentry, Abdulhamid actually produces a prosthetic leg for Onnik. Onnik's mother's grateful response, in turn, reveals how Western powers circulate fake news about the Empire's fundamentally compassionate multicultural politics:

ONNIK'S MOM: We are grateful to you. We are Christians but are your subjects. For centuries, we've known that the Ottoman Empire has justly ruled over its subjects. Excuse me, but what we read these days on newspapers …

ABDULHAMID II: What do the newspapers write?

ONNIK'S MOM: They call you a murderer. But now, you've made a leg for my child with your own hands. It seems that you are person who have a big heart that transcends state borders and one that cares about the health of even a disabled kid.

In sum, Abdulhamid's paternal multiculturalism is completely altruistic, concerned with justice, didactic, and informed with historical knowledge. It is not his own political future but rather the future of the Empire that actually concerns him. The country is alive only when its multicultural subjects are with them in an intact fashion.

## Conclusion

Contemporary debates regarding the rise of right-wing populism mostly focused on electoral results or the role of digital media and algorithms, producing somewhat presentist accounts (Chakravartty and Roy 2017). Our examination of *Payitaht Abdulhamid* demonstrates that television, especially in the Global South, still plays a central role in governments' desire to reconstruct history and establish cultural hegemony. This is particularly important as Turkey is going through a crisis of hegemony since the public is completely divided in its support for the government. Within the context of this hegemonic crisis, televised popular culture is vital, perhaps more than ever. Specifically, the show reduces a complicated history into easily understandable dichotomies and projects them on to contemporary politics in order to consolidate support for the government. Through televised popular culture, the government mobilizes history for purposes of cultural hegemony and populist politics flavored with nationalist, Islamist, and anti-Western motifs. Ultimately, the TV show presents yet another moment for the mediated nature of 21st-century politics. Yet, how pro-government populations engage with these shows or the extent to which these shows are successful in maintaining government support is open to further ethnographic analysis.

## Notes

1 The show *Magnificent Century* (2011–2014) was a highly popular historical drama with major international success. It features the life of one of the greatest figures in Ottoman history, namely Sultan Suleiman. The government became quite critical of the show, arguing that it misrepresented the ancestors of the Turkish citizens by simply focusing on political and emotional rivalries in the Ottoman Dynasty, rather than foregrounding Sultan Suleiman's conquests and the Empire's power.

2 The struggle over the historical iconography of Sultan Abdulhamid II has mostly been overcome in the sense that new studies on the period have moved away from dualist interpretations of Abdulhamid's personality as either "the Red Sultan" or "the Almighty Sultan" (Özbek 2002; Zürcher 2008). Such studies reject the modernization paradigm, which posits the Ottoman Empire into a backward position compared to Europe. Rather, this new revisionist historiography emphasizes the diversities of the Abdulhamid period within the context of a global temporality. Scholars foreground Abdulhamid II's policies on industrial transformation, education, public health, and security and make comparisons between Europe and the Ottoman Empire without reproducing the discourse of "development" or "belatedness" (Alkan 2000; Deringil 2009; Fortna 2002; Özbek 2002).

3 For a detailed analysis on the dynamics of Ottoman Bank Raid, please see Edhem Eldem (2011) and Necdet Sakaoğlu (1994).

4 We were unable to locate this phrase in the relevant literature on the Abdulhamid II era. In that regard, it has more to do with the work of the screenwriters rather than being a historically accurate phrase or a phrase that is meant to evoke contemporary post-Cold War ideas of globalization. Having said that, we have come across this phrase via the Paris Peace Conference following World War I (Ertan and Özgün 2016; Kasalak 2016). So, it might be the case that the writers of the show may be doing a backward interpretation of history in order to highlight the colonial goals of England and other European powers.

## References

Acikel, Fethi. 1996. "'Kutsal Mazlumlugun' psikopatolojisi." *Toplum ve Bilim* 70:153–198.

Akyol, Taha. 2017. "Abdülhamid." *Hürriyet*. 7 March. www.hurriyet.com.tr/yazarlar/taha-akyol/abdulhamid-40386695.

Alkan, Mehmet Ö. 2000. "Modernization form Empire to Republic and Education in the Process of Nationalism." In *Ottoman Past and Today's Turkey*, edited by Kemal Karpat, 47–133. Leiden: Brill.

Arsan, Esra and Yasemin Yıldırım. 2014. "Reflections of Neo-Ottomanist Discourse in Turkish News Media: The Case of The Magnificent Century." *Journal of Applied Journalism and Media Studies* 3 (3):315–334.

Baştürk, Efe. 2014. "Mağduriyet ile İktidar Arzusu Arasında Türk Sağı: Komplocu Bir Söylemin Anatomisi." *Teorik Bakış* 5 (September):117–150.

Batuman, Elif. 2014. "Ottomania." *The New Yorker*. www.newyorker.com/magazine/2014/02/17/ottomania.

Bulut, Ergin. 2013. Role Model Turkey's Bankrupt News Media: Lessons from Gezi Parkı Protests. *The Political Economy of Communication*,.1(2): 102–113.

Bulut, Ergin. 2016a. "Social Media and the Nation State: Of Revolution and Collaboration." *Media, Culture & Society* 38 (4):606–618.

Bulut, Ergin. 2016b. "Dramın Ardındaki Emek: Dizi Sektöründe Reyting Sistemi, Çalışma Koşulları ve Sendikalaşma Faaliyetleri." *Galatasary İletişim Dergisi* 24:79–100.

Bulut, Ergin and Erdem Yoruk. 2017. "Digital Populism: Trolls and Political Polarization of Twitter in Turkey." *International Journal of Communication*, 11:4093–4117.

Canovan, Margaret. 2005. *The People*. Cambridge. Malden, MA: Polity.

Carney, Josh. 2014. "Re-Creating History and Recreating Publics: The Success and Failure of Recent Ottoman Costume Dramas in Turkish Media." *European Journal of Turkish Studies* 19 1-25.

Celik Rappas, Ipek A., and Sezen Kayhan. 2016. "TV Series Production and the Urban Restructuring of Istanbul." *Television & New Media* 1–21. doi:10.1177/1527476416681500

Chakravartty, Paula and Srirupa Roy. 2017. "Mediated Populisms: Inter-Asian Lineages –Introduction." *International Journal of Communication* 11:4073–4092.

Dağlıoğlu, Emre Can. 2017. "Osmanlı tahtında bir Polat Alemdar." *Agos*. 9 March. www.agos.com.tr/tr/yazi/17910/osmanli-tahtinda-bir-polat-alemdar.

Deringil, Selim. 2009. *The Well-Protected Domains: Ideology and the Legitimation of Power in the Ottoman Empire, 1876-1909*. London; New York: I.B. Tauris.

DHA. 2017. "Erdoğan: Biz 14 Yıldır Kesintisiz Siyasi Iktidarız Ama Hala Sosyal ve Kültürel Iktidarımız Konusunda Sıkıntılarımız Var." *Hürriyet*. May 28. www.hurriyet.com.tr/erdogan-biz-14-yildir-kesintisiz-siyasi-iktida-40472521.

Diken. 2017. "Erdoğan'dan 'Özeleştiri': Son 14 Yılda En Zayıf Halka Eğitim ve Kültür." *Diken.com.tr*. February 9. www.diken.com.tr/erdogandan-ozelestiri-son-14-yilda-en-zayif-halka-egitim-ve-kultur/.

Eldem, Edhem. 2011. "26 Ağustos 1896 'Banka Vakası' ve 1896 'Ermeni Olayları'." In Fahri Aral *İmparatorluğun Çöküş Döneminde Osmanlı Ermenileri, Bilimsel Sorumluluk ve Demokrasi Sorunları Konferansı*, 125–152. İstanbul: Bilgi Universitesi Yayınları.

Ertan, Ali Erhan and Tevfik Orçun Özgün. 2016. "I. Dünya Savaşı'nın Ardından Yeni Dünya Düzenine Doğru: 1919 Paris Barış Konferansı." *Türkiye Sosyal Araştırmalar Dergisi*, 3:551–572.

Fortna, Benjamin. 2002. *Imperial Classroom: Islam, the State, and Education in the Late Ottoman Empire*. Oxford: Oxford University Press.

Haber10. 2017. "Erdoğan, Diriliş Ertuğrul Ekibiyle Aynı Uçakta." May 10. www.haber10.com/cumhurbaskani/erdogan_dirilis_ertugrul_ekibiyle_ayni_ucakta-703399.

Haberler.com. 2017. "Payitaht Ekibi Cumhurbaşkanı İftar Yemeğinde!" *Haberler.com*. June 13. www.haberler.com/payitaht-ekibi-cumhurbaskani-iftar-yemeginde-9724952-haberi/?utm_source=facebook&utm_campaign=tavsiye_et&utm_medium=detay.

Hanioğlu, Şükrü. 2016. "Sultan Abdülhamid'e Yeni Bakışlar." *Derin Tarih*. December.

Kasalak, Kadir. 2016. "İngilizlerin Filistin Politikası ve Filistin Mandası." *Süleyman Demirel Üniversitesi Sosyal Bilimler Enstitüsü Dergisi* 25:65–78.

Koçer, Suncem. 2015. "From the 'Worst Menace to Societies' to the 'Robot Lobby': A Semantic View of Turkish Political Rhetoric on Social Media." In *New Media Politics: Rethinking Activism and National Security in Cyberspace*, edited by Banu Baybars-Hawks, 229–244. Cambridge, UK: Cambridge Scholars Publishing.

Kraidy, Marwan M. and Omar Al-Ghazzi. 2013. "Neo-Ottoman Cool: Turkish Popular Culture in the Arab Public Sphere." *The International Journal of Media and Culture* 11 (1):17–29.

Laclau, Ernesto. 2005. *On Populist Reason*. London: Verso.

Moffitt, Benjamin. 2016. *The Global Rise of Populism: Performance, Political Style, and Representation*. Stanford, CA: Stanford University Press.

Mudde, Cas. 2004. "The Populist Zeitgeist." *Government and Opposition* 39 (4):541–563.
Müller, Jan-Werner.. 2016. *What Is Populism?* Philadelphia, PA: University of Pennsylvania Press.
Onaran, Burak. 2017. "Payitaht Abdülhamid vesilesiyle: Tarih-kurgu-propoganda." *Birikim* 336:76–81.
Özbek, Nadir. 2002. *Osmanlı İmparatorluğu'nda Sosyal Devlet, Siyaset,İktidar ve Meşruiyet, 1876–1914*. İstanbul: İletişim Yayınları.
Özbek, Nadir. 2004. "Modernite, Tarih ve İdeoloji: İkinci Abdülhamid Dönemi Tarihçiliği Üzerine Bir Değerlendirme." *Türkiye Araştırmaları Literatür Dergisi* vol 2 1:71–90.
Saka, Erkan. 2018. "Social Media in Turkey as a Space for Political Battles: AKTrolls and Other Politically Motivated Trolling." *Middle East Critique*. doi:10.1080/19436149.2018.1439271
Sakaoğlu, Necdet. 1994. "Osmanlı Bankası Olayı." In *Dünden Bugüne İstanbul Ansiklopedisi*, Ankara, İstanbul: Kültür Bakanlığı ve Tarih Vakfi Ortak Yayını vol 6, 166–168.
Shaw, Stanford. 1989. "Sultan Abdülhamid II: Last Man of the Tanzimat." In *Tanzimat'ın 150. Yıldönümü Uluslarası Sempozyumu*, Ankara: Millî Kütüphane 179–197.
Somer, Murat. 2016. "Understanding Turkey's Democratic Breakdown: Old vs. New and Indigenous vs. Global Authoritarianism." *Southeast European and Black Sea Studies* 16 (4):481–503.
TRTHaber. 2015. "Cumhurbaşkanı Filinta'nın Yönetmen Koltuğunda!" April 19. www.trthaber.com/haber/turkiye/cumhurbaskani-filintanin-yonetmen-koltugunda-179662.html.
Tunç, Aslı. 2015. "Quest for Democracy: Internet Freedom and Politics in Contemporary Turkey." In *Digital Transformations in Turkey: Current Perspectives in Communication Studies*, edited by Banu Akdenizli, 207–220. Lanham, MD: Rowman & Littlefield.
White, Jenny. 2002. *Islamist Mobilization in Turkey: A Study in Vernacular Politics*. Seattle, WA: University of Washington Press.
Yavuz, M. Hakan. 1998. "Turkish Identity and Foreign Policy in Flux: The Rise of Neo-Ottomanism." *Critique: Critical Middle Eastern Studies* 7 (12):19–41.
Yesil, Bilge. 2016. *Media in New Turkey: The Origins of an Authoritarian Neoliberal State*. Geopolitics of Information. Urbana, IL: University of Illinois Press.
Yesilyurt, Adem. 2016. "Gündelik Hayat, Karnaval ve Direniş: Gezi'de Mizah." *Praksis* 42:635–656.
Yörük, Zafer and Pantelis Vatikiotis. 2013. "Soft Power or Illusion of Hegemony: The Case of the Turkish Soap Opera Colonialism." *International Journal of Communication* 7:2361–2385.
Zürcher, Erik Jan. 2008. *Modernleşen Türkiye'nin Tarihi*. İstanbul: İletişim Yayınları.

# 22

# TRANSNATIONAL SCREEN NAVIGATIONS

## Priyanka Chopra's Televisual Mobility in Hollywood

*Pawan Singh*

There was a time in Indian television when actors who had limited luck in the Hindi film industry would migrate to Indian television. By the 1990s, with the beginning of television's transformation in India owing to economic liberalization, the converse was also occasionally true with former TV actors such as Shahrukh Khan and Vidya Balan becoming successful in Hindi cinema. The boundaries between Indian film and television were slowly blurring toward the end of the twentieth century. Yet, the migration of stars from the big screen to the small was still considered a "failure" and the less common movement from television to film was deemed more successful. In the twenty-first century, however, television is no longer considered a consolation medium. The Hollywood television debut of the hugely successful Bollywood star Priyanka Chopra in a leading role on the TV show *Quantico* (U.S., ABC, 2015–2018) and her subsequent numerous appearances on American television talk and award shows, including the Oscars, offer a prime example of television as a competitive medium for established stars. However, Chopra's case is noteworthy for exemplifying not just star mobility between film and TV but also across national industries. Her move to American TV testifies to the increasing transnational viability of established Bollywood stars in the twenty-first century. Importantly, the uptake of her rise on the American TV screen has also been seen as part of the broader arrival of South Asian performers on American TV. But her success differs from American actors of South Asian origin whose trajectory recapitulates the immigrant narrative of breaking free of stereotypical roles to play realistic, meaningful characters on the screen. Chopra's representational currency and her "global" Indian English accent instantiate the logic of televisual mobility – her transnational screen navigations speak, not to a teleological moment of arrival on the American screen, but rather to her ability to make professional choices that enable her to represent Indians everywhere.

### Introduction

The transnational turn in media studies has astutely demonstrated the various ways in which television has crossed national borders within both a political economy and a cultural globalization framework (Chalaby 2005). Within this infrastructural flexibility of television systems across borders and content localization for national and diasporic audiences, the star figure as a mobile embodiment of national culture and values and their travel along international circuits of media visibility has also substantially concretized the meaning of the transnational. While historically, stars served to stage the national on an international scale, in the contemporary moment, this function is tied up with the stars acting "as agents of the global media systems that create them while also acting as sources of pleasures and identification for their audiences" (Meeuf and Raphael 2013: 5).

The dramatically transforming political economy of television in the twenty-first century has resulted in greater screen access to a diversity of talent in terms of race, nationality, sexual orientation and gender identity. American television, in particular, has emerged to be more inclusive of immigrant stories and characters with the advent of digital content delivery platforms such as Hulu, Netflix and Amazon Prime. Within this set of developments, American actors of South Asian origin in the twenty first century are touted as having arrived from the margins to the center of the American television screen.

This chapter develops the concept of televisual mobility to examine the success of established Bollywood actors such as Priyanka Chopra in American television and film. The chapter proposes that the recent success of Indian actors of American or British origin in American television reprises a teleological narrative that moves against and away from a history of stereotypical representations of South Asian performers in American media. On the other hand, the case of Indian film star Priyanka Chopra demonstrates that her debut on American television in ABC's show *Quantico* instantiates cultural and economic logics of televisual mobility, a mode of professional diversification through which she has successfully navigated transnational screens. Televisual mobility may be conceptualized as a media commodity logic through which established Bollywood stars amplify their stardom on television screens in a transnational context. These screen affordances speak to transnational television's competitiveness on the one hand while enabling established stars to make professional choices in a manner that continues to instantiate their televisual mobility. Priyanka Chopra's transnational screen navigations are indicative of this nascent trend in the twenty-first century; however, whether Hollywood stardom may be amplified by the same logic of televisual mobility is a question I bracket for now. Historically, the ambition of Hollywood stardom has been prevalent in Bollywood whose actors on various occasions have vied for the former's attention.

Even though it is television's dramatically changing political economy in the digital age that has fostered a diverse representational canvas allowing non-white actors to play leading roles, Chopra's televisual mobility has enabled her to represent Asia in global terms on American television, transcending the teleological narrative of on-screen arrival and representational parity of parts. In this representation, Chopra's screen presence on mainstream American television edifies a much longer narrative of the Hindi film industry and its modern-day avatar, Bollywood, grabbing American media's attention. Examples of prominent Indian film stars in American film and TV abound – from Om Puri, Amrish Puri and Irrfan Khan in mainstream productions to Aamir Khan's Oscar nomination for *Lagaan* (India, 2001) and many more occasions on which declarations of Bollywood going global (meaning Hollywood) have been made (Power 2000; Parthasarathy 2006; Ramesh 2008). But Chopra's success as a mainstream Indian actor in Hollywood is distinct from this pedigree and is attributable to the rapid proliferation of media platforms globally as well as new circuits of star visibility enabled by television, the Internet and the historical conjuncture of globalization (Mazumdar 2012). Chopra's success has been reported in Indian media as bringing great pride to India (Iyer 2017), a claim that speaks to the Hollywood ambitions of Bollywood actors. Hollywood as a media destination is often regarded in Bollywood as highly prestigious and affording global visibility to Indian film stars. For instance, Indian actor Aamir Khan who has consistently nixed Indian film awards as biased and unfair expended a great deal of effort to take his 2001 film *Lagaan* to the Oscars. The film's Oscar nomination in the best foreign film category generated great national excitement as well as disappointment when the film lost to Bosnia's *No Man's Land* (2001).

At a personal level, Chopra claims to have defied conventional wisdom advising her not to foray into the international music industry in her mid-thirties (India Today 2018), a professional move that subsequently set the stage for a leading role in American Broadcasting Corporation's (ABC) *Quantico* (2015–2018, USA).

In both cases – American actors of Indian origin and actors from India – national and diasporic audiences have staked racial pride in the star images of South Asian actors. However, the teleological narrative of South Asian actors of American origin remains inscribed within the racial history of American TV on which until very recently, screen portrayals by South Asians were predominantly stereotypical and marginal (Dave 2013) without the opportunity to play ordinary American characters in realistic parts (Mahdawi 2017). However, the digital diversification of television beyond the TV set and the proliferation of television content through shows of diverse genres has led to diversification in casting. According to media studies scholar, Mary Beltrán, viewership has become more diverse and television messages are no longer geared toward predominantly white viewers (Ramanathan 2015). This development could suggest a demand for ethnically diverse portrayals from audiences within the broader changes around racial justice social movements in the United States. Consequently, non-white performers and South Asian actors in particular, are no longer bound by the vapid visibility of marginal characters. Their success has in part created the conditions for an already established Indian film star like Priyanka Chopra to succeed on the American television screen.

The chapter analyzes Priyanka Chopra's global mobility and transnational stardom as an instance of the logic of televisual mobility that enables an actor of Chopra's stature to make professional choices beyond Bollywood. Her numerous appearances on American television talk and award shows have allowed her fans in the South Asian diaspora to stake a racial pride in her popularity. Chopra's transnational screen navigations are distinct from American actors of South Asian origin even as she partakes of the teleological narrative of arrival and progress by minority actors on the American mediascape. Importantly, Chopra's Hollywood forays may be understood as a professional strategy enabled by access to multiple media industries in contradistinction to American origin South Asian actors whose success recapitulates the struggles for racial recognition and mainstreaming of previously marginal parts.

The mainstream media discourse on the teleological narrative of South Asian arrival on the Hollywood screen tends to conflate diverse trajectories of professional struggle into a singular image of Hollywood's racism that is understood to be waning gradually. Take for instance, the 2018 story in the VICE magazine entitled 'South Asian Actors Are Fighting Hollywood's Racism', featuring juxtaposed shots of Priyanka Chopra, Mindy Kaling, Hari Kondabolu and Utkarsh Ambudkar. The story goes on to detail the typical predicament American actors of Indian origin find themselves in – dearth of parts, simplistic and stereotypical roles, too many actors auditioning in addition to preconceived notions about the range of character possibilities in casting a South Asian actor. As South Asian music and film scholar Nilanjana Bhattacharjya has observed, the path for the South Asian arrival on the Hollywood screens was paved by actors such as Kal Penn, Parminder Nagra, Maulik Pancholi and Aasif Mandvi who often held small roles in the 1990s and early 2000s (Rao 2018). The stand-up comedy by American comedians of Indian origin, for instance, Hari Kondabolu, Aziz Ansari and Hasan Minhaj has humorously critiqued racial stereotyping in Hollywood. This critique may be read as diasporic actors leading the charge toward demanding more complex and diverse representations of South Asians in America.

Chopra, who grew up in India and spent her teenage years in America, did experience racial discrimination. But her professional trajectory follows the trials and tribulations of any outsider (not a star's son or daughter) trying to make it big in Bollywood. The conflation of two distinct experiences of professional struggles overlooks the commodity logics of celebrity circulation (Govil 2007) within the international mediascape of TV, cinema, award shows, television talks shows, film festivals and other cultural production. Chopra's visibility may be seen as a sublimation of the long history of Bollywood actors appearing on Hollywood screens, a kind of access that is not necessarily afforded to American actors of South Asian origin in Bollywood, or anywhere else outside America. The latter's success reinscribes the standard trajectory from the margins to central screen space, a movement that celebrates the telos of racial recognition within the United States. The success of Indian American and British actors is often framed in linear terms of shattering cultural stereotypes through stories that are predominantly told as comedic critique of American race relations. Further, their success can be ascribed to the privileging of comedy as the preferred genre through which a number of non-white actors have claimed central screen space. Yet, their emergence on American TV continues to be predominantly limited to comedic parts even when the actors play leading roles.

The role of the American television talk show is particularly noteworthy for the transnational visibility of Bollywood actors. Some of the most notable appearances include Aishwarya Rai and Abhishek Bachchan on the *Oprah Winfrey Show*, Deepika Padukone on *The Ellen DeGeneres Show*, Sridevi on the *George Stroumboulopoulos Tonight Show* and Priyanka Chopra on multiple shows including Stephen Colbert's, Jimmy Fallon's, and Jimmy Kimmel's shows. The television talk show has been understood by scholars as setting social, political and cultural agendas, helping define the common sense for the viewing public (Timberg 2002), as constructing a hierarchy of behaviour for an active, gendered audience in the American context (Shattuc 1997) and as providing a template for grounding the cultural assumptions of the audience as well as their identity formation (Mittell 2003). While scholars have observed how celebrities as media texts express values in a modern society toward the reconfiguration of the public and private aspects of social life (Marshall 1997) as well as individuality and class marker in consumer culture (Marshall and Redmond 2016), the transnational television celebrity has come to represent a screen mobility that negotiates stardom in terms, not always of immigrant status or racial assimilation but also in economic terms of strategic professional decisions. In numerous appearances on American TV talk shows, Chopra has debunked stereotypes about Indian actors' ability to speak English, played Holi (the Hindu festival of colors) and related her experiences of struggles in the Hindi film industry (Ajay 2017). Such television appearances constitute a kind of curation of the transnational celebrity's stardom that augment their celebrity while bringing them closer to audiences through the revelations of an actor's journey to stardom.

To reiterate, while this arrival of the South Asian actors of American origin on the center of the American television screen has brought racial pride to the South Asian diaspora, their struggles to carve out this space differ from the trajectory of Indian actors such as Chopra with an already celebrated body of work in Bollywood, a long list of national and international brand endorsements, and humanitarian work that includes her appointment as the United Nations International Children's Emergency Fund (UNICEF) goodwill ambassador. Chopra's foray into American television and Hollywood cinema demonstrates a form of transnational screen navigation constitutive of her global celebrity.

In order to offer an account of Chopra's transnational screen navigations, I begin with a brief history of the movements of actors between film and television in the Indian context during the 1990s, a time of media liberalization and the arrival of cable television in India. This history gives way to the contemporary moment of television's digital proliferation, a set of developments that undergird transnational stardom across screens.

## Indian Television during the 1990s and Beyond

During the 1990s, a dramatically changing media landscape in India, owing to economic liberalization, generated an influx of Western media content that appealed to the burgeoning Indian middle classes (Thussu 1999). In significant measure, Murdoch's Star TV implemented the strategy of localizing Western content for Indian audiences, leading to the proliferation of English and Hindi news channels as well as an expansion of television genres, such as reality TV and game shows, under the banner of an indigenous globalization (Thussu 2007). In documenting the history of television in India, scholars have paid attention to important aspects of the medium's evolution including the appeal of state-sponsored national programming to foster feelings of nationalism among diverse audiences (Rajagopal 1993), ethnography of television audience in terms of class and gender in postcolonial India (Mankekar 1999) and the popularity of Indian soap operas among middle- and lower middle-class audiences, during the early years of the twenty-first century, which focused on social issues in everyday life, emphasizing the importance of family values, and hence traditional notions of Indian culture (Munshi 2010).

The rise of transnational television channels as a form of electronic capitalism fostered communities of television viewers that were unimaginable (Kumar 2006) in contrast to Benedict Anderson's notion of bounded imagined communities brought about by print capitalism in Europe. These television viewers formed unimaginable communities in the sense that they were "infinite, limitless, and unbounded in the worldwide flows of national, transnational, and translocal networks" (Kumar 2006: 15). A focus on the negotiation of cultural boundaries and changing youth identities in the post-liberalization Indian context with the arrival of transnational television has remained a central thematic in the study of Indian television (McMillin 2001; Butcher 2003). These scholars have made valuable contributions to the study of Indian television through their analyses of television's political economy, globalization and negotiation of cultural identities and the role of TV programming in mobilizing national and diasporic audiences.

Despite these important contributions to the study of Indian television, the role of the television actor in the Indian national and diasporic context, and their significance as celebrity or star has remained relatively under-studied. The lack of attention to the role of television actors and their movements between television and the film industry during the period of economic liberalization, roughly between 1992 and 1999, may be owing to the paradigmatic shift in television's political economy that impelled, in part, the urgency of studying such profound economic and political transformations. Audience ethnographies and research on the localization of global content on Indian television and attendant negotiations of cultural identities in national and diasporic contexts may have similarly been inspired by the burgeoning scholarship on the study of culture under the aegis of cultural studies. This research direction reshaped the study of media significantly during the 1990s and early 2000s. Yet, a closer look at the TV star or celebrity in the Indian and transnational context has been generally overlooked in the wake of these scholarly investments to understand the dramatically changing landscape of Indian television.

Bombay, which is home to India's entertainment media industries, has become synonymous with Bollywood, a neologism that designates a new culture industry that prominently markets high-budget Hindi films to diasporic audiences (Rajadhyaksha 2003). It began by marketing a kind of traditional Indian "national culture" through its focus on themes of joint family values, weddings, romance and nationalism in relation to the lives of upwardly mobile, global Indians. But it is also home to India's television industry, which as noted above underwent significant economic and cultural changes in the 1990s. During this period, television often served as a consolation medium for actors who had little to moderate success as Hindi film actors. Given the dynastic model of the Hindi film industry – the trend of star kids having easy access to debut ventures to begin their careers as actors or directors – the barriers to entry into the movie industry have historically tended to be higher for those without family connections in the film industry (Iyer 2014). While most aspiring film actors would disappear from the screen altogether in the face of failure, a few with moderate success would migrate to television. For instance, the 1980s actor Shekhar Suman, who acted in B-grade films subsequently moving to marginal characters, went on to act in notable TV sitcoms such as *Dekh Bhai Dekh* (*Look Brother Look*, Doordarshan, 1993). With the arrival of cable TV around the same time, Suman hosted a television talk show, *Movers and Shakers*, modeled on American talk shows like *The Tonight Show with Jay Leno*, to which he invited other film and television celebrities. Interestingly, most of the guests who appeared on the show weren't necessarily eminent film actors but lesser known media celebrities including other TV actors, winners of beauty pageants, models and musicians/singers. Aired on Sony Entertainment Television (1997, SET, India), *Movers and Shakers* was scheduled as a late-night show, possibly the only one of its kind with a music band and risqué humor. Its American-style format and gossipy banter in English made it appealing to niche, English-speaking audiences.

In fact, part of this emergent picture of television globalization in the 1990s had to do with the boom in diverse programming with the proliferation of television channels such as MTV, Sony, Channel V and Star TV. Initially, many TV producers largely followed Western formats, which served as a benchmark (Mehta 2008). With increasing competition, as Nalin Mehta has noted, these TV channels began to tap into "Indian oral traditions and traditional patterns of social communication" that pertained to a distinct "Indian" identity (ibid.: 6). Despite the commercially viable deployment of the cultural identity logic in television programming, talk shows in English featuring elite hosts remained popular. If *Movers and Shakers* gave the erstwhile film star Shekhar Suman the opportunity to stage his comic sensibilities on television talk shows through clever banter, Simi Garewal, an eminent Hindi film actress during the 1970s and the 80s, wooed elite Indian audiences through her celebrity chat show, *Rendezvous with Simi Garewal* (1997). On her chat show, she invited more eminent and popular Hindi film stars who opened up to her intimate style of interviewing, offering the viewers personal insights into the lives of the stars.

While television offered a new professional lease to some erstwhile film actors, there were also those who began their acting careers on the small screen but migrated to Hindi films successfully. The two most notable examples are Shahrukh Khan and Vidya Balan who both acted in television series before attaining celluloid stardom in the Hindi film industry. Within the Indian mediascape, these professional moves to the big screen have been indicative of a general trend among aspiring actors to graduate to the cinema screen in due course of time. However, both Khan and Balan, who gained television visibility in their respective roles in *Fauji* (*Soldier*, Doordarshan, 1989) and *Hum Paanch* (*Us Five*, Zee TV 1995), went on to become viable, commercial stars through a series of critically acclaimed performances in blockbuster films. The trajectory of migration from TV to film during the early to mid-1990s must be understood in the early days of globalization of television channels and programming in India when television and film represented distinct modes of stardom. This distinction began to blur during the early years of the twenty-first century, as television channels proliferated and offered a range of programming including commercial Hindi films to the Indian viewers. Television thus became as competitive an entertainment medium as the movies. In fact, a common colloquial refrain among the movie-going middle classes who owned television sets with a cable connection would be, *why go to the movies when the film would be shown on TV in a few days in any case?* During this period, India emerged to be the third largest television market after China and the United States (Mehta 2014). Historically, Hindi films have always been viewed as a more desirable medium for actors to showcase their talent as well as embody a superior stardom. The recognition of film awards, and the near-absence of television awards historically further reinforces the desirability of Hindi films over television roles. Although television awards in India only began with the diversification in television programming (e.g. Indian Television Academy Awards, Indian Telly Awards) in 2001, Indian TV has also emerged as a higher paying medium according to a few TV actors who believe that an actor's shelf life and commercial viability changes with every film. Television, on the contrary, offers long-standing parts, with more money over the long run (Ramasubramanian 2017).

In considering Indian television's relationship to Indian films,Kalyani Chadha and Anandam Kavoori (2012) have described the latter as an invasive species to television. They demonstrate that given Indian television's high upfront costs and an inability to definitely identify products assured of success in the marketplace, films and film-based programming constitute 20% of television programming schedules nationally and film channels attract 30–35% of advertisement revenue in a crowded market. Add to this the popularity of talent shows, reality TV programmes and film music channels among audiences, all of which rely on the iconic power of the film stars as hosts and guests on these programmes.

While Chadha and Kavoori's argument regarding films as invading television screen space seeks to illustrate the inability of Indian television to articulate a non-film identity, and at a conceptual level "break free of film-related personalities and content" (115), it is premised on Indian television's economic viability as an indicator of the medium's success among mass audiences. This, according to the authors, is also true of regional television programming that offers linguistic variation with content homogeneity through the same format of programmes. The political economy argument, while instructive in understanding Indian film's invasive influence on television programming, may be reformulated from a televisual mobility perspective that decenters films as the ultimate destination of stardom and accounts for television's ongoing liveliness in a screen-saturated culture as well as malleability of stardom across film and television. It is to this dynamic that the chapter turns in the next section in its discussion of televisual mobility in a transnational context.

## Televisual Mobility and Transnational Screen Navigations

It wouldn't be wholly inaccurate to suggest that the transnational movement of stars predates media and cultural globalization. After all, both Hollywood and the Hindi film industry, before it became a global entity under the aegis of Bollywood (Kavoori and Punathambekar 2008), have had a long history of exchange through the traffic of stars, audiences and finance (Govil 2015). Part of this long history of exchange are the Hollywood parts played by veteran Indian actors such as the late Amrish Puri in *Indiana Jones and the Temple of Doom* (1984) and the late Om Puri in *Gandhi* (1982), *City of Joy* (1992) and *The Hundred Foot Journey* (2014) as well as Aishwarya Rai in transnational productions such as *Bride and Prejudice* (2004) and *The Last Legion* (2007). These Hollywood film parts and appearances on American TV talk shows had, in large part, defined a standard paradigm of global visibility for Hindi cinema and Bollywood actors. Bollywood's Oscar aspirations have also for a long time shaped the discourse on star mobility. India's own national film awards, the Filmfare awards, have held an allure among Bollywood actors but, in some instances, big names like Aamir Khan and Ajay Devgan have nixed the Filmfare awards on grounds of their lack of credibility. In the actors' view, the awards are either biased toward particular actors or given to those who are able to attend the ceremony, implying that Filmfare awards are promotional events for the sponsors rather than a recognition of an actor's craft (Jha 2014). Aamir Khan is among the few Indian actors who attended the Oscars in 2002 when his film *Lagaan* (Taxation, 2001, India) was nominated in the best foreign-language film category in 2002. But the global mobility of Bollywood actors generally remained confined to cinematic occasions whether in the form of film parts or television chat show appearances.

American television until the late 2000s too, had only managed to cast South Asian and non-white actors in marginal, stereotypical parts. American actors of South Asian origin, in particular, were, however, gradually becoming visible on American television in non-stereotypical parts. For instance, Mindy Kaling's character Kelly Kapoor in *The Office* (NBC, 2005, USA) and Aziz Ansari's character Tom Haverford in *Parks and Recreation* (NBC, 2009, USA) are examples of prominent parts for South Asian actors on major network television in the U.S. These appearances further led to the two actors launching their own shows, *The Mindy Project* (Fox, 2012, USA) and the *Master of None* (Netflix, 2015, USA).

However, when the ABC approached Priyanka Chopra with a talent holding deal that required the channel to develop a starring project for her or cast her in an existing project, a new trend began. Chopra's leading role, amid a racially diverse cast in *Quantico*, as an FBI agent who gets framed for a terrorist conspiracy, is the first of its kind for a South Asian performer from India on American television. Chopra's debut on the American television screen needs to be understood against the long history of stereotypical roles of South Asian actors in Hollywood film and television. For instance, up until the rise to prominence of South Asian American actors such as Mindy Kaling, Kal Penn and Aziz Ansari and British Indian actors like Riz Ahmed and Archie Punjabi, stereotypical characters such as Apu on *The Simpsons* generally defined the representation of South Asians on American TV (Mahdawi 2017). The proliferation of television content distributed via online platforms has brought about significant changes in casting practices that increasingly cast non-white actors owing to the success of shows like *Empire*, which focuses on a black family (2015, Fox, USA) and *Fresh Off the Boat* about an Asian American family (2015, ABC, USA) as well as the rising albeit not particularly significant number of non-white writers, actors and directors working in TV (D'Addario 2016).

The shift from Apu of *The Simpsons* (Fox, 1989, USA) to diverse roles for many South Asian actors of American origin has been an important, albeit gradual, one. The character Apu offers a stereotypical representation of an Indian immigrant who has a PhD in computer science but speaks with a thick accent and runs the Kwik-E-Mart convenience store in the fictional American city of Springfield. Voiced by the American actor Hank Azaria, the character is among the most iconic "racist" images of Indian immigrants. In a documentary, *The Problem with Apu* (dir. Melamedoff, truTV, 2017), American comedian of Indian origin, Hari Kondabolu, calls out the racist portrayal of Indian characters, played by white actors such as Hank Azaria. In his appearance on *The Late Show with Stephen Colbert* (CBS, 2015, USA), Hank Azaria said that it wasn't his intention to offend South Asians with his voice of Apu; he expressed sadness at South Asian kids' experiences of being bullied and teased in school because of the popularity of the character Apu. The changing norms around South Asian representations on American TV may also be attributed to the South Asian community's success and prominence in American society, particularly in science, technology, business and politics, as well as a higher median household income (Bhattacharjee 2018). The rise to prominence of South Asian American and in some cases, British actors has dramatically altered the terms on which their representations have gone from being the object of humor to actors, writers, directors and stand-up comedians taking control of the stories that are told about Indian culture, society and the diasporic experience. At the same time, it would be remiss to regard South Asians as one homogeneous group. India often serves as the synecdoche for South Asia as Indian stories dominate the South Asian representational canvas. Actors of Pakistani origin, such as Kumail Nanjiani who starred in a leading role on the sitcom *Silicon Valley* (HBO, 2014, USA) and British actor (half Indian, half Pakistani), Jameela Jamil in a major role in the sitcom *The Good Place* (NBC, 2016, USA), are equally a part of the South Asian wave on American TV. The on-screen arrival of these actors speaks, in part, to the general upward immigrant mobility in diasporic contexts. Within the social media landscape of cultural politics, it also speaks to the ability of actors to control the narrative and representation, and to collectively author a counter-narrative that questions established institutional structures of power, just as Hari Kondabolu did with his documentary, *The Problem with Apu*. The documentary narrates the experience of South Asian diasporic actors, every one of whom had an Apu story of being bullied, being laughed at and made fun of in random public places.

Kal Penn's decision to make public, in a series of tweets, old Hollywood scripts given to him during his early years in the industry is another noteworthy moment in this counter-narrative. The scripts contained the parts of Indian characters that required a more "authentic" accent or conform to other stereotypes about nerdy intelligence (Neelakantani 2017). In one particular script, when asked to audition for the role of a "quirky Indian lab buddy", the actor tweeted, "Jeez I remember this one! They were awful. 'Can you make his accent a little more AUTHENTIC?' That usually meant they wanted Apu" (Hoffman 2017). Shilpa Dave, in her analysis of the Apu character in *The Simpsons*, has argued that such portrayals represent a racialized performance of South Asians in the United States that reinforces rather than challenges the stereotypes associated with them (Dave 2013). This argument about what Dave calls "brown voice", certainly rings true in the light of Kal Penn's tweet exposing Hollywood scripts that required stereotypical performances of ethnicity through a stereotypical Indian English accent.

But here, the recent screen arrival of American actors of South Asian origin in primetime, non-stereotypical, "realistic" parts notwithstanding, it is clear that actors of South Asian origin who grew up in the USA and the UK face a different set of challenges than those who are established actors in India. This is because even as the film and TV parts historically available to American actors of South Asian origin have been limited in range, so too has the narrative through which they are beginning to claim a more realistic representational screen space in meaningful roles. Inevitably, they are confined by a narrative of progress through which diasporic actors find their contributions to the American mediascape framed, always achieving only in terms of the moment of arrival after the shattering of stereotypes. There is a subtle difference, therefore, between the experiences of diasporic actors who grew up in the USA or the UK and Bollywood actors who appear on Hollywood screens. American and British actors succeeded after battling cultural stereotypes of immigrants in TV characters like Apu, a character that typifies an accent, profession and personality and reduces an entire culture to a humorous object. On the contrary, Bollywood stars arrive on the Hollywood screen with an already extensive body of work constituting their transnational legibility as viable performers, now saleable on American TV and film.

Priyanka Chopra, who spent three years as a teenager in the United States, became a Hindi film actress after winning the Miss World title in 2000. Chopra's professional trajectory spans a well-established career in Hindi films followed by a brief 2012 stint in music in the United States as well as international visibility through global brand endorsements before her lead role in *Quantico* earned her central screen space on American TV. In contrast to the teleological struggles of Indian American actors whose success is defined in terms of "breaking the racial stereotype" through meaningful and central characters (e.g. *The Mindy Project*, *Master of None*, *The Night of*…), Chopra's television debut in America instantiates the strategy of professional diversification within the framework of transnational screen navigation. Rather than struggling to make it to the American film or television screen, as most Indian American actors do, Chopra's case demonstrates a form of screen access afforded to well-established stars whose ability to choose from scripted parts defines the logic of televisual mobility. I propose that this televisual mobility also marks a departure from, as well as extends, the Hollywood forays of Indian actors and stars from Bollywood. My usage of the terms actor and star is deliberate in trying to elucidate the distinctions that I frame as defining particular celebrity trajectories of Bollywood figures in Hollywood. The appearances of veteran Hindi film/Bollywood actors such as Amrish Puri, Om Puri, Anupam Kher and Naseeruddin Shah in Hollywood films precedes the circulation of the Bollywood star figure in the American mediascape including social media. Actresses such as Aishwarya Rai who appeared on the *Oprah Winfrey Show* multiple times, Priyanka Chopra and Deepika Padukone circulate as star figures in multiple promotional media economies such as the television talk show, film festivals, award shows and social media. This specificity of the distinction between an actor and a star in relation to Bollywood figures in Hollywood further clarifies the logic of televisual mobility.

American television has begun to illuminate the Bollywood star in a different light. It is television's specificity as an audio-visual medium in the twenty-first century that needs to be examined for the new omnivisible mobility afforded to transnational figures. TV is no longer confined to the TV set. It has diversified across screens and content delivery platforms in the digital age, which has further led to a proliferation of content and multiplication of TV genres.

Consequently, the space to showcase a diversity of talent from different cultural backgrounds has expanded. The changing portrayal of Indian Americans on television in leading and non-stereotypical parts has also opened up screen access to Bollywood stars like Chopra. Her televisual mobility must be situated within the context of American television's political economy as well as its rapidly diversifying racial politics of casting. The political economy of television in the twenty-first century – of which online streaming is a major part – has brought about greater inclusion of diverse talent and allowed comedians such as Azir Ansari and Mindy Kaling to have their own shows in which they played the lead. The convergence of proliferating TV platforms and democratization of talent also parallels the progressive changes in American culture, in part impelled by Obama's presidency from 2008 to 2016 according to a Pew Research Center report (Dimock 2017). The success of American actors of South Asian origin fits snugly within this set of cultural, political and economic developments. It has allowed diasporic actors to take charge of the narrative and the stories that are told about them. But again, these stories may be broadly understood as humorous snippets of South Asian American culture as a critique of white America that historically relied on immigrant stereotypes as a gesture of inclusion of the Other within the media representational canvas. These stories default to comedy as the privileged genre of reversing damaging representations. If, historically, American actors of South Asian origins had limited roles available to them given the choice of marginal, stereotypical immigrant parts, then in the contemporary moment, the success of these performers on mainstream screens owes predominantly to comedic portrayals. In other words, their success appears to still remain limited to formulaic portrayals that reverse the stereotypes. Of course, there may be exceptions to this trend but the majority of primetime performers such as Aziz Ansari, Mindy Kailing, Kumail Nanjiani, and Hasan Minaj have gained popularity through their humorous snippets of immigrant lives.

The rise of Priyanka Chopra as a global star who is often positioned prominently in the narrative of a South Asian takeover of the American screen, can neither be subsumed within this story of arrival nor reduced to the trend of a stereotype-fighting performer. The logic of televisual mobility enables her to diversify her stardom, partake of the teleological emergence of South Asians and yet move on to other ventures, a quintessential attribute of her star mobility. Whether the diasporic actors can chart a similar course is debatable.

In an interview, Chopra revealed that she was asked by the vice president of casting at an American broadcast television network about her interest in doing television. She clearly stated to the television network to not come to her with a show that showcases a big, fat Punjabi wedding. She was given 26 scripts of which she selected the part of a half-Indian FBI agent (Kaushik 2018). The ability to dictate terms and choose from a selection of scripts would not likely be part of the experience for diasporic actors who must first tell their stories of the immigrant experience through humorous portrayals and stand-up comedy.

Chopra also played a villain in the movie *Baywatch* (2017) and has appeared on numerous American talks shows including *The Ellen DeGeneres Show*, *The Late Show with Stephen Colbert* and *The Tonight Show Starring Jimmy Fallon* aside from interviews on all major news channels. Where Indian American actors like Kal Penn were offered parts that required an "authentic" Indian accent, or what Indians stereotypically sound like to American ears, Chopra's "global" English accent, in addition to her status as a sex symbol and beauty queen, is an integral part of her televisual mobility. Dave's examination of cultural citizenship of American immigrants – the idea of "brown voice" that points to a successful performance of model minority – demonstrates that Indian Americans can't be movers and shakers or anything more than as assimilated immigrants. In contrast, Chopra complicates that conception. Her global English accent has been a subject of curiosity in some of her interviews. In an interview on Fox, Priyanka was asked about her accent in the Disney movie *Planes* (2013) in which she lent her voice to a character who is an Asian racing champion. Alisyn Camerota complimented Chopra on her "sultry" beautiful accent, asking where her accent came from and whether she grew up speaking English, to which Chopra responded that her accent was Indian (2013). In another story in the online publication, *Brown Girl* magazine, Chopra talked about her global accent for the character Victoria Leeds in *Baywatch* (2017), and also recalled talking to American comedian of Indian origin, Aziz Ansari, about the increasing size of the South Asian presence at the MET Gala. The story describes Chopra as a "true trailblazer of South Asian Representation in media" (Arora 2017).

It is likely that Chopra's global success is unique and the audiences may not see another star rising on the transnational media horizon. However, the logic of televisual mobility that undergirds her stardom must be seen as flexible and evolving, a mode of professional diversification that holds the potential to accommodate and remold celebrity figures' screen navigations. Televisual mobility may even be understood as a form of transnational screen affordances to established Bollywood stars – a set of professional opportunities that give them access to Hollywood screens. But the potential success of their ongoing and future professional ventures may continue to validate the evolving logic of televisual mobility.

Chopra's multiple stints of a leading role in a primetime drama series, music artist, appearances at the MET Gala, film festivals and the Oscars illustrate her screen mobility, which, rather than being formulaic, may be as strategic as serendipitous. By this reasoning, we can also regard her attendance at the televised, high-profile royal wedding of Prince Harry and Meghan Markle – Chopra's close friend – as yet another instance of her screen navigation. The televisual mobility instantiated by Chopra's transnational visibility may manifest as another set of opportunities through which emerging Bollywood stars may appear on Hollywood film and television screens to expand the logic in their own unique manner. Chopra's global stardom is unprecedented in terms of scale and popularity but it isn't necessarily an exception given that prominent Bollywood actors such as Aamir Khan, Aishwarya Rai, Deepika Padukone and Amitabh Bachchan have all made their way to Hollywood in some measure.

Chopra's 2016 appearance at the Oscars, where she presented an award, was yet another illustration of her televisual mobility, a representational synecdoche where Chopra stood in for the rest of Asia. When the interviewer (a woman of East Asian origin) on the Oscar red carpet asked Chopra if she had any advice for women or Asians in general who would like to achieve similar success, Chopra's response literally seized the representational moment with: "I just think *represent* … own who you are and own your roots and we are amazing! We come from Asia!" The interviewer concluded by saying, "we are so proud of you … as Asian women, as Asians in general, you are really representing us!" ("Priyanka Chopra at the Oscars …" 2016). Having won the People's Choice Award for best actress in a drama series consecutively in 2016 and 2017, Chopra has also presented at the Emmy's and Golden Globe Awards.

A combination of leading parts in American TV and notable, non-stereotypical roles in Hollywood has also brought her India's eminent civilian honor, *Padma Shri*, in 2016. But it was her Oscar appearance that made her the pride of India and its community according to a write-up about Chopra's accomplishments on the Indian diaspora website. The website describes Priyanka's 2016 Oscar appearance as an honor that many international stars have only dreamt about (Datt 2016). Indian mainstream newspaper outlets such as the *Indian Express* and *Time of India* and other news media have also come to the same foregone conclusion that Priyanka Chopra is making India proud (Banerjee 2016; Sahadevan 2016; Iyer 2017).

## Conclusion

Television in the early twenty-first century has undergone profound changes in its political economy in terms of content and delivery platforms in the digital age. Such developments have also diversified television's politics of racial representation given the success of many young American actors of South Asian origin on American television in shows such as *The Mindy Project*, *The Night of …* and

*Master of None*. Consequently, the representational canvas for South Asian actors has expanded, making television a mode of entertainment in the same league as Hollywood. This expansive screen space has not only enabled American actors of South Asian origins to break out of accepting stereotypical parts and marginal characters in film and TV, but also opened up transnational screen access to well-established Bollywood stars to appear in leading roles in American TV shows.

Priyanka Chopra's media stint in Hollywood, first as a singer, and then as the protagonist of the American TV thriller *Quantico*, as well as her portrayal of a villain in the movie *Baywatch*, has brought her critical international exposure as a global celebrity and an actor not bound by cultural borders and unfamiliar work cultures. With her celebrated status as a beauty queen and sex symbol as well as her "global" Indian English accent, Chopra's malleable stardom demonstrates the logic of televisual mobility, which moves beyond the narrative of stereotypical roles for non-white actors. Televisual mobility allows Chopra to take control of her star image, enabling her to diversify her talent across American film and TV. Her numerous American TV talk show appearances and her presentations at the Emmy's, Golden Globes and the Oscars further testifies to the logic of televisual mobility that, while reinscribing in part the narrative of racial progress, pride and representational recognition in the media, transcends the teleological conclusion of such a narrative.

Chopra's international stardom, even when it represents Asia, is a key function of her forays into music, American television and Hollywood films. She made these choices, not simply to break the stereotypes to move from representational margins to the screen's center. Despite being a newcomer, she seems to have chosen the terms of her own portrayal. For instance, while ABC offered her more than a dozen scripts and roles to choose from, which ultimately led to her appearing as the lead in *Quantico*, her role of Victoria Leeds as the villain in *Baywatch* was initially written for a male character. Chopra brought her own creative ideas to the table about what she wanted the character to be. This ability to choose is a reflection of her viability as an international performer and an illustration of how the televisual mobility comprising a star's transnational screen navigations may be both strategic and serendipitous. The scale of Chopra's stardom and success partakes of, in some measure, the American immigrant narrative represented by diasporic actors who emerged on to the center of mainstream American screen by battling cultural immigrant stereotypes, most notable in the character of Apu on the show *The Simpsons*. But Chopra's presence on American television screens instantiates much more than the arrival of immigrant performers on the screen. It signals the constantly shifting visibility of a star across media industries.

## Bibliography

Ajay, Namrata. "7 Times Sharp-Witted Priyanka Chopra Left International Chat Show Hosts Tongue-tied." *News18*, June 2017.

Arora, Priya. "Priyanka Chopra on 'Baywatch' and the Importance of South Asian Representation." *Brown Girl Magazine*, May 2017.

Banerjee, Madhuri. "Why Every Indian Should be Proud of Priyanka Chopra." *Daily O*, May 2016.

Bhattacharjee, Yudhijit. "How South Asian Americans are Building a New American Dream." *National Geographic*. September 2018.

Butcher, Melissa. *Transnational Television, Cultural Identity and Change: When STAR Came to India*. New Delhi: Sage, 2003.

Chadha, Kalyani and Anandam Kavoori. "Mapping India's Television Landscape: Consitutive Dimensions and Emerging Issues." *South Asian History and Culture*, 3 No. 4 (2012): 591-602.

Chalaby, Jean. ed., *Transnational Television Worldwide: Towards a New Media Order*. London: I.B. Tauris, 2005.

D'Addario, Daniel. "There Is No Such Thing as 'Too Much Diversity' on TV." *TIME*, September 2016.

Datt, Kavita Bajeli. "Priyanka's Oscar Moment: Pride of India and Community." *The Indian Diaspora*, February 2016.

Dave, Shilpa. *Indian Accents: Brown Voice and Racial Performance in American Television and Film*. Urbana, Chicago and Springfield: University of Illinois Press, 2013.

Dimock, Michael. "How America Changed During Barack Obama's Presidency." *Pew Research Center*, January, 2017.

Govil, Nitin. "Bollywood and the Frictions of Global Mobility." In *Media on the Move: Global Flow and Contra-flow*, eds. Daya Kishan Thussu. London, New York: Routledge, 2007, 84–98.

Govil, Nitin. *Orienting Hollywood: A Century of Film Culture Between Los Angeles and Bombay*. New York: New York University Press, 2015.

Hoffman, Ashley. "Kal Penn Criticizes Racist Stereotypes in Audition Scripts From His First Years Acting." *TIME*. March, 2017.

Iyer, Meena. "Bachchans, Kapoors, Khans Continue Dynasty Rule in Bollywood." *Times of India*, November 2014.

Iyer, Meena. "Priyanka Chopra: India's First Truly Global Superstar." *Times of India*, January, 2017.

*India Today*. "Priyanka Hits Back at Naysayers: I Was told Indians aren't Cut Out for Hollywood." *India Today*, October, 2018.

Jha, Subhash K. "Bollywood Insider: Why B-town Stars Like Ajay, Aamir say no to Award Shows." *Firstpost*, August, 2014.

Kaushik, Divya. "Priyanka Chopra: My Dad Isn't Sitting in Hollywood, Making a Film For me. I Have to Choose from What I Get." *Times of India*, August, 2018.

Kavoori, Anandam P. and Aswin Punathambekar. eds., *Global Bollywood*. New York and London: New York University Press, 2008.

Kumar, Shanti. *Gandhi Meets Primetime: Globalization and Nationalism in Indian Television*. Urbana and Chicago: University of Illinois Press, 2006.

Mahdawi, Arwa. "From Apu to Master of None: How US Pop Culture Tuned into the South Asian Experience." *The Guardian*, May 2017.

Mankekar, Purnima. *Screening Culture, Viewing Politics: An Ethnography of Television, Womanhood and Nation in Postcolonial India*. Durham and London, Duke University Press, 1999.

Marshall, P. David. *Celebrity and Power: Fame in Contemporary Culture*. Minneapolis and London: University of Minnesota Press, 1997.

Marshall, P. David and Sean Redmond. *A Companion to Celebrity*. Malden and Oxford: Wiley Blackwell, 2016.

Mazumdar, Ranjani. "Film Stardom after Liveness." *Continuum: Journal of Media & Cultural Studies*, 26, no. 6 (2012): 833–833.

McMillin, Divya C. "Localizing the Global: Television and Hybrid Programming in India." *International Journal of Cultural Studies*, 4, no. 1 (2001): 45–68.

Meeuf, Russell and Raphael Raphael. eds., *Transnational Stardom: International Celebrity in Film and Popular Culture*. New York: Palgrave Macmillan, 2013, 5.

Mehta, Nalin."Ravana's Airforce: A Report on the State of Indian Television." In *Television at Large in South Asia*, eds. Aswin Punathambekar and Shanti Kumar. Abingdon and New York: Routledge, 2014.132-143.

Mittell, Jason. "Audiences Talking Genre: Television Talk Shows and Cultural Hierarchies." *Journal of Popular Film and Television*, 31, no. 1 (2003): 36–46.

Munshi, Shoma. *Prime Time Soap Operas on Indian Television*. Oxford and New Delhi: Routledge, 2010.

Mehta, Nalin. ed., *Television in India: Satellites, Politics and Cultural Change*. Abingdon and New York: Routledge, 2008.

Neelakantani, Shailaja. "Actor Kal Penn blasts 'racist, stereotypical' Hollywood in Series of Tweets." *Times of India*, March 2017.

Parthasarathy, Anand. "Bollywood Goes Global." *The Hindu*. March 2006.

Power, Carla. "Bollywood Goes Global." *Newsweek*. February 2000.

"Priyanka Chopra and Jimmy Fallon Celebrate Holi with a Messy Paint Fight." *The Tonight Show Starring Jimmy Fallon* on *Youtube*. www.youtube.com/watch?v=NYvO9m0zoMQ. March 14, 2017.

"Priyanka Chopra at the Oscars 2016 Talks about Jamshedpur, Jharkhand." YouTube. www.youtube.com/watch?v=pqLsO9WJWJc. March 1, 2016.

Rajadhyaksha, Ashish. "The Bollywoodization of the Indian Cinema: Cultural Nationalism in a Global Arena." *Inter-Asia Cultural Studies*, 4, no. 1 (2003): 25–39.

Rajagopal, Arvind. "The Rise of National Programming: The Case of Indian Television." *Media, Culture and Society*, 15 (1993): 91–111.

Ramanathan, Lavanya. "Television is Getting More Diverse. Here's Why it's Good for the Business." *The Washington Post*, November 2015.

Ramasubramanian, Uma. "Why Actors Move from TV to Films." *Deccan Chronicle*, February 2017.

Ramesh, Randeep. "Bollywood Goes Global with Hollywood Movie Deal." *The Guardian*, May 2008.

Rao, Ankita. "South Asian Actors are Fighting Hollywood' Racism." *VICE* April 2018.

Sahadevan, Sonup. "Very Proud of Priyanka Chopra, She is Making India Proud, Says Shahid Kapoor." *The Indian Express*, February 2016.

Shattuc, Jane M. *The Talking Cure: TV Talk Shows and Women*. London and New York: Routledge, 1997.

Thussu, Daya Kishan. "Privatizing the Airwaves: The Impact of Globalization on Broadcasting in India." *Media, Cultures and Society*, 21 (1999): 125–131.

Thussu, Daya Kishan. "The `Murdochization' of News? The Case of Star TV in India." *Media, Culture and Society*, 29, no. 4 (2007): 593–611.

Timberg, Bernard M. *Television Talk: A History of the TV Talk Show*. Austin, The University of Texas Press, 2002.

"Watch Alisyn Camerota be Surprised as she finds out Miss World 2000 Is from India and Speaks English." *YouTube*. www.youtube.com/watch?v=oYCBNMEAaN4 .August 10, 2013.

Williams, Raymond. *Marxism and Literature*. Oxford: Oxford University Press, 1977.

# 23

# MEDIA SPECTACLE AND DONALD TRUMP'S AMERICAN HORROR SHOW

*Douglas Kellner*

The political rise of Donald Trump will challenge historians and critical theorists of US politics for years to come, but will also occupy the work of media studies scholars. That is because the Trump phenomenon is in no small part the result of his particular mastery of media spectacle, a concept that I've been developing and applying to US politics and media since the mid-1990s. In this chapter, I will discuss Trump's developing use of media spectacle in his business career, in his effort to become a celebrity and reality-TV superstar, and ultimately in his political campaign and presidency.

I have argued elsewhere that the Trump phenomenon may be approached as a teachable moment that helps us understand the changes and contour of US politics in the contemporary global era and the role of broadcast media, new media and social networking, and the politics of the spectacle (Kellner 2016: 1). Trump's rise to political power is of a moment with the growing global threat of authoritarian populism, a result in part of unparalleled global economic inequality[1] leading to widespread dissatisfaction with current political leadership and thus leaving political processes open to various forms of manipulation. In the case of Donald Trump, this means a form of political leadership premised on a reputation for wealth and business prowess that seemingly paradoxically masquerades as a "voice of the forgotten man." In practice his policies, such as they are, have economically benefitted primarily the already extremely wealthy on the one hand, while cementing a base of voter support through a symbolic politics meant to stir misogynist, xenophobic, and racist discord much more than address actual economic concerns. As *The Atlantic* magazine observed after the first year of Trump's presidency,

> we have never had a president so ill-informed about the nature of his office, so openly mendacious, so self-destructive, or so brazen in his abusive attacks on the courts, the press, Congress (including members of his own party), and even senior officials within his own administration.
>
> *(Goldsmith 2017)*

These characteristics of Trump's rise to global celebrity and now political power are in fact bound up with his use of, and immersion in, twenty-first century media spectacle, so I begin my study with analysis of Trump and the politics of the spectacle.

## The Politics of the Spectacle

I first came up with the concept of media spectacle to describe a key phenomenon of US media and politics in the mid-1990s. This was the era of the rise of cable news networks such as Fox, CNN, and MSNBC and the 24/7 news cycle that has dominated US politics and media since, with detailed, ongoing televisual coverage and commentary on events such as the televisually orchestrated first Persian Gulf War,[2] the O.J. Simpson murder case and trial—from slow motion car chase through shocking (for some) verdict (and resulting eventually and unexpectedly in the media phenomenon we now know as "the Kardashians")—, and President Bill Clinton's sex scandals and impeachment, to name a few.[3] The 1990s was also the period when the internet and new media began to take off, facilitating direct distribution for anyone who desired to be a political commentator, player, or participant in the spectacle.

By "media spectacles" I am referring to media constructs that present events which disrupt ordinary and habitual flows of information, and which become popular stories that capture the attention of the media and the public, and circulate through broadcasting networks, the internet, social networking, smart phones, and other new media and communication technologies. In a global networked society, media spectacles proliferate instantaneously, become virtual and viral, and in some cases become tools of socio-political transformation, while in other cases simply become mere moments of media hype and tabloidized sensationalism. Media spectacles insistently gain possession of two of media's most precious commodities: time and attention. Media spectacle in this sense may be thought of as the further commodification and instrumentalizing of what Mary Ann Doane has described as television's "crisis of temporality which signifies *urgency* and which is attached to the information itself as its single most compelling attribute" (Doane 2015: 253).

In a twenty-first-century media environment characterized by excessive access, thousands of channels of distribution, always-on coverage of the globe, and immense competition for audiences and revenue, such interruptions in the form of media spectacle are to be highly desired precisely for their ability to alarm and attract viewers by being outside the rhythms of programming and reception, snatching their attention from the otherwise potentially banal. Media spectacles stand out dramatically from otherwise banal events, in order to disrupt normal expectations and attract increasingly elusive audience attention. The spectacle of a white Ford Bronco driving slowly down California freeways, police equally as slowly pursuing, interrupted television programming around the country, elevating the seeming significance of O.J. Simpson's actions in this event from the individual level to the mass level and ultimately extending a few minutes across hours of television time and into history books. What better way for would-be celebrities or campaigning politicians to gain air time and media exposure than through the construction of events and activities designed to fit the model of media spectacle?—fame through orchestrated, interrupting, attention seeking "spectacle." The media spectacle is harnessed regularly now by media programmers and those seeking media attention, alike.

The proliferation and scope of media spectacle has increased in the past decades with the growth of accessible new media and social networking like Facebook, YouTube, Twitter, Instagram, Snapchat, Whatsapp, and the like that increase the scope of and participation in the spectacle. "Spectacle" has also proliferated as global corporate media companies have faced increased competition for audiences, advertisers, and other forms of revenue generation, even as several have emerged to become global superpowers. In the hypercompetitive global, regional, national, and local media markets, sensationalistic spectacle becomes a tool for attracting audiences and advertising—a relatively inexpensive addition to programming schedules that stands apart from the clutter of so much media content by its very nature as an interruption in regular media flows. To be successful, celebrities, politicians, and news and information programmers are virtually obliged to create spectacles that attract mass audiences and hold attention—at least fleetingly.

As I have argued since 2008, a crucial key to Barack Obama's success in two presidential elections was his talent as a master of media spectacle, blending politics and performance in carefully orchestrated events designed as much for how they would play on screens and appear in various media as attractions and spectacle (Kellner 2005, 2012). Prior to Obama's success, Ronald Reagan had been the model of the mastery of presidential media spectacle, every day performing his presidency in a well-scripted and orchestrated daily spectacle. Reagan was trained as an actor and every night Ron and Nancy reportedly practiced his lines for the next day's performance like they had done in their Hollywood days. Reagan breezed through the day and well-orchestrated media events, scripted with a teleprompter, smiling frequently, and pausing to punctuate the soundbite—the line of the day.

Donald Trump has now emerged as a major instigator of media spectacle, regularly making outrageous claims, attacking allies and institutions, utilizing the various affordances and convergences of social media, television, and journalism in apparently orchestrated grabs for headlines that will saturate media coverage, marginalizing other news and other voices. In fact he has long been a celebrity and honed his use of media spectacle since the 1980s as he promoted his early real estate endeavors in New York, right through multiple movie and TV cameos (usually as himself), a reality-TV competition program, and finally his presidential campaign, election, and administration. Hence, Trump's ability and success in running for the presidency is attributable in part to his lengthy experience gaining mastery of media spectacle as media spectacle has become a major force in US politics, helping to determine elections and more broadly the ethos of our culture and political sphere. His political success, building on and significantly expanding the models of previous presidents, would seem to suggest that a dominance of the media—by any means of attraction, spectacle, or crisis—remains among the surest ingredients for obtaining the White House.

Throughout his presidential campaign, Trump was the master of bringing media spectacle into his followers' lives through his use of social media in a manner more pronounced, dominant, apparently uncensored, and certainly more effective than previous US politicians. Trump's frequent tweets were part of the way that he helped shape news cycles of both broadcast and social media throughout his campaign and into his presidency. Through these he claimed to bring his message directly to his Twitter (and political) followers, circumventing the editorial practices and contextualization of more traditional forms of media-ated political communication. He was not the first to use social media along these lines, Obama allegedly had over one million Facebook "friends," yet Trump's seemingly unguarded Twitter feed gave the appearance of emanating directly from his mind (or his id), unguarded and unrehearsed (and often misspelled), allowing his followers to feel like insiders and participants with seemingly direct access to his every thought and mood. Broadcast media, in turn, were left to cover these Twitter pronouncements, often rendering Trump's frequently controversial tweets major story-lines of the day and therefore helping him dominate news cycle after cycle.

Trump rose to prominence in New York during the 1980s, an embodiment of wild, entrepreneurial cowboy capitalism in an era of deregulation, the celebration of wealth, and the "greed is good" ethos of Wall Street, enabled by the Reagan administration. Trump's biographies characterize him as driven by a need to compete and win,[4] and entering the highly competitive real estate business in New York in the 1980s, his language and behavior developed from within this highly competitive and ruthless New York business culture. Trump soon saw the utility in media and publicity to promote his celebrity and image and gained an appreciation of the importance of media and celebrity, even (or perhaps especially) when rendered in adversarial terms, to succeed in a media-centric hypercapitalism. Trump has been particularly persistent in branding (and indeed commodifying) the Trump name (in 2015 considering it his company's most valuable asset) and selling himself as a businessman, a celebrity, and now as a national politician (see Klein 2017; Nasirpour 2019).

In a time of tabloid culture and media-driven celebrity, Trump at one point even adopted a pseudonym, "John Barron," which he used to give the media gossip items that touted Trump's (i.e. his own) successes in businesses, as a Lothario, and as a rising man about town. He gossiped about himself to a New York City media environment eager to increase readership by revealing semi-scandalous details about a prominent local celebrity, which in turn helped promote Trump's name, brand, and presumably business practices (or, at least, ego). This symbiotic—both mutually beneficial, yet seemingly adversarial—relationship to sensationalistic media has continued ever since. It is perhaps nowhere better encapsulated than in the origin of his odd moniker, "The Donald." According to the *Washington Post*, "The Donald" was so named by *Spy* magazine, the snide and irreverent 1980s New York insider publication most responsible for building Trump's fame by ruthlessly mocking him. A May 1989 article in *Spy* written to deflate claims Trump had made about his then-wife, Ivana Trump, revealed that—with English not her first language—she referred to her husband always with that definite article, the Donald. Although mentioned almost as an aside in the *Spy* article, "The Donald" soon replaced "short-fingered vulgarian" as the magazine's most lasting contribution to the growing fame it was helping him obtain as other media sources began to adopt it as well. Thus, Trump gained fame, a reliably distinctive if odd moniker, and increasing media presence from a publication intent on growing its readership by mocking him. This mutually symbiotic, seemingly adversarial relationship with so-called mainstream media has continued in much the same way ever since (Argetsinger 2015; Van Meter 1989).

What propelled Trump's fame beyond New York novelty toward national and then global recognition was his embrace of three less "mainstream" forms of media and publicity. First, he appeared on a number of reality-TV programs, eventually finding some success with NBC's *The Apprentice* (2004–2017). Second, he began to generate a grass roots political following by frequently calling in, unbidden, to conservative and far-right talk radio programs and then the conservative Fox News chat shows—he was a popular phone guest on alt-right radio and TV. Third, he started an account on the social media microblogger, Twitter.

Of these, *The Apprentice* (later *The Celebrity Apprentice*) most certainly popularized him into a supercelebrity and made "The Donald" a major public figure for a national and then global audience. While previous, nationally recognized politicians have first gained success as entertainment celebrities, Trump is the first reality-TV candidate who ran his campaign like a reality-TV series. While he had been profiled on early reality programs like *Lifestyles of the Rich and Famous* and had made numerous cameos in television programs and feature films, usually playing himself as confident, successful and wealthy (recall his assisting young Macaulay Culkin's character in *Home Alone 2: Lost in New York* (1992; set inside the Plaza Hotel, at the time owned by a Trump company), or telling young Waldo in *The Little Rascals* (1994) reboot, "you're the best son money can buy"), alongside a lengthy string of sitcom cameos and fast food commercials, Trump was not a Hollywood celebrity. Rather than a movie actor like Ronald Reagan, or California governor Arnold Schwarzenegger, or a professional wrestler like Minnesota Governor Jesse Ventura, Trump's celebrity derived from the logics of New York tabloid culture and was refined on reality-TV competition programming, wherein the most outrageous behavior and most outlandish statements are rewarded with attention, increased viewership, and program renewal. Thus, Trump's use of language, sophomoric name calling, apparently off-the-cuff and (initially) surprisingly nonsensical asides produced chaos and disruption he could use as part of his campaign appearances to more surely gain and extend vast media coverage. During the most chaotic episodes in his campaign, he would boast that his rallies and media appearances were the most entertaining (and suggest that they produced the highest ratings for channels covering them). Hence, Trump is the first celebrity candidate whose use of the media and, specifically, celebrity star power to disrupt media flows and exceed the normalized political coverage and discourse into the realm of disruptive media spectacle has proven to be a most potent weapon in a presidential campaign.[5]

## Reality-TV and Social Media Political Spectacle

Already known to New Yorkers, Trump's national celebrity perhaps began with publication of his book *The Art of the Deal* (Trump, coauthored with Tony Schwartz 2005 [1987]), but it most certainly derived in largest part from his role in the reality-TV series *The Apprentice*.[6] We therefore need to interrogate Trump's reality-TV show to help explain the Trump phenomenon. *The Apprentice*'s opening theme music "For the Love of Money," a 1973 R&B song by The O'Jays, established the capitalist ethos of the competition show in which the winning contestant was to get a job with the Trump organization. Obviously then, money is the key to Trump's business and celebrity reputation.[7] Such a premise, however, also recalls Laurie Ouellette's reminder that the ways in which Trump's wealth has been represented to the public resonate with larger questions of political economics. Reality-TV's participation in neoliberal reinvention, for example, means that "for nearly two decades, reality-TV has resonated with critiques of 'big government' and has enacted private alternatives to state oversight and responsibility for public welfare" (Ouellette 2016: 648). Whether demonstrating savvy, wit, and a competitive spirit to win a job with Trump, flip some residential real estate, or improve one's life through weight loss or elective surgery, reality-TV has been demonstrating the ways in which individuals are meant to improve themselves outside of safety nets and government support.

In the original format to *The Apprentice*, several contestants formed teams to carry out a task dictated by Trump, rendering him boss, judge, and sage, adjudicating each "contest" which resulted in a winner and Trump notoriously barking "you're fired" to a member of the losing "team." The structural logic of this program pitted contestants against each other in a cut-throat, capitalist competition to ultimately win Trump's approval. This program produced Trump as an approval-withholding father figure, a boss with whom the buck stops, and an icon of excessive wealth—although a growing body of evidence is emboldening Trump skeptics, who do not believe he is as wealthy and successful in business as he claims, to contest this image of Trump that he successfully sold to, first, his TV fans, and then his political base, and implicitly the press covering him and the public viewing that coverage (see Buettner and Craig 2019).[8] Curiously, some commentators believe in the 2012 presidential election that Barack Obama beat Mitt Romney handily because he early on characterized Romney as a billionaire who liked to fire people. This initially appears ironic since firing people is Trump's prime managerial move in his business, on TV, and now in his political career, which saw him fire two campaign managers and more advisers and made dramatic firings of key officials in his administration a defining feature of his first years as president. Yet reality-TV can again help us understand this apparent shift, as in the intervening few years the reality-TV premise that anyone can "'maximize' themselves out of precarious labor conditions and a shrinking welfare state" has become increasingly difficult to sustain and instead "reality TV shows about pawn shop sharks and right-wing patriarchs have emerged, highlighting harsher entrepreneurial subjectivities and new forms of 'bare enterprise' that openly exploit the poor, people of color, and the dispossessed" (Ouellette 2016: 649). These tendencies parallel Trump's political persona (political outsider with wealth and business savvy barking "you're fired") as it "intersects with xenophobia, racism, and sexism" on the campaign trail, in presidential rhetoric, and in official policy (Ouellette 2016: 649).

*The Apprentice*'s producer, Mark Burnett, broke into national consciousness with his reality-TV show *Survivor*, a scantily-clad, neo-Darwinian epic of alliances, pandering, backstabbing, and nastiness, which provides an allegory of how one succeeds in the dog-eat-dog business world in which Donald Trump has thrived, and spectacularly failed (see Buettner and Craig 2019). Both Burnett and Trump share the ethos of nineteenth-century ultracompetitive capitalism, with some of Donald Trump's famous witticisms proclaiming: "I think everyone's a threat to me" and "When somebody challenges you unfairly, fight back—be brutal, be tough—don't take it. It is always important to WIN!" as well as "Everyone that's hit me so far has gone down. They've gone down big league" (Trump in Pogash 2016: 30, 152, 153).

As *The Apprentice* faltered in the ratings and was revamped as a showcase for C-list celebrities to attempt career resurrections as *The Celebrity Apprentice*, Trump emerged as a legitimate national celebrity. Apparently incensed by the election of Barrack Obama, Trump began calling talk radio shows and later Fox News chat shows questioning the validity of the president's birth certificate and whether the president was actually born outside the US and thus constitutionally ineligible to hold the position. This bizarre, so-called "birtherism" functioned as thinly veiled racism leveled at the first African American president and attracted a rabid, loyal, and not insubstantial base of conservative radio listening voters. Although apparently interested in the idea since at least 1987, Trump had become well known enough in 2015 to announce a run for president with fanfare and media attention and then lurch through a campaign in which he leveraged this celebrity to gain media attention and dominate air time. Building on his experience in New York tabloid culture, reality-TV, and talk radio, Trump produced himself as a media spectacle, suggesting to at least one reporter that his provocative and outrageous campaign was deliberately designed, "If I were totally presidential, I'd be one of the many people who are already out of the race," opting instead for a strategy that, the article noted, "openly stoked racial tensions and appealed to latent misogyny" (Sherman 2016).

In addition to his campaign's ability to dominate broadcast media through outrageous spectacle, Trump is also a heavy user of Twitter and tweets out his messages throughout the day and night. Indeed, Trump may be the first major Twitter candidate, and certainly he is the one using it most personally, aggressively and frequently right into his actual presidency. Twitter, a social media microblogger, limited to messages of 140 characters (until late 2017 when that limit was doubled) was launched in 2006. During the 2008 election it was not yet widely used enough to be considered a significant factor. While Presidential candidates utilized Twitter as one of an array of social media and online forms of communication in the 2012 campaign (see Kreiss 2016), Donald Trump has taken its use to new levels so that this private media company is seemingly inseparable from Trump's apparently spontaneous public political messaging and communication with the public and especially his followers.

Twitter is, in many ways, a perfect vehicle for Trump as it has been defined by three key features: "simplicity, impulsivity, and incivility" (Ott 2017: 60). Its 140-character framework, in other words, seemed designed for Trump's short, incendiary style of emotional attack ("Very Sad!"), bragging ("... to President of the United States (on my first try). I think that would qualify as not smart, but genius ... and a very stable genius at that!"), and simple messaging ("I WILL FIX IT"). His "tweets" also engage receivers who feel they are in the know and involved in Trump's minute by minute thoughts and directives. In this way Trump has intuitively utilized social media in the way that Paolo Gerbaudo describes as offering "a channel for the populist yearning to 'represent the unrepresented', providing a voice to the voiceless and unifying a divided people" (Gerbaudo 2018: 746). This populist use of social media is the twenty-first-century equivalent of Bill Clinton's "I feel your pain" moment, but extended a million-fold, entering into people's individual phones, computers, and devices, even when the unifying message is organized around difference, otherness, fear, anger, and hate and even when the voiceless are given voice through a wealthy media celebrity. As Gerbaudo notes, after all,

> if social media has come to provide a suitable channel for populist appeals, it is first and foremost because of the way in which it has come to be understood as a platform for the voice of people in opposition to the mainstream news media, accused of being in cahoots with the financial and political establishment.
>
> *(Gerbaudo 2018: 749)*

It is this strain of populism that Trump effectively tapped into with his use of Twitter and that has informed his attacks on journalists as well as politicians (even as it ignores the global corporate ownership structure of social media, mobile phone, and internet companies that facilitate the use of social media).

Hence, Trump has mastered the "hybrid" media system (traditional *as well as* new social media). His approach during the campaign was characterized by regularly scheduled appearances in conventional media through standard public relations strategies utilized by most politicians: press conferences, rallies, occasional interviews, and in what became a "Trump specialty, uninvited call-ins to radio and television programs." In addition to these tactics, however (each often orchestrated to produce spectacle for media coverage), Trump would also release "tweetstorms" in order to "encourage supporters to extend his narratives and create *new* stories about what is happening in social media" (Wells et al. 2016: 670). This led at least one early, quantitative study to conclude that

> what we can say with some certainty, and without being overly dramatic, is that Trump's use of conventional information subsidies in the form of press conferences and scheduled interviews and his triggering of social media activity in the form of retweets of his messages, were key factors in explaining his coverage in print news outlets and online blog posts at legacy media. His mastery of conventional and digital media—hybrid campaigning—helped drive his coverage to the nomination.
>
> *(Wells et al. 2016: 675)*

For example, from the start of Trump's presidential campaign kickoff speech on June 16, 2015, when he announced he was running for president, Trump and his wife, Melania, dramatically descended down a golden escalator at Trump Tower as a crowd of spectators applauded him and cameras clicked. It was later reported that at least part of the crowd in attendance consisted of paid actors, hired to "wear t-shirts and carry signs and help cheer him in support of his announcement" (Couch and McDermott 2015), thus using the political campaign practice of "astroturfing" to assure the right setting for the media spectacle that was being orchestrated. The Donald strode up to a podium and henceforth dominated media attention for days with his dramatic pronouncements and the controversy they provoked. The opening speech of his campaign made inflammatory remarks about Mexicans crossing the border, whom he called "rapists, murderers, and drug dealers," that dominated subsequent news cycles for days and excited his base who had never seen mainstream politicians announce their candidacy in such a (faux) glamorous site, or heard such racially charged and xenophobic statements from a candidate for president uttered so flamboyantly, explicitly, and confidently.

> The U.S. has become a dumping ground for everybody else's problems. [Applause] Thank you. It's true, and these are the best and the finest. When Mexico sends its people, they're not sending their best. They're not sending you. They're not sending you. They're sending people that have lots of problems, and they're bringing those problems with us. They're bringing drugs. They're bringing crime. They're rapists. And some, I assume, are good people.[9]

This initial assault on undocumented immigrants began a cycle in which outrageous and offensive missives would dominate daily news cycles during the Republican primaries, leaving little air for the many other candidates also in the race. Hence, Trump's orchestration of media spectacle was irresistible for traditional and legacy media coverage. The sight of an empty podium, awaiting Trump's next appearance, became commonplace on cable news channels as crawlers at the bottom of the screen announced "Waiting for Trump" and newscasters anticipated the next outrage. The speeches that followed were usually broadcast live, interrupting regular programming for their entire duration. This was treatment no other candidate was awarded. In the lead-up to the first Republican primary debate in Fall 2015, Donald Trump got the majority of media time, and his daily campaign appearances and the Republican primary debates became media spectacles dominated by Trump. An early quantitative study of news media coverage of major candidates indicated that from mid-June 2015, after Trump announced he was running, through mid-July 2015, Trump appeared in 46% of the news media coverage of the

Republican field, based on Google News hits; he also received 60% of Google News searches. Another found that across the four major US television networks in 2015, Trump held 31% of airtime, more than all the other presidential candidates combined (Tyndal Report 2015).

The media spectacle and discussion that would inevitably follow were so relentlessly covered —to the near exclusion of others' campaign events—simply because it seemed to bolster viewership for channels that covered Trump, an assertion Trump himself frequently made. Indeed, such coverage has seemingly continued to do so into his presidency. The *Los Angeles Times* proclaimed in headline that "President Trump Turned Cable News into Must-See TV in 2017" while industry trade journals continue to run columns about "How Americans Became Addicted to Cable News in the Trump Era" and articles on how the "Mueller Report Release Draws 11 Million Total Viewers Across TV News" (Battaglio 2017b; D'Addario 2018; Thorne 2019). Trump's orchestration of media spectacle in order to pursue politics as a reality-TV show, and thus to collapse the distinction between entertainment, news, and politics, has proven impossible to ignore and for commercial media, impossible to resist. Trump's media spectacles are good for the business of the media business.

## Fake News

If Trump ran his presidential campaign like a reality-TV show, so too has he run his presidency as media spectacle, playing for approval, attention, and ratings. Yet the Trump presidency has also been one of the most controversial since its very start. This is in part, because, from the beginning, the Trump presidency has been a media spectacle orchestrated like a reality-TV show that has rapidly devolved into an American Horror Show (Kellner 2017). This characterization refers less to the similarly titled, popular television program *American Horror Story*, and more to the tropes of anxious anticipation, shock, fear, spectacle, thrill, and lingering threat of dread and death that an ongoing series of largely negative media spectacle orchestrations have produced. It seems now at times as if there is only interruption, with the next crisis orchestrated as spectacle to distract from the last. In concrete terms, *Vanity Fair* characterized a sample Monday afternoon amid this American Horror Show, or "Next-Level Horror Show" as the magazine characterized it:

> after a weekend spent mentally decomposing that reportedly included telling G.O. P. donors "Democrats hate Jewish people"; insisting a thing he said on-camera never happened; and claiming he could easily become prime minister of Israel if he wanted to; Donald Trump released his 2020 budget proposal.
>
> *(Levin 2019)*

The accumulation of never-ending media spectacle interruptions produces nothing so much as horror for those who cannot avoid paying attention.

Antics such as these, accompanied by unrelenting media coverage on television, in print, and all across online sources, has characterized this media spectacle presidency. Indeed, Trump began his presidency with a lie concerning the numbers of people attending his inauguration, claiming that his inauguration was the biggest ever. When TV pictures, taking the bait, showed that there were many more people at the 2008 Obama inauguration, Trump dispatched his hapless press secretary Sean Spicer to read an unpleasant attack on the media for misrepresenting the number of people who had attended Trump's inauguration, and threatened that the media would be held responsible for their lies and distortions.

Following a misleading tweet about television ratings for his inauguration, these misrepresentations of inauguration numbers were cited on the inside-Washington television news program *Meet the Press* as "a provable falsehood," prompting Kellyanne Conway, counselor to the president, to respond: "Don't be so overly dramatic about it, [...]. You're saying it's a falsehood, and they're giving—our press secretary, Sean Spicer, gave alternative facts to that. But the point really is ..."—at which point the host jumped in to retort: "Wait a minute. *Alternative facts!?* Alternative facts!? Four of the five facts he uttered ... were just not true. Alternative facts are not facts; they're falsehoods." Alternative facts, as a concept, have since been paired with "fake news" and "false news" to characterize negative or embarrassing reporting about Trump and his administration by his administration, even as they ignore demonstrable online trolling, fake social media accounts, and an apparently systematic production of falsehoods released to influence voter decisions. The media spectacle of this intentional crisis of reality and truth, repeated over and over, spread across media sources, debated, and repeated again, has produced an unavoidable and ongoing "American Horror Show" characterizing media's relationship to political actions (Kellner 2017). Even as their very premise seems to fall apart under these attacks, the journalists and media accused of fake news promulgation continue to cover the administration and continue to benefit from enlarged viewerships. Meanwhile, the adversarial relationship to media Trump learned in New York remains symbiotic, improving ratings for the media under attack and producing Trump as the dominant center of all news, media, and spectacle while confirming his ideological bona fides with his political base.

Perhaps one of the most explosive media spectacles of the first two years of the Trump administration erupted in August 2017, as a media uproar exploded over Trump's failure to condemn neo-Nazi, Klan, and white supremacist groups until more than two days after deadly alt-right demonstrations in Charlottesville, Virginia over the weekend of August 11–13, 2017, which produced three deaths, many injuries, and tremendous outrage over the extremist demonstrations—much of which was televised and played back repeatedly, often in slow motion, as events continued to unfold. In a now infamous news conference on August 15 (that was strongly denounced by the media and political establishment), Trump symmetrized the white supremacist forces that had gathered in Charlottesville with protestors against neofascism and white supremacism, arguing they were equally responsible for violence, while he also expressed admiration for Robert E. Lee and the confederacy! As Trump hardened his support of far-right groups, executives from major corporations began resigning from advisory panels, leading Trump to cancel the panels, while major military leaders and some Republican Congressmen and Senators denounced white supremacism and the president's failure to more sharply criticize extremist groups.

The media responded with sharp critique of an increasingly embattled Trump who was finding out that war against the media cut two ways. Bowing to pressure, and perhaps wanting to change the media narrative, Trump fired Steve Bannon, the central figure of the alt-right in the White House on August 18, 2017. Yet Trump continued on his hard-right course and has continued his war against the media into the present moment.

Another ongoing media spectacle centered on this administration has been the investigation into US–Russia–Trump relations. It soon became the center of congressional investigations, multiple state investigations, as well as Special Counsel Robert Mueller's investigation of Trump and the election. In addition, the Trump–Russia connection was a major focus of liberal segments of the US media doing its job investigating shady government actions, with mainstream media like *The New York Times, The Washington Post, The Wall Street Journal* and some of the cable news networks, notably MSNBC, working feverishly (and, yes, spectacularly) to break new and startling daily revelations of the Trump–Russia saga, which emerged as the most astonishing and jaw-dropping political-media spectacle of my lifetime—although by mid-2019 it had not eroded Trump's base.

Indeed, the release of the Mueller Report in April 2019 became a bizarre media spectacle, as it wound its way through the Justice Department, apparently the White House had its own press conference in which a hastily written, four-page memo was reported to summarize its findings (and implied that Trump had been exonerated on both "collusion/conspiracy" charges and obstruction of justice). This outcome was widely reported and reproduced on social media, only to have this memo contested by the report's author, followed by another press conference that prefigured and attempted to frame the release of a redacted version of the full report. The release of this longer report will be remembered for the televisual spectacle of legal correspondents reading the lengthy volume, to themselves, live on air, while reporters nearby speculated about the possible contents. That this report was still redacted led to a series of stand offs between Congress and the executive branch of government, still unresolved at the time of this writing.[10] The released version of the Mueller Report indicated in the first half that multiple interactions between the Trump organization and campaign and Russian operatives came close to collusion or conspiracy, and the second half of the report documented at least ten occasions in which Trump obstructed justice by attempting to interfere in the investigation, and even to shut it down. An outraged Congress demanded to see the redacted parts of the Mueller Report, the underlying evidence, and called for Mueller and others mentioned in the report to testify to Congress. Trump immediately demanded that none of his administration should testify, declaring executive privilege, and as of May 2019 a battle is unfolding between the executive, legislative, and judicial branches of the US government that could constitute one of the most intense and consequential sets of media spectacle in history, in which the very fate of US democracy is at stake.

Part of Trump's appeal for his devoted followers (who Trump infamously claimed would vote for him even if he shot someone on Fifth Avenue), was that, at least in his campaign rhetoric, Trump presented himself as anti-political correctness, anti-establishment, and anti-globalist. This nationalist populist rhetoric linked Trump with a global movement of authoritarian populism that was capitalizing on extreme economic disparity by manipulating fears over immigration, racism, Islamophobia, and other *bêtes noirs* of the global right (see Kellner 2016, 2017). Trump's movement and rhetoric played out daily on television, in print, and online, all corporate media interconnected with global multinational corporations now rendered as vehicles for purveying anti-globalist rhetoric and political positions.

On one hand, Trump presenting himself as anti-globalist and anti-establishment is simply false, as Trump is part of the global establishment with business connections throughout the world and since running for and capturing the presidency has become a global media spectacle, promoting and circulating his brand and businesses throughout the world. Moreover, Trump's administration is full of billionaires and "Swamp Creatures" (he promised during his campaign to remove careerists from DC, using the metaphor of "draining the swamp") from the globalist political and economic establishment, and his major policies, such as his tax "reform," have largely helped the 1% while he has done little for his working-class supporters angry about globalization and an establishment apparently hostile or blind to their interests. Yet, it appears that Trump is seriously against the US political establishment and institutions of US democracy. So, for example, while The Donald loaded his cabinet and administration with establishment businessmen, generals, and Republican politicians, he has acted from the beginning as a wrecking ball for the political institutions of liberal democracy and has arguably diminished severely the US role and standing in the world and the global community, and reconfigured the relationship between politics and media spectacle.

Throughout his presidency, Trump has continued to hold campaign rallies and provide an anxious public with frequent tweetstorms. In these ways he appears to speak directly to the public, or at least his adamant supporters. A frequent topic in both venues are his attacks on the press and legacy media institutions, at one point referring to journalists as "the enemy of the American people," a phrase pilfered from Comrade Stalin who used to use it against bourgeois media. Trump's equation of the media as a site of "fake news" is an attack on the truth, but just as significant, it is meant to foster distrust for long-established institutions of US democratic practices.

Hence, while one cannot foresee the final trajectory of the Trump presidency, one can be certain that it will involve continuing war with the media and media spectacles of huge intensity and consequence. Trump's seemingly paradoxical attacks on the very media that has facilitated his startling rise to political power make sense when you realize that the media has long favored nothing so much as reporting on itself, particularly when such reporting generates increased revenue. The symbiotic relationship between media reportage and Donald Trump is fueled by mutually held self-admiration. Therefore, it is highly likely that the fate of the Trump presidency will be determined by media spectacles that will either continue or destroy his presidency. The continuing Trump spectacle might be very entertaining, indeed spectacular, were it not for the way these spectacles have material implications, which may have destructive effects on US democracy and hurt many real, embodied people. Indeed, the fate of US democracy hangs in the balance and the outcome is far from certain.

## Notes

1 For a concise summary of current global economic inequality, see, for example, the Oxfam International report: www.oxfam.org/en/even-it/5-shocking-facts-about-extreme-global-inequality-and-how-even-it-davos
2 See my "The Persian Gulf TV War Revisited" here for more details on modern war and television spectacle: https://pages.gseis.ucla.edu/faculty/kellner/essays/gulftvwarrevisited.pdf
3 I provide accounts of the O.J. Simpson Trial and the Clinton sex/impeachment scandal in the mid-1990s in Kellner (2003b); engage the stolen election of 2000 in the Bush/Gore presidential campaign in Kellner (2001); and describe the 9/11 terrorist attacks and their aftermath in *From 9/11 to Terror War* in Kellner (2003a).
4 See Blair (2000); D'Antonio (2015); Kranish and Fisher (2016). Blair's chapter on "Born to Compete," op. cit., pp. 223ff., documents Trump's competitiveness and drive for success at an early age.
5 See Wheeler (2013). On Trump, the media, and his long cultivation and exploitation of celebrity, see O'Brien (2016 [2005]).
6 On *The Apprentice*, see (Klein 2017), which documents how the series replaces the ethos of a dog-eat-dog predatory capitalism and fixed Trump's brand/image as a high-powered billionaire. Trump's book *The Art of the Deal*, co-written with Tony Schwartz (2005 [1987]), helped introduce him to a national audience and is a key source of the Trump mythology; see Blair (2000), pp. 380ff.
7 Although there is much controversy over how rich Trump is, how much he contributes to charity, and whether he pays all, little or no taxes
8 For the story of Trump's financial down-fall and near collapse in the 1980s and 1990s, see the detailed and well-documented narratives in Barrett (2016); D'Antonio (2015); O'Donnell and Rutherford (1991).
9 For a complete transcription and full video, see: http://time.com/3923128/donald-trump-announcement-speech/
10 The Mueller Report can be found at www.cnn.com/2019/04/18/politics/full-mueller-report-pdf/index.html (accessed May 6, 2019).

## References

Argetsinger, A. (2015) "Why Does Everyone Call Donald Trump 'The Donald'? It's an Interesting Story," *The Washington Post*, September 1, at: www.washingtonpost.com/news/arts-and-entertainment/wp/2015/09/01/why-does-everyone-call-donald-trump-the-donald-its-an-interesting-story/?noredirect=on&utm_term=.63b2c29ff2c3

Barrett, W. (2016) *Trump: The Greatest Show on Earth: The Deals, the Downfall, the Reinvention*. New York: Regan Books.

Battaglio, S. (2017b) "President Trump Turned Cable News Into Must-See TV in 2017," *Los Angeles Times*, December 30, at: www.latimes.com/business/hollywood/la-fi-ct-cable-ratings-yearend-20171229-story.html

Blair, G. (2000) *The Trumps*. New York: Simon and Schuster.

Buettner, R. and S. Craig (2019) "Decade in the Red: Trump Tax Figures Show Over $1 Billion in Business Losses," *The New York Times*, May 8, www.nytimes.com/interactive/2019/05/07/us/politics/donald-trump-taxes.html

Carol Pogash (Editor) (2016) *Quotations from Chairman Trump* New York: RosettaBooks (it is referenced on page 9).

Couch, A. and E. McDermott (2015) "Donald Trump Campaign Offered Actors $50 to Cheer for Him at Presidential Announcement," *The Hollywood Reporter*, July 17, at: www.hollywoodreporter.com/news/donald-trump-campaign-offered-actors-803161

D'Addario, D. (2018) "How Americans Became Addicted to Cable News in the Trump Era (Column)," *Variety*, https://variety.com/2018/tv/opinion/cable-news-must-see-tv-trump-1202980753/

D'Antonio, M. (2015) *Never Enough. Donald Trump and the Pursuit of Success*. New York: Thomas Dunne Books.

Doane, M.A. (1990, 2015) "Information, Crisis, Catastrophe" in *New Media, Old Media* edited by W.H. K. Chun, A.W. Fisher and T. Keenan (2nd edition) New York: Routledge, 251–264.

Gerbaudo, P. (2018) Social Media and Populism: An Elective Affinity? *Media, Culture & Society*, 40(5), 745–753.

Goldsmith, J. (2017) "Will Donald Trump Destroy the Presidency?" *The Atlantic*, October, at: www.theatlantic.com/magazine/archive/2017/10/will-donald-trump-destroy-the-presidency/537921/

Kellner, D. (2001) *Grand Theft 2000. Media Spectacle and a Stolen Election*. Lanham, MD: Rowman and Littlefield.

Kellner, D. (2003a) *From September 11 to Terror War: The Dangers of the Bush Legacy*. Lanham, MD: Rowman and Littlefield.

Kellner, D. (2003b) *Media Spectacle*. London and New York: Routledge.

Kellner, D. (2005) *Media Spectacle and the Crisis of Democracy*. Boulder, CO: Paradigm Press.

Kellner, D. (2012) *Media Spectacle and Insurrection, 2011: From the Arab Uprisings to Occupy Everywhere*. London and New York: Continuum/Bloomsbury.

Kellner, D. (2016) *American Nightmare: Donald Trump, Media Spectacle, and Authoritarian Populism*. Rotterdam, The Netherlands: Sense Publishers.

Kellner, D. (2017) *The American Horror Show: Election 2016 and the Ascendency of Donald J. Trump*. Rotterdam, The Netherlands: Sense Publishers.

Klein, N. (2017) *No is Not Enough. Resisting Trump's Shock Politics and Winning the World We Need*. Chicago, IL: Haymarket Books.

Kranish, M. and M. Fisher (2016) *Trump Revealed. An American Journey of Ambition, Ego, Money and Power*. New York: Scribner.

Kreiss, D. (2016) Seizing the moment: The presidential campaigns' use of Twitter during the 2012 electoral cycle. *New Media & Society*, 18(8), 1473–1490. doi:10.1177/1461444814562445 (accessed May 5, 2019).

Levin, B. (2019) "Trump's Fantasy Budget is a Next-Level Horror Show For Everyone But the Wall," *Vanity Fair*, March 11, at: www.vanityfair.com/news/2019/03/trump-budget-is-a-next-level-horror-show

Nance, M. (2016) *The Plot to Hack America. How Putin's Cyberspies and WikiLeaks Tried to Steal the 2016 Election*. New York: Skyhorse Publishing.

Nasirpour, S. (2019) "Trump's Political Success Comes with a Cost for His Business," *Bloomberg*, February 15, at: www.bloomberg.com/news/articles/2019-02-15/trump-s-political-success-comes-with-a-cost-for-his-business

O'Brien, T.L. (2016 [2005]) *TrumpNation: The Art of Being the Donald*. New York: Grand Central Publishing.

O'Donnell, J. and J. Rutherford (1991) *Trumped!: The Inside Story of the Real Donald Trump-His Cunning Rise and Spectacular Fall*. New York: Simon and Schuster.

Ott, B. (2017) The Age of Twitter: Donald J. Trump and the Politics of Debasement. *Critical Studies in Media Communication*, 34(1), 59–68.

Ouellette, L. (2016) The Trump Show. *Television & New Media*, 17(7), 647–650.

Sherman, G. (2016) "Operation Trump: Inside the Most Unorthodox Campaign in Political History," *New York Magazine*, April 3, at: http://nymag.com/intelligencer/2016/04/inside-the-donald-trump-presidential-campaign.html?gtm=bottom

Thorne, W. (2019) "Mueller Report Release Draws 11 Million Total Viewers Across TV News," *Variety*, April 19, at: https://variety.com/2019/tv/news/mueller-report-release-draws-11-million-total-viewers-across-tv-news-1203193710/

Trump, D.J. with T. Schwartz (2005 [1987]) *The Art of the Deal*. New York: Ballantine Books.

Tyndal Report (2015). "Campaign 2016 Coverage: Annual Totals for 2015," December 20, at: http://tyndallreport.com/comment/20/5773/ (accessed April 15, 2016).

Van Meter, J. (1989) "That's Why the Lady is a Trump: A Special Investigative Tribute to Ivana," *Spy*, May: 86-97 Archived here, at: https://books.google.com/books?id=xaCC4gI-zJkC&lpg=PA98&dq=trump&pg=PA86#v=onepage&q=trump&f=false

Wells, C. D. Shah, J. Pevehouse, J. Yang, A. Pelled, F. Boehm, J. Lukito, S. Ghosh and J. Schmidt (2016) How Trump Drove Coverage to the Nomination: Hybrid Media Campaigning. *Political Communication*, 33, 669–676.

Wheeler, M. (2013) *Celebrity Politics*. Cambridge, UK: Polity.

# PART IV

# Cultures and Communities

Part IV is interested in television's circulation and the social practices associated with television. Television may do things to us, but we also do things to and with television. From a variety of perspectives, approaches, and methods, authors in this part of the book inquire into the ways in which television is made a meaningful part of everyday life through culture, engagement, participation, and community. This section, therefore, considers television and citizenship (Graeme Turner), television and identity (Alexander Dhoest), television and diversity (Ana-Christina Ramón and Darnell Hunt), television as local (Frederic Chaume), television as art (Nomusa Makhubu), television and gaming (Divya McMillin), and television and online communities of shared affinity (Ruoyun Bai). Through these cultural categories, this part guides us through the programming of a small European nation (Dhoest), it looks at assumptions about and formations of community produced through the multiple practices of translating and localizing television (Chaume), and the ways that television is adapted to accommodate the growth of online gaming communities (McMillan) as well as online social and cultural sites (Bai). Communities and cultural practices here range from the white room of the art gallery and the black box of the television, contrasted, compared, and brought together in the context of post-colonial African intersections of art and television (Makhubu), through to the facilitation and obstacles television meanwhile presents for transforming increasingly global renderings of cultural and political citizenship (Turner). Globally, television remains a lively and important catalyst for multiple forms of social engagement, cultural practice, and the formation of community.

# 24
# TV CITIZENSHIP

*Graeme Turner*

In the broadcasting era, the era of mass media, it was customary to regard media institutions, and especially television, as structural components of the democratic state. To function effectively in their capacity as citizens, and thus for democracy to work, it was understood that people required accurate and reliable sources of news and information. The media were foremost among these sources, and while it was commonly held that newspapers may have set the news agenda, television was the primary platform through which that agenda was prosecuted and disseminated. Acknowledging the civic importance of the practice of journalism, regimes of regulation were established; they varied in their details and policy orientation but they shared an interest in ensuring that the media functioned in ways that served the national interest while protecting the ethical and democratic principles of press freedom and citizens' rights. Regulation of this kind was seen as essential to the successful functioning of a democratic state. Indeed, notwithstanding the enthusiasm for deregulation that affected many national systems from the 1980s onwards, aspects of these regulatory systems were normatized and exported into developing countries and 'transitional democracies' (Voltmer, 2013), with varying success, as guideposts for the desired relation between the media, the citizen and the state. It is important, however, to recognize that the role television plays in constructing citizenship is not at all confined to stable democratic states; indeed, its political role is necessarily more extensive, intrusive and explicit in the emerging democracies Voltmer examines, as well as in the nation-states with more authoritarian political structures. In such jurisdictions, the citizen is the final ground upon which the state must construct its legitimacy and over which it must maintain its power by controlling the access to information and the circulation of opinion. Typically, television is an integral part of such programmes of control, especially where, for varying reasons, the literacies required for print media are not widespread.

In the post-broadcast era, there have been many developments which have substantially affected the relation between TV and citizenship – in widely varying ways, it is true, but in just about every location. There are numerous factors involved in what I have described elsewhere as the post-digital 'reinvention of the media' (Turner, 2016) which has transformed the political economy of the media industries, how they conceive of and address their audiences, and how the media's platforms, affordances and content are used by those audiences. The expansion of the online environment into entertainment and retail, the uneven but widespread globalization of media industries, and the evolution of social media are all relevant here, and they are discussed in a number of other chapters in this volume. There are some changes, however, which are of particular relevance to this chapter: the decline of the mass media paradigm; the privatization of public service television in Europe; the shift away from the common culture created by appointment viewing and linear scheduling as the audience for broadcast media in some countries declines; the emergence of the multiplatform environment; the shift toward mobile and out-of-home devices; and what has been described as the 'crisis in journalism' (McChesney, 2007; Dahlgren, 2009; Meikle and Young, 2012). In relation to this last issue, the key elements involved there include the collapse of the advertising-led business model – especially for print media; the proliferation of alternative modes of aggregating and distributing journalistic content online; and commercial journalism's reorientation around its identification with the entertainment rather than the information industries. Greater commercialization has been assisted by programmes of deregulation responding in part to the increasing globalization of the media industries and, consequently, the competition among media outlets overall, and perhaps most comprehensively in television, has focused more intensely upon entertainment than previously. The widespread tendencies toward increased choice in content and sources, coupled with greater capacities for customization and the enhanced potentials for transnational connectivity, have focused upon a highly individualized consumer with specific preferences and a personalized menu of choices for news, entertainment and social media. Within such a context, the citizen disappears from view, the notion of media publics loses some of its relevance, and the freely choosing consumer takes centre stage. In train since well before the arrival of the digital, and as so often noted within the contemporary media studies literature, the addressee of the contemporary media message is now, most of the time, the consumer rather than the citizen.

It needs to be acknowledged here that not all of these things happen in the same way everywhere – or in some cases at all. Indeed, what a contemporary television studies needs to accommodate is not the irresistible roll-out of a set of universal tendencies but rather the increasing diversity of media systems around the world, and the varied and contingent manner in which they have evolved in response to their framing historical, political, economic and cultural conditions. As the range of possible components of the audience's media consumption expands, and as consumer choices within these components multiply, the experience of television depends, more radically now than ever, on where you are (Turner and Tay, 2009, 8). The relationship between television and citizenship has become more contingent and in some cases more compromised than it has been in the past; in a contradiction which further complicates matters, it has become both more transnational *and* more located. All of that acknowledged, in what follows I discuss some of the shifts in this relationship with regard to four sets of overdetermining influences that are, while not universal, sufficiently widespread to require attention: technological change, the information explosion or what Andrejevic (2013) calls 'infoglut', globalization, and television's participation in the production of 'cultural citizenship'.

## Technological Change

Over the last two decades the traditional or so-called legacy model of television production and distribution – a mix, in most places, of broadcast and cable, commercial and public service – has been hit by a technological perfect storm. The arrival of the digital brought us into a multiplatform post-broadcast era, enabled catch-up and time-shifting services, radically enhanced the capacity for copying and sharing content, and integrated television into the online environment and consequently with the multiplicity of affordances it provided and the range of devices to which it connected. New modes of production and delivery were developed in response to the emerging (or projected) patterns of consumption; in many cases, these broke significantly with the modes of production and delivery that had preceded them. What Lotz usefully categorizes as 'internet-distributed' television employed strategies and practices that were 'unavailable' to broadcast and cable distribution; as a result, industrial and audience practices that had been in place for many years, she says, required significant 'reconceptualization' – notwithstanding 'the important similarities that persisted' (Lotz, 2017, 6). Non-linear access, for example, distinguishes 'portals from their channel brethren by freeing them from the task of scheduling' (ibid.). Streaming services became serious competitors and in some locations their flexibility and menu of content choices have been increasingly preferred over the linear schedules of the broadcasters and cable companies. What Lotz calls the 'portals', or channels, of internet-distributed television do something very different to the broadcast and cable channels; she suggests they perform what might be best described as a task of 'curation – of curating a library of content based on the identity, vision and strategy that drives [their] business model' (Lotz, 2017, 8). The different business models of internet-distributed television have troubled the longstanding models of television production and consumption to such an extent that Tim Havens' chapter in this volume is entitled, provocatively but with good reason, 'what *was* television'.

Most of these elements of the 'perfect storm' share a shaping principle: they offer greater choice, convenience and flexibility for the consumer, and in instances such as streaming services like Netflix and the free catch-up services offered by broadcasters they are taken up by consumers at little or no cost and without the limitation of being tied to a linear schedule or to viewing on a fixed device. The expansion of choice, customization and mobility constitute the key difference between the various platforms of internet-distributed television and those of its broadcast and cable predecessors. As noted earlier, this difference is commonly described as a shift from addressing the audience as a public and its individual members as citizens, to addressing them primarily as a market and consumers. This is not just a minor adjustment to the social function of television. If most people were to shift from watching a mass medium such as free-to-air television, and select in its place a narrowcast service such as cable television, or an even more customizable platform such as an online streaming service, it would constitute a significant change in the social meaning of the experience of television. The withdrawal of a sense of the imagined co-presence of the shared, usually national, audience watching along with us, that is one of the core components of the experience of watching live broadcast television, would be a withdrawal from a practice that in one way or another situates the individual as not just a member of an audience, or as a consumer identified with a taste niche, but as a member of a community – and in the broadest sense as a citizen. There are political considerations here, to which we will turn in the next section, but for the moment I just want to make the point that, for many individuals around the globe, their experience of the relation between television and their sense of themselves as a citizen is fundamentally different as a result of the technological changes in the television industry.

## Infoglut

If successful democracies depend upon the citizen's access to information, one might consider that democracy should be in exceptional health at the moment. We have never had such access to the volume and varieties of information we now enjoy online. The problem for us now is that there is too much of it. It is proving extremely difficult to sort out the accurate from the inaccurate, the legitimate from the fake, the useful from the irrelevant. And there is always more to find out. As Mark Andrejevic puts it at the beginning of *Infoglut: How too much information is changing the way we think and know* (2013, 2):

> at the very moment when we have the technology to inform ourselves as never before, we are simultaneously and compellingly confronted with the impossibility of ever being *fully* informed. Even more disturbingly, we are confronted with this impossibility at the very moment when we are being told that being informed is more important than ever before to our livelihood, our security, and our social lives.

More worryingly, the task of sorting out what is and what is not important to us is largely an automated process driven by algorithms in search engines such as Google. The proliferation of news sites, blogs and various forms of commentary and opinion, in addition to the extraordinary capacities introduced with the 'culture of search' (Hillis et al., 2013) were initially hailed as an empowering and even emancipating development for ordinary people. It has become increasingly clear, however, that the principles that drive these platforms and their search engines can have highly contradictory social consequences and effects. Democratic participation is only one of many competing drivers here, and so we are finding some of the same crew of commentators who had hailed the democratizing potential of the digital world are now starting to wonder aloud if the internet is 'destroying democracy' (Edsall, 2017).

The news media have become oriented around entertainment to a greater extent than previously, and this is influencing what news is gathered, what is made available, how it is represented, and where it appears in the hierarchy of results from an online search. Since the 1980s, there has been growing sense of a crisis in relation to the future of journalism, and of the news media in general. In more recent years, concern has accelerated in response to the declining commercial viability of news organizations which provide investigative journalism, fact-oriented journalism, and hard news, as they strive to compete with those more commercially and entertainment-oriented outlets which nonetheless still operate within the discursive and institutional framework of journalism. The line between news and entertainment has blurred, as commentary and opinion overtake reporting, and social media's circulation of fictional stories presented as news – so-called 'fake news' – plays its part in undermining the ethical standards and cultural authority of journalism. Furthermore, the availability of news sites that are targeted to particular patterns of belief, and the decline in the numbers of people consuming what we might call a common or public diet of news that is based upon a consensus not only of opinion but also on facts and values, has raised concerns about the political and cultural fragmentation of the public. Cass Sunstein (2009) has critiqued what he has called 'cyberbalkanisation' – the fracturing of community into smaller, self-contained and inward-looking mini-public spheres increasingly disinterested in, intolerant of, or simply unacquainted with, different points of view.

Notwithstanding the increasing volume of news now available via the information aggregators, very few of them actually gather news themselves. What may seem like an explosion of news and information is to some extent an illusion. While aggregating news sites and bloggers flourish, and opinion and sensationalized 'clickbait' displaces news as means of attracting their audiences, the organizations charged with actually producing the news are under commercial threat. The production of 'new' news, particularly so-called 'hard news' to do with politics, the economy and society, is barely viable in many locations today.

This has to have an effect upon the political structures that depend upon the free flow of accurate information. A large comparative programme of research led by James Curran (2011) has charted the consequences of a declining level of consumption of political information in a number of countries. The results suggest that the decline of a common culture of media consumption – effectively that generated by the mass media and the public service models which are gradually being displaced – has deleterious effects on the capacity for the citizen of some, perhaps many, states to make the kinds of informed judgements upon which a successful democracy depends. The survey examined the levels of political knowledge in audiences from the US, the UK, Germany, Denmark and Finland, and paid particular attention to those differences that may have been attributable to the different roles played by commercial and public service broadcasters. They found considerable variations in the levels of public knowledge of both 'hard' and 'soft' news; the US, with its highly commercialized system and the most fragmented and narrowcast market, and with its network coverage of international news at comparatively low levels, returned the lowest scores for public knowledge. Importantly, however, the survey also established that there were more factors involved in this than just the structure of the local industry; inequities of income, the role of gender, and the existence, or lack of, mechanisms for social inclusion of the disadvantaged in each country are among examples cited. Interestingly, one of the key areas where this variable was most active was in the proportion of viewers regularly watching the national evening news bulletins. Since that is the location within which one would expect information to be most directly framed around an address to the viewer as citizen, and organized around a consensus on facts and values, this is a significant finding. This research has been dealt with at some length in many publications and I am not able to do it justice in the space available here. However, it is worth citing one of Curran's concluding remarks. As an increasing number of countries are 'converging toward the entertainment-centred model of American television', he says, the evidence from the survey would suggest that this fosters 'an impoverished public life characterized by declining exposure to serious journalism and by reduced levels of public knowledge' (2011, 60).

## Globalization

One of the individual indicators of the levels of public knowledge that Curran's survey examines is to do with respondents' knowledge about the rest of the world. The US scored particularly poorly against this indicator, and this was connected to the comparatively low levels of international news on American television screens. Paradoxically, this occurs at a time when America is routinely offered as the leading edge of the contemporary media's evolution, and when there is constant talk about media globalization and the declining relevance of the nation-state in the face of expanding transnational connectivity. The survey results highlight just how complex, contingent and contradictory the processes and effects of globalization have turned out to be. On the one hand, there is a global political economy that is shaped by transnational concentrations of media power which is overwhelmingly organized around commercial interests rather than the interests of the state or of the citizen. The processes of deregulation that gathered force toward the end of the 20th century have served an agenda of marketization which has sought legitimacy by referring to the forces of globalization – held to make a single market regulatory framework anachronistic – and to the imputed impossibility of unilaterally regulating digital media and the internet, which also makes the nation-state look like an anachronism. Programmes of deregulation have reduced the presence of, or have privatized, publicly funded television in many markets; typically, it has been claimed that sources of news and information have proliferated online to the point where their regulation or preservation on broadcast television was no longer necessary. Whatever the merits of such claims might be, it is hard to deny that, in general and for the most part, deregulation has worked to effectively subordinate the rights of the citizen of the nation-state to the imperatives of the market.

On the other hand, if we were to approach globalization from other angles of inspection, we would find powerful countervailing tendencies toward localism and regionalism at various levels and on various scales: we have the cities such as Hong Kong or Seoul which Michael Curtin (2004) describes as the new 'media capitals'; we have the establishment of dynamic geo-linguistic television markets in Asia and Latin America; and we have the persistent overdetermining influence of local, regional, and national political and cultural structures such as the historical specificities affecting how television in each location is experienced, how it is structured as an industry, and how it has become embedded in the political formation of the state. Where strong publicly funded television broadcasting persists, it is precisely as policy-driven mechanisms for informing the citizens of the nation-state – as correctives, that is, to the forces of the market.

There are, then, a number of contrasting tendencies in play. Globalization and transnationalism have introduced new patterns of affiliation and community for television audiences, as the spread of transnational streaming services is turning large parts of the world into what is, in effect, an aggregated market for entertainment. That market often articulates itself by way of the networks of social media rather than through the corporate structures provided by the producers – contributing to shifts in distribution and release patterns such as the simultaneous global release of traditional television network entertainment content. On the other hand, we have the developments described by those dealing with television, the citizen and the state in post-Communist central Europe (Kaneva, 2012; Volcic and Andrejevic, 2016), where new representations of nationalism are constructed, prosecuted and captured by commercial interests within the state.

In general, however, we have moved to a point where the national is no longer necessarily always the default point of reference for the TV citizen, nor is it necessarily always the key location for power. Social media networks seem now to have the potential to challenge how we might think of the national TV citizen, just as they have challenged how we might think of community (Turner, 2016). That said, and keeping in mind notions such as Straubhaar's (2007) analysis of cultural proximity as a means of understanding the development of regional television markets, there is also evidence of just how stubborn and resilient the identification with the nation can be. When Anna Cristina Pertierra and I were examining television consumption in Mexico some years ago (Pertierra and Turner, 2013), our expectation was that the Mexican audiences would recognize and perhaps distance themselves from the external sources of much of their programming – the telenovelas from Colombia, for example. However, we found that many of our informants were not even aware that these programmes were made elsewhere; if it was on Mexican television, and if the dialogue was in Mexican-accented Spanish (the telenovelas concerned had been dubbed with Mexican accents), the standard perception was that it was Mexican. That provided some insight into the power of the cultural preconditions to the local or national consumption of content from a globalizing television market.

This was also a reminder of the point that Curran makes, when he emphasizes how important it is not solely to concentrate on television news and current affairs when we are examining how television constructs forms of citizenship:

> The great bulk of content produced by contemporary media systems … has nothing to do with public affairs. Indeed, even *new* media devote increasing proportions of their output to soft news and entertainment … In other words, most media output consumed most of the time is unrelated to conventional understandings of politics. Any reconceptualization of the democratic role of the media needs to think through the implications of this transformation.
>
> *(2011, 63)*

Curran argues that we need to more seriously consider the 'democratic meaning of entertainment', and that it would be mistaken to assume that entertainment programming plays no part in educating and informing its audience:

> In brief, entertainment connects to the democratic life of society in four ways. It provides a space for exploring and debating social values, which occupy a central place in contemporary politics. It offers a means of defining and refashioning social identity, something that is inextricably linked to a sense of self-interest. It affords alternative

> frameworks of understanding, which inform public debate. And it provides a way of assessing, strengthening, weakening and revising public norms that are an integral part of the way we govern ourselves.
>
> *(2011, 75)*

This takes us into territory where television has been considered as the medium most comprehensively engaged with the production of cultural citizenship.

## Cultural Citizenship

There is, of course, much television entertainment that is directly concerned with politics: situation comedies such as *Veep* (USA), *The Thick of It* (UK), and *Utopia* (Australia); dramas such as *The Newsroom* (USA), and *Borgen* (Denmark); and the longstanding tradition for late-night satire and talk in the US which has created *Saturday Night Live, The Daily Show, Last Week Tonight with John Oliver, Full Frontal with Samantha Bee* and the raft of contemporary 'late shows' hosted by Jimmy Fallon, Seth Myers and Stephen Colbert. However, the point that Curran is making does not take us to these. The connection between mainstream and 'non-political' television entertainment and the production of cultural citizenship has a long history in television studies. It goes back as least as far as Fiske and Hartley's *Reading Television* (1978) which was one of the earliest attempts to explain the cultural politics of 'light entertainment' genres such as quiz shows and ballroom dancing competitions. Fiske and Hartley engaged positively with television as 'culture', approaching it in as close to its own terms as could be managed at that point. As Hartley puts it in his foreword to the 2nd edition, their focus upon the broad and everyday cultural function of television, recognized that 'citizenship flows from a sense of community as much as from obligations to a state' (2003, xv). That sense of community was generated in diverse ways through the audiences' engagement and modes of identification with, as well as the pleasures derived from, the performances, spectacles, images and narratives on the screen. Initially, television studies focused on how elements of the social or the cultural were represented in the text: the tradition of textual analysis of television. Such a focus gradually gave way to a more complicated and nuanced examination of the audience's active and contingent engagement with the versions of living they saw on their television screens. The tendency here was toward investigating how television participated in the production of cultural identities, evident in Hartley's own later work, where television as 'democratainment' leads to the formation of what he describes as 'DIY citizenship' – the 'citizenship of the future', with its focus on 'identity and selfhood' (Hartley, 1999, 161).

The ultimate provocation for the analysis of this mode of television citizenship was reality TV, specifically *Big Brother*; its determination to stage and replicate everyday life, and the 'ordinariness' of its participants, combined to make its various formats the perfect playing field for the performance (and the witnessing) of cultural identities. Some decades later, there is now an enormous television studies literature dealing with reality TV – and especially with the format's capacities for addressing values and identities through its display of diverse modes of performing the self. There are many examples of highly sophisticated elaborations of how particular reality-TV formats have focused upon the cultural self-fashioning of the individual: for instance, by inculcating rules for a self-regulated approach to 'better living' (Hay and Ouelette, 2008), by eliciting both identification with and resistance to recommended class identities (Skeggs and Wood, 2012), and by interrogating the protocols of self-presentation and appearance used to legitimate 'makeover TV' (Weber, 2009). As such analyses have demonstrated, the gap between television and everyday life has begun seriously to shrink, and this has been most dramatically evident in the formats of entertainment television. There is considerable consonance, then, between such analyses and the points that Curran makes about how entertainment television 'relates to politics' or, for our purposes here, to citizenship, through its 'representation of values, identities, cognition and norms' (2011, 63). We know from the work that has been done over the last twenty years that the citizen/consumer watching *Wife Swap* is just as likely to engage in what Skeggs and Wood (2012) describe as a 'tournament of values' over what they see on the screen as is the citizen/consumer watching the evening news bulletin.

Finally, it is worth noting, in this connection, that one consequence of the heightened competition for television audiences in the post-broadcast era has been changes in the range of content required and made available. On the one hand, such competition has been responsible for a wearying pattern of repetition and imitation that can make the standard cable schedule look at best unpromising. On the other hand, it is possible to argue that new levels of originality and creativity (Mittell, 2015) in certain areas of television content have been facilitated and made viable by the multiplatform environment. I am thinking here of, for instance, the influence of HBO and Netflix with high-end drama series such as *House of Cards* or *Orange is the New Black*, but also of the transnational success of the so-called Scandi-noir dramas such as *The Bridge* and the global phenomenon of *Game of Thrones*. Such programmes have taken advantage of the fragmented and globalized market by producing what might once have been considered 'niche' content that is free to be more narratively and morally complex, more socially and politically nuanced than might have been possible with content aimed at a single national mass market. At this point, we know almost nothing about how such content might play into the production of cultural citizenship in the current conjuncture. However, we know enough – as the comments from Curran and the resonance of the work on reality TV would indicate – to predict that it will almost inevitably figure in any future accounting of the changing relationship between TV and the construction of the citizen.

## Conclusion

The commercialization of the media, as noted a number of times in this chapter, does enforce those practices which are most responsive to the pressures of the market, and the transformations of television in recent years have been shaped more by these pressures than by earlier obligations to the national interest or the interests of its citizens. Crucially, it has become commonplace for contemporary state policy-makers to claim an equivalence between the interests of the market and the interests of the citizen, and the way television has largely dealt with the citizen, around the world, reflects this. Commercial interests have expanded their influence on policy, upon regimes of state-based regulation, and upon what television offers to the average consumer in most locations. While the influence of technological change, of globalization, and the opening of access to information all contain within them potentials that could well enhance television's responsiveness of the needs of the citizen, there are few tendencies that could be held to have actually enhanced and exploited such potentials so far. The decline in the number of publicly funded television systems around the world, and the redefinition of the news media, generally but most actively in television, as part of the entertainment industry (Turner, 2016), certainly contribute to what Miller describes as the 'democratic deficit run up by the contemporary media' (2007, 17). That said, such key indicators as the decline in the audience for television's evening news in the US are not automatically replicated elsewhere; there are plenty of places where the television news still leads the ratings and serves a large mass audience – Australia, where I am based, is one such location. The massively differentiated levels of public knowledge recorded in Curran's research cited earlier, demonstrates, too, that there remains considerable capacity for delivering better results for the citizen in this area of television's performance. However, it might be argued that any balance sheet prepared in order to provide an account of what has been lost and what has been gained for the television citizen in the post-broadcast era would have to look elsewhere than news and information for positive returns. For the moment, it would appear that it is global television's development of the quality niche for drama, as well as a wider range of formats across its entertainment content, that is likely to provide the most productive location for such an enquiry.

## References


Andrejevic, M. (2013) *Infoglut: How the Internet Has Changed the Way We Think and Know*, New York and London: Routledge.

Curran, J. (2011) *Media and Democracy*, London and New York: Routledge.

Curtin, M. (2004) 'Media Capitals' in L. Spigel and J. Olsen (eds.) *Television after TV: Essays on a Medium in Transition*, Durham, NC: Duke University Press, 270–302.

Dahlgren, P. (2009) *Media and Political Engagement: Citizens, Communication and Democracy*, Cambridge: Cambridge University Press.

Edsall, T. (2017) 'Democracy, Disrupted', *New York Times*, March 2. Accessed 22 August, 2017. www.nytimes.com/2017/03/02/opinion/how-the-internet-threatens-democracy.html?mcubz=3

Fiske, J. and Hartley, J. (1978) *Reading Television*, London: Methuen.

Hartley, J. (1999) *Uses of Television*, London and New York: Routledge.

Hartley, J. (2003) 'Foreword' in J. Fiske and J. Hartley (eds.) *Reading Television*, 2nd edition, London and New York: Routledge.ix-xxi.

Hay, J. and Ouelette, L. (2008) *Better Living through Reality TV: Television and Post-Welfare Citizenship*, Malden, MA: Blackwell.

Hillis, J., Petit, M. and Jarret, J. (2013) *Google and the Culture of Search*, London and New York: Routledge.

Kaneva, N. (2012) *Branding Post-Communist Nations: Making National Identities in the 'New' Europe*, London and New York: Routledge.

Lotz, A. (2017) *Portals: A Treatise on Internet-Distributed Television*, Ann Arbor, Michigan MI: Maize Books and Michigan Publishing.

McChesney, R. (2007) *Communication Revolution: Critical Junctures and the Future of Media*, New York and London: The New Press.

Meikle, G. and Young, S. (2012) *Media Convergence: Networked Digital Media in Everyday Life*, London: Palgrave Macmillan.

Miller, T. (2007) *Cultural Citizenship: Cosmopolitanism, Consumerism, and Television in a Neoliberal Age*, Philadelphia, PA: Temple University Press.

Mittell, J. (2015) *Complex TV: The Poetics of Contemporary Television Storytelling*, New York and London: New York University Press.

Pertierra, A.C. and Turner, G. (2013) *Locating Television: Zones of Consumption*, London and New York: Routledge.

Skeggs, B. and Wood, H. (2012) *Reacting to Reality Television: Performance, Audience and Value*, London: Routledge.

Straubhaar, J. (2007) *World Television: From Global to Local*, Thousand Oaks, CA: Sage.

Sunstein, C. (2009) *Republic.com 2.0*, Princeton, NJ: Princeton University Press.

Turner, G. (2016) *Re-Inventing the Media*, London and New York: Routledge.

Turner, G. and Tay, J. (eds.) (2009) *Television Studies after TV: Understanding Television in the Post-Broadcast Era*, London and NewYork: Routledge.

Volcic, Z. and Andrejevic, M. (eds.) (2016) *Commercial Nationalism: Selling the Nation and Nationalizing the Sell*, London: Palgrave.

Voltmer. I. (2013) *The Media in Transitional Democracies*, London: Polity.

Weber, B. (2009) *Makeover TV: Selfhood, Citizenship and Celebrity*, Durham, NC: Duke University Press.

# 25
# TELEVISUAL IDENTITIES
## The Case of Flemish TV Drama

*Alexander Dhoest*

This chapter explores the multiple connections between television and issues of identity, focusing on the interplay between the global and the local. Despite undeniable processes of globalization, contemporary television is often strongly anchored in local and particularly national contexts. In a context of ever easier access to television content across national borders, thanks to neoliberal broadcasting policies and digitization, television productions worldwide are not only inspired by global (often, but not always, American) examples, but also aim to distinguish themselves through a strong anchoring in local culture. As such, television continues to contribute to the discursive construction of cultural and national identities, capitalizing upon audience preferences and affinities while also reconfirming them.

The current chapter develops these insights by referring to research on television drama, particularly in the context of a small European country, Belgium. While the choice of Belgium is arbitrary and inspired mostly by the author's knowledge of this particular case, it does offer a valid example. First, as a smaller, non-Anglophone TV market it helps to pinpoint the continued relevance of such national contexts for an understanding of local television. Second, the focus on a particular case supports this chapter's main point: that the national remains key to understand television cultures, but that it may take on very different guises. There are large and small nations, old and new ones, unitary and fragmented ones. Each of these differences, and many more, determine to what degree and how the national matters for any specific case.

Throughout the chapter, theoretical insights will be illustrated with concrete examples drawn from Belgian television. Beside a broad discussion of the continued importance of national and cultural identities in and for television, the implications of this national focus for the representation of other identities will also be explored, focusing in particular on ethnic and sexual identities. Discussing the entire complexity of such broad issues in a single chapter is, of course, impossible, so instead of aiming for a complete account, the chapter reflects on some aspects of these issues, in dialogue with concrete examples drawn from the Belgian context.

### Television Drama and National Identity

Television drama has always been a central television genre and it continues to be a key asset in the battle for audience attention, be it on classical network television or on subscription services such as Netflix. In the past, television in general and drama in particular were often tied to national identities. Certainly in the era of "scarcity" (Ellis, 2000), television was seen as a medium for the nation, offering a shared televisual culture. This gradually changed in the era of "availability", with its ever growing and diversifying offer of channels through cable and satellite, in which national productions increasingly competed with a global television offer. This has only strengthened in the current age of "plenty", with a multitude of digital channels and content. However, even today television – and particularly public service television – remains an important medium to "work through" current issues: "No longer the agent of a standardizing notion of national unity, public service broadcasting can provide the forum within which the emerging culture of multiple identities can negotiate its antagonisms" (Ellis, 2000: 87).

Waisbord (2004) equally stresses the continued relevance of the national for media in a context of globalization, which he situates mostly in the everyday: "The power of the media lies in making national feelings normal on an everyday basis" (p. 386). To him, media regularly make available cultural forms and representations identified with the nation; they offer opportunities for shared experiences; and they institutionalize national cultures. In short, mass media such as television bring members of the nation together: "The media still manage to nurture a sense of home, collectivity, and community linked to nationhood" (p. 389). More recently, Flew and Waisbord (2015) defended the continued relevance of national media systems in a globalized media context, also pointing at the need for local adaptation in order for global companies and formats to be successful in specific countries. Similarly, in his work on format adaptation, Albert Moran (1998) has repeatedly observed the strong and continued audience preference for "own" programmes, either locally developed or (if based on global formats) indigenized through the use of national languages and incorporated in national television culture.

In the same vein, Straubhaar (2007) states that "national cultures, national markets supported by national governments, and national television networks still dominate the television viewing reality of most audiences" (p. 7). He attributes this to cultural identification and proximity: "This is a seemingly common attraction audiences feel for cultural products, such as television or music, that are close in cultural content and style to the audience's own culture(s)" (p. 26). Nelson (2007) makes a similar observation in relation to drama, where he notices a textual influence of American television style but also resistances and an increasing focus on cultural specificity. All of this is not to discount the irrefutable and increasing importance of the global, nor the growing multiplicity and hybridity of cultural identifications; the point is, quite simply, that the national does also remain a key source of identification.

Belgium offers a good case in point, albeit a complicated one. To start, it is important to note that television is organized separately for both linguistic communities, the northern Dutch-speaking Flanders and the southern French-speaking Wallonia (including the mostly Francophone capital of Brussels). As a consequence, "the national" in relation to Belgian television refers mostly to one of these regions. This is most strongly the case in Flanders, the region with the strongest national and independentist aspirations. The Dutch-language television market in Flanders is dominated by two domestic companies and their channels, the public service broadcaster VRT and the Flemish-owned commercial broadcaster Medialaan. In 2017, VRT had a market share of 35.5% with its channels Eén and Canvas, and Medialaan 29% with its core channels vtm, Q2 and Vitaya (CIM, 2018). Even if these channels do broadcast international drama, it is Flemish drama that gets the highest ratings. For instance, in December 2016 four of the five top-rated shows were domestic dramas, while the top imported show (the British *The Missing*) only reached 19th place (CIM, 2017a). While these shows are influenced by global (American, but also British and increasingly Scandinavian) examples, they deliberately look for local topics and approaches, "domesticating" genres and formats by capitalizing on pre-existing ideas of Flemishness. The latter is not conceptualized as a fixed cultural or national identity but rather as a changeable historical construction, which is discursively constructed, among other things in (media) representations and narratives (Hall, 1992).

One of the key instances where this interplay of globalization with strong local anchoring can be witnessed, is that of "quality drama". While this label may be questioned for a number of reasons, for instance for legitimating a small segment of television while continuing to denigrate most of it (Newman & Levine, 2012), it does help to classify a number of generically innovative shows that have appeared on Flemish television from the 2000s. Prior to that, in the 1990s, the liberalization of the broadcasting market and the introduction of commercial broadcasting led to a steep rise in Flemish drama productions, which, however, closely followed international genre conventions: soaps, sitcoms and crime dramas (Dhoest, 2007). The growing output and budgets (through government support and tax shelter) led to the gradual maturing and professionalization of Flemish TV drama production, which from the 2000s led to a growing number of more original, high-quality dramas (Dhoest, 2014).

On the one hand, recent Flemish "quality" dramas are clearly inspired by the rise of HBO-style "quality TV", shows with a clear visual identity and thematically combining established genres. Robert J. Thompson, writing on the Second Golden Age of American television, characterizes it as follows:

> American quality TV breaks the rules of established television; it is produced by people of quality aesthetic ancestry outside the field of television; it attracts a blue-chip audience; uses ensemble casts and multiple, overlapping plot lines that indicate literary values; includes social and cultural criticism; and creates a new genre by combining old ones.
>
> *(In Fricker, 2007: 13)*

However, rather than simply copying these shows, Flemish writers and showrunners started to dig into "typically Flemish" themes and settings, such as the cycling Tour of Flanders (*De Ronde*, 2011) or the local amateur theater company (*Amateurs*, 2014). Partly this is due to the wish to distinguish these shows from their international competitors, which generally have much higher budgets. Partly this also reflects the requisites for government funding, which stress the "Flemishness" of productions. For instance, the rules of the VAF (Flemish Audiovisual Fund), which sponsors quality series through its "Mediafonds" (Media Fund), stipulate that productions should be Flemish in at least two of three possible ways: their artistic crew, production context, or content. The latter is linked to the topic, source of adaptation, heritage, history, social relevance, political context, arts, etc. (VAF, 2016). Ultimately, however, audience figures are essential in understanding the focus on Flemish issues: these shows simply get the best ratings.

One recent example is *Beau Séjour* (2017), a "quality show" broadcast by public broadcaster VRT on its main channel Eén, in prime time on Sunday evening, the key drama slot. At the time of writing, this show ranked second in the viewing charts, with a market share of 46.7% (about 1.4 million viewers), while another local drama, the daily soap *Thuis*, led the charts with a market share of 49.4% (over 1.5 million viewers; CIM, 2017b). *Beau Séjour* (the name of a hotel) is a 10-episode crime drama firmly set in the Flemish province of Limburg. In line with American quality drama, it has a strong visual style – dark and slightly eccentric, set in the present but with stylishly retro settings – and it is generically hybrid. Beside being a classical "whodunit", where the police investigate the murder of a girl, Kato, it also draws on fantasy: Kato is still visible to five people, and she helps them investigate her murder, of which she has no memories. As such, it mixes genres, a crucial characteristic of American quality drama: "These shows were generic mongrels, often scrambling and recombining traditional TV formulas in unexpected ways; they had literary and cinematic ambitions beyond what we had seen before, and they employed complex and sophisticated serialised narratives and inter-series 'mythologies'" (Thompson, 2007: xix).

Beside combining the police procedural genre with fantasy elements, *Beau Séjour* has a large ensemble cast with some 18 key protagonists in a number of complexly intertwined storylines. The show also reflects the auteurism which is so central to quality TV (Newman & Levine, 2012), in this case not in the person of a single author or producer as showrunner, but through a couple (professionally and romantically) of female directors who also developed and co-wrote the show, Nathalie Basteyns and Kaat Beels, who previously directed a number of shows with a clear directorial identity as well as strong female protagonists.

*Beau Séjour* is also reminiscent of the "Nordic Noir" crime shows which conquered the world since the mid-2000s. These shows tend to be dark, with a strong but understated visual and acting style. Creeber (2015) describes them as follows: "These dramas are typified by a dimly lit aesthetic (hence its implicit reference to *film noir*) that is matched by a slow and melancholic pace, multi-layered storylines and an interest in uncovering the dark underbelly of contemporary society" (p. 22). The murder mystery, he continues, is often attached to a number of other storylines and themes, with enigmatic and barren landscapes often seen as reflecting the psychological mood. *Beau Séjour* equally has a rather slow pace, multiple storylines and a great number of dirty secrets such as mental problems, alcoholism, drug dealing, etc. Both the opening titles and the episodes are interlaced with dark, mysterious images of woods and desolate landscapes. Like most of the Nordic Noir shows, it also has a couple of investigators forced to work together; in the case of *Beau Séjour*, these are two female inspectors who have to work together with incompetent and fraudulent male local police officers. Often, Nordic Noir shows have a strong female protagonist with a "quirk" (such as the emotionally distant Sarah Lund in *The Killing*, 2007–2012), a feature that is taken one step further with Kato – who is a ghost.

While similar to American quality TV as well as Nordic Noir, *Beau Séjour* is nothing like a copy, as it develops its own, local style. For instance, the actors speak with a local Limburg accent, which strongly anchors it in a particular Flemish region. Similarly, the key "arena" of the show is a local "motocross" track, a sport that is very popular in Limburg, the home region of tenfold world champion Stefan Everts. Several of the characters either race on motorcycles, they train racers, or run the motorcycle club. The show also occasionally makes fun of the stereotype of Limburg's inhabitants as hospitable (they continuously offer each other pie) yet slow-witted, as exemplified by the naive cop Bart. Testifying to its successful mixture of global and local elements, *Beau Séjour* was purchased for international distribution by Netflix in 2017.

Such local interpretation of international tendencies is not limited to "quality drama" in Flanders. The public broadcaster's soap *Thuis* (1995–), while closely following international genre conventions (in particular as developed in British soaps such as *EastEnders*, 1985–), is also strongly anchored in Flemish society and culture. Franco (2001), comparing *Thuis* to *EastEnders*, describes the link between soaps and national identity as follows:

> By addressing the national/regional audience in its own language, speaking about domestic situations/matters and communicating values/commonsense assumptions and understanding, soaps engage in a process of enculturation that promotes a sense of community and identity among a particular group of viewers within a national territory.
>
> *(p. 453)*

From a distance, the characters and storylines in *Thuis* may look very similar to those in any long-running soap: there are fathers and mothers, colleagues and lovers, doctors and taxi drivers, bar-tenders and students, entangled in ever-evolving dynamics of love and betrayal, illness and crime. Upon closer inspection, however, all the tropes and clichés get a very Flemish slant. Rather than the more working-class culture of *EastEnders*, *Thuis* focuses on the middle class which takes up a more central position in the Flemish imagination (Franco, 2001). The characters speak contemporary Flemish Dutch (with a particular accent, sometimes bordering on dialect), live in recognizable houses and visit familiar settings such as cafés and restaurants. Small businesses abound, such as a plumbing company, a taxi company and a café. Thematically, contemporary issues are dealt with in a way that corresponds to mainstream Flemish beliefs and values. Franco (2001) ties the show's Flemishness to its

> naturalistic language/traditions of performance, the familiar personages of local stars (…) established throughout their television career, a sense of nostalgia for a rural way of life and a defensive/conservative worldview tempered by a mild anarchism emblematized by the characters of the rebellious working-class underdog as the locus of emotional investment.
>
> *(p. 468)*

All of this leads to a particular take on all kinds of issues, including the representation of ethnic and sexual diversity.

## Ethnicity and Sexuality

As indicated above, the national continues to matter for television in a global context, leading to local inflections of international genres and formats. Visually and thematically, productions are rooted in the local, in the process (re-)constructing national identities. The world that is represented, and the way it is represented, contribute to a sense of identification. A key question, however, is who is included in this television nation, and how they are represented. Nations tend to be presented as homogeneous, defining a central "us" against marginalized others (Hall, 1992). Minorities are generally under-represented and relayed to the periphery of the world as shown on television. While it is impossible to discuss all the ramifications of this issue, I will focus on two identities and refer to the Belgian example to illustrate how the way they are dealt with is always nationally inflected.

A first identity of relevance to issues of in- and exclusion in the television nation is ethnicity, often discussed in conjunction with race. Regardless of the ethnic and racial constitution of the particular country Western television is made in, ethnic minorities tend to take up a marginal position, as do non-White people (Hall, 1997; Haynes, 2007). In the US, this has led to a long tradition of quantitative and qualitative analyses of television (and other media), generally criticizing media's whiteness and stereotypical representation of minorities (e.g. Tuckachinsky, Mastro & Yarchi, 2015). Most attention has gone to the media representation of African Americans, but increasingly the attention also turns to Latinos and Asians. It is important to keep in mind, however, that the particular groups and issues do vary across nations.

For instance, on Flemish television ethnicity is a more central issue than race, reflecting the ethno-cultural context of a majority White population still struggling with a growing but relatively small and recent (compared to the US) influx of non-European migrants, which do not identify as Black, Latino or Asian (as in the US) but are Northern African or Middle-Eastern and increasingly identify as Muslim. In the US, categorization by race is commonplace in government statistics and policies, and the population is very racially diverse; in 2015, the US population was 61.6% White, 17.6% Hispanic or Latino, 13.3% Black or African American and 5.6% Asian (United States Census Bureau, 2017). While race is an important issue in Belgium, too, it is hardly addressed, and discussions of on-screen representation rather focus on the sufficient and balanced portrayal of ethno-culturally defined minorities. Although representations capitalizing on the fear of Muslims, as exemplified by American shows such as *24* (Halse, 2012), do not appear in Flemish drama, there is a lack of diversity in racial and ethno-cultural terms.

Public broadcaster VRT is regulated by five year government contracts, which stipulate among other things that VRT has to adequately represent social diversity, including ethno-cultural minorities which are called "new Flemings" and defined as anyone with at least one parent not born in the EU15. The last government agreement stipulates that 7.5% of all characters on screen should be "new Flemings", with a minimum of 5% (VRT, 2016: 52). An annual "Diversity monitor" quantitatively measures the presence of (visible) ethnic diversity on TV, and the numbers for 2016 show an overall percentage of 8.0% "new Flemings", but only 6.3% in drama (De Swert et al., 2016). It is hard to relate these numbers to actual numbers of new Flemings in society, as there are no reliable statistics on people's origins, only on nationality, but it should be noted that the numbers are substantially lower than those for the non-White population in the US. The last available numbers, for 2013, estimate the number of new Flemings at 10% (Van den Broucke et al., 2015).

Zooming in on the way particular ethnic minority characters are represented, it is clear that the public broadcaster does make an effort to get it right, but due to the limited number of representations individual characters carry a heavy "burden of representation", echoing a long-standing critique of ethnic and racial minority representation in TV drama: "the black character as an individual disappears under the responsibility of carrying the 'race' issue" (Geraghty, 1991: 142). Nevertheless, one striking evolution is that these characters' ethnic and racial "otherness" is less and less an issue in Flemish TV drama. For instance, in crime drama *De Ridder* (2013–2016) the protagonist, public prosecutor Helena De Ridder, has a sidekick of Moroccan descent, police inspector Zohra El Zarkaoui. Most of the time, her roots go completely unnoticed: neither her dress (mostly jeans and T-shirt) nor her language (accentless) or behaviour refer to her cultural background and Muslim religion. Occasionally, her ethnicity is explicitly addressed, for instance in an episode about gay-bashing where a Turkish Muslim is questioned by Zohra, who suspects he had a relationship with the victim but doesn't dare to admit it to her because she is also Muslim. Overall, however, Zohra is represented in a counterstereotypical way, consciously avoiding and even contradicting all stereotypes about Moroccans and Muslims as represented in Western media.

Similarly, the public broadcasting soap *Thuis* has, over the years, introduced a number of ethnic minority characters in its cast, moving to decreasingly ethnically marked representations. From 1999 to 2011, the key minority character was Mo, a plumber of Moroccan descent. His Dutch was heavily accented and he was characterized as quite conservative in relation to cultural values and gender roles, which led to criticism. Gradually, more characters were introduced, such as his daughter Aisha, who was more Westernized and modern, which led to disputes with her father, for instance about premarital sex. Both characters have since disappeared and new ethnic minority characters tend be less ethnically marked. For instance, Adil Bakkal was introduced in 2013 as yet another Moroccan plumber, whose background is hardly an issue: he is Muslim and doesn't drink alcohol, but beyond that he is smoothly integrated in the existing social networks and his origin or religion are only very occasionally referenced. At the same time, he is only one of a few minority characters in the current cast, which remains predominantly White and of Flemish origin. To Franco (2001) *Thuis*, like most Flemish TV, represses difference and represents a small, insular and family-based community confirming White ideology. Clearly, in terms of ethnicity, White ethno-cultural majority characters are the norm and ethnic minorities, if present at all, take up a marginal and increasingly unmarked position in the homogenizing representation of the Flemish nation.

Sexuality is another contentious issue in relation to diversity within the televised nation. In Western media, LGBTQs have a history moving from television underrepresentation, to negative representation and stereotyping, to a greater and more balanced representation (Fejes & Petrich, 1993; Gross, 2001). However, even if gay and lesbian characters are more frequently included, they are generally not "queer" (understood here as resisting and subverting sexual and gender categories and norms; Hall & Jagose, 2013) but rather heteronormative, as they confirm the centrality of heterosexual norms which they mostly live up to (Chambers, 2009; Provencher, 2005). Moreover, diversity remains an issue, with transgender characters only recently appearing on screen while bisexual or queer characters are still very rare; visibility is still most evident for "white, affluent, gay males who hold traditional family values" (Avila-Saavedra, 2009: 18).

Within these general contours, again, the national context impacts representations. First of all, in many parts of the world LGBTQs are excluded from the nation, as they are considered as "outsiders" whereas heterosexual families are seen as the building blocks of the nation (Pryke, 1998), also on television where they remain invisible or strongly stigmatized. Second, in regions where same-sex sexuality is visible on television, representations vary according to the legal, cultural and social context. For instance, Belgium presents a very welcome environment for LGBTQs, where same-sex marriage was legalized quite quickly (in 2003) and without much controversy (Eeckhout & Paternotte, 2011). Moreover, there is an antidiscrimination law, adoption by same-sex couples is possible, reproductive technologies are available to them, transgender rights are quite well protected and there are a number of equal opportunity policies.

As a consequence, television representations of sexual minorities and gay marriage tend to be rather mainstream and uneventful, unlike the American context where same-sex marriage has been such a controversial issue over the past decade. However, because of the small scale of the domestic market, Flemish productions tend to be oriented toward a broad, mainstream audience (in order to reach a sufficient number of viewers), which partly explains the lack of diversity and the strong normativity of lesbian and gay characters; rather than questioning heterosexual norms, they conform to them, for instance by forming long-lasting monogamous couples. While sexual minority characters appear in many shows, what is lacking is a broader range of representations also exploring the intersections of gender, sexuality and race, as presented for instance in the Netflix show *Orange is the New Black* (2013–2019).

Most lesbian and gay characters on contemporary Flemish TV are caught in this dilemma. On the one hand, they are "positive" as stereotypes and negative portrayals are carefully avoided. Particularly in the last few years, their sexuality tends to be presented in a matter of fact way, without drawing much attention to it, nor giving it a central narrative role. For instance, in *Marsman* (2014), a "quality" show about Nico Marsman who takes care of his brother with autism, Nico's daughter Femke turns out to be a lesbian. At first, we don't know this and her girlfriend Dominique is presented as "just a friend". Later on we see a scene in the park where Femke and Dominique kiss, and not much is made of this until Femke's mother (Nico's ex-wife) mistakes Dominique for her ex-husband's lover and hides in a closet, sobbing – until she's asked by her daughter to (literally) come out of the closet. So, all in all, Femke's same-sex sexuality is the object of a single silly joke, after which it is, again, taken for granted by all characters. On the face of it, this is a positive, inclusive portrayal. However, as in most Flemish TV drama, Femke and Dominique are exemplars of White, middle-class acceptability.

Soap opera *Thuis*, similarly, has a long-standing commitment to including lesbian and gay characters and giving them a central position in the storylines. For instance Ann, one of the key characters, is a lesbian and was in a relationship with Mayra. After Ann came out to her conservative mother, her sexuality was hardly a theme, although her relationship troubles were. From 2011 to 2017, she was together with Mayra and both were presented as feminine, attractive middle-class women. For Franky, the son of the central couple Frank and Simone and the first gay character on *Thuis*, sexuality was a key source of storylines. When he came out, in 2011, this led to a huge clash with his father; later, he was the victim of gay-bashing; but in 2012, the wedding of Franky with his boyfriend Tibo was the season finale and theme of the annual fan day. Despite the strong narrative focus on his sexuality, Franky was broadly considered as a role model: a core character in a daily soap, who comes out and is ultimately accepted into the community. At the same time, again, he and his boyfriend aimed for acceptability; as a plumber, Franky epitomized normative masculinity, so his portrayal, while countering certain stereotypes of gay men, was hardly subversive.

In 2013, the character of Franky left the show, travelling to the US for his honeymoon and staying there; in 2016, the character suddenly reappeared as Kaat and it is revealed that she is transgender and underwent surgery in the US. On the one hand, it is problematic that the one gay character is recycled to address transgender issues, and it is equally problematic that the whole transition takes place out of sight, Kaat appearing after the transition as a "passable" woman played by a cisgender female actress. On the other hand, Kaat is the first transgender character in Flemish television drama, and her portrayal is sympathetic and understanding. As with Franky, she manages to make a whole set of issues palatable for a broad audience; the character was created after consulting with transgender activists, and the appearance of Kaat led to a lot of media attention for transgender issues. Again, however, because of the limited number of portrayals (in this case: one) and the broad family audience they are oriented to, the representation of transgender people in Flemish drama is very normative, only presenting White, middle-class characters adjusting to mainstream gender norms – quite unlike the more daring representations of gender and sexuality in American shows such as *Orange is the New Black*, which do not have to aim for the greatest possible mainstream audience but manage, through the global streaming model of Netflix, to find a large niche audience.

## Conclusion

Globalization and the erosion of national boundaries may have been a key theme in academic research over the past decades, but the persistence and even resurgence of the national merits equal attention. Even in television, which thanks to digitization and convergence is able to cross national boundaries to an unprecedented degree, producers and audiences remain attached to national contexts. National TV markets as well as cultural values continue to be important and to strongly influence the representation and discursive construction of national and other social identities. To understand these processes, rather than generalizing accounts we need localized understandings of television in particular contexts, as this chapter aims to offer by focusing on the small European case of Flanders, in comparison with the more widely known American context which is, however, equally specific and should not be universalized.

Flemish TV epitomizes the persistent interest of audiences in content that is culturally "close to home". While the TV offer has become much larger and more diverse over the past decades, I would argue that television still retains some of the power it had in its classical age, in offering viewers identity and security:

> With one anchor, one accent, one look, one message at a time, and one viewpoint of "the way things are", the evening news gave a large majority of otherwise diverse Americans a unified, seamless, and clear-cut image of their nation, its central players, and its agenda.
>
> *(Blondheim & Liebes, 2009: 183)*

The notion of "ontological security", as discussed by Silverstone (1994) helps to understand this function of television; television, to him, is full of routines, rituals, traditions and myths, giving the viewers a sense of reassurance by "providing in its genres and its narratives stimulation and disturbance, peace and reassurance, and offering within their own order an expression and a reinforcement of the containing temporalities of the everyday" (p. 19). It is, indeed, in the representation of the everyday that the contemporary national power of television lies – in the daily reinforcement of the national as the natural order of society, echoing Billig's (1995) notion of "banal nationalism". Inclusion in this taken-for-granted portrayal of the world is empowering, as the discussion of ethnic and sexual minority characters disclosed. At the same time, the absence of less normative minority characters is problematic; the price for being included in the TV nation, certainly in Flanders, seems to be a sanitized portrayal focusing on acceptability. As a consequence, in Flanders as elsewhere, the homogenization of TV nations remains a problem and the struggle for more diversified and inclusive portrayals is an ongoing one.

## References


Avila-Saavedra, G. (2009) "Nothing *queer* about *queer* television: Televised construction of gay masculinities", *Media, Culture & Society*, 31 (1): 5–21.

Billig, M. (1995) *Banal Nationalism*. London: Sage.

Blondheim, M. & Liebes, T. (2009) "Television news and the nation: The end?", in E. Katz & P. Scannell (eds.)) Special issue: The end of television? Its impact on the world (so far). *The Annals of the American Academy of Political and Social Science*, 625 (1): 182–195.

Chambers, S. A. (2009) *The Queer Politics of Television*. London: I.B.Tauris.

CIM. (2017a) "Top 20 Programma's Noord 12/2016", Brussels: CIM (available online on www.cim.be/nl/televisie/openbare-resultaten, consulted 1 February 2017).

CIM. (2017b) "Top 20 Programma's Noord van 16/01/2017 tot 22/ 01/2017", Brussels: CIM (available online on www.cim.be/nl/televisie/openbare-resultaten, consulted 1 February 2017).

CIM. (2018) "CIM TV marktaandelen 2017", Brussels: CIM (available online on www.cim.be/nl/televisie/openbare-resultaten, consulted 6 June 2018).

Creeber, G. (2015) "Killing us softly: Investigating the aesthetics, philosophy and influence of *Nordic Noir* television", *The Journal of Popular Television*, 3 (1): 21–35.

De Swert, K., Kuypers, I., De Smedt, J., Hardy, A. & Walgrave, S. (2016) Monitor diversiteit 2016: Een kwantitatieve studie naar de zichtbaarheid van diversiteit op het scherm in Vlaanderen (unpublished internal report). Brussel: VRT.

Dhoest, A. (2007) "The national everyday in contemporary European television fiction: The Flemish case", *Critical Studies in Television*, 2 (2): 60–76.

Dhoest, A. (2014) "It's not HBO, it's TV: The view of critics and producers on Flemish "quality TV"", *Critical Studies in Television*, 9 (1): 1–22. doi: 10.7227/CST.9'1.1.2.

Eeckhout, B. & Paternotte, D. (2011) "A paradise for LGBT rights? The paradox of Belgium", *Journal of Homosexuality*, 58: 1058–1084.

Ellis, J. (2000) *Seeing Things: Television in the Age of Uncertainty*. London: I.B.Tauris.

Fejes, F. & Petrich, K. (1993) "Invisibility, homophobia and heterosexism: Lesbians, gays and the media", *Critical Studies in Mass Communication*, 10 (4): 396–422.

Flew, T. & Waisbord, S. (2015) "The ongoing significance of national media systems in the context of media globalization", *Media, Culture & Society*, 37 (4): 620–636.

Franco, J. (2001) "Cultural identity in the community soap: A comparative analysis of *Thuis* (At Home) and *Eastenders*", *European Journal of Cultural Studies*, 4 (4): 449–472.

Fricker, K. (2007) ""Quality TV" on show: The Irish Times' review of 'Quality American Television international conference, held at Trinity College, Dublin in 2004", pp. 13–16 in J. McCabe & K. Akass (eds.) *Quality TV: Contemporary American Television and Beyond*. London: I.B. Tauris.

Geraghty, C. (1991) *Women and Soap Opera: A Study of Prime Time Soaps*. Cambridge: Polity Press.

Gross, L. (2001) *Up from Invisibility: Lesbians, Gay Men, and the Media in America*. New York: Columbia University Press.

Hall, D. E. & Jagose, A. (eds.) (2013) *The Routledge Queer Studies Reader*. London: Routledge.

Hall, S. (1992) "The question of cultural identity", pp. 273–325 in S. Hall, D. Held & T. McGrew (eds.) *Modernity and its Futures*. Cambridge: Polity Press. Open University.

Hall, S. (1997) *Representation: Cultural Representation and Signifying Practices*. London: Sage.

Halse, R. (2012) "The Muslim-American neighbour as terrorist: The representation of a Muslim family in 24", *Journal of Arab & Muslim Media Research*, 5 (1): 3–18.

Haynes, A. (2007) "Mass media re-presentations of the social world: Ethnicity and "race"", pp. 162–190 in E. Devereux (ed.) *Media Studies: Key Issues and Debates*. London: Sage.

Moran, A. (1998) *Copycat TV. Globalisation, Program Formats and Cultural Identity*. Luton: University of Luton Press.

Nelson, R. (2007) *State of Play: Contemporary "High End" TV drama*. Manchester: Manchester University Press.

Newman, M. & Levine, E. (2012) *Legitimating Television: Media Convergence and Cultural Status.* New York & London: Routledge.

Provencher, D. (2005) "Sealed with a kiss: Heteronormative narrative strategies in NBC's will & grace", pp. 177–189 in M. Dalton & L. Linder (eds.) *The Sitcom Reader: America Viewed and Skewed.* Albany: SUNY Press.

Pryke, S. (1998) "Nationalism and sexuality, what are the issues?", *Nations and Nationalism*, 4 (4): 529–546.

Silverstone, R. (1994) *Television and Everyday Life.* London: Routledge.

Straubhaar, J. D. (2007) *World Television: From Global to Local.* Thousand Oaks: Sage.

Thompson, R. J. (2007) "Preface", pp. xvi–xx in J. McCabe & K. Akass (eds.) *Quality TV: Contemporary American television and beyond.* London: I.B. Tauris.

Tuckachinsky, R., Mastro, D. & Yarchi, M. (2015) "Documenting portrayals of race/ethnicity on primetime television over a 20-year span and their association with national-level racial/ethnic attitudes", *Journal of Social Issues*, 71 (1): 17–38. doi: 10.1111/josi.12094.

United States Census Bureau. (2017) *Quick Facts.* (Available online on www.census.gov/quickfacts/table/PST045216/00, consulted 15 February 2017).

VAF. (2016) *VAF/Mediafonds: Reglement, In uitvoering van de beheersovereenkomst met de Vlaamse Gemeenschap.* Brussel: VAF. (available online at www.vaf.be/downloads-mediafonds, consulted 3 February 2017).

Van den Broucke, S., Noppe, J., Stuyck, K., Buysschaert, P., Doyen, G. & Wets, J. (2015) *Vlaamse migratie- en integratiemonitor 2015.* Antwerpen: Steunpunt Inburgering en Ingratie.

VRT. (2016) *Beheersovereenkomst 2016–2020 tussen de Vlaamse gemeenschap & VRT.* Brussel: VRT.

Waisbord, S. (2004) "Media and the reinvention of the nation", pp. 375–392 in J. Downing (ed.) *The Sage Handbook of Media Studies.* Thousand Oaks: Sage.

# 26

# THE FUTURE IS NOW

## Evolving Technology, Shifting Demographics, and Diverse TV Content

*Ana-Christina Ramón, and Darnell Hunt*

For the past 80 years, television as a viewing experience has evolved with changes in technology. The small screen has gone from offering limited content through a single viewing option to offering almost unlimited content through various viewing options. The post-network era of television now offers television programming on virtually any device with a screen that can be made available on demand and not constrained by a traditional broadcast schedule. Lotz (2014) describes how television in the U.S. has changed through the decades from being solely focused on producing advertiser-supported programming with broad appeal to a newer, subscriber-based model capable of delivering high production value programming for niche audiences. In order to survive in today's market, both television networks and subscription services must increasingly appeal to diverse audiences with varied interests and viewing habits.

Indeed, while television technology has evolved over the years, the United States has also morphed demographically. In the 1930s, non-Hispanic whites constituted about 88 percent of the population. By 2016, they accounted for only about 61 percent of the population according to the U.S. Census (2017). Today's TV audiences are more racially and ethnically diverse than ever before, challenging traditional notions of the "mainstream" audience, which certainly can no longer be considered simply "white."

In this chapter, we consider findings from our *2017 Hollywood Diversity Report* (Hunt et al. 2017) in order to explore some of the ways in which the combination of evolving television technology and shifting American demographics may be creating an opening for television content options not considered viable in the past. The *2017 Hollywood Diversity Report* is part of an annual UCLA study series[1] focused on exploring the relationships between diversity and the bottom line in the Hollywood entertainment industry. The report stems from the Division of Social Science's larger Hollywood Advancement Project, which has three primary goals: (1) to generate comprehensive research analyses of the inclusion of diverse groups in film and television, including lead roles, writing, directing, producing, and talent representation; (2) to identify and disseminate best practices for increasing the pipeline of underrepresented groups into the Hollywood entertainment industry; and (3) to consider the broader implications of diverse industry access and media images for society as a whole.

While the trend in television appears to be toward increasing diversity in front of the camera, and to a lesser degree behind it, the 2017 report revealed that the television industry still has a long way to go before anything resembling proportionate representation for people of color is achieved—despite the finding replicated throughout the study series that television shows with casts reflective of America's diversity receive, on average, the highest viewer ratings. People of color account for about 40 percent of the nation's population today, a share that has been increasing by about half a percent each year in recent years. According to Census Bureau projections, America will become majority-minority by 2044 (Colby and Ortman 2014). In light of the profound technological changes reshaping the television landscape, how are content creators responding to the demographic shifts reshaping the nation? What are the possibilities for diverse programming content?

## The Evolving TV Industry

Television today can be broadly classified into two different business models—advertiser-supported broadcast and cable TV and paid subscription services. The first model relies heavily on viewer ratings as eyeballs are sold to advertisers who seek to raise awareness of and increase demand for their consumer products. The higher the ratings for desired viewer demographics, the more broadcast and cable networks can charge advertisers for a given amount of airtime for commercials to promote their products. The second model is based more on the portfolio of viewing options a given media platform makes available to its subscribers. Here, the value-added for viewers is access to a relatively diverse collection of provocative, often critically acclaimed programs that are unencumbered by commercial interruptions. Originally, this latter model was rooted in "premium" cable networks such as HBO and Showtime. More recently the trend has been toward digital streaming video-on-demand services (SVOD) like Netflix and Hulu, which have popularized a new modality of television consumption by allowing viewers to binge view entire seasons of a series on demand. It's worth noting that these two basic business models are "heavily intertwined by retransmission, syndication, digital distribution and myriad other factors" (Masnick et al. 2014, 12). TV series now can gain viewers in ways not traditionally measured and still be considered successful through DVD box set purchases, DVR-delayed viewing, on-demand viewing (Masnick et al. 2014), and/or their contributions to the overall portfolio packaged and sold to subscribers. This fact raises interesting questions about the possibilities for diverse content, given America's shifting demographics and Hollywood's tendency to privilege traditional notions of a (white) mainstream in the programming it develops and green-lights.

Because traditional television networks eventually will have to adjust to recent increases in the original programming produced by other, newer television platforms, some scholars believe this will ultimately work to the benefit of the viewer in one way or another. For example, Einav and Lipson (2015) argue that the increase in distribution models will benefit viewers by creating competition for their attention and thus requiring more high-quality programming. Lotz (2014) argues that because subscription services are not constrained by advertisers and the dictates of ratings for desired demographics, they are free to produce TV series with original stories that appeal to "niche" audiences often considered too narrow for traditional television platforms. Indeed, the most innovation and potential for change on the diversity front may rest with subscription streaming services. As of late 2017, there were over 20 video-on-demand (VOD) services. The major services that also produced original programming were Netflix, Amazon, and Hulu. With over $6 billion in revenue, Netflix was expanding at a fast rate and producing more content than any other platform (traditional or subscriber-based), with over 120 original programs in 2016 alone (Masters 2016). Netflix had a decided advantage over traditional outlets in 2017 due to its 104 million digital subscribers (Ng 2017) in 190 countries, distributed across 20 languages (Masters 2016).

Despite the dramatic rise of SVOD platforms in recent years, Nielsen reports that traditional, advertiser-supported television still reached the largest number of users at 226 million in early 2016, second only to AM/FM radio among popular media platforms. These measures also indicate that black Americans tend to watch more live TV than other groups, while Latinos tend to spend more time using digital devices (PC, smartphone, and tablet) than most Americans (Nielsen 2016). Meanwhile, SVOD penetration continues to grow exponentially. From 2015 to 2016, it increased by 19 percent to 50 percent penetration in American households (Nielsen 2016). As of the first quarter of 2016, 41 percent of black households, 51 percent of Latino households, and 66 percent of Asian American households reported subscribing to a VOD service (Nielsen 2016). If traditional television does not provide programming that is representative of or appealing to diverse audiences, is it possible that these audiences are searching for and finding other types of programming content on SVOD platforms? Is it possible that the threat posed by SVOD platforms' ability to produce relatively diverse television content has placed pressure on traditional television platforms to also develop more diverse content? We address these questions by exploring the prevalence of people of color in show casts and corresponding audience engagement data for broadcast, cable, and digital television shows.

## The Data

The *2017 Hollywood Diversity Report* (Hunt et al. 2017) examined 1,206 television shows airing or streaming during the 2014–15 season in the U.S. The 2014–15 season is defined as television programming that originally aired between September 1, 2014 and August 31, 2015. The television shows were distributed across six broadcast networks, 64 cable networks, 13 digital platforms, and through syndication. They were sorted into the following categories for analysis: 125 broadcast scripted shows; 87 broadcast reality and other shows; 204 cable scripted shows; 688 cable reality and other shows; 56 digital platform scripted shows; 39 digital platform reality and other shows; and seven syndicated shows.[2] For this chapter, we focus on the scripted shows airing or streaming on broadcast, cable, and digital platforms during the 2014–15 season. These data were compared to those from three prior television seasons (2011–12, 2012–13, and 2013–14) in order to identify any apparent trends in content diversity.

Variables considered in the analyses for the study include the following:[3]

- Racial status of lead talent
- Gender of lead talent
- Overall cast diversity
- Show creator racial status
- Show creator gender
- Show locations
- Writer diversity
- Director diversity
- Genres
- Talent agency representation
- Emmy awards
- Types of diversity initiatives
- Nielsen viewer and social media ratings[4]

## Cast Diversity and Audience Engagement

Diverse television content, we suggest, is rooted in storytelling that reflects and resonates with the experiences of a wide range of different groups in society. It is inclusive content that balances the needs, hopes, and fears of dominant groups—expressed through the struggles and triumphs of featured characters in the narrative—with the issues salient to groups traditionally subordinated in "mainstream" television content. In other words, it is content that moves the stories of women and racial and ethnic minorities from the margins to the center, exploring them alongside the stories of white men that have typically organized television narratives. To truly get at the degree to which television content can be said to be "diverse," one would have to engage in careful textual analyses that consider the relationships between the stories constituting the narratives and the relative positioning of differently raced and gendered characters. Whose stories are being told? How inclusive is this mix of storytelling with respect to demographic realities? This, of course, is a qualitative, nuance-dependent project that necessarily sacrifices generalizability for specification and interpretation.

With this realization in mind, we rely in this chapter upon objective, quantitative measures of content diversity as a proxy for the admittedly more qualitative concept of actual content diversity. That is, we employ cast diversity, which can be measured reliably over time, as a proxy that provides us with a bird's-eye view of differences in content diversity for hundreds of television shows across the platforms of interest, over time. The first question we ask is what relationships, if any, exist between cast diversity and audience engagement? Moreover, to what degree might any relationships pertaining to audience engagement vary by the ethnic and racial makeup of viewers?

Table 26.1 presents median ratings (aged 18–49, white households, black households, Latino households, and Asian American households) by minority cast share[5] for broadcast scripted shows from the 2014–15 season. For the purpose of analysis, we partitioned broadcast scripted shows into six distinct categories: (1) shows with casts that are 10 percent minority or less, (2) shows with casts that are from 11 percent to 20 percent minority, (3) shows with casts that are from 21 percent to 30 percent minority, (4) shows with casts that are from 31 percent to 40 percent minority, (5) shows with casts that are from 41 percent to 50 percent minority, and (6) shows with majority-minority casts. The general upward trajectory of the lines suggests that cast diversity has a meaningful, positive relationship to audience engagement and the ratings bottom line. (It should be noted that previous reports in the *Hollywood Diversity Report* series found similar relationships for the three previous seasons.)

*Table 26.1* Median ratings by minority cast share, 18–49 and HH race, cable scripted shows, 2014–15 season. (n=198)

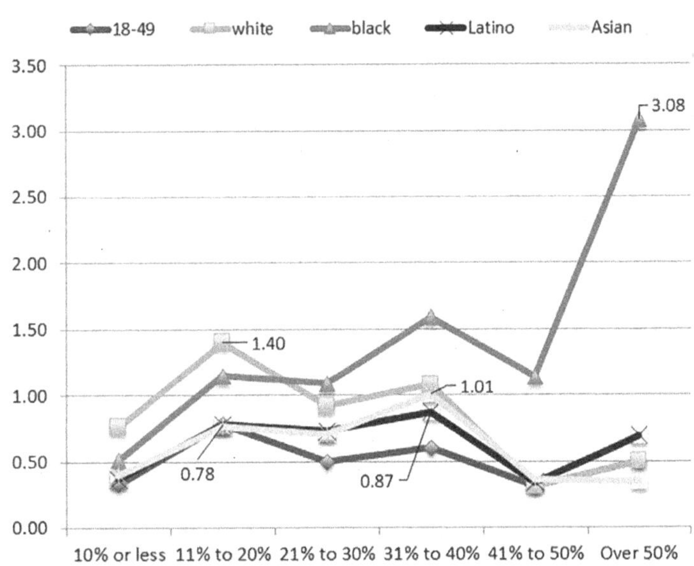

*Table 26.2* Media tweets and authors, by minority cast share, broadcast scripted shows, 2014–15 season. (n=112)

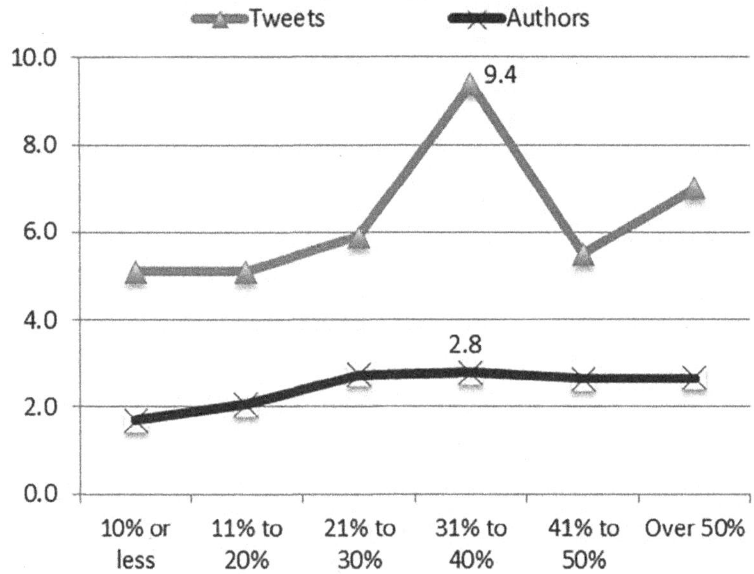


For white households (4.98 ratings points), black households (6.24 ratings points), and Latino households (2.77 ratings points), median ratings peaked for shows featuring casts that were majority-minority. Eleven shows fell into this cast diversity interval during the 2014–15 season. Examples include: *Empire* (Fox), *Black-ish* (ABC), *Law and Order: Special Victims Unit* (CBS), *Hawaii Five-O* (CBS), and *The Carmichael Show* (NBC). Meanwhile, for viewers 18–49 (2.37 ratings points) and Asian American households (2.96 ratings points), median ratings peaked for shows with casts that were from 41 to 50 percent minority. Examples include: *Brooklyn Nine-Nine* (Fox), *Elementary* (CBS), *New Girl* (Fox), and *Sleepy Hollow* (Fox). In short, the most popular broadcast scripted shows for the 2014–15 season were typically shows with casts that were at least as diverse as the U.S. population in terms of race and ethnicity.

Another way to measure audience engagement with television content is to consider how often people tweet about a show in the aftermath of its airing. Table 26.2 charts the median volume of tweets for and unique authors for broadcast scripted shows by cast diversity interval for the 2014–15 season. It shows that the median volume of tweets (9,400) peaked for shows with casts that were from 31 to 40 percent minority—the cast diversity interval that contained the minority share of the population in 2014–15. The median number of unique Twitter authors (2,800) also peaked in this interval, though the differences across the intervals were less pronounced than with the volume of tweets.Table 26.3 applies the ratings analyses presented above for broadcast scripted shows to the case of cable scripted shows from the 2014–15 season. Though the trajectory of the lines reveal that relatively diverse cable scripted shows excelled with most audience segments, the picture for cable is considerably more complex than the one we present above for broadcast. This is largely due to the niche marketing strategy adopted by key cable networks that almost exclusively target specific ethnic and/or

*Table 26.3* Median ratings by minority cast share, 18–49 and HH race, broadcast scripted, 2014–15 seasons. (n=127)

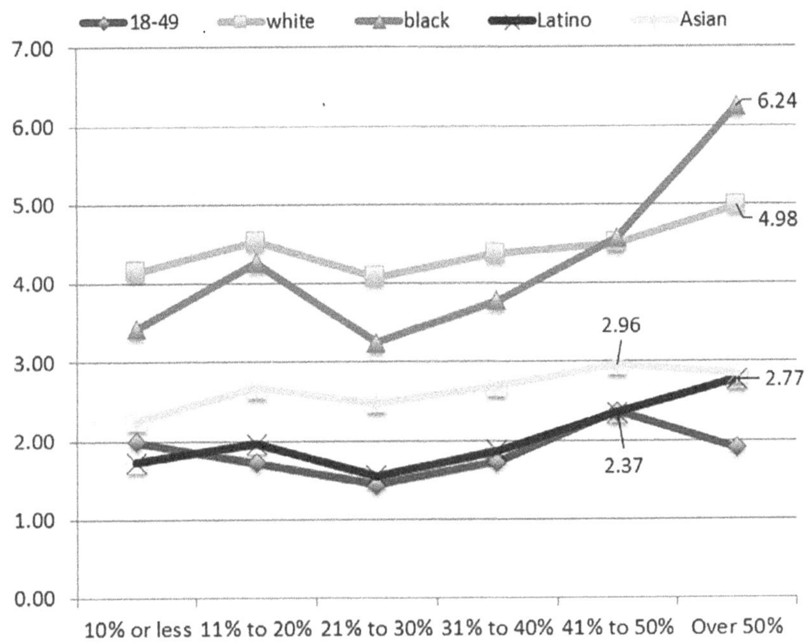


racial groups (e.g., BET, TV One, El Rey, etc.). Indeed, median ratings peaked among black households (3.08 ratings points) for cable scripted shows that had majority-minority casts. Twenty shows fell into this diversity interval during the 2014–15 season, including several black-themed sitcoms and dramas. Examples include: *Being Mary Jane* (BET), *The Game* (BET), *Survivor's Remorse* (Starz), *Power* (Starz), *The Haves and the Have Nots* (OWN), and *Love Thy Neighbor* (OWN). Meanwhile, among Asian American households (1.01 ratings points) and Latino households (.87 ratings points), median ratings peaked for cable scripted shows featuring casts that were from 31 to 40 percent minority—the cast diversity interval encompassing the minority share of the U.S. population in 2015 (i.e., 38.4 percent). This interval contained 19 shows during the 2014–15 season. Examples include: *Fear the Walking Dead* (AMC), *Major Crimes* (TNT), *Suits* (USA), and *Jessie* (Disney). By contrast, median ratings for white households (1.40 ratings points) and viewers 18–49 (.78 ratings points) peaked for cable scripted shows with casts that were from 11 to 20 percent minority during the 2014–15 season. Thirty-six shows fell into this cast diversity interval. Examples include: *American Horror Story* (FX), *Salem* (WGN), *Sons of Anarchy* (FX), and *The Librarians* (TNT).

Table 26.4 charts the median volume of tweets for cable scripted shows by cast diversity interval for the 2014–15 season. It shows that the median volume of tweets (3,400) peaked for cable scripted shows with casts that were from 21 to 30 percent minority. Majority-minority cast shows and those with casts that were from 11 to 20 percent minority followed closely behind, both with median volume of tweets of 3,200. Meanwhile, the median number of unique Twitter authors (1,500) peaked for cable scripted shows with majority-minority casts.

*Table 26.4* Median tweets and authors, by minority cast share, cable scripted shows, 2014–15 season. (n=187)

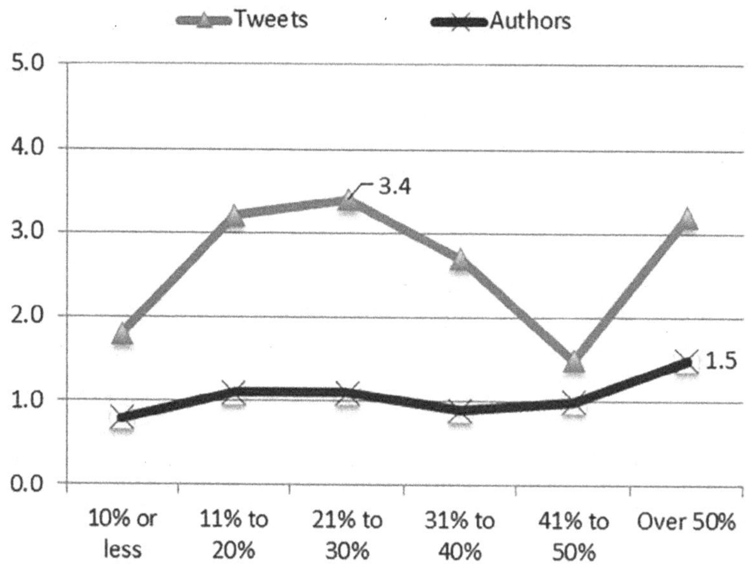


In short, ratings data from the 2014–15 season—like those from the three previous seasons examined in the *Hollywood Diversity Report* series—suggest that cast diversity had a meaningful, positive relationship with audience engagement. Of course, Nielsen did not report audience ratings for digital scripted shows over the period considered in this study, and thus the type of analysis we conducted for broadcast scripted and cable scripted shows could not be completed for this arena. Nonetheless, there is every reason to expect that audiences' apparent preference for scripted programming populated by characters who looked like them, whose stories resonated with their own experiences in a diversifying America, would also be a prominent feature of audience engagement with content in the digital scripted arena.

## Increasing Diversity in Television Content

As we argue above, the combination of shifting audience demographics and new television technologies such as SVOD creates, in theory, a golden opportunity for moving beyond some of the barriers that have traditionally stood in the way of more diverse television content. In this section, we begin to explore this possibility by considering recent trends in the prevalence of diverse television show leads, overall cast diversity, and minorities' overall share of roles across the three major types of television platforms (broadcast, cable, and digital). To what extent have show development practices across the different types of platforms reflected audiences' demonstrated preference for content that reflects America's diversity?

## *Lead Actors*

Lead actors[6] are significant figures in the television arena because, more often than not, the storytelling from week to week revolves around the characters they play. These central characters are usually the show's protagonists, and it is their hopes, fears, and achievements that drive their respective show narratives. As discussed throughout the *Hollywood Diversity Report* series, people of color and women traditionally have been underrepresented among television leads, which raises questions about audience engagement with the images of American society in circulation. Indeed, some research suggests that audience members are more likely to engage with characters with whom they identify and can relate (Weaver 2011). Race, ethnicity, and gender are core dimensions of identity that thus become highly salient in the optics of the on-screen television environment.

Table 26.5 shows that in the broadcast television arena, the trend has been toward increasing diversification among leads in recent years. Minority actors claimed 11.4 percent of the lead roles in broadcast scripted programming during the 2014–15 season—more than double the 5.1 percent figure posted for the 2011–12 season. Though the trend line for minority leads has a clear and steady upward trajectory, minorities remained seriously underrepresented in the broadcast scripted arena. That is, given that minorities accounted for 38.4 percent of the population in 2015, their 2014–15 share of broadcast scripted lead roles corresponded to underrepresentation by a factor of more than 3 to 1.

The picture was less clear for cable television. Table 26.6 shows that minority actors accounted for 15.8 percent of the lead roles in cable scripted shows during the 2014–15 season, down about a percentage point from the 2013–14 figure (16.6 percent). The minority share of leads in cable had peaked at 19.3 during the 2012–13 season, suggesting that progress on the diversity front for leads in cable was more complicated than in broadcast. Still, consistent with earlier reports in the series, the minority share of leads in cable remained considerably larger than the corresponding figures in broadcast, most likely due to the fact that several cable networks market primarily to minority (or "urban") audiences. Minorities were underrepresented by a factor of more than 2 to 1 among cable scripted leads during the 2014–15 season.

*Table 26.5* Leads by race, broadcast scripted, 2011–12 to 2014–15 seasons

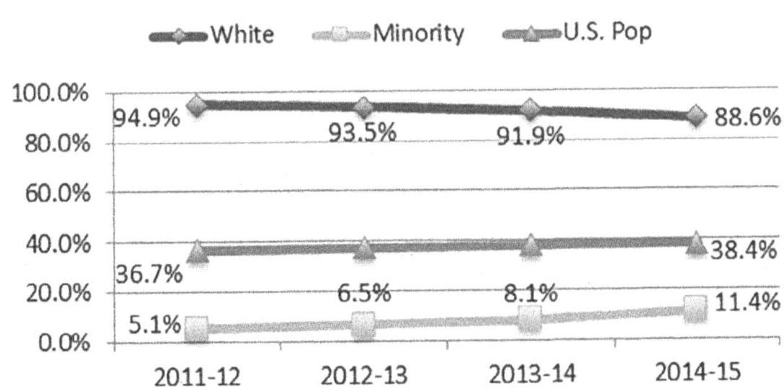

*Table 26.6* Leads by race, cable scripted, 2011–12 to 2014–15 seasons

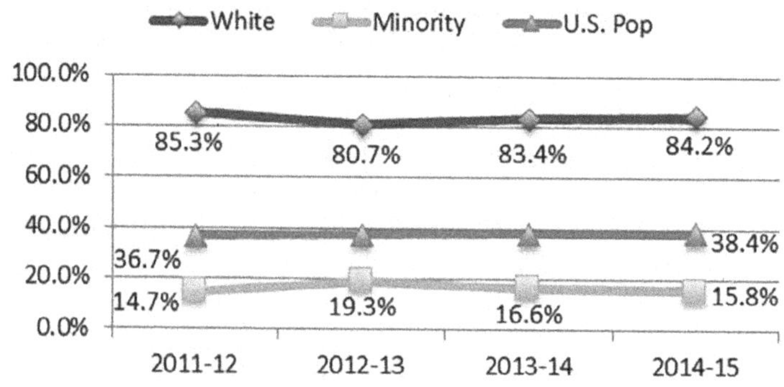


*Table 26.7* Leads by race, digital scripted shows, 2013–14 and 2014–15 seasons

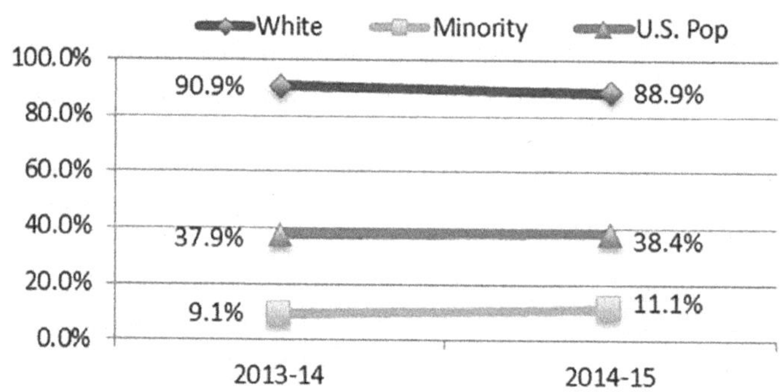


Though only two years of complete data were available for digital platforms, it appears as if both the number of digital shows and the diversification of digital leads may be on upward trajectories. Table 26.7 reveals that the number of scripted digital shows increased 60.6 percent between the 2013–14 and 2014–15 seasons, from 33 to 54 shows. Meanwhile, minorities accounted for 11.1 percent of lead roles in digital scripted shows for the 2014–15 season, up from 9.1 percent a season earlier. This figure is comparable to the minority share of leads in broadcast scripted programming for the season (11.4 percent) but significantly lower than the group's share in cable scripted programming (15.8 percent). As in the broadcast scripted arena, minorities were underrepresented by a factor of more than 3 to 1 among leads on digital scripted shows.

*Table 26.8* Minority cast share, by share of broadcast scripted shows, 2011–12 to 2013–14 seasons

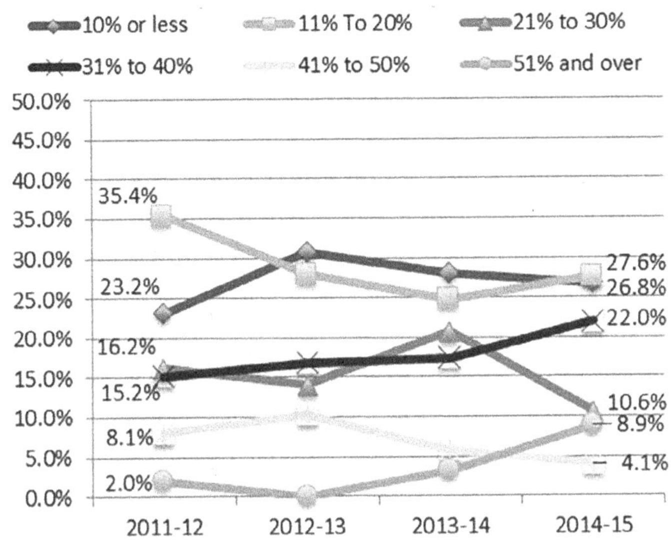


## *Overall Cast Diversity*

The overall racial and ethnic makeup of a television show's regular cast presents us with another measure of the degree to which television content may be diversifying. Table 26.8 presents findings for the minority cast share in broadcast scripted shows from the four seasons, 2011–12 through 2014–15. Like the racial and ethnic makeup of show leads, this measure also suggests that diverse content was on the rise in the broadcast scripted arena. Most notably, the share of broadcast scripted shows with majority-minority casts more than quadrupled between 2011–12 and 2014–15, from just 2 percent to 8.9 percent of all shows. Meanwhile, broadcast scripted shows with casts from 31 percent to 40 percent minority—the interval in which the 38.4 percent minority share of the population fell in 2015—also increased significantly, from 15.2 percent of all shows in 2011–12 to 22 percent in 2014–15. While the share of the least diverse broadcast scripted shows also increased over the four seasons (those with casts 10 percent minority or less), the increase was smaller and the trend has been downward since the 2012–13 season.

Table 26.9 presents the same analysis for cable scripted television shows over the four seasons. Here the pattern is altogether different from what we see for broadcast scripted shows. The least diverse cable scripted shows—those with casts that were 10 percent minority or less—accounted for lion's share of the total in 2014–15 (45.1 percent), a share that had increased significantly since the 2011–12 season's 37 percent figure. By comparison, the share of shows with majority-minority casts increased only marginally over the four seasons, from 8.4 percent of all shows in 2011–12 to 9.9 percent in 2014–15. Meanwhile, the share of cable scripted shows with casts from 31 percent to 40 percent minority actually declined, from 13 percent of all shows in the 2011–12 season to just 9.4 percent in 2014–15.

*Table 26.9* Minority cast share, by share of cable scripted shows, 2011–12 to 2014–15 seasons

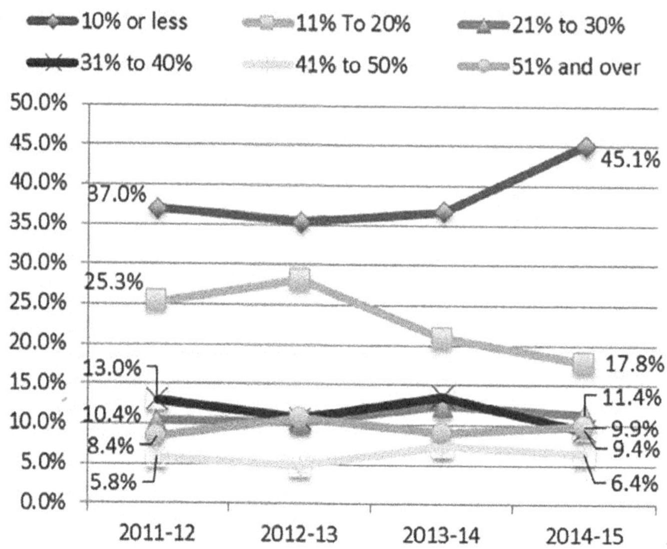


For digital scripted shows, Table 26.10 reveals that minority cast share increased as the number of original shows skyrocketed over the three seasons for which data are available. There were 54 digital scripted shows streaming during the 2014–15 season, compared to just 11 two seasons earlier—a 391 percent increase. Notably, there were no digital scripted shows for the 2012–13 season with casts from 31–40 percent minority, the interval containing the 38.4 percent minority share of the population. But by the 2014–15 season, 13 percent of digital scripted shows fell into this diversity interval. Meanwhile, digital scripted shows with majority-minority casts increased from 9.1 percent of all shows in the 2012–13 season to 11.1 percent in 2014–15. By contrast, the least diverse digital scripted shows (i.e., with casts 10 percent minority or less), though still the largest category of digital scripted shows, declined markedly over the three seasons—from 45.5 percent of all shows in 2012–13 to 38.9 percent in 2014–15.

## *Overall Share of Roles*

A final measure of content diversity considered in this study is the overall share of roles in broadcast, cable, or digital scripted programming that is portrayed by minority actors versus white ones. This measure differs from the overall cast diversity measure discussed above, which is based on individual shows as the unit of analysis as opposed to the arena (broadcast, cable, or digital) as a whole. Table 26.11 reveals that white actors claimed 76 percent of the 806 roles we examined in broadcast scripted programming for the 2014–15 season, down a bit from the 80 percent share the group posted a season earlier. Meanwhile, minorities combined took 24 percent of all roles, up from the 20 percent share evident in 2013–14. Breakdowns for specific minority group shares

*Table 26.10* Minority cast share, by share of digital scripted shows, 2012–13 to 2014–15 seasons

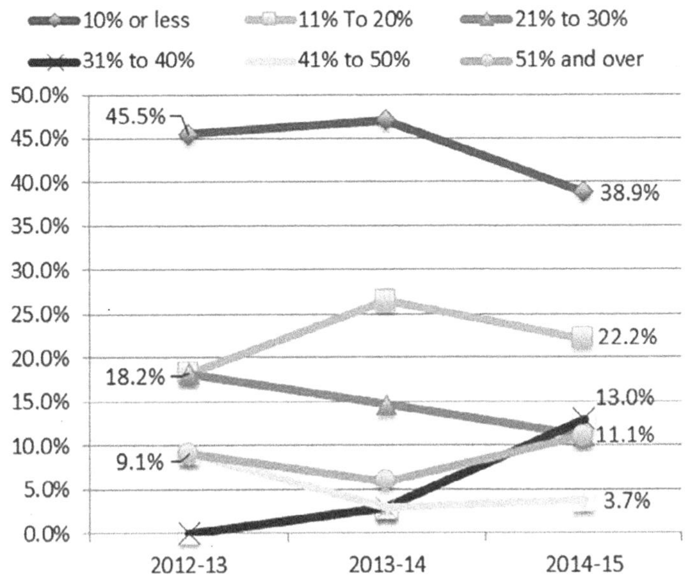


*Table 26.11* Share of roles, by race, broadcast scripted shows, 2014–15 season

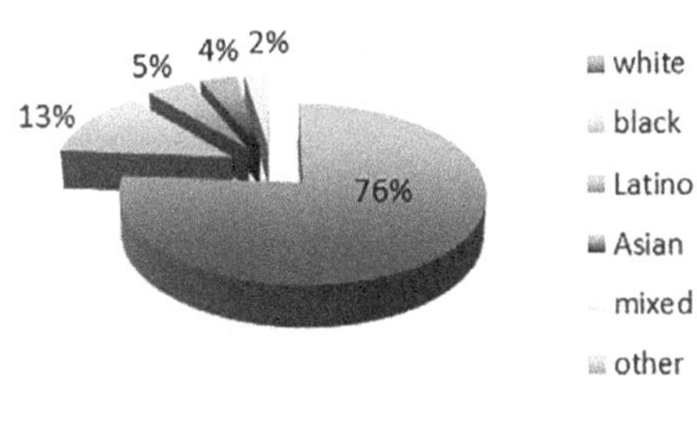


that season include: black, 13 percent; Latino, 5 percent; Asian American, 4 percent; Native Americans, 0 percent; and mixed, 2 percent.

For the cable arena, there was no change over the four seasons. Table 26.12 shows that white actors claimed 79 percent of the 1,141 cable scripted roles, while minorities combined took 21 percent. These white and minority shares were nearly identical to those we documented for

*Table 26.12* Share of roles, by race, cable scripted shows, 2014–15 season

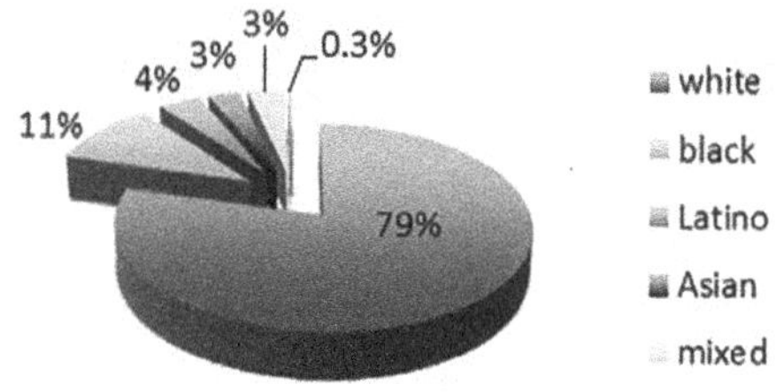


*Table 26.13* Share of roles, by race, digital scripted shows, 2014–15 season

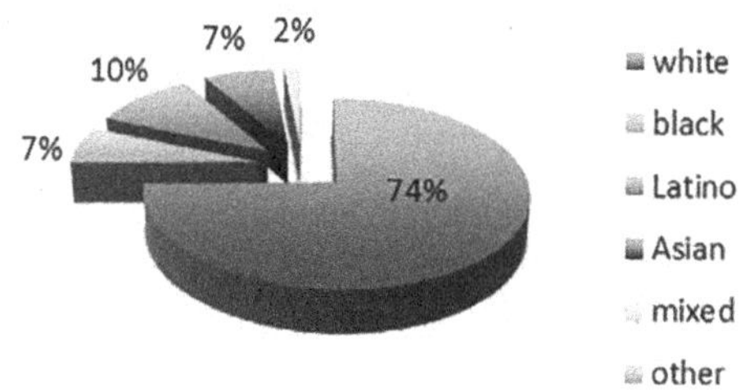


the 2013–14 season. African American actors accounted for more than half of all minority roles in scripted cable shows in 2014–15. Claiming 11 percent of the roles in this arena, the group nearly matched its share of the U.S. population. The shares of cable scripted roles claimed by other minority groups were similar to the small shares observed in broadcast scripted programming (4 percent for Latinos, 3 percent for Asian Americans, 3 percent for mixed actors, and 0 percent for Native Americans[7]).

Consistent with earlier findings, we see the largest change in the digital scripted arena. Table 26.13 shows that white actors claimed 74 percent of the 301 scripted roles examined in digital platform programming during the 2014–15 season, down from the 80 percent figure posted a season earlier. Accordingly, minorities combined took 26 percent of the roles in 2014–15, up from the 20 percent share they claimed in 2013–14. The white/minority breakdown in digital scripted programming is similar to that for both broadcast scripted and cable scripted programming, though minority actors fared slightly better overall in digital during the 2014–15 season. Latino actors claimed the largest share of all roles among the minority groups,

10 percent, while African Americans accounted for 7 percent of all roles, Asian Americans another 7 percent, and mixed actors 2 percent.

Taken together, these findings support the idea that the combination of technological change and shifting demographics may present unique opportunities for the development of diverse television content. Previous reports in the *Hollywood Diversity Report* series found that most digital scripted programs had casts that were considerably less diverse than those on broadcast or cable television (Hunt et al. 2014, 2016, 2017; Hunt and Ramón 2015). In an effort to cement their place among the more prominent content options for television viewers, the decision makers behind early digital platforms relied heavily on "name" (i.e., mostly white and male) producers to develop new scripted programs. But as SVOD became more integrated into consumers' viewing choices and the production of original, digital scripted programs increased, it appears as if digital platforms also expanded their portfolios of content options to feature shows with more diverse casts and storylines (and in some cases, behind-the-scenes talent)—the types of shows our earlier findings suggest that diverse audiences engage with the most.

## Conclusion

Longtime critics of Hollywood's response to its diversity problem are often firm in their cynicism: meaningful progress has been mostly illusory, as the industry traditionally has greeted pressures for change with momentary appeasements that dissolved the instant pressures were relaxed (e.g., see Hunt 2005; Cruz 2016). This is a situation where diversity may be "in" now but business-as-usual will soon return. The resulting cycle of progress and retreat on the industry diversity front, they argue, has resulted in an industry that continues to be dominated by white men and to privilege their stories. The question we consider in this chapter is whether the twin factors of recent technological changes and radical shifts in audience demographics may have fundamentally disrupted this cycle for television.

The combined buying power of people of color in America approached $3.5 trillion in 2015 (Humphreys 2015), and these growing, diverse audience segments watched more television on a per capita basis than their white counterparts. Not appealing to Latinos and Asian Americans, the two fastest growing demographic groups in particular, is done at the industry's own peril. Latinos and Asian Americans are already attuned to using new technologies to meet their entertainment needs and are thus ready to wield their considerable buying power to demonstrate a preference for content that resonates with their experiences. With digital networks such as Netflix investing in increasingly large amounts of original content, there is more opportunity for the development of diverse programming that caters to a receptive audience of diverse subscribers. SVOD seems poised to benefit the most from these trends, given its business model and the huge observed increases in digital platforms' diverse programming in the past few seasons. However, in the race to remain competitive, we believe, broadcast and cable networks will likely follow suit. Indeed, our data suggest that *both* digital networks/services *and* broadcast networks are trending upward in terms of racially diverse programming.

In the final analysis, the concept of "niche" audience, which traditionally has been defined in terms of particular demographic groups, will soon be (and in some places, already is) conceptualized as part of the "mainstream." Younger and more diverse audiences, in particular, are demanding content that is relevant and relatable to their lives. This present reality will become only more engrained in the future. If television is to remain relevant for America's increasingly diverse audiences, business-as-usual approaches to content development will have to be abandoned for good.

## Notes

1 The annual report was previously housed at the Ralph J. Bunche Center for African American Studies at UCLA. It is now housed at the Institute for Research on Labor and Employment (IRLE) at UCLA. The most recent report was released February 2018.
2 Total numbers for the various analyses in this report may deviate from the overall total for the number of films or television shows considered due to focused analyses on a subset of the data (noted below) or missing values for some cases (e.g., many reality shows do not identify a cast, directors, or writers).
3 These data were compiled from a variety of sources that industry stakeholders rely upon for day-to-day updates on industry developments. These sources include The Studio System, Variety Insight, the Internet Movie Database (IMDb), and Nielsen.
4 Household ratings (HH) are defined as the percentage of the universe of households tuned to a particular TV program during the average minute of the program. This includes incremental viewing to programs watched at the time of the telecast as well as watched in DVR playback that occurs within seven days of the original telecast. The HH ratings presented by race are based on the race of the head of household, while 18–49 ratings are based on individual viewers. Social media ratings consist of tweets and unique authors, which are measures of relevant U.S. Twitter activity from three hours before through three hours after a broadcast, local time. "Tweets" are ascribed to a linear TV episode, while "unique authors" refer to unique Twitter accounts that have sent at least one tweet ascribed to a specific TV episode.
5 The cast diversity measure used in this study is based on the first eight credited actors for a given television show.
6 "Lead role" is defined in this report as the first credited actor/performer for a given project's list of cast members.
7 The number for native actors was so low (just three) that the Native American share of cable scripted roles rounded to 0 percent.

## References

Colby, Sandra, and Jennifer M. Ortman. 2014. "Projections of the Size and Composition of the U.S. Population: 2014 to 2060." *Current Population Reports*. P25-1143. U.S. Census Bureau, Washington, DC. Accessed December 3. www.census.gov/content/dam/Census/library/publications/2015/demo/p25-1143.pdf.

Cruz, Lenika. February 26, 2016. "The Missing Piece of the Oscars' Diversity Conversation." *The Atlantic*. Accessed December 4. www.theatlantic.com/entertainment/archive/2016/02/hollywoods-black-and-white-filter/460326/.

Einav, Gali, and Nathan Lipson. 2015. "The Times They Are A' Changin' … From Newspapers to TV, Traditional Shifts to Digital." In *The New World of Transitioned Media. The Economics of Information, Communication, and Entertainment (The Impacts of Digital Technology in the 21st Century)*, edited by Gali Einav, 81–101. Switzerland: Springer, Cham.

Humphreys, Jeffrey M. 2015. *2015 Multicultural Economy Report*. Athens, GA: Selig Center for Economic Growth, University of Georgia.

Hunt, Darnell M. 2005. *Channeling Blackness: Studies on Television and Race in America*. New York: Oxford University Press.

Hunt, Darnell M., and Ana-Christina Ramón. 2015. *2015 Hollywood Diversity Report: Flipping the Script*. Ralph J. Bunche Center for African American Studies at UCLA. https://socialsciences.ucla.edu/deans-initiatives/hollywood-diversity-report/.

Hunt, Darnell M., Ana-Christina Ramón, and Zachary Price. 2014. *2014 Hollywood Diversity Report: Making Sense of the Disconnect*. Ralph J. Bunche Center for African American Studies at UCLA. https://socialsciences.ucla.edu/deans-initiatives/hollywood-diversity-report/.

Hunt, Darnell M., Ana-Christina Ramón, and Michael Tran. 2016. *2016 Hollywood Diversity Report: Busine$$ as Usual?* Ralph J. Bunche Center for African American Studies at UCLA. https://socialsciences.ucla.edu/deans-initiatives/hollywood-diversity-report/.

Hunt, Darnell M., Ana-Christina Ramón, Michael Tran, Amberia Sargent, and Vanessa Díaz. 2017. *2017 Hollywood Diversity Report: Setting the Record Straight*. Ralph J. Bunche Center for African American Studies at UCLA. https://socialsciences.ucla.edu/deans-initiatives/hollywood-diversity-report/.

Lotz, Amanda D. 2014. *The Television Will Be Revolutionized*. New York: New York University Press.

Masnick, Michael, Michael Ho, Joyce Hung, and Leigh Beadon. October 2014. *The Sky Is Rising: 2014 Report*. CCIA and floor 64. www.ccianet.org/wp-content/uploads/2014/10/Sky-Is-Rising-2014.pdf.

Masters, Kim. September 14, 2016. "The Netflix Backlash: Why Hollywood Fears a Content Monopoly." *The Hollywood Reporter.* Accessed December 4. www.hollywoodreporter.com/features/netflix-backlash-why-hollywood-fears-928428.

Ng, David. July 29, 2017. "Netflix Is on the Hook for $20 Billion. Can It Keep Spending Its Way to Success?" *Los Angeles Times.* Accessed December 4. www.latimes.com/business/hollywood/la-fi-ct-netflix-debt-spending-20170729-story.html.

The Nielsen Company. 2016. *The Nielsen Total Audience Report: Q1 2016.* New York: The Nielsen Company.

U.S. Census. 2017. "Quick Facts: United States." *U.S. Census.* Accessed December 4. www.census.gov/quickfacts/fact/table/US#viewtop.

Weaver, Andrew J. April 2011. "The Role of Actors' Race in White Audiences' Selective Exposure to Movies," *Journal of Communication* 61 (2): 369–385. doi:10.1111/j.1460-2466.2011.01544.x

# 27

# LOCALIZING MEDIA CONTENTS

## Technological Shifts, Global and Social Differences and Activism in Audiovisual Translation

*Frederic Chaume*

### Localizing Media Contents

Three key developments are radically changing the media world: technological change, global and social difference, and the new active role of audiences. These three major changes have a huge impact not only on media production and distribution, but also on media localization and consumption. Regarding the first, technological change, digital transformation has brought about the widespread introduction of information technology (IT) in all markets. In the field of media communication, digitalization facilitates the creation, production, distribution and potential manipulation of new audiovisual contents. Digitalization has also led to the multiplication of audiovisual distribution platforms and devices (TV stations, video-on-demand platforms, streaming services, mobile telephones, etc.), which, broadly speaking, are cheap and user-friendly. For more people all the time, we are living in the so-called age of convergence, where almost everything can be accessed via smartphones, tablets and laptop computers, where an audiovisual product can be broadcast across multiple platforms and formats through current digital technologies.

Digital technology is also playing a pivotal role in shifts in the booming market of localization practices. Localization is an overall term used to describe any kind of *audiovisual translation* practice, namely, the transfer of audiovisual texts either interlingually or intralingually, such as dubbing, subtitling, voice-over, subtitling for the deaf and the hard of hearing, audiodescription, surtitling, respeaking, videogame localization, etc., *and* any kind of *media adaptation*, such as format licensing, adaptations, transcreations and remakes – in fact, all types of translation understood in the broadest sense of the concept. The staggering number of hours of audiovisual content being localized – translated and adapted – every single day, and the speed with which this is occurring, as well as giving audiences a wider and better choice has led to a growing diversity in audiovisual content consumption and in the use of different translation practices. This decisive role of digital technology is the main reason behind the present state of great activity in new ways of producing and consuming audiovisual products, in optimizing use of new devices (laptops, tablets, smartphones), and also in new forms of communication (social networks, crowdsourcing).

All these technological innovations entail the use of new localization practices, in the form of new audiovisual transfer modes, through a combination of traditional audiovisual transfer modes, and in the incipient use of machine translation (MT) in traditional audiovisual translation modes. Audiovisual content transcends all borders thanks to audiovisual translation; at the same time audiovisual translation continues to stretch its own limits to accommodate new ways of localizing products.

## Global Difference: Different Localization Practices

The history of Translation Studies has taught us that, broadly speaking, there are two major translation methods: either the target text – the actual final translation – keeps the flavor and characteristics of the original text; or it is accommodated to the customs and tastes of the target culture that will consume and enjoy it, regardless of the characteristics of the original text. These two methods have been labeled differently over time, but Venuti's (1998) foreignization–domestication dichotomy now seems to be the most widely accepted term. According to the author, translators should convey the cultural norms and conventions of the source language, preserving their meaning and their foreignness, to the target-language text. Every strategy and technique used in the translation process – from the selection of foreign texts to the implementation of particular translation techniques and including the review process – should reflect the original values embedded in the original text. However, instead, it is usually mediated by the multifarious cultural values that govern the target receiving culture and circulate in the target language. There is therefore always a battle in translation, a kind of arm wrestling match, between preserving the original flavor and characteristics of the original text, what Toury (1995) terms adequacy, and bowing to the demands of the target recipients, conventions, tastes and habits. Although Venuti advocates foreignization, not to forcefully erase the cultural values of the original text, but instead to highlight the foreignness of the source text and thus prevent the dominant target culture from assimilating the differences of the source culture, in the real world the opposite seems to occur, especially in the world of media communication. The romantic, sometimes pamphleteering claims he makes clash with how real markets work and how economic flows determine the industry of localization.

Audiovisual translation has always been dominated by a balance between respect for the original contents and, at the same time, fluent domestication. Venuti heavily criticized translators who, in order to minimize the foreignness of the target text, reduce the foreign cultural norms to target-language cultural values. However, one of the premises of most media localization practices is precisely to create a text that appears to have been written (and sometimes shot) in the target language and culture, and that follows the cultural norms of the target audience. The author unrealistically advocates the foreignization strategy, in order to highlight the linguistic and cultural differences belonging to the foreign text, a utopia achieved only by some fansubs and fandubs, precisely those translation modes that run parallel to real markets. This approach contrasts sharply with the market-based approach more likely adopted in media localization, whereby choices are customized and determined by target culture needs and preferences.

In fact, different target cultures use different audiovisual transfer modes to bring foreign texts closer to their audiences. Audiovisual translation modes respond to the way a particular culture wants to consume a foreign audiovisual text. And these modes are deeply rooted in cultures, since habits and tastes are not easily changed in a short space of time. Modes of audiovisual translation are understood as all types of audiovisual text transfer between two languages and cultures (interlingual) or within the same language and culture (intralingual, such as the accessibility modes: subtitling for the deaf and the hard of hearing, audiodescription for the blind and visually impaired, respeaking, audiosubtitling, etc.). Essentially, translations of audiovisual texts entail introducing a target text with the translation or reproduction of the dialogues and inserts (captioning) on or next to the screen, or by inserting a new soundtrack in a different language and either cancelling out the original soundtrack of the source language dialogues (dubbing) or leaving it in place (voice-over). In other words, the audiovisual text is either subtitled or revoiced (Chaume, 2012). What follows is a description of the most relevant forms of localization used all over the world, which shows how global difference shapes media contents in distinct ways.

## Dubbing

Dubbing consists of translating and lip-syncing the script of an audiovisual text, which is then performed by actors directed by a dubbing director and, where available, with advice from a proofreader and sometimes from a dubbing assistant. This complex process is linguistic and cultural, but also technical and artistic, in which teamwork is vital to produce a high-quality end product. Dubbing is a deeply rooted practice in some countries all over the world, especially in the case of TV and video-on-demand (VoD) platforms, where it tends to be more popular than in cinema. For historical, economic and cultural reasons (Chaume, 2013), it has become the common mode of localizing foreign audiovisual products for TV and VoD. A detailed map of the dubbing countries can be found in some publications (Chaume, 2012). However, for the purpose of this research, it is interesting to look at an example of how digitalization has dramatically changed the world of dubbing. Digital technology has changed the way voices are recorded; for example, dedicated software is now available that reproduces the sound waves of the original actors' voices, so isochrony – when the original and target dialogue lines are the same length, a standard of quality in dubbing – no longer presents the problems it used to. Bad isochrony obliged actors to redub their lines, perhaps on a different day, if the mistake was noticed after the dubbing session; nowadays, however, the engineer simply makes the recorded lines in the target language fit the sound waves of the original voice to ensure perfect isochrony. But technology has also helped to delocalize dubbing, to the point where voice talents are able to dub in their own homes and send the audio files to the dubbing company, as in the case of *collaborative dubbings* – a process whereby voice talents join in live dubbing sessions from anywhere in the world, at any given time. This ground-breaking step allows companies to work on the same project in many different places. Technology also allows translators to work in teams, or in the cloud, and so on.

Collaborative dubbings also reveal a hitherto unknown feature: voice talents living in different countries have different accents, and although they may mask them in favor of a more neutral, standard accent, the dubbing might show indications of their respective origins, as is the case in Latin-American collaborative dubbings (Scandura, 2015). In these cases, actors can dub their lines at home in different cities and send the audio files to the dubbing company, where the engineer will mix them up and send the final edited dubbing to their client (usually TV stations or VoD platforms).

## Voice-overs

Voice-overs, another revoicing mode to localize foreign products, are edited by broadcasting the audio track with the recording of the original dialogue at the same time as the track with the translated version. The sound engineer reduces the volume of the original soundtrack and raises the volume of the dubbed track, such that the original text can hardly be heard in the background below the translated text. This mode is widely used for documentaries in most countries, and to translate fictional film texts (films, TV series) in Poland, Russia and other former Soviet Union countries such as Estonia, Latvia and Belarus. The now hugely popular reality show genre prefers this audiovisual translation mode for localization across borders. A new audiovisual transfer mode, called *simil-sync*, has also appeared in Italy; this is half way between dubbing and voice-over: like dubbing but unlike voice-over, simil-sync respects isochrony (same length of utterances), and unlike dubbing but like voice-over it does not respect proper lip-sync (matching the translation with the bilabial and labiodental phonemes, and open vowels in close-ups and extreme close-ups). New genres, then, also create new audiovisual transfer modes, new ways of becoming localized.

## *Simultaneous Interpretation of Film*

Simultaneous interpretation of films is a little used audiovisual translation mode that is in decline. In this process, an interpreter, who is present in the cinema or location where the film is being screened, interprets the film and translates and voices-over the screen actors' voices through a microphone connected to headphones. In TV and VoD this is done by respeaking (see below).

## *Free Commentary*

The determining feature of free commentary is that it is not a faithful rendering of the source text; rather, commentators are free to create and give their own impressions and opinions, to narrate and describe what they see in their own words, and to add any further details and information they deem necessary. The translation can also be adapted to the program's potential audience. This mode usually complements dubbing or subtitling, especially in the case of gags from film excerpts, but it can also substitute them; in Europe, for example, it is frequently used in comedy or sports videos and programs. Fundubs, spelled with a *u*, are a kind of extremely domesticating free commentary (see below).

## *Fandubs*

Fandubs – amateur dubbing – and fundubs – gag dubbing – are home-made dubbings of television series, cartoons (particularly the anime genre) and trailers for films that have not yet been released in the target-language country or region. Fandubs are usually translated and recorded by fans of these genres; they download the film texts from Internet and, using a digital sound editing program, manipulate or eliminate the soundtrack of the original text and record and insert the dubbed track they record at home using a microphone (Chaume, 2013). They are sometimes spelled *fundubs* but only when the real function of the "creative translation" is to make fun, also called *gag dubbing*, because of the witty and humorous nature of this type of home-grown dubbing. Literal video versions, also known as literal music videos, are parodies of official music video clips in which the original lyrics are replaced by new ones that provide a literal description of the visuals. This technique is normally applied to music videos in which the imagery seems to be illogical and disconnected from the lyrics. Usually, this is done intralingually, meaning that there is no *proper* translation as such, though it can also be done interlingually. These are clear cases of creative writing for the media.

## *Subtitling*

Subtitling consists of incorporating a written text (subtitles) in the target language on the screen where the original version film is shown, such that the subtitles coincide approximately with the screen actors' dialogues. Díaz Cintas and Remael (2007) provide a detailed account of subtitling in terms of formats, condensation strategies, reading speeds, semiotic features of subtitling, and so on. For the purposes of this chapter, several examples illustrate how technology has changed the practice and industrial process of subtitling. Subtitling is so tightly linked to technology that any technical advancement generally has a direct impact on this professional practice (Díaz Cintas, 2013). For example, during the past decade the introduction of DVD and Blu-Ray as standard formats for watching films and television series at home transformed subtitling into an established professional practice all over the world, including traditionally dubbing countries. Nowadays, thanks to technology, DTV and VoD platforms also broadcast most of their products subtitled into many different languages. But digitalization in the translation process and product has further implications. For instance, technological advances in subtitling can be easily identified in the new typographical formats and fonts available to translators (the newer PTSans, Avenir or Tiresias font types now coexist with the more traditional Arial, Verdana and Helvetica), in the production and design of creative subtitles (for example, Netflix now allows subtitles to match the content on screen with new fonts, colors and text effects depending on the audiovisual genre being watched,) in translators' working conditions (especially the possibility of working in the cloud), in the audience's perception of subtitles (enhanced visibility), and in the provided access to several languages (source and target languages), etc. Digitalization of images has signified a major advancement in two directions: first, it has spawned new distribution channels (DTV, DVD, Blue-Ray, VoD platforms, streaming services, etc.), and second, it has led to the development of professional and amateur subtitling programs. An original audiovisual production in any language can now simultaneously have multiple translations in various languages and in several modes (dubbing, subtitling, SDH, AD, etc.); an achievement that, only two decades ago, would have been regarded as science fiction.

In academic circles, research has focused on the possibilities offered by digitalization and, particularly, on the development of useful software and applications for professional translators and lecturers. The literature describes the audiovisual translator's new workstation and the resources available to carry out their work, which can now integrate different programs and applications for the professional practice (and teaching) of subtitling. New software tools are being developed in universities, for example Black Box (González Iglesias, 2014), created to analyze and edit subtitles, or the tool designed by Martí Ferriol (2014) to calculate reading speeds. There is also a growing body of research on the impact of MT in subtitling, incorporating translation memories into the subtitling process in the cloud.

## *Surtitling*

Interlingual and intralingual surtitling is a specific form of subtitling for theater and opera. It allows spectators to understand the characters' dialogues, and follow the storyline of the play or opera. The subtitles are usually projected on a screen located above the stage in the proscenium so the audience in the boxes can follow events on stage and at the same time read the subtitles – thence called *surtitles* – with a translation (interlingual) or transcription (intralingual) of the dialogues. Conventions are similar to those of commercial subtitling (Mele, 2016), although they differ in some ways, such as not repeating opera chorus lines, in order to let the audience enjoy the music without being constantly shown the same repeated translation.

## *Fansubbing*

Fansubs are home-made subtitles for television series, cartoons (particularly the anime genre) and films that have not yet been released in the target-language country. As in fandubbing, they are usually prepared by fans who download the clips from the Internet and use computer editing programs to insert new subtitles they create at home using free software. Fansubbing, or subbing, is usually less orthodox than conventional subtitling. Colors can be used; subtitles may appear anywhere on the screen (above or below the speaker, or at the side); they use more characters than conventional subtitles; fonts may vary throughout the film; translations are frequently highly foreignizing; higher reading speeds are demanded than for conventional subtitles, etc. (Chaume, 2013). As in fundubbing, the aim of fansubs can be to make fun of a popular character, a particular film/TV series scene or a sketch or clip, thus becoming funsubs, that is, home-made subtitles that do not translate the original dialogues, but create new witty humorous lines unconnected to the meaning of the original text.

## *Videogame Localization*

The process of manipulation of video game software and hardware for distribution in a new region or country is known as videogame localization. This process encapsulates all changes made to a game, not only translating the text, but also modifying art assets, creating new packaging and manuals, dubbing dialogue lines for cinematic scenes – especially in the most modern ones, and those classified as "AAA", i.e. a classification used for videogames produced and distributed by a major company, typically having higher development and marketing budgets – transforming parts of the hardware, editing and sometimes deleting whole scenes of the game to accommodate differing cultural values and taboos, and even inserting new sections to replace cut content. Videogame localizers use memory tools to speed up the process, since many utterances and instructions are repeated in several games. Dubbing in videogames differs from canonical dubbing for the cinema and TV, especially regarding synchrony norms: the videogame industry acknowledges up to five different syncing types as opposed to the three traditional ones (Mangiron and O'Hagan, 2013; Bernal-Merino, 2014; Pujol Tubau, 2015). Videogame subtitling also differs from canonical subtitling for TV and cinema, in that the use of creative subtitles is far more extended, subtitles are longer in terms of characters and lines, and reading speed ratios are also lower than in commercial subtitling.

Mangiron and O'Hagan (2006, 20) recommend the term *transcreation* to describe what takes place in videogame localization, because "localisers are granted *quasi* absolute freedom to modify, omit, and even add any elements which they deem necessary to bring the game closer to the players and to convey the original feel of gameplay." This insightful link to transcreation (see below) helps us understand that the borders of localization (and hence of audiovisual translation) go beyond canonical audiovisual transfer modes, because the elements added to the translation the authors refer to are icons and written text included in the images. This gives us leeway to also connect transcreation with the different adaptation modes (see below).

## *Transcreations*

Transcreations are all forms of semiotic adaptation and manipulation where some or most – if not all – semiotic layers of the original (audio)visual product are localized, that is, manipulated. Traditionally, only signs belonging to the five acoustic codes of the audiovisual text (linguistic, paralinguistic, musical, special effects and sound provenance) were localized using dubbing/voice-over and subtitling. However, nowadays, images (icons, indices and symbols), lighting, movement (kinesic signs) and types of shots can also be manipulated in order to produce a domesticated product that, allegedly, satisfies a specific target audience. This process is called transcreation. Yahiaoui's (2016) study of the ideological constraints operating in audiovisual translation from English into Arabic in 52 dubbed episodes of *The Simpsons* (Matt Groening, 1989–) revealed an ideological process of censorship in the dubbing into Arabic, implemented by transcreation, namely, via the manipulation of images. *The Simpsons* addresses many sensitive issues, such as sex, drugs, religion, politics, racial and gender stereotypes, with a bluntness and boldness rarely seen elsewhere, and goes beyond passive entertainment. The author reports that the series is closely monitored and regarded with suspicion in the Arab world. One of the interesting examples of manipulated images and content he gives is the way Homer Simpson's customary beer is replaced with juice. These manipulations are not always a result of censorship, but to tie in with the habits and tastes of the target culture. In the same vein, Zabalbeascoa (2008, 169) describes how in the dubbing of two French TV series for the Catalan teenage audience, proper names were changed and the music was replaced by Catalan contemporary rock. Such examples abound in the localization of commercials for TV and Internet, and also in cartoons (Chaume, 2016).

## Adaptations

However, localization, in its broader and contemporary meaning, is not constrained to the canonical modes of audiovisual translation described above. Audiovisual translation should be considered more broadly in order to encapsulate modern practices which, due to digitalization, are merging with, or complementing, traditional audiovisual translation modes. Research on TV formats, for instance, shows how the origins, flow and pattern of internationally distributed formats are systematically localized for other markets, thus becoming another kind of "translation" (Esser, Bernal-Merino and Smith, 2016). The concurrent global and local essence of these localized products creates a tug-of-war, a kind of compromise between source culture and target culture ideologies and ideological agendas. Research on TV formats examines the extent to which an adapted franchise differs from its original counterpart, how several adaptations differ from one another, and also how far these differences influence the audience to watch and consume a different product. For example, the Spanish TV series *Cuéntame Cómo Pasó* (Miguel Ángel Bernardeau, 2001–) has been adapted in other countries, but both plots and historical backgrounds have also been changed to accommodate recent local history and historical figures. This is the case of the adaptation for Italy (*Raccontami*, 2006–2008), Portugal (*Conta-me Como Foi*, 2007–2011), Argentina (*Cuéntame Cómo Pasó*, 2017–), and the United States (*Remember When*, in progress). Different national regulations and obvious cultural differences, which cannot be hidden or looked down on in any globalization process, as well as the (un)conscious agendas of the agents involved in the localization process, are the fundamental reasons for all forms of localization, and subsequently, for any clashes and confrontations between original and domesticated products. There are hundreds of examples of adaptation, especially on TV, and they fall into two major types: (1) the format and the contents are adapted, but the original plots and some original characteristics of the main characters are kept, like in the Spanish TV series *Médico de Familia* (Daniel Écija, 1995–1999) adapted in Italy as *Medico in famiglia* (Carlo Bixio et al. 1998–); and (2) the format is bought but everything else is changed, as in the case of *Masterchef* (Francis George Roddam, 1990–), *Kitchen Hell* (Arthur Smith, Kent Weed and Gordon Ramsey, 2005–), etc.

Adaptations are not restricted to TV, however. The theater also has its own examples of adaptation, as is the case of the plays written in Catalan by Pau Miró and localized in Barcelona and adapted into Italian (Neapolitan), for example *Plou en Barcelona* into *Chiove*, and *Els Jugadors* into *i Giocatori*, among many other examples. In cinema, they may also be known as remakes, discussed below.

## Remakes

Transnational remakes can also be considered as a kind of localization. Evans (2014) argues that film remakes have often been neglected by Translation Studies in favor of other forms of audiovisual translation such as subtitling and dubbing. He considers that remakes should also fall under the umbrella of film translation, since they reveal a multimodal, semiotic and adaptive process of translation: the semiotic codes of the original film are translated or adapted, and the source text reworked; they can therefore be studied under this new broad prism of audiovisual translation. Esser, Bernal-Merino and Smith (2016, 2) also share this view, considering transnational remakes as "the most enhanced form of localization" – clearly an illustration of global cultural borrowing. There are hundreds of examples of remakes, not only the most well-known instances of films from local film industries and traditions being taken to Hollywood, but also between local markets. For example, *Bienvenue chez les Ch'tis* (Dany Boon, 2008) has been adapted in Italy with a remake entitled *Benvenuti al Sud* (Luca Miniero, 2010); in turn,

*Perfetti Sconosciuti* (Paolo Genovese, 2016) has been remade in Greece as *Teleioi Ksenoi* (Thodoris Atheridis, 2016) and in Spain as *Perfectos Desconocidos* (Álex de la Iglesia, 2017).

### *The Localization of Transmedia Projects*

The term *transmedia* refers to content served to consumers across a wide range of locales and platforms (Singh, 2012), including movies, books, comics, games, apps, toys, etc. Transmedia content creates a narrative that expands into several media forms: generally audiovisual (born-digital), multimedia, interactive, episodic, and shaped in multiple formats. Modern transmedia makes use of multiple devices to involve audiences by creating stories that are linked together in different platforms and formats. Thus transmedia storytelling, also called transmedia narrative, cross-media seriality, and multiplatform storytelling, is the general technique of telling a story using several formats and platforms, ideally with some audience or end user involvement, and every adaptation to each format uses text (linguistic codes), images (iconographic codes), music (musical codes), sound effects (sound codes), puzzles and games to illustrate and enhance the narrative (Pujol Tubau, 2015; Ferrer Simó, 2016).

## Social Difference: Localizing for Disabled Communities and Migrants

The Media Accessibility Platform (MAP) project, funded by the European Union, seeks to promote all kinds of accessibility in the media, a term understood as a tool for social and cultural inclusion of people with specific needs, such as deaf and blind users, people with cognitive disabilities, foreign audiences, children, and the elderly. All over the world similar projects, institutions and organizations are pursuing the same objectives. The ethos of these initiatives is that everyone, regardless of their sensorial and linguistic barriers, must have access to media. Therefore, audiovisual contents should also be localized – in a way, adapted – for all these groups. In this case, the localization is not regional but social, an ethical and supportive localization. It is not only the hard of hearing, the visually impaired and other groups with sensorial barriers who have exactly the same right to access information and culture as the rest of the population; migrants, students of foreign languages, children and the elderly also have the same right to be part of the society to which they belong or wish to belong.

Although the so-called accessibility modes of audiovisual translation are now on the increase (subtitling for the deaf and the hard of hearing, respeaking, audiodescription for the blind and visually impaired, sign language interpreting, audiosubtitling, etc.), in this section only the three main modes are dealt with. The section ends with a discussion of a modern concept that invites film producers and distributors to create new audiovisual products that are accessible from the start: the concept of accessible filmmaking.

### *Subtitling for the Deaf and Hard of Hearing*

Subtitling for the deaf and hard of hearing (closed captions) is another mode of translation used to enhance accessibility for people with hearing difficulties, or for older or foreign audiences. It consists of an intralingual translation (it might also be interlingual, but this is not so usual) or rendition that reproduces the characters' dialogues – verbatim or edited – in such a way that the subtitles are shown on the screen at the same time as they are spoken. In contrast to live subtitling, this mode is generally used for recorded programs, and colors are used to discern characters in crowd scenes or those with their back to the camera or off field, sound effects are represented using symbols or onomatopoeias, subtitles for sounds or songs are sometimes placed at the top of the screen, etc.

## *Respeaking*

Respeaking or live subtitling is another mode that falls somewhere between simultaneous interpreting and audiovisual translation. This technique shows subtitles at the bottom of the screen during a live broadcast (Romero Fresco, 2011). The interpreters use a computer with voice recognition software, previously trained to recognize their voice, and listen to the live broadcast through headphones. This method can also be used with recorded broadcasts. The interpreter hears the characters' dialogues or the narrator's voice and respeaks them in his or her own words, usually summarizing the original dialogues quite substantially so their re-translation fits into the subtitle space that the software program generates when it processes the spoken sentences (Chaume, 2013).

## *Audiodescription*

In audiodescription (AD) for the blind and visually impaired, the scenes, locations, characters, costumes, objects, and action of a film are described and inserted between dialogues. These descriptions allow blind or visually impaired audiences to follow the story line, while lending coherence to the dialogues throughout the film. There is an intense debate on whether audiodescriptions should be objective or whether they should convey the emotions and feelings of the audiovisual product to the blind audience. Most authors and guidelines advocate extremely realistic and objective descriptions; however, subjective values in ADs have recently been analyzed from the receiver's point of view, showing that despite potential fierce rejection, metaphors and other literary devices are naturally accepted by the audience (Chaume, 2019).

## *The Notion of Accessible Filmmaking*

New technological developments are spawning new access services. Most access services, including here access services to overcome sensorial and linguistic barriers, have been traditionally added as part of the film or audiovisual post-production process. Once the audiovisual content is finished, and only as an afterthought, the product is dubbed, subtitled, subtitled for the deaf, respoken, audiodescribed, voiced-over, surtitled, etc. The way audiovisual products are currently made accessible no longer involves communication between translators, mediators, localizers and the creative team. Romero-Fresco (2013) first came up with the concept of accessible filmmaking, referring to a way of making audiovisual products that incorporates linguistic and sensorial access as an integral part of the production, through close collaboration between localizers and creators. Accessible filmmaking, therefore, has the potential to integrate AVT and accessibility during the filmmaking process, by including experts in translation, localization and accessibility in the crew, all of whom have their say in how the scenes are shot, and who can foresee typical localization problems, such as the future position of subtitles against colors, certain translation issues, or problems of reading speed when too many characters speak at a time or speak fast, of lip-syncing in close-ups and extreme close-ups, etc.

## Activism: The Audience's Turn

New technological developments are also giving rise to new ways of consuming and interacting with audiovisual products. Almost four decades ago, Toffler (1980) coined the term *prosumer* to refer to individuals who consume and produce media, meaning production by consumers. In the world of media localization, passive consumers sitting on their comfortable sofas at home watching TV have radically evolved into *interactive and proactive consumers or prosumers*, who now can make their own domestic dubbing, subtitling, voice-over, etc., interact with all kinds of localizations on the Web, revoice and caption from across the world, use apps to revoice and caption, and instantly upload any audiovisual clip available on the Web, create new dubbing and subtitling genres, such as free-commentaries, literal music videos or literal video versions (see above) and literal dub versions (a parody of an official music video clip in which the original lyrics are replaced by translated lyrics that describe the visuals in the video in the target language) and gag dubbings, creative subtitles, etc. Drawing on the notion of prosumers, Romero-Fresco (2013) shows us how films made by users with dual sensory impairments can potentially change our understanding of cinema, moving away from ocular-centrism to a more haptic notion of film.

The advent of digitalization has indeed brought about a new culture, which could be called *digital culture*. Consumers' participation as co-creators in the audiovisual production process has increased significantly in the past ten years (Díaz Cintas and Baños-Piñero, 2015). The concept of Web 2.0, an inclusive notion for a new, more interactive and dynamic Internet use, also attracts users to participate in the (creation and) translation of audiovisual content, the generation of new ideas and the interaction of collective intelligence. This invites us to ponder the concepts of *empowerment* and intervention. Pérez González (2014) demonstrates how passive consumers have become active consumers or prosumers, since these changes have enabled them to partially take on a fraction of the power and responsibility traditionally held by producers. Orrego-Carmona (2015) focuses on non-professional subtitling use as an efficient way to overcome linguistic barriers, and provides a general picture of users' engagement with audiovisual content, their attitude toward subtitling and non-professional subtitling, and how they manage their expectations and adapt to new conditions.

Both digitalization and empowerment converge toward a new way of carrying out subtitling, known as *creative subtitling* – both inter- and intralinguistic – which grants subtitles functions other than the simple communicative task of standard commercial subtitling. Creative subtitles feature as footnotes or captions placed on any part the screen, creative titles or words emerging from coffee machines or chimneys, dialogue placed next to the onscreen characters' mouths, etc. This clearly brings other aspects into the picture, such as multilingual movies and translation, or the concept of accessible filmmaking (Romero Fresco, see above). Creative subtitles have their roots in fansubs (Ferrer Simó, 2005; Díaz Cintas and Muñoz, 2006; Orrego-Carmona, 2015), an increasingly popular phenomenon, both because of the growing communities of people who enjoy foreign, particularly Japanese, products, and because of the free, easy-to-use computer software for home subtitling readily available to fans across the world. Finally, another localization modus operandi of note is voluntary subtitling, commissioned by certain non-profit associations or crowdsourcing platforms (Díaz Cintas, 2013), such as TED, Universal Subtitles and Khan Academy.

However, the most interesting form of current active localization is political dubbing and subbing. Sometimes in the form of gag dubbing, active viewers create fake dubbings of political speeches by famous presidents, leaders and people's representatives in order to denounce and sometimes ridicule and deride their agendas. The same process is seen in the realm of fansubbing, where fansubbers ridicule politicians and leaders by creating fake subtitles that either expose their hidden agenda or simply make them sound ludicrous through absurd utterances. This new political use of audiovisual translation is changing the traditional way of localizing audiovisual content and is stretching the boundaries of the old notions of equivalence, faithfulness and ethics of translation.

## Conclusions

When one reflects on the intense process of colonization by American culture across the world through its cinema and TV productions, we often disregard the fact that translation, adaptation, and localization processes in general, intentionally or unintentionally, make changes to and diverge from the original product, recreating and modifying the original contents and intentions. Dubbings and subtitlings from all over the world offer some glorious examples of subtle changes where the original message is substantially subverted (El Halli Obeid, 2012). As Franco Aixelá states (1996, 54),

> faced with the difference implied by the other, with a whole series of cultural signs capable of denying and/or questioning our own way of life, translation provides the receiving society with a wide range of strategies, ranging from conservation (acceptance of the difference by means of the reproduction of the cultural signs in the source text), to naturalization (transformation of the other into a cultural replica). The choice between these strategies will show, among other factors, the degree of tolerance of the receiving society and its own solidity.

The reality is even more complex than this simple binomial. Localization practices tend to show a balance between foreignization and domestication, depending on the audiovisual genre, the target culture norms, the audiovisual translation mode chosen to localize the product, etc. Technology, through digitalization, new localization practices, social differences, regional differences and the new active role of audiences, has dramatically changed the world of localization. Figures for the domestic and overseas revenue from American blockbusters reveal that virtually half the total income derives from their dubbed and subtitled versions, despite a mere 0.1–1% of their budget being devoted to translation and accessibility (Romero-Fresco, 2013). However, in spite of their pivotal role in the process of localization, translators are still among the least valued professionals in the film and videogame industry. The time has come, therefore, for localization to form part of the original conception of the audiovisual product, and for cinema and TV stakeholders to change the way they produce and distribute audiovisual contents, taking into account this social shift in audiovisual consumption and translation.

## Further Reading

Romero Fresco, Pablo (2018) *Accessible Filmmaking. Making Films Accessible to Foreign and Sensory-Impaired Audiences*, London and New York: Routledge. The first chapters provide filmmakers, film scholars, professional translators and translation scholars with the main concepts and notions that must be considered to show how audiovisual translation and accessibility modes can have an impact on the distribution and reception of a film. The second part offers a well-informed proposal about how translation and accessibility can be integrated into the different stages of the filmmaking process.

## References


Bernal-Merino, Miguel Ángel. 2014. *Translation and Localisation in Video Games: Making Entertainment Software Global.* London and New York: Routledge.

Chaume, Frederic. 2012. *Audiovisual Translation: Dubbing.* London and New York: Routledge.

Chaume, Frederic. 2013. "The turn of audiovisual translation: New audiences and new technologies." *Translation Spaces* 2: 105–123.

Chaume, Frederic. 2016. "Audiovisual Translation Trends: Growing Diversity, Choice and Enhanced Localization." In *Media Across Borders*, edited by Andrea Esser, Miguel-Ángel Bernal-Merino and Iain Robert Smith, 68–84. London and New York: Routledge.

Chaume, Frederic. 2019. "Audiovisual Translation in the Age of Digital Transformation: Industrial and Social Implications." In *Reassessing Dubbing: Historical Approaches and Current Trends*, edited by Irene Ranzato and Serenella Zanotti. Amsterdam and Philadelphia: John Benjamins.
Díaz Cintas, Jorge. 2013. "The technology turn in subtitling." *Translation and Meaning* 9: 119–132.
Díaz Cintas, Jorge and Aline Remael. 2007. *Audiovisual Translation: Subtitling*. London and New York: Routledge.
Díaz Cintas, Jorge and Pablo Muñoz. 2006. "Fansubs: Audiovisual translation in an amateur environment." *JoSTrans, The Journal of Specialised Translation* 6: 37–52.
Díaz Cintas, Jorge and Rocío Baños-Piñero. 2015. *Audiovisual Translation in a Global Context. Mapping an Ever-Changing Landscape*. Basingstoke and New York: Palgrave McMillan.
El Halli Obeid, Leticia. 2012. "La normalidad de Frankenstein: El español neutro y el doblaje." *Nexos*, August 1, 2012. www.nexos.com.mx/?cat=3173.
Esser, Andrea, Miguel-Ángel Bernal-Merino and Iain R. Smith 2016. "Introduction." In *Media across Borders*, edited by Andrea Esser, Miguel-Ángel Bernal-Merino and Iain R. Smith, 1–18. London and New York: Routledge.
Evans, Jonathan. 2014. "Film remakes, the black sheep of translation." *Translation Studies* 7 (3): 300–314.
Ferrer Simó, María. 2005. "Fansubs y scanlations: la influencia del aficionado en los criterios profesionales." *Puentes* 6: 27–43.
Ferrer Simó, María. 2016. "La gestión de los proyectos de traducción audiovisual en España. Seis estudios de caso." PhD diss., Universitat Jaume I.
Franco Aixelá, Javier. 1996. "Culture-Specific Items in Translation." In *Translation, Power, Subversion*, edited by Román Álvarez and África Vidal-Claramonte, 52–78. Clevedon: Multilingual Matters.
González Iglesias, David. 2014. *Desarrollo de una herramienta de análisis de los parámetros técnicos de los subtítulos y estudio diacrónico de series estadounidenses de televisión en DVD*. Salamanca: Ediciones de la Universidad de Salamanca. Colección Vítor.
Mangiron, Carme and Minako O'Hagan. 2006. "Game localisation: Unleashing imagination with 'restricted' translation." *Jostrans: The Journal of Specialised Translation* 6: 10–21.
Mangiron, Carme and Minako O'Hagan. 2013. *Game Localization: Translating for the Global Digital Entertainment Industry*. Amsterdam and Philadelphia: John Benjamins.
Martí Ferriol, José Luis. 2014. "Subtitle reading speed: A new tool for its estimation." *Babel* 59 (4): 406–420.
Mele, Antonia. 2016. "La sobretitulación teatral: Estudio descriptivo de las convenciones profesionales en España." PhD diss., Universitat Jaume I.
Orrego-Carmona, David. 2015. "The reception of (non) professional subtitling." PhD diss., Universitat Rovira i Virgili.
Pérez González, Luis. 2014. *Audiovisual Translation. Theories, Methods and Issues*. Abingdon: Routledge.
Pujol Tubau, Miquel. 2015. "La representació de personatges a través del doblatge en narratives transmèdia. Estudi descriptiu de pel·lícules i videojocs basats en *El senyor dels anells*." PhD diss., Universitat de Vic-Universitat Central de Catalunya.
Romero Fresco, Pablo. 2011. *Subtitling through Speech Recognition: Respeaking*. London and New York: Routledge.
Romero-Fresco, Pablo. 2013. "Accessible filmmaking: Joining the dots between audiovisual translation, accessibility and filmmaking." *JoSTrans 20*: 201–223.
Scandura, Gabriela. 2015. "El español neutro del doblaje de programas infantiles y juveniles: ¿un caso de estandarización o de censura?" MA diss., Universitat Jaume I.
Singh, Nitish. 2012. *Localization Strategies for Global E-Business*. New York: Cambridge University Press.
Toffler, Alvin. 1980. *The Third Wave*. New York: Bantam Books.
Toury, Gideon. 1995. *Descriptive Translation Studies and Beyond*. Amsterdam and Philadelphia: John Benjamins.
Venuti, Lawrence. 1998. *The Scandals of Translation: Towards an Ethics of Difference*. London and New York: Routledge.
Yahiaoui, Rashid. 2016. "Ideological constraints in dubbing *The Simpsons* into Arabic." *Altre Modernità, Numero speciale: Ideological Manipulation in Audiovisual Translation* 182–200. doi:10.13130/2035-7680/6855
Zabalbeascoa, Patrick. 2008. "La credibilidad de los diálogos traducidos para audiovisuales." In *La oralidad fingida: descripción y traducción*, edited by Jenny Brumme, 157–175. Madrid: Iberoamericana.

# 28

# CURATING LIFE, STAGING ART

## Modernisms and the Art Practices of Television

*Nomusa Makhubu*

### Introduction

In 2017, Samsung announced a new product called *The Frame*: television that "is literally a hanging wall art."[1] Designed to blend in with other artworks hanging on lounge walls in the home, *The Frame* TV could switch to Art Mode "when you're done watching Netflix." One can "[cycle] through over 100 different artworks curated by Samsung, from oil paintings to architectural photography" and subscribe to *The Frame* Art Store to purchase "[c]urated artwork from world renowned institutions and collections" such as Saatchi Art, Albertina Museum, Art Space, Museo del Prado and other major national and private collections.[2] An advertisement shows *The Frame* TV installed in the Albertina Museum in Austria, displaying canonical art by artists such as Claude Monet to astonish gallery goers as they "spot the difference" between the televised version and the original painting.

Merging television and art, *The Frame* is meant to create "a new way of seeing exceptional art in everyday life," and to facilitate art collecting, consumption and the curation of a virtual gallery in one's home "in just one click."[3] What is striking about this is that the art is not referred to as a copy, representation or simulacrum on TV but can be purchased as a collectible *original* artwork purchased from a museum or a collector, to be appreciated on TV. That is, not only does the television become a virtual gallery for "rendering modern art as an everyday lifestyle for progressive publics," as Lynn Spigel (2008: 22) aptly points out, but it also becomes, I argue, an extraordinary site of exhibition, fusing the white cube of the conventional art gallery with the black box of the television.[4]

Art discourse in general has relegated television to philistinism. However, Lynn Spigel (2008: 20, 22) asserts that the set piece "Hail! Modern Art" in television programs such as *Admiral Broadway Revue* during postwar America, buoyed modern art, promoting art consumption and gallery-going. By constructing the national "cultured American" as one who "displaye[d] one's progressive, au currant, tastes" (Spigel 2008: 20, 22), television in postwar America cultivated an interest in art and the art buying of "'first class reproductions' at a cost many can afford" (Spigel 2008: 20). That is, the popularization of modern art as "high art" on TV was not necessarily treated as trivialization but reconstituted communities of spectatorship among the bourgeoisie, middle classes and lower-middle classes who aspired to the sense of "bourgeois publics." Within those communities, it also shifted the gendered notion that art connoisseurship is the prerogative of the male spectator whereas television is in the domain of the feminine domestic "restless hermit" (Spigel 2008: 154). Although elitist and sexist perceptions about taste and sensibility pervaded communities of spectatorship, television became the gallery space through which modern art could be accessible. In the context of postwar America, popular culture buttressed and elevated, if not created, cultures of consuming art.

As an exhibition site in the curated public domains of the private home (such as lounges), television not only represents the curated lifestyles of bourgeois publics but it is also a site where cultural values can be contested and class hierarchies subverted. While lifestyle television programs, dramas, soaps and films display artworks as wealth, power, cultural finesse and class, these programs can also conversely parody and subvert bourgeois publics. The collapsed boundary between mass culture and art as high culture, the ubiquity of television in cultures of viewing, makes art, television and power inseparable.

This relationship is not always conceived this way, however. Television is predisposed to the perception that it is the opposite of art and politics and, as the epitome of adulterated thinking, it is a time-consuming, superficial, hypnotic means for "dumbing down," placating, disempowering and benumbing the citizenry. To that effect Pierre Bourdieu (1996: 10) affirmed "television poses a serious danger for all the various areas of cultural production – for art, for literature, for science, for philosophy and for law." Television, it would seem, would corrupt the minds of free thinkers and dominate people's social practices and values through cultural imperialism.

Nevertheless, Bourdieu (1996: 12, 14) recognized the power that lies in television but warned against "the threat of instrumentalization" and argued that there needed to be "tools or weapons" for "image professions who are struggling to keep what could have become an extraordinary instrument of direct democracy from turning into an instrument of symbolic oppression." For Marshall McLuhan television had transformed politics. He and Quentin Fiore argued that "[a] new form of politics is emerging [...] the living room has become a voting booth. Participation via television is changing everything" (McLuhan, Fiore and Agel 2001: 22). Since television reaches people on an emotional and psychological level, the "danger" Bourdieu foresaw, is manifest in the manipulability of perception or its influence on the social imaginary.

Given the criticism levelled against television and its inherent visual politics, it has become the site in which politics are played out and where identities, cultural conventions, and canons are created, reinforced but also subverted. In this chapter, I focus on how these manifest in postcolonial African contexts. It did not, as with postwar America, create a sense of new communities of spectatorship that would be brought closer to art appreciation through television. Rather, television, as the political, social and cultural phenomenon of modernization in various African countries, seemed to corrupt what was regarded as "pristine" cultural practices. However, since it coincided with colonial independence and the emergence of postcolonial nations, it became intertwined with various modes and mediums through which new forms of social consciousness aimed at self-definition and self-representation could be disseminated.

On its arrival on the African continent, television was a state-owned enterprise (Ziegler and Asante 1992). At the time, it was seen as a medium that had "been used to promote the political aims of the controllers of power, particularly the heads of state" (Bourgault 1995: 103). However, by the 1990s, a number of independent private television stations emerged in most African regions. More recently, locally produced content and access to online television streaming on the internet means that there is now a complex network in which power in the politics of representation is constantly shifting.

While television represented being global and being modern, African art struggled to escape the colonial grasp which sought to contain it as indigenous, traditional "tribal" art and from being "corrupted" by modernity (Kasfir 1992; Steiner 1994; Oguibe 2004). In a classic essay, Sidney Littlefield Kasfir (1992: 41) lambasts the assumption that pre-colonial art "occurring in most places from the mid-nineteenth to early twentieth century, exhibited qualities that made it authentic (in the sense of untainted by Western intervention)" while colonial and postcolonial art "is relegated to an awkward binary opposition: it is inauthentic because it was created after the advent of a cash economy and patronage from missionaries, colonial administrators, and more recently, tourists and the new African elite." The hypervisibility of modern and contemporary African art on the global stage comes much later than European art. The former is generally dated to the 1960s, coincidentally when television first arrived. Even then, the category modern African art remains less "settled" than Euro-American modern art, whose sense of being "modern" is not continuously contested and questioned.

In a paper calling for more scholarship on what then seemed to be a nascent African modern art, Rasheed Araeen (2005: 411) stated that "Africa has now acquired everything necessary to be part of the modern world," since "it has modern factories, aeroplanes, airlines, the latest cars, buses and lorries, radios and colour TVs." His concern was that all these creature comforts were essential to a "modern consumer society," but indicated passive consumption rather than active creation. In Araeen's (2005: 412) argument, at least, this is what separates things like television from art, the one is passive while the other is a way to "think, innovate and produce." The idea of television as a virtual art gallery for progressive publics, in this case, seems incongruous.

Television and art in postcolonial Africa mark very distinct politics of seeing and being seen implicated in what Andrew Apter (2002: 564) refers to as the "imperial spectacle" – the extension of empire through *fields of spectatorship* to mediate colonial relations. In this chapter, I use, as a conceptual framework, the convergence of the white cube and the black box to interrogate dynamics of postcoloniality, vision and power and how this forged divergent modernisms. I also discuss the way in which television could give a glimpse into actively creative modernities in Africa that preceded and were disrupted by colonialism. My use of the term African art considers art and popular culture as correlative, rather than separate and incompatible, creative practices in Africa.

## Spectacular Modernisms, Spatial Politics and the Televisual Gallery

With rising nationalist and Pan-Africanist political and cultural ideologies in the 1960s, television marked, for some countries, the transition from pernicious colonial spectacle to postcolonial globalism. Since television was an expensive feat, presidents such as Tanzania's Julius Nyerere considered it a "luxury" and not a priority, but it appealed to African elites who saw it as a medium that "goes with being a modern state" (Ziegler and Asante 1992: 56). In Senegal, television was called by one of Senghor's Director-Generals, "a jewel for the spent and tired bourgeoisie" (Katz and Wedell 1977: 87). Nevertheless, TV and other forms of modern technology, would re-assert Africa's place in a globalizing world.

However, James Ferguson (2006: 26) points out that "[a]gain and again, it seems, when it comes to globalization, Africa just doesn't fit the story line [...] Defenders of neoliberal structural adjustment programs naturally find Africa an inconvenient example." Africa's extraction from the globalization narrative was reinforced through the rhetoric that it is among the "remote" parts of the world where television would contaminate and erode "tribal" arts and cultures.

Media theorist Marshall McLuhan, for example, saw the simultaneous global consumption of TV and radio as traumatic detribalization. He states:

> Detribalization by literacy and its traumatic effects on tribal man is the theme of a book by the psychiatrist J. C. Carothers, *The African Mind in Health and Disease* (World Health

> Organization, Geneva, 1953). [...] Again, it is electric speed that has revealed the lines of force operating from Western technology in the remotest areas of bush, savannah, and desert. One example is the Bedouin with his battery radio on board the camel. Submerging natives with floods of concepts for which nothing has prepared them is the normal action of all of our technology. But with electric media Western man himself experiences exactly the same inundation as the remote native. We are no more prepared to encounter radio and TV in our literate milieu than the native of Ghana is able to cope with the literacy that takes him out of his collective tribal world and beaches him in individual isolation. We are as numb in our new electric world as the native involved in our literate and mechanical culture.
>
> *(McLuhan 2006: 288)*

The psychosocial effects of detribalization caused by these technologies (such as radio and television), argues McLuhan, would individuate and isolate "Western man" and "the remote native" equally, in a global mediatized world. Assuming that so-called tribal man is illiterate and assuming that Africans are homogenously tribal and in "remotest areas of bush, savannah, and desert," television was seen as something that would have adverse effects on the psychological health of Africans.

In the case of South Africa, television was considered a medium that would have adverse effects on black South Africans and would also interrupt the racial state of affairs. Being culturally "remote" from the globalizing world was convenient for apartheid. Television was only introduced in 1975, during the height of apartheid. From the onset, the introduction of television was contested. For conservative Afrikaaner nationalists, as Terry Bell and Dumisa Ntsebeza (2003: 35) show, television was regarded by the Afrikaaner Broederbond (AB) as "a work of the devil" while "modernisers were exerting pressure within the AB advocated for the introduction of a television service." Since it was regarded as "the devil in the black box," South Africa "was among the last countries in Africa, and in fact the world, to receive television" (Orgeret 2015: 188). The apartheid government feared that television would interfere with its guarded separatist ideology. They feared that it would weaken internal government control and that even though it was not aimed at black South Africans, they could by chance see images on television while at work, cleaning white homes. Arguing against television in parliament, Dr. Albert Herzog used an anecdote to demonstrate the "danger" of TV:

> It is afternoon and the bantu house boy is in the living room cleaning the carpet. Someone has left the television set on. The house boy looks up at the screen, sees a chorus line of white girls in scanty costumes. Suddenly seized by lust, he runs upstairs and rapes the madam.
>
> *(cited in Orgeret 2015: 189)*

Television represented apartheid bigotry and constructed psychosocial fear of free thinking among black people who were imagined as unthinking, savage and violent. More apparent however, was that for the government, television not only undermined the regulation of what black people see, think and do but, significantly, the restriction or "protection" of the minds of white South Africans, who might be exposed to international liberal and leftist thought and realize that apartheid is a global anomaly.

TV in South Africa had the potential to politically undermine the apartheid government. For this reason, it became guarded as a manipulable political tool. Orgeret (2015: 191) argues that "the possibility of using television as a government propaganda tool was clear from the beginning." At the launch of the South African Broadcasting Corporation Television (SABC TV) in

1976, Prime Minister B. J. Vorster noted that "what is clear is that we are dealing with a medium that, as it has already been experienced by all other countries, can have a powerful influence whether for good or for bad" (cited in Orgeret 2015: 191). Once parliament approved the introduction of TV, broadcasts would only cater to white South Africans since it was deemed dangerous for black South Africans to be exposed to the liberal global world.

It was only in 1982 that a channel catering for black South Africans was broadcast owing to, among other things, the "view that the apartheid government did not want black South Africans to see any interracial relationships on television" (Muller 2004: 34). The bulk of the content was imported American programs that were dubbed into Afrikaans and later, black languages such as Zulu (Muller 2004: 35). People could imagine something as rare in South Africa as white people speaking indigenous languages. In other words, for a black African in those days, watching TV could be an act of resistance against the isolation of racial groups. TV, in some ways, presented the possibility of re-fashioning transgressive racial identities in a racist South Africa, through creating the tension between the containment of the internal "world" and the "threat" posed by external cultural contaminations.

Beyond South Africa, there is a general view that TV signifies the "threat" of cultural imperialism that is eroding "authentic" cultures. The concern with the "flood of Western programming" in Africa – a phenomenon "of media imperialism" (Fair 2003: 189), reflects apprehension toward ominous cultural dissolution. The idea that "Western man" was so far ahead in historical time that he could not possibly be in the same community of spectatorship, consuming similar things at the same time, with the "native" who is imagined to be remote in place and time, would be destabilized by the advent of television in Africa in the 1960s. The possibility of diverse cultures of spectatorship and agency engendered by global television consumption contradicts the apathetic colonial gaze on people in "remote" countries.

This spatio-temporal distanciation and differencing is also suggested in John Tomlinson's thesis on cultural imperialism, which is initiated through an analysis of a British Television Christmas card portraying an aboriginal family watching television. He states:

> a very ordinary domestic scene. Yet there is clearly something extraordinary about the image, something that immediately strikes an exotic note. The family is watching in the open air; instead of armchairs, they are seated on blankets and oil-drums or on the desert sand. These people are obviously not Westerners, and the starkness of the setting seems to concentrate our attention on the Western technology absorbing them. The presence of the television is made strange in this context by the lack of the usual trappings of western affluence. This picture can thus quickly be grasped as representing cultural imperialism.
>
> *(Tomlinson 1991: 1)*

In this analysis, Tomlinson (1991: 2) critiques the ways in which the discourse of cultural imperialism that is propagated in the "imperialist West" marks the spectatorship of aborigines as "strange" and detrimental. Even with the prolonged physical proximity between so-called Western man and the "remote native" through infiltration, colonial terrorism and settlement, the communities of spectatorship were imagined as divided and isolated.

Likewise, in the visual arts, exhibitions were technologies of "seeing the world out there" and constructed socially differentiated communities of spectatorship. The gallery or museum enabled a privileged view of the world for the "Western man" where the "remote native" was looked at but never did the looking. With TV, however, the concept of the "Western man" as distanced subjective spectator *par excellence* is subverted. Television meant that spectators could be anywhere looking in. That is, television could be seen as the exhibitionary site for global communities of spectatorship, whether bourgeois or not. Therefore, even what Araeen saw as passive consumption could be an active challenge to power within the politics of seeing and being seen (how art exhibitions also operate).

Through the convergence between the television screen and the art gallery – the black box and the white cube – the discourse of imperialist modernism gains a visual currency through the spectacle of distance and difference, through *tele*casting. Constructing the imaginary of exotic "worlds" out there, ironically involved a view of the self as distanced. In the spectacle of bourgeois publics, art, specifically twentieth-century modern art in public museums and private collections, was not just about "seeing the world out there" through appropriating African objects as its raw material, it marked the dissimulation of colonial spatio-temporal proximity as fundamental social distance.

Writing about the early museums in modern European societies, Tony Bennett (1995) points out that it was through spectacular exhibitions that the bourgeoisie could render itself not only visible but, more significantly, visibly different. European modernization relied on grand scale spectacle. The "reorganization of the social space of the museum," Bennet observes "occurred alongside the emerging role of museums in the formation of the bourgeois public sphere" (1995: 25). Public museums, to which major royal collections where re-located, would create a visible and visibly different bourgeois public. In juxtaposition to Foucault's "institutional articulations of power and knowledge relations," institutions of exhibition

> were involved in transfer of objects and bodies from the enclosed and private domains in which they had previously been displayed into progressively more open and public arenas where, through the representations to which they were subjected, they formed vehicles for inscribing and broadcasting messages of power throughout society.
>
> *(Bennett 1995: 60–61)*

Exhibition spaces such as museums and markets, in this case, the *Bon Marche* in Paris, "offered a vision of a bourgeois lifestyle that became a model for others to follow" (Bennett 1995: 30).

In colonial contexts, this spectacularization became a strategy for displaying power and constructing racial and class difference. Moreover, it simulated an image of Africa that Apter (2002: 585) calls an "inversion of simulacrum and original [that] characterizes the imperial ontology of the world-as-exhibition, in which the framing devices of models and plans became political realities with perceived truth-effects." Imperial power entailed *showing* difference. The depiction of Africans through colonial photography were transformed into "objects of imperial vision and knowledge" while "the pomp and pageantry of empire" solidified defined racial hierarchies, turning the illusion of fundamental difference into reality (Apter 2002: 566, 572). Through exhibitions, performances and sports, Apter (2002: 572) argues, "the African-European opposition formed an important visual axis that would emerge and recede under different conditions as one among several dominant relations of visual engagement." Television in postcolonial Africa is therefore haunted by the colonial spectacle, where programming seems to be dominated by the broadcasting of Euro-American splendor and power (for example, the perpetual broadcasting of Britain's Royal Weddings at a time when the private affairs of former colonial powers are no longer relevant) alongside the *showing* of Africa's debasement.

Even after colonialism, and from the 1980s and 1990s, to the outside world, Africa was turned into a spectacle, plagued by poverty, dreaded diseases and civil war. Television images of children with visible ribs, pot-bellies and flies all about, became the spectacle that the world would come to associate the continent with. Cameroonian media critic Simon Inou (cited in Udeze 2009: 32) remarks that "the usual pictures we see on European and US TV's are those of Africans who can hardly hold up their skeletal bodies, brawling for food and gifts thrown out at them from trucks owned by Western aid organizations." In a similar vein, the singer, Miriam Makeba rejected the "Tarzan movie view of Africa" that "people in the United States still have" and proclaimed "we [too] watch television and listen to the radio, and go to dances and fall in love" (cited in Udeze 2009: 32).

This was during the same historical period when the Nigerian Television Authority (NTA) was screening Adiela Onyedibia's *Things Fall Apart* adapted from Chinua Achebe's classic novel in modern literature in 1987. The tragic figure of Okonkwo, faced with the contradictions of modernity and the disruptive tyranny of colonial presence, kills himself. This act of radical resistance and refusal to accept the Western colonial court and its judiciary system is regarded by his people as a denigration of their own set of modernizing moral codes and judiciary systems. While television elsewhere was broadcasting poverty and war in Africa, in Nigeria, people were watching images on television depicting paradoxes of intersecting but divergent modernities on the continent.

If one also considers, for example, Ken Saro-Wiwa's television program called *Basi and Company* (1986–1990), the ironies in the display or exhibition of postcolonial social class are illuminated.[5] Basi, nicknamed Mr B, is a street-smart tenant in a home owned by a wealthy landlady. He goes by the slogan "To be a millionaire, think like a millionaire." The madam's house and Basi's room have a display of paintings and from this site, the ironies of class are played out. Madam who keeps her purse behind her television set sometimes has parties to celebrate her lavish cars (in one of the episodes, a party is held for the arrival of her Lamborghini). While Basi uses any trick to make money and become a millionaire, including pawning a multi-purpose machine he refers to as the "million-naira baby," which "makes money, produces ice-cream, flashes images and drowns poverty," he remains a fleecer. The television box-like machine is technology that Basi says he built himself. Parodying the assumption that technology only comes from outside Africa and that modernity is expressed in exhibiting accumulated commodities, Basi represented to African TV spectators an innovative character navigating a complex globally connected world.

Also around the same time in South Africa, people were watching the TV mini-series of William Faure's *Shaka Zulu*, which was broadcast in South Africa in 1986. Not only did it generate visual images of the historical moments where the colonial encounter disrupted polymorphic modernities in which trade networks, mobility and technologies, military prowess and complex governance structures were already established, but it also created a lasting visual memory of a tragic but heroic Shaka. The television show was so impactful that statues of Shaka Zulu (in South Africa and in London, such as that at the Shaka Zulu restaurant, Camden market, in London) resemble the television actor, Henry Cele, who played the lead role of Shaka Zulu in the television series. As Gary Minkley and Phindezwa Mnyaka (2015: 59) also point out, this mini-series "re-configured a very particular Africanized image as 'Zulu.'"

Television is therefore a gallery of historical intricacy, of cultural and political intersection and resistance. It is not about only passively looking at the European or American other but also looking at the internal proximity and entanglement of the so-called first world in the continent. Television in Africa is a process of engaging with cultural conditions of modernity and with modernism as "a constellation of formal and conceptual strategies adopted by artists everywhere who, exploiting ideas and expressive forms from within and outside of their cultural contexts, fashioned new ways of seeing and experiencing the modern" (Okeke-Agulu 2006: 15). Arguably, television is a way of negotiating historically colonially circumscribed alienation and ways of being at home in the postcolonial global world.

I have previously argued that television is embedded in the profound space of home, is an interspace providing "multiple connections to spaces at various scales (national and global)" (Makhubu 2016: 64). Using Zina Saro-Wiwa's (Ken Saro-Wiwa's daughter) artworks and curatorial approaches, I referred to television as "a frame through which the distinction between interior and exterior is destabilized" and a way of "replicating the act of looking through boundless representation." In her work, Zina Saro-Wiwa uses the cathode ray tube TV as an artistic medium. In art installations such as *The Mourning Class* and *Table Manners*, each TV screen displays a woman crying, and in the latter work, each TV shows a person, looking directly into the camera and eating. The television also features in her video-work, *Phyllis*, which is about a woman living alone who watches Nollywood and televangelist programs on her TV set in her room. While I will not discuss these works in detail as I do elsewhere, they are examples of the ways in which art on TV and TV in art conversely circumscribe two distinct but inter-related exhibition spaces of the black box and the white cube as a negotiation of Africa's political geography of pre-colonial, colonial and postcolonial modernities.

Television becomes an interspace, and this spatio-temporal matrix of television can enable a dialogic politics of vision, power and global modernism. By this proposition, I am not implying that television, and other technologies of seeing, are examples of the "arrival" and exhibition of modernity and cultural modernism to Africa, through which Africans could imagine, exhibit and see themselves as "modern". On the contrary, I propose that it became a way of negotiating power and transcending the colonial aura in the politics of seeing and being seen in relation to the rest of the world. It is a mode of scrutinizing the distortion of time and space through the primitivism constructed through colonial modernity. Television is an exhibition of the same historical events from multiple perspectives, both documentary and fictional. The televisual gallery as curation and as a collage of time and place confronts the politics of seeing, of looking in and looking back.

## Politicizing the Televisual Archive: Television as Collage in Art

While television has become culturally ubiquitous, as a form of self-defining and "telling African stories" as in, for example, the rise of local video film industries like Nollywood over the years, it seems to remain entangled with the burden of deleterious colonial televisual imagery. Imperialist nostalgia, as Renato Rosaldo (1993) terms it, haunts the televisual archive and, to an extent, the visual arts.[6] In this section, I shift my attention to the ways in which the archived televisual image is politicized through collage in the arts.

The intangible nature of television delayed the process in which television could be recognized as archival. Archival discourse often illuminates the Foucauldian relationship between the archive, knowledge and power. Therefore, materials worth archiving are selected, symbolizing authority and the conveyance of specific cultural values. Marita Sturken (1990: 103) observes that "television is coded as the immediate – the live image transmitted to many locations at once [and] has never been conceived, either culturally or industrially as an archival medium." Archives, as Achille Mbembe (2002: 19) asserts are "a collection of documents" which derive power and status "from this entanglement of building and documents." In this definition, it would be difficult to see the fleeting image as a historical document to be stored and preserved in a building. The first television archives in the United States of America necessitated the "amassing of space [...] in order to preserve the ethereal TV form as an embodied thing-like medium" (Spigel 2010: 56). In this way, the TV medium could become archival, requiring a place to preserve its otherwise ephemeral past. This spatial dimension, while it seems to create a taxonomic proximity between TV, archival documentation and art museum, is also faced with multiple locations of the televisual archive in people's homes. That is, the convergence of the white cube and the black box intersect with concepts of home-place.

With VHS, many families recorded television programs. Owing to its immateriality, however, the televisual archive refuses the cumulative, selective and often inaccessible yet systemically institutionalizing process of archiving. Unlike other forms of archives, historic television programs are prone to being replayed on contemporary retrospective television shows. Furthermore, recorded television broadcasts, reflect the ideas and values of a particular time and place that can be looked at retrospectively and analytically. This contributes to the notion that the televisual archive, in tangible and intangible forms, has time-space ubiquity that "original art" does not possess. It is an archive of content that could be copied, replayed or repeated, imitated, remixed and can be represented on mass produced, reproduced and recycled objects (tapes). Furthermore, that archive can be located in accessible, uncategorized virtual space platforms such as YouTube, which can be accessed on portable devices.

Contemporary art practices illuminate this ubiquity through quoting and replaying. The televisual archive, for modern and contemporary artists, is a trans-disciplinary discursive field. Take, for example, the re-fashioning of photographic and televisual images by the Malawi-born artist Samson Kambalu, whose work borrows from the sepia toned televisual aesthetic. Through what he terms Nyau cinema, Kambalu produces imagery that evokes historical documentary footage and archival material to illustrate what can be seen as political circumlocution.

For example, in a set of sepia videos shot in Connecticut, USA, in 2014, Kambalu's *Runner, Dancer in the Woods* and *A Road in the Countryside* interrogate slavery and racism in America. In *Runner*, Kambalu runs on the same spot – the same action being rewound and replayed – but he never goes anywhere. Alluding to a fugitive slave running from the plantation, this artwork is reminiscent of television documentaries and fictional broadcast films in which the narratives of fugitive slaves are shown and re-enacted. In this loop of re-enactment by Kambalu, the running fugitive is as at once an historical figure and a contemporary one. Through the repeated act of futile running, not only does the archive of the late 1700s runaway slave surface, but also the broadcast imagery of young African American men being killed by police, often while running away (for example, the case of Walter Scott among others) is evoked. More significantly, the televisual archive in this instance casts global modernism (built on slave labor) as a recurrent phenomenon of being "stuck" in time and place. It interprets the sense of modern progress as, in Vladimir Lenin's (1951) words, "one step forward, two steps back."

In *Dancer in the Woods* Kambalu looks up at a tree and walks toward it. This artwork is reminiscent of the way in which a camera pans over a photograph in television documentaries about slavery and lynching in America. Kambalu's repetitive movements elicit the eerie suggestion of the ghost-like presence of a lynched body. Consonant with the notion of the archive as "the place that marks an emptiness and loneliness that indicates what was once animated and filled – a memory space that becomes a place," Kambalu's work operates through inference (Carolyn Steedman, cited in Spigel 2010: 53). In other words, it relates to visual memories from which we can surmise a narrative from a single image based on the historical luggage of imagery that has been screened on TV. This produces a telepresence of history and fiction that is interpreted by Spigel (2010: 54) as "television's textual aesthetics, especially its 'live-ness' and sense of the here and now, [that] open onto a series of uncanny reversals that take place in the archive." Kambalu's other works, such as *A Road in the Countryside, Amistad* and the recent *Laughing at the Big Tree on a Slave Plantation* (2017), directly make references to the contemporary ghostly presence of slavery writ-large in current socio-economic conditions. Here, the historical, or the archival is repeated, re-enacted and re-inscribed into the present. This tautology in history and political circumlocution in Kambalu's work is characteristic of the ways in which the televisual archive can be seen as a discursive field, re-producing old meanings while generating new ones in the re-fashioning of recurrent, disruptive and disrupted modernisms.

Television programs and news broadcasts that illustrate political circumlocution and tautology in history but refer to memory are utilized by artists such as South African-born Francois Knoetze. Through a collage of television programs, home videos and recorded performance in

*Cape Mongo* (2014–2015), Knoetze presents the televisual archive as nostalgic, sometimes poignant. Knoetze's work shows how television gave a visual form to dogmatic apartheid and normalized segregated, politically apathetic domestic lifestyles. Television programs showing the daily lives of white farmers, the glorification of the Voortrekker monument as a symbol of Afrikaaner nationalism, the celebration of Cecil John Rhodes and other didactic shows were ways of framing apartheid as a moral and just cause. Using archival television footage that was broadcast in South Africa since the 1970s, Knoetze's set of five video artworks are based on five characters each made out of a specific material: unwound and broken VHS cassettes, glass, metal, paper and plastic. Each signifies a commodity that sustained the economic growth of South Africa through racist labor practices. The characters in each work traverse the segregated spaces in Cape Town and all perish at the end.

*Cape Mongo* begins with a television color bar. The color bar in early television broadcasting was used as a color test pattern and would show when there was no programming. In Knoetze's art, however, there is a conscious play on the image of "the color bar" and the silhouette of the Cape Town landscape as indicative of color/racial segregation, in the sense used by W.E.B. du Bois. With reference to South Africa, anthropologist Max Gluckman (1955: 164) also refers to "the bonds in the color bar" to emphasize that "discriminatory custom against the Africans and other colored groups chains the dominant white group." Similarly, Knoetze's work demonstrates the tragic impact of racial segregation on black and colored communities but, reciprocally, the ways in which the apartheid government used television for the mental bondage and benumbing of white communities as well.

In *Cape Mongo*, Knoetze uses home video film footage, recorded in 1994 – the dawn of a democratic post-apartheid South Africa. The main character, made out of VHS tapes, realizes he is now obsolete and resorts to bingeing on alcohol and begging on the streets. Feeling dejected, he walks up to the mountain, looks down on the Cape Town landscape and raises a television antenna with magnetic tapes streaming in the wind. As he does this, one hears broadcasting noise and sees television clips of rugby games, children's cartoons such as the South African Afrikaans language *Liewe heksie* (which translates as *Dear Witch*), the English children's show *Mr Chinwag*, a puppet donkey, and American shows such as *Sesame Street* and *Planet Man*. In this work, there is also an aerial image of the Voortrekker monument complemented by the words: "it's a God-given right." As Robert Weisbord (2017: 161) observes, "[i]nspired by their Old Testament faith, Boers believed that they had a God-given right to their South African land, and a God-given right to enslave the bantu." The juxtaposition of TV shows representing childhood innocence with those perpetuating the horrors of apartheid through religion, politicizes the televisual archive.

In another related artwork, *Cape Mongo: Glass*, the main character explores the exploitative wine industry in Cape Town. Throughout, one sees images from old agricultural television shows of black people laboring on the land while a white farm owner inspects it. This is paired with black-and-white television footage of black farmworkers, happily singing and stomping on grapes (an older form of wine making) in a wine barrel. Over this representation of farms is the voice of a Xhosa-speaking man who says

> I have been living here for fifty years, my first daughter died here, my second daughter died here as well. I'm left with my two sons. When these white people bought this place I was already here. Now they are trying to evict us. I told them I'm not going anywhere.

This experience of land dispossession is juxtaposed with images of leisurely wine drinking in white communities contrasted with the adverse effects of alcohol on colored communities.[7] In

Cape Town, colored laborers working on farms were paid in alcohol through a system generally referred to as "the dop system." The detrimental generational impact this has had on colored communities is juxtaposed with wine drinking among the wealthy, or as the television footage announces, "the more capable citizens of the day," as a "cultured" practice. Through images of red wine flowing in factory pipes as though it were blood flowing in veins, the violence of apartheid characterized as bloody and ruthless is interpreted as recurring in the structural violence of contemporary South Africa. Interlaced with this is the footage of news broadcasts depicting wine farm protests over the minimum wage of R150.

This work is punctuated by an image of the Coca-Cola bottle landing in the desert sand in the South African film, *The Gods Must Be Crazy*. This film, made in Botswana in 1980, utilizes the ethnographic lens that is typical of television nature documentaries where the Khoi-San people are portrayed as simple, "primitive" people who know nothing about globalization. This skewed historical view of a primitive land so far away, is also contrasted with a scene from the TV show, *Megaman* in *Cape Mongo: Metal*. A voice proclaims: "Welcome to Mingo city," and as this is announced one hears the classic words of Gil Scott-Heron boldly stating "you will not be able to plug in, turn on, and cop out, the revolution will not make you look five pounds thinner" and although we do not hear it, here we are reminded: "the revolution will not be televised!"

In relation to the spatio-temporal matrix of the black box and the white cube, the collage of the televisual footage circumscribes modernity and modernism as a particular kind of dislocation or displacement and alienation. It is not only the heroic traversing of "the world out there" displayed within the exhibitionary cube or box, but it is also colonial settlement, colonial raids, land theft, and systematic dispossession that signals a sense of dislocated-ness. In the work *Cape Mongo: Paper*, for example, current housing crises in South Africa are portrayed as the recurrence of colonial displacement. A cartoon image in which a woman says: "look there, it's [...], it's a city! [...] A city [...] in the clouds!" is contrasted to a statement made by a local Cape Town man who says: "I call this a dump place. [...] that's what the government do." After he says this, we see footage of a news broadcast showing a protest against Helen Zille, the Western Cape Premier's perpetuation of evictions and the destruction of informal homes in Cape Town. This housing protest is contrasted by snippets from a lifestyle television program, *Top Billing*, which is "targeted at successful South Africans" where homes of wealthy South Africans are profiled. In the *Top Billing* footage used by Knoetze is the description of a home with space for African sculptures, which are defined on the show as "African figures inhabiting the house ranging in age from thirty to a hundred [years]." Immediately, a cartoon advert image follows, showing a "house of tomorrow" represented as a small wrapped gift. This gift box is juxtaposed to the box-like architecture of tin shacks. The cartoon advert continues to state that in the "box" is a "completely pre-fabricated and ready to set up" house, which then pops up into a mansion: "and there it is: modern in every respect."

Through the collage of the archival television sequence, Knoetze exhibits the neurosis of a traumatized country, still haunted by the incoherence and chaos of apartheid. The televisual archive, in this artwork, reflects the psychosocial conditions of a given time and place, demonstrating the recurrent nature of the modernity discourse. In pasting bits and pieces from pre- and post-1994 television programs, the "whole" becomes perceptible. The racialized and racializing nature of television programs is illuminated through collage. Seemingly innocent children's television shows and programs displaying bourgeois lifestyles are shown to be reinforcing race, gender and class hierarchies. In this work, television, like art, becomes a useful anthropological lens for unveiling and scrutinizing underlying social inequities.

## Participatory Strategies: Live Art, Video and Television in Africa

In an artwork entitled *Back to Me* (2003), the artist Thando Mama sits naked in a dark room, leaning in, looking at a television screen. In this simple yet potent image, as the artist gazes at the static noise of the television screen, he is the viewer and the one who is viewed. In this way, he contravenes perceived communities of spectatorship. Artworks, like Mama's work, use TV as an allegory of self-image on the global image-saturated stage. Artists use television to "think, innovate and produce," to appropriate Araeen's words.

In art, television as medium is often associated with conceptual, video and live art. While it is generally seen to function as a device for transmitting video in multimedia art installations, or a discursive instrument in performance or live art, television in art represents the critique of ideological manipulation. It is, as artists have shown, a mode of interrogating socio-political conditions in modern and contemporary societies. The use of cathode ray tube (CRT) television sets as well as television archival footage also lends a particular kind of aesthetic. Television gained a specific cultural currency in conceptual art, not as a broadcasting medium but as a methodological approach to institutional critique, self-reflexivity, the critique of authenticity and the conditions of production, as well as the questioning of ideological dynamics in the politics of representation.

For example, the Korean American artist Naim June Paik, a Fluxus artist for whom the CRT television set would become key in his oeuvre, is generally seen as the pioneer of video art. The CRT became, for Paik, an artistic medium or, as he states, a way to "shape the TV screen canvas as precisely as Leonardo, as freely as Picasso, as colorfully as Renoir, as profoundly as Mondrian, as violently as Pollock and as lyrically as Jasper Johns." Ken Hakuta, his nephew, observes that Paik "foresaw the globalization of everything."[8] Paik's work was not only about what is transmitted but most significantly, the very nature of television itself.

Artists such as Adrian Piper, took the medium to be confrontational. If one considers, for example, the work *Cornered* (1988), Piper is on the television screen placed in a corner of a room behind, but not concealed by, an upturned table and a set of empty audience chairs facing the television. On the television, she boldly asserts:

> I'm black. Now, let's deal with the social fact of me stating it together. Maybe you don't see why we have to deal with it together. Maybe you think this is just my problem and I should figure it out by myself. But it's not just my problem, it's our problem.

Television, in this case becomes an agonistic medium, through which political struggles about racial and gender representation are waged.

In these artistic strategies, the spectator is meant to participate, to be implicated in the act of looking and to be conscious of how being looked at operates in exhibitionary spaces. These participatory strategies are also used in artwork of artists such Dineo Seshee Bopape. In her installations, for example, *lešobana!! lešobana! lešobana!! (le bulegile); lešobana! lešobana! lešobana!! (go phunyegile)*, the virtual space, both in the gallery and the opened up box of a CRT TV set, is not just the "world out there" but the internal imaginary. She asks "if there is space within the virtual, how deep does it go? How far can a sound or an image recede? [This is also] a question of metaphysics, and trying to draw a type of parallel timeline."[9] Likening the colonial moment as a break-in, a rupture, or a puncture, the words of the title, *lešobana* and *go phunyegile* which translates to "the hole, the hole, it's punctured." The double coding of the hole as a point of destruction (something punctured) and the hole as something to look into, as a telescopic or camera hole, elicits a dialogic, participatory, yet implicated, notion of spectatorship.

## Conclusion: The Speculative Image

In the Nguni language, Zulu, television is almost directly translated as *umabonakude*: "the one who can see far" or the "one who can see into the future." It is very similar to tele (far) vision (sight). However, in Nguni languages, the pronoun "u" refers to a person of any gender and "i" refers to an object. The slight difference is that in Nguni languages television is not just the object, or the instrument that enables one to see far, but also a subject. This distinction between object(ive) and subject(ive) may seem minor but establishes a hermeneutic approach to television, at least in some African contexts. Since television coincided with colonialism, cultural imperialism, the spring of postcolonial independence coupled with neo-colonial neoliberal compulsion, the pronoun "u" signals the strange historical relationship with television. Positioning it as a subject, rather than object, and as a subjective projection of perceptions illustrates a conscious recognition of the machinations of ideology.

*Umabonakude* is also speculative, capturing the sense of a spatio-temporal matrix ("far" as distance or space and "future" as time). Television, especially when considered as art, engages with the geopolitics of modernity and modernism. It is, to a certain extent, a form of participating in the politics of seeing and being seen. The black box and the white cube have been agonistic spaces in the postcolonial context, not only showing the "world out there" but drawing resistances against certain forms of representation. Thinking of these exhibitionary spaces, as a convergence, enables a critical engagement with the menace of colonially segregated communities of spectatorship.

## Notes

1 (Pinkus, D. 2017). "Samsung's new framed TV is literally a hanging wall art," https://homes.nine.com.au/2017/03/15/10/13/samsugs-fram-tv-is-literally-a-piece-of-wall-art [accessed 12/03/2018]
2 www.samsung.com/za/curation-contents/theframe/curated-collections/ [accessed 12/03/2018]
3 www.samsung.com/global/tv/blog/the-frame-art-store-bring-worldclass-galleries-into-your-home/[accessed 12/03/2018]
4 Although the term "black box" is associated with theater, early black box theaters were based on basic TV studio sets.
5 Ken Saro-Wiwa was a Nigerian writer and activist known for his role in the Movement of the Survival of the Ogoni People (MOSOP), which campaigned against big oil corporations such as Shell and their exploitative and polluting acts. Saro-Wiwa and other MOSOP members were hanged by the Nigerian government in 1995 for their activist work.
6 Also see Andrews, K. "Colonial nostalgia is back in fashion, blinding us to the horrors of empire," *The Guardian*, 24 August 2016, www.theguardian.com/commentisfree/2016/aug/24/colonial-nostalgia-horrors-of-empire-britain-olympic [accessed 12/03/2018]
7 In South Africa, colored refers to the mixed-race population. As an apartheid-constructed racial category, it is different from black or black African.
8 www.sfmoma.org/ken-hakuta-my-uncle-nam-june-paik/ [accessed 12/03/2018]
9 www.oneart.org/dineo-seshee-bopape [accessed 13/05/2018]

## References

Apter, A. 2002. "On Imperial Spectacle: The Dialectics of Seeing in Colonial Nigeria," *Comparative Studies in Society and History*, 44(3): 564–596.

Araeen, R. 2005. "Modernity, Modernism and Africa's Place in the History of Art of Our Age," *Third Text*, 19(4): 411–417.

Bell, T. and Ntsebeza, D. B. 2003. *Unfinished Business: South Africa, Apartheid, and Truth*. London and New York: Verso.

Bennett, T. 1995. *The Birth of the Museum: History, Theory, Politics*. London and New York: Routledge.

Bourdieu, P. 1996. *On Television*. Priscilla Parkhurst Ferguson [transl.]. New York: The New Press.

Bourgault, L. M. 1995. *Mass Media in Sub-Saharan Africa*. Bloomington and Indianapolis: Indiana University Press.

Fair, E. J. 2003. "Francophonie and the National Airwaves: A History of Television in Senegal," in Parks, L. and Kumar, S. (eds.), *Planet TV: A Global Television Reader*, New York and London: New York University Press. pp. 189–210.

Ferguson, J. 2006. *Global Shadows: Africa in the Neoliberal World Order*. Durham and London: Duke University Press.

Gluckman, M. 1955. *Custom and Conflict in Africa*. Oxford: Blackwell Publishers.

Kasfir, S. L. 1992. "African Art and Authenticity: A Text with a Shadow," *African Arts*, 25(2): 40–53+96–97.

Katz, E. and Wedell, G. with Pilsworth, M. and Shinar, D. 1977. *Broadcasting in the Third World: Promise and Performance*. Cambridge, MA: Harvard University Press.

Lenin, V. 1951. *One Step Forward, Two Steps Back: The Crisis in Our Party*. Moscow: Progress Publishers.

Makhubu, N. 2016. "'This House Is Not for Sale': Nollywood's Spatial Politics and Concepts of 'Home' in Zina Saro-Wiwa's Art," *African Arts*, 49(4): 58–69.

Mbembe, A. 2002. "The Power of the Archive and its Limits," in Hamilton, C., Harris, V., Taylor, J., Pickover, M., Reid, G., and Saleh,R. (eds.), *Refiguring the Archive*, Dordrecht, Boston and London: Kluwer Academic Publishers. pp. 19–26.

McLuhan, M. Fiore, Q. Agel, J. 2001. *The Medium is the Message: An Inventory of Effects*. Corte Madera, CA: Gingko Press.

McLuhan, M. and Fiore, Q. 2006. "The Medium is the Message: An Inventory of Effects," in Morra, J. and Smith, M. (eds.), *Visual Culture: Critical Concepts in Media and Cultural Studies*, Volume 4, London and New York: Routledge. pp.136–292.

Minkley, G. and Mnyaka, P. 2015. "Seeing Beyond the Official and the Vernacular: The Duncan Village Massacre Memorial and the Politics of Heritage in South Africa," in Peterson, D. R., Gavua, K., and Rassool, C. (Eds.), *The Politics of Heritage in Africa*, New York: Cambridge University Press. pp. 50–69.

Muller, C. 2004. *South African Music: A Century of Traditions in Transformation*. Santa Barbara, Denver and Oxford: ABC Clio.

Oguibe, O. 2004. *The Culture Game*. Minneapolis and London: University of Minnesota Press.

Okeke-Agulu, C. 2006. "The Challenge of the Modern: An Introduction," *African Arts*, 39(1): 14–15, 91.

Orgeret, K. O. 2015. "From 'The Devil in the Black Box' to a Nation Building Tool: Early TV in South Africa a New Medium for a New Nation," in Anderson, S. and Chakars, M. (eds.), *Modernization, Nation-Building, and Television History*, New York and London: Routledge. pp. 187–206.

Rosaldo, R. 1993. *Culture & Truth: The Remaking of Social Analysis*. Boston: Beacon Press.

Spigel, L. 2008. *TV by Design: Modern Art and the Rise of Network Television*. Chicago: University of Chicago Press.

Spigel, L. 2010. "Housing Television: Architectures of the Archive," *The Communication Review*, 13(1): 52–74.

Steiner, C. B. 1994. *African art in Transit*. Cambridge and New York: Cambridge University Press.

Sturken, M. 1990. "Paradox in the Evolution of an Art Form: Great Expectations and the Making of a History," in Hall, D. and Fifer, S. J. (Eds.), *Illuminating Video: An Essential Guide to Video Art*, New York: Aperture. pp. 101–124.

Tomlinson, J. 1991. *Cultural Imperialism: A Critical Introduction*. London and New York: Continuum.

Udeze, B. 2009. *Why Africa?: A Continent in a Dilemma of Unanswered Questions*. Bloomington: Xlibris Corporation.

Weisbord, R. G. 2017. *Racism and the Olympics*. New York and Oxford: Routledge.

Ziegler, D. and Asante, M. K. 1992. *Thunder and Silence: The Mass Media in Africa*. Trenton: Africa World Press, Inc.

# 29

# IN THE BIG LEAGUE

## Television and Gaming in India

*Divya McMillin*

### Introduction

The darkened café and gaming areas with black-painted floor, ceiling, and walls, and lit with green LED pipes throughout, offer a welcome relief from the heat and dust outside. The League of Extraordinary Gamers (LXG) in the Indiranagar neighborhood of Bangalore City (now Bengaluru), in India, is a prime location, part of a relatively new neighborhood that has only in the past decade expanded to upscale shopping strips and gleaming high-tech offices. You could almost miss it, packed as it is in a mostly residential neighborhood, on a road that connects suburb to highway. On August 13, 2017, the place was crowded, pizza stacks and Coca Cola pallets almost reached the ceiling as the FIFA 17 AFGC (Asian Football Gaming Championship) Offline Qualifier was on, the biggest ESports tournament in the country. In separate rows, Dota2 contenders played furiously, live streaming to potential viewers around the world through Twitch and YouTube. With a global grand prize of USD500k for FIFA 17, emotions were running high; fans crowded into a viewing room as teams played in another section. The darkened rooms, high-action content, and collective attention to one screen, could very easily be analyzed through a conventional neoliberal North American framework of television studies along the levels of political economy, textual analysis, and audience studies. Yet there is something quite new in the multiplayer capabilities that demands new ways of seeing and interpreting. Convergence technologies and diversification of products require new methodologies and theoretical frameworks. In nonwestern contexts, different points of entry are necessary and attention to the postcolonial condition is critical. Specifically, the immersive and interactive capabilities mark the difference between television and gaming. While some of the critiques and concerns of television studies apply, the agency engendered in gaming and the limitations placed on such agency in the postcolonial labor conditions in urban India, offer us new directions to consider in media studies within globalizing contexts.

How then may we place gaming in the larger media landscape in India? Is this an extension of television viewing? How may we expand television studies through an exploration of gaming? Over-the-top (OTT) platforms facilitating digital content are on the rise, increasing users for streaming programming, gaming, music, and video-on-demand (VOD) services. Television studies in India has to necessarily include the burgeoning gaming industry, and it is this industry that is the focus of this chapter. The users constitute a minority urban elite, but they coalesce into a significant high-paying population, making a real impact on the industry nationally and globally. The chapter situates gaming within the larger Indian media landscape and explores how this extends television studies, through the case study of LXG as well as through observations and surveys of young television audiences in Bengaluru, India, over four phases of fieldwork spanning 2013–17.

The imperative to digitize across the nation's metros was issued by the Ministry of Information and Broadcasting in 2011. With slogans such as Digital India, Skill India, Startup India, and Make in India, the government led by Prime Minister Narendra Modi, embarked on an aggressive digital campaign, with its controversial biometric identity system, Aadhar,[1] receiving global recognition. By the end of 2017, India was the third largest mobile internet and smartphone market in the world with rural use widespread and rapidly expanding. With government approved cashless, paperless, digital services offered by companies such as IndiaStack, users can access a range of services securely and personalize their entertainment needs (Balaji, 2017). Such a move, which paralleled the government's demonetization initiative, created confusion and panic, with inadequate attention to the services necessary to equip the nation's vast majority of the poor and illiterate in the transition. For a country that is barely behind China in population at 1.34 billion (in January 2018), despite a landmass three times smaller, digitization for those who can afford it, seems a magical escape from the insurmountable barriers to resources created by overpopulation.

The first phase of fieldwork for this study, in summer 2013, revealed probably the most significant upheavals in the city with the installation of fiber optic cables throughout. Pubs, breweries, cyber cafés, lifestyle malls, and gaming centers were cropping up everywhere, providing much needed air-conditioned relief from crawling traffic and packed streets. It was that summer that LXG was established, necessitating fieldwork in summer 2015 when the gaming industry achieved a dramatic spike. Analysis of LXG campaign strategies and marketing materials, as well as interviews continued in early 2016, both with shoutcasters[2] and marketers in Bangalore as well as developers in Seattle and Bellevue, Washington, where the most popular games are developed. The final phase of fieldwork was conducted in summer 2017, where visits to four private universities in the city, surveys on television usage, and interviews and participant observation at LXG, filled in the gaps on television preferences and gaming habits.

The Indian Media and Entertainment (M&E) industry is the fastest growing in the nation, largely due to the pivotal deregulation effected by the Indian government in 1991 at the dawn of satellite television. The M&E 2017 Report notes that while television and box office sales dropped considerably that year, the digital market surged ahead. The gaming industry reported a revenue of INR 30 billion (USD 470 million) in 2016. As Andre Cesaro, Shanghai-based Directive Games Founder and CEO observed in a February 6, 2017 interview:

> The most popular in Asia are free-to-play games, it's a way of distributing and selling that is being perfected in China and Japan. The game is free, later on it is monetized heftily. The number of games downloaded is incredibly high.

Indeed, the number of games downloaded in India in 2016 was 1.6 billion, the most popular being Farmville, Mafia Wars, Rummy, and Teen Patti, on various social media platforms.

The city of Bengaluru, one of the first to be digitized, provides a rich site for the exploration of gaming as an extreme version of the individualized mediated experience. For a theoretical framework, we turn to the concept of Third space. Third space allows us to explore more fully the process by which young media users in India leverage their connectivities and technology options to produce a "place" they can inhabit meaningfully and powerfully, subverting lived realities which offer, oftentimes, limited mobility and even threats to personal safety (see McMillin (2009) for a full treatment of teens and media in India). As developed by postcolonial scholars, this concept allows us a relevant and timely framework for explorations of youth agency and identity in the digital age.

## Third Space as "Place"

In the complex transmedial environment of Bengaluru we use the concept of Third space from critical geography studies as a point of entry to understand the unique ways in which gaming offers a realm of identity production for its users. We begin with a quick overview of the media landscape in urban India.

With a population of over 1.3 billion, India possesses the largest youth population in the world; over two-thirds is under the age of 35. Television itself has been declining worldwide in the past decade as a medium of choice for youth, with handheld technologies and gaming far outpacing it as a vehicle for leisure. A 2017 report of the Swiss cooperative Klynveld, Peat, Marwick, and Goerdeler (KPMG) commissioned by the Federation of Indian Chambers of Commerce and Industry (FICCI) of mobile phone and internet usage in India reveals that social networking, instant messaging, and video watching, particularly on YouTube, accounted for the highest traffic, around 80%. Top social platforms in the country are Whatsapp (11%), Facebook Messenger (9%), Facebook (8%) and Skype (8%). Developers are working to extend applications to on-the-ground services. Subsequently Uber (for example) has seen an exponential boost in usage. There are 2,500 wifi spots throughout the country under the Bharat Sanchar Nigam Limited's (or BSNL – the national government's telecom service) initiative. Internet users were estimated at 500 million in June 2018, with around 700 million using smartphones.[3] These numbers outrank users in the United States and will soon overtake those in China. The roll out of 4G by private telecommunication companies and the government support for public wifi have further increased consumption of digital media, with a reported 1.1 billion wireless subscribers in July 2018.[4]

It is no wonder that "television" has transitioned from a stationary box to a fluid medium, accessed through devices connected to the internet. Streaming services such as YouTube, Netflix, Amazon Prime, and Hotstar, to name just four of the nine in India, are algorithmically informed, integrating social media and television habits of the consumer.[5] In addition, recording capabilities through DVR, and streaming services such as Roku, Chromecast, and Apple TV, have significantly changed when exactly television is watched and how episodes are viewed (i.e. with or without commercials). Groshek and Krongard (2016) discuss the trend in differentiated social media use as well as increased personalization of content. Television is only part of a larger dialogue, with users demanding media that are interactive and that provide opportunities for sharing and control.

In *Transmedia Television: Audiences, New Media, and Daily Life*, Evans (2015) writes that great similarities exist between television and gaming, most notably in audiences' engagement with the text, especially as the narrative demands interactivity with puzzles (as in mystery serials such as *Lost*) or immersive reality (as in shows such as *Big Brother*). Some theorists emphasize differences between the two in terms of the technology and medium itself (e.g. Gisprud, 1998), while most others address the content that flows across such media and instead seek to understand what audiences do with it (see Ang, 1985, for example). In the current transmedial environment across urban centers worldwide, shared content across technologies retains cross-media fandom and spawns franchises. Dramatic film-gaming examples have emerged in the Indian transmedia environment in just the past few years. The 2015 big screen blockbuster *Bahubali* grossed $20 million in the United States alone[6] (Raghunathan, 2017) and was almost immediately available as a game download for Androids and iPhones.[7] It is no wonder that ESports are advertised across platforms and capitalize on winning strategies of real world sports arenas as well as televised sports for its streaming tournaments.

Third space is extremely relevant for our study of gaming in India. The concept developed most cogently by postcolonial theorist Homi Bhabha (2009) and critical geographer Edward Soja (2009) is useful in its imperative to understand experience over representation. It opens up the concept of the postcolonial subject and is productive for our understanding of how media globalization is experienced in postcolonial countries. In India, with over 200 years of British colonialism and increasingly aggressive deregulation and privatization policies since independence in 1947, hybridity is everywhere, in cuisine, clothing, media, and general lifestyle. Bhabha (2009) notes that in such hybridity is also a loss, it reveals the postcolonial subject, the "third" in "Third space." The postcolonial subject is always in that liminal space negotiating among identities, feeling tensions and riding the interbred global and local. English as a colonial language is also simultaneously the language of urban transformation, a space for resistance and appropriation where the large English-speaking population in Indian cities have made it their own and productively and profitably adapted it to their own needs.

Briefly, First space encompasses direct spatial configurations that can be mapped, such as physical buildings and streets of a city center. Second space refers to spatial representations which are the cognitive constructions related to the purpose or representation of the place. For example, a coffee shop would represent a space for getting coffee, meeting a friend, or a quick spot to get work done informally. Third space is a "transcending composite," an "also/and" of First space and Second space, simultaneously real and imagined. To extend our previous example, the coffee shop becomes a space of possibilities, where friends meeting over steaming lattes create a vision for a safe space for the city's homeless children. The desires of its inhabitants take flight through its physical structure and location as well as its purpose of connectivity, comfort, and convenience. Ikas (2009) discusses the cultural production that is possible, with Third space allowing for multiple identities to emerge, with no one conclusion or positivist framework determining outcomes. Methodology has to be open ended and immersive, allowing for a discussion that recognizes that within agency is limitation and within limitation is also agency. The case study of LXG allows us to draw out the intricacies and effectiveness of gaming as Third space. It is a combination of real and imagined spaces, hospitable to an enduring sense of self amid the rapidly changing First space of the globalizing city of Bengaluru.

## The League of Extraordinary Gamers: Third Space Relief in the Globalizing City

LXG was founded in 2013 by aeronautical engineer Aravind Ananth and, its website states, was rated the best gaming lounge in the country within six months. Collaborations with Logitech and NVIDIA were instrumental in boosting its technical capabilities and ambience. Within barely two years, the location had registered around 9,000 users. Through ingenious home-made ventilation systems, Ananth was able to build liquid cooling units in his office as well as harness solar power for energy efficiency. LXG Indiranagar has its in-house café, but is surrounded by a wide variety of eateries as well, open through the night. It is quite common for the place to be packed around the clock, with day customers leaving only in the wee hours of the morning to get ready for a full day of work. As has been discussed elsewhere in the case of call centers (see McMillin, 2008), the presence and subsequent emergence of a host of businesses such as eateries, tech shops, malls, and arcades, support and build the young clientele and workers at such gaming centers. It docks in a most hospitable First space of modern buildings and retail indulgence, providing the construction of Second space in what it represents, a very progressive space for young, urban leisure. And then within, together with physical environs and representational aspects, it produces a Third space where imaginations take flight, where fantasy architecture engenders the formation of new identities and allows the expression of ephemeral selves and agency through virtual avatars.

A fundamental understanding of the richness of "what lies beneath" is the prerogative of creative teams. Their business is the production of the most vivid, interactive Second spaces where user and content can come together in creating the Third space of fantasy. Cesaro, (Directive Games Founder and CEO) explained in a February 2, 2017 interview, that the best content that keeps users wanting more is that which plays on time-tested themes, such as:

> Archetypes. Stories told through puzzles, riddles, challenges. These are the best games which have a narrative. Without that we cannot build a franchise. If you want to become successful, you have to have a franchise with hero, heroine, and *always* a villain.

He continued on to discuss the ubiquity of online relationships. The stigma of connecting online is long gone. Whether real or virtual, relationship building online is the norm. Gaming as an immersive environment does not spark as much concern about social isolation as it did about a decade ago. He said,

> In our generation—when online dating started, (that) was considered only for losers. These days when we look at younger demographics, they may use that as a tool. What is normal these days? Being on Tinder is normal. Same with gaming, it is part of everyday life. Because of mobiles and smartphones, what started with Nintendo is now at their fingertips. Everybody is a gamer. Whether it is candy crush at the office or at home.

Directive Games' storylines are time-tested, tapping into the deepest desires and anxieties of its users. The creators, whose credits include the VR Super Kaiju as well as The Machines, one of the first cross-platform VR/Mobile games, stay within the classic archetypes of good and evil. Cesaro noted that his initial team wanted to work with franchises in gaming, and to bring the PC and console quality to VR and Mobile. They recognized the importance of high-quality graphics, even over the storyline, which itself had to be strongly iterative:

> For people to go back into a game over and over again, you need to build a social fabric and human interaction. Multiple (story) layers speak to the new generation where social and internet connectivity is inherent to what they do. The way we see the new generation play and going forward ... they are always interactive, always social, always connected. Even the most immersive VR has the problem of isolation. You are isolated by goggles in front of your face, so for us the real nut to crack, is how to bring multiplayers to VR. This is what we are working on right now.

Cesaro's comments directly connect to the immersion/interaction possibilities of gaming, as Evans (2015) describes in her analysis of the BBC television serial *Spooks* that crossed over to online gaming. While the televised version allowed for immersion, for "losing oneself" in a fictionalized world, gaming heightened this through interactivity. The both/and possibilities of gaming, of immersion *and* interactivity, set it apart as a very different user experience compared to that of a film goer or television viewer, and connect to the both/and environment of Third space. Despite the growing popularity of mobile games, the steady increase in the clientele at LXG is indicative of the undeniable draw of Third space exploration with infinite possibilities for mastery and conquest. Uninterrupted and high-speed internet are critical for the total immersion experience, and LXG was savvy to this right from its early days.

With 40 high-end PCs and 10 PS4s with 60–120 fps, the gaming experience is relatively seamless at LXG, allowing for such fast-paced games as DOTA 2 to be played at maximum settings. Backup internet connection with 1:1 bandwidth through Airtel means that there is hardly any downtime or disruption in service. Pricing is usually 70 Rupees per hour, with package deals available through the Stampact card that provides reward points. The liquid cooling systems efficiently sustain the high-end PCs, all loaded with the latest graphic cards. Mechanical keyboards and ergonomic headsets together with bean bags for comfort make this a very popular place of leisure. The venue has a dedicated PS4 section and an onsite cafeteria. Originally just a make-shift room sharing storage space with wires and old PCs, the café used to serve basic junk food at high prices. A frequent customer recognized the potential of the cafeteria space and invested in it to transform it into the Dire Café, with DOTA2-themed snacks and beverages. In 2014, LXG launched its own LXG India TV where gamers could stream their games for free on platforms such as YouTube Gaming, Hitbox, and Twitch. Over time, it moved into the ESports realm with considerable resources invested in teams and tournaments, as well as corporate relationships for sponsorship. Academic research on gaming in India is less than scarce, and trends worldwide are connected to the case study here with the caution that while there may be some relevance, effects and impact cannot be assumed from western to non western contexts, as was the case in early television studies. We turn now to the ESports profile of LXG, beginning with a quick overview of ESports research worldwide.

## The Ethics of "Labor" in Entertainment

Hollist (2015) provides a comprehensive analysis of ESports, which, simply defined, are professional video game matches. Players could be one-on-one as in tennis, or five-member teams as in the globally successful League of Legends. LXG provides a remarkable snapshot of the journey from a small video gaming center to a major host of ESports. In the paragraphs that follow, selected interviews with personnel from LXG and game developers from their biggest suppliers in Seattle and Bellevue Washington will be interwoven with critical legal and social scholarship on the respective ramifications for the industry's vast numbers of players and employees. The attempt here is to acknowledge that some global trends apply in terms of software and connectivity, yet the First space and Second space are uniquely Indian, with gaming centers mirroring the video parlors and illegally wired cable centers of the early 1980s and 1990s. The spaces begin as not much more than garages or one-room offices in a strip mall. Labor is cheap as fascination with the technology and its capabilities far overrides concerns for equitable pay or sustainable employment, as we see below.

The early history of LXG and its journey to exponential growth emerged in conversations in 2014 and 2017 with one of its earliest employees, Shaurya Singh. Typical of start-up environments, a multipurpose core team of just three to four people (including the owner) managed everything, from machine installation and maintenance, to marketing and promotion. Singh himself moved quickly from an initial position of battery re-charger and systems maintenance tech to social media manager and general marketer/campaign manager. The small core team worked practically around the clock with below average pay, fulfilling a variety of duties as each day demanded. Initially the place was threadbare, with all capital invested in the computers. Ananth astutely focused on LXG's social media presence, with Singh taking on its Facebook, Instagram, and YouTube accounts. Entirely self-taught, Singh produced a stream of fast-paced videos edited to catchy music and with almost no voiceovers. Once its online presence grew with actual increase in users, attention turned to hosting events. Tournaments in high schools as an

Annual Day feature as well as in-house competitions brought in a lot of capital, and more importantly, attracted investors and co-sponsors. Staffing included talented gamers who were paid in game time in part and in some cases, entirely. Bean bags served as sleeping bags, with some staying overnight to complete various duties and squeeze in play time. A social media optimizer in the mix was pivotal in boosting LXG's online media presence, with the name receiving priority status in Google searches on gaming in India. Using the example of Dota 2, the incredibly popular multiplayer online battle arena (MOBA) game where two teams consisting of five players each try to destroy the "Ancient," a large structure, while in opposition with each other, to explain the hypnotic immersive attraction, Singh said:

> Dota (Defense of the Ancients) 2 is a lifestyle. People want to come for free, just work, sleep on the bean bags at night, work just to play. Many come in on a Friday evening and leave on a Monday morning. We needed rechargers, designers, coders, web developers, back end workers, even guys to man the cafeteria, we needed ushers, security, odd jobs men, lots of them. Some were willing to take a break in salary, some were hired for games in lieu of their entire salary.

Two big partnerships were with Logitech and NVIDIA. While Logitech provided hardware in return for co-branding at all events, NVIDIA invested in an entirely new interior. The walls were blacked out while floors and ceiling were lit through LED pipes, mirroring the black and neon green colors of NVIDIA. The company refurbished the administrative offices and supported the development of the in-house cafeteria. In return, NVIDIA sought a shared name itself, resulting in the "LXG Experience Zone." Most popularly played were ESports and FIFA. High-rated PC games were Counterstrike Global Offensive (CS:GO) and Dota 2, with the latter growing into the biggest tournament sport with teams practicing over years to compete and win big on regional, national, and even international circuits. Consistent with the global gamer trends,[8] users of the more action-oriented games and of ESports were predominantly male. In the urban Indian context where safety for women is a real concern, the crowded and competitive Third space of the gaming center is much less amenable for the female user.

To draw the crowds, Logitech embarked on an ingenious strategy. LXG was given the capital to secure a venue, host a free-to-play CS:GO and Dota 2 tournament, manage its marketing and promotion entirely, and even include shoutcasting. In return, Logitech provided merchandising, game jockeys, prize money, and promotional materials. With a format that stretched over three years with one tournament per year, LXG saw a phenomenal surge in players, with over 770 14–24-year olds in two days. This was a brilliant strategy as it led to notoriety for the smaller tournaments in between, built up energy and excitement for the major annual ones, and drew the attention of big Chinese investors and North American developers. LXG now enters big conventions such as Comic Con, which is moving Indian comics to a digital platform, with such gimmicks as a free-to-play Just Dance which was so hugely successful that it had to shut down its stage so people could pay attention to Comic Con's mainstage.

Shoutcaster Kiran Noojibail described his work as similar to that of a cricket commentator on televised matches, providing running commentary as well as serving as streamer and observer. This comparison is worth deconstructing as it draws out the dual worlds gaming occupies that propel it to success. Drawing from narrative analysis in the deconstruction of televisual texts (see Kozloff, 1992), we see that gaming contains existents, that is, as with a fictional show on television, it contains characters and setting. Realism is heightened when the narrator is heterodigetic and omniscient, that is, external to the world constructed by the story, and situated at a distance in space and time from the story world. The shoutcaster's commentary carries with it the air of authority, running in real time and place as opposed to the fantasy world playing out on the screen. His omniscience is heightened beyond that of a commentator for televised sports in that the shoutcaster can sometimes also control the streaming on such platforms as YouTube and Twitch.

Together with the players who control movement and destiny to a certain extent, the shoutcaster's role is easily recognized as essential to the experience. Tournaments build on the mass excitement already drummed up through cricket and tennis, and are organized through already popular formats such as intercollegiate cultural and literary competitions. The deft interplay of television codes and of real-life sports and varsity competition rules result in an audience in-the-know and ready to play. Noojibail described that he often shoutcasted over the streams as well as provided live shoutcasting for gaming tournaments. With an aptitude for public speaking since high school, combined with a passion for gaming since his days as an engineering student, Noojibail said he easily transitioned to a shoutcaster, finding it far more interesting and lucrative as a profession compared to being an engineer. His early professional experience as a customization engineer meant he had to work directly with corporate clients, understanding their needs and customizing engineering services for them. He chanced upon LXG when he was invited to a birthday party there and was hooked, having watched streams on Twitch for at least three years before that. He was thrown right into it after meeting Ananth the CEO, when the more experienced shoutcaster he was teamed with for the weekend tournament fell sick. The play-by-play commentary was an intense experience, but now he is a pro, working regularly at weekend tournaments, including international ones such as the eXtremesland Zowie Asia 2016 which had 128 teams in all participating from the Asia Pacific region, with the winner receiving an all-expense paid trip to China. Popular games in the café were FIFA 16, League of Legends, Dota 2, and CSGO Counterstrike, developed and marketed by Valve located in Seattle. Audiences were either lucky enough to be in the coveted tournament space (in some cases an entire football field was commissioned as a venue), or streamed the games through YouTube and Twitch.

As can be expected, engagement around the game is diverse, with a range of opportunities for marketing and consumption. Audiences are drawn to the events not just for the competition itself but for the merchandising, opportunity to learn techniques and strategies from shoutcasters and pre- and post-game interviews, and the freebies thrown in liberally. Free-to-play opportunities are also available, in exchange for user information which then of course becomes part of a larger targeted clientele universe. The League of Legends World Championship finals attracted more than 8 million television and online viewers, with 8,000 in attendance at the game itself. By 2013, television and online viewership jumped to 32 million worldwide. The pay-as-you-go or purchase-for-play access point also situates the gamer as a willing and decision-making participant, a point we will come back to later as we evaluate the ethics of game labor.

Acutely aware of the commercial potential of the immersion-interaction environment, game developers are increasingly serving as game sponsors and organizers as well, taking control of the wrap-around experience from software to actual event logistics and promotion. Valve, a Seattle developer of games such as Dota 2, has made significant investments in broadcasting and prize money, with a prize pool of $22.5 million at the time of this writing.[9] Software giants such as Microsoft and Sony, meanwhile, are actually latecomers to this playing field, announcing their launch of ESports leagues only at the end of 2014. It is clear that there is significant overlap between televised sports and streamed gaming tournaments, with the big prize money, the ethos of competition, the players signing big ticket contracts, the merchandising, the commentary, and all the accompanying concerns of addiction and health issues. Particularly relevant and perhaps different in the gaming environment is the age of players and audience members. Hollist (2015) draws out these concerns with the example of games producers Riot and Valve.

Both promote their games only, with players signing no-compete contracts against other video game competitions, preventing them also from participating in any other activity such as livestreaming, which can actually be far more lucrative than professional gaming. In one instance, Riot's contracts with players were leaked, resulting in improvement of player conditions. However, its most stringent conditions of limitations on claims, restrictions on class actions, and a right of publicity to use player image and voice indefinitely were questionable given the high school and college-age players, who may be inadequately informed or represented. While this scenario, however problematic, is regulated, it is specific to the United States. It is here that we pause and examine the "agency" of gaming workers and players, especially in unregulated postcolonial conditions with all its uneven modernization, as in urban India.

As interviews with LXG employees demonstrated, worker terms are negotiated over handshakes, rewards and pay are nebulous with game time serving as salary in many cases, and where real addiction is encouraged as a cost-efficient way to run the business. As Singh stated, it was not uncommon, in fact, for parents to barge into the venue and hustle their children out, or for LXG to receive threats from parents to shut down the place because it was enslaving their children to the games.

Labor considerations extend to that of developers as well. Perhaps these could be explained as a mere feature of the start-up environment where "laborers" are generally privileged to be a part of the high-cost tech environment at all, and where the start-up phase is certainly a very temporary first stop to a lucrative career. For example, interviews with developers for Sony provided a fascinating picture of the architects of the First spaces of games. Origin stories of developers were not unlike LXG employees, where opportunities to work directly with code for real salaries well before the usual milestones of graduation and formal employment, were plentiful and revered. Self-taught, innovative, and entrepreneurial problem-solvers defined core teams at both gaming site as well as developer companies. Brian Fleming, CEO of Sucker Punch, a developer for Sony Computer Entertainment America, discussed the early years of the company, in a January 31, 2017 interview. As a programmer for Microsoft, he and a core team decided to branch out on their own into the video game business, just at the time when people were leaving to hitch on to the dot.com boom. As described in a study of the rise of the hi-tech entrepreneurial class in India (McMillin, 2008), the dot.com bust left many out of a job, but the demand for video games was only just beginning to explode. Sucker Punch grew to a relatively small team of around 90 (as compared to cross-national teams of over 1,500), first contracting with Sony and then completely acquired in 2011, and now developing games exclusively for PlayStation 4, most notably the inFamous series. The goal is to build "visually arresting and emotionally compelling games"[10] and now, celebrating its 20th anniversary, Sucker Punch is an example of stability in what is often seen as a capricious and volatile industry. With Sony's sophisticated transnational market research, Sucker Punch could rely on the fact that this giant publisher's assessment of the viability of their product, meant that it would be a global success. Pitches then were to Sony's team, a relatively low-barrier task compared to the complex one of assessing varying demand in diverse parts of the world.

Contrary to the conventional process of developing a show for television where the pilot is tested after it is complete, the process of developing a game is an iterative and consequently more arduous one. Audience feedback is solicited even in the early stages, requiring deft filtering among suggestions and thick skins to weather continuous critique. The process often involved over 200–300 hours of labor with a team of close to a 100 people, just to work out a few details in a game. The process is decentralized with individual artists making independent decisions on aspects of the product and with creative directors deciding at each level, what to sustain. The ability to discard ideas for the overall success of the product is a heady and humbling experience, Fleming explained. Good ideas were abundant. The ability to discern the good from the really good, and then move with the really good ones that would actually work through from product to market success, required emotional stamina and trust in the team process. Sony itself was deeply involved in content development and the relationship between developer and publisher is one of deep respect and trust. Focus group testing with players, discussions of what worked and what didn't and then feedback on what to do better, are frustrating and difficult processes. Fleming explained:

> Incredibly important is what they (focus group members) liked, what they enjoyed, what they didn't enjoy. The flipside is that they will tell you ten things to fix, how and why, then they will tell you what worked in their favorite game and that you should change it like that. Their job is to react to what is. Your job is to decide what to do about it. If you get caught up in the suggestions, in their expectations, you start resisting, it steals you emotionally.

Juxtaposing the creative process of game development with the creative strategies of the architects and marketers of gaming centers allows us to see the correspondence and similarities in the producers and consumers of this fantasy product. In these comments we can see the construction of First space and its dynamic interaction with the purpose-producing Second space. Driving the energy is the desire for Third space, that untethered zone from the material realities of daily life.

The enduring popularity of tournaments may be attributed to their resonance with other urban youth mass events, such as college cultural festivals, pop and rock music shows, and college competitions. Tournaments with the familiar food stalls, product sampling kiosks, quick raffles, and pre-games were pivotal venues to draw large crowds. At LXG, shoutcaster Noojibail's corporate customization experience had oriented him toward what was *not* efficiently working at the LXG tournaments. Tournament execution then became an extension of his work, with an average pay of around 1,000–2,000 rupees per hour for the duration of the tournament, both online and offsite. Customers ranged in age from 10–33 years, with casual gamers even in their late 40s. Professional or serious gamers were typically between 16–22 years old, with reflexes declining significantly after around 23 years. Kiran himself noticed considerable slowing down after 23 and now was a full time shoutcaster and tournament organizer.

Gearing up for the ILG.LXG India tournament, Noojibail discussed the behind-the-scenes preparation necessary. Just the scale was formidable: a series of seven tournaments held in seven cafés in seven top gaming cities in the country: Bangalore, Chennai, Hyderabad, Mumbai, New Delhi, Pune, and Kolkata. The tournament website lists the multiplayer games played: Dota 2, CS:GO, and Overwatch, with a total prize pool of 20 lakhs. This is billed as probably the biggest ESports tournament, with many smaller-scale ones spinning out because of it. During January to July, 21 preliminary minor events were scheduled with six major ones during September to October, after which eight champion teams would move up to the finals. Prizes for major events were at 1.5 lakh rupees per player for each of the two-member winners of Dota 2, Overwatch, and CS:GO, for a total of 9 lakh rupees, while minor events were at 28,500 rupees per player for seven-member winning team for each of the games, for a total of 6 lakh rupees.

It was only a matter of time before entrepreneurs latched on to LXG to expand its services and co-opt its best features for theirs. One prominent example was the six-floor extravagant microbrewery Big Pitcher, which partnered with LXG to boost traffic in its first floor. Singh oversaw this new venture, purchasing television sets, controllers, and bean bags to outfit the place with a micro gaming venue for six months. The tournaments worked beautifully, with free food and beer for players. Both Big Pitcher and LXG gained, with an online signup sheet tracking guests for both businesses. Similarly, BenQ (established in India in 2001), entered a partnership with LXG to provide monitors and help sponsor small events. As LXG grew its capacity as a gaming venue, a well-equipped studio became a necessity, as well as specialized technicians, game developers, and shoutcasters. With all the ancillary businesses the gaming center generated, it served as a very lucrative venue for youth, particularly young men, who could earn six-figure salaries per year for playing, streaming, and compiling videos for YouTube.

## Gaming Life Worlds

Edward Soja (2009) urges us to think beyond conventional modes of space and place to address the representational strategies of both real as well as imagined spaces. Individuals are inherently spatial beings. Social spatiality, he argues, includes form or physical and material conditions in that space, and the thoughts about that space. Taken together, social spatiality is dynamic, produced and reproduced in relation to society. As described earlier in the chapter, Third space is beyond the material, physical First space and the representative Second space. It is what is experienced in the combination of First and Second space. As we have seen, interviews with developers revealed a powerful correspondence between creator and consumer. Developers tap into the high need for an iterative story line with complex layers and sophisticated all-encompassing landscapes.

Of course, the protectionist rhetoric and "moral panic" that accompanies each new technology is at its height with the new media environment of personalized and interactive technologies. Bochner, Sorenson, and Belamarich (2015) summarize that in the United States, between 2009 and 2010, almost one-third of the 2–18-year-old population was obese, with television watching and video game playing identified as prime culprits. Their 2014 meta-analysis of 7 to 19-year-olds' use of exergaming, that is video games that require exercise through dance or sports, showed that no significant health benefits such as loss of weight, had been recorded. In Singapore, Liau et al. (2015) report from two related studies including almost 2,500 youth, that pathological video gaming resulted in impulsive behavior and low self-regulation, raising concerns for social well-being with implications for anti-social behavior. Patriarchal and determinist stances that have plagued the study of youth and media in general are even more so when turned to those in developing economies. Postcolonial scholars call for a measured re-insertion of Marxism in the analysis of anything global and "new." Of particular scrutiny should be the continued privileging of capital accumulation in globalization, the ever expanding proletariat, without equitable distribution of profits in tandem with distribution of labor and value (Chowdhury, 2006). Youth should be understood as mobile and dynamic, *as well as market-led*, with limited range of expression and employment (see also Nuttall, 2008).

This chapter has demonstrated that gaming is on the continuum of an increasingly individualized media experience and blurs boundaries among television, PC, and mobile phones. Multi-layered story lines are critical, as in soaps, they have to get the user to return. Like reality TV, they have to offer opportunities for a grand prize. The chapter extends television studies in strongly arguing for analysis of the lifeworlds created through use, and using those as an entry point, a unit of analysis as it were. Focus on medium or audience response to a single text no longer accommodates the affordances of a digital media environment. It calls for analysis of patterns in audiences'/users' Third spaces to understand what gaps in reality are filled through the First spaces offered by media texts. The immersion possible in fantasy worlds, as well as the various demands placed on the user and worker in such centers as they connect to their social and material lives outside of the center, is presented through the lens of Third space.

It is true that gaming narratives reaffirm hierarchies of gender, class, religion, and nation and that stereotypical fantasy characters may actually provoke the strong criticism that games are hardly the resistant Third space that Bhabha (2009) and Lefebvre (1974) theorize. However, in the context of limited agency offered to urban Indian youth through an English-language education system that still for the most part lumbers through a colonial lens with meager job opportunities after graduation, gaming centers offer selfhood as well as employment. Thinking through gaming centers and streaming programming as Third space advances our understanding of confluences between global and local by urging us to think through new configurations of space as well as the reordering of imagination they engender and the translation of such imaginations into agency.

## Notes

1 Privacy was just one of the issues surrounding the roll out of the Aadhar system, most pressing were the ways in which the very poor and illiterate would be documented and their income and property accessed. See www.businesstoday.in/current/economy-politics/aadhaar-privacy-key-issues-that-all-aadhaar-card-holders-should-bear-in-mind/story/256723.html and https://economictimes.indiatimes.com/news/economy/policy/big-data-to-address-problems-of-jobs-growth-nandan-nilekani/articleshow/59972379.cms

2 Quite simply, shoutcasters provide running play-by-play commentary and analysis of online and in-person ESports. Various descriptions abound, but a useful synopsis may be found here: https://medium.com/broadcastgg/introduction-to-shoutcasting-definitions-and-goals-e0c2e571782b

3 "500 million Internet users in India by June, say experts." (April 27, 2018). *Financial Express*, Retrieved August 12, 2018 from: www.financialexpress.com/industry/technology/500-million-internet-users-in-india-by-june-say-experts/1147709/

4 "India's wireless user base increases by 6 mn in May, top four telcos hold 87.2% market share: ICRA." (July 31, 2018). Retrieved August 12, 2018 from: https://telecom.economictimes.indiatimes.com/news/indias-wireless-user-base-increases-by-6-mn-in-may-top-four-telcos-hold-87-2-market-share-icra/65212332

5 Top nine streaming services in India are: YouTube, Netflix, Amazon Prime Videos (with mostly English-language programming and films for monthly fees around $10), as well as Hotstar, Voot, Sony Liv, Spuul, and Eros Now (with mostly Hindi language films and serials for around $5 per month).

6 Raghunathan, A. (June 27, 2017). "India's Most Expensive Film Franchise—Bahubali—Wows Audiences," Forbes Asia. Retrieved September 5, 2018 from: www.forbes.com/sites/anuraghunathan/2017/06/27/indias-most-expensive-film-franchise-bahubali-wows-audiences/#6435ca14f89f

7 See for example, https://baahubali-the-game.en.uptodown.com/android

8 "Distribution of gamers playing selected game genres worldwide as of January 2017, by gender," (The Statistics Portal, nd). Retrieved August 12, 2018 from: www.statista.com/statistics/694381/gamer-share-world-genre-and-gender/

9 http://dota2.prizetrac.kr/international2016

10 www.suckerpunch.com/

## References

Balaji, S. (December10, 2017). "Life after Demonetization: Meet the Local Startups Driving Digitization in India's New Economy." *Forbes*. Retrieved December 21, 2017: www.forbes.com/sites/sindhujabalaji/2017/12/10/life-after-demonetization-meet-the-local-startups-driving-digitization-in-indias-new-economy/#1d590974b01a

Bhabha, H. (2009). "In the Cave of Making." In K.Ikas and G. Wagner (Eds.), *Communicating in the Third Space*. New York: Routledge. ix-xiv.

Bochner, R. E., Sorensen, K. M., and Belmarich, P. (2015). "The Impact of Active Video Gaming on Weight in Youth: A Meta-Analysis." *Clinical Pediatrics*, *54*(7), pp. 620–628.

Chowdhury, K. (2006). "Interrogating "Newness": Globalization and Postcolonial Theory in the Age of Endless War." *Cultural Critique*, *62*, pp. 126–160.

Evans, E. (2015). *Transmedia Television: Audiences, New Media, and Daily Life*. New York: Routledge.

Groshek, J. and Krongard, S. (2016). "Netflix and Engage? Implications for Streaming Television on Political Participation during the 2016 US Presidential Campaign." *Social Sciences*, *5*(65), pp. 1–18.

Hollist, K. (2015). "Time to be Grown-Ups about Vido Gaming: The Rising ESports Industry and the Need for Regulation." *Arizona Law Review*, *57*(823), pp. 824–847.

Ikas, K. and Wagner, G. (2009). "Postcolonial Subjectivity and the Transclassical Logic of the Third." In K. Ikas and G. Wagner (Eds.), *Communicating in the Third Space*. New York: Routledge. pp. 96–106.

KPMG. (2017). "Media for the Masses: The Promise Unfolds." Indian Media and Entertainment Industry Report. Retrieved December 30, 2017: https://assets.kpmg.com/content/dam/kpmg/in/pdf/2017/04/FICCI-Frames-2017.pdf

Lefebvre, H. (1974). *The Production of Space* . Malden, MA: Blackwell.

Liau, A. K., Neo, E. C., Gentile, D. A., Choo, H., Sim, T., Li, D., and Khoo, A. (2015). "Impulsivity, Self-Regulation, and Pathological Video Gaming among Youth: Testing a Mediation Model." *Asia-Pacific Journal of Public Health*, 27(2), pp. 188–196.

Kozloff, S. (1992 [1987]). Narrative Theory and Television. In *Channels of Discourse Reassembled. Television and Contemporary Criticism*, ed. Robert C. Allen, London: Routledge 52–76.

McMillin, D. C. (2008). "'Around Sourcing': Peripheral Centers in the Global Office." In R. Gajjala (Ed.), *South Asian Digital Diasporas and Global Technocpaces: Culture and Economics in Cyberspace*, New York: Peter Lang Publishers. (pp. 249–263).

McMillin, D. C. (2009). *Mediated Identities: Television, Youth, and Globalization*. New York: Peter Lang Publishing.

Nuttall, S. (2008). "Youth Cultures of Consumption in Johannesburg." In N. Dolby and F. Rizvi (Eds.), *Youth Moves: Identities and Education in Global Perspective*. New York: Taylor and Francis. pp. 151–179.

Soja, E. (2009). "Third Space: Toward a New Consciousness of Space and Spatiality." In K. Ikas and G. Wagner (Eds.), *Communicating in the Third Space*. New York: Routledge. pp. 49–61.

# 30
# REFASHIONING CHINESE TELEVISION THROUGH DIGITAL FUN

*Ruoyun Bai*

The challenges and upheavals that television industries around the world are experiencing in the digital media ecology affect China as well (Bai, 2016; Keane, 2015; Zhao, 2016). Television content is everywhere via proliferating digital platforms and devices; viewers increasingly access their favorite shows through online video-on-demand services so television stations are facing the dim prospect of a dwindling, graying viewing population; digital media giants, e.g. Tencent, Alibaba, and Baidu, are bringing new industrial players and practices to television production. All these developments are sharply contrasted with the clearly defined boundaries of Chinese television not that long ago, when the production and distribution of television content were strictly organized in accordance with the political mandate of broadcasting as propaganda and economic interests of commercialized public broadcasters. Although the commercial transformation of the broadcasting system into a media industry has made its boundaries porous through advertising, content outsourcing, joint ventures, and publicly listed subsidiaries, the role of television stations as the central player in organizing the field has remained largely intact, that is, until very recently. Against the new reality of media convergence, the entire terrain of television production and consumption is refashioned by the mutually shaping digital entertainment businesses and online youth cultures. At the same time that television stations are diminishing in both quantity and status, television entertainment, now de-territorialized from the broadcast medium, has continued to thrive and dominate the mainstream culture. In fact, the remediation of popular television, especially drama serials and reality shows, by the Internet and social media has given it a new lease on life. This chapter argues that as television production, distribution, and consumption are redefined by the digital entertainment industry, new genres, identities, and representations have emerged in recognition of youths as the most valuable and desirable category of audience.

In about a decade, China's digital entertainment industry evolved from an array of digital entrepreneurs offering free, unlicensed audiovisual content to the dominance of a few digital oligopolies that provide online video-on-demand (VOD) services for hundreds of millions of users. In its early days in 2004 and 2005, many budding online businesses thrived on pirated content; by 2010, such content had been largely eliminated through economic and legal sanctions, smaller firms were either closed down or merged into larger ones, resulting in a rapid wave of consolidation and concentration (see Montgomery and Priest, 2016; Zhao and Keane, 2013). In the meantime, the rollout of 3G smartphones and tablets in 2009 and the state-facilitated launch of the 4G network services in 2014 ushered in the era of mobile and connected viewing (Zhao, 2016). Online VOD services began to prosper in the new media ecology characterized by proliferating viewing platforms. From 2013, the advertising revenue for these services grew about 40% each year to 462 million yuan in 2017. By July 2017, there are 565 million people using online VOD services, which means that 75.2% of Internet users in China watch online video. Of these, 170 million have paid subscriptions – they are either subscribers paying membership fees to enjoy premium online viewing packages or watch their favorite shows on the pay-per-view basis (China Netcasting Services Association, 2017).

As the industry grew, it became a game changer for Chinese television. Competing for subscribers, web traffic, and user engagement, the appetite of these VOD firms for content is seemingly without limit. While user-generated short-form videos are rapidly gaining popularity, professionally created entertainment such as drama serials and variety shows remains unparalleled in its ability to create massive viewership and fandom bases. As the era of streaming pirated content ended in 2009, competition for copyrighted content intensified and online VOD firms paying hefty prices for online streaming rights became a new industrial norm (Bai, 2016). This reality gives an incentive to the online video firms to reduce dependence on drama suppliers by becoming investors in drama productions, including exclusively made-for-web dramas. It should be noted that these firms are able to pay or invest not only because of the robust growth of advertising for online audiovisual content and a continually expanding subscription base, but more importantly because they each have a digital media empire as a parent company and financial backer, so even though these firms have been fighting losses, they are here to stay as shakers and movers of the television industry. Thus, new players and new money are now supporting the continually expanding base of television production. As production companies now expect to earn more from selling to online VOD services than to television stations, the notion of audience is re-conceptualized as well to include both traditional television audiences, the prototypical of whom are middle-aged women, or the "mama generation" (see Schneider, 2012, p. 139), and younger audiences watching TV online. The broadcast-era audience ratings system continues to guide advertising transactions while services of social TV analytics proliferate to provide all sorts of data about audience engagement online. Online audiences are becoming increasingly important both as viewers and as reputation amplifiers that can positively impact broadcast ratings. This is so because a hot drama property is usually simultaneously shown via broadcast and online; it frequently happens that when a drama becomes a hit on the Internet and social media, its television ratings would rise subsequently. Given the importance of online distribution and consumption, interests, preferences, and sentiments specific to online youth audiences have become a significant influence on what kinds of television get produced. In what follows, I will illustrate three ways in which Chinese television is re-programmed by digitally driven media convergence.

## Internet Literature and TV Fantasies

Chinese media officials have always treated fantastic, improbable elements in television dramas with caution and distrust, with the exception of televisual adaptations of some pre-modern popular literary works said to represent the best of Chinese traditional culture. Stories that defy natural laws (time travels, ghosts, supernatural forces, etc.) are censured similarly to those that fabricate histories – they are both considered as examples of "excessive entertainment" providing a distorted worldview in a rush to pander to "vulgar" tastes. Yet despite official censuring, such dramas have continued to be made in large numbers because first, they are not in the strictly forbidden zone of political censorship; second, they provide entertainment that sells well in the Chinese television drama market; and last but not least, the Internet and social media are making younger audiences' preferences not only visible but also increasingly central to the development of television dramas. This section reflects on the symbiotic relationship between the Internet and the remaking of television culture by focusing on how Internet literature has contributed to the fantastical turn of Chinese television.

When compared with other Internet cultures, the Chinese Internet distinguishes itself in the massive formations of online literary communities, a phenomenon that has been observed and extensively analyzed (Chao, 2016; Feng, 2013; Hockx, 2015; Kong, 2005). The first literary portal website in China, "Under the Banyan Tree," was established in Shanghai in 1997 and became "an unprecedented social platform for readers and amateur writers to consume and produce original literary work easily" (Chao, 2016, p. 229). A large number of literary websites and online forums appeared in the next few years, including the most influential ones – Qidian (Starting Point), Jinjiang Literature City, and Hongxiu Tianxiang. Reading, writing, and sharing novels online became a popular pastime for Chinese youths, for whom Internet literature was "an inexpensive and relatively safe venue for self-expression, spontaneity, and exchange" (Linder, 2005, p. 896). The online reading population has been growing consistently. According to the most recent survey done by the China Internet Network Information Center, online literary readership comprised 48.9% of China's Internet population by the end of 2017. More than 377 million people read literature online, more than 90% of whom read on their mobile phones (CNNIC, 2018). Corresponding to the vast readership of Internet literature is the formation of a vast pool of creative labor producing immense quantities of multi-chapter fictions online and the spectacular rise of a handful of young writers as celebrities-cum-entrepreneurs.

Internet literature not only opened up a new literary space (Kong, 2005), but also impacted the much wider field of cultural production encompassing film, television, videogame, and most recently comic book publishing. An early television drama that drew inspiration from Internet literature was *My Own Swordsmen* (2006), a nonsense-style comedy "packed with witty jokes, sophisticated satire, and numerous intertextual references to contemporary popular culture and Internet youth culture" (Liu, 2013). Though this drama was not directly based on any online novel, its writer, known as Ning Cai Shen, belonged to the first generation of Internet literature writers before leaving his online career to work on television screenplays. *My Own Swordsmen* was unlike any other television drama of the time and became an event TV drama drawing a large following of youthful fans. In the 2010s, when battles for original content intensified, Internet literature became the most coveted source of original works for transmedia adaptations.

For television companies, Internet literature gives the following advantages. One is the sheer size of online communities that facilitates massive production and consumption of online novels. While shallow, mediocre works abound, the more popular works are generally believed to reflect the tastes and preferences of a vast number of young people, and therefore possess the unique ability of translating into popular television dramas. Second, is the depth of engagement that readers develop with their favorite works and authors. As readers are actively involved in the writing process (Chao, 2016), their enthusiasm tends to develop into long-term fandom that would cause them to follow the adaptations of their favorite novels in other media. In other words, born-digital novels are assumed to "bring their own traffics," which is a highly attractive quality in the attention economy. Third, is the wide availability of online writers, which provides much needed cheap, informal, precarious labor for the industry (Zhao, 2017). Despite the fame and wealth of a handful of online writers/celebrities, most remain highly vulnerable to exploitation. For these reasons, Internet literature is regarded by many in the industry as a desirable incubator of IP or Super IP.

IP is understood as copyrighted content that has the potential to generate a universe of franchised visual and material commodities, including but not limited to film, television, comic books, and videogames (see Qin, 2016). The idea of media synergy is not new to Chinese media businesses; for example, when a television drama is broadcast, its spin-off novel or the original novel from which the drama was adapted is usually simultaneously released or reprinted. The newly energized discourses surrounding IP indicate a more systemized approach to content production. It seems that what Chinese media companies are aspiring to is no less than the level of media convergence that super IPs in the West such as Marvel Comics, *Harry Potter, Game of Thrones*, and *Hunger Games* achieved, or the "media mix" (Steinberg, 2012) that characterizes Japanese anime ecology. What is unique about the IP-centered entertainment industry in China is its heavy dependence on and exploitation of youth-oriented online literary communities.

The majority of online novels hosted by commercial literary websites belong to such genres as Fantasy, Immortal Swordsmen, Historical, Urban, Sci-Fi, Military, Romance, Matriarchal, Boys' Love, Online Games, and Fan Fiction. Maintaining boundaries of genre makes good economic sense and may be welcomed by loyal fans of very specific genres. However, they are hardly fixed in actual reading and writing practices, as each genre is open to further breakdown, intertextual allusion, and hybridization. Also, as they are translated into television dramas, the industry's instinct for catering to diverse audience tastes generally favors bringing different generic features together in one drama. Take Fantasy and Immortal Swordsmen for example. Fantasy, or *Xuanhuan*, frequently portrays a world populated by "fire-spitting dragons, treasure-hoarding dwarves, blonde fairies, long-bearded elderly wizards wearing pointed hats, Gothic-style castles occupied by pale, red-eyed vampire" (Chao, 2012, p. 114). Clearly influenced by *The Lord of the Rings* and the *A Song of Ice and Fire* series, Chinese fantasy novels are similar to their Western counterparts "in every way except for the [Chinese] identity of the protagonists" (ibid.). *Xianxia*, or Immortal Swordsmen, meanwhile, is more inspired by the Chinese-language fiction about martial arts masters and clans popularized by bestselling Hong Kong and Taiwanese writers. But whereas martial arts masters may die in works of the previous generation, in *Xianxia* novels, the protagonists are immortal gods living in a fairy world. Any fan of *Xuanhuan* or *Xianxia* would be able to enumerate all the differences between the two genres; however, their television drama adaptations are often lumped together in the category of *Xuanhuan Xianxia*, a designation that downplays the differences between the two genres and highlights their common ground constituted by the fantastical and the magical, i.e. the other-worldly. Furthermore, the fantastical is not limited to *Xuanhuan* or *Xianxia*, but is increasingly incorporated in other online fictional genres such as romance, (quasi)historical, and online game drama. The protagonists may find themselves endowed with the ability to time-travel from modern times to an ancient dynasty; or they fall in love with immortal beings in the human form traveling all the way to the present from a thousand years ago; or they encounter supernatural powers on tomb-raiding expeditions; or they may be online gamers whose identities and experiences in the virtual space are depicted and whose travel between the virtual and the offline is facilitated by the gaming interface. In other words, while *Xuanhuan* and *Xianxia* fictional works are frequently adapted into television dramas, the other-worldly fantasy is by no means limited to these genres, but immanent in the online literary space. Their transmedia effects are strongly felt in the large number of fantasy dramas and the increasing normalization of incorporating fantastical elements into contemporary dramas.

As for why Chinese Internet literature is suffused with other-worldly fantasies, it is beyond the scope of the chapter to fully investigate. Suffice it to say that China's online space is extremely porous and allows the relatively free circulation of global popular culture, especially that from the US, Britain, Japan, and Korea. What we see in Chinese television dramas may well be local versions of the global genre of fantasy bolstered by computer-generated images. More specific to the Internet culture, online fictional fantasies co-exist with online games. *Xuanhuan* and *Xianxia* narratives render themselves easily to the gaming world because in them, trainings, battles, powers, and magical items are often essential to the protagonists' endeavors to eliminate monsters of increasing ferocity or evolve from a lower being to a god in the hierarchical other-world. The affinity between online fiction and games can be seen in how most online literary websites are owned and controlled by digital entertainment businesses with an intention to attract gaming youths. In 2004, Shanda, a famous online gaming company at that time, purchased a leading online literary website, Qidian (Starting Point), which hosted a large number of fantasy novels. In 2008, Shanda established its Internet literature subsidiary, merging the most influential online literary portals under its name and dominating 70% of the Internet literature market in its heyday. The co-evolution and interlocked development of Internet literature and online games is further evidenced in the digital entertainment giant Tencent's interest in the online literary market. Tencent became a dominant force in Internet literature by purchasing Shanda Literature and merging it with its own Internet literature arm, Tencent Literature, to form China Literature Inc. (Yuewen Group) in 2015. Through such mergers and acquisitions, Internet literature and online gaming are being brought under the same roof, which it is hoped would facilitate the incubation and exploitations of IPs, and all these developments are also giving rise to new filmic and drama productions.

## Female Power and Idol Consumption

Another way of exploring the fantastical turn of Chinese television is to examine it as a playground of fantasies of, by, and for women. Television has always been a site where fantasies about women (and men) actively participate in the production of gender ideologies (Huang, 2008; Song, 2010; Zhong, 2010). While priorities of a patriarchal society continue to dictate gender representations in Chinese media, some recent developments seem to indicate female empowerment as a result of female writers carving out an online literary space for themselves, and the increased economic importance of female fans/consumers to the entertainment industry. This section interrogates two new discursive categories – strong female leads and sweet males.

Television dramas featuring strong female leads are loosely and alternately labeled in the Chinese press and social media as "Mary Sue [*ma li su*] Drama," "Supreme Heroine [*da nü zhu*] Drama," or "Women's Inspirational [*nü xing li zhi*] Drama." A Mary Sue drama typically features a perfectly beautiful and virtuous heroine whose enchanting power is such that she causes practically all males whose eyes set upon her to suffer in love, to unconditionally serve as stepping stones for her ascendency in the social or power hierarchy, or even to die for her. As such she is absolutely hated by her jealous female arch-enemies, whose sole objective is no less than to destroy her, an objective forever thwarted. A romance like this can be seen as a 21st-century version of the Cinderella story. It is mockingly referred to as "Mary Sue," a nod from Chinese popular culture to Paula Smith's 1973 "A Trekkie's Tale" and its seemingly perfect character Mary Sue. If in a Mary Sue drama, the heroine does not need to be specially endowed except with good looks, a golden heart, and an army of male admirers, a "Supreme Heroine" drama adds power to the heroine, power associated with intelligence, competence, assertiveness, and a strong will. The heroine is not just beautiful but also ambitious. She is no longer contented with being an object of male love; instead she has the desire to leave the domestic sphere and seek meanings of life in male-controlled domains. In the drama, she is shown to possess, and be admired for, a supreme ability to rule in politics, extraordinary business acumen, or excellence in professional practices such as medicine. These Supreme Heroine dramas are also known as "women's inspirational dramas," because of the success the heroines achieve due to their hard work, confidence, and perseverance.

To echo the previous section on the impact of Internet literature on television production, the proliferation of the female success fantasies may largely be attributed to the influence of women's writing on the Internet. Internet romances written by and for women represent a new space where women reinvent gender and cultural identities (Feng, 2013). Aside from traditional romances that celebrate heterosexual love, many novels targeting online female readers actively explore gender boundaries. The Boys' Love genre (known as *Danmei*), for example, allows readers to imagine male-to-male romances. Another example is the so-called "matriarchal novel" (*nüzun* or *nüqiang*), which

> describes a woman's ascent to power in the public arena, or her success at establishing and heading a happy domicile including one or more male sexual partners, or both. Utilizing a strategy of excess and conjuring up a fantastical world, matriarchal narratives delineate ideal femininity, or at least unrestrained superwomanhood, in a more vivid fashion than run-of-the-mill heterosexual romances.
>
> *(Feng, 2013, p. 85)*

Short of celebrating the possession of multiple male partners as in Matriarchal novels, many "Supreme Heroine" dramas share the structure of feeling with the former by positively portraying the heroine's pursuit of power and agency.

Aside from women's online literature, the independent and confident womanhood on Chinese television are due to larger socio-cultural forces. As television increasingly turned to younger and more affluent urban female audiences since the early 2000s, it was only a matter of time before the "virtuous wife, good mother" type of womanhood gave way to female characters who are not saddled with family responsibilities and are more concerned with self-actualization, whether through love, sex, or career. The so-called women's inspirational dramas must be seen as one piece with and a continuation of contemporary Chinese, and indeed East Asian, popular culture inflected by an embrace of more assertive girl/womanhood. They result from the expanded social space for well-educated urban middle-class women and speak to their desire for more economic and social power. Of course, these "inspirational" dramas inspire in a rather limited and problematic manner. The successful heroine has to embody perfection, an unattainable standard for women in real life; she is often pitted against other women in her pursuit of success, casting the notion of sisterly solidarity into doubt; during all her endeavors, men's love, support, and sacrifice are indispensable; she has to think and act like a man and play by the rules of a patriarchal society; and above all, most Supreme Heroine dramas are in historical settings, for the distance from reality makes female power less threatening to the social status quo. All these are powerful reminders that these inspirational dramas are, at most, fantasies of female power that brush aside real social struggles.

A parallel development to the redefinition of ideal womanhood is the emergence of good-looking, well-trimmed young male idols adored by massive numbers of predominantly female fans. Starting from 2014 or so, a sexually evocative term that underscores the consumptive side of female fandom came into being, "small fresh meat." At first, it refers to male Korean entertainers who have "trim physiques, flawless skin and often can be seen wearing make-up" (Mok, 2017). But soon it came to designate and hail into being an entire fleet of boyish-looking Chinese idols, including Wu Yifan, Lu Han (both used to be members of the Korean boy group EXO), Li Yifeng, Ma Tianyu, Chen Weiting, among others. In 2015, *Forbes China* published its first list of "hottest small fresh meat stars." In 2017, members of a boy band, known as TFBoys (or The Fighting Boys), became the new sweeties in the world of idols. Each of these idols has a huge number of followers on Weibo ranging between 25 and 28 million. The overwhelming majority (somewhere between 70% and 85%) of their Weibo fans are women (Mok, 2017). These idols are truly transmedia commodities, starring in film, television, commercials, and music videos. Their ability to sing, dance, and act is of secondary importance; first and foremost, they exist as idols, as projections of their fans' desires and fantasies.

This is not the first time in history when "men become a commodity for female consumption" (Baranovitch, 2003, p. 143). In Baranovitch's study of popular music in mainland China under the influence of Hong Kong and Taiwanese popular culture, he notes that the enthusiastic reception of soft male images on the musical stage was the result of women being empowered in the consumer society. On television, softer male images increased as a result of the rise of the so-called "youth idol dramas" in East Asia since the end of 1990s. Made to appeal to young viewers, idol dramas feature handsome youths living a modern, fashionable, urban life. Male protagonists in these idol dramas are invariably gentle, considerate, and committed to pure love, although they might initially appear full of themselves and unapproachable due to the wealth and high social status of their families. While idol dramas are said to originate in Japan, it was the Korean Wave that established soft masculinity as the new norm of gender representation, whether via television dramas or boy band performances. By the mid-2010s, public discourses about the new type of masculinity came to congeal around the discussions and debates on "small fresh meats." Whereas defenders of rugged masculinity see "small fresh meats" as symptomatic of cultural degradation, the widespread usage of the phrase itself suggests a kind of female subjectivity becoming accepted by the mainstream culture. Compared with earlier phases of cultural transformation, "small fresh meats" represents a continuity and at the same time points to the more direct, forceful impact that female fandom has on public discourses of ideal masculinities.

The impact of female fans upon gender representation was seen in an early instance in the stardom of Li Yuchun, a female singer who won Super Girl in 2005, China's first reality TV hit that achieved massive fan mobilization and participation. Mainstream cultural commentators were mystified by Li Yuchun's victory because her gender-neutral appearance and nonchalant attitude on the stage defied every received predictor of female popularity. A most recent example that underscores the discursive power of female fans is the 2018 idol-training reality show, Creation 101. A chubby girl named Wang Ju distinguished herself from multitudes of good-looking contestants and captured the hearts of millions of fans because of her assertiveness and self-confidence. These anecdotal success stories of alternative female role models might not indicate any major change in the standard of beauty in Chinese society; however, their message echoes that disseminated by popular women's inspirational dramas that celebrate female power. When it comes to male idols, female fans have consistently demonstrated a preference for "small fresh meats." Idol-training reality shows empower fans to select, vote, and canvass on behalf of their favorite candidates, thereby raising the level of fandom involvement and engagement to a new high. In the process of nurturance and training, fans are extraordinarily invested in their boy idols and root for them as their adopted mothers and sisters, as it were. They are described as

> the most loyal of followers, with the strong feelings and long-term relationships with their celebrity idols. They may even see themselves as a part of an artist's extended family, and believe that they have the right to protect and support their idols as if they were supporting their own family members.
>
> *(Liang and Shen, 2016, p. 333)*

## How Social-Media-Oriented Marketing Transforms a Political Drama

Chinese youths and their cultural consumption not only provide a clue to the emergence of new television genres and gender representations, but also shed light on how political communication can be reconstructed in the context of digital culture. Broadcasting has always been an essential component of the state's propaganda system. In the post-broadcast era, Party leaders and media officials are evidently anxious about the increasing irrelevance of the broadcast platform. State broadcasters have acted in response to such anxiety (and of course to their own existential concerns as well), with more or less success. CCTV, the Party's leading mouthpiece, launched CNTV (China Network Television) in 2009, which was to handle all forms of digital distribution of its programs in and outside of China in the coming years, including the flagship website (cctv.com), foreign-language channels targeting overseas Chinese, and various mobile applications for its dramas and newscasts. Almost all CCTV channels, programs, and emcees have presence on social media through Weibo and WeChat official accounts. Interactive features are embedded in its programs to navigate its viewers through multi-screen experiences. CCTV's present director, Shen Haixiong, stated the broadcaster's mission in the digital era aptly – "to expand and enrich our new media offerings so netizens cannot have enough of them" (SAPPRFT News, 2018). These digital strategies are intended to bring official discourses to the digital sphere and Chinese millennials; in the process, official discourses are recontextualized, and often repackaged for youth consumption by media entrepreneurs, young people themselves, or both. In this section, I will introduce a mainstream television drama and its encounter with the online world. Through this example, I hope to start making the case that television, when circulated online, undergoes deconstruction and reconstruction by the logic of digital media culture.

In 2017, an anticorruption drama, *In the Name of the People* (hereafter referred to as *People*) became the most watched and discussed drama of the year. Set in a fictional province, it portrays the interpenetration of power and money that involves corrupt government officials and businesspeople, and a densely woven fabric of influence among several prominent political families. It is also a feel-good melodrama that positively portrays a variety of state heroes, some of whom dedicate themselves to bringing the bad guys to justice and others to solving subsistence and environmental problems facing local residents. The drama is explicitly anticorruption propaganda,[1] and promoted as such.

The 55-episode *People* was simultaneously broadcast on Hunan TV's Satellite Channel and on the Internet. The TV broadcast lasted an entire month from March 28 to April 28, while the complete drama series has remained available on several online video platforms. The broadcast scored an average of 3.661% of national ratings, which made it the leading drama of 2017, and reached a peak of 8% across an aggregate of 52 cities on the day of the finale (Liang and Qiu, 2018; Sohu Entertainment, 2017). On authorized online video platforms – those that had the legal right to play the show, as opposed to the many platforms that ran the pirated version – the drama was viewed billions of times by the end of the television broadcast (Juhong Data, 2017). Young people aged 15 to 34 constituted the bulk of the online audience (see Chen, 2017; Chen and Chen, 2017; Jiang, 2017; Wang, 2017). For several weeks, all major social media platforms, online forums, and portals were saturated with news, behind-the-scene stories, commentaries, interviews, discussions, emojis, gifs, memes, and video remixes related to this drama. Watching *People* or at least just a part of it became virtually compulsory for those who did not want to feel "left out."

This media event presents a textbook example of digital marketing that targets younger audiences, and a case that underscores the transformation of a mainstream text by the youth culture. PPTV, a second-tier online video distributor, purchased the drama's exclusive online distribution rights (before it struck last-minute rights-sharing deals with the largest online VOD service providers to reduce financial risks) at the end of 2016. It then implemented a high-budget digital marketing strategy, a core objective of which was to encourage social media discussions and fandom activities through creating and orchestrating trending topics and memes (see Ding, 2017; Huang, 2017; "PPTV", 2017; Shao and Zeng, 2017; Wu, 2017). First, it positioned *People* explicitly as an anti-idol drama, a quality drama that does not depend on idol-status stars for success. From the very first PR release, *People* was promoted as a "drama with a conscience" and "a fresh breeze" in the tainted waters of commercial culture ("PPTV", 2017). The term *lao xigu* (literally translated as "old acting bones") was used to refer to the leading male actors who move the audience with their performance instead of good looks, an implicit reference to the "small fresh meat" phenomenon. The focus on quality was soon amplified in subsequent PR releases and media coverage, much of which compared the cast favorably with idol actors and contrasted their modest earnings to the much higher cost of idols. Castigating the idol culture as a way to attract the attention of the youths immersed in this culture seemed to have worked well. Instead of turning youths away, it actually helped distinguish the drama from the rest and sent out an invitation to youths to experiment with something different. Reputation-building also involved getting key online opinion leaders on board. PPTV created topics for discussion and invited reputed writers to review on Douban and Zhihu, two leading platforms where large numbers of educated youths and intellectuals concentrate. The strategy of targeting the more sophisticated strata of online youths was one of creating topics and images that could trigger quick responses from fun-seeking occupants of the digital world. Part of PPTV's marketing team's job was to find memefiable elements in the drama that youths were likely to identify with. Then the team of millennials created potential memes to be released at appropriate times during the broadcast. These memes recast the Party cadres and government officials in the drama into cute figures with rabbit's ears, or as members of a boy band (i.e. HD Boys with reference to the TF Boys band mentioned earlier). In addition, PPTV threw out potential trending topics to online publics. In the first episode, a corruption-busting team raids a suspect official's house to find stacks of cash neatly piled from bottom to top along an entire wall; this scene was followed by team members' manual counting of the numerous one-hundred-yuan bills. Government officials hoarding cash-cum-corrupt-money in their secret abodes has become an urban legend in China. This scene provides visual affirmation and satisfaction for people's curiosity. Nonetheless, spectacles of corruption are by no means rare and the attention this scene garnered was disproportionately huge and would be hard to understand if we did not consider the fact that marketers deliberately made this a topic for trending in online and mainstream media. Even more inscrutable would be the viral communication of the bank tellers' demonstration of their cash-counting skills. Not incidentally, these trending topics were drawn from the first episode when the drama was still building audiences, a time when deliberate marketing was essential before ordinary audiences could be counted upon to spread the word. Furthermore, PPTV created and disseminated dozens of digital shorts – trailers, behind-the-scene videos, exclusive interviews, and light-hearted remixes. To reach the large ACG-based (animation, comics, and games) youth subculture, PPTV even disseminated remix videos to the famous *bilibili*, the largest video-sharing website for ACG fans.

Firms such as PPTV are the key nodes of distribution and marketing of television content in the age of social media. As such, they take the lead in the production of paratexts about the original content, bringing *People* to online publics in bits of fun. It encouraged a particular way of watching the drama that resonates with the way young people consume popular culture online. PPTV's marketing team, consisting of people in their 20s and 30s, demonstrated a solid understanding of online youth cultures, and had a direct impact on how *People* was received among youths and the general population. Social media platforms, a central locale of PPTV's marketing efforts, were absolutely essential to the success of *People* in generating audiences, facilitating connected viewing, and guiding public discussions. Platform-based online communities were formed around the drama, and a participatory culture ensued. All major platforms, online forums, and online news portals turned into voluntary marketing machines in order to stay relevant and not to be left out of the hyper transmedia event.

The astounding marketing success of this blockbuster demonstrates that social media platforms can be effectively utilized by the Chinese state to reach younger audiences and that media convergence may present opportunities for the state to reconsolidate its ideological power. As Chinese television repositions itself as a youth's media in the context of the rise of digital entertainment, the Chinese state's approach to ideological work will continue to evolve. However, what remains less clear is the efficacy of the anticorruption message *People* carried for the country's leaders. After all, selling a cute, cartoonish figure based on a Party chief character is not the same as selling the idea that the Party can and will resolve the problems of rampant corruption and power abuse through its anticorruption campaigns. It is highly plausible that when *People* was delivered to online viewers in bits of fun, it remains as bits of fun, no more and no less than that.

## Conclusion

Changes in Chinese television in the 2010s must be understood as a result of industry, text, and audience all undergoing transformations under the current conditions. The production and distribution of television programs increasingly involve digital and social media firms, big and small. This newly formed digital entertainment industry is exercising a profound impact on the financial, creative, and meaning-making processes surrounding television programs. Post-broadcast televisual texts have taken on distinctive features of born-digital popular culture. Attitudes and sentiments that first achieved cultural and subcultural expressions on the Internet and social media are now made not just visible but prominent in mainstream television dramas and reality shows. I highlighted how Internet literature and participatory culture of fandom are used to market media goods to the youths. In this sense, new genres and expressions on Chinese television reflect the desire of its producers, marketers, and advertising clients to seek out younger populations whose primary mode of access to entertainment is through digital mobile devices; they are the product of negotiations between the power of television producers and the agency of online youth cultures.

Television drama genres are reconstituted, with old genres being either replaced by new ones or spiced up with elements from the web. As online literature and online games, and increasingly comic books, become the major inspirations for drama production, dramas for middle-to-old-aged audiences are taking a hit, such as serious historical dramas and dramas that represent contemporary social concerns. Dramas dealing with rural themes or issues of urban poverty, already rare in commercial media, are facing extinction. Increasingly, Chinese television dramas are constructing a dreamland where fantasies reign. Fantasies, of course, are not always devoid of social significance. They can be vehicles for imagining and inventing identities that are suppressed in real life. For young women in particular, television consumption is a means of negotiating meanings of femininity and masculinity. Through supporting their dramas and idols, their wishes and desires are affirmed and fulfilled, if only temporarily. In this chapter I hope the importance of female fandom to the study of contemporary gender representations in China has been made clear.

Television is being refashioned not only through genres and representations, but also through marketing and distribution. *In the Name of the People* presents a case of how traditional television content may be recreated by online distributors and influencers so as to be aligned with the habits, attitudes, and preferences of the younger audiences. In this case, a drama that deals with a serious social issue and carries a strong anticorruption message from the Party opens up an unexpected space for digital fun, which in turn transforms the drama into an ensemble of digital (para)texts. It could be argued that to understand today's commercial popular culture, the Internet and social media have become an integral component of inquiry because of their powerful remediating role in the public communication of any cultural text.

## Note

1 Throughout the history of the PRC, the Chinese Communist Party had periodically launched anticorruption campaigns aimed at curbing the abuse of political power for personal gain among party and state officials. China's current leader, Xi Jinping, went further than his predecessors to make anticorruption a centerpiece of his rule and political brand. Among the toppled officials was a former member of the Politburo Standing Committee, Zhou Yongkang, and former top military leaders, Xu Caihou and Guo Boxiong. Xi's anticorruption campaign, while popular among ordinary Chinese, has been interpreted by both his detractors and impassioned observers as a tool of power consolidation. It was against this backdrop that this political drama was given the green light by the Party's Propaganda Department after an extended period of ban on any television drama dealing with the issue of corruption.

## References

Bai, Ruoyun. 2016. "Commercial and Digital Transformation of Chinese Television." In Michael Keane (ed.) *Handbook of Cultural and Creative Industries in China*. Cheltenham, UK: Edward Elgar Publishing, pp. 360–376.

Baranovitch, Nimrod Author. 2003. *China's New Voices: Popular Music, Ethnicity, Gender, and Politics, 1978–1997*. Berkeley, CA: University of California Press.

Chao, Shih-chen. 2012. *Desire and Fantasy On-line: A Sociological and Psychoanalytical Approach to the Prosumption of Chinese Internet Fiction*. Dissertation. Manchester, UK: University of Manchester, School of Arts, Languages and Cultures.

Chao, Shih-chen. 2016. "Production of Consumption, Consumption of Production: Readers Empowered, Authors Enabled and Digital Prosumption Facilitated in the Landscape of Popular Literature in China." In Daria Berg and Giorgio Strafella (eds.) *Transforming Book Culture in China, 1600–2016*. Wiesbaden: Harrassowitz, pp. 225–247. Kode. Jahrbuch de International Buchwissenschaftliche Gesellschaft.

Chen, Linjing. November 2017. "Cutification of Serious Drama Sheds Light on the Evolution of Youth Discourses and Reversed Cultural Influence – A Look at *In the Name of the People*" [cong zhengju menghua kan qingnian huayu bianqian he wenhua fanbu – yi renmin de mingyi weili]. *New Media* [xin chuanmei], No. 303: 69–71.

Chen, Xiao and Qixin Chen. December 2017. "Survey on University Students Watching the Television Drama *In the Name of the People*" [daxuesheng guankan renmin de mingyi qingkuang diaocha baogao]. Jishou University Bulletin, No. 38: 74–78.

China Internet Network Information Center (CNNIC). January 2018. *The 41th Statistical Report on Internet Development in China*. Available at: www.cac.gov.cn/2018-01/31/c_1122347026.htm.

China Netcasting Services Association 2017 Summary of *Report on Chinese Online Video Development: 2017*. Available at: www.199it.com/archives/663363.html.

Ding, Daoshi. 2017. "From Exclusive to Shared Control of the Copyright of in the Name of the People" [cong dubo dao fenxiao renmin de mingyi]. *Zhihu Column*, May 5. Available at: https://zhuanlan.zhihu.com/p/26731303

Feng, Jin. 2013. *Romancing the Internet: Producing and Consuming Chinese Web Romance*. Leiden; Boston: Brill.

Hockx, Michel. 2015. *Internet Literature in China*. New York: Columbia University Press.

Huang, Jue. 2017. "How did *In the Name of the People* Become a Hit?" [renmin de mingyi daodi zenme huoqilai de?] Available at: www.sohu.com/a/133565544_663268

Huang, Ya-chien. 2008. "Pink Dramas: Reconciling Consumer Modernity and Confucian Womanhood." In Ying Zhu, Michael Keane, and Ruoyun Bai (eds.) *TV Drama in China*. Hong Kong: Hong Kong University Press, pp. 103–114.

Jiang, Ying. 2017. "Big Data on Public Opinion Interprets *In the Name of the People*" [dashuju yuqing fenxi jiedu renmin de mingyi]. *Zhuhu Column*. Available at: https://zhuanlan.zhihu.com/p/26884724

Juhong Data. 2017. "Big Data Shed Light on How In the Name of the People Became an Online Hit" [*dashuju jiemi fanfu zhengju renmin de mingyi "wangluo zouhong" lichen*] Available at: www.juhongbigdata.com/news_detail.php?id=51.

Keane, Michael. 2015. *The Chinese Television Industry*. London: British Film Institute.

Kong, Shuyu. 2005. *Consuming Literature: Best Sellers and the Commercialization of Literary Production in Contemporary China*. Stanford, CA: Stanford University Press.

Liang, Yilu and Wanqi Shen. 2016. "Fan Economy in the Chinese Media and Entertainment Industry: How Feedback from Super Fans Can Propel Creative Industries' Revenue." *Global Media and China*, 1 (4): 331–349.

Liang, Yingrui and Chaohan Qiu. 2018. "In the Name of the People Tops the Ratings List" [renmin de mingyi 2017 shoushi duoguan]. *Pheonix Weekly*. No. 645. Available at: www.ifengweekly.com/detil.php?id=5467

Linder, Birgit. 2005. "Web Literature." In Edward L. Davis (ed.) *Encyclopedia of Contemporary Chinese Culture*. London: Routledge, p. 896.

Liu, Jin. 2013. *Signifying the Local: Media Productions Rendered in Local Languages in Mainland China in the New Millennium*. Leiden and Boston: Brill.

Mok, Laramie. 2017. "Have China's 'little fresh meat' actors Lu Han, Li Yifeng, Ma Tianyu, Yang Yang and Zhang Yixing passed their sell-by date?" South China Morning Post Online, September 22. Available at: www.scmp.com/magazines/style/people-events/article/2112413/have-chinas-little-fresh-meat-male-actors-lu-han-li.

Montgomery, Lucy and Eric Priest. 2016. "Copyright in China's digital cultural industries." In Michael Keane (ed.) *Handbook of Cultural and Creative Industries in China*, Cheltenham, UK; Northampton, MA, USA: Edward Elgar Publishing, pp. 339–359.

"PPTV". 2017. "PPTV Provides Support; the Broadcast of *In the Name of the People* Begins" [pptv juli zhuli renmin de mingyi huore kaibo]. *Pheonix Finance*. Available at: http://finance.ifeng.com/a/20170328/15268133_0.shtml

Qin, Amy. 2016. "Craving a Hot TV Show in China? Start Scouring the Web." *New York Times*, October 28. Available at: www.nytimes.com/2016/10/29/arts/television/craving-a-hot-tv-show-in-china-start-scouring-the-web.html?_ga=2.127706824.496718447.1527840248-435711176.1446575592

SAPPRFT News. 2018. "Shen Haixiong Visits CCTV Cadres and Workers on the Frontline." February 28. Available at: www.sapprft.gov.cn/sapprft/govpublic/11078/359770.shtml.

Schneider, Florian. 2012. *Visual Political Communication in Popular Chinese Television Series*. Leiden and Boston: Brill.

Shao, Deng and Zeng, Ni. 2017. "Marketing Strategies of Drama Hits: Cost Increased by 100 times; Post-90s Lead Trending Topics" [reboju yingxiao da qidi: feiyong 5nian fan baibei 90hou yindao huati]. *Tencent Entertainment*, April 27. Available at: http://ent.qq.com/a/20170427/052208.htm

Sohu Entertainment. 2017. "Another New Record! In the Name of the People Surpassed 8%; Finale Expected" [zaichuang xingao! Renmin de mingyi shoushi po8, jieju shou qidai]. Available at: http://yule.sohu.com/20170427/n490874002.shtml.

Song, Geng. 2010. "Chinese Masculinities Revisited: Male Images in Contemporary Television Drama Serials." *Modern China*, 36 (4): 404–434.

Steinberg, Marc. 2012. *Anime's Media Mix: Franchising Toys and Characters in Japan*. Minneapolis: University of Minnesota Press.

Wang, Wei. 2017. "Foreign Media: *In the Name of the People* is a Chinese Anticorruption Propaganda Drama" [waimei: renmin de mingyi shi zhongguo fanfu xuanchuan pian]. *Elite Reference* [qingnian cankao], April. 19. Available at: http://qnck.cyol.com/html/2017-04/19/nw.D110000qnck_20170419_1-12.htm

Wu, Yanyu. 2017. "Online Marketing of In the Name of the People: Audience Engagement that Originated from 'Li Dakang'" [renmin de mingyi wangluo tuiguang lujing: yichang shiyu li dakang de guanzhong hudong]. *21st Century Economic News*, April 26. Available at: www.21jingji.com/2017/4-26/0NMDEzODFfMTQwNzc0Nw.html

Zhao, Elaine Jing. 2016. "Getting Connected in China: Taming the Mobile Screen." Michael Keane (ed.) *Handbook of Cultural and Creative Industries in China*. Cheltenham, UK: Edward Elgar Publishing, pp. 396–411.

Zhao, Elaine Jing. 2017. "Writing on the Assembly Line: Informal Labour in the Formalised Online Literature Market in China." *New Media and Society*, 19: 1236–1252.

Zhao, Elaine Jing and Michael Keane. 2013. "Between Formal and Informal: The Shakeout in China's Online Video Industry." *Media, Culture & Society*, 35 (6): 724–741.

Zhong, Xueping. 2010. *Mainstream Culture Refocused: Television Drama, Society, and the Production of Meaning in Reform-Era China*. Honolulu: University of Hawai'i Press.

# PART V

# Systems, Structures, and Industries

This final part considers the implications of political, economic, and technological change for the experience and the meaning of television. This section tracks how governance, changing and expanding ideas of markets, and institutions—both public and private—contribute to an infrastructure that allows for television to be made and experienced. Thus, Part V begins with an analysis of media industries as a global system, wherein value is both structural and structuring in chains of transnational production networking (Jean K. Chalaby). The section then considers regional, rather than national, formations of television systems and industries, beginning with a historical survey of Eastern European, Soviet-style socialist television during the Cold War (Aniko Imre) and continuing with a current analysis of television industries and emerging alternatives in the Arab region (Joe F. Khalil), structural changes in the television connections between the Iberian peninsula and Latin America (Guillermo Mastrini and María Trinidad García Leiva), and the complexities of digital television convergence in Africa (Lyombe Eko). Specific national issues of the political economy, politics, and infrastructure of television are drawn from examples in China (Ying Zhu), Turkey (Ece Algan), and South Africa (Ruth Teer-Tomaselli). The section concludes with considerations of industries and technologies examined from informal, extra-legal, and alternative perspectives. A history of peer-to-peer sharing is recounted in the context of pirate radio, privatization, and the limits of ownership (Martin Fredriksson), while the recent shift from file sharing to streaming as the preferred technology for superseding geo-blocking, copyright, and distribution restrictions are considered by Ramon Lobato. Finally, Aymar Jean Christian recounts web series production of the 1990s by way of introducing the potentials and the limits of independent television production and distribution outside of the mainstream industry. Together, the authors in this part engage with critical approaches to understanding the infrastructure and conditions of possibility for our encounters with television, its production, circulation, and technological transformations.

# 31

# UNDERSTANDING MEDIA GLOBALIZATION

## A Global Value Chain Analysis

*Jean K. Chalaby*

### Introduction

The research problem this chapter seeks to elucidate is as follows: in which sense do the media industries form a global system? And if the media work that way, which forces bind companies together across borders? The consensus among political economists is that there is a global media system insofar as there is a global industry. International media conglomerates occupy a dominant position in this system and claim an ever growing share of the industry's worldwide revenues, it is claimed. The chief exponents of this theory are Herman and McChesney who state that 'the global market is being constructed by these and other large firms seeking commercial gain in a congenial and political and economic environment' (Herman and McChesney, 1997: 69).

Terry Flew identified flaws in this opinion, arguing, *inter alia*, that the global reach of the largest conglomerates can be overestimated. Thus, while acknowledging the presence of these firms, 'in referring to the global creative industries,' he writes, 'we need to apply some caution. It should not be taken to mean that a small number of global media and entertainment conglomerates now dominate the world's cultural markets' (Flew, 2013: 24).

Flew makes a valid point: while the international footprint of the US-based media conglomerations is vast, they derive a large proportion of their revenue from their home market (Table 31.1).

It is undeniable, however, that the close relationship between media and nation has been unravelling over the past two decades. Globalization and digital technologies are remapping media spaces and contribute to the emergence of new products, practices and experiences. They are not only intensifying trade flows, but reshaping national media systems from within by embedding them in production/consumption transnational networks. This new order has several distinctive features.

### Media Globalization

First, national markets are far less self-contained than they used to be: borders have become porous, media flows have expanded, and trade has grown. Data from the United Nations Conference on Trade and Development (UNCTAD) show a sharp rise in the total value of creative goods exports, passing from US$ 208.5 billion to US$ 509.8 billion between 2002 and 2015, an average growth rate of 6.6 per cent per annum (Table 31.2).

*Table 31.1* Leading media conglomerates' geographic segments in terms of revenues (percentages), 2016

| | *North America* | *Europe* | *Asia Pacific* | *Latin America* | *Japan* | *Other* |
|---|---|---|---|---|---|---|
| *Alphabet (incl. Google)* | 47 | – | – | – | – | 53 |
| *21st Century Fox* | 71 | 12 | – | – | – | 17 |
| *CBS Corporation* | 86 | – | – | – | – | 14 |
| *Discovery* | 53 | – | – | – | – | 47 |
| *Disney* | 77 | 12 | 8 | 3 | – | – |
| *Facebook* | 46 | – | – | – | – | 54 |
| *Sony* | 21 | 23 | 19 | – | 29 | 8 |
| *Time Warner* | 72 | 16 | 7 | 5 | – | 1 |
| *Viacom* | 75 | 17 | – | – | – | 8 |

Source: Thomson ONE Banker database, 2017

*Table 31.2* Values and shares of creative good exports, 2002-2015, in US$ at current prices in millions

| | *2002* | *2015* | *Annual growth rate* |
|---|---|---|---|
| *World* | 208,493 | 509,753 | 6.6% |
| *Developed economies* | 122,911 | 241,624 | 5.0% |
| *Percentage of world exports* | 58.9.% | 47.4% | – |
| *Developing economies* | 84,365 | 265,001 | 8.5% |
| *Percentage of world exports* | 40.5% | 52.0% | – |
| *Transition economies* | 1,217 | 3,048 | 6.8% |
| *Percentage of world exports* | 0.6% | 0.6% | – |

Source: adapted from UNCTAD, 2017

In the television industry alone, international traffic has not only intensified but has become complex and multi-faceted. The first TV series began to travel in the 1950s, with international audiences' appetite for US drama growing steadily in the ensuing decades (Bielby and Harrington, 2008). In this period, US export landmarks include *Star Trek*, *Dallas* and *Baywatch*. The first TV formats – defined as the structure of a show that can generate a distinctive narrative and is licensed outside its country of origin in order to be adapted to local audiences – were also exchanged in the 1950s. This market developed slowly and, until the 1990s, consisted mostly of US game shows (e.g. *The Dating Game*, *The Price is Right*, *Wheel of Fortune*) adapted in foreign markets (Chalaby, 2016a). The TV industry's *global shift* came later, with the internationalization of TV networks and content platforms, and the expansion of the TV series and format trade.

In Europe, TV channels began to cross borders when the first communications satellites were launched in the early 1980s, but the planets aligned the following decade when barriers and obstacles receded and new technologies broke through: digitization facilitated the formation of large-capacity international communications networks, a viable international copyright regime was created (not least through the 'SatCab' EU Directive adopted in 2003), regulation evolved (e.g. the Television Without Frontiers EU Directive voted in Paris in 1989), the number of cross-border subscribers expanded and advertisers began to pay attention. This led to a rapid growth of transnational TV channels across a wide variety of genres, including news and business news (Bloomberg, CNBC, Euronews, etc.), factual entertainment (e.g. the Discovery and National Geographic suite of channels), entertainment (AXN, HBO and Fox-branded channels), children's television (such as Cartoon Network, the Disney brands and Nickelodeon), sports (Eurosport, ESPN), music television (MTV, Viva, etc.) and films (e.g. Studio Universal, Turner Classic Movies) (Chalaby, 2009).

Today, transnational TV networks count among TV's most prestigious brands and can be found on every cable and satellite platform across the world. Discovery Networks International, for instance, operates networks in more than 220 territories, distributing an average of ten channels in every market and reaching 3 billion cumulative viewers worldwide (company source). In Europe alone, the ten leading international channel operators earned US$ 4.3 billion in advertising and carriage fees income in 2013, and are predicted to generate US$ 5.5 billion by 2018 (Chalaby, 2016a).

The advent of the Internet as a distribution platform has only accentuated this trend. Entertainment platforms (e.g. Netflix) and video-sharing sites (e.g. YouTube) rely on a technology that is intrinsically international. These platforms may not necessarily be 'born global' (Gabrielsson and Kirpalani, 2012), but they cross borders at lightning speed. Netflix, originally a California-based DVD-by-post company, began streaming films in its domestic market in 2009. Eight years later, its number of international subscribers (57.8 million) is higher than the domestic figure (52.8 million), and its revenues from international streaming is fast approaching US revenues (US$ 5.1 billion against US$ 6.1 billion as of 31 December 2017; Netflix, 2018).

The historical evolution of the format trade echoes the development of cross-border channels. TV formats remained at the fringe of the TV industry until the late 1990s, when a combination of factors made the trade explode (Chalaby, 2016a; Moran, 2013). In mature markets, television was becoming more competitive and broadcasters welcomed the opportunity to acquire TV shows that had been tested in other territories and which enabled them to de-risk their schedules. It was also in the 1990s that commercial broadcasting developed in Central and Eastern Europe, the Middle East, across Asia and Latin America. Fledgling broadcasters quickly realized that their audiences preferred local shows to imported ones, but local programming requires know-how and resources, both in short supply at the time. They realized that it would be more convenient and efficient to rely on concepts that came with a pre-established method of production and guidelines.

Furthermore, new TV genres emerged providing a rich source of TV formats. All reality and factual entertainment shows come with a clear structure and narrative blocks that are easily identifiable, described and reproduced. By the end of the millennium, four shows (*Who Wants to Be a Millionaire?*, *Survivor*, *Big Brother* and *Idols*), later to become known as the four super-formats, swept audiences off their feet and announced TV formats to the world. They helped turn a cottage industry into a multi-billion dollar business, multiplying the number of formats, the number of countries they travelled to and the number of companies that distributed and produced them (Chalaby, 2011; Moran, 2013).

TV format trading was worth in excess of €3.0 billion per year by the late 2000s. Formats have become part of the daily diet on broadcasters' schedules. A sample of 84 European channels aired an average of 338 hours of formats in 2013 (FRAPA, 2009: 8–13; TBI Formats, 2014: 23). The majority of top-rated programmes worldwide are formatted. It is the case that those ubiquitous talent competitions (*Idols, Dancing with the Stars, Got Talent* and *MasterChef*) have all been produced in more than 50 territories, where they have often become cultural phenomena (Chalaby, 2016a).

The trade is no longer limited to game shows and reality TV; fiction has caught up, as dramas and comedies have all joined the format revolution and are being re-made across borders. *The Bridge, Homeland* (based on Keshet's *Prisoners of War*) and Netflix's *House of Cards* (originally a BBC One political thriller) are all scripted formats (Chalaby, 2016a).

TV formats have not only contributed to the expansion of cross-border flows but have aided the internationalization of production companies. Formats hold the great advantage of enabling these firms to reach scale with a minimum investment. When adaptation rights are sold, the format rights holder receives a licence fee. Production companies, however, have realized that there is more money to be made if they hold on to those rights and produce the format themselves. Although many format licences are still exchanged, producers try to adapt their own shows in as many markets as possible, a strategy that has led to their international expansion (Chalaby, 2012). As a result, there are now 11 global TV content production majors – TV content production companies with deep multinational capabilities – supplying programmes to broadcasters around the world. The largest, Endemol Shine, brings together in excess of 100 companies with production facilities spread across more than 20 territories (Chalaby, 2016a; Esser, 2016).

The synchronous growth of cross-border TV channels, platforms and formats constitutes an important shift in the TV industry – it has increased the number of international businesses and expanded cross-border media flows.

The second key aspect of media globalization is the increasing *poly-centric* nature of this trade. Although it remains dominated by Western companies, media businesses from the Global South are rapidly extending their reach. As seen in Table 31.2, the annual growth rate of exports in creative goods from developing economies stands at 8.5 per cent between 2002 and 2015, compared to 5.0 per cent for developed markets. As a result, developing countries now hold 52.0 per cent of the trade, compared to 47.4 for developed economies.

Most of this progress is down to the sharp growth of South–South trade, as many companies from developing countries have built a strong regional presence. MultiChoice, South Africa's leading video entertainment and Internet provider, has millions of subscribers across the region. Zee Entertainment, based in India, operates tens of channels in the subcontinent. To the ire of Qatar's neighbours, Al Jazeera Media Network broadcasts the Middle East's most influential news channel. Latin American companies have extensive regional activities, including Brazil's media powerhouse Organizações Globo, Mexico's media giant Grupo Televisa and Venezuela's Cisneros. South–South traffic includes a wide range of products, such as news channels, telenovelas, and dramas from Turkey, India or Pakistan. The South–South format trade is also developing strongly, with strong intellectual property generators emerging in the Middle East, Asia and Latin America (Chalaby and Esser, 2017).

The progress of South–North trade – known as the counter-flow of communication (Thussu, 2007) – remains sketchier. Many of these firms have global activities but they have yet to make solid inroads into Western markets: Bollywood movies, telenovelas and South Korean formats are yet to be widely distributed in the West.

## Media Transnationalization

The growing importance of trade is transforming the nature and fabric itself of the TV industry. It is becoming increasingly transnational, denoting a constant interplay between the local, the regional and the global. Media products, services and companies must adapt to local cultures.

As broadcasters expanded across frontiers they learned to deal with a multinational universe and adapt their channels to local taste. They progressively broke up the pan-regional satellite feeds they had launched initially and turned them into *transnational networks of local channels*. Providing a prime example of 'the globalization of the business model of television and the efforts of international and domestic companies to deal with the resilience of national cultures' (Waisbord, 2004: 360), these channels share a concept, a brand, resources and infrastructures and much of the programming, but adapt to their cultural and commercial environment. The depth and type of localization, however, varies from genre to genre (Chalaby, 2009). Today, media owners apply this principle across a portfolio of brands, which they deploy territory by territory. For instance, Discovery operates 24 brands in Europe, representing in excess of 200 channels in total when all the local versions are taken into account.

Entertainment platforms operate on a similar basis. Video-sharing websites provide language and content preferences (or interface localization) in the territories they launch (Lobato, 2016: 12–13), and film subscription services increasingly adapt their catalogue, as there is 'a strong relationship between the degree to which SVoD [subscription video-on-demand] catalogues have been localized and overall subscriber uptake' (Bisson, 2017: 9). Among Netflix's latest commissions, almost half (48 per cent) comes from outside the USA, and originations from non-English-language territories are in sharp rise (e.g. Mexico, 6 per cent, Brazil and India, 5 per cent each) (Bisson, 2017: 13). Amazon Prime has already ordered 18 original Indian titles barely a year after starting in this market (Bylykbashi, 2017). All told, global platforms 'currently have around 5% to 8% local content, depending on market' (Bisson, 2017: 8).

A process of adaptation lies at the heart of the TV format trade. The format combines a kernel of rules and principles that are immutable and elements that are adaptable as it travels. Albert Moran equates a format to a 'cooking recipe', that is, a 'set of invariable elements in a programme out of which the variable elements of an individual episode are produced' (Moran, 2006: 20). Format rules might be global but their adaptation is local. Formats acknowledge both the universality of great ideas and the perennial nature of local cultures and languages.

The explosion of the format trade alone illustrates well the progressive transnationalization of media markets. Until the 1990s, most entertainment programmes were purely domestic productions. Such was the case with the old variety genre, and even a good proportion of game shows were local creations (as opposed to licensed adaptations) (Cooper-Chen, 1994: 270–89). Today, factual entertainment or talent competition formats, among others, are the storytelling tool kits which have become de rigueur for broadcasters around the world.

## Media Firms' New Organizational Structure

Operating across borders has prompted several media conglomerates to re-think their organizational structure and integrate elements of an ideal type called the *transnational organization model*. This architecture is labelled 'transnational' because it enables companies to achieve simultaneous global efficiency, local responsiveness and knowledge transfer (Bartlett and Ghoshal, 1998: 65). It moves a corporation to go beyond a 'hub and spokes' anatomy and adopt an *integrated network configuration*. Local subsidiaries break the pattern of dependence/independence toward the headquarters to become units in a web of 'relationships based on mutual independence' (Bartlett and Ghoshal, 1998: 148).

The key organizational characteristics of the transnational company are network integration and agility. Some assets and decisions are centralized, while others are localized. Some subsidiaries are closely controlled from headquarters, others are given full autonomy. The headquarters is not the custodian of all centralized resources and responsibilities, which are dispersed throughout the network. The role of subsidiaries is differentiated and the most important among them can make a distinct contribution to the company's international operations. Local units continuously collaborate and share knowledge, sometimes with one another, sometimes at network level.

This organizational structure enables firms to break down the opposition between the local and the global to make each term an extension of the other. It gives them the ability to generate knowledge about both dimensions and to combine global efficiency with responsiveness to local and regional markets and to move anywhere within the *local/global matrix* (Figure 31.1). This transnational generation of knowledge and interaction between the two terms often becomes a driver for growth and creativity. In the process, these firms acquire their unique transnational quality that makes them more than the sum of their parts

Globalization is clearly a driver of change in the media industries. As the number of international companies and products expands, so does trade, as reflected in the UNCTAD data. This intensification of cross-border exchange is transforming the nature of media businesses and goods, and the fabric itself of national media markets. The question, however, remains: do the media industries work as a global *system*? And if so, what is the dynamic that binds firms and markets together? The answer lies with global value chains (GVCs).

## Global Value Chains and the Media Industries

Over past decades, a trend known as the *vertical disintegration of production*, which corresponds to the slicing and dicing of production processes across multiple firms and countries, has thoroughly transformed the structure of the world economy (Baldwin, 2013; Feenstra, 1998). In the earlier periods of capitalism, corporations were vertically integrated and took charge of most, if not all, of the tasks involved in the making of a product or service – a model known as the Chandlerian firm. These corporations have found it more efficient to break up the value chain and outsource – and often offshore – non-core tasks. They now tend to specialize in the most profitable segments of the chain, such as branding, marketing and finance, leaving other duties to subcontractors. Today, many industries, from fashion to electronics, are dominated by 'fabless manufacturers' that sell goods they do not manufacture. The best illustration of this strategy is Apple, which reported US $ 229.2 billion annual revenue in 2017 and became the world's first trillion-dollar company by market value in August 2018 (Davies, 2018), owns no plant and offshores the entirety of its manufacturing operations to Far East subcontractors (Langlois, 2003; Milberg and Winkler, 2013).

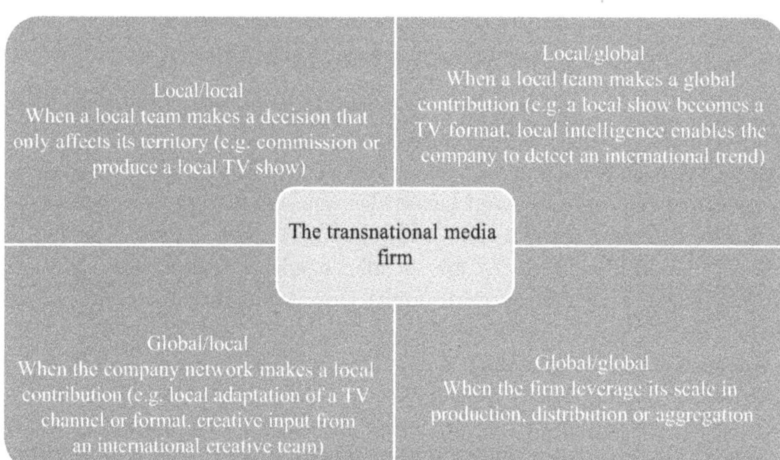


*Figure 31.1* Managing transnational media firms: the local/global matrix
Source: Author

As a result, Chandlerian industries characterized by large vertically integrated corporations have been progressively replaced by global value chains (GVCs), or transnational networks of firms collaborating on the tasks involved in design, manufacturing, marketing and delivery of products and services (Gereffi and Fernandez-Stark, 2016). GVCs have several dimensions including the input-output structure, which designates the 'sequence of value adding economic activities' involved from the creation to the distribution of manufactured goods or services (Gereffi, 1994: 97). Segments vary from one chain to another but often include inputs (e.g. agrochemicals in the fruit and vegetable value chain), research, design, manufacturing, distribution, marketing, sales, consumption and recycling. The shape and nature of these sequential stages gives a chain its anatomy, or structure.

Vertical disintegration has led to the globalization of production for multiple reasons. Although the geographical configuration of value chains varies, all GVCs cross borders and involve firms spread across multiple territories either regionally or globally. In addition, as companies tend to vertically specialize within their segments, their best option for growth is often to grow internationally. In today's economy suppliers are often as global as the firms they serve (Gereffi and Lee, 2012; Sturgeon and Kawakami, 2010).

GVCs have also led to a growth in international trade which in recent decades has increased dramatically both as a percentage of the world GDP and in absolute value: world merchandise exports have risen from US$ 59 billion in 1948 to US$ 18,301 billion in 2013 (Feenstra, 1998: 31; WTO, 2014: 23). Some of this growth can be ascribed to the expansion of 'trade in tasks' (or 'trade in intermediate goods') within value chains, which is routinely assessed as representing more than half the total value of (non-fuel) global exports (Milberg and Winkler, 2013: 37–48; WTO, 2013: 182–83).

This evolution is relevant to the media industries where outsourcing has become as prevalent as in other industrial sectors (Caldwell, 2008: 154–67), leading to vertical disintegration and the formation of transnational production networks (Chalaby, 2016b, 2019). Some markets have been more resistant to these changes than others due to cultural protectionism, but in many countries media companies are far less vertically integrated than they used to be.

The disassembling of the old production model occurred when a combination of factors, including economic growth, rising industrial complexity, deregulation, trade liberalization, new technologies, consumer demand and preference, progressively led media firms to concentrate on those activities in which they retained a competitive advantage and/or could capture the most added value. This strategy created distinct segments that form the value chains through which creative goods travel from inception to consumption.

The TV industry constitutes an interesting case study, where two value chains have recently formed. The first, the TV content GVC, encompasses four segments: production, facilities, distribution and aggregation (Chalaby, 2016b):

- *TV production* consists of the production of content that is sold or licensed to aggregators. Independent TV production sectors developed initially in the United Kingdom and the United States but are now growing across Europe and all other world regions. TV producers serve a content market that is estimated at US$ 400 billion globally (Chalaby, 2017). Today, TV production companies serve a wide variety of scheduling windows, platforms and off-screen markets.
- TV and film producers have their own suppliers which form the *facilities sector*. It comprises numerous specialist providers, including film and TV studios, outside broadcast firms, post-production houses (video and audio editing, etc.), and a multitude of businesses offering the back-up services integral to a shoot, from prop and film equipment rental firms to special effects and prosthetics experts. Most of these suppliers regroup in specific areas that become

clusters for film and TV production, known as Hollywood in LA and Soho in London (Pratt, 2009).

- Further down the supply chain, both the facilities and production segments have their own suppliers, which include all hardware and software infrastructure providers (e.g. equipment manufacturers, audio and video production technology specialists).
- The role of *distributors* is to coordinate content production and content aggregation, if and when content is not directly commissioned by aggregators and acquired on the international market. A distributor's core remit is to wring every last drop out of the intellectual property they represent, while nurturing relationships with buyers and understanding their needs. With the multiplication of genres, rights, windows and platforms, distribution is an increasingly complex activity requiring legal, financial and marketing skills.
- The final segment is *aggregation*, which consists of bringing content together under the umbrella of a single brand and marketing it to audiences and/or advertisers. Digitization has multiplied the delivery modes and business models broadly fall into four categories (Figure 31.2).

The fastest growing business model is to be found in the fourth quadrant: on-demand pay-to-view which encompasses all types of digital content delivery services that charge for access. Several distinctions exist in this market segment. First, there is a difference between *transactional video-on-demand* (TVoD) services that charge for each request (or transaction, hence the name) and *subscription video-on-demand* (SVoD) which involves a recurring billing model. Technically, content can be streamed over the Internet, as most OTT (over-the-top) content providers do (e.g. Netflix), or it can be downloaded by customers to rent or own (Apple's iTunes and some broadcasters' on-demand websites). The SVoD market is dominated by two types of players: OTT providers that are solely Internet companies and the on-demand services of pay-TV platforms (EAO, 2014: 137).

Once aggregators have the master tape, a second inter-firm production network comes into play, whose task is to deliver the signal and/or files to consumers. The three segments of the media delivery GVC are as follows:

- *Publication* consists of preparing the files for transmission. In today's global market, a Hollywood movie or a documentary like the BBC's *Planet Earth* will be issued in hundreds

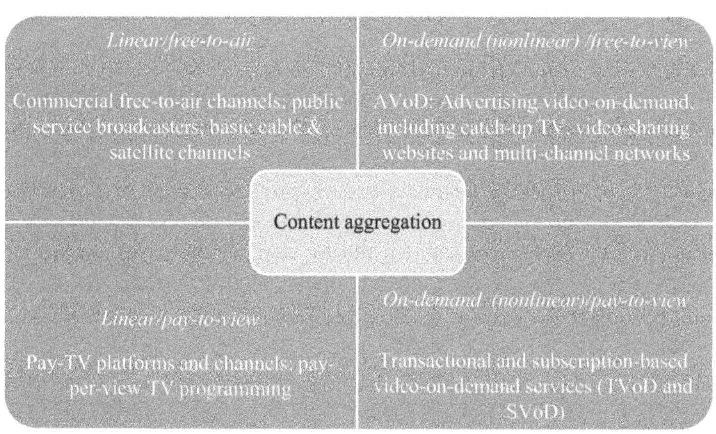


*Figure 31.2* Content aggregation business models
Source: Adapted from Chalaby (2016b)

of versions to suit a variety of transmission paths, platforms, electronic devices, languages and cultures. The tasks involved in this segment are multiple and include validation, versioning and storage (i.e. content management), and playout. Businesses specializing in this segment include media asset management and archive systems providers, channel management firms and playout specialists.

- *Transmission* ensures content distribution to customers. Infrastructure and media service companies, CDN providers, data centres and satellite operators are among the segment's specialist businesses. The Internet has revolutionized transmission paths. In a classic broadcast configuration – or *multicast* – a piece of content is transmitted over a radio wave that can be received by anyone wishing to tune in. The transmission mechanism with Internet protocol-based delivery differs, as every single client is sent their own copy of the material (*unicast*). IP-based transport networks have brought a host of newcomers (Akamai, Cisco, etc.) to the media delivery chain offering their services to old and new aggregators that use the Internet as a delivery mechanism.
- The *reception* segment represents the moment when the signal or file reaches its final destination. It brings together broadband providers and consumer electronics companies, including TV set and set-top box manufacturers.

## GVCs as a Globalizing Force

Such developments are significant because these chains are the engines behind the globalization of the media industries: not only do the chains' dynamics stimulate media firms in order to expand internationally but they also increase their transnational interdependence.

Vertical integration is a noted trend because merging conglomerates grab headlines. Three recent takeovers (21st Century Fox by Disney, Time Warner by AT&T, and Sky by Comcast) show that when faced with uncertainty, media giants strive to integrate content and production capabilities with large (and international) distribution networks. This phenomenon, however, occurs in the wider context of industry segmentation.

GVCs have formed in the media and entertainment sector because the tasks involved in the creation, production and distribution of programming are increasingly complex and require a substantial amount of expertise. A single company cannot occupy all the segments, and the making of a programme inevitably involves the collaboration of many firms. Even media conglomerates will work with hundreds of suppliers and outsource the tasks they deem marginal to their business plan (Chalaby, 2019).

Most suppliers stick to one segment and opt to expand within it. Thus, the best opportunity for growth is global expansion and, as a result, one of the most significant developments in the industry is the *internationalization of suppliers*. Examples abound in every segment of the industry's production networks, as illustrated by the recent evolution of TV content production majors (see above).

The same phenomenon can be observed in the facilities sector. Pinewood Studios, for instance, the film and TV studio near London, offers shooting facilities in locations across Asia, North America and the Caribbean. In the media delivery chain, many firms have gained the ability to serve clients almost irrespective of location. Ericson, a media management and playout business, originates more than 500 TV channels through eight hubs across the Middle East, Europe and the USA. The company also has the ability to operate channels remotely, with its team located in a different country than the broadcaster's channels. As seen above, the pace of internationalization in the aggregation segment has accelerated since the advent of the Internet as a video distribution medium.

International scale is no longer a premium for media firms but a necessity that often makes the difference between success and failure. In the aggregation segment, scale brings economies in all operational aspects (management, marketing, procurement, etc.) and allows for larger commissioning budgets. Netflix's budget, for instance, exceeds those of the largest conglomerates, and some reports estimate the platform's spending on original programming to be as high as US $ 13 billion in 2018 (Morris, 2018). By the end of 2018, the FAANGs (Facebook, Apple, Amazon, Netflix and Alphabet's Google) had, collectively, 424 original titles in the pipeline (Cooper, 2018). Smaller aggregators are becoming aware of the implications: 'Nobody else will be able to compete with these guys [Amazon, Apple, Google and Netflix] in terms of bringing scale entertainment. The scale and structural advantages that they have cannot be replicated', says Efe Cakarel, chief executive of film subscription service Mubi (in Farber, 2017: 13).

In the production segment, scale also brings a host of benefits, including enhanced creativity and innovation (notably through central development teams and funds, creativity events, career incentives), access to better market intelligence, and ability to take larger bets and deficit-fund large-scale scripted projects (Chalaby, 2012, 2016a).

Furthermore, being embedded in cross-border production networks has increased the *transnational interdependence* among businesses. While the content value chain used to be local, aggregators have become adept at sourcing content further afield. For instance, US networks have recently adapted series from South Korea and Turkey. In the media delivery chain, aggregators must also look beyond their borders to find the expertise they need. This entails a *deterritorialization of competition*: if programme suppliers can, today, fill the schedules of broadcasters and platforms located anywhere, the reverse side of the coin is that they compete for these slots against producers from other countries.

The same trend is characterizing the aggregation segment. While the markets of broadcasters used to be delineated by national borders, it is far less the case with entertainment platforms. Access can be blocked, notably in China and Russia, progress in certain markets can be slow, but OTT platforms are essentially global and often reach 100+ territories within a few years of existence. Beyond the well-known brands, smaller platforms, such as HBO Go, Mubi (see above), Discovery's Eurosport Player, or Walter Presents also reach audiences across borders (Blair, 2017).

Finally, production processes themselves have become transnational: films and TV series, from *Lord of the Rings* to *Game of Thrones*, are increasingly cast, shot, edited and post-produced in multiple locations. High-capacity Internet networks and high-speed data transfer technology enable teams in multiple production sites to work simultaneously on the same project.

## Conclusion

We set out to answer the following question: in which sense are the media industries globalizing? The answer lies in the formation of GVCs. Outsourcing and vertical disintegration are shaping the media industries that increasingly function as transnational production networks, bringing together sets of businesses collaborating on the creation, production and commercialization of cultural products. GVCs are transforming business strategies, pushing firms to specialize within a chosen segment and to grow internationally within it. The cross-border partnerships of companies have created transnational interdependence, whereby firms are embedded in cross-border production networks, competition is deterritorialized and production processes are transnationalized (Figure 31.3).

The 'global' in global media system stems from the *coordinating effect* of GVCs: they are at once a *structural* reality – media industries are globalizing because value-adding sequences have become international – and a *structuring* reality – as corporations and national markets are increasingly coordinated by GVCs (Chalaby, 2016b: 54). A handful of global firms and products do not make a system global. What does are the GVCs that bind together national markets and connect interdependent businesses, institutions, places and networks.

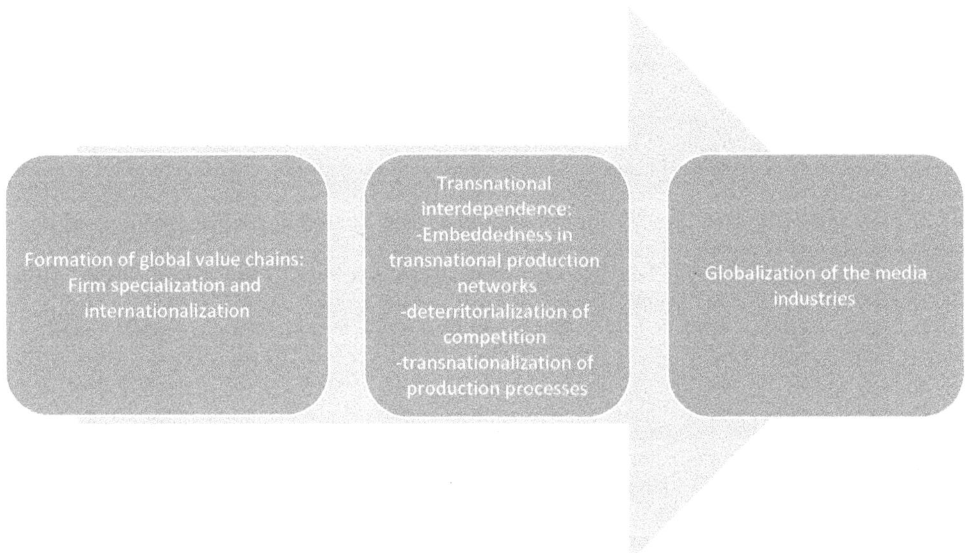


*Figure 31.3* Globalizing processes in the media industries
Source: Author

## References

Baldwin, R. (2013) 'Global supply chains: why they emerged, why they matter, and where they are going', pp. 13–59 In: D. K. Elms and P. Low (eds), *Global Value Chains in a Changing World*. Geneva: WTO.

Bartlett, C. A. and S. Ghoshal (1998) *Managing across Borders: The Transnational Solution*. Boston, MA: Harvard Business School Press.

Bielby, D. D. and C. L. Harrington (2008) *Global TV: Exporting Television and Culture in the World Market*. New York: New York University Press.

Bisson, G. (2017) *SVoD Content Strategies: The Drive for Originality (Thinking Local, Going Global)*. London: Ampere Analysis.

Blair, M. (2017) 'Determining the right distribution model for going OTT overseas', *Rapid TV Everywhere*, September: 8–10.

Bylykbashi, K. (2017) 'Making Amazon Tick', *TBI Vision*, August/September: 4–5.

Caldwell, J. T. (2008) *Production Culture: Industrial Reflexivity and Critical Practice in Film and Television*. Durham, NC: Duke University Press.

Chalaby, J. K. (2009) *Transnational Television in Europe: Reconfiguring Global Communication Networks*. London: I.B. Tauris.

Chalaby, J. K. (2011) 'The Making of an Entertainment Revolution: How the TV Format Trade became a Global Industry', *European Journal of Communication* 26(4): 293–309.

Chalaby, J. K. (2012) 'Television and the Globalization of the Intellectual Property Market: The Emergence of the International Production Model', *Journal of Media Business Studies* 9(3): 19–40.

Chalaby, J. K. (2016a) *The Format Age: Television's Entertainment Revolution*. Cambridge: Polity.

Chalaby, J. K. (2016b) 'Television and Globalization: The TV Content Global Value Chain', *Journal of Communication* 66(1): 35–59.

Chalaby, J. K. (2017) 'Can a GVC-Oriented Policy Mitigate the Inequalities of the World Media System? Strategies for Economic Upgrading in the TV Format Global Value Chain', *International Journal of Digital Television* 8(1): 9–28.

Chalaby, J. K. (2019) 'Outsourcing in the U.K. Television Industry: A Global Value Chain Analysis', *Communication Theory* 29(2): 169–190.

Chalaby, J. K. and A. Esser (eds) (2017) 'Special Issue: The TV Format Trade and the World Media System', *International Journal of Digital Television* 8: 1.

Cooper, R. (2018) 'FAANGs Fight First-Run Battle', *C21Media*, 29 September. www.c21media.net/perspective/faangs-fight-first-run-battle/, accessed 15 October 2018.

Cooper-Chen, A. (1994) *Games in the Global Village: A 50-Nation Study of Entertainment Television*. Bowling Green: Bowling Green State University Popular Press.

Davies, R. (2018) 'Apple becomes World's First Trillion-Dollar Company', *The Guardian*, 2 August. www.theguardian.com/technology/2018/aug/02/apple-becomes-worlds-first-trillion-dollar-company, accessed 16 October 2018.

EAO. (2014) *On-Demand Audiovisual Markets in the European Union*. Strasbourg: European Audiovisual Observatory and Council of Europe.

Esser, A. (2016) "Challenging U.S. Leadership in Entertainment Television? The Rise and Sale of Europe's International TV Production Groups', *International Journal of Communication* 10: 3585–3614.

Farber, A. (2017) 'Big Four to Battle it Out: Content Producers told they Must "Super-Serve" Audiences to Succeed', *Broadcast*, 19 May: 13.

Feenstra, R. C. (1998) 'Integration of Trade and Disintegration of Production in the Global Economy', *Journal of Economic Perspectives* 12(4): 31–50.

Flew, T. (2013) *Global Creative Industries*. Cambridge: Polity.

FRAPA. (2009) *The FRAPA Report 2009: TV Formats to the World*. Cologne: FRAPA.

Gabrielsson, M. and Kirpalani, V. H. M. (2012) *Handbook of Research on Born Globals*. Cheltenham: Edward Elgar.

Gereffi, G. (1994) 'The Organization of Buyer-Driven Global Commodity Chains: How U.S. Retailers Shape Overseas Production Networks', pp. 95–122 in G. Gereffi and M. Korzeniewicz (eds), *Commodity Chains and Global Capitalism*. Westport, CT: Praeger.

Gereffi, G. and K. Fernandez-Stark (2016) *Global Value Chain Analysis: A Primer*. Durham, NC: Duke University, Center on Globalization, Governance & Competitiveness.

Gereffi, G. and J. Lee (2012) 'Why the World Suddenly Cares about Global Supply Chains', *Journal of Supply Chain Management* 48(3): 24–32.

Herman, E. S. and R. W. McChesney (1997) *The Global Media: The New Missionaries of Corporate Capitalism*. London: Cassell.

Langlois, R. N. (2003) 'The Vanishing Hand: The Changing Dynamics of Industrial Capitalism', *Industrial and Corporate Change* 12(2): 351–385.

Lobato, R. (2016) 'Introduction: The New Video Geography', pp. 10–22 in R. Lobato and J. Meese (eds), *Geoblocking and Global Video Culture*. Amsterdam: Institute of Network Cultures.

Milberg, W. and D. Winkler (2013) *Outsourcing Economics: Global Value Chains in Capitalist Development*. Cambridge: Cambridge University Press.

Moran, A. (2006) *Understanding the Global TV Format*. Bristol: Intellect.

Moran, A. (2013) 'Global Television Formats: Genesis and Growth', *Critical Studies in Television* 8(2): 1–19.

Morris, D. Z. (2018) 'Netflix is Expected to Spend up to $13 Billion on Original Programming This Year', *Fortune.com*, 8 July. http://fortune.com/2018/07/08/netflix-original-programming-13-billion/, accessed on 16 October 2018.

Netflix. (2018) *Annual Report 2017*. Los Gatos, CA: Netflix.

Pratt, A. and G. Gornostaeva (2009) 'The Governance of Innovation in the Film and Television Industry: A Case Study of London, UK', pp. 119–136 in A. Pratt and P. Jeffcut (eds), *Creativity, Innovation and the Cultural Economy*. London: Routledge.

Sturgeon, T. J. and M. Kawakami (2010) 'Global Value Chains in the Electronics Industry: Was the Crisis a Window of Opportunity for Developing Countries?', *Policy Research Working Paper 5417*. Washington, DC: The World Bank.

TBI Formats. (2014) 'Europe's most valuable formats', April/May: 22–25.

Thussu, D. K. (2007) *Media on the Move: Global Flow and Contra-Flow*. London: Routledge.

UNCTAD. (2017) 'Values and shares of creative goods exports, annual, 2002–2015', http://unctadstat.unctad.org/wds/TableViewer/tableView.aspx?ReportId=14772, accessed 7 September 2017.

Waisbord, S. (2004) 'McTV: Understanding the Global Popularity of Television Formats', *Television & New Media* 5(4): 359–383.

WTO. (2013) *International Trade Statistics 2013*. Geneva: WTO.

WTO. (2014) *International Trade Statistics 2014*. Geneva: WTO.

# 32

# THE OTHER KIND OF COLD WAR TV (NOT SO DIFFERENT AFTER ALL)

*Aniko Imre*

## Introduction

Much of television's development as a mass medium took place during the decades following World War II, under the influence of the Cold War. The ideological conflict and competition between the two superpowers impacted every aspect of television, from building its technological and institutional infrastructure to its aesthetic forms and ideological influences. Thanks to television studies, we know quite a bit about a part of this history, that of postwar US TV and, to a lesser extent, other television systems on the Western side of the Iron Curtain. Television studies has evolved from and revolved around the commercial broadcasting model that had come to dominate the American TV industry, whose only registered alternative was the public broadcasting systems developed in Western Europe.

What we know far less about is how television worked on the other side of the Cold War competition, in Soviet-type socialist countries. To some extent, this absence of socialist television from the global history of the medium is due to shortage of information about socialist cultures. But there are also other important reasons for the omission: a long-term, motivated disinterest in television as a "lowly" popular medium within studies of socialist cultures; and television studies' own focus on Cold War American TV as the dominant and normative path of the medium's development. These trends have been further supported by the common-sense normalization of a stereotyped view of life on the other side of the Cold War divide: that of scarcity, uniformity, housing shortages, censorship, propaganda, isolation from global flows, and a resulting, persistent longing for a superior way of life offered on Western screens. In other words, socialist TV has long been imagined as a version of contemporary footage from North Korean TV that occasionally pops up on the news or, just as frequently, on late-night news comedy.

Thankfully, in the past decade, a burgeoning interest in the histories and everyday lives of socialism across the social sciences and humanities has also fermented historical research on television in the Soviet empire. Recently published work has not only challenged the lingering stereotypes about socialist systems but has also introduced a different model of television into the historical and theoretical study of media (see Mihelj and Huxtable 2018; Imre 2016; Gumbert 2014a; Imre et al. 2012; Bren 2010; Roth-Ey 2011; Evans 2016; Čulik 2013; see also the special issue of the *Historical Journal of Film, Radio and Television* (Vol. 24, No. 3, 2014) dedicated to television in the former GDR). Just how different is this model from either the Cold War stereotypes imposed on communism or from the commercial version that television studies has embraced as the default model of television's development?

I answer this question here by offering a compressed overview of the most recognizable and distinctive features of television systems in the Soviet Union and its Eastern and Southern European satellite states. While I focus on what distinguished television in the Eastern bloc from contemporaneous Western developments, I will also emphasize a surprising extent of similarity. In fact, socialist television's extensive international connections, borrowings, exchanges and networks undermine some of the very received notions about socialist systems themselves that have remained unchallenged since the West "won" the Cold War and "socialism" turned from global threat to a dirty word. My purview is decidedly regional, identifying shared features and resisting the traditional country-by-country approach that has dominated studies of socialist cultures in the humanities and social sciences. This is not to impose an artificial homogeneity on a region of great diversity. I also point out how some of the differences in terms of language, cultural tradition, political attitudes, economic development and ideological distance from the Soviet Union were reflected in the television systems of the various nation-states and sub-regions.

## Histories

Television in the socialist region went through stages of transformation that were largely aligned with those in the rest of Europe. The technology itself was first introduced in most countries of the region in the pre-war period. As elsewhere in Europe, small-scale state television broadcasting began in the inter-war years, to be interrupted by World War II.[1] Socialist TV's postwar history roughly corresponds with John Ellis's three-stage West European chronology, from the 1950s of "scarcity" through the 60s and 70s of "availability" to the era of "plenty" since the 1980s. Raymond Williams's periodization from "emergent" through "dominant" to "residual" cultures, adapted by Charlotte Brunsdon to Western European broadcasting, also maps onto television's changes in the East.

With the exception of Britain, the Soviet Union and Germany, where regular broadcasting began before the war and was restarted immediately after, and France, where state broadcasting launched its first limited service as early as 1945–46, most Western European countries began broadcasting in the 1950s–60s.[2] In the Soviet-ruled region of Europe, state service also started up in the 1950s, with sporadic broadcasts received by a few thousand subscribers in each country. The first program types were similar across the continent: live theatrical and sports events, news programming and feature films. By the late 1960s, the proliferation of television sets in the home and the quality and quantity of programming transformed socialist television into a national form of entertainment.[3] In most countries, the mid-60s saw the launch of a second channel and the extension of broadcast time to five, then six, and eventually seven days a week.

By the mid-1960s, all Soviet satellite governments faced pressure to revise their ideological positions and programming policies to adjust to the opportunities and challenges presented by the new home-based mass medium. The launch of communication satellites, beginning with Sputnik-1 in 1957, the first Earth-orbiting artificial satellite and a key component of the Soviet space and communication strategy, increased the regimes' anxiety over their populations' access to Western programming. This challenge could only be minimized by re-channeling desires for capitalist lifestyles toward fostering national cohesion on the party leadership's terms. Socialist governments therefore began a strategic domestic production of scripted programming in the 1960s. Similar to Western European public broadcasting, television's shift to the center of public culture in the 1960s allowed Eastern European socialist governments to expand and solidify their nationalistic and Eurocentric educational-propaganda directives by wrapping those in increasingly entertaining forms.

Media and communication reforms in the 1960s–70s focused on television as the main institution for implanting socialist democratic values within entertainment. Television had to provide carefully selected information. It was also charged with shaping citizens' tastes so they understand and value Eurocentric art and culture and resist what were widely perceived as the detrimental effects of television: reducing appreciation for cultural quality and promoting mental and physical inertia.[4] But the party leadership was never entirely successful in deploying television toward these lofty goals. Rather, socialist parties played catch-up in their effort to understand and harness the potential of the mass medium of the Cold War. A gap consistently remained between top-down party policies and their translation into programs that became embedded into citizens' daily lives.

By the late 1960s–70s of the political thaw, as it became increasingly clear that the Soviet bloc was losing out in the economic competition with the West, party leaders shifted the arena of Cold War competition to "lifestyle". Television became a crucial venue for communicating this rhetorical shift. Soviet Central Television was charged with displaying what Leonid Brezhnev, at the Twenty Fifth Party Congress in 1976, defined as the "socialist way of life": "an atmosphere of genuine collectivism and comradeship, solidarity, the friendship of all the nations and peoples of our country, which grows stronger from day to day, and moral health, which makes us strong and steadfast" (Evans 2015). This new direction in Soviet ideology identified a moral, spiritual and emotional mode of existence unique to socialism and superior to capitalism. Television was to break up the boredom and ineffectuality of formulaic news and documentary programs and serve as a form of passionate counterpropaganda that would bring ordinary people and their stories of everyday heroism to the screen.[5]

In a similar vein, Paulina Bren describes the post-1968 normalization period's televisual turn in Czechoslovakia in terms of pursuing "a more qualitative socialist lifestyle" (2010: 207). But Bren also characterizes this turn as part of a European and even global shift in the 1970s and 80s toward moving the exercise of politics into the sphere of private, emotional relations. In Czechoslovakia, Hungary, Poland and Yugoslavia, where the opposition to Soviet occupation and cultural colonization was strong throughout the period, the affective turn within television took on less sentimental and more ironic and sarcastic tones than in the Soviet Union. During the thaw, socialist parties in these countries consolidated their leadership on the platform of cultural nationalism, often subtly positioned in opposition to the imperial Soviet regime.

The 1980s brought more economic and ideological differentiation among televisions in the countries of the region. At one extreme, Yugoslavia, Hungary and Czechoslovakia steadily increased both their import of Western programming and their domestic production of fictional programs and more liberal current affairs and news programs. This was done to satisfy viewer demand and to prove the regimes' tolerance toward a controlled plurality of viewpoints and willingness to usher in new economic measures that mixed socialist and capitalist features. In Poland, by contrast, the declaration of martial law in 1981, in response to the influence of Solidarity, reduced broadcast hours for a few years, until restrictions were lifted in 1983 (Mihelj and Huxtable 2018: 82). At the extreme of television scarcity, we find Albania, with scant and controlled domestic production but unable to prevent citizens' access to Italian broadcasting from across the Adriatic; and Romania, where broadcasting was reduced to 2–4 hours a day to save electricity, and featured mostly footage of dictator Ceauşescu and his family.

## Propaganda and Censorship

Those are probably the first ideas that spring to mind when one thinks about broadcasting under state socialism. And with good reason: "Agit-prop" departments and censors were active components of all socialist institutions. However, as I explain below, these activities did not necessarily translate into strict control or prohibition. Neither were such practices unique to Soviet-type societies. The Nazi government in Germany was already interested in the propaganda value of television, partly to compete with large US corporations. The 1936 Berlin Olympics was a stimulus to the development of German TV, with viewing rooms established in cities to hold audiences as large as 400 (Fickers and Bignell 2008a: 24). Television was formed under military dictatorship in Greece in 1967; and the second channel remained under direct army control and government censorship until 1982. Even the French government monitored and censored TV programs from the 1950s onwards. Under de Gaulle, this was direct political control, impacting every type of program, which also provoked direct resistance during the 1968 strikes, when some TV personnel resigned rather than agreeing to submit scripts ahead of time.[6] Using television as an instrument of positive propaganda also united Eastern and Western European broadcasters around educational initiatives, from teaching literacy in rural regions during the pioneering years of broadcasting to educating the population about a wide range of subjects once television matured into a mass medium.

The actual extent of censorship varied greatly within the socialist region, from almost no direct censorship over TV in Hungary to the heavily censored television of Ceauşescu's Romania. In much of the region, by the 1970s, drama programming was recognized as a more effective vehicle for affirming the regimes' ideological directives than the news and other factual programming, which were viewed with suspicion by the public. Comedic programs, in particular, regularly issued anti-Soviet criticism; although these could be seen as a strategically tolerated form of expression that, in fact, coopted both artists and audiences and ultimately served the parties' ideological interests as much as it undermined them.

Rather than being reduced to propaganda, socialist programming was highly influenced by the ethos of public service broadcasting (PSB), modeled on a common European ethical and ideological ideal, whose roots reach back to the pre-war era. As Jonathan Bignell and Andreas Fickers explain, the identity of European TV crystallized in the 1930s around what was to become a shared European public mission at the broadcasting technology exhibits of the 1937 World Fair in Paris. In this European model, spearheaded by German discourse, nation and education already appeared as twin key terms, in clear distinction from commercialization and entertainment, which characterized American ideas around television (Fickers and Bignell 2008b: 233). The main features, successes and difficulties of PSB were, indeed, embraced by all European television cultures. These included the government-led goal to inform and educate while promoting nationalism – a mission that was permanently challenged by the imperative to entertain (Keinonen 2012; Hickethier 2008) – and the nationalistic cultural hierarchy that assigned low value to television (see Adamou et al. 2008; Bourdon et al. 2008).

In addition, Soviet-type and Western European versions of socialism both drew on the values of cultural nationalism and a top-down intention to educate and enlighten all social classes. The differences between the two systems were less evident in the principles than in the degree of dogmatism with which they were put into televisual practice. More precisely, while the letter of Marxist-Leninist ideology continued to be reiterated in the Soviet-controlled region into the 1980s, its spirit operated much more closely to the ideological principles of Western European socialist democracies. This discrepancy between letter and spirit enacted an ongoing, performative repetition of the socialist order in the East. Around television, the discrepancy was particularly striking. The more elitist, austere, realistic and educational television attempted to be, the more it was mocked and abandoned by viewers, who wanted fiction, humor and entertainment. In essence, Eastern European socialist TV reproduced in more pure form and preserved well into the 1980s the educational principles and realistic aesthetic that Western European public broadcasters had already gradually abandoned under pressure from competition with commercial broadcasters.

## Financing and Advertising

To speak of advertising in state-controlled economies might strike one as a paradox. Indeed, advertising did not constitute a significant source of revenue in most countries. Television was mostly funded via license fees. The Soviet Union replaced these with a levy on TV sets in 1962 (Mihelj and Huxtable 2018: 82). Yet, television advertising did exist in every socialist country, although often in a contradictory, hybrid ideological dimension between centrally planned and market economies. The presence of advertising in itself is not that surprising given that, as Heather Gumbert (2013) succinctly put it, "the entire system was geared towards advertising itself." Competition was inherent to the structure of Soviet-type socialism since the system was built on utopian principles that needed constant fortification, justification and adjustment on the domestic front, particularly in light of ongoing comparisons with Western European social democracies.

TV ads bore the traces of these ideological negotiations in a range of aesthetic forms that would be characterized as awkward, funny or camp today. Much of earlier television advertising resembled public service announcements. These were often fairly lengthy and tedious short films about products or services or informational videos that warned citizens of the dangers of smoking or littering. DEFA, the East German film studios, began making TV commercials in the 1960s but suspended their production in the 70s because there simply wasn't enough stuff to advertise. However, GDR television continued to serve the economy by showcasing product placements that helped manage the inadequacies of socialist economic planning. For instance, they would feature herring recipes on cooking shows in order to move unsold herring out of the warehouses (Gumbert 2013; Beutelschmidt 2013; Oehmig 2013).

The largest supplier of ads for Soviet TV was Eesti Reklaamfilm (Estonian Film Advertising), which began making TV commercials in 1967 and continued to produce over 5,000 of these, advertising a variety of organizations, services and products over the company's 24-year run.[7] Soviet ads were broadcast in five-minute blocks that aired three times a day to minimize interruption. They became one of the highlights of the TV day, televisual events that gathered a massive cult following. As elsewhere, they displayed symptoms of central economic planning's weaknesses. Often they advertised products that were in short supply or nonexistent by the time the ad aired; or they were aired preemptively, in the hope that the product would be available in time. In a way that was true all through the region, what these ads lacked in production values and marketing savvy, they often made up in creativity.[8] Polish ads, in particular, bordered on avant-garde creations. They were often abstract and even absurd, especially when they promoted products that were not available in stores. In Poland, TV advertising began as early as 1956, initially limited to 15 minutes a week. The first ads were modeled after promotional segments developed for Polish Radio. It was not until the early 1980s, however, that an advertising market developed.

By the 1970s–80s, the most liberalized states had their own more or less developed marketing departments, which embedded TV advertising in an extensive set of commercial activities. Yugoslavia's relative ideological independence from the Warsaw Pact under Tito's leadership opened up greater financial independence for media elites. However, in exchange they had to supplement decreasing state revenue. As early as 1968, 20.8% of Slovenian Ljubljana TV's revenue came from advertising. First this meant advertising blocks; later, commercials began to interrupt programs, which provoked some backlash from viewers. Typical products advertised were food, cosmetics and services. Commercially funded programming also existed in Yugoslavia, in the form of sports and entertainment sponsorship. Slovenian TV Ljubljana and Croatian TV Zagreb established an official cooperation with the Italian RAI (Radiotelevisione Italiana) in the early 1960s. This happened despite protests from communist party authorities, who were anxious about the influx of Western news programs and the consumerist values transmitted by fictional programming (Pusnik and Starc 2008). Significantly, such unprecedented openness toward

Western-type entertainment was finally approved because it strengthened the national leadership's own strategy of championing Slovenian national autonomy against the encroachment of Yugoslavian federalism, which was pushed by the Yugoslav state broadcaster RTV (Pusnik and Starc 2008: 786). In Yugoslavia, federal revenue from advertising grew steadily in the 1960s and reached a peak of 23.1% of the total television budget by 1971 before it began falling throughout the 1970s (Mihelj and Huxtable 2018: 77). The deployment of advertising to harness market competition went hand in hand with the emergence of niche markets and audience research. The first studies of commercials' audience ratings were conducted in the 1960s, with the purpose of making commercials more effective (Mihelj 2013).

While Yugoslavia represented the most market-oriented variant, supplementing commercial state support with increasing commercial revenue was not uncommon practice among socialist broadcasters in the East. Hungarian TV and Radio established their joint marketing department in 1968 and engaged in extensive domestic and international marketing activities, including their participation in the European Group of Television Advertising (EGTA). The marketing department created an institutional framework around the commercial activities that had been features of Hungarian broadcasting from the start, what the institution's longtime commercial director Ilona Pócsik calls a "manager" type of thinking. The first advertisements appeared in the late 1950s, just a few years after television's launch in 1956. These first ads came out of an agreement between Hungarian Television and the advertising agency Magyar Hirdető ("Hungarian Advertiser"), which had been in the business of movie advertising since the 1940s. In the beginning, TV ads were run as text columns interspersed with pictures. Television's marketing arm also produced ads for other large state institutions, including labor unions (interview with Ilona Pócsik, December 2013).

Hungarian Television's marketing activities went well beyond the production of commercials. They encompassed concert organization, film, video and record production and distribution, book publishing, and trading film rights. As in the case of Yugoslavian TV, Pócsik and her team placed significant emphasis on interactions with viewers. For this purpose, they established a "Public Information" department in the 1970s, which was in charge of organizing venues to collect feedback from the viewing public across the country and responding to viewer letters and calls. Even though such outreach continued to run under the label of "propaganda," it more closely resembled public relations. In fact, such hybrid activities required the creation of hybrid categories such as "socially oriented propaganda," which referred to providing free or inexpensive informational outlets for socialist institutions that catered to public health or education. From the 1970s, the sponsorship of programs was also allowed.[9]

While its extensive marketing activity sat uncomfortably with socialist ideology, it was also clear that Hungarian Television and Radio increasingly relied on the revenue it produced. As Pócsik put it, "We searched for every possible source of commercial revenue. We took advantage of every opportunity that made profit." According to a report on Hungarian Radio-TV's marketing activities, created in 1988 for the 20th anniversary of the Marketing Department, during its two decades the department increased its annual revenue 40-fold, from 3.9 million to 200 million Hungarian forint; and the number of its staff from 5 to 80 (ibid.).

Socialist advertising *is* an oxymoron. As one of the many articles that responded to the boom in TV ads in 1970s Hungary cautions:

> We need strictly to adhere to the codes of socialist advertising ethics. Advertising, a tool of market competition, cannot lead to actions that hurt the interests of workers and companies. It's not allowed to mislead consumers and it's not allowed to favor or disfavor companies.
>
> *(Varga 1970)*

Clearly, socialist commercials were situated at contradictory ideological crossings. They were awkward reminders of a market-based economy operating within the matrix of centralized planning. They sold services that had no competition, or sold products that were often scarce, nonexistent, or in excess. Socialist ads thus capture the contradictions of an era in a way that most histories are not able to. This is most likely why they triggered so much nostalgic remembrance since the end of socialism. It is precisely due to the contradictions that they embody that they act as "authentic" manifestations of lived socialism.

## International Exchanges and Networks

Far from being sharply divided by the Iron Curtain, Europe's communication space during the Cold War was interwoven by broadcast cooperations, exchanges of technological expertise and knowhow, program trade and unofficial borrowing. In addition, broadcast signals that reached across borders allowed sizeable populations to watch neighboring countries' programming.

The history of European broadcasting cooperation goes back to the interwar years. It was first necessitated by technical issues: before World War II, there were different technical standards for the number of lines and images per second as a way to protect the national industries of TV set makers (Adamou et al. 2008: 91). The International Broadcasting Union (IBU), founded in 1925 to regulate international broadcasting, was the first international, non-governmental broadcasting institution to harness the cross-border nature of radio waves. It was broken up after the war partly due to the abuse of its technical facilities by the German army and partly under challenge by the establishment of the Organisation Internationale de Radiodiffusion et de Télévision (OIRT) in 1946, which became the dedicated "eastern" network (Fickers and Bignell 2008: 27). The Western response to the OIRT was the European Broadcasting Union (EBU), founded in 1950, in an effort led by the BBC to form a union for Western European broadcasters. The EBU initially included 23 broadcasters, but soon expanded to include some Mediterranean and Middle Eastern companies. The OIRT and the EBU eventually merged in 1993 with the goal of fostering a shared European consciousness (ibid.: 78–100).

Despite the existence of two dedicated broadcast organizations, the interconnectedness of television systems across the Iron Curtain was ensured by ongoing exchanges between the two from the start (see Gumbert 2014b; Beutelschmidt and Oehmig 2014). Both institutions created sub-organizations in charge of program exchanges in 1954. The OIRT's sub-division was called Intervision and the EBU's Eurovision. Both organizations fostered intra-regional cooperation but also exchanges between the two regions, in a spirit of friendly competition (Mihelj and Huxtable 2018: 71). In addition, some countries actively disrupted the political alliances, or, from another point of view, bridged the two networks. Yugoslavia chose to join the EBU and Eurovision to assert its independence from the Soviet Union (ibid.: 75), and Finland, balancing between Eastern and Western interests throughout the Cold War, was a member of both the EBU and the OIRT (Keinonen 2012). The OIRT also included some non-European allies such as Mongolia (joined in 1972), Cuba (1979), Afghanistan and Vietnam (both 1982).

Collaboration between the two organizations became increasingly active as the Cold War thawed: The first discussion of exchanges happened in 1956; the first Eurovision-Intervision transmission, from the Rome Olympics, in 1960 (Fickers and Bignell 2008b). Program exchanges were particularly extensive in children's television (Adamou et al. 2008: 94), where Eastern European broadcasters made a very significant although still rarely acknowledged contribution. The program exchanges ensured that most of Europe was watching many of the same programs, at times simultaneously.

This trend only intensified in the 1970s–80s. Programs popular in the East and West ranged from German crime dramas such as *Tatort* through American drama and soap serials such as *Colombo, Star Trek, The Fugitive, Kojak, Charlie's Angels, Starsky and Hutch, Roots, Isaura, Dynasty* and *Dallas* to more high-brow fictions such as the German *Heimat* and the US series *Holocaust*. In most Eastern and Western European countries, the gap between domestic supply and audience thirst for American-style drama caused a great deal of consternation about US media imperialism's threat to national cultures and to public broadcasting. The exception was Britain, where a cross-pollination of imports and format exchanges with the US had been ingrained in television from the beginning (Bjondebjerg et al. 2008).

As Sabina Mihelj (2012) shows, as early as the 1960s, the proportion of foreign programs in Eastern Europe (excluding the Soviet Union) ranged from 17% in Poland to 45% in Bulgaria. In the early 1970s, 12% of all imported programming on Hungarian television came from the UK, 10% from France, and 10% from Western Germany. In non-aligned Yugoslavia, a full 80% of all imported programs came from outside of the socialist bloc and 40% from the US alone. In the early 1980s, an average of 43% of imported programming in Eastern Europe came from Western Europe, almost equaling the 45% from the Soviet Union and other Eastern European countries (ibid.). Even in dictator Enver Hoxha's Albania, a period of relative liberalization began in the 1970s (Idrizi 2013). In Romania, where television also experienced a regression under dictatorial control, selected US content and genre models were nevertheless welcome – not least thanks to Ceauşescu's attempt to demonstrate his Western orientation and independence from Moscow.[10] It was only in 1985, as the regime became especially isolated and paranoid that the temporary second channel set up in 1968 ceased to operate and broadcasting hours were reduced from 10–11 to 4–5 hours a day on weekends and to 2 hours on weekdays. *Dallas* was one of the last programs to go. It was broadcast on Saturdays into the early 1980s, although severely censored and shortened (Bjondebjerg et al. 2008: 177–178).

In a scenario that was more typical, in 1985 the rate of imported programs in Hungary was 40%, 80% of which came from the US. It was hard to compete with the price of these imports: a 60-minute imported drama episode cost an average of $1,000, or 60,000 Hungarian forint, while it would have cost 4–5 million forint to produce domestically. About a quarter of the population had a choice of programming beyond Hungarian TV's offerings through access either to a VCR or to foreign channels. By the mid-80s, the estimated number of VCR devices was 100,000 – that is, one attached to about every fifth TV set (Horvát 1986). By the late 80s, Hungarian TV was woefully unequipped to meet the demand for TV entertainment. Its technical facilities did not allow for transmitting more than 95–100 hours of programming a week between the two channels.[11] In 1986, the first satellite channel, Sky Channel, a holding of Rupert Murdoch's News International, premiered at the Budapest Hilton before it was made available to households. A total of 7.1 million Western European households were able to access Sky Channel at this time.[12]

The greatest political risk involved in the expansion of television broadcasting was that, unlike feature films or print publications, broadcast signals could not be confined to state borders. Inhabitants of large regions in Yugoslavia, East Germany, Czechoslovakia, Hungary and Albania received either Austrian, Italian or West German programming. Shared TV signals had the most profound effect in East Germany, where viewers – with the exception of the "Valley of the Clueless" near Dresden, where foreign signals did not reach – were able to view West German broadcasting in a shared language, often specially directed at East German viewers. But border-crossing signals literally disrupted communism even in the most isolated corners of the communist empire. In Albania, while the communist elite retained to itself the privilege to watch foreign broadcasts in the 1960s, the brief period of relative liberalization in the early 1970s removed restrictions on the general population when it came to consuming Italian TV. By 1973, the party leadership realized that the pleasures of Italian programs far outweighed their potential to demonize capitalism; so they installed signal jammers. These proved ineffective, however, especially in border areas where one did not need antennae to receive signals (Idrizi 2013; see also Carelli 2014).

In a similar vein, TV-deprived Romanians took advantage of the overspill of terrestrial broadcast signals to watch more liberalized Hungarian, Bulgarian and Yugoslav programming in the 1970s–80s (Mustata 2012). Annemarie Sorescu-Marinkovic's ethnographic research conducted in the Banat region (bordering Yugoslavia and Hungary) shows that it was not only the significant bilingual and trilingual populations who watched foreign TV. Romanian viewers went through the trouble of learning foreign languages just to access a slice of the outside world. Ironically, the nationalistic restriction on physical mobility across borders motivated a TV-mediated flow of exchange that recreated, at least virtually, the multilingual, multiethnic culture the region inherited from the Habsburg Empire (Sorescu-Marinkovic 2013).

Romanian viewers' reactions to Yugoslav and Hungarian TV echo Annika Lepp and Mervi Pantti's interviews with Estonian viewers, who were fixated on Finnish programming in the 1970s and 80s, after the Finnish broadcaster YLE built a new TV broadcast transmitter in Espoo, which brought Finnish TV to households in northern Estonia. Even though only two Finnish public broadcast channels existed during this period, these carried American and Western European series. This opportunity even created a weekend "TV tourism" from other parts of Estonia to the northern areas so fans could follow their favorite serials. Many people who lived in the north were also able to learn the Finnish language from TV, which lent them cultural and economic capital as trade with Finland expanded in the 1980s and 90s (Lepp and Pantti 2012). Television thus facilitated cultural identification with Nordic Europe, re-establishing the Baltic region's older imperial ties with Scandinavia in a quest for independence from the Soviet Union.

## Program Types

One of the main strategies socialist authorities employed for message control was to insist on a line of separation and strict value hierarchy between factual and fictional programming. The only exception to the lower value assigned to TV fiction was art. But what qualified as art was also determined by ideological principles: although it did not have to be realistic, it was assumed to be educational, contributing to the central mission of "taste training." "Art" programming had to demonstrate social commitment to the cause of the nation, to European high culture and to the future of socialism – for instance, in the form of moralizing historical teleplays and theatrical broadcasts of European classics.

This value order was so marked that, browsing print sources about television programs from the 1960s–70s, one would hardly know that non-fiction and non-artistic programming even existed – except for the occasional short piece lamenting the popularity of imported dramatic serials, or expressing hope for improving taste education in the future. TV as a public educational ideal and TV as it actually functioned in the living rooms were clearly distinct during these decades.

In its early postwar years, while party authorities were trying to figure out the uses and potentials of the new medium, television was imagined as a massive school for the masses. Every program was imbued with the intention to educate. Tele-education was seen as a key to the citizenry's erudition, from academic and ideological training through learning a variety of practical skills to the all-important but most contradictory goal of "taste cultivation" – something rooted in the Kantian idea of aesthetic education. Programs varied as to how explicit this educational intent was. In its most direct educational form, television actually functioned as a people's classroom in the various national installments of School TV. School TV was a Europe-wide movement, rooted in the model of collective viewing organized around the first, scattered TV sets (Fickers and Bignell 2008a: 24). In Italy, the program *Non é mai troppo tardi* ("It's Never Too Late") was financed by the Italian Ministry of Education to promote literacy in rural regions. It was broadcast by RAI (Radiotelevisione Italiana) from 1959 to 1968 for a mostly older audience, who watched it weekly in classrooms or municipal buildings. The French télé-clubs are perhaps the most influential examples of education in the context of collective viewing. They began in 1951 and operated mostly in rural primary schools, where people could attend TV transmissions by the state broadcaster RTF once or twice a week (Wagman 2012). School TV also drew on antecedents in radio such as the BBC's Listening Groups (1927–47), which offered early adult education; farm radio programming based in Canada for adult education produced by the Canadian Broadcasting Corporation; and farm radio produced in cooperation between UNESCO and All India Radio in Poona, which was followed by similar radio experiments in the developing world. Educational radio broadcasting aimed at school children also preceded public service educational television in Belgium, Switzerland, Germany, Portugal and other countries throughout the 1920s, 30s and 40s (ibid.: 123).

In the socialist region, School TV was launched in the 1960s. In most countries, it complemented and then replaced School Radio. It is important to note that, even in the Eastern bloc, School TV was not seen as simply, or even primarily, a propaganda instrument. Rather, it was considered a tool of social transformation that could bring knowledge and cultural sophistication to the masses by opening a window to the world in tiny schoolhouses, community centers and eventually family homes. The knowledge to be disseminated was to be a grand equalizer among different social classes, professions and educational levels. However, socialist TV soon outgrew the confines of the TV-school model and embarked on other forms of education better suited for the medium. In fact, much of postwar Eastern European public television was meant to teach its viewers how to read and write, to understand math, physics and geography, to appreciate fine national and European literature, film and music, and Marxist-Leninist philosophical principles. But it also offered a broader educational program in how to raise children and navigate legal issues, in cooking, sewing, gardening, agricultural work, mining and operating heavy machinery. It prioritized genres most closely associated with Western European PSB, which also happened to be in the center of socialist programming. These are genres that valued politically committed documentary realism above fictional representations. The aesthetic of socialist realism they adopted was supposed to support an overarching educational intention to teach viewers how to be good socialist citizens.

Their pairing of realist aesthetic with pedagogical intent is not unlike the self-professed profile of contemporary reality programs. The difference is that the latter are in the business of neoliberal citizenship education. They teach us how to behave, dress, build, decorate, raise children, lose weight and gain confidence. Socialist TV was driven by noble initiatives to democratize access to education, to create a level playing field among people of different class and educational backgrounds, and to socialize the individual as always primarily a community member. In this regard it was the opposite of contemporary reality shows, which tend to isolate the individual and teach the virtues of self-help. Socialist television's contribution to the prehistory of reality-based programming is not only significant in itself but also offers a critical counterdiscourse to reality TV's ideological presumptions and academic assessments.

By the 1970s, more and more educational programs began to solicit viewers' emotional engagement and participation, employing a playful or humorous tone, embedding their lessons in competitive formats, mixing live action with animation and studio conversations with dramatic re-enactments, and employing well-liked, entertaining personalities as guides and moderators instead of scholars. In the long run, these proved to be the most enduring formats, many of which survived socialism. Such was the Romanian *Teleenciclopedia*, a general interest magazine, which mixed fiction films with short documentaries featuring the voices of popular actors, singers and other celebrities as narrators. It took a thematic approach to topics of general knowledge. Since it was neither elitist nor exclusivist, *Teleenciclopedia* actually popularized scientific knowledge in Romania. The balance it found between informative, educational programming and popular, general topics of interest guaranteed its survival through the 1980s, when programming was reduced to two hours a day (Mustata 2011). The Hungarian program *Delta*, launched in 1964, was similar to *Teleenciclopedia*. A long-running popular TV magazine that covered mostly scientific and technological topics of popular interest, *Delta* was hosted by the charming

Júlia Kudlik, one of the most universally liked announcers and celebrities of the entire period. Polish TV launched its first current affairs program, *Tele-Echo*, in 1956. It remained on the air until 1981. Like *Delta*, it featured a beloved hostess, Irena Dziedzic, who was also the creator and writer of the program. *Tele-Echo* covered current affairs, arts and culture in an interview format. During its 25-year history, *Tele-Echo* hosted over 12,000 guests. While Dziedzic – much like Kudlik – was often and rightly accused of political conformism, she was also a phenomenal TV personality, who was able to evoke intense emotional reactions (Pikulski 2002: 54). Her role was similar to that of legendary Soviet Central Television hostess Valentina Leonteva, who helmed the melodramatic talk show *Ot vsei dushi* ("From the Bottom of Our Hearts") between 1972 and 1987, a program that played a central role in visualizing a "Soviet way of life" through the emotional display of a "Soviet way of feeling" (Evans 2015).

Such programs successfully combined documentary realism with fiction to reach into viewers' actual lives, making television the most vivid and organic platform for public engagement. The desire to use television as a live public forum brought forth other, more directly politicized, region-wide formats. The tellingly titled Hungarian program *Fórum* (1969) bypassed lectures and propaganda and invited artists, intellectuals and politicians to a live town hall meeting held in varying locations. It put party leaders in front of the cameras and connected them with actual viewers, who asked questions about the economy and its reforms, political issues and foreign relations. In a similar vein, *Current Debates* in Yugoslavia (TV Belgrade 1965–69) was a participatory discussion program that revolved around current issues based on audience suggestions, including unemployment, living standards and political reforms. The East German *Prisma*, which ran from 1963 to 1991, was modeled after the West German current affairs magazine *Panorama*. Unlike *Panorama*, however, which focused on large-scale political issues, *Prisma* was positioned as a liaison between the party leadership and ordinary citizens, inviting viewers to contribute questions, comments and complaints about a variety of issues that affected everyday socialist lives (Gumbert 2014a: 146).

Other program types that successfully balanced socialist TV's educational mandate with an engaging format were quiz, game and variety shows. The generic boundaries among these types tended to be unclear; but they had competition and participation as their emotional engines. Much like on Western European public service television, quiz shows were some of the earliest genres on socialist TV, introduced as part of live broadcasting in the late 1950s. If we trace their history, we see a transformation from the early, open-ended formats of the 1950s to more or less centralized attempts at instituting more rules and controls, which were to adjust the genre to serving the mass-educational policies of the 1960s. The earliest formats arose in an era of fairly low regulation and high confusion among socialist parties as to the purposes, potentials and dangers of the new medium. This uncertainty gave TV professionals some leeway to experiment with the genre, which was cheap to produce. As entertainment increasingly came to define television in the 1970s and 80s, quiz and game formats bore more and more of the pressure from capitalist competition and viewer demand. Direct, codified format borrowings began on a large scale in the 1990s (Dankovics 2012).

Self-directed, lifelong learning is a very contemporary idea, as is the notion that competitive play increases motivation and participation. In the Soviet Union of the 1960s, hundreds of groups were formed to create local versions of popular quiz and talent shows around the famous Soviet sketch comedy competition show, Soviet television's most successful game show of all time, *Club of the Merry and Quick-Witted* (*Klub veselykh I nakhodchivykh* or *KVN*, 1961–), described by its co-creator Sergei Muratov as "intellectual soccer" (Evans 2016). It inspired regional competitions among KVN teams in factories, schools, agricultural collectives, the armed forces and many other groups. Thus, it significantly fostered youth mobilization and mass recreation. Some *KVN* stars became celebrities. The program was discontinued in 1972 but was resurrected during *glasnost* and is still broadcast in Russia.

Such competitive game formats also highlight fundamental contradictions embedded in the structure of socialism. In the first place, since they were not domestically developed for the most part, they reveal socialism's intricate international connections with capitalism. They were often formats more or less directly borrowed from Western European public service broadcasters and, like those broadcasters, also incorporated elements of American commercial variants. Others, such as musical talent shows, were also seen by governments as prime export products. These competitive formats contribute to an alternative prehistory of reality-based programs and pose a challenge to our ideas about socialism.

When it comes to programs of fiction, two signature genres stand out: historical dramas and dramatic serials that I elsewhere call "socialist soaps" (Imre 2016). From the late 1960s through the 1970s, during their heyday, domestically produced historical adventure serials allowed socialist regimes to teach selective history lessons and foster national identifications that also appeared to conform to ideological prescriptions demanded by the Soviet occupiers. These serials often ostentatiously demonstrated adherence to Soviet socialist dogma in formal features such as the glorification of folk culture or plotlines that rewarded peasant characters at the expense of the wealthy and powerful. Socialist historical drama serials participated in a European circulation that reached across the Iron Curtain. From the beginning, most East European viewers were exposed to a number of successful Western European historical serials, which were safe to import for socialist TV. Many were literary adaptations with a strong cultural pedigree, telling stories about the nation, often dealing with the cultural roots and formative moments of history.[13] They tended to have high production values to match their cultural quality (De Leeuw et al. 2008: 134).

While the historical adventure series of the 1960s–70s revolved around patriotic men's heroic deeds in the distant past, "socialist soap operas" took place in the present of 1970s–80s late socialism, and featured the family as a microcosm of the socialist nation. They reflected socialist parties' increased attention to women, consumerism and the ideological power of emotional engagement. Similar to historical adventure shows, domestic serials continued the project of political education in less didactic, more entertaining ways.[14] They also took for granted viewers' familiarity with and yearning for imported drama serials. Much like Western soap operas, socialist soaps showcased communities that were metaphors for modern society, offering fictional "feminine" pleasures of identification (Ang and Stratton 1995). In other ways, however, the "socialist soap" was a peculiar hybrid specific to the conditions of late socialism. Rather than addressing women only as consumers, it addressed them first and foremost as citizens, whose biological features simply assigned them to unique roles within the socialist collective. It was also distinct in tone and aesthetic as it absorbed the influence of other socialist TV genres, most prominently the didacticism of public affairs programs and the satirical tone of comedy shows. While socialist soaps were not overtly politicized, they all modeled ideal socialist lifestyles in ensemble dramas that encompassed the workplace and the family. They featured central female characters who acted as problem-solvers, linchpins between the public and private worlds.

Nowhere was the production of soap operas as prolific and profitable as in Czechoslovakia. Dramatic serials date back to 1959 when *Rodina Blahova* ("Family Blaha") began broadcasting live once a month. The most popular of these were created by the writer-director duo Jaroslav Dietl and Jaroslav Dudek during the thaw period and circulated within and beyond the region. By the 1970s, the genre settled into a one-hour format, with four 12–13-episode serials covering the entire year (Bren 2010: 126). Dietl's most popular soap, *Hospital at the Edge of Town* (*Nemocnice na kraji města*, 1977) drew an average viewership of 88% and was also a roaring success in many other Soviet bloc countries as well as West Germany (ibid.: 144), the Soviet Union, Poland, Romania, Hungary, Bulgaria, Yugoslavia and Cuba. It was also broadcast in Austria, Switzerland, Finland, Cyprus, Australia, China and Afghanistan (Bednařík 2013). While the program's original purpose was to promote socialist healthcare, its popularity and translatability were, ironically, due to its utopian depiction of the socialist hospital: the well-equipped institution with kind doctors who care about their patients more closely resembled American medical soaps.

The soap opera genre has been described as escapist or even utopian, depicting a universe of privileged consumption (Geraghty 1991). The Dietl serial with the most direct gendered address, *Žena za pultem* (*Women Behind the Counter*, 1977), fully indulges in this utopianism. The protagonist, Anna Holubova, works in the delicatessen section of a grocery store, behind a stack of canned caviar and other luxury items that hardly do justice to the more austere shopping conditions of the 1970s. But the show stands out in the first place because it was designed to demonstrate the success of state feminism (Machek 2010). The feminized workplace setting allowed the 12-episode program to focus on a female collective and thus model women's central roles in socialist society.

As elsewhere in the region, the post-68 period of normalization had brought about a crisis of masculinity and foregrounded women's role as a major anchoring power for nationalism in Czechoslovakia (ibid.: 169). One would see men bumbling and constantly in need of female help in domestic serials. However, one would *not* see them cleaning or taking care of children. Ultimately, the socialist soap confirmed the gendered status quo of state feminism, which further naturalized the traditional division of roles while it vastly expanded women's workload. Instead of recognizing the enduring patriarchal scaffolding of nationalism as the problem, women tended to blame their extraordinary burden on the socialist system and its failed feminism (ibid.: 47), in a tragic misrecognition that helped demonize feminism as an alien construct.

## Conclusion

Recent work on the history of socialist TV has been removing the Cold War filter that has occluded alternative histories of European television. Instead of a clear-cut East–West divide, a much more significant difference emerges between what Bourdon et al. (2008) call the "courteous" European model that reigned through the 1980s and the "competitive" American model that subsequently challenged European broadcasters to shift to a dual broadcasting mode in a commercial, multi-channel, deregulated global environment. In this revised historical view, Eastern European broadcasters were aligned with the European model but were simply a little slower to join the competitive model. Nevertheless, competition always underscored their operations. Socialist state broadcasters always had to compete for eyeballs distracted by foreign broadcasts and had continually to increase the volume of imports needed to fill schedules on state channels (see Mihelj and Huxtable 2018). Socialist TV developed in synchronicity with European television from interwar experimental broadcasts through a postwar relaunch in the 1950s, to adopting the principles of educational-nationalistic broadcasting from European public service media and gradually shifting to an entertainment-focused model by the 1970s of the political thaw. Throughout this history, socialist TV consistently participated in European media flows. The history of socialist TV in Europe is an integral part of global television history.

## Notes

1 1936 in Hungary and 1937 in Poland.

2 Danish and Belgian broadcasting in 1953, Spain in 1956, Sweden and Portugal in 1957, Finland in 1958, Norway in 1960, Switzerland in 1958, Ireland in 1961, Gibraltar and Malta in 1962, and Greece in 1966 (Hickethier 2008).

3 In Hungary, where regular broadcasting began in 1957, the number of programming hours a week jumped from 22 to 40 between 1960 and 1965. In Czechoslovakia, trial public broadcasts began in 1953. The rapid increase in television access in the 1960s played a central role in the liberalization of the country's political climate. This liberalization was arrested following the Prague Spring of 1968, which was brutally crushed by the Soviet Union. Television Romania was established in 1956 and added a second channel in 1968, which was then suspended in 1985 due to dictator Ceauşescu's energy-saving program until after 1989.

4 For instance, to conform to these directives, in 1968 Hungarian Television (MTV) divided its programming among different departments this way: art films and programs that promoted cultural appreciation made up 30% of all programs; 9% of broadcast time went to literary and dramatic programming; news programs, responsible for political agitation, comprised 29%; youth and children's programming made up 11.5%; and informational programming such as nature documentaries took up 2.5%. In addition, a daily morning programming block called "School Television" made up 11%. With the addition of Friday, a sixth day of weekly programming, that year (Monday remained a non-broadcast day devoted to work), programs were reorganized so that each weekday had a distinct educational profile. Entertainment programs were concentrated on the weekend. See Horváth, Edina. "A magyar televízió müsorpolitikája – 1968." Hungarian Television (MTV) Archives. www.tvarchivum.hu/?id=279930

5 Evans (2015) analyzes the popular Soviet program *Ot vsei dushi* (*From the Bottom of My Heart*), launched in 1972 and hosted by legendary Central Television hostess Valentina Leont'eva, as the most evocative showcase of how television's emotional realism came to be deployed to represent the Soviet socialist way of life as a way of feeling.

6 At the same time, commercial radio and the press were free to criticize the government, which ensured some pluralism of opinion (Adamou et al. 2008).

7 "Soviet TV Advertisements From the 1970s and 1980s," http://boingboing.net/2013/02/17/soviet-tv-advertisements-from.html.

8 "The Only Anthology of Retro Soviet TV Commercials," www.retrosovietads.com/.

9 "Az RTV Belkereskedelmi Igazgatóság tevékenysége 1968–88 között" ("Report on the Activities of Hungarian Television and Radio's Commercial Department 1968–88"). Internal departmental correspondence, courtesy of Ilona Pócsik.

10 As is widely known, Ceauşescu built his legitimacy on currying favor with Western countries. He visited with Nixon, and even gained most favored nation status in the US in 1975.

11 L.G., "Másfélből kettő," *Magyar Nemzet*, 20 September 1988.

12 "Beszélgetés a műholdas televíziózásról. Megnyíló égi csatorna," *HVG*, 8 November 1986.

13 For a detailed analysis of Eastern European historical drama serials such as the Polish *Janosik* and the Hungarian *The Captain of Tenkes*, see Imre 2012.

14 As Paulina Bren (2010: 125) points out, this was not an isolated strategy. Under Brazil's military rule, Globo teamed up with the government to showcase upscale lives in telenovelas as a sign of modernization and upward mobility.

## References

Adamou, Christina, Isabelle Gaillard, and Dana Mustata (2008) "Institutionalizing European Television: The Shaping of European Television Institutions and Infrastructures," in *A European Television History*, eds. J. Bignell and A. Fickers (New York: Blackwell).

Ang, Ien, and Jon Stratton (1995) "The End of Civilization As We Knew It: *Chances* and the Postrealist Soap Opera," in *To Be Continued … Soap Operas around the World*, ed. Robert C. Allen (London and New York: Routledge), 122–144.

Bednařík, Petr (2013) "The Production of Czechoslovakia's Most Popular Television Serial The Hospital On the Outskirts and Its post-1989 Repeats," *VIEW Journal* 1(3). Available at: http://ojs.viewjournal.eu/index.php/view/article/view/jethc029/54

Beutelschmidt, Thomas (2013) "East German TV and Global Transfers." Paper presented at the conference "Television in Europe beyond the Iron Curtain – National and Transnational Perspectives since the 1950s," Friedrich-Alexander-Universität Erlangen-Nürnberg, December 5–7.

Beutelschmidt, Thomas, and Richard Oehmig (2014) "Connected Enemies? Programming Transfer Between East and West during the Cold War and the Example of East German Television," *VIEW: Journal of European Television History and Culture* 3(5): 60–67.

Bjondebjerg, Ib, Tomasz Goban-Klas, Michele Hilmes, Dana Mustata, Helle Strandgaard-Jensen, Isabelle Veyrat-Masson, and Susanne Vollberg (2008) "American Television: Point of Reference or European Nightmare?" in *A European Television History*, eds. J. Bignell and A. Fickers (New York: Blackwell), 154–183.

Bourdon, Jerome, Juan Carlos Ibanez, Catherine Johnson, and Eggo Mueller (2008) "Searching for an Identity for Television: Programmes, Genres, Formats," in *A European Television History*, eds. J. Bignell and A. Fickers (New York: Blackwell), 101–126.

Bren, Paulina (2010) *The Greengrocer and His TV: The Culture of Communism after the 1968 Prague Spring* (Ithaca, NY: Cornell University Press).

Brunsdon, Charlotte (2008) "Is Television Studies History?" *Cinema Journal* 47(3): 129.

Carelli, Paolo (2014) "Italianization Accomplished: Forms and Structures of Albanian Television's Dependency on Italian Media and Culture," *VIEW: Journal of European Television History and Culture* 3(5): 68–78.

Čulik, Jan (ed.) (2013) *National Mythologies in Central European TV Series: How J.R. Won the Cold War* (Eastbourne: Sussex Academic Press).

Dankovics, Noémi Dóra (2012) "Játékmesterek és vetélkedőműsorok a magyar televíziózás történetében" ("Quiz Show Hosts and Game Shows in the History of Hungarian Television." PhD diss., Budapest College of Communication and Business.

De Leeuw, Sonja, Alexander Dhoest, Juan Francisco Gutierrez Lozano, Francois Heinderyckx, Anu Koivunen, and Jamie Medhurst (2008) "TV Nations or Global Medium? European Television Between National Institution and Window on the World," in *A European Television History*, eds. J. Bignell and A. Fickers (New York: Blackwell), 127–153.

Evans, Christine (2015) "The 'Soviet Way of Life' as Way of Feeling: Emotion and Influence on Soviet Central Television in the Brezhnev Era," *Cahiers du Monde Russe* 56(2–3).

Evans, Christine (2016) *Between Truth and Time: A History of Soviet Central Television* (New Haven, CT: Yale University Press).

Fickers, A., and C. Bignell (2008a) "Introduction: Comparative European Perspectives on Television History," in *A European Television History*, eds. J. Bignell and A. Fickers (New York: Blackwell), 1–54.

Fickers, A., and C. Bignell (2008b) "Conclusion: Reflections on Doing European Television History," in *A European Television History*, eds. J. Bignell and A. Fickers (New York: Blackwell), 229–256.

Geraghty, Christine (1991) "Soap Opera and Utopia," in *Cultural Theory and Popular Culture: A Reader*, ed. John Storey (Harlow: Pearson Press), 246–254.

Gumbert, Heather (2013) Paper presented at the conference "Television in Europe beyond the Iron Curtain – National and Transnational Perspectives since the 1950s," Friedrich-Alexander-Universität Erlangen-Nürnberg, December 5–7.

Gumbert, Heather (2014a) *Envisioning Socialism: Television and the Cold War in the German Democratic Republic* (Ann Arbor: University of Michigan Press).

Gumbert, Heather (2014b) "Exploring Transnational Media Exchange in the 1960s," *VIEW: Journal of European Television History and Culture* 3(5): 50–59.

Hickethier, Knut (2008) "Early TV: Imagining and Realising Television," in *A European Television History*, eds. J. Bignell and A. Fickers (New York: Blackwell), 55–78.

Horvát, János (1986) "Külföldi műsorok a magyar képernyőn," *Pártélet* (August–September): 89–93.

Idrizi, Idrit (2013) "Das magische Gerät. Die Bedeutung des Fernsehers im isolierten Albanien und für die Erforschung des albanischen Kommunismus." Paper presented at the conference "Television in Europe beyond the Iron Curtain – National and Transnational Perspectives since the 1950s," Friedrich-Alexander-Universität Erlangen-Nürnberg, December 5–7.

Imre, Aniko (2012) "National History and Cross-National Television Edutainment," *Journal of Popular Film and Television* 40(3): 119–130.

Imre, Aniko (2016) *TV Socialism* (Durham, NC: Duke University Press).

Imre, Aniko, Timothy Havens, and Katalin Lustyik (eds) (2012) *Popular Television in Eastern Europe during and since Socialism* (London: Routledge).

Keinonen, Heidi (2012) "Early Commercial Television in Finland: Balancing between East and West," *Media History* 18(2): 177–189.

Lepp, Annika, and Mervi Pantti (2012) "Window to the West: Memories of Watching Finnish Television in Estonia during the Soviet Period," *VIEW: Journal of European Television History and Culture* 3(2): 76–86.

Machek, Jakub (2010) "'The Counter Lady' as a Female Prototype: Prime Time Popular Culture in 1970s and 1980s Czechoslovakia," *Media Research* 16(1): 31–52.

Mihelj, Sabina (2012) "Television Entertainment in Socialist Eastern Europe: Between Cold War Politics and Global Developments," in *Popular Television in Eastern Europe during and since Socialism*, eds. Aniko Imre, Timothy Havens, and Katalin Lustyik (London: Routledge), 13–29.

Mihelj, Sabina (2013) Keynote address. Paper presented at the symposium "Television Histories in (Post) Socialist Europe," Stockholm, November.

Mihelj, Sabina, and Simon Huxtable (2018) *From Media Systems to Media Cultures: Understanding Socialist Television* (Cambridge: Cambridge University Press).

Mustata, Daniela (2011) "The Power of Television: Including the Historicizing of the Live Romanian Revolution." PhD diss., Utrecht University.

Mustata, Daniela (2012) "Television in the Age of (Post)Communism: The Case of Romania," *Journal of Popular Film and Television* 40(3): 131–140.

Oehmig, Richard (2013) Paper presented at the conference "Television in Europe beyond the Iron Curtain – National and Transnational Perspectives since the 1950s," Friedrich-Alexander-Universität Erlangen-Nürnberg, December 5–7.

Pikulski, Tadeusz (2012) *Private History of Public Television* (Warsaw: Muza).

Pusnik, Marusa and Gregor Starc (2008) "An Entertaining (R)evolution: The Rise of Television in Socialist Slovenia," *Media Culture and Society* 30(6): 777–793.

Roth-Ey, Kristin (2011) *Moscow Prime Time: How the Soviet Union Built the Media Empire that Lost the Cultural Cold War* (Ithaca, NY: Cornell University Press).

Sorescu-Marinkovic, Annemarie (2013) "We Didn't Have Anything, They Had It All. Watching Yugoslav Television in Communist Romania." Paper presented at the conference "Television in Europe beyond the Iron Curtain – National and Transnational Perspectives since the 1950s," Friedrich-Alexander-Universität Erlangen-Nürnberg, December 5–7.

Szostak, Sylwia (2012) "Poland's Return to Europe: Polish Terrestrial Broadcasters and TV Fiction," *Journal of European Television History and Culture* 1(2):80. Available at: http://ojs.viewjournal.eu/index.php/view/article/view/jethc021/43

Varga, György (1970) "Műsoron a hirdetés," *Figyelő*, 30 September.

Wagman, Ira (2012) "Tele-Clubs and European Television History beyond the Screen," *VIEW: Journal of European Television History and Culture* 1(2): 118. Available at: www.viewjournal.eu/articles/113/

# 33
# ARAB TELEVISION INDUSTRIES
## Enduring Players and Emerging Alternatives

*Joe F. Khalil*

When compared to Arab press, radio, and film, television has been the most resilient form of mass media throughout the rise of digital platforms. With more than 300 million Arabs in an estimated 50 million households with TV sets, Arab television is at the crossroads of national, regional, and global state and business interests (Dubai Press Club 2017). Between January 2004 and August 2017, free-to-air satellite television channels grew by 848% to more than 1,000 (Al Atiyat and Khreisha 2017). In 2017, Pay TV made more than USD 1 billion in profit driven by sports and movies (Digital TV Research n.d.). But for many in the industry, television today seems very similar to that of the 1990s: major production activities in hubs, a very demanding young viewership, and a competition between free and paid distribution services. Despite these shared continuities, emerging players, content forms, and platforms offer contradictory directions.

So what, exactly, is Arab television today? In this chapter, I am careful not to define the term too narrowly. Historically, Arab television referred to state-owned channels with their inward-looking content and their limited broadcasting coverage. It should also mean satellite television as technological, economic, and political regional mass media in transition, and must consider the change in television culture from free, often pirated, to paid content. Today, the term also denotes a multi-screen experience no longer bound by time and space: "anytime, anywhere!" The chapter traces the development of Arab television production, programming, and distribution, and offers a glimpse into the current alternatives. Arab television today is a state affair, a regional industry, and an increasingly global enterprise.

This chapter is subdivided into three sections. The first provides a context for understanding the development of television in the Arab region. The subsequent sections provide a detailed analysis of the current structure of Arab television's production, programming, and distribution. As a principal method, this chapter is the result of institutional research designed to be a structured review of case studies in which surveys, ratings, press releases, and other information are cross-analyzed with insights from 13 industry professionals in production and programming. Such an undertaking presents a difficult set of problems. The first is the absence of official or public information about Arab television: private or state-owned channels are not mandated to present any data to the public. The second is the reluctance of television producers, programmers, and other executives to participate in research. The third is that much of the available data is restricted by sampling methods or political or social practices (Khalil 2014).

Traditionally, Arab television has been treated as a uniform field with a focus on nation-states (Boyd 1993, 1999). Some recent studies have approached satellite television as a regional or transnational industry (Sakr 2007; Kraidy and Khalil 2009; Barry 2011; Mellor 2011; Sakr, Skovgaard-Petersen, and Della Ratta 2015). A significant number of publications have tried to interpret television coverage and practices regarding specific issues including, for example, Islam (Ferjani 2010), women (Al Malki 2012; Kharroub and Weaver 2014; De Poli 2017), children (Sakr and Steemers 2017), politics (Hafez 2007), and regulation (Shukrallah 2009). Factors influencing the production, programming, and distribution of specific television content have been explored in several studies: news (Zayani 2005; Rinnawi 2006), music (Frishkopf 2010), reality television (Kraidy 2010); drama series (Abu-Lughod 2005), and format television (Khalil 2017a).

## Culture, Politics, and Economics of Arab Television

This section is dedicated to outlining the context within which Arab television industries have emerged over the last 50 years. The focus is on the various cultural, political, and economic vectors that have contributed to charting specific paths for Arab television. In order to assess the importance of culture in television programming in the Arab region, one must look at the relationship between geography and culture as well as the movement of media products that have developed over a long period of time. For many Arabs, media infrastructure, institutions, and products have come to define Arab cultural and national identity during the modern era. Arab culture assumes a special significance because the idea of an Arab television is founded on the premise that the people living in the region share similar customs, ideas, and social behaviors. However, the region inhabited by Arab-speaking people is itself divided into several geo-cultural regions: the Levant (Lebanon, Syria, etc.), the Gulf (Saudi Arabia, Qatar, etc.), and the Maghreb (Algeria, Morocco, etc.). Each of these regions has its own set of linguistic, cultural, and socio-political practices.

Many media scholars argue that Arab cultures are under threat from foreign, mostly Western, media content. As a result, researchers often evaluate television industries on their ability to protect and reflect national identity and social norms. Anchored in studies of national media content, these researchers evaluate state practices around questions of culture (Abu-Lughod 2005; Hafez 2007; Mellor 2011), while other Arab media scholars offer a nuanced perspective on the relationship between culture and Arab television industries. Culture, then, is negotiated at the nexus of the local, regional, and global. Television becomes a site where debates about culture are manifested in talk shows, reality television, and other content (Zayani 2005; Rinnawi 2006; Kraidy 2010; Khalil 2017a).

At this stage, it is important to recognize that Arab culture is not a collection of set ideas, customs, and social behaviors; rather, it is interactive, evolving, and constantly challenging. What is changing about Arab culture? Foremost is the demographic youth bulge at the core of moral panics about television content and, increasingly, social media practices. With almost 50% of the Arab population under the age of 24, screen content—whether produced for television or for social media—increasingly reflects the challenge of cultural preservation and openness, an integration in global affairs or reclusion to national aspirations.

With the exception of Lebanon, television in the Arab world has been an instrumental part of the state formation since the first broadcast in the 1950s (Boyd 1993). The state was the founder, regulator, financier, and primary controller of television in the Arab world. Although the Arab region is composed of 21 different nation-states, politics is often imagined as a singular space dominated by an autocratic Islamic rule unwilling to modernize its political systems (Khiabany and Sreberny 2007). In practice, these nation-states have different local, regional, and global aspirations that are reflected in the organization of their media, particularly television. Because it is still the most popular medium with Arab audiences, television is treated as a symbol of independence, a mouthpiece for regimes, and an instrument to exert soft power across the region (Rugh 2004; Kraidy and Khalil 2009). Internally, television was instrumental in forging a national identity and fostering socioeconomic development. These roles persisted even after the introduction of satellite channels in the 1990s. However, some regional and international political forces became influential players in the era of transnational, pan-Arab television.

In television news, the competing agendas of Arab regimes and foreign states led to the emergence of a cacophony of news channels. The regional broadcasting that Egyptian president Nasser pioneered in the 1960s continues today with multiple players such as Qatari-owned Al Jazeera and Saudi-owned al Arabiya. Meanwhile, state-sponsored international broadcasters such the US-channel Al Hurra, Russia Today (RT), China Central Television (CCTV), Deutsche Welle (DW), France 24, and the British Broadcasting Corporation (BBC) have all introduced television services in Arabic. At the same time, entertainment programs, particularly reality TV shows, are at the center of political discourses and tensions. In these shows, participants are representatives of their countries and their personal success is that of their nations. Entertainment as politics is well documented in several cases including the canceling of *Big Brother* (Kraidy 2010) or *Al Maleeka* (Khalil 2017a).

It is within this cultural and political context that the current economic landscape of Arab television industries is reflective of the region's macroeconomic indicators and television's microeconomic trends. While almost 80% of the region's GDP is dependent on oil, the map of Arab economies includes oil-exporting countries (e.g., Saudi Arabia) and large importers (e.g., Morocco). However, the dependency between Arab television industries and the oil-exporting countries of the Gulf is addictive: Arab television industries rely on petrodollars for production and on the Gulf's large population and high per-capita spending for distribution. At the core of this relationship are media moguls who in production, distribution, and advertising have managed to developed a strong ecosystem (Khalil 2015, 2017b).

An emphasis on the economy is central to this study; the question to be asked is, to what extent can we refer to Arab television as industries when its funding relies on a combination of state or private parties, commercial and advertising sources, and international states or NGOs? In the absence of public service media, Arab governments continue to fully or partially fund state-owned television stations. As a symbol of the state, these channels have a relatively secure financial stance. However, the past two decades have seen attempts to make these channels more financially prudent and to seek commercial funding sources. Similar to state-owned channels are television stations that act as media platforms to support the party or individuals' political ambitions. While lacking the financial stability of state-owned channels, these channels rarely venture into securing other funding sources. Finally, channels that offer popular programs—and consequently demonstrate high viewership ratings—can command the advertising market. Valued at USD 4 billion, this advertising market has been shrinking as budgets are diverted toward more targeted platforms such as social media (Señal International n.d.). However, subscription-based channels, and over-the-top (OTT) platforms, are enjoying growth, particularly in the educated middle-class sectors across the region (Deloitte and Touche 2016).

## Production

The development of Arab television production is rooted in the era of post-independence, as newly formed nation-states began to assert their control over the means of producing information and entertainment. Similar to most international histories of television, the first producers of Arab television came from radio, print, and literary backgrounds and had to develop their audio-visual skills. The majority of Arab television channels focused on building their capacities for producing local programs, often depending on foreign or other Arab technical or production expertise. Some countries, such as Algeria, inherited the technical and organizational infrastructure of television production and needed to develop the mission and content strategy for a truly national television channel. Other countries, such as Lebanon, developed partnerships with American and French channels to develop the technical and editorial skills of their own first generation of television producers. Still other countries, such as the UAE, relied on regional expertise (especially from Kuwait) to develop their own technical and programming expertise. What started as local television with the aims of asserting a sociopolitical, cultural, and economic independence slowly became a platform for the promotion of a pan-Arab regional culture with the introduction of satellite broadcasting in the 1990s. Of particular interest is inter-Arab collaborations, which have given rise to regional productions and distribution outlets.

Unlike regional productions that have evolved organically in traditional media hubs with little government involvement, media cities are the result of active government engagement in the development of media "ecosystems." Often referred to as special economic zones, these media cities offer unique opportunities for governments to diversify their economies and to regulate media (Khalil 2017b). They are organized based on geographic location, vertical integration, and sectoral groupings, such as film or television production, television broadcasting, print, design, and so on. Briefly, these media cities operate as "one interdependent trans-territorial production, distribution, and consumption site" (Khalil 2017b: 143).

Building on its media legacy, the Egyptian government established a quasi-public company to offer film and television studios as well as facilities for broadcasting. Established in 1997, the Egypt Media Production City started operating as a free zone in 2001. It operates six studio complexes with a total of 64 fully equipped studios for film and television productions. In a bid to attract the repatriation of pay TV Arab Radio and Television (ART) from Italy to the Arab region, the Jordanian government offered to establish Jordan Media City. With limited production facilities, this media city soon became a central node in satellite broadcasting for ART as well as numerous small Arab channels. Owing to its geographic proximity, Jordan Media City played an instrumental role in the growth of the Iraqi television scene.

Dubai has been the most proactive state in developing media cities. Under the umbrella of Dubai Creative Clusters Authority (DCCA), 11 specific cities (or zones) are home to 5,100 businesses employing more than 76,000 people (Deloitte and Touche 2016). In 1999, Dubai Internet City (DIC) was established to provide a hub in the expanding knowledge economy with a focus on Information and Communication Technology (ICT). Building on the success of DIC, Dubai Media City (DMC) became "home" for regional and international media companies. One of the first three buildings at DMC was dedicated to the Middle East Broadcasting Center (MBC) Group as it relocated from London in 2002. The success of DMC and the increased demand from various sectors of the media industries led to the establishment of Dubai Production City and Dubai Studio City to focus on specific sectors of the industry. The attractiveness of Dubai for media entrepreneurs rests in its simple and streamlined approach for attracting investments, licensing, visa regulations, zoning regulations, and developing talent pool.

In the last decade, the media cities model has spread in the UAE and across the Arab world. Abu Dhabi opted for an integrated production facilities approach with the establishment of Twofour54, while Fujeirah targeted a broad spectrum of media activities. Saudi Arabia, Bahrain, Qatar, Iraq, Morocco, Lebanon, and Syria are also planning and developing media cities.

## Programming

Acknowledging specific variations between state-owned channels and privately owned channels in the acquisition, censorship, and scheduling structures, the following are two distinctive features about general viewership channels.

Often compared to a combination of soap operas and telenovelas, *Musalsalat* are melodramatic series built around historical, social, political, or religious narratives that often include love stories. These series are associated with Ramadan, the Islamic month when Muslims fast from dusk until dawn. The episodes air daily, often repeating several times during a 24-hour period, culminating with a grand finale to coincide with the end of Ramadan. This television genre is anchored in an earlier version of TV series inspired by Arab and world theater and literature. During the 1970s and 1980s, Egyptian TV dominated Ramadan television programming with *Fawazeer* (Riddles), singing-dancing-comedic sketches that had their start on radio. Due to a growing conservative thrust primarily in the Gulf markets, *Fawazeer* gave way to the *Musalsalat*. During the 1990s, Egyptian series dominated the programming grids and established a trend that relied on studio shooting with budgets consumed by famous stars, writers, and directors. Because of the growing content demand of multiplying satellite channels, dubbed Latin American telenovelas became a regular feature until 2008, when series were imported from Turkey, South Korea, Iran, and elsewhere (Elouardaoui 2013). Simultaneously, the languages in these dubbed series changed from the classical Arabic, spoken in religious and elite circles, to local dialects. By promoting its dialect through dubbing, the nascent Syrian television drama slowly gained government support, foreign financing, and increasing popular appeal. Despite the ongoing war, Syrian *Musalsalat* became a staple of Ramadan schedules. The increasing popularity of the genre has led to the establishment of dedicated free-to-air and pay TV channels offering current and old *Musalsalat*.

A second distinctive television genre is format adaptation, which emerged in force after the *Idol* adaptation (*Superstar*) on Lebanese Future TV in 2003. This ushered a trend of reality TV format adaptation that included global formats (e.g., *Star Academy*, *Fear Factor*, *Perfect Bride*, etc.) and the development of original Arab formats (e.g., *Stars of Science*, *Million's Poet*, etc.) (Arab Advisors Group 2017). Both processes involved a negotiated adaptation of format rules, television requirements, institutional structures, and cultural practices to execute or develop these programs in the Arab world. Kraidy (2010) pointed to specific adaptation to Arab political circumstances, particularly in the case of *Star Academy* and *Big Brother*. Ayish (2011) noted the priority for localized programming to promote both religious morality and scientific creativity. Despite budget restraints, reality television formats continue to hold prime-time space on general television—except during Ramadan. Channels had to diversify and expand their revenue models to recognize young audiences' uptake of social media. The current formats include big variety shows such as *Dancing with the Stars* and *The Voice* but also talk show formats such as *The Doctors* and news magazines such as *Entertainment Tonight* in Arabic. Even formats for television drama series (*ER*, *Ugly Betty*, and *Everybody Loves Raymond*) were adapted to the Arab market.

Taking a cue from the success of pay TV channels, general viewership channels have managed to fragment their audience by creating multi-channels based on niche programming. This audience fragmentation forces pay TV channels to develop different programming strategies to survive in a highly competitive market (Khalil 2016). The following is a discussion of two distinctive pay TV offerings.

Over the past 20 years, the field of sports programming has changed dramatically. At the core of this change are global and local forces affecting the rights to broadcast specific games. State-owned channels are relatively invested in the coverage of local sports as a way to attract eroding audiences and much-needed advertising funds. The free-to-air satellite channels now compete with pay TV channels for rights to major regional sports leagues and events. It has been increasingly challenging for these free-to-air channels to meet the requirements of sports leagues that are capitalizing on their broadcast license fees. Regional and international rights sales treat the region as a single market and thus sell rights to one broadcaster, who in turn could sell rights to specific languages, countries, and so on to other broadcasters. To illustrate, we can trace the course of soccer's World Cup television rights in the region.

Until 2006, the broadcasting rights for the World Cup were negotiated on a country-by-country basis between FIFA, the games host broadcaster, and national broadcasters in the Arab region. The pay TV satellite network, Arab Radio & Television (ART), acquired rights for the entire region in order to offer the games exclusively on its channels for three consecutive World Cups (2006, 2010, and 2014). As expected, a battle ensued between several governments, under pressure from their publics, claiming their right to broadcast the games for free; ART responded that it was trying to protect its right to recoup its investments. A combination of uncontrolled piracy, an immature pay TV market, and audience rejection led to ART's failure to market the World Cup as pay TV programming. However, the Qatar-based Al Jazeera Sports seized the opportunity to acquire ART's rights for the 2010 and 2014 World Cups. Armed with a range of exclusive rights for European and American sports, it was only a matter of time before Al Jazeera Sports was rebranded as beIN Sports and became an independent global leader in sports programming in the region. In 2014, beIN Sports expanded into beIN Media Group, acquiring Miramax Studios, establishing a number of sports channels (USA, France, Canada, Australia, etc.), and offering a dedicated service for the Arab world with a range of exclusive pay TV drama and lifestyle channels, as well as the 2018 World Cup.

From the start, television programming for children has been closely aligned with the mission of state-owned television. Children's programs were developed to foster national identity, education, and social cohesion. After some experimentation, a format developed across state-owned channels consisting of an older person, often a woman, acting out stories, interviewing children, and allowing them to show their talents including singing, poetry, puppetry, and more. Starting in the 1980s, cheaper programming in the form of dubbed Japanese and Eastern European cartoons occupied the prime-time lead-in segments. Acknowledging a great many notable exceptions, cartoons have become the primary programming option for parents and children. The advent of satellite television offered several additional opportunities for children's programming. In the mid-1990s, pay TV operator Orbit Network partnered with Disney to offer a dedicated service to Arab children featuring classic original and dubbed Disney programs as well as exclusive shows in Arabic. Similarly, an exclusive agent for children's toys invested in a channel, Spacetoons, as a platform to market its products while offering a range of children's programs. In contrast, the State of Qatar decided to invest in the development of a dedicated service in Arabic for the region under the umbrella of its Al Jazeera brand. Between 2005 and 2013, Al Jazeera Children's Channel (JCC) offered Arab children and parents across the region and the diaspora a blend of quality education and entertainment (Sakr and Steemers 2017). Succumbing to financial and logistical pressures, the two Al Jazeera Children's channels Jeem and Baraem folded under beIN Media as a paid service. The fate of JCC clearly reveals the pressures and challenges of sustaining quality programming for children in the Arab region.

## Distribution

Arab audiences may view television as content, entertainment, or news; paid or free; local regional, or international. In contrast, television executives perceive it as distribution of television content. There are three legally established models of distribution for Arab television content: terrestrial national broadcast, regional direct-to-home satellite broadcast, and Internet Protocol TV (IPTV).

With few exceptions (such as Lebanon, Iraq, and the Palestinian territories), terrestrial television is limited to national state-owned channels. The control of the VHF and UHF spectrums is considered a matter of national security in many countries, since the revolutionary *coups d'état* that swept the Arab world in the 1960s and 1970s used state television and radio to broadcast their communiqués. Even with the introduction of satellite broadcasting, the regulatory framework remained resistant to the allocation of terrestrial spectrums to private channels. For instance, most of the Egyptian and Kuwaiti channels are dedicated to serving their respective nationals, yet they are only licensed to broadcast on satellite. This policy of spectrum allocation and economic concerns have limited the transition to digital terrestrial broadcasting (DTB), which was expected in 2015. To expand their transnational reach, state-owned channels were the first to occupy the spectrum provided by ArabSat, a satellite consortium owned by Arab states. By the mid-1980s, state-owned channels were available across the region and some provided their signal to Europe and the United States.

Given the shared governance between the state-owned channels and the authorities supervising the telecommunications sector, satellite connectivity is then under the jurisdiction of the state (Kraidy and Khalil 2009). In the 1990s, the Lebanese government banned satellite connectivity to local channels, which were then forced to send their tape-recorded content to Italy to be broadcast on satellite. During the Egyptian January 25, 2011 revolution, the state limited the access of many satellite broadcasters in an attempt to control the message.

Direct-to-home satellite broadcasting can be divided to two broad categories: free-to-air (FTA) and pay TV. The satellite providers are state and commercial services: Arabsat, Nilesat, Noorsat, Yahlive, Eutelsat, Gulfsat, and Es'hailsat. The Arab region is largely an FTA market with more than 1,000 channels often operating on one or more of these satellites. While more than 30% of the channels provide programming for a general viewership, the remaining channels offer niche programming such as news, music, movies, and shopping. The FTA model has gained popularity particularly because audiences consider television as a service originally provided by the government. The success of the commercial model of FTA television built primarily on the general appeal of entertainment and news underwent a process of audience fragmentation. This strategy was pioneered by pay TV, which made music videos, movies from the West, and even reality shows both accessible and tempting to watch. Multi-channels dominated the FTA channels with a main service dedicated to a general audience and several other channels dedicated to a specific genre, audience, geography, or other niche.

Two Saudi-owned groups emerged as leaders in multi-channel offerings. After 11 years in London, the MBC relocated to Dubai in 2002. Rebranded as the MBC Group, this multi-channel service offers 18 channels and a number of ancillary services. The second group is the Rotana Media Group, which transformed from a music distributor to the major producer and broadcaster of music videos in 2003. With a focus on Arab music, film, and television series, Rotana has acquired rights to more than 100 Arab artists, 2,000 Arab films, and hundreds of dubbed and original series. In November 2017, the media moguls behind MBC and Rotana, Sheikh Walid Al Ibrahim and Prince Al Waleed Bin Tala, were part of a group of Saudi princes and businessmen detained on alleged corruption charges. The details of their release remain undisclosed.

In response to increased competition, pay TV benefited from opportunities afforded by the introduction of Internet Protocol TV (IPTV). This technology allows the transmission of television content through data networks offering regular TV programs and video on demand. The next section discusses how this distribution method is allowing pay TV to expand its offerings to become a player in OTT services.

The last eight years have been marked by the introduction of OTT services to the region. The growth of these applications is closely linked to the widespread development of internet connectivity and the increase in internet access and speed. Whether through 4G or fiber optics, Arab audiences, particularly the younger demographics, are driving the adoption of OTT services (Northwestern University in Qatar n.d.). The landscape is becoming increasingly complex with regional players, global players, and regional independent producers.

Local players involve both traditional broadcasters and OTT entrepreneurs. When MBC Group started its Shahid.net in 2010, it was designed as a "catch-up" TV platform particularly for *Musalsalat* followers. While major offerings of Shahid are still free, MBC has made significant revenues from exclusive content and a competitive subscription rate. In an attempt to maintain their limited audiences and to curb churn, pay TV channels have started offering OTT platforms with the benefit of achieving a multi-screen universe. For the two main players, beIN Media and OSN's Wado, the financialization of pay TV confronts audiences' limited ability to afford the added premium for receiving these services. It remains to be seen if such services could become an additional source of revenue for pay TV.

Numerous regional OTT entrepreneurs are trying to claim a piece of this emerging alternative television market. Each has a different approach to developing content, devising revenue models, and curbing competition. Cinemoz has developed a large archive of Arab films and depends on advertising to sustain its more than two million average views per month. Amman-based Istikana provides a variety of content including Arab and dubbed series, documentaries, and cartoons for USD 2.82 per month. Using ads on satellite television to lure subscribers, Icflix focuses on Arabic films and series as well as dubbed Hollywood and Bollywood content. Established in 2012, Icflix has amassed more than 8,000 hours of Arabic movies and TV shows and has been sourcing content from young talents. As first movers in content streaming, these local OTTs negotiated deals with telecom operators to provide a single bill to their customer base due to limited credit card penetration. These local OTTs deliver significant valued traffic for their telecom partners.

In 2016, Netflix announced its arrival in the Middle East; it was quickly followed by Amazon Prime's limited access. But even before Netflix, Starz Play Arabia had succeeded in attracting more than 700,000 subscribers. These global players focus on the Gulf countries as a key market comprised of a large expatriate population, higher disposable income, limited entertainment opportunities, high-end infrastructure, and a strict anti-piracy legal framework. In order to facilitate and guarantee payments, Starz Play partnered with the IPTV and telecommunications sectors to enlarge their customer base and offer a unified bill. Netflix, which originally relied on credit card payments, then partnered up with pay TV operator OSN to expand its customer base and facilitate payment processes. In less than two years, Netflix has managed to increase its subscribers nearly threefold, from 660,000 to 1.88 million; it also acquired and developed a number of projects in the region including series, films, and stand-up comedies.

Even in news, Al Jazeera Network has embraced the appeal of short-form videos. Linked to the popularity of its coverage of the Arab Uprisings, AJ+ offers news and current affairs content directly to social media platforms in English (2014), Spanish and Arabic (2015), and French (2017). It offers brief news clips of up to 30 seconds summarizing the day's most salient stories. In addition, Al Jazeera Frames is an online portal offering a selection of short (3–5-minute) documentaries. It specifically targets the online community and offers a platform for documentary filmmakers and animators to promote their productions.

For local independent producers, YouTube remains an important platform to offer content and develop alternative channels. While there are several noteworthy examples, two will serve as illustrations. Named as one of YouTube's top ten channels in 2013, Jordan's Kharabeesh ("scribblings") produced animated satirical shorts. Kharabeesh has amassed more than 60 million views and continues to offer timely commentary on the sociopolitical and economic conditions in the Arab world. Similarly, Saudi-based U-Turn became a collective channel for a number of comedians, social commentators, and filmmakers. Bypassing restrictions on access to other means of distributing their content, U-Turn used YouTube to develop a channel that successfully attracted more than a million subscribers. Both channels survive on an advertising/product-placement model and self-production. A noteworthy dramatic series, *Takki*, written and directed by Mohammed Makki and featuring an ensemble cast of young Saudi female and male actors, had its first season on U-Turn, in 2016, and a year later was picked up by London-based Al Arabi Television.

## Concluding Observations

Arab television has always been a regional and global endeavor. With few exceptions, television has been about the intersection of state and business. Arab television today is a medium in evolution: its production, programming, and distribution have been mutating steadily for the past three decades.

Since 2010, the Arab region has been experiencing a shifting cultural, political, and economic landscape that signals contradictory trends in television production, programming, and distribution. The Arab Uprisings of 2010–2011 and the enduring conflicts in the region have contributed significantly to the emergence of a primarily young, proactive audience demanding content that reflects its multifarious sociocultural conditions. At the same time, Arab states have recognized the need to diversify their economies, recognize the changing public mood, and offer young people venues for self-expression. Television industries themselves are at the center of these contradictory forces, the increasing popularity of local content, the growth in short-form production and popularity, and the multi-screen environment, all of which challenge established modes of production, programming, and distribution.

Yet, television remains a buoyant and adaptable source for news and entertainment in the region. What this chapter speaks to is the need for scholarship to historically ground analysis and recognize that television production, programming, and distribution practices are not born out of a vacuum, but of cultural, political, and economic contexts, and their academic inquiry is dialectically related to studies of global television.

## References


Abu-Lughod, Lila. 2005. *Dramas of Nationhood: The Politics of Television in Egypt*. The Lewis Henry Morgan Lectures. Chicago, IL: University of Chicago Press.

Arab Advisors Group. 2017. *Arabic Reality TV Shows. Amman*. Jordan: Arab Advisors Group.

Al Atiyat, Hiba, and Dina Khreisha. 2017. *Satellite TV in the Arab World*. Amman, Jordan: Arab Advisors Group.

Al Malki, Amal Mohammed. 2012. *Arab Women in Arab News: Old Stereotypes and New Media*. London, New York: Bloomsbury Publishing.

Ayish, Muhammad. 2011. "Television Reality Shows in the Arab World: The Case for a 'Glocalized' Media Ethics." *Journalism Studies* 12 (6):768–779. doi: 10.1080/1461670x.2011.614816.

Barry, Wail A. 2011. "Global Communication Models and Formats: A Case Study of the Sustenance of the Media Industry in the Arab World." *Journal of Arab & Muslim Media Research* 4 (1): 23–39. doi: 10.1386/jammr.4.1.23_1.

Boyd, Douglas A. 1993. *Broadcasting in the Arab World: A Survey of the Electronic Media in the Middle East*. 2nd ed Ames: Iowa State University Press.

Boyd, Douglas A. 1999. *Broadcasting in the Arab World: A Survey of the Electronic Media in the Middle East*. 3rd ed Ames, IA: Iowa State University Press.

De Poli, Barbara. 2017. "Arab Television Channels: The Image of Women Between Local and Global Trends." *Arab Women and the Media in Changing Landscapes*, Eds. Elena Maestri and Annemarie Profanter, 13–33. Cham, Switzerland: Palgrave Macmillan.

Deloitte and Touche. 2016. *Technology, Media & Telecommunications Predictions 2016:* Middle East. Dubai, UAE: Deloitte and Touche (M.E.).

Digital TV Research. n.d. "Pay TV Revenue in the Middle East and North Africa from 2010 to 2022, by Platform (in Million U.S. Dollars)." In *Statista – The Statistics Portal.* Retrieved January 21, 2018, from www.statista.com/statistics/252170/pay-tv-revenue-in-the-mena-region-by-platform/.

Dubai Press Club. 2017. *Arab Media Outlook 2016–2018: Youth Content Digital Media.* Dubai, UAE: Dubai Press Club.

Elouardaoui, Ouidyane. 2013. "Spanish-Language Telenovelas and Turkish Soap Operas on Arab Television: Cultural Adaptations and Social Effects." PhD diss., University of California, Santa Barbara, CA. https://alexandria.ucsb.edu/lib/ark:/48907/f3bc3wh9.

Ferjani, Riadh. 2010. "Religion and Television in the Arab World: Towards a Communication Studies Approach." *Middle East Journal of Culture and Communication* 3 (1): 82–100. doi: 10.1163/187398609x12584657078367.

Frishkopf, Michael Aaron. 2010. *Music and Media in the Arab World.* Cairo, Egypt: The American University in Cairo Press.

Hafez, Kai. 2007. "Arab Mass Media: Newspapers, Radio, and Television in Arab Politics." *Political Communication* 24 (1): 96–98. doi: 10.1080/10584600601130537.

Khalil, Joe F. 2013. "Towards a Supranational Analysis of Arab Media: The Role of Cities." In *National Broadcasting and State Policy in Arab Countries*, edited by T. Guaaybess, 188–208. New York: Palgrave Macmillan.

Khalil, Joe F. 2014. "'The Mass Wants This!': How Politics, Religion, and Media Industries Shape Discourses about Audiences in the Arab World." In *Meanings of Audiences: Comparative Discourses*, edited by Richard Butsch and Sonia M. Livingstone, 111–122. London: Routledge.

Khalil, Joe F. 2015. Modalities of Media Governance in the Arab World. In N. Sakr, J. Skovgaard-Petersen, & D. Della Ratta (Eds.), *Arab Media Moguls.* (pp. 13-30). IBTauris.

Khalil, Joe F. 2016. "The Business Push and Audience Pull in Arab Entertainment Television." *International Journal of Communication* 10: 3632–3646.

Khalil, Joe F. 2017a. "From Big Brother to Al Maleka: The Growing Pains of TV Format Trade in the Arab Region." *International Journal of Digital Television* 8 (1): 29–46. doi: 10.1386/jdtv.8.1.29_1.

Khalil, Joe F. 2017b. "Change and Continuity in Arab Media: A Political Economy of Media Cities." In *Bullets and Bulletins: Media and Politics in the Wake of the Arab Uprisings*, edited by Mohamed Zayani and Suzi Mirgani. New York: Oxford University Press. doi: 10.1093/acprof:oso/9780190491550.003.0007

Kharroub, Tamara and Andrew Weaver. 2014. "Portrayals of Women in Transnational Arab Television Drama Series." *Journal of Broadcasting & Electronic Media* 58 (2): 179. doi: 10.1080/08838151.2014.906434.

Khiabany, Gholam and Annabelle Sreberny. 2007. "Introduction: Mediated Politics in the Middle East." *Comparative Studies of South Asia, Africa and the Middle East* 27 (3): 509–512.

Kraidy, Marwan M. 2010. *Reality Television and Arab Politics: Contention in Public Life Communication, Society and Politics.* New York: Cambridge University Press.

Kraidy, Marwan M. and Joe F. Khalil. 2009. *Arab Television Industries International Screen Industries.* New York: Palgrave Macmillan.

Mellor, Noha. 2011. *Arab Media: Globalization and Emerging Media Industries.* Malden, MA: Polity.

Northwestern University in Qatar. n.d. Share of National Respondents Who Watched News or Comedies on TV or Online in the Middle East and North Africa as of March 2017. In *Statista – The Statistics Portal.* Retrieved January 21, 2018 from, www.statista.com/statistics/780505/mena-nationals-watching-online-or-tv-news-or-comedies/.

Rinnawi, Khalil. 2006. *Instant Nationalism: McArabism, Al-Jazeera, and Transnational Media in the Arab World.* Lanham, MD: University Press of America.

Rugh, William A. 2004. *Arab Mass Media: Newspapers, Radio, and Television in Arab Politics.* Westport, CT: Praeger.

Sakr, Naomi. 2007. *Arab Television Today.* London: I.B. Tauris.

Sakr, Naomi, Jakob Skovgaard-Petersen, and Donatella Della Ratta. 2015. *Arab Media Moguls.* Library of Modern Middle East Studies. London: I.B. Tauris.

Sakr, Naomi and Jeanette Steemers. 2017. *Children's TV and Digital Media in the Arab World: Childhood, Screen Culture and Education.* London: I.B. Tauris.

Señal International. n.d. "Television Market Value in the Middle East and Africa from 2010 to 2021, by Region (in billion euros)." In *Statista – The Statistics Portal.* Retrieved January 21, 2018, from www.statista.com/statistics/707136/tv-market-value-middle-east-africa/.

Shukrallah, Hani. 2009. "The Arab Charter for Satellite Broadcasting: A Reply to Hussein Amin." *Westminster Papers in Communication & Culture* 6 (1): 118–124.

Zayani, Mohamed. 2005. *The Al Jazeera Phenomenon: Critical Perspectives on New Arab Media.* Boulder, CO: Paradigm Publishers.

# 34

# STRUCTURAL CHANGES IN THE IBERO-AMERICAN TV MARKET

## Concentration and Convergence against Diversity?

*Guillermo Mastrini and María Trinidad García Leiva*

### Introduction

Several studies have empirically demonstrated, both in Latin America and in Spain and Portugal, that the pace of business concentration and technological convergence of info-communication industries accelerated during the first decade of the new millennium (Mastrini and Becerra, 2009; Nikoltchev, 2012; Becerra and Mastrini, 2012, 2017; Noam, 2016; Cappello, 2017). The audio-visual landscape is no exception to this norm; in fact, the TV industry emerges as one of the most strongly concentrated sectors.

This work analyzes the structural changes observed in Ibero-American TV in the first 15 years of the 21st century in terms of concentration and convergence. It focuses on the major TV markets in the region (Mexico, Brazil, Argentina and Spain), for two main reasons: the size of the domestic markets (the largest in the region) and their political influence at regional level. It provides an overview of their characteristics, potentials and limitations, by reference to their distinctive historical evolutions. To conclude, policies and recommendations for action are given, with the aim to promote and protect audiovisual diversity, having recognized the issues, determinations and challenges that the future development of TV in the region must face.

It should be noted that the concept of Ibero-America, which in this chapter refers to Latin America and the Iberian Peninsula, points not so much to a common identity, distinguishable due to the existence of two main languages and certain cultural proximity, but rather to a shared space that is under construction. As García Canclini and Martinell (2009) point out, there is no biological base or single common tradition that would enable a definition of "being Ibero-American." On the contrary, there is a history of convergence and conflict that makes it more convenient to refer to a shared space – a cultural, political and socioeconomic arena bringing together many identities, languages and practices. From this perspective, the Ibero-American audiovisual sector in general, and the TV market in particular, are part of a common space of enormous diversity but also of shared cultural traits.

## Background and Context

The audiovisual industries had an early development in Ibero-America: first the radio and then the TV took over the free time of consumers widely, without allowing newspapers and paperbacks to catch up. This explains the high penetration rates that TV has always had in the region, both in terms of households and audience. However, while in Latin America television emerged in the 1950s with a clear business drive and consolidated as a mass media in the following decade, in both Spain and Portugal TV was born in a contemporaneous manner, under dictatorial regimes, and remained linked to the state until the 1990s, when private competition was admitted. In any case, in the region as a whole, TV has prevailed among audiovisual media since the 1960s.

Television in Latin America has predominantly followed the US commercial licensing model, financed by advertising and based on the granting of few high-power licenses to commercial operators. Public service and not-for-profit broadcasters have always had a secondary role and the market has been controlled in each country by few companies. This historically remarkable commercial structure, with no real public service broadcaster (despite most other countries having had state-run media, closely aligned with the ruling party), has also shown a varying degree of intervention of political power in news content (Fox and Waisbord, 2002; Waisbord, 2010).

Conditioned by its strong ties with foreign capital, particularly with US sources, TV became an everyday life activity for Latin Americans right from the start, with average consumption above that of other parts of the world. This has garnered the attention of several researchers, including Jesús Martín Barbero, Omar Rincón or, more recently, Orozco and Miller (2017), who have depicted the ability of this medium to become pervasive in the everyday lives of people.

Additionally, since the origins of Latin American TV, tacit agreements have taken place between media owners and political and economic power. Latin American TV has seldom been critical of the administrations in office, and governments have rarely questioned the economic growth of TV groups. This has led to the creation of networks such as Globo (Brazil) and Televisa (Mexico), which have become integrated in the international trading of TV programs. Even though early on TV typically depended on foreign content (almost exclusively from the USA), since the 1990s there has been an increase in the production capacity, and national programs have displaced foreign ones from prime-time slots (OBITEL, 2017). The structure of the industry, however, continues to be highly concentrated (Becerra and Mastrini, 2017).

As Becerra, Mastrini and Waisbord (2015: 37) succinctly note,

> Before commercialization and globalization became dominant forces in television systems worldwide starting in the 1980s, television in the region was already structured around commercial principles and linked to the dynamics of internationalization. Latin America as a region represented the first case of internationalization and commercialization of the global television industry.

Local politics and the globalization of media markets are two forces that have shaped the development of Latin American media since the 1990s (Fox and Waisbord, 2002).

The evolution of TV in the Iberian Peninsula was quite different until the mid-1990s. There are differences between the Latin American model (Muraro, 1985) – with the prevalence of a commercial, advertisement and entertainment sector right from the start – and the cases of Portugal and Spain, where the state monopoly remained for several decades, with the addition of censorship. However, there are also common traits. To begin with, in the Peninsula TV was not born out of the logic of the public service of northern and central European countries since it was launched in the Francoist Spain and the Portuguese authoritarian Estado Novo. It can also be pointed out that commercialization came early on, hand in hand with national broadcasters being almost completely open to hegemonic advertiser-funded programming. The liberalization of the TV market and the advent of commercial private operators, in the early 1990s, reinforced the entertainment side of Iberian TV, undermining its informational and cultural functions (Bustamante, 2007).

In this historical context, cooperation and exchange of audiovisual products among Ibero-American countries were confined to testimonial expressions. While in recent decades certain actions designed to achieve increased interaction were promoted, so far any achievements in this area represent a catalog of good intentions (Moreno Domínguez, 2008; Sánchez Ruiz, 2011; Fuertes and Marenghi, 2016; García Leiva, 2017). As Bustamante (2007) points out, Ibero-American cooperation in the TV field has never been seriously addressed in any of its fundamental components: either in relations with Ibero-American cinema or in the co-production or exchange of TV programs, or as a strategic platform for all other areas of culture. Having said that, the circulation of TV products in Ibero-America has been much larger than in the cinema, due to the importance of soap operas in channel programming. In step with market needs and based on popular consumption, fiction producers such as Mexico and Brazil have managed to export their production to the entire region.

## Structural Changes: Concentration and Convergence

> Several Latin American countries have approved new media laws that have been perceived by some as an opportunity to make the media landscape more pluralistic and less concentrated, and by others as an opportunity for the governments to act against media outlets that have been critical of their administrations. The same debate has applied to steps to revise out-of-date media laws, including those left over from military dictatorships. There has been a trend of public officials initiating criminal legal actions against journalists and media outlets, although in the majority of cases these do not move forward. Countries that have typically maintained international standards on freedom of expression and access to information have continued to do so.
>
> *(UNESCO, 2014: 7)*

UNESCO's 2014 Freedom of Expression Report lays the groundwork for analyzing the situation of television in Latin America. The quote raises some fundamental issues that are also present in this chapter: a high level of concentration, the development of new regulations intended to reconfigure the media scenario away from the traces remaining from the times of military rule, and the role of growing intervention played by governments that strive in their relations with media corporations and journalists. To describe the full picture, one should add convergence between the audiovisual and the telecommunication sectors. Fueled by digitization, a new pattern emerged in the early 2000s: telephone companies began to build dominant positions in the media sector, particularly in the cable TV segment, turning into audiovisual players by means of service integration (e.g. offering bundles and packages usually known as triple-, quadruple- or quintuple-play services).

Multimedia groups in Latin America can be broken down into two categories: the main groups of Brazil (Globo and Abril), Mexico (Televisa) and Argentina (Clarín),[1] which well exceed the revenue threshold of one thousand billion dollars per year, and the rest of the companies in other countries, which are far from reaching the size of the abovementioned groups. This is partly due to the fact that, considering the demographic and economic differences in the region, Latin America has three large information and communication markets, namely Brazil, Mexico and Argentina, in which the major groups have greater magnitude. Their ability to expand into the USA (Mexican groups) or export content (Globo) also counts when analyzing their strengths.

As Becerra, Mastrini and Waisbord (2015) explain, the large communication groups in the region, for which audiovisual is part of their main business, have been adjusting to the new convergent environment since the turn of the century. Internally, they have been completing the transformation process from family business to conglomerate structures. Externally, some have taken advantage of globalization, diversifying their interests in other countries (particularly Televisa and Globo). This has also been the case of Telmex, the Mexican telecommunications company that expanded toward Central and South America, as well as toward IPTV, in the second half of the 2000s.

To be precise, Latin American groups have had to face the challenges posed both by telecommunication companies and their expansive corporate strategies in the audiovisual industry, as well as by political sectors that seek to redefine the regulatory framework. One should recall that during the first years of the 21st century, several "populist" or "new leftist" administrations (today there is no clearly agreed definition) encouraged policies to limit the concentration of ownership of communication media, with the declared goal of "democratizing communication." These processes led to an interesting social debate regarding the role of the media, coupled with strong opposition by media owners. Outcomes, in terms of changes in the structure of the audiovisual system, have been, nonetheless, mostly imperceptible (Badillo, Mastrini, and Marenghi, 2015).

As far as pan-European audiovisual groups are concerned, both Spain and Portugal evidence significant levels of concentration of the broadcast and the pay TV/distribution markets (Cappello, 2017). More specifically, according to the 2016 Media Pluralism Monitor (Masip et al., 2016), Spain scores medium risk regarding the market plurality area, although one indicator of this area is at high risk (71 percent): cross-media concentration of ownership and competition enforcement. Even though Spanish legislation covers ownership restrictions in the audiovisual sector, based on audience share and number of licenses, market concentration is high in both.

Among Spanish audiovisual groups, Atresmedia and Mediaset hold prominent positions. PRISA, a multimedia group established in the 1970s, with major assets in Latin America (radio and television), has just reduced its activities in Spain to radio, newspapers and book publishing, given that due to its substantial debt it has sold almost all its TV-related assets. Those assets ended up in the hands of Mediaset and Telefónica; furthermore, the telco giant Telefónica became a shareholder of PRISA by acquiring 13 percent of the company (the rest of the shares are held by financial investors).

Similar to Latin America, the Spanish groups must face the challenges posed by the telcos that offer audiovisual services: in addition to Movistar (owned by Telefónica), Vodafone and Orange are worth mentioning.

In Spain, the media market is characterized by an overall dominance of television, which attracts about 40 percent of the total advertising expenditure in the country. Television also remains the prevailing source of information (88 percent), followed by the Internet (68 percent) (AIMC, 2016). The point is that TV is controlled by Atresmedia and Mediaset, together reaching 55 percent of the audience and acquiring 88 percent of advertising revenues, and net profits of 129 and 171 million euros, respectively, in 2016. TVE, the public service broadcaster, is no longer a leader of the Spanish TV system in general terms or as a provider of basic audiovisual news for citizens, live TV or online channels, as it has been subject to a process of regression and deterioration since 2012 on all fronts: management, financing, programming, etc. (Bustamante and García Leiva, 2017).

A strong concentration of revenues is also seen in the TV sector in Latin America. According to survey data (see Table 34.1), this market is an oligopoly. In Southern Cone countries in particular, the four largest TV channels of each country have a share of at least 80 percent of total industry revenues. Likewise, it can be claimed that very high concentration levels are found in terms of audience. Despite the marked variation in the number of licenses existing in the different countries (over 300 in Brazil, less than 50 in Argentina), concentration levels are high in both cases. This finding points to the fact that those players who gain dominant positions in terms of audience also succeed in capturing the most substantial share of the market.

*Table 34.1* Evolution of Concentration in the TV Sector

| *Chart 1: Evolution of Concentration in the TV Sector (CR4 index – largest markets)* | | | | |
|---|---|---|---|---|
| Year | 2000 | 2004 | 2008 | 2014 |
| Argentina | 51 | 82 | 89 | 92 |
| Brazil | | 91 | 91 | 97 |
| Chile | 76 | 97 | 83 | 91 |
| Colombia | 75 | 68 | 63 | 88 |
| Mexico | 67 | 80 | 100 | 100 |

Source: Becerra and Mastrini (2017)

In Latin America, the deployment of digital terrestrial television (DTT) has occurred with little success so far. Three phenomena explain this situation: several countries have started a technological migration recently; the high penetration of pay TV entails a major barrier; and incumbent operators have little interest in carrying out a reconversion or upgrade for which they do not find many incentives for the time being.

In contrast to many Latin American countries, where there is a high penetration of pay services in the TV market (around 80 percent in Argentina and 40 percent in Brazil, via cable TV, and 60 percent in Mexico, via satellite TV), in Spain DTT is the most widespread platform in terms of audience and revenue. After the implementation of aggressive digitization policies that secured the switch-off of analog television in 2010, DTT was developed to saturation point and, as a result, subscription services have had a hard time expanding in a multichannel free-to-air environment. The 2008 economic crisis has also played its part.

Nevertheless, pay TV in Spain has been experiencing an important rise since 2015 with the extensive deployment of fiber optics infrastructure in homes and, with it, the increase in the purchase of bundles of audiovisual and telecommunication services. According to the regulator, *Comisión Nacional de los Mercados y la Competencia* (National Commission on Markets and Competition, CNMC), at the end of the first quarter of 2017 there were 6.09 million pay TV subscribers, 86 percent of whom had hired some bundled offering (the quintuple-play option being the one with the largest growth). As regards market share, pay TV achieved 7.8 percent in 2017 (growing from 5.6 percent in 2013). It is worth mentioning that in any of its distribution platforms, pay TV in Spain is highly concentrated, and that since the acquisition of PRISA assets by Telefónica, the market leader is the IPTV service offered by Movistar.

Finally, in the context of digitization and globalization of the audiovisual industry, over the past 15 years the TV market on both sides of the Atlantic has shown, even on different paths, the strengthening of two trends that already existed. They can be summarized in the concepts of concentration and convergence.

In a recent analysis of the new challenges faced by the audiovisual sector in Latin America, Becerra and Mastrini (2017) identified a number of challenges that are easily applied to the Ibero-American set:

- The relationship existing between the concentration of the industry and the progressive appearance of new pay offerings.

- The technical and economic evolution of the different TV platforms, whose borders have become blurred in the form of bundles of audiovisual and telecom services.
- The recent emergence of over-the-top (OTT) services with global reach, including Netflix or YouTube, which not only place the groups of traditional media at stake, but also jeopardize the audiovisual business of telcos.

For the above reasons, TV as a media outlet is faced with the challenges of more sophisticated competition, in some cases unplanned and ubiquitous in terms of distribution, marketing and consumption. One can only wait to see whether these trends will lead to a strengthening of convergence processes (Telefónica announced in January 2018, for instance, that it will make available the entire Netflix content on its Movistar+ content platform) and/or to a leap in industry concentration (as of January 2018, Netflix had nearly 118 million streaming subscribers globally).

## Future Outlook

Audiovisual industries face the challenge of digitization, which has called into question the traditional models of production, distribution and consumption of the analogic environment. The rapid pace of change is set in an unprecedented marketing and globalization framework, in which nations find that their capacity for intervention is at stake. In this context, Ibero-American television should adapt to a situation that presents both threats and opportunities.

In certain countries, the situation does not appear to be worrisome. Free-to-air TV continues to create revenues and pay TV has gained subscribers in recent years, particularly in Brazil, Spain and Mexico, the markets with the largest numbers of inhabitants. The phenomenon of substitution of pay TV with OTT streaming services has not been significant yet in the region, but it is a worldwide trend that should be a clear sign for traditional operators. In Spain, for instance, CNMC is already monitoring the rapid increase in the pay platforms to view contents online: in mid-2017, almost one out of four homes with Internet access was subscribed to some of these video streaming services. Additionally, TV audiences have become increasingly fragmented among diverse platforms, affecting advertiser-based funding. Though this phenomenon is softer in Latin America than in other countries, the latest report from OBITEL (2017) indicates that it is already noticeable in Spain, Brazil and Mexico.

Globalization encourages the production of contents for the global market. In this context, a real threat has to do with the need to adjust to these conditions to continue producing, with the consequent loss of cultural diversity that the region has historically shown. As Néstor García Canclini (2004) points out, cultures that are historically more numerous are confined to minority spots in global markets. Conversely, an opportunity that has come about is taking advantage of the creativity shown in terms of producing transmedia contents. The OBITEL (2017) annual report underlines that such strategy is still in the experimentation phase in most countries. But the popularity of contents such as soap operas and their local/regional nature open up interesting avenues.

Orozco and Miller (2016: 103) claim that Latin American TV has changed to keep its privileged function due to its particular role in society:

> Everyday existence becomes mixed up with *telenovelas* as per *realismo mágico*, making both programs and experiences into an inter-calculation of the fictional and the factual, with the dividing lines cosmically blurred. Watching becomes a safe place for many Latin American viewers to emote, to cry and laugh, without social consequences, and to ponder the inequality that so discolors the supposed togetherness of *mestizaje*.

As explained in the first section, even though the Ibero-American audiovisual sector in general, and the TV market in particular, are part of a common space of enormous diversity but also of shared cultural traits, regional integration has been minimal in this regard. The circulation of audiovisual products has, in fact, been rare outside large media groups. In this regard, Moreno Domínguez (2008) notes that building an Ibero-American space has to do with leveraging the different cultural industries in a region which, despite its potential as a space for diversity and cultural wealth, in many cases is relegated to a minority standing by the occupation of large transnational companies that control recording, publishing or audiovisual circuits. Cooperation strategies should be strengthened. In this regard, regional platforms for viewing audiovisual contents are a must. National initiatives such as Cinear in Argentina or Retina Latina in Colombia demonstrate it is possible. If a joint effort was undertaken among several countries, opportunities and scale would be maximized. According to Sánchez Ruiz (2011), public policies are needed, both at national and regional levels, that are conducive to a greater number of exchanges, which are in turn expected to enable greater diversity in the audiovisual offering for the sake of Ibero-American audiences.

Last, but not least, the levels of concentration observed do not contribute to create a space that encourages audiovisual diversity. The large groups understand the difficulties created by digitization and globalization and start to demand less regulation, arguing an asymmetry with global corporations such as Netflix. While historically those groups had shown similarities with transnationals, which, for instance, provided them with content and in some cases with capital, today they perceive that their positioning is under threat, and they ask for state protection to keep their privileged place. Therefore, the idea of having "national audiovisual champions" to compete in international markets with transnational players, that gained strength in Europe in the 1990s, could regain momentum in the region.

## Conclusion

As Albornoz and García Leiva (2017: 57) have pointed out, in the context of global digital networks and re-intermediation of Internet giants in the field of culture and communication,

> achieving "equitable access to a rich and diversified range of cultural expressions from all over the world" (article 2, principle of equitable access of the 2005 Convention on the Protection and Promotion of the Diversity of Cultural Expressions) today remains a formidable challenge not only for developing nations but also for the more technologically advanced countries.

The design of public policies for the audiovisual sector in general, and for the TV sector in particular, has been and still is a challenge that the governments of Ibero-America must tackle if they are to stimulate access to culture and the promotion of production.

The market-centered focus of public policies seems to persist, following the failure of the initiatives of so-called "progressive" administrations. The advent of new governments with a rightist or center-rightist orientation has taken place in a context of de-legitimation of state intervention, which has favored the reorientation of audiovisual policies. However, we are convinced of the need for public intervention when culture is concerned.

To start with, public policy should take as a starting point the restructuring that is taking place in the audiovisual sector based on digitization and globalization, and on that basis create actions to ensure a diverse and democratic cultural environment. In this regard, we would like to propose a set of recommendations, which are aimed to serve as orientation and do not constitute an all-inclusive list.

1. Full access to culture should be ensured.

The shift from consumption to pay-focused contents and the restriction of access to relevant contents such as soccer, are trends in the audiovisual sector that should be reversed. There should be no room for exclusions of symbolic goods due to unaffordability. Digitization has caused a considerable drop in production costs, but this does not automatically result in an improvement of conditions to access; i.e., it does not necessarily resolve, and it may even aggravate the difficulties facing consumption of domestic cultural production.

2. More social participation in audiovisual policymaking and design should be enabled.

Democratization of culture should not be limited to supporting or instigating actions to ensure access, but rather, it should be informed from the bottom up. If cultural production is rich in grassroots expressions, and in fact cultural industries are nourished by such production, policies cannot exclude society from taking part in their making.

3. Excessive concentration of ownership in the info-communicational sector and in the TV market in particular is a problem that should be tackled as a priority.

In the current globalizing context, there might be discourses which, arguing themes such as competition, may promote even more concentrated environments. One should not forget that the theory of "national champions" has already shown major limitations in creating a context of more diversity both in European and Latin American countries.

4. More attention should be paid to consumption.

The trend to focus the problem only on the production side should be revised, investing, for instance, in the always necessary but systematically forgotten formation of audiences.

5. Measures should be refocused to stimulate production of regional and local contents.

It is necessary to design initiatives of a federal nature to offset the centralizing trends of the market. All stories deserve to be told. Faced with the homogenizing globalization, the action of the state to promote diversity is a fundamental tenet.

6. The growing importance of online audiovisual distribution demands the urgent upgrading of public policies.

The inclusion of local contents in online catalogs should be promoted, investment in national production should be proposed, and taxes should be levied and redirected to local and regional production. It should be considered that platforms such as Netflix contribute to the development and maintenance of Internet connectivity infrastructures, as these types of companies take a major share of the transport capacity of these infrastructures.

7. The integration and development of Ibero-American distribution platforms are fundamental to leverage the new technological environment in a productive manner.

If connectivity is increased and no platforms are created for Ibero-American audiovisual consumption, the bridging of the digital divide will only contribute to expand the gap in cultural consumption. Cooperation is therefore a key concept.

In sum, the new context requires rethinking the key elements of a democratic cultural policy in the Ibero-American audiovisual sector. It cannot be limited to protecting market niches or productive sectors; it should rather integrate fully the whole range of producers, distributors and, especially, citizens/consumers. Without a policy enabling access of all social classes to a diversified range of cultural assets, in which all voices, perspectives and stories are represented, little will be contributed to the development of a democratic culture in society.

## Funding Disclosure

This work has received funds from the ICEP Research Project at Universidad Nacional de Quilmes and the Ubacyt Project at Universidad de Buenos Aires.

This work is partly based on research undertaken for the project "Audiovisual Diversity and Online Platforms: Netflix as a case study" [CSO2017-83539-R], supported by the Spanish State Research Agency (AEI), within the National RDI Program Aimed at the Challenges of Society of the Spanish Ministry of Economy and Competitiveness, and the European Regional Development Fund (ERDF) of the European Union.

## Note

1 The Venezuelan group Cisneros could also be added to this group, although in this case activities in the USA should be taken into account.

## References

AIMC (2016). *Marco General de los Medios en España.* Madrid: Asociación para la Investigación de Medios de Comunicación.

Albornoz, Luis and García Leiva, Mª Trinidad (2017). *Diversidad e industria audiovisual. El desafío cultural del siglo XXI.* Mexico: Fondo de Cultura Económica.

Badillo, Ángel, Mastrini, Guillermo, and Marenghi, Patricia (2015). "Teoría crítica, izquierda y políticas públicas de comunicación: el caso de América Latina y los gobiernos progresistas", *Comunicación y Sociedad,* 24, 95–126.

Becerra, Martín and Mastrini, Guillermo (2012). "Structure of the Info-Communications Sector in Latin America", in C. Bolaño, G. Mastrini, and F. Sierra (eds.). *Political Economy, Communication and Knowledge.* New York: Hampton Press, 221–247.

Becerra, Martín and Mastrini, Guillermo (2017). *La concentración infocomunicacional en América Latina: nuevos medios y tecnologías, menos actores.* Quilmes: Universidad Nacional de Quilmes & Observacom.

Becerra, Martín, Mastrini, Guillermo, and Waisbord, Silvio (2015). "Television in Latin America: from commercialism to reform?", in M. Alvarado, M. Buonanno, H. Gray, and T. Miller. *The SAGE Handbook of Television Studies.* London: SAGE, 36–49.

Bustamante, Enrique (ed.) (2007). *La cooperación cultura-comunicación en Iberoamérica.* Madrid: AECID.

Bustamante, Enrique and García Leiva, Mª Trinidad (2017). "La TDT de servicio público estatal y RTVE: balance y futuro", in E. Gutiérrez (ed.). *La televisión digital terrestre. Situación actual y tendencias de futuro.* Madrid: COIT, 19–36.

García Canclini, Néstor (2004). *Diferentes, desiguales y desconectados: mapas de la interculturalidad.* Barcelona: Gedisa.

García Canclini, Néstor, and Martinell, Alfons (2009). "El poder de la diversidad cultural", *Pensamiento Iberoamericano,* Vol 4, XIII–XXI.

Cappello, Maja (ed.) (2017). *Media ownership - Market realities and regulatory responses.* IRIS Special. Strasbourg: European Audiovisual Observatory.

Domínguez, Moreno (2008).Diversidad audiovisual e integración cultural: analizando el programa Ibermedia. *Comunicación y Sociedad,* Vol 9, 95–118.

Fox, Elizabeth and Waisbord, Silvio (eds.) (2002). *Latin Politics, Global Media.* Austin: University of Texas Press.

Fuertes, Marta and Marenghi, Patricia (2016). "Cooperación y cultura: veinte años de coproducción cinematográfica entre España e Iberoamérica 1995/2014", *Alcance,* 5 (10), 100–123.

García Leiva, Mª Trinidad (2017). "Las culturas latinoamericanas en España y Europa: entre la invisibilidad del mercado y los vaivenes de la cooperación", in E. Bustamante (coord.), *Informe sobre el estado de la cultura en España 2017. Igualdad y diversidad en la era digital.* Madrid: Fundación Alternativas, 87–101.

Masip, Pere, Ruiz, Carlos, Suau, Jaume, and García Castillejo, A. (2016). *Media Pluralism Monitor 2016. Monitoring Risks for Media Pluralism in the EU and Beyond. Country report: Spain.* Florence: European University Institute.

Mastrini, G. and Becerra, M. (2009). *Los dueños de la palabra. Acceso, estructura y concentración de los medios en la América latina del siglo XXI.* Buenos Aires: Prometeo.

Muraro, Heriberto (1985). "El modelo latinoamericano de televisión", *Revista Telos*, 3, 78–82.

Nikoltchev, Susanne (ed.) (2012). *Converged Markets – Converged Power? Regulation and Case Law.* IRIS Special. Strasbourg: European Audiovisual Observatory.

Noam, Eli (2016). *Who Owns the World's Media? Media Concentration and Ownership around the World.* Oxford: Oxford University Press.

OBITEL (2017). *Una década de ficción televisiva en Iberoamérica. Análisis de diez años de Obitel (2007-2016).* Porto Alegre: Ediciones Globo.

Orozco, Guillermo and Miller, Toby (2016). "Television in Latin America Is "Everywhere": Not Dead, Not Dying, but Converging and Thriving", *Media and Communication*, 4 (3) 99–108.

Sánchez Ruiz, Enrique (2011). "El espacio audiovisual iberoamericano: entre el mercado y las políticas públicas", *Revista ALAIC*, VIII, 7 (4), 54–67.

UNESCO (2014). *World Trends in Freedom of Expression and Media Development.* Paris: UNESCO.

Waisbord, Silvio (2010). "The pragmatic politics of media reform: Media movements and coalition-building in Latin America", *Global Media and Communication*, 6 (2), 133–153.

# 35

# AFRICAN TELEVISION IN THE AGE OF GLOBALIZATION, DIGITIZATION, AND MEDIA CONVERGENCE

*Lyombe Eko*

African television systems have always been ensconced in diverse political economies, their geographies of freedom of expression, and the respective journalistic cultures they have engendered. The continent essentially has 54 different television systems that correspond to the latitudes of freedom that the regimes allow them. These systems have come a long way since the height of governmental monopoly broadcasting in the 1980s and 1990s under the perspective of development communication theory. This perspective was based on the view that African countries were fragile, fledgling societies with many internal and external problems. Striving for national unity and cohesion was considered more important than freedom of the press. The mass media, usually a government newspaper and government-controlled broadcasting facilities, were conceptualized as tools for national development. Most politicians argued that in situations of dire poverty, where the bare necessities were absent, having mass media that concentrated on checking government action and criticizing it was a misuse of resources (Eko, 2012; Eribo and Jong-Ebot, 1997). The typical African television system was mainly a terrestrial TV system with the governmental public broadcaster as the monopoly actor. In addition to analogue terrestrial TV, a number of countries allowed free-to-air digital satellite TV services provided by international broadcast entities such as French-owned Canal+ Afrique, Arabsat (Tunisia), and others.

Things changed dramatically in the 1990s as a result of two major forces: (1) The end of the bi-polar, East/West geo-strategic system triggered by the fall of the Berlin Wall, the collapse of communism in Eastern Europe, and the disintegration of the Soviet Union in 1991; and (2) revolutionary innovations in information and communication technologies, and media convergence on the Internet. This was accompanied by the diffusion of digital technologies, as well as standards and policies, from the International Telecommunications Union (ITU) and centers of innovation such as Silicon Valley, China, and Japan to the multiple broadcast systems across Africa.

## Aim of This Chapter

The aim of this chapter is to survey, describe, analyze, and explain the major changes that have taken place in the African television landscape since the liberalization of the 1990s. It focuses on two major developments that have marked the African television landscape: (1) the political-economic liberalization of the television industry to keep abreast of international developments, and (2) the analogue to digital switchover shepherded by the ITU. Perhaps the most significant technological development in television since the invention of color television, the transition to digital terrestrial broadcast television programming followed pursuant to policies agreed upon in the context of the regulatory and developmental activities of the ITU. Digital television encoding allows stations to offer higher definition video and better sound quality than analogue TV. In the United States, the digital switchover was mandated by law in the 1990s. The Federal Communications Commission phased in the transition and allowed for delays and postponements where necessary, and the National Telecommunications and Information Administration subsidized digital-to-analogue set-top converter boxes for low-income citizens who applied for a subsidy voucher. In Africa, the switchover presented several economic and regulatory challenges to television stations. These included: the role of the ITU in the digital transition, the impact on the structure of African television regimes, and television regulation in different national contexts.

## Liberalization of African Television in the Post-Cold War Era

The end of the bi-polar, East/West geo-strategic system triggered by the fall of the Berlin Wall, the collapse of communism in Eastern Europe, and the disintegration of the Soviet Union in 1991 had a liberalizing effect in Africa. The reason is that during the Cold War, Western powers, such as the United States and France, supported authoritarian African regimes like Zaire/Congo and Cameroon that were aligned with the West, while the Soviet Union and its Warsaw Pact allies supported Marxist-Leninist regimes such as Ethiopia and Congo-Brazzaville. These alliances ended with the Cold War. Africans took to the streets and demonstrated for more political freedoms. Embattled governments granted a modicum of political reforms and freedom of expression. As a result of the political liberalization of the 1990s most African governments ended state monopolies on television broadcasting through the licensing of private radio and television stations that had various degrees of independence from the government.

Many countries also created regulatory agencies that served as legal and ethical watchdogs of the changing media environments, although not necessarily of the public interest. These new regulatory agencies included the Tanzania Communications Regulatory Authority, the Communications Authority of Kenya, the *Conseil National de Communication* (National Communication Council) of Cameroon, and the *Haute Autorité de la Communication* (High Authority on Communication) in Gabon. This liberalization process has included privatization of segments of the airwaves, termination of government's monopoly on television broadcasting in most parts of the continent, licensing private, commercial television stations with different degrees of independence from the authorities, and the regulation of national broadcasting as convergent "Content Services Providers" operating in the sphere of global communication in real space and cyberspace. The term "content service providers," which originated from online communication, includes television, Internet Service Providers, and interactive computer service providers. These changes in the African television scene were necessitated by changing political geographies of freedom of expression – a relative increase in freedom of speech – innovations in the field of information and communication technologies, media convergence, social networking sites (social media), a boom in mobile telephony, and policy transfers from the ITU.

The liberalization of the political economy of the television industry in Africa in the post-Cold War era was carried out under the impetus of the neo-liberal deregulation and privatization programs of the International Monetary Fund, the World Bank, and the International Telecommunication Union. These market-oriented reforms triggered acute tensions between: (1) global information and communication technology companies eager to take advantage of new openings in Africa, (2) newly authorized private or independent African TV stations of diverse ideological persuasions intent on exercising newly acquired rights of freedom of expression, (3) entrenched "legacy" governmental public television regimes, who, with their monopolistic and censorious reflexes, tried and true methods of information control and manipulation, were intent on maintaining specific political regimes, (4) the public diplomacy or soft power initiatives of the major Western powers, and (5) the international dimensions of China's state capitalism, and Russia's emerging global communication regime. Generally speaking, most African countries modified their radio and television systems in line with the changing geo-politics and the realities of their countries. The following configuration was typical.

## *Free-to-Air Governmental Public Service Broadcasters*

These are the heirs and custodians of the old government monopoly broadcasters that were inherited from the colonial powers – France, the United Kingdom, Portugal – and evolved to become the backbone of entrenched governmental communication. They inherited the nation-wide broadcasting infrastructures of the legacy governmental public broadcasters, as well as their universal public service mandate. They are funded by tax-payers and or license fees and have different levels of editorial independence. Countries around the continent regulate these television organizations differently. In Francophone countries, these entities continue to be the mouthpiece of the government and serve as the disseminators of official government policies and ideologies. They do not, therefore, enjoy editorial independence in terms of the content they broadcast about politics, religion, sexuality, and other controversial issues. In Tanzania, the Tanzanian Broadcasting Corporation, the governmental broadcaster, is classified as part of "Content Services Providers" in real space and cyberspace. The law in Tanzania defines governmental public service broadcasting as a "broadcasting service that is made for the public, financed by the public and non-profit with obligations prescribed under regulation" (Electronic and Postal Communication Act, 2018). In Kenya, the Information and Communications (Amendment) Act, 2013 vests the Kenya Broadcasting Corporation with the national public broadcasting service. The KBC is authorized to operate as a mixed public service and commercial service, with universal service obligations, and with government funding, in order to sustain its universal service obligations

## *Private or Commercial Free-to-Air Content Service Providers*

One of the novelties of African television broadcasting is its particular structures of commercialization. Most African countries have licensed "commercial service" broadcasting stations that are largely funded by advertising revenue. These hundreds of "independent" for-profit stations compete against entrenched national governmental public broadcasters who use their extensive nation-wide infrastructures, platforms, and government funding to attract advertisers who desire to reach large national or regional audiences. These stations are also required to abide by content-based regulations. This situation gives government broadcasters many advantages in the marketplace. For example, in Tanzania, commercial radio and television stations are required to "promote the use of standard Kiswahili and English languages; (e) avoid racial and religious hatred; (f) protect minors from harmful program content; (g) avoid programs related to nakedness, gambling, violence, superstition and astrology" (Tanzania Communications Regulatory Authority, 2018). As we see in the case studies below, the independence and freedom of expression of these private commercial stations are not always guaranteed. Nevertheless, these independent stations can have branches in multiple African countries, thereby developing an international audience base that entrenched government stations cannot match.

### *Community Broadcasting Services or Free-to-Air Community Content Service Providers*

Community broadcasting refers to licensees who are authorized to broadcast or provide content services in specific regions or communities or societies. Community broadcasting services are classified as non-profit content providers. Since they have very little to no government funding, community broadcasting services are generally limited to radio broadcasting in regional languages or dialects. Due to political and economic reasons, community television is a rarity in most parts of the continent. The closest thing to community television in Africa is a Chinese initiative known as "Access to Satellite TV for 10,000 African Villages." This initiative, which was kicked off in South Africa and Nigeria, is intended to link millions of Africans in rural areas, many of whom have no electricity, clean-drinking water, health centers, or schools, with satellite television courtesy of the government of the Peoples' Republic of China (Naik, 2018; Saliu and Baoping, 2017). This development aid initiative, which is spearheaded by Chinese government-backed media and electronics company, StarTimes Group, will provide African villagers with global news, sports, educational, and entertainment programs free of any charge. The project involves equipping every village with giant solar-powered digital TV sets, satellite antennae (dishes), and decoders (Naik, 2018; Saliu and Baoping, 2017). Critics of this initiative see it as a project to reduce Chinese production overcapacity and keep the electronics industry humming.

### *Subscription Broadcasting Services or Pay-TV*

One of the major developments in African broadcasting in the 21st century is the emergence of Pay-TV. This subscription television targets the growing number of Africans who can afford to pay subscription fees for programming supplied by continent-wide satellite broadcasters. This important sector of African broadcasting is a domain that is highly contested by major international satellite broadcasters: DSTV, part of the Multichoice Group of companies (South Africa), Canalsat Afrique, a subsidiary of French-owned Canal+ and Vivendi, that covers Francophone Africa, StarTimes a Chinese media company that competes across the continent, and ZAP, an entity designed to serve the Portuguese-speaking countries of Africa: Angola, Mozambique, São Tomé & Príncipe, and Guinea Bissau. Africa is the fastest-growing Pay-TV market in the world, and this competition has led to lower prices and, especially, the production of local content (Monnier and Hoije, 2018). Indeed, the most interesting development in subscription television on the African continent is the increased amount of African content broadcast, in response to local demand. In fact, it is interesting that countries have rules requiring Pay-TV services to reserve a certain percentage of their content to local content. Pay-TV is no longer a matter of importing foreign programming for the upper classes and foreign expatriates. The focus is now on creating programming with local content that satisfies the demands of local or regional African audiences (Monnier and Hoije, 2018). In addition to local content quotas, Pay-TV services in many countries are required to observe "must-carry" rules that mandate carriage of the programming of the governmental Public Service Broadcaster (PSB) in their subscription TV services, except in situations where the content requires copyright clearance or is exclusive in nature. Must-carry rules are designed to ensure that Pay-TV subscribers are presented an opportunity to watch local programming. In order to meet this requirement and present fresh African content, StarTimes, the Chinese government-backed electronics and media corporation, has a significant footprint on the African subscription television scene as a content generator and distributor. It has signed exclusive agreements with international and African national soccer teams and soccer clubs to secure broadcasting rights for its Pay-TV service. These rights agreements provide a lot of African content for StarTimes, making it a major competitor in most parts of the continent. Taken together with China's other television program distribution initiatives (discussed below), China is now a significant player in subscription TV in Africa. Due to the popularity of sports on the continent, many Pay-TV channels have been licensed as the exclusive distributors of international sports signals from the Olympics and FIFA World Cup championships to European, Latin American, Asian, and African soccer leagues.

## The Analogue–Digital TV Switchover: The African Experience

The digital revolution, rapid development, and global diffusion of information and communication technologies around the world represented what Kuhn (1962) called a "paradigm shift," a revolutionary change in conceptualizations and practices in the sciences and in technology. As innovative information and communication technologies, networks, databases, platforms, social networking sites (social media), and algorithm-based linguistic capitalism diffuse from centers of innovation in the United States to the rest of the world, one of the major challenges facing African countries in the 21st century is keeping abreast of these dynamic changes. Accordingly, Africa has gone through a wave of ICT-inspired technologies that have revolutionized communications within the continent, and between the continent and the rest of the world. In 2001, a - $600 million, 28,000 kilometer (17,500 mile) underwater, fiber-optic cable system linking Southern and Western Africa with Europe (SAT-3/WASC), and Southern Africa with Asia and the Far East (SAFE), was successfully launched. The system linked the African continent to the major telecommunications hubs of Europe, North America, the Middle East, and Asia for the first time (Eko, 2010). Seacom, another fiber-optic project funded mostly by African corporate investors, finalized an East African fiber-optic link in 2009. These technological changes have had an impact on African television. African television networks, the majority of which were no more than unwieldy, governmental bureaucratic mouthpieces, were faced with the daunting task of switching their television infrastructures and networks from analogue to digital in order to keep abreast of new digital technologies and of television systems in North America. In keeping with its role as the multilateral catalyst that sets standards, transfers telecommunications polities, and facilitates the diffusion of new information and communication technologies around the world, the ITU launched a "Project on the digital broadcasting transition roadmap in Africa" in 2008 to guide African countries to make the analogue–digital switchover. This project was aimed at assisting African countries in creating their nation-specific roadmaps to facilitate a smooth transition from analogue to digital terrestrial TV (DTTB) as well as good practices in mobile TV (MTV) implementation. MTV is television that can be watched on hand-held or mobile devices, Pay-TV services delivered via mobile telephone networks, or received via free-to-air terrestrial stations. MTV includes features that enable time-shifting – downloading TV programs and podcasts, and storing them for later viewing. The ITU guidelines were designed to provide information and recommendations on policy, regulation, technologies, network planning, customer awareness, and business planning for the smooth transfer to DTTB and MTV (ITU, 2010).

Digital television has multiple advantages over analogue TV. It allows more national and regional channels, and has potential for commercial television services, multiple Pay-TV services, as well as a spectrum for community TV services, all at lower costs and better image quality than analogue TV. Ultimately the ITU analogue to digital roadmap project was aimed at spurring African broadcasters "to promote investment in digital and mobile TV broadcasting to narrow the digital divide in the region" (ITU, 2010). The premise was that digital and mobile television would revolutionize television broadcasting in Africa, open up spaces for independent broadcasters, and bypass governmental gatekeepers who traditionally control TV content. Interestingly, only 22 of Africa's 54 countries responded to an ITU questionnaire about their plans to introduce DTTB services and terrestrial mobile digital broadcasting (MTV). Of these, only 12 were planning to make the switch over from analogue to digital TV (ITU, 2018). Nevertheless, 18 countries indicated that they would participate in policy transfer assistance from the ITU. They wanted the ITU to assist them in drafting national legislation that would create enabling regulatory environments for implementation of the new digital technologies and services.

The analogue–digital crossover is a complex, laborious, and on-going process. It is predicted to continue for the next several years. The process was complicated by technological, political, and economic realities. Once terrestrial television stations switch from analogue to digital broadcasts, analogue TV sets essentially become obsolete because they are incapable of receiving the new, digital over-the-air broadcasts without purchasing digital television sets or paying for a set-top converter box (sometimes called a digital TV adapter or a "digital set-top box" (DSTB)). This is a digital-to-analogue converter, an electronic device that connects to the viewer's analogue television in order to allow the television to receive the digitized broadcast. The alternative is to purchase a new digital television and discard analogue TV sets that are currently in use. This was an additional cost to African television viewers, an expense that policy makers in many countries did not factor into the switchover. It also turned out to be a huge source of electronic waste, in situations where there were plans put in place to recycle discarded analogue TV sets. The analogue–digital crossover triggered an international competition between the StarTimes group of China, which has positioned itself as a multi-purpose supplier of digital technologies, television receivers, and even content, on the one hand, and French companies such as Thomson and Sagem, which partnered with Canal+ to provide digital technologies and content (Akdim, 2015).

In practice, the analogue to digital switchover was carried out differently in different jurisdictions. The political geography of freedom of expression and market conditions for the analogue to digital switchover varied across the continent. Countries like Egypt, Nigeria, and South Africa, which have high terrestrial television penetration, had the more complex task of preparing for the switch and carrying it out than smaller countries, such as Malawi, or countries with low terrestrial television penetration such as Tanzania or Mozambique (ITU, 2015). South Africa, which has the most advanced television market in Africa, with about 10 million households, decided to give away more than five million digital set-top converters in recognition of the fact that the cost of the transition would be onerous to its low-income citizens (Gqirana, 2015). This policy initiative is identical to the initiative in the United States, where the government subsidized the set-top conversion boxes for certain classes of citizens. In South Africa, beneficiaries had to show identification cards as proof of citizenship, as well as a valid TV license to show proof of ownership of a TV set (Gqirana, 2015; South Africa to spend Sh23bn on set top boxes for the poor, 2015).

Kenya's transition from analogue to digital television was not without its problems. The Kenyan communications sector regulatory body, the Communications Authority of Kenya – whose mandate includes regulation of broadcasting, multimedia, telecommunications and telephony, electronic commerce, postal and courier services – insisted that the transition had to be carried out by June 2015. Television audiences had two choices: buy digital television sets or acquire set-top conversion boxes. The only other option was for them to subscribe to direct broadcast satellite Pay-TV services, which thousands did. Since purchasing new digital TV sets was out of the question for millions of Kenyans, the Communications Authority of Kenya licensed tens of vendors to import a total of 3.2 million set-top converters (Digital switch-off leaves 1.3 million homes in TV blackout, 2015). This was essentially a stop-gap measure to give Kenyans time to purchase digital television sets in the long run. In view of the complexities of the situation, media owners and the Consumer Federation of Kenya asked the Communications Authority of Kenya to delay implementation of the switchover. The response was negative. Ultimately, when the transition took place, it was not hitch-free. More than 1.3 million households with television had a TV blackout (Digital switch-off leaves 1.3 million homes in TV blackout, 2015). This loss of service essentially forced many viewers to resort to Pay-TV channels, which are required to include governmental public broadcasting channels in their "bouquet" of programming, under "must-carry" provisions of the law. Nevertheless, due to its subscription fees, Pay-TV was not considered a sustainable option for ordinary Kenyans (Digital switch-off leaves 1.3 million homes in TV blackout, 2015).

Some countries, such as Burkina Faso, that have a liberalized television broadcast market, welcomed the additional spectrum space and channels that the crossover would enable (Télévision: Le Burkina faso lance officiellement le TNT, 2017). Autocratic countries, such as Cameroon, Gabon, Ethiopia, and Eritrea, wanted to ensure that the transition did not reduce government power and control over the new television landscape. Nigeria, Tanzania, and other countries worked with China StarTimes, to facilitate the transition and provide the digital set-top boxes. Other countries opted for hybrid options. The general picture that the analogue to digital transition brings into relief is that African television is caught in a tension between the centrifugal, decentralizing forces of innovative digital information and communication technologies, and the centripetal (centralizing) forces of governmental instrumentalization of television as a central component of the architecture of repression.

## Television Regulation and Control in the "Liberalized" Environment of Africa: Case Studies of Controversies in Kenya, Gabon, and Cameroon

Despite liberalization and the licensing of private and independent broadcasters, most governments maintained strict control of the media through their judiciary and extra-judicial regulatory power. While governmental public television stations are still an integral part of the architectures of repression, controversies in Kenya, Gabon, and Cameroon exemplify the tightrope that even private television stations have to walk within the parameters of authoritarian regimes.

## Kenya Television: Between the Pull of the State and the Push of the Market

Kenya has been the media hub of East Africa since the 1960s, the leader in matters of freedom of expression in the region. However, the narrative of the development of television in Kenya reflects the turbulent politico-cultural history of the country in the post-Cold War era. While the airwaves were progressively opened up in the first decade of the 21st century, the process was subject to the vicissitudes of the political environment, which demonstrated that the end of governmental monopoly on television broadcasting, and political (ministerial) licensing of broadcasters did not mean an end to authoritarian abuse of power.

The most talked about happening involved the First Lady of Kenya. On Monday, May 2, 2005, just after 11 p.m., Mrs. Lucy Kibaki, wife of President Mwai Kibaki, stormed the Nation Centre, headquarters of the largest media company in Kenya, the Nation Media Group. Mrs. Kibaki, who was accompanied by her armed, official bodyguards, marched up to the NTV multimedia newsroom to complain about the negative coverage she claimed her husband was receiving at the hands of the Nation Group's flagship newspaper, *The Daily Nation* of Nairobi, and the group's television station, NTV. While Mrs. Kibaki was berating staff in the newsroom, she suddenly noticed that a camera operator, Clifford Derrick Otieno, was recording the unusual event. Mrs. Kibaki suddenly attacked Otieno, slapped him in the face, knocked down the camera, and damaged it. Mrs. Kibaki and her bodyguards sat in the newsroom for five hours, while the media workers watched helplessly, trying to do their work under the intense pressure put on them by the First Lady. Ironically this incident occurred on International World Press Day.

Clifford Derrick Otieno, the camera operator Mrs. Kibaki had attacked, filed a formal criminal complaint with the Attorney General of Kenya, accusing Mrs. Kibaki of assault in his workplace. When the Kenyan Attorney General and the Director of Public Prosecutions declined to prosecute Mrs. Kibaki, Mr. Otieno filed a private prosecution against the First Lady, along with a motion asking the court to summon Mrs. Kibaki to appear in court and answer charges of assault and malicious destruction of property. The Attorney General promptly terminated Mr. Otieno's suit against Mrs. Kibaki, claiming that it was an abuse of the court process (Court Fails to Hear Case, 2005). The Nation Media Group did not take legal action against Mrs. Kibaki for destroying its camera. Doing so would clearly have endangered the station's license.

The second attack on the media occurred in 2007. After contentious elections that resulted in contested results – the re-election and swearing in of President Mwai Kibaki – riots broke out around the country, resulting in more than 600 deaths and 250,000 internationally displaced persons, who fled their homes after political attacks. In order to bring the situation under control, the Kenyan Minister of Internal Security banned live radio and television news coverage of political events, based on the premise that such broadcasts had the potential to contain inciting or alarming material that could further inflame the situation. Many human rights and freedom of expression organizations in Kenya and abroad described the ban as extreme and unnecessary (CPJ, 2008).

In 2008, Kenya enacted the Kenya Communications (Amendment) Bill 2008 whose provisions include heavy penalties and fines for press offenses. Many international freedom of expression organizations described the bill as repressive (president deals major blow to press freedom, 2009). The law gives the information minister and the Communications Authority of Kenya, powers to regulate television and radio content and schedules. The powers of the Kenyan government and the Communications Authority of Kenya to regulate media content were again evident in the contested presidential election of 2017. President Uhuru Kenyatta was declared the winner of elections in August 2017 but the Kenyan Supreme Court annulled the elections on the grounds that they were "neither transparent nor verifiable" and ordered a repeat of the whole exercise. Kenyatta was declared the winner of the repeat election, which the opposition had boycotted. He was sworn in as president in November 2017. However, the losing opposition candidate, Raila Odinga, did not recognize the results. Claiming that the elections had been rigged, he organized his own swearing-in ceremony in Nairobi on January 30, 2018 (Kenya's Raila inaugurates himself, 2018). Since this was a major development in the sad and often tragic drama of Kenyan politics, the media in Kenya and abroad covered the event. While Kenyan television channels were broadcasting the event, officials of the Communications Authority of Kenya, accompanied by armed police officers, marched into the broadcasting facilities of the main independent television stations of Kenya – Citizen TV, NTV, and KTN – and ordered engineers to switch off their transmitters, leaving their viewers with nothing but blank screens (Cherono, 2018; kenya cuts tv transmissions over live coverage of opposition's Odinga, 2018). This unprecedented, censorious action followed warnings from President Uhuru Kenyatta and other government officials against coverage of the event. In effect, four days earlier, the president had summoned media managers and editors to a meeting and warned them that their TV stations would be shut down and their licenses revoked if they broadcast live coverage of Odinga's "treasonous," symbolic, "swearing in" (Cherono, 2018). The Kenyan government indicated that the shutdown was punishment for the TV stations' defiance of the government ban on the broadcast of live events coverage of the Odinga swearing in (Agutu, 2018; Kenya cuts TV transmissions over live coverage of opposition's Odinga, 2018). The vague and overbroad television shutdown and ban on coverage of live events was lifted five days later after a hue and cry from the Kenyan media and international human rights organizations, and legal action against the Kenyan government.

## Politico-cultural Context of Television in Gabon and Cameroon: Between the Pull of the "Republican Monarchy" and the Push of Technological Innovations

Gabon and Cameroon are Francophone countries that can be described as highly centralized, authoritarian, "*chefferies républicaines*" (African republican chiefdoms). The ideological nomenclature "republican," refers to French "secular republicanism" the Jacobine, politico-cultural dogmatic ideology that emerged during the French Revolution of 1789, and replaced the monarchical *ancien régime*. This secular republican ideology, with its strong, centralized, Napoleonic governmentality, was transferred to French African colonies. Both countries can be described as French "post-colonies," to use Mbembe's term (2001). Their media regulatory systems are grounded in the Jacobine/Napoleonic governmentalities inherited from French colonialism. Gabon has a record of corrupt, authoritarian, nepotistic, kleptocratic governance under the rulership of the Bongo Ondimba "clan." President Omar Bongo Ondimba governed the oil-rich nation of Gabon with an iron fist for 42 years (from 1967 until his death in 2009). His son, Ali Bongo Ondimba succeeded him in 2009 and has stayed in power after two hotly contested elections, which the opposition claimed were rigged.

Gabon and Cameroon exemplify clearly the tensions that exist between the entrenched, tightly controlled, governmental television channels and the new generation of private commercial television stations struggling to compete in difficult censorious environments, while dealing with the challenges of professionalization. On October 27, 2018, Vision 4, a private television station in neighboring Cameroon that is part of the bouquet of programming available in Gabon, announced in its evening news broadcast that President Ali Bongo Ondimba was dead. Citing anonymous sources in Saudi Arabia, where Ondimba was attending a summit meeting, Vision 4 reported that President Bongo Ondimba had died of a massive stroke. The nation was in shock, and the opposition started preparing for the aftermath of this situation. Ike Ngouoni Aila Oyouomi, spokesman of the Presidency of the Republic of Gabon, told reporters on October 28, 2018, that President Bongo Ondimba was resting in a hospital in Saudi Arabia after suffering from "*une fatigue passagère*" (temporary fatigue). The world would soon find out that the president had had a massive stroke during a Summit in Saudi Arabia, and had undergone brain surgery. Neither the government nor Gabon Télévision, the national television station, disclosed the nature of the president's health and exaggerated the rapidity of his recovery. Oyouomi and Gabon Télévision essentially played the role of national dissimulator and mis-communicator. On November 11, 2018, Oyouomi appeared again on Gabonese national television to "provide an update on the state of health" of President Ali Bongo Ondimba, who, the spokesman said, had been admitted to the King Faisal Hospital in Riyadh, Saudi Arabia, after experiencing "discomfort and persistent dizziness." Oyouomi stated that information on the state of health of the president was "extremely reassuring," after the president had undergone a "medico-surgical intervention." This was the first time the Gabonese government had hinted at the gravity of Bongo Ondimba's health situation.

In the meantime, the Gabonese media regulator, the *Haute Autorité de la Communication* (High Authority on Communication) banned the Cameroonian TV channel, Vision 4, which had erroneously announced the death of President Bongo Ondimba, from the Gabonese airwaves for six months. The charge was "publishing fake news." Pay-TV and satellite companies that carried Vision 4 in their bouquets were ordered to remove its signal with immediate effect. The High Authority on Communication also banned French TV channel, France2, a BBC Radio French language reporter, and summoned the correspondent of the French News Agency (AFP) to report to its offices for a hearing due to the content of their reporting. After close to one month in the King Faisal Hospital in Riyadh, Saudi Arabia, a partially paralyzed President Ali Bongo

Ondimba was transferred to a hospital in Rabat, Morocco, for physical therapy. Gabon Télévision showed the visibly ailing President Ali Bongo Ondimba meeting with King Mohammed of Morocco. Despite being hospitalized outside the country for months, the government of Gabon did not officially inform the people of the nature of the illness Bongo was suffering from. Only unofficial reports in the French press indicated that he had indeed had a stroke while in Saudi Arabia and that he was undergoing physical therapy in Morocco (Le Monde & AFP, 2018). On January 1, 2019, the Gabonese government broadcast a pre-recorded New Year's message by a visibly ill President Ali Bongo Ondimba on Gabonese national TV. On January 7, 2019, junior officers of the Gabonese Presidential guard, led by Lieutenant Kelly Ondo Obiang, seized the national TV station, Gabon Télévision, the symbol of governmental communicational power and control, in the capital, Libreville, and announced that they had overthrown President Bongo Ondimba in a *coup d'état*. French troops that have been based in Gabon since independence in 1960, crushed the coup attempt, and President Ali Bongo Ondimba returned to Gabon shortly thereafter (Gabon president back home a week after failed coup, 2019).

## Cameroon: Using the Sledge Hammer of the Law to Crush Freedom of Expression

The situation in Cameroon demonstrates a slightly different dynamic between the national regulatory agency and a private television station that has close ties to the ruling elite. We saw earlier that Cameroonian private television channel, Vision 4, was banned for six months by the Gabonese government for falsely announcing that President Ali Bongo Ondimba had died while he was recovering from a stroke in Saudi Arabia. In order to assuage the hurt feelings of the Gabonese government, the Cameroonian media regulator, the *Conseil National de la Communication* (National Communication Council) also imposed a one-month ban on Vision 4 in Cameroon for publishing fake news. CNC issued a warning to Vision 4 CEO, Jean Pierre Amougou Belinga, and news reporter Nadine Patricia Mengue, for presenting the fake news to the general public. In another matter before the *Conseil National de la Communication*, Vision 4 news director and commentator, Ernest Obama, was banned from practicing journalism in Cameroon for one month for utterances that the CNC said could be construed as "incitement to tribal hatred and violence."

The Vision 4 affair once more put the spotlight on the National Communication Council of Cameroon, an "autonomous" media regulator created by presidential decree in 1991 to serve as the ethical watchdog that promotes the professionalization of media. Its other roles include promotion of governmental ideologies of national unity, peace, and cultures. The NCC's main task is extra-judicial content regulation. It is empowered to issue warnings, temporary suspension of activities, and permanent interdictions against unprofessional media organizations and journalists. Over the years, the NCC has become a censorious agency that powerful members of the government and the ruling elite routinely use to suppress newspapers, intimidate and ban journalists, and silence their owners. Interestingly, the NCC has no enforcement mechanism at its disposal so journalists and media outlets close to the regime routinely ignore the decisions of the NCC.

One of the most controversial decisions of the National Communications Commission was the suspension, in 2015, of Afrique Média TV, the largest, multilingual and multi-cultural television channel in Africa. Afrique Média TV is a self-identified "Pan-African" broadcaster, which reports the news from an African perspective. Pan-Africanism is an ideology that emphasizes the brotherhood of all peoples of African descent on the African content and in the diaspora. It is very critical of Africa's former colonial rulers – France, Belgium, and Portugal. When the NCC banned Afrique Média TV Cameroon police and soldiers sealed off its offices in Yaoundé and Douala. The station moved to nearby Equatorial Guinea and Chad where it continued to broadcast via a Russian satellite and cable to its audiences across the continent. It ultimately returned to Cameroon. During the 2018 elections in the Democratic

Republic of Congo, Afrique Média TV showed its pan-Africanist colors when Congolese Opposition leader, Felix Tsihesekedi Tshilombo, was proclaimed the winner. France and Belgium were unhappy with the outcome and called an emergency meeting of the United Nations Security Council, claiming that the elections had not been free and fair. Panelists on Afrique Média TV spent hours denouncing French, Belgian, and Catholic Church neo-colonial interference in African politics. Due to the continued influence of France in the African continent, no governmental public station criticized French and Belgian interference in the Congolese elections.

In Cameroon there is also a phenomenon whereby the government and its political allies outside government seek to neutralize the power of private, independent TV stations. They create and license private, pro-government radio and television stations that echo governmental religious, ethnic, or tribal ideologies. We have seen the case of Vision 4 TV, whose news director and commentator was banned for one month for utterances that the NCC construed as promoting tribal hate and violence. In effect, the journalist Ernest Obama had said in his program on Vision 4: "Hypocrisy, wickedness, and bitterness are hereditary in some tribes in Cameroon. Guess which ones?" This was not the first time the Vision 4 commentator, whose owner and CEO is politically and ethnically close to the regime of President Paul Biya, has been the subject of criticism for his utterances. When the Cameroon government arrested lawyers, professors, teachers, students, journalists, and other members of civil society who were demonstrating peacefully in the English-speaking regions of Cameroon in 2016 and 2017 (O'Grady, 2018), Jean-Jacques Ze of Vision 4 said during the program he jointly hosts with Ernest Obama, that the arrest of Anglophone civil society leaders in Cameroon was just a process of "*deraticisation*" (elimination of rats). These pronouncements led critics of Vision 4 to nickname the television channel, "Vision 4 Mille Collines," an allusion to Radio Mille Collines, whose broadcasts incited genocide in Rwanda in 1993–1994.

These African television case studies demonstrate that political, technological, and political economy transformations of television in Africa have given rise to tensions that have brought to the fore the censorious and authoritarian practices used by various governments to maintain monopoly control over broadcasting. The multiple transitions and tensions African nations have to negotiate – from European colonies to post-colonies – the ideological conflicts and contradictions of the Cold War, Western transfer of economic globalization and neo-liberalism, have all had an impact on the African continent and especially radio and television systems. Despite governmental efforts to maintain broadcast monopolies and censorious information control, the new phenomenon of private and independent television stations is a breath of fresh air on the African broadcast landscape.

## Trends in the Development of African Television: The Role of Chinese Soft Power Diplomacy and State Capitalism

We will close this chapter with a discussion of the greatest single influence on African television in the post-Cold War era: China. So far we have seen that China has made its presence felt on the African television scene in the sub-sectors of community television (particularly in the two largest economies on the continent, South Africa and Nigeria), subscription or Pay-TV (particularly by generating African content and bringing international sports and other content to the continent), and partnerships with a number of African countries in making the transition from analogue to digital television. The Western powers have taken notice and have begun to express their concerns. On December 13, 2018, Ambassador John Bolton, the United States National Security Adviser, presented the Trump Administration's New Africa Strategy. It was aimed at countering America's rivals in the continent. Bolton said that: "China and Russia are rapidly expanding their financial and political influence across Africa ... China uses bribes, opaque agreements, and the strategic use of debt to hold states in Africa captive to Beijing's wishes and demands" (Gearan, 2018). The Trump administration's Africa strategy acknowledged the fact that China has become the biggest player on the African continent in the fields of sports infrastructure, road and railway construction, transport, communications, information and communication technology, and broadcasting. China even built the headquarters complex of the African Union. China is now a major player in the economic and entertainment landscape in virtually all parts of the continent.

One of the earliest developments in the African cinematic scene was the diffusion of cheap Chinese martial arts films produced in Hong Kong, across the continent. These films, which have traditionally been popular in the urban areas of Africa, had very little dialogue. Their action-packed sequences compensated for the lack of dialogue and bridged the linguistic and cultural gaps between Africa and Asia. That early "Chinese" movie presence in Africa facilitated the Chinese government policy of globalizing Chinese culture and audiovisual productions in keeping with its status as an emerging global power (Keane, 2008). In effect, as part of its "Going Out" globalization policy, China Radio International (CRI), a broadcaster that was founded in1941 by Mao Zedong's Communist Party of China as "Xinhua New Chinese Radio," to spread Maoist communist propaganda and resist Japanese aggression and occupation of Manchuria, has now become a multilingual, multimedia, state-owned global media conglomerate that is the cutting edge of Chinese soft power diplomacy and a state-driven commercial, entertainment complex (Huang, 2018). China Radio International broadcasts in more than 70 languages. China's full-fledged venture into the African continent was intended to protect its interests there, increase its influence among Africans, create a semblance of South–South flow of communication to compete with and reduce the dominant North–South (US and Western Europe to African countries) paradigm of global communication. China also wants to challenge Western ideological rhetoric against China, assert China's place in the global economy, and highlight Chinese economic development projects in Africa (Huang, 2018). The African broadcasting sphere has been a major sector of Chinese interest and influence. Chinese soft power and public diplomacy in Africa include developing digital broadcast infrastructures and television program content with African partners in real space and cyberspace. As early as 2012, China Radio International set up a center to translate, dub (voice synchronize), subtitle, and post-produce specially selected Chinese television programs and TV series for distribution to specific countries and markets around the world.

This policy was shaped by the unlikely success of a 36-episode Chinese TV series in Africa. When Chinese President Xi Jinping paid a state visit to Tanzania in 2013, he declared that the "Chinese Dream will be more easily accepted by foreign audiences when it is manifested in small and touching stories" (cited in Huang, 2018). Xi Jinping was alluding to the success of the Chinese TV series, *Doudou and Her Mother-in-law*, that had been translated and dubbed into Swahili, using Tanzanian actors, and broadcast on Tanzania's national TV (Tanzania Broadcasting Corporation) in 2011. The series was very successful not only because of its storyline, which dealt with the universal theme of family and generational conflict, but also because of the novelty of Chinese actors "speaking" Swahili with perfect African accents. English and other African language versions of the series aired in other countries. China decided to make the successful Tanzanian experiment a model for its television and film distribution in Africa and other parts of the world. The Propaganda Department of the Chinese Communist Party soon became the most important sponsor of the translation of Chinese television programs and movies into different languages. In 2014, Chinese Radio International and the State Administration of Press, Publication, Radio, Film and Television, launched a multilingual television and film translation and post-production center to post-produce Chinese televisions programs and films for the global market (Ming, 2013). Several Chinese TV series and films have been dubbed into English, French, Swahili, and Hausa, major languages used in Africa. It was the first time a government-funded, non-sports series of this nature had been translated and dubbed for mass audiences in free-to air government television channels on the African continent. These popular series, which were generally provided free of charge to African governmental public television stations as "national gifts" from the Chinese people, have been broadcast on national television stations in Tanzania, Uganda, Kenya, Comoros,

Senegal, Gabon, Burundi, Namibia, and Togo, where they have had very high ratings. Other Chinese series are being specially post-produced for the African market. The ultimate goal of the Chinese "going out" policy is building relationships between Chinese and African television organizations for purposes of marketing Chinese TV and movie content, and exerting Chinese "soft power" in Africa (Huang, 2018). This is consistent with China's policy of making its voice heard internationally (Thusu et al., 2018).

Since the Chinese government engages in "cultural governance" (Schneider, 2012, p. 6), its political apparatuses maintain control over television broadcasting infrastructures and content. The state uses TV programming to further its ideological objectives, and provide mass entertainment that is consistent with the Chinese Communist Party's ideological posture, as well as legitimizing the state and its actions. Television programming is thus highly didactical and homogeneous in terms of discourse (Schneider, 2012). Nevertheless, during the process of translating Chinese programs for the African and other international markets, the Chinese government ensures that stridently ideological scenes designed to serve didactic purposes for Chinese audiences, are neutralized for the international market (Schneider, 2012; Yun & Wu, 2018). Nevertheless, when Chinese state-driven distribution and marketization of programming takes place, countries on the receiving end also receive Chinese politico-cultural values as part of the package of soft power.

## Conclusion

African television is at the crossroads. Information and communication technologies and mobile telephony mean it has newly converged platforms to work with. This has led to tensions between entrenched governmental political television, private commercial stations, and international players such as China. African television is no longer merely an extension of the sphere of state governments who made monopoly broadcasting part of the architecture of repression and monolithic communication. The on-going analogue–digital transition means more spectrum space and channels for public service, commercial, community, and subscription television, as well greater presence in cyberspace. The Internet and social media platforms have eliminated the traditional gatekeeping and agenda-setting roles of governmental PSBs and given African viewers choices. However, African television is still constrained by restrictive regulations on freedom of expression, and administrative agencies that allow authoritarian regimes to maintain systems in which the law trumps rights. In some regimes, such as Cameroon, Nigeria, and Egypt, religious rites and ethnic solidarity trump human rights and freedom of expression. The massive presence of China at all levels of African television – from infrastructural development, the analogue to digital switchover, to rural community television and subscription and commercial television – provides more competition, more choices, and better television experiences for African viewers. However, some political scientists see China's concerted effort to enter the major sectors of the African television market as governmental public diplomacy intended to exert Chinese soft power (Huang, 2018). To others, it is part of Chinese political and cultural imperialism.

## References

Agutu, N. (2018). CA delivered KTN shutdown letter to Bamba an hour after. *Star* (Nairobi). January 30. Retrieved from: www.the-star.co.ke/news/2018/01/30/ca-delivered-ktn-shutdown-letter-to-bamba-an-hour-after_c1705999.

Akdim, Y. (2015). *Télévision: Qui est prêt pour la TNT en Afrique? Jeune Afrique.* September 22. Retrieved from: www.jeuneafrique.com/mag/238185/economie/television-qui-est-pret-pour-la-tnt-en-afrique/.

Cherono, S. (2018). Matiang'I says 3 TV stations to remain shut pending probe. *Daily Nation,* (Nairobi, Kenya) January 31. Retrieved from: www.nation.co.ke/news/Matiangi-TV-stations-remain-shut/1056-4286150-10l49h0/index.html.

Digital switch-off leaves 1.3 million homes in TV blackout (2015). *Business Daily* (Nairobi, Kenya), June 18. Retrieved from: www.businessdailyafrica.com/corporate/Digital-switch-off-leaves-1-3m-homes-in-TV-blackout/539550-2755674-sakw9jz/index.html.

Eko, L. (2010). "New Technologies, Ancient Archetypes: The *Boston Globe's* Discursive Construction of Internet Connectivity in Africa." *Howard Journal of Communications 21(2),* 182-198.

Eko, L. (2012). *New Media, Old Regimes: Case Studies in Comparative Communication Law and Policy.* Lanham, MD: Lexington Books.

Electronic and Postal Communication (Radio and Television Content) Act (2018). CAP 301, Tanzania Government Notice 134, published 16/3/2018.

Eribo, Festus and Jong-Ebot, William eds. (1997). *Press Freedom and Communication in Africa.* Trenton, NJ: Africa World Press.

Gabon President Bongo Back Home a Week After Failed Coup (2019). *BBC News.* January 15. Retrieved from: www.bbc.com/news/world-africa-46885467

Gearan, Anne (2018). Bolton Says Predatory China is out-pacing the U.S. in Africa. *The Washington Post* (December, 13). Retrieved from: https://washingtonpost.com/politics/bolton-says-predatory-china-is-out pacing-the-us-in-africa/2018/12/13/7b906626-fee2-11e8-83c0-b06139e540e5_story.html?noredirect=on

Gqirana, T. (2015). Cabinet sets aside R3-billion for set-top box rollout. *Mail and Globe* (South Africa). Retrieved from: https://mg.co.za/article/2015-03-05-cabinet-sets-aside-r3-billion-for-set-top-box-rollout

Huang, K. (2018). The "Going Out" of China Radio International. In *China's Media Go Global* Daya Thussu, Hugo de Burgh, and Anbin Shi (eds.). pp. 141–152 New York: Routledge.

ITU (2010). Guidelines for the transition from analogue to digital broadcasting. Retrieved from: www.itu.int/ITU-D/tech/digital_broadcasting/docs/TransitionAD_Leaflet.pdf

ITU (2018). Project on the digital broadcasting transition roadmap in Africa" International Telecommunications Union Retrieved from:www.itu.int/en/ITU-D/Technology/Pages/Projectonthedigitalbroadcasting transitionroadmapinAfrica.aspx

Keane, M. (2008). From National Preoccupation to overseas aspiration. In Ying Zhu et al (ed.). *TV Drama in China* (pp. 145–156). Hong Kong: Hong Kong University Press.

Kenya Cuts TV Transmissions over live coverage of Opposition's Odinga (2018). Committee to Protect Journalists, 2018.

Mbembe, A. (2001). *On the Post-Colony.* Berkeley, CA: UCLA Press.

Kenya's Raila Inaugurates Himself (2018). BBC News. Retrieved from: www.bbc.com/news/world-africa-42870292

Kuhn, T. (1962). *The Structure of Scientific Revolutions.* Second Edition, Chicago: The University of Chicago Press.

Le Monde & AFP (2018). Le président gabonais Ali Bongo poursuit sa convalescence au Maroc. *Le Monde,* December 4. Retrieved from: www.lemonde.fr/afrique/article/2018/12/04/au-maroc-premieres-images-du-president-gabonais-ali-bongo-en-convalescence_5392365_3212.html

Ming, X. (2013). Dramatically-Liked: Daily Life of a Chinese Family is TV Hit in Africa. *Global Times* (Beijing). April 11. Retrieved from: www.globaltimes.cn/content/774307.shtml

Monnier, O. and Hoije, K. (2018). In West African TV, It's France vs. China as StarTimes Enters. *Bloomberg,* June 27. Retrieved from: www.bloomberg.com/news/articles/2018-06-28/in-west-african-pay-tv-it-s-france-vs-china-as-startimes-enters

Naik, S. (2018). Satellite TV for Africa Courtesy of the People's Republic of China. *Independent Online (*IOL, South Africa), July 10. Retrieved from: www.iol.co.za/saturday-star/news/satellite-tv-for-africa-courtesy-peoples-republic-of-china-15934797

O'Grady, S. (2018). In Cameroon, Journalists are being Jailed on Charges of "fake news." *The Washington Post,* December 16. Retrieved from: www.washingtonpost.com/world/africa/in-cameroon-journalists-are-being-jailed-on-charges-of-fake-news/2018/12/15/80bcb5c6-f9ad-11e8-8642-c9718a256cbd_story.html?utm_term=.f465a533fa1d

President Deals Major Blow to Press Freedom (2009). Reporters without Borders. Retrieved from: https://rsf.org/en/news/president-deals-major-blow-press-freedom

Saliu, O. and Baoping, Z. (2017). Feature: China initiates satellite TV project in rural Africa. *Xinhuanet.* August 11. Retrieved from: www.xinhuanet.com/english/2017-08/11/c_136517844.htm

Schneider, F. (2012). *Visual Political Communication in Popular Chinese Television Series.* Leiden: Brill.

South Africa to Spend Sh23bn on set top boxes for the poor (2015). *Business Daily* (Nairobi, Kenya). March 7. Retrieved from: www.businessdailyafrica.com/South-Africa-to-spend-Sh23bn-on-set-top-boxes-for-the-poor/-/539546/2645352/-/5edj20z/-/index.html.

Tanzania Communications Regulatory Authority (2018). The Electronic and Postal Communications (Radio and Television Broadcasting Content) Regulations 2018. Retrieved from: www.tcra.go.tz/images/docu

ments/regulations/SUPP_GN_NO_134_16_03_2018_EPOCA_RADIO_AND_TELEVISION_BROAD CASTING_CONTENT_REGULATIONS_2018.pdf

*Télévision: Le Burkina faso lance officiellement le TNT* (2017). *Jeune Afrique*. December 29. Retrieved from: www.jeuneafrique.com/505888/economie/television-le-burkina-faso-lance-officiellement-la-tnt/.

Thusu, D., de Burgh, H. & Shi, A. eds. (2017). *China's Media Go Global*, pp. 1–14. New York: Routledge.

Yun, G. and Wu, J. (2018). Internationalization of Chinese New Documentary. In *China's Media Go Global* Daya Thussu, Hugo de Burgh, and Anbin Shi (eds.). pp. 153–166 New York: Routledge.

# 36

# TV CHINA

## Control and Expansion

*Ying Zhu*

Initiated in 1958, television in China has transformed itself over the past several decades from a state propaganda organ to a profit-generating media juggernaut. Atop China's TV pyramid is China Central Television (CCTV), the only network TV in China, which was funded entirely by the state from 1958 to 1978 and served at the party's pleasure. TV became a popular medium in China when advertising was introduced as a revenue source in the late 1970s amid the post-Mao marketization and commercialization drive, which propelled Chinese TV to create party-sanctioned programs that would nevertheless appeal to the public and thus advertisers. The first commercial appeared on Shanghai TV in January 1979. CCTV cautiously jumped onto the market wagon and experimented with commercial spots in 1979. With TV's popularity on the rise and advertising bringing in revenue, the state began to reduce its annual budget to CCTV in 1984. Development of the TV industry took off in leaps and bounds in the 1980s following a series of central government decisions to boost the production of television sets and expand the number of channels and broadcast hours. Fewer than one television receiver per hundred people existed in China back in 1978, and only 10 million had access to television, but by 1996 there was a television receiver for every four people, and 1 billion had access (Zhu and Berry, 2009, 2–3). By 2000, this figure had risen to 1.19 billion, representing 92 percent of the population and making TV the dominant medium in China. As China sped up its overall marketization and decentralization drive in the 1990s, state funding gradually diminished, and was dropped entirely by 1997.

As state subsidies to CCTV dwindled during the early 1990s, the station's then chief, Yang Weiguang made the radical decision to attach commercials to the "National News Bulletin," the prime-time news program carried by all stations in the country (Zhu, 2012, 29–30). Though commercial spots had been introduced since the late 1970s, National News Bulletin, as party's major propaganda vehicle, remained a sacred cow in national TV broadcasting, being shielded from the market "contamination." It was a risky proposition because, though television commercials during light entertainment programs were well established by then, the flagship prime-time network news program was the program most scrutinized and controlled by CCTV's overseer, the State Administration of Radio-Film-Television (SARFT), so attaching commercials to it might trigger negative reactions. The station started out conservatively, running a 30-second spot after the news program. The 30-second spot was stretched into 60 seconds six months later. Sponsors began to compete for the spot and lobby for longer airtime for the spot. CCTV had to hold an auction, awarding the time slot to the highest bidder. Two Chinese liquor brands, Confucius Family Wine and Confucius Banquet Wine became the finalists. In the end, the Family Wine secured the first-year contract, leaving the Banquet Wine for the second-year contract. With the blessing of the Confucius branded Chinese liquors, CCTV came out of the state subsidy by 1997 an enviable winner of China's new market economy.

Indeed, economic reform in the form of commercialization and marketization played a major role in the rapid development of the Chinese television industry. In November 2002, the 16th Party Congress of the Chinese Communist Party officially designated broadcasting as both a cultural industry and institution, with the latter attending to the public good and the former following market incentives. The official endorsement of market mechanism allowed the Chinese television industry to become more diverse in its content and pushed the industry to adopt international management and programming practices, which led to professionalization, commercialization, and internationalization of a previously draconian Soviet-style bureaucratic system, though centralization in terms of the party's political control has continued, and has even been enhanced under the leadership of Xi Jinping, the party chief and president of China since 2012.

## China Central Television and China's TV System

Though CCTV now operates as a commercial enterprise abiding by market principles, its fundamental function as party mouthpiece remains. As such, it retains its protected status as China's monopoly national television broadcaster. CCTV now operates over 50 channels, each broadcasting different programs, and is accessible to more than one billion viewers. Day-to-day supervision of CCTV falls to SARFT, which coordinates and evaluates the network's key propaganda efforts, regulates its signal coverage, controls its senior appointments, and decides on its organizational structure and all of its programming. Meanwhile, the Ministry of Central Propaganda under the Chinese Communist Party (CCP) manages the broad propaganda mission and provides ideological guidance. These bodies exist to "strengthen the structural management of the media through specialized government agencies" (Zhao, 2008, 26). The structure of these two is replicated at the network level: CCTV's internal leadership is comprised of a party committee and a senior management team. The former is responsible for ideological control, while the latter is in charge of the station's daily operation. The memberships of these two groups closely overlap, though the party committee ultimately trumps the management team. Formally, the state controls broadcasting through administrative regulations, which include administrative statutes promulgated directly by the State Council and department rules enacted by SARFT (Zhu, 2012, 23).

The relationship between this national CCTV operation and the many local stations is a flawed one, partly as a result of the cumbersome structure of China's TV system. China's overall commercialization and decentralization drive led to the initiation in 1983 of a four-tier TV system that had TV stations set up at the national, provincial, county, and city levels. Both national and local regulators operated their own TV stations and served audiences within their own administrative boundaries. The policy was aimed at getting local communities and authorities involved in local broadcasting so as to lessen the central control. The four-tier policy came into existence further because the new science of audience research pointed toward localized consumption preferences (Keane, 2015). The change coincided with the upgrading of China's Central Broadcasting Administrative Bureau to the Ministry of Radio and Television in 1982. The agency was later upgraded to the Ministry of Radio, Film and Television (MRFT) in 1985. Decentralization granted local stations the autonomy for financial management and program development and specified that a television station could be established if a certain financial threshold was achieved, leading to growth in the number of television stations in the ensuing years. From 1984 to 1990 the number of terrestrial television stations increased from 93 to 509

(Huang, 1994, 236). By 1990 there were over 1,000 cable television stations (Liu, 1994, 218). By the end of 1995, a total of 2,740 stations including terrestrial broadcasters, cable stations and university television channels were beaming out signals (Tu, 1997, 4). Such an ecosystem ensures that television stations, broadcasting bureaus, and governments at the same administrative level are closely linked in economic and political patronage. TV stations seek respective government and broadcasting bureaus' protection to maintain their grip on regional markets while government and broadcasting bureaus count on television stations to ensure their political influence and to also generate revenue. Since the early 1990s, TV stations have been important economic engines to local governments and their broadcasting bureaus. This is particularly evident in affluent regions where local channels were able to leverage enough local resources to compete with CCTV-1's terrestrial channel for ad revenue.

The four-tier policy did not dismantle CCTV's monopoly in national news carried by CCTV-1, which was to be integrated into local program flow. CCTV-1 is CCTV's flagship channel and carries national news programs that are beamed to the entire nation via local station relays. SARFT emphasized that guaranteeing CCTV-1's national coverage is a political mission, an "undeniable" obligation and responsibility of local broadcasting bureaus and television stations. CCTV is further granted exclusive coverage rights to major national and international events, and the CCP regularly leaks exclusive information to CCTV, making it the go-to source for insight on the party. In addition, CCTV has the exclusive coverage rights to national and international sports events, including the Olympics and the World Cup, which bring in huge profits. SARFT's favorable treatment of CCTV caused widespread discontent among local stations. In the 1990s, although MRFT reiterated the "must carry" policy, many local stations ignored these directives. To ensure local compliance, MRFT issued several policy documents, demanding that local terrestrial, cable, and retransmitting stations carry CCTV-1's programs in full, including commercials. The monopoly of news programs has turned CCTV news into a lucrative profit-generating vehicle, with massive annual auctions for advertising spots. Advertising is not allowed during the news bulletin itself, so spots immediately before and after are highly sought after. Another monopolistic and ad revenue generating program of CCTV is the annual Spring Festival Gala, which was watched by hundreds of millions of domestic Chinese audiences and thus commands big price tags for ads. Spring Festival Gala was accessible by Chinese diaspora around the world with eager inserts of local ads. TV drama is another area where stations can fetch sizable price tags for commercial spots. However, recent rules from the State Administration of Press Publications Radio Film and Television (SAPPRFT) restricting advertising spots within television dramas have compelled advertisers to seek less restrictive online platforms for dramas since 2012. Overall, SARFT has been consistently criticized for suppressing attempts by provincial TV stations to expand regionally and nationally, thereby securing the network's national monopoly. As the financial stakes grew huge, discontent over SARFT's preferential policies mounted among the local broadcasters, especially as provincial stations began launching local satellite TV channels.

In China, each provincial TV station is allowed to operate one satellite TV channel with signal coverage technologically capable of reaching the entire nation; yet because of the administrative boundaries and local protectionism, each provincial TV station must negotiate with other provinces to bring its satellite channel to their local cable networks. Most provincial broadcasters have managed to extend their regional reach via independent satellite and cable distribution deals with other provincial broadcasters. The youth and entertainment-oriented Hunan Satellite TV (HSTV), in particular, has become a formidable CCTV challenger in recent years. HSTV's *Super Girls*, a singing competition show modelled on *American Idol* with mobile phone voting, became an overnight rating sensation when it debuted in 2004. Feeling the heat, CCTV launched a campaign to attack HSTV, calling it a rogue broadcaster with vulgar taste. As a local station not on an equal footing with the state-run CCTV, HSTV took great pains to drive home the point that *Super Girls* was not aimed at competing with CCTV (Zhu, 2012, 201–204), thus preserving the party-mandated political and institutional hierarchy.

## Chinese Television at the Service of the Party-State

Between May 1994 and February 1995, following party orders, the MRFT undertook a comprehensive analysis of broadcasting in China, culminating in the "Report to Further Strengthen and Improve Radio, Film and TV Work," which stipulated four major objectives for Chinese media: retaining correct opinion guidance, improving program quality, enhancing government oversight, and promoting technological development (Zhu, 2012, 24–25). In anticipation of increasing social tension caused by inflation and cuts in social safety net and housing systems as a result of economic reform, Li Tieying, a member of the Central Committee of the CCP, spoke at a national broadcasting and film working conference in 1995, emphasizing that the imperative of leading people with correct opinion, thus providing a positive environment for reform, development, and stability, and letting "the whole Party and entire people of the country recognize the socialist market economic system and understand its basic elements," should be the "top priority" for broadcasters (Li, 2000, 254–259). The second MRFT objective was to improve the quality and quantity of radio and TV programs. "Quality" was defined as content that would "promote the main melody and advocate diversity" (Li, 1995, 255). The "main melody," according to the then President Jiang Zemin, was "patriotism, socialism and collectivism." "Diversity" promoted "richer," more "colorful," and "creative" content to compete with the rising tide of East Asian popular culture for viewership. Xu Guangchun, vice minister of the Ministry of Central Propaganda laid down the "four no changes" in China's media industry at the Fifteenth National Congress in 1997: "no change in nature as the Party, Government, and people's mouthpiece and the Party's important public opinion apparatus and battlefield"; "no change in glorious duty of serving the Party's national work"; "no change in responsibilities of persisting in correct public opinion guidance, creating a good public sphere, and providing strong public support for reform, liberalization, economic, and social development"; and "no change in the leadership of the Party over broadcasting" (Xu, 1998, 262). The directives were reiterated in 2011 as the party-state reined in reality-based programs such as dating shows and game shows that were deemed to have deviated from its propaganda and control mandate (Zhu, 2012). SARFT and the General Administration of Press and Publications merged into SAPPRFT in 2013 to streamline regulatory control and content oversight. SAPPRFT is responsible for drafting the laws and regulations that provide "supervision and management" of radio and television programs. Such laws and regulations are to be ratified by State Council. SAPPRFT can suspend or cancel programs and revoke the licenses of stations/companies that do not adhere to the party directives.

As I discussed in *Two Billion Eyes: The Story of China Central Television* (Zhu, 2012), cultural policy in China follows directives issued by the party's Propaganda Department, which is embedded at all layers of government, from the national level all the way down to the local. These party propaganda departments, along with the party committees within media institutions, act as censors and set the policy and propaganda tone according to the directives of the Ministry of Central Propaganda. The local Propaganda Department operates mainly behind the scenes and in broad strokes, except for editorial content (news and information), which it manages very closely via secret weekly instructions to the media. In the past, the Propaganda Department also actively directed the variety, quantity, and content of cultural products. In the reform era, however, self-censorship by media professionals has been the principal mode of control. While the party-state must now attend to markets and their systems of rules, it still exercises what often amounts to ad hoc control over China's entertainment industry. The state government establishes laws, but because the party apparatus ranks above the state government, party directives trump everything else. For the entertainment industry, this amounts to an informal regulatory practice that results in often disruptive regulatory fiddling. For entertainment industry professionals, it creates a climate of constant uncertainty and self-censorship. Meanwhile, economic logic continues to encourage entertainment programs of popular appeal, which will inevitably exert pressure for censors to loosen their grip. These contradictory mission statements and swings—from forays into market reform to retreats to conservatism, from initiatives to commercialize to those attacking it—have been typical of media regulation in the post-Mao era, and they have contributed to the on-the-job training of a new generation of media professionals thoroughly steeped in the fine art of intuiting what their overseers want. It may involve self-censorship, the primary mechanism of cultural control in contemporary China, or at times out-and-out censorship. Under this environment, the development of Chinese television must reconcile the party's demand for loyalty and stability and the market's demand for profits and popular appeal, and the end result has been the elimination of ideologically and artistically provocative programs that do not conform to either the party's or the market's demand. As I argued elsewhere (Zhu, 2015), the battle against the West initially launched on political and ideological grounds has been complicated by an all-encompassing market force. The patronage between the party and the state-run enterprises means that they more often than not act in unison to protect their mutual interest in keeping the masses and profits on a short leash. Meanwhile, the West continues to serve as a punching bag and a convenient archenemy for the CCP to rally popular support in the name of patriotism. The economic dimension of China's battle against Western and entertainment cultures adds complexity to the political and ideological dimension of control and censorship. While Western reports of Chinese prime-time TV's cultural cleansing focused mostly on political and ideological control, a close look at the unique structure and control mechanism of the Chinese television industry brings to light a more intricate economic tussle between the national and local broadcasters. The CCP has maintained that the biggest threat to China is Western cultural erosion. China's cultural regulators and guardians have always been vigilant against Western cultural encroachments such as the spreading of political ideas associated with liberal democracies and popularity of Hollywood-led US pop culture. The on and off cultural war against the US entertainment industries notwithstanding, Mao's Cultural Revolution was said to have been launched as a counter-punch to the strategy of "peaceful evolution" promoted by the US Secretary of State John Dulles in the 1950s that aimed to turn Communist China into a US-style liberal democracy (Ma, 2012).

The campaign against Western cultural "infiltration" only intensified under China's new President, Xi Jinping. A leaked party memo issued in April 2013 titled "Briefing on the Present Ideological Situation" listed a number of ideological "perils" that included constitutional democracy, universal values of human rights, press freedom, and complaints about the undisclosed personal wealth and financial ties of government officials. The memo warns that Western anti-China forces and internal "dissidents" are still actively trying to infiltrate China's ideological sphere, and have made a fuss over asset disclosure by officials, fighting corruption with the Internet, media supervision of government, and other sensitive hot-button issues, all of which stoke dissatisfaction with the party and government. By 2014, the Western "cultural threat" escalated to be one of several types of "unconventional security threats" under China's new National Security Committee's surveillance. The senior colonel of the People's Liberation Army, Gong Fangbin, singled out the US as a particularly dangerous source of Western culture infiltration and Hollywood movies and the Internet as lethal tools for disseminating Western cultural values. Chinese youth were perceived as particularly vulnerable to such Western cultural influence.

## Popular Programs

The most popular genre on Chinese TV is serial drama, though the rise of reality shows since the mid-2000s has threatened to challenge its dominance. China broadcast its first television drama, a black and white single-episode anthology drama *A Mouthful of Vegetable Pancake* (*yi kou caibingzi*) in 1958. From 1958 to 1966, roughly 200 television dramas were broadcast. The Cultural Revolution interrupted much of TV production and broadcasting. Beijing TV Station, the predecessor of CCTV announced in 1966 that starting from January 3, 1967, the station would "suspend normal broadcasting service except on important matters and occasions" (Huang, 1997). Between 1967 and 1976, the only "entertainment programs" shown on Chinese television were eight revolutionary model operas. During the period, TV broadcasts typically started at 7 p.m. with the sound of "The East Is Red" in praise of Chairman Mao with Mao's image on Screen (Hong, 1998, 55). The opening would be followed by newscasts of political stories ranging from commemoration of heroes to the work of educated youth in a remote village, etc. The news program would then be followed by reruns of revolutionary ballet and old Chinese movies about the anti-Japanese or anti-Nationalists war. The station signed off promptly at 10:30 p.m. Beijing Television was renamed China Central Television in 1978, one year after the end of the Cultural Revolution in 1977. China had less than one television set per 100 people when CCTV officially started its operation in 1978.

Drama production recommenced in 1978. The first serial drama, *Eighteen Years in the Enemy Camp*, was produced by CCTV in 1980 and screened in 1981. The production of a highly acclaimed serial adaptation of the Chinese classic *The Water Margin* (*shui hu*) in 1982 by Shandong TV station led to a slew of tele-adaptations of popular classics, making costume serial drama PRC's niche export program (Zhu, 2008). The Provisional Rules for the Censorship of Television Serials was promulgated in April 1999 to provide guidance on what was permitted in TV dramas. The regulations state that a television drama must be modified or edited where it contains scenes, language or images that are overtly sexual, violent, of "low taste" or that promote activities such as gambling, drug taking, fortune telling, religious zealotry or intentionally damaging the environment. For imports, programs that incite ethnic separatism, ethnic hatred or discrimination were forbidden. So were programs that destroy national solidarity or denigrate the customs or habits of the Chinese people.

Television serial drama dominated local programming until the late 1990s, when entertainment programming such as talk shows and reality TV became fashionable. The popularity of entertainment shows, especially dating shows such as Jiangsu Satellite TV's *If You Are the One*, led SARFT to issue an edict in 2010, "Strengthening the Regulation of TV Dating Shows," stipulating that dating shows "shall not insult or slander participants or discuss sex in the name of love and marriage, shall not display or sensationalize unhealthy, incorrect outlooks on love and marriage such as money worship" (Bai, 2014). In October 2011, Chinese authorities banned scores of racy and overtly materialistic entertainment shows on prime-time television. *If You Are the One*, which showcased provocatively dressed young women publicly embracing materialism and opting for wealth and affluence over love and romance, is frequently cited as an example of "excessive entertainment." Beginning on January 1, 2012, SARFT enacted a mandatory cap on the number and duration of entertainment shows that would be allowed during prime time (7:30 to 10:00 p.m.). The order restricted the number of entertainment programs that satellite channels can broadcast to two per week, with a maximum of 90 minutes of content defined as entertainment. Programs promoting "traditional virtues and socialist core values" were encouraged. As Michael Keane notes (2015), a new edict was issued the same year by SAPPRFT advising TV serial production companies to rectify six contentious areas. The first was to ensure that the depiction of characters in historical dramas are "unambiguous," which meant that the party-sanctioned heroes should be flawless. The second was to avoid portraying fraught family relations in times of unparalleled social change. The third was to stop production of satirical costume dramas. The fourth emphasized that dramas depicting business rivalries must be guided by positive values. The fifth warned producers not to be corrupted by overseas programs. The sixth advised against using online stories and online games as sources for drama adaptations.

Single-episode anthology dramas dominated prime-time television in China throughout the 1980s. However, CCTV began to import serial dramas from the United States, Japan, Hong Kong, and South America in the early 1980s (Zhu, 2008). *Iron-Armed Atongmu* and *Doubtful Blood Type* from Japan, *Dynasty* and *Dallas* from the United States, *Huo Yuanjia* from Hong Kong, *Woman Slave* from Brazil, and *Slander* from Mexico became popular hits. Between 1982 and 1984, CCTV signed contracts with six American television companies. The deal with CBS in 1984 imported 64 hours of "off-the-shelf" programming for 320 minutes of advertising time, which CBS then sold in ten packages of 32 minutes to nine foreign sponsors (Hong, 1993). As part of the agreement, CCTV shared in half the revenue paid by the foreign advertisers. A similar programs-for-advertising deal was negotiated in 1986 between Lorimar Telepictures and Shanghai Television, this time for 500 hours of American programming (Huang, 1994, 230). Between 1985 and 1986, national and local stations were said to be getting as many as 750 television dramas and telefilms from overseas. The government eventually stepped in to rein in the explosion of foreign programs on Chinese TV. In 1995, SARFT issued a circular to limit imported TV dramas to 25 percent of overall broadcast time and no more than 15 percent of prime time, although the rules were not strictly enforced and many local stations applied the quota on a channel by channel basis. In practice, foreign programming might account for as much as 50 percent of programming (Wolf, 2012). The popularity of the imports nonetheless helped to re-define both the story structure and the program format of China's domestic television drama, encouraging the experimentation with multiple-episode dramas.

Serial dramas did not begin to outnumber anthology dramas in prime-time programming until after the debut of China's first long serial drama, *Yearnings* in 1990, which marked the turning point for multi-episode, hour-long drama. *Yearnings*' sweeping success firmly established the multi-episode serial as a commercially viable format in Chinese prime-time television. The early 1990s thus became the pivotal moment when television drama in China broke its tie with feature-film-style narrative and became aware of its unique narrative structure and commercial potential. Serial dramas became the default prime-time genre. China now produces the largest number of TV dramas in the world, though with limited global fanfare beyond the confines of diasporic Chinese communities.

To produce programs and broadcast in China, foreign entities must obtain a license from SAPPRFT. Phoenix TV is a Hong Kong-based satellite channel originally a 55/45 joint venture between a Chinese businessman. Liu Changle. and Rupert Murdoch. The elite news talk show-oriented Phoenix is the most successful non-mainland satellite channel with limited access to the mainland market and with loyal followers among China's elite viewers. Starry Skies Satellite station (Xing Kong weishi) is a Mandarin-language satellite channel specializing in light entertainment fares. It was formerly connected with Murdoch's STAR TV China based in Hong Kong. Murdoch gradually withdrew its investment and eventually sold all his remaining shares in January 2014 to concentrate on regions with less strict government control. Censorship remains the most serious obstacle to competitiveness. In order for a program to be approved it must run the gauntlet of "content management." First, full scripts are submitted to provincial-level examination and re-examination committees, and then, after initial approval, to dedicated committees in SAPPRFT. Receiving clearance from SAPPRFT does not guarantee distribution. While more evident in the film industry, co-productions have taken off in the TV industry as well, with Chinese companies collaborating with Hong Kong, Taiwanese and Korean companies to produce serial dramas. Co-productions benefited local producers by bringing in knowledge and skills and at the same time allowing their East Asian counterparts a foothold in the Chinese market.

## An Evolving TV System

In recent years, China's TV industry has witnessed the rise of private media companies, many of which are spinoffs of state broadcasters. A number of owners of these private companies are former state employees with seasoned knowledge on how the system works. In China, any person or enterprise can form a media production company as long as SAPPRFT greenlights the license. Approximately 80 percent of television production companies are now classified as private. As the industry matured, in house production was replaced by productions from private companies with contracts to stations. Private companies produce shows on a contractual basis for the broadcaster, which holds the rights to the program and can distribute it in whatever way benefits the broadcaster. High-rating shows now command exclusive release on the network that invested in the show before being released on the network affiliates and finally on online video streaming sites.

The rapid expansion of digital media and the proliferation of over the top (OTT) content in recent years has transformed China's entertainment industry. Many young viewers now access entertainment content via online devices. A slew of Chinese digital media companies, such as Youku-Tudou, Sohu, Baidu, IQiYi, LeTV, BesTV, PPS, and PPTV, have joined the Chinese media market both in content creation and distribution. The presence of online players has expanded the industry from "network television" dominated by the big commercial networks to "networked television" where viewing is connected to devices and networks other than television (Holt and Sanson, 2014). With the rise of revenues from online distribution, the Ministry of Industry and Information Technology (MIIT) is getting involved in the regulation of the television sector as MIIT must remit profits to SAPPRFT, thus ensuing turf wars between SAPPRFT and MIIT over control and profit.

A mixture of state-run and private media companies with an influx of online delivery mechanisms has become the new norm of China's television and media system, with fierce competition persisting between the established media and the new media, both vying for state-sanctioned contents with popular appeal. As domestic competition intensified, the Chinese state called for a united domestic front in exporting China's culture abroad. Chinese television under President Xi Jinping is to aid "national rejuvenation" by exerting China's influence globally and exercising political and ideological control domestically. To this end, China has invested heavily in the television industry. State subsidies in the form of low taxes and direct investment have allowed Chinese television to grow at a rapid pace, and with an eye on the global market, though there is little demand for Chinese entertainment contents in the rest of the world, a "cultural deficit" that China is eager to overcome.

## References


Bai, Ruoyun. 2014. "Curbing entertainment: television regulation and censorship in China's disjunctive media order", in *Chinese Television in the Twenty first Century: Entertaining the Nation*, edited by Ruoyun Bai and Geng Song, 67–68. London: Routledge.

Holt, Jennifer and Kevin Sanson. 2014. "Introduction: getting connected", in: *Connected Viewing: Selling, Sharing and Streaming Media in the Digital Era*, edited by Jennifer Holt and Kevin Sanson, 1–16. London: Routledge.

Hong, Junhao. 1993. "China's TV program import 1958–1988: towards the internationalization of television?", *Gazette* 52 (1), 1–23.

Hong, Junhao. 1998. *The Internationalization of Television in China: The Evolution of Ideology, Society and Media since the Reform*. Westport, CT: Praeger.

Huang, Yu. 1994. "Peaceful evolution: the case of television reform in post-Mao China", *Media, Culture and Society* 16 (2), 236.

Huang, Yu and Xu Xu. 1997. "Broadcasting and Politics: Chinese television in the Mao Era, 1958–1976", *Historical Journal of Film, Radio and Television* 17 (4), 563–574.

Keane, Michael. 2015. *The Chinese Television Industry*. London: BFI.

Li, Tieying. 2000. "Speech Given in the Meeting with the CCTV's Staffs", in *The Important Documents of Broadcasting and Film Work 1995*, edited by General Office of SARFT. 254-259. Beijing: SARFT.
Liu, Yu-li. 1994. "The growth of cable television in China", *Telecommunications Policy* 18 (3), 216–228.
Ma, Damin. 2012. "Beijing's 'Culture War' Isn't About the U.S.—It's About China's Future", *The Atlantic* (January 5) www.theatlantic.com/international/archive/2012/01/beijings-culture-war-isnt-about-the-us-its-about-chinas-future/250900/.
Tu, Chuangbo. 1997. "The development and legal policies of China's broadcasting network (woguodianshi wang de fazhan jiqi falu zhengce)," *Television Arts* (*dianshi yanjiu*) 6.
Wolf, David. 2012. "Television Regulations: New Bottle, Same Wine (With Corrections)", *Silicon Hutong* (Febraury 15) http://siliconhutong.com/2012/02/15/television-regulations-new-bottle-same-wine/.
Xu, Guangchun. 2000. "Speech Given in the National Meeting of the Chiefs of Provincial TV and Radio Stations" in *The Important Documents of Broadcasting and Film Work 1998*, edited by General Office of SARFT. 262. Beijing: SARFT.
Zhao, Yuezhi. 2008. *Communication in China, Political Economy, Power, and Conflict*. Lanham, MD: Rowman & Littlefield.
Zhu, Ying. 2008. *Television in Post-Reform China: Serial Dramas, Confucian Leadership and Global Television Market*. London: Routledge.
Zhu, Ying. 2012. *Two Billion Eyes: The Story of China Central Television*. New York: New Press.
Zhu, Ying. 2015. "China's Cultural War Against the West", in *Media at Work in China and India: Discovery and Dissection*, edited by Robin Jeffrey and Ronojoy Sen. 69-88. London: Sage.
Zhu, Ying and Chris Berry. 2009. *TV China*. Bloomington: Indiana University Press.

# 37

# TACTICS OF THE INDUSTRY AGAINST THE STRATEGIES OF THE GOVERNMENT

## The Transnationalization of Turkey's Television Industry

*Ece Algan*

Within two decades, Turkey's television industry went from producing content only for the domestic market to exporting content to over 100 countries from the Balkans to the Middle East and from Asia to the Americas. In 2014, the Turkish television industry became the second-largest exporter of scripted TV content after the US ("Turkey world's second" 2014; Vivarelli, 2018), and in 2016 it became the fifth-largest TV program exporter overall in the world ("France and Turkey" 2016). In 2016 alone, Turkish television series exports brought over US$350 million in revenue and reached over 500 million viewers, with a number of TV series breaking viewership records both inside and outside of Turkey (Sofuoglu, 2017). Reactions to Turkey's TV becoming a global growth industry have been mixed both within and outside the country. On the one hand, its global success and popularity has brought both TV industry professionals and government officials together to opportunistically unite under the claim that these exports strengthen Turkey's soft power in the world, particularly in the Middle East, and thus, it has been welcomed as a source of national pride. On the other hand, its fast-paced, flashy and sometimes edgy content has caused concern among government officials, policymakers and interest groups that it might misrepresent Turkey's history, culture and everyday life and thus negatively affect the public's sensibilities and the country's image abroad. As a result, Turkish TV producers and the well-known actors in these dramas have often found themselves trying to balance the demands of staying competitive in the global market with those of an increasingly unstable domestic socio-political climate with widening freedom of speech violations and direct verbal attacks from senior government officials, including President Erdoğan.

Indeed, during the height of the Turkish dramas' rise in the global markets, Turkey has faced major challenges, such as the change from a parliamentary system to a presidential one in 2017, preceded and followed by a number of elections, Turkey's involvement with the Syrian war, its military actions against Kurdish insurgent groups on the border, and its declining relations with Europe, the US and the Middle East, all of which contributed to the financial turmoil of 2018 to some extent. In addition, there was an attempted military coup in 2016, armed conflict in 2015 in its Kurdish cities with ongoing curfews, and a major uprising (Gezi) in 2013 that shook the country deeply, to name only a few. This environment of conflict and instability, both within and outside of Turkey's borders, coupled with the government's rhetorical strategy of suggesting foreign threats to Turkey aimed at dividing the country, created anxieties for the people, and fueled patriotism and a need for security. In response to this and the need to produce content based on local audience preferences due to high domestic competition driven by ratings and demand for quality, Turkey's TV industry catered its programming to appeal to people's nationalistic sentiments on the one hand, and their desire for a modern urban middle-class lifestyle, on the other.

Stuck between the political economy of the larger domestic television production industry and global market imperatives, I argue that Turkey's TV industry executives and professionals had to develop and implement a number of tactics to achieve a locally based transnational cultural industry able to withstand both global and domestic pressures. In this chapter I identify three main tactics employed by Turkey's TV industry executives and professionals to combat the socio-economic and political challenges they face: These tactics are: (1) carefully managing the content to skirt government restrictions; (2) adopting the government's soft power discourse and public diplomacy aspirations by cooperating with government officials and businesses in their cultural promotion and nation-branding efforts; and (3) adapting to global TV trends by undertaking rigorous marketing and branding campaigns.

I believe that a discussion of these tactics in the Turkish case can help us understand how culture industries in the developing world, which had to integrate into a neoliberal media environment after the 1980s due to market- and state-driven policies propelled mostly by US-based global media giants, negotiate being "local-based transnational cultural industries" (Jin, 2016) in the face of increasingly authoritarian and right-wing domestic political climates, as in the cases of India, Brazil and Korea. Before I discuss the tactics implemented by the Turkish TV industry executives, I will first provide a brief overview of the development of the Turkish television industry during the era of economic liberalization in light of recent transformations in the global television business. I then discuss the tactics that Turkish TV professionals have developed in order to operate within a highly unstable domestic political and economic environment while continuing to appeal to both domestic and global audiences.

In order to explain the context within which the Turkish TV industry flourished and became a transnational player, and to identify the tactics of the industry, I conduct a political economy analysis. This approach can illustrate how power between different groups and individuals is distributed and contested. My analysis is based on reports from industry trade associations such as RATEM (Radio and TV Broadcasters Association in Turkey) and corporate entities such as Deloitte, as well as news stories and op-ed columns from mainstream newspapers and industry trade magazines such as *Variety*. I also draw upon five unstructured interviews I conducted with television executives, producers and distributors in Istanbul in July 2015[1] and MIPCOM[2] newsletters which also feature interviews with a number of production and distribution company executives that provide Turkish TV content.

## Development of the Turkish TV Industry in the Era of Economic Liberalization

The first exported Turkish TV drama was *Deli Yürek* (*Crazy Heart*), which its producer sold to Kazakhstan for US$30 per episode in 1997 with the goal of getting the foreign markets into the habit of purchasing Turkish content legally instead of pirating it via the internet, as had been the case (Birsel Çıkıncı, July 2015. Interview by author). It was not until the mid-2000s, with the success of *Gümüş* (*Silver*, *Noor* in Arabic) and later *Muhteşem Yüzyıl* (*Magnificent Century*), that the Turkish TV industry started to become a visible global player. While it may seem like a late blooming industry, it should be noted that broadcasting was a state monopoly until the early 1990s, with only the state-owned and controlled network TRT dominating the airwaves (Algan, 2003), so the development in less than three decades of 196 TV channels[3] (Radyo Televizyon Yayıncıları Meslek Birliği, 2018) and around 85 TV production companies (Deloitte, 2014) is quite significant. Not surprisingly, there is much speculation about the reasons for this success.

Based on the feedback they gathered from their distributors, the TV production and distribution executives I talked to attributed the Turkish TV industry's success and expansion in the international markets to the high production values and universal themes[4] of Turkish soaps, along with audience boredom with American programming. The high production quality is a consequence of the industry's initial emergence and development within a statist economy, and later due to fierce domestic competition for ratings[5] and revenues. The competition is so fierce, in fact, that more than half of TV series are cancelled before the end of their first season (Deloitte Touche Tohmatsu Limited, 2014). With Turkish TV series costing up to US$800,000 per episode and actors demanding six-figure salaries (Vivarelli, 2017), the industry increasingly turns to foreign markets to wring profits out of shows that were domestically disappointing. However, compared to the overall number of TV dramas being produced, only a small percentage of those are seen by international viewers. For instance, only ten out of 70 different dramas produced in 2017 were exported, and only four or five of those became international hits (Vivarelli, 2017).

A number of scholars (Alankuş and Yanardağoğlu, 2016; Özalpman, 2017; Tunç, 2012; Yanardağoğlu, 2013; Yesil, 2015; Yörük and Vatikiotis, 2013) have criticized assertions that the success of Turkish dramas rested on cultural proximity or Turkey's soft power only. Studying Turkish TV dramas' reception by Iranians in Vienna, Özalpman (2017) argues that the transnational popularity of these dramas cannot easily be explained by simple proximity or through soft power theses as Iranian viewers' perceptions of Turkey or Turkish people do not necessarily correlate with their consumption practices surrounding these soaps, such as the increase in Iranian tourists to Turkey. She found that "the series' melodramatic, humorous and Cinderella storyline" (38) typical of the soap opera and telenovela genres are responsible for the Iranian diaspora's enjoyment of Turkish dramas. Her finding indicates the importance of genre proximity in transnational audiences' reception preferences as opposed to cultural proximity.

While Yesil (2015) also believes that the concept of "multiple proximities" might explain better how both the shared cross-cultural genre structures (melodramas) and narrative themes (family, romance, power, etc.) contribute to the transnational popularity of Turkish dramas, she also asserts that the proximity theses still fail to address Turkish dramas' success in the global market (45). According to Yesil (2015), the Turkish TV industry's transnational success should be seen as a consequence of various socio-political and economic factors in the mid–late 2000s that increased drama production in Turkey, while also creating a favorable environment for their distribution. Yesil (2015, 45) argues that a number of shifts in both global and local markets, such as

> the burgeoning of the production sector and the growth of the advertising market in Turkey, the search for additional revenue streams in foreign markets by Turkish producers, the integration of these actors into the global networks of [the] television trade and the governmental support given to producers and distributors

are all responsible for creating the export opportunities for Turkish dramas. Indeed, Turkish TV advertising revenues grew to US$1.3 billion in 2017 domestically (Vivarelli, 2017), enabling expensive, high-quality productions.

Similar to Yesil (2015), Alankuş and Yanardağoğlu (2016) also illustrate that cultural attraction abroad to Turkish dramas may not always be "admiration and consumption driven, but they may be contingent on a number of factors such as national and global television market dynamics as well as entrepreneurial initiatives" (3619). Nevertheless, Tunç (2012) has observed how larger economic dynamics, such as the Greek economic crisis, played a role in Greeks consuming Turkish dramas when Greek broadcasters found it cheaper to purchase Turkish serials for 7,000–8,000 euros per episode while domestic productions cost around 70,000–80,000 euros (339). Greek broadcasters' decision to risk featuring Turkish programs, despite existing Greek nationalism resulting from the long history of war and disputes over land and population between the two countries, illustrates the role of broadcasters in making decisions over purchasing content. Similarly, Chalaby (2015) also found that broadcasters have a big influence in TV programming trading and shaping trade flows. Chalaby (2015) argues that an emphasis on the cultural rationale explaining what content transcends national and cultural boundaries "tends to ignore the influence of capitalist power structures on the world TV industry and in particular its embeddedness in international trade" (468). Defining a trading system as "a singular transnational space that brings together interdependent economic agents, institutions, places, networks and commodities" (461), Chalaby (2015) illustrates how the TV business – format sales, in particular – became a transnational trading system in the 2000s in which the Turkish TV industry has actively participated.

Indeed, the Turkish TV industry's rise in the international markets should be evaluated within the larger developments in the global television business, such as the rise in cross-border media flow since the late 1990s (Chalaby, 2015, 468), increased opportunities for producers in small-market nations to sell their programs globally and the concomitant intense competition for content (Tinic, 2015), the replacement of national institutional structures supporting the media industry by global ones and the transformation of national TV production practices as a result (Keinonen, 2018, 165), the advent and rise of streaming technology (Burroughs, 2018), and the incorporation of big data technologies by media organizations (Arsenault, 2017; Havens, 2014). However, that the global TV business is operating increasingly as a "series of interlocking international production networks" (Chalaby, 2016, 38) does not necessarily make the role that nation-states play irrelevant. As Becerra et al. (2014, 47) have argued, globalized media does not undermine the power of governments. On the contrary, governments can shape the transnational TV trade actively via incentives and regulations, as in the example of Korea, and "retain power through patrimonial and quid pro quo practices given their control of decisions affecting business calculations, programming and news" (Becerra, Mastrini, and Waisbord, 2014, 47), as is the case in Turkey. Stressing the importance that nation-states have in purchasing K-pop products, Leung (2008) explains how in strictly regulated and controlled media environments such as China, transnational media have to be politically de-odored – meaning depoliticized – and "culturally correct" in order to be successful. She illustrates how Korean dramas exported to China had to underline and acknowledge Chineseness in Korean culture in order for them to be sanctioned (64–5).

Discussing the role of the Korean state and its cultural policies in the worldwide popularity of Korean audiovisual cultural products, Ryoo and Jin (2018, 4) argued that despite the conflicts between state-developmentalism and neoliberal globalization, the Korean state supported and even initiated the growth of the cultural sector through the development of a unique cultural policy, which was crucial to the global success of the Korean Wave. The Korean state's cultural policy aimed at developing and promoting the Korean Wave included a number of economic incentives and legal protections such as financial subsidies, low corporate taxes and low interest rates, increasing the cultural ministry's budget, supporting film festivals, developing domestic audiovisual industries by increasing the number of television programs produced by independent producers, and actively helping broadcast Korean soap operas abroad while continuing to deregulate and liberalize the culture industries.

Similar to Korea and other countries, Turkey has also deregulated and liberalized its culture industries based on global neoliberal imperatives since the mid-1980s, allowing commercial TV channels to operate legally after 1993 by changing the old broadcasting laws that mandated a state monopoly in broadcasting (Algan, 2003). However, according to the TV executives and producers I interviewed, the Turkish state has never developed a systematic set of policies to aid the growth of its private cultural sector abroad, except to offer some financial aid toward the promotion of TV series during international trade fairs and festivals. On the contrary, the Turkish government chooses to reward TV content that reflects a conservative worldview and nationalistic storyline by promoting them within and outside the country, as in the example of President

Erdoğan visiting the set of *Resurrection: Ertugrul* (Sofuoglu, 2017) and having its actors accompany him during his state visits abroad ("Diriliş dizisi oyuncuları" 2017), while cutting support to the ones that contradict or pose a challenge to official ideologies either by directly targeting them, as in the example of Erdoğan's public criticisms of *Magnificent Century*, or indirectly. For instance, during our interview, the president and CEO of Tims Productions, Timur Savcı, suggested that they had a hard time obtaining permits to film on location at historical sites (Timur Savcı, July 2015. Interview by author). Governmental favoritism, clientelism, exercise of patrimonial state power and quid pro quo practices, coupled with the lack of state policies aimed at helping the industry, such as tax incentives, are all challenges that the Turkish TV industry negotiates on a daily basis as it strives to grow transnationally.

## TV Industry Tactics for Overcoming Domestic Challenges

While I believe that the recent developments in the global TV business, the generic similarities of Turkish dramas to telenovelas, and the political economy of domestic TV production can all help us understand the processes behind the transnationalization of the Turkish TV industry, they fail to address how industry professionals working in smaller and semi-authoritarian national media systems, such as Turkey, deal with the domestic challenges that affect its transnational expansion. Exploring tactics developed and implemented by industry professionals to counteract domestic challenges can illustrate how a locally based transnational culture industry manages tensions between global and local demands in a neoliberal media environment. It can also help us understand the role of politics in shaping cultural policies and determining the audiovisual content produced in that country. In the next three subsections, I will argue and illustrate that the Turkish TV industry has developed a number of tactics in three main areas, namely: (1) the management of content creation in a way that withstands competition among a high number of domestic channels and new platforms to access content while minimizing government control and censorship; (2) industry assuming a "cultural envoy" role to take advantage of its cultural and economic influence abroad and prevent possible political retribution; and (3) the adoption of competitive promotional business strategies that respond to new trends in global TV, such as producing content for new platforms, launching new platforms and making international agreements to trade TV formats and adaptations.

### *Managing Content as a Tactic*

Emre Cetin (2014) states that the transformation that Turkish television has undergone since the early 2000s was propelled by "the expansion of the television production market, political challenges to the freedom of the press, and the eventual penetration of political content into dramas" (2464). Indeed, after the 2000s, television serials reflecting real-life problems became a very popular genre. The portrayal of ethnic identities in drama serials created an opportunity to discuss the existence of other groups, alternative ways of living and other invisible, previously ignored features of social life (Işık, 2013). Especially in the past decade when press freedoms were significantly curtailed, society was politically polarized, and mainstream newspapers, TV channels and the state news agency increasingly became mouthpieces for the government, Turkish dramas with political content became appealing for viewers who wanted to safely ponder and discuss the socio-political changes through imaginary plotlines that nevertheless were close to real life. For precisely that reason, as Emre Cetin (2014) has observed, Turkish TV dramas react to various daily socio-cultural and political developments by "tak(ing) sides in an overtly politicized media environment, while at the same time politicians and government institutions intervene in drama production" (2464).

Due to the industry's interest in exporting fresh, edgy and relatable content without attracting government scrutiny, this content often gets disguised as normal power relations taking place in historical or ahistorical settings, which in part explains why TV executives who are not closely aligned with the AKP government claim that they are only featuring universal themes. Despite their efforts, the industry continues to feature content that can be deemed culturally and/or politically sensitive by the Turkish state. As a result, numerous mainstream TV channels have faced repeated fines and requests for content modification, at times through RTÜK (Radio TV Regulatory Council) for violating the family or traditional values clause in the telecommunications law, and other times directly from high government officials, including President Erdoğan, for misrepresenting Turkish history and traditions. RTÜK's fines on the TV channels for showing certain groups of people in a negative light or encouraging lewd behavior, drinking, gambling, etc., encourages the industry to censor itself. Birsel Çıkıncı, who was the production designer and supervisor of Focus Film when I interviewed her, expressed how they began refraining from including passionate kissing scenes in their scenarios while making the characters in despair drink a glass of water instead of alcohol in an effort to prevent a possible RTÜK warning (Birsel Çıkıncı, July 2015. Interview by author). In some instances, the TV production companies even avoid casting actors who have attracted the attention of politicians due to activism or political comments on social media. Focusing on three popular Turkish TV series' actors – Mehmet Ali Alabora, Halit Ergenç and Beren Saat – Vitrinel (2017, 8–9) argues that "the characters played by celebrities can be challenging in the political debates and confusion between the 'role play' and the reality is not so rare."

Struggling to self-censor can create frustrations and even cause a creative backlash from industry professionals. For instance, an absurd comedy series broadcast in 2011 on TRT called *Leyla ile Mecnun* (*Leyla and Mecnun*) became popular in part for using clever plotlines to bypass RTÜK's restrictions on cigarettes and alcohol.[6] The show used certain objects as signifiers of banned substances, such as eating grapes to stand in for drinking wine, hazelnuts for vodka and chewing gum for smoking. The absurdity of the plotlines that treated these substitute substances as if they were the actual ones – such as characters made to wear Russian-style fur hats and goatee beards just for that scene in which they ate hazelnuts out of a bottle while speaking in a drunken manner about how one could go to the moon eating those hazelnuts, or one character asking another if he would be interested in eating Bordeaux grapes from the 1940s – not only created a comedic effect but also criticized the censorship foisted on the industry. While fans of the program suspected for some time that TRT would not renew the series due to its edgy content – which became a phenomenon among young populations' discontented with the conservative AKP government – TRT did not wait for renewal and instead managed to cancel the program immediately when some of its cast members openly supported the Gezi uprising that criticized Erdoğan's increasing authoritarianism.

Erdoğan's speeches criticizing Turkish dramas had a direct effect on several series' content. There are numerous examples, but a well-known one about *Magnificent Century*, a TV drama on the life of the Ottoman Sultan, Suleiman the Magnificent, is worth recalling here. Publicly claiming in a speech that the series depicts the Sultan as a leader who never leaves his palace and harem (meaning the Sultan's family, concubines and children) instead of a successful, expansionist ruler who fought wars, President Erdoğan threatened to file a formal complaint against the series' producers, director and the owners of the TV channel for misrepresenting Turkish history. There were even public protests and threats against the president and CEO of Tims Productions, Timur Savcı, and the series' scriptwriter Meral Okay, after Erdoğan targeted the series ("Meral Okay: Bana dinsiz" 2011). In such a climate, Tims Productions chose to immediately comply with the government's demands for content change by including numerous battle scenes and depicting the

Sultan's mother and concubines praying with a headscarf, instead of featuring extended harem scenes depicting love and lust, as was previously the case.

In order to attract less official attention to its content, TV industry professionals have opted to craft their storylines and imagery more in line with the government's conservative cultural policies. For instance, Tunç (2012, 338) illustrates how the emphasis on family values and the honor of women in the popular TV dramas *Aşk ve Ceza* (*Love and Punishment*) and *Ezel* coincide with the AKP government's efforts to strengthen family values and the party's ideology of embracing a modernity that does not reject tradition. Kocamaner (2017) observes that in the last decade, Islamic TV channels have increasingly focused on family entertainment programs such as dramas as opposed to religious teachings, and that "Islamic TV professionals define their 'social responsibility' as assisting or complementing the state in fighting threats and risks to the family" (705). Similarly, Coşar and Gençoğlu Onbaşi's (2016) feminist analysis of the TV series *Yeşeren Düşler* (*Blossoming Dreams*) that is broadcast on an Islamic TV channel, illustrates that such family dramas reproduce the AKP's neoliberal-conservative hegemony, which is a "blend of faithful capitalism, Islamic moralism, and conservative patriarchy" (229). So, the tactics the Turkish TV industry has resorted to include crafting storylines in line with the AKP government's conservative ideologies, self-censorship in anticipation of possible RTÜK warnings, and complying with requests for content modification and/or demands for new content that reflects conservative ideology.

## *"Secret Cultural Envoys": Soft Power as a Discursive Tactic*

Focusing on the concurrence of Turkey's foreign policy initiatives to improve political and economic relations with its neighbors and the growing popularity of Turkish TV series in the Middle East, Tunç (2012, 339) points out that

> perhaps coincidentally, the spread of Turkish TV to Arab countries and the Balkans falls neatly into the soft power strategy of Turkey's Foreign Minister, Ahmet Davutoglu, as outlined in his 2001 seminal book, The Strategic Depth, where he outlines his neo-Ottoman ambitions in an area extending from the Middle East to the Balkans.

Indeed, the popular excitement surrounding the new "Ottoman-cool image" (Kraidy and Al-Ghazzi, 2013) of Turkish TV series encouraged many governmental, non-governmental and corporate entities and actors, including TV executives, to rally around the idea that Turkish dramas have soft power abroad.

To illustrate how different entities and actors adopted a similar "soft power" and "nation-branding" discourse about the impact and utility of Turkish TV series, I will give a few examples – among many – from their publicly made statements quoted in mainstream newspapers. For instance, Abdurrahman Çelik, who manages copyrights at the Culture and Tourism Ministry in Ankara, stated that "Turkey's growing soap opera exports are increasingly important for the promotion of our national brand" (quoted in Candemir, 2013). Another government official, Deputy Minister of Culture and Tourism Hüseyin Yayman, similarly suggested that "When we visited Spain, they also asked about Turkish series. Turkish series are shown in Latin America as well … In fact, that tells us that Turkey's soft power has also risen" (quoted in Kale, 2016, 86). Birol Güven, the head of the Turkish Producers Association, said, "I expect the series industry, which is a great contribution on Turkey's promotion, will grow about 12 per cent this year" ("Astonishing Revenues for Turkish Soap Operas" 2014). While the industry's growth was measured at 10.9% the following year in 2016 (Radyo Televizyon Yayıncıları Meslek Birliği, 2018, 37), it is worth noting the parallels he draws between the industry's growth, its contribution to the economy, and its possible role in promoting the country. Similarly, İzzet Pinto, the CEO of one of Turkey's biggest distribution companies, Global Agency, stated "We are showing our country to millions of viewers. We are showing the beautiful scenery, our lifestyle and traditions. So we have a great influence on people through soft power" (quoted in Williams, 2013).

It is dubious to assume that the success of a particular nation's cultural products in global media markets results in a gain of soft power. As Nye (2014) has cautioned us, we should not confuse the international appeal of media products with soft power, and that soft power can only be enhanced if foreign policy and democratic values are also adequately developed. Nevertheless, increases in trade and other collaborations with Turkey's allies in the Middle East are the result of AKP policies that go back to the early 2000s before these soap operas became popular there. Moreover, as Yörük and Vatikiotis (2013, 2374) stated:

> While Turkish political and economic influence coincides with the improving exports of Turkish TV series, the rhetoric of the 'Turkish model' and 'soft power' do not convincingly demonstrate the link between these phenomena, given that cultural popularity and power of any type (be it soft or hard) do not automatically follow one another.

However, my goal here is not to prove or disprove whether Turkish dramas have soft power but rather to draw attention to the fact that Turkish TV industry executives use the same "soft power" discourse as government officials, and to illustrate why the government and industry discourses coincide so readily.

I argue that the overwhelming agreement about and use of the discourse of Turkish series' soft power is beneficial for all stakeholders, but especially for the Turkish TV industry, because it helps their efforts of both marketing their own productions and gaining more hegemonic power through collaborations with government agencies so they can have more leverage against any potential government retribution. For instance, embracing their new role of cultural promotion, the Turkish TV industry has cooperated with the Ministry of Economy and the Ministry of Culture and Tourism to help businesses that are eager to expand internationally and draw tourists to Turkey. As a result, the popular actors from the Turkish dramas were enlisted to market business ventures abroad and take part in the nation-branding and promotional efforts of the Ministry of Culture and Tourism in a variety of ways. They are cast in ad campaigns, commercials and videos that promoted Turkey's historical sites or Istanbul's vibrant shopping life, such as the Turkish Airlines commercial featuring Kıvanç Tatlıtuğ – the well-known actor in *Gümüş* (*Noor*) – visiting Istanbul's landmarks (Artan, 2010). Businessmen and government officials also take TV celebrities to various promotional events abroad, such as launch parties, openings and press conferences. For instance, in May 2017, President Erdoğan took the well-known stars of *Resurrection: Ertugrul* to Kuwait for the groundbreaking of a US-$4.5 billion international airport being built by a Turkish construction business ("Diriliş Dizisi Oyuncuları Cumhurbaşkanı Erdoğan'ın Uçağında" 2017).

While the actors are well compensated for these trips and appearances and they are good opportunities for self-promotion, the trips are mainly seen as cultural promotion and nation-branding exercises by government entities, and thus actors are expected to deliver certain messages about Turkish businesses and products as well as Turkish culture, with the assumption that Turkish dramas' assumed soft power would translate into other realms of Turkish economy and socio-political relations. During these trips abroad, actors usually present themselves as "cultural envoys" – a phrase used by a special magazine issue published by the PR arm of AKP to refer to the TV actors (Kale, 2016). For instance, actor Kıvanç Tatlıtuğ said during a press conference in Dubai for the promotion of Istanbul Shopping Fest that as an actor who is well known abroad, he is happy to contribute to the promotion of Turkey and attract more interest in the country ("Kıvanç Tatlıtuğ ve Songül Öden Dubai'de" 2012). During the same press conference, his co-star Songül Öden said that they take upon themselves the mission of promoting and merging cultures ("Kıvanç Tatlıtuğ ve Songül Öden Dubai'de" 2012).

Perhaps the most interesting example of how the industry's cooperation with the government and embrace of the cultural envoys' role help the industry to gain and maintain power comes from the forced resignation of RTÜK President İlhan Yerlikaya in January of 2019. The Broadcasters Association complained to the Ministry of Culture and Tourism that while the ministry claimed that Turkish TV series play a big role in promoting Turkey and bringing in export revenues, it had become impossible for them to continue producing TV series due to RTÜK's increased and arbitrary censorship, fines and favoritism under Yerlikaya's leadership (Çelik, 2019; *Yavuz Oğhan'dan Bidebunudinle*, 2019). Until recently, TV production companies included in their contracts with TV channels that they would be responsible for any RTÜK fines, but they had to stop including this clause because of the fines' unpredictability, frequency and amount.

## Global Marketing and Branding as Tactics

While the Turkish TV industry still makes content decisions predominantly according to the needs of the domestic market, it has reorganized its marketing, branding and distribution to respond to the innovations occurring in the international TV markets. As a result, the Turkish TV industry has ended up following global business trends, such as developing original content for over-the-top (OTT) platforms, getting involved in joint projects with numerous TV markets abroad, selling the remake rights of its content, creating new TV formats and concepts for TV shows that are licensed for adaptation in local markets, and purchasing TV formats from various markets abroad.

The industry has also invested heavily in promotion and marketing campaigns, which include participating in global TV trade shows, attending festivals with popular actors, organizing parties and press conferences at these festivals, advertising on many platforms, including international trade journals, and thus, following all the strategies for "creating a buzz" as identified by Havens (2003). For instance, İzzet Pinto, who is the founder and CEO of Global Agency, told me that his distribution agency participates in 12 international TV trade fairs annually spending millions of dollars, and that today some Turkish TV series are sold for 4–5 times more than their American counterparts while others are 2–3 times more expensive (İzzet Pinto, July 2015. Interview by author). Some of the distributors and producers in the Turkish TV industry adapt to global markets by doing more than just marketing their content abroad, they also globalize their business practices more broadly, such as by creating new TV formats and adaptations of their existing content for specific markets, and by employing international personnel or opening offices abroad. For instance, Tims Productions opened an office in Los Angeles and also teamed up with Sony TV and Universal TV to co-produce an American remake of a Turkish thriller-drama called *Suskunlar* (*Game of Silence*) for NBC (Timur Savcı, July 2015. Interview by author). Global Agency distributes content for not only Turkish producers but also a number of producers from abroad, and thus operates as a global agency buying 10–15 TV formats from various national markets and distributing them to the international markets while also creating their own formats for music, game and reality show programs (İzzet Pinto, July 2015. Interview by author).

After the success of *Magnificent Century*, which was broadcast in 75 countries and watched by over 200 million people ("Turkish soap opera sector …" 2016), the industry realized that the Turkishness of the content was a welcome novelty for the international market, and the tactic fit well with the cultural envoy role assigned to the industry by the government. So, the industry branded its products as Turkish, especially for their scripted content. A recent example of nation branding is the promotion of Turkish TV series as "*Novellas Turcas*" in Latin America. While the nation branding of Turkish soaps as "*Novellas Turcas*" has been adopted as a strategy to stay competitive and distinct in Spanish-language markets, it also functions as a tactic aimed at minimizing domestic political interference and maximizing government cooperation in the long run.

## Conclusion

The transformation of Turkish television to a locally based transnational cultural industry was influenced by numerous supportive local and global forces, as detailed in this chapter. However, it was never a process without domestic challenges posed by the highly unstable political and economic environment and deepening crony capitalism, and Turkish TV's success as a global growth industry has not mitigated these challenges. I argue that the Turkish TV industry has responded with three main tactics. The first is focused on the careful management of content, which usually means either complying with the government's demand to modify their content or producing content in line with its conservative ideology, but it sometimes means resisting government bodies, such as RTÜK, whose arbitrary enforcement hurts the industry. While the Turkish state has never had a policy of enhancing the nation's image through television exports, it has made direct or indirect impositions on the TV industry from time to time to encourage the representation of the country in a certain way, which has resulted in an increase in self-censorship on the one hand and the growth of conservative and nationalistic storylines on the other. The presentation through TV series of a jingoistic and homogeneous ethnic nationalism based on Turkishness and patriarchal family values, in turn, further reinforces the right-wing and statist agenda of the AKP government.

The second tactic involves the industry adopting the government's soft power discourse and embracing the role of "cultural envoys" while cooperating with officials and businesses in their cultural promotion and nation-branding efforts. The industry's third tactic is to incorporate global TV business strategies by undertaking rigorous marketing and branding campaigns abroad in order to gain economic power and influence to withstand any possible domestic political interference or retribution for content. While I have invoked in this chapter the distinction that de Certeau (1984) drew between "strategies" and "tactics," I do so with a caveat. De Certeau sees strategies as produced by institutions and structures of power, and tactics as developed by individuals, consumers or users, so the tactics of the weak then constitute a response to the strategies of the powerful. The AKP's increasingly authoritative regime clearly represents the structure of power against which the TV industry develops its tactics. However, in the highly polarized and concentrated media environment of Turkey, characterized by political oligopoly where "several different actors ... are affiliated to the same network created by the government" (Irak, 2015, 16), not all industry actors are equally weak against the government. As my chapter has illustrated, the TV production and distribution companies that are close to the government are more immune to the aforementioned challenges that face the industry and are supported in various ways, such as direct praise, production support, immunity from RTÜK fines or government tenders issued to their holding companies, while smaller independent production firms that are outside of the government-created network are more vulnerable and, therefore, strive more for transnational growth as a means to stay strong.

## Notes

1 Of the nine TV industry executives I asked, seven of them agreed to an interview. However, I was only able to interview five of them due to their busy schedules and my narrow time frame for conducting the interviews. My interviewees were either high-level producers, such as production designers and production supervisors, or CEOs and presidents of TV production and distribution companies in Istanbul. Also, in this chapter I do not necessarily use direct quotes from or cite all the interviews I conducted.

2 MIPCOM (Marché International des Programmes de Communication, a.k.a. International Market of Communication Programs) is an annual trade fair and content market for the international television industry. In

2015, Turkey was MIPCOM's country of honor, and its annual meeting and newsletters featured interviews with Turkish TV executives and government officials.

3 Out of these 196 TV channels, 19 are national, 165 are local and 12 are regional. This is still not as high as it was prior to the 2016 military coup attempt, when the total number of TV channels was 246 (Radyo Televizyon Yayıncıları Meslek Birliği, 2018). In 2016 alone, 68 channels were either shut down via emergency state decree or sold immediately following the coup attempt. However, the industry's revenues continued to show growth.

4 As Havens (2002, 383) has suggested, "Often, distributors market their programming as 'universal' in an attempt to convince buyers that it can surmount cultural barriers." However, based on the MIPCOM newsletter interviews with TV industry executives I have reviewed and what I gathered in my own interviews, while the producers who are close to the AKP government boast that their content reflects Turkish culture, history and values, those who are not aligned with the government use the universality discourse as a tactic to prevent any possible retribution from their representation of "Turkishness."

5 Turkish viewers spend an average of four hours per day watching TV and they mostly prefer romantic series (Deloitte Touche Tohmatsu Limited, 2014).

6 I thank Şahika Erkonan for bringing this example to my attention.

## References

Alankuş, Sevda and Eylem Yanardağoğlu. 2016. "Vacillation in Turkey's Popular Global TV Exports: Toward a More Complex Understanding of Distribution." *International Journal of Communication* 10: 3615–3631. Accessed September 16 2018, https://ijoc.org/index.php/ijoc/article/view/4808

Algan, Ece. 2003. "Privatization of Media in Turkey and the Question of Media Hegemony in the Era of Globalization." In *The Globalization of Corporate Media Hegemony*, Eds. Lee Artz and Yahya R. Kamalipour, 169–192. New York: SUNY Press.

Arsenault, Amelia H. 2017. "The Datafication of Media: Big Data and the Media Industries." *International Journal of Media & Cultural Politics* 13 (1/2): 7–24. https://doi.org/10.1386/macp.13.1-2.7_1

Artan, Gökhan. 2010. "Tatlıtuğ'lu reklam filmi Ortadoğu'da gösterilecek." *HaberTurk*, June 7, 2010. www.haberturk.com/haber/haber/521478-tatlituglu-reklam-filmi-ortadoguda-gosterilecek

"Astonishing Revenues for Turkish Soap Operas." 2014. *Yeni Safak*, September 5, 2014. www.yenisafak.com/en/economy/astonishing-revenues-for-turkish-soap-operas-2021470

Becerra, Martin, Mastrini, Guillermo, and Waisbord, Silvio. 2014. "Television in Latin America: From Commercialism to Reform?" In *The Sage Handbook of Television Studies*, Eds. Manuel Alvarado, Milly Buonanno, Herman Gray, and Toby Miller, 36–49. London: Sage.

Burroughs, Benjamin. 2018. "House of Netflix: Streaming Media and Digital Lore." *Popular Communication* 1–17. https://doi.org/10.1080/15405702.2017.1343948

Candemir, Yeliz. 2013. "Turkish Soap Operas: The Unstoppable Boom." *Wall Street Journal*, blogs, April 29, 2013. Accessed September 12, 2017. https://blogs.wsj.com/middleeast/2013/04/29/turkish-soap-operas-the-unstoppable-boom/

Çelik, Özlem Akarsu. 2019. 'RTÜK Başkanı istifa ettirildi'. *Gazate Duvar*, January 22, 2019. www.gazeteduvar.com.tr/gundem/2019/01/22/RTÜK-baskani-istifa-ettirildi/

Chalaby, Jean K. 2015. The advent of the transnational TV format trading system: a global commodity chain analysis. *Media, Culture & Society* 37 (3): 460–478.

Chalaby, Jean K. 2016. Television and Globalization: The TV Content Global Value Chain. *Journal of Communication* 66: 35–59.

Coşar, Simten and Funda Gençoğlu Onbaşi. 2016. "Moralism, Hegemony, and Political Islam in Turkey: Gendered Portrayals in a TV Series." *Journal of Mediterranean Studies* 25 (2): 217–234.

de Certeau, Michel. 1984. *The Practice of Everyday Life*. Translated by Steven Rendall. Berkeley: University of California Press.

Deloitte Touche Tohmatsu Limited. 2014. "World's Most Colorful Screen TV Series Sector in Turkey Report." *Deloitte*, August 2014. Accessed December 8, 2018. www2.deloitte.com/content/dam/Deloitte/tr/Documents/technology-media-telecommunications/tr-web-tv-series-sector-in-turkey.pdf

"Diriliş Dizisi Oyuncuları Cumhurbaşkanı Erdoğan'ın Uçağında." 2017. *Posta*, May 10, 2017 www.posta.com.tr/dirilis-dizisi-oyunculari-cumhurbaskani-Erdoğan-in-ucaginda-1294877

Emre Cetin, Kumru B. 2014. "The 'Politicization' of Turkish Television Dramas." *International Journal of Communication* 8 (January): 2462–2483. Accessed September 5, 2018. https://ijoc.org/index.php/ijoc/article/view/2749/1265

"France and Turkey join biggest programme exporters." 2016. *Digital TV Europe*, April 15, 2016. www.digitaltveurope.com/2016/04/15/france-and-turkey-join-biggest-programme-exporters/
Havens, Timothy. 2014. "Media Programming in an Era of Big Data." *Media Industries Journal* 1 (2): 5–9. Accessed November 28, 2018. http://dx.doi.org/10.3998/mij.15031809.0001.202
Havens, Timothy J. 2002. "'It's Still a White World Out There': The Interplay of Culture and Economics in International Television Trade." *Critical Studies in Media Communication* 19 (4): 377–397.
Havens, Timothy J. 2003. "Exhibiting Global Television: On the Business and Cultural Functions of Global Television Fairs." *Journal of Broadcasting & Electronic Media* 47 (1): 18–35.
"Turkish soap opera sector plans to hit $1 billion in exports by 2023." 2016. *Hurriyet Daily News*, October 19, 2016. www.hurriyetdailynews.com/turkish-soap-opera-sector-plans-to-hit-1-billion-in-exports-by-2023-105133
Irak, Dağhan. 2015. "Reading political clientelism and concentration through new media: The case of Anadolu Agency." *Yeni Medya Çalışmaları* 3: 13–30.
Işık, Nuran E. 2013. "Parables as Indicators of Popular Wisdom: The Making of Piety Culture in Turkish Television Dramas." *European Journal of Cultural Studies* 16 (5): 565–581. 10.1177/1367549413491722.
Jin, Dal Y. 2016. *New Korean Wave: Transnational Popular Culture in the Age of Social Media*. Urbana: The University of Illinois Press.
Kale, İsmail. 2016. "Secret Cultural Envoys: Turkish Series." *Magazine of Culture of Turkey* 2 (July 15th Exclusive Issue): 82–88.
Keinonen, Heidi. 2018. "The Power of the Format Catalogue." *Critical Studies in Television* 13 (2): 153–169.
"Kıvanç Tatlıtuğ ve Songül Öden Dubai'de." 2012. *Haber7*, May 10, 2012. www.haber7.com/yasam/haber/878541-kivanc-tatlitug-ve-songul-oden-dubaide
Kocamaner, Hikmet. 2017. "Strengthening the Family through Television: Islamic Broadcasting, Secularism, and the Politics of Responsibility in Turkey." *Anthropological Quarterly* 90 (3): 675–715.
Kraidy, Marwan M. and Al-Ghazzi, Omar. 2013. "Neo-Ottoman cool: Turkish popular culture in the Arab public sphere." *Popular Communication 11* (1): 17–29.
Leung, Lisa. 2008. "Mediating nationalism and modernity: The transnationalization of Korean dramas on Chinese (satellite) TV." In: *East Asian Pop Culture: Analyzing the Korean wave*, Eds. Beng Huat Chua and Koichi Iwabuchi, 53–70. Hong Kong: Hong Kong University Press.
Nye, Joseph S. 2014. "The Information Revolution and Power." *Current History* 113 (759): 19–22.
Özalpman, Deniz. 2017. "Transnational consumers of Turkish Television drama series." *Transnational Marketing Journal* 5 (1): 25–43.
Radyo Televizyon Yayıncıları Meslek Birliği. 2018. "Türkiye Radyo ve Televizyon Yayıncılığı Sektör Raporu." RATEM. Accessed December 8, 2018. www.ratem.org/h99sektor-raporu
Ryoo, Woongjae and Jin, Dal Yong. 2018. "Cultural Politics in the South Korean Cultural Industries: Confrontations Between State-developmentalism and Neoliberalism." *International Journal of Cultural Policy* (February): 1–15. Accessed September 6, 2018. https://doi.org/10.1080/10286632.2018.1429422
Sofuoglu, Murat. 2017. "The Giddying Rise of Turkish Television Series." *TRT World*, January 27, 2017. https://medium.com/trt-world/the-giddying-rise-of-turkish-television-series-5f6c5b456a0c
"Meral Okay: Bana dinsiz Allah'sız dediler ama ben yara almam." 2011. *T24*, January 29, 2011. https://t24.com.tr/haber/meral-okay-bana-dinsiz-allahsiz-dediler-ama-ben-yara-almam,124774
Tinic, Serra. 2015. "Where in the World Is Orphan Black? Change and Continuity in Global TV Production and Distribution." *Media Industries Journal* 1 (3): 54–59. Accessed November 28, 2018. http://dx.doi.org/10.3998/mij.15031809.0001.310
Tunç, Asli. 2012. "Missing Byzantium: Explaining Greeks' Love for Turkish TV Serials." *International Journal of Media & Cultural Politics* 8 (2/3): 335–341.
"Turkey world's second highest TV series exporter after US." 2014. *Hurriyet Daily News*, October 26, 2014. www.hurriyetdailynews.com/turkey-worlds-second-highest-tv-series-exporter-after-us-73478
Vitrinel, Ece. 2017. "Forced Politicization of Television Celebrities in Turkey." *Journal of Balkan and Near Eastern Studies*: 1–12. Published September 18, 2017 online. https://doi.org/10.1080/19448953.2017.1367585
Vivarelli, Nick. 2017. Turkish TV Reaches More Viewers Abroad. *Variety*, October 16, 2017. https://variety.com/2017/tv/markets-festivals/turkish-tv-reaches-more-viewers-abroad-1202591287/
Vivarelli, Nick. 2018. "Turkey Experiences Its Own Wave of Peak TV." *Variety*, April 8, 2018. https://variety.com/2018/tv/features/turkey-experiences-its-own-wave-of-peak-tv-1202746009/
Williams, Nathan. 2013. "The rise of Turkish soap power." *BBC News*, June 28, 2013. www.bbc.co.uk/news/magazine-22282563
Yanardağoğlu, Eylem. 2013. "The Fever That Hit Arab Satellite Television: Audience Perceptions of Turkish TV Series." *Identities* 20 (5): 561–580.

*Yavuz Oğhan'dan Bidebunudinle*. 2019. "Yavuz Oğhan'dan Bidebunudinle 22-01-2019 Yayını." *Sputnik*. Aired January 22 on YouTube. Accessed January 22, 2019. www.youtube.com/watch?v=HJWX92YDFM4

Yesil, Bilge. 2015. Transnationalization of Turkish dramas: Exploring the convergence of local and global market imperatives. *Global Media and Communication* 11 (1): 43–60.

Yörük, Zafer and Vatikiotis, Pantelis. 2013. "Soft Power or Illusion of Hegemony: The Case of the Turkish Soap Opera '*Colonialism*'." *International Journal of Communication* 7: 2361–2385. Accessed September 10, 2018. https://ijoc.org/index.php/ijoc/article/view/1880

# 38

# SOUTH AFRICAN TELEVISION MOVES INTO THE GLOBAL AGE

*Ruth Teer-Tomaselli*

## History of Television in South Africa

National broadcasting in South Africa was inaugurated in 1936, when the South African Broadcasting Corporation (SABC) was formed. Radio services were divided along language and racial lines. Television was introduced experimentally in the early 1970s (Teer-Tomaselli, 2015), enabled by the Broadcasting Act (73/1976), underpinning the legislative and regulatory framework for the next 30 years. The SABC reported directly to the relevant minister rather than to parliament. On January 6, 1976, television was officially inaugurated and, by the opening night, only 300,000 television sets had been sold. Initially, a single channel – SABC-TV, transmitted for five hours every evening and seven on Saturday. Equal time was accorded to English and Afrikaans. Advertising was permitted after two years, and by 1980 constituted a third of the Corporation's revenue. The remainder came from television – as opposed to radio – licenses (SABC, 1980). As was the custom in many countries with a public service broadcaster, license fees were the initial manner in which revenue was raised. In order to own a radio set, one had to produce a valid radio license for which one paid an annual fee. With the introduction of television, this fell away, and only households with a television were required to pay the annual license fee. Over the years, approximately 18% of the SABC's income has been derived from license payments.

## Introduction of Television for African Audiences and Network Expansion

Initially no television programming occurred in any African language. In 1982, SABC-TV was renamed TV1 and two additional channels, TV2 and TV3 were introduced. These channels shared a single frequency but were beamed to different parts of the country, characterized by the predominance of different African languages spoken in those areas. TV2 catered for the Nguni languages, specifically isiZulu and isiXhosa. The Sotho family of languages – Tswana, SeSotho and Pedi found a home on TV3 (SABC, 1982: 8).

During this period, spectrum scarcity limited television programming expansion. The language needs of black viewers were too diverse to be catered to by merely a single channel; however, there was only sufficient bandwidth for one more channel. The SABC's technical solution was to use the affordances of the very high frequency (VHF) radio waves to 'split' the channel geographically. Because the VHS signal is propagated as a near-line-of-sight phenomenon, it was possible to divide it into two different carriers broadcasting in different directions at the same time. (Similar use had been made of the Frequency Modulation (FM) radio service in the early 1960s, where the line-of-sight properties had been put to use to reinforce the idea of different ethnic (and therefore language) 'groups' living in different geographic areas (Hayman and Tomaselli, 1989).)

The example here is important, since despite the scarcity of spectrum, it was possible to introduce a 'double value' that was considered necessary. It is also exemplary of the way in which technology served the interests of ideology, and not the other way around. Apartheid held that different ethnic groups, and therefore different language groups, constituted different political entities. This was the basis of the apartheid divide-and-rule philosophy. Thus, providing different language services to specific areas reinforced the notion of ethnically specific 'bantustans'.

The programming on both TV2 and TV3 finished at 21:30, and to make the most use of the spectrum available, a fourth service, TV4, was introduced in 1985, scheduled after 21:30, but using the same spectrum allocation. This late-night channel differed from the previous models, based as they were on language. In contrast, TV4 was entirely made up of sports and entertainment programming, most of which was English or Afrikaans. In 1992, just before the SABC became the first of the state-owned enterprises to come under the transitional government structures that dismantled apartheid and introduced the first democratic elections, the channels were once again rationalized. TV2/3/4 were amalgamated into a single channel, and named Contemporary Community Values (CCV). At the same time, a third channel was introduced, TopSport Surplus, which transmitted the overflow of sporting programmes produced by the SABC's sports' production arm, Topsport. This was later changed to National Network Television (NNTV), an educational channel without advertising that was envisaged as a pristinely 'public service' arm of the SABC. These changes in the SABC's channel make-up gloss over profound and deeply rooted ideological, political, economic and aesthetic power struggles within the SABC itself, moving between the poles of political support for the dying body of apartheid to the energetic embrace of a new and democratic political order; and from rampant commercialism (seen at its zenith in the establishment of TV4) to an idealistic enclosure of public service values in the establishment of NNTV (Duncan and Glenn, 2010; Jacobs, 2013; Teer-Tomaselli, 2004).

Thus, by 1993, which was to prove a watershed with the appointment of the first democratically appointed SABC Board, three distinct channels were being broadcast: TV1, an all-purpose channel with a classic public broadcasting mix of news and informational programming, entertainment, sports and children's programming, equally divided by time into English and Afrikaans; CCV, with a similar mix, but with multi-lingual broadcasts of English, Nguni and Sotho; and NNTV, an educational mix that was financially cross-subsidized by the revenue derived from the other two channels (Teer-Tomaselli, 2011).

## Introduction of M-Net

The introduction of a new television service is usually the result of protracted policy debate, followed by the application of a regulatory framework. Not so with the introduction of M-Net in 1986. M-Net arose from political negotiation between the four major newspaper consortia and the government, in order to compensate for the loss of advertising revenue resulting from the introduction of television. Within a short period, three of the four groups sold their shares to Nasionale Pers (Naspers), which became the sole owner of M-Net.

Technically, the signal was delivered as an encrypted terrestrial feed, and unencrypted, or 'decoded' with a set-top box enabled by a monthly subscription. Advertising supplemented the subscription fees. Despite being an encrypted service, M-Net was allowed an 'open time' slot from 17:00 to 19:00, which constituted the peak 'family viewing' period. During this period, the signal was unencrypted and available to all households with a television set. The open time enabled considerable amassment of advertising revenue and served as a very useful showcase to promote M-Net offerings. This concession was to have been temporary in order for the new service to achieve financial viability, but in fact it remained until 2007, a neglectful policy implication that raised both eyebrows and ire within the broadcasting community (News24, 2007).

Subscription rates rose rapidly, from 50,000 in 1986 to 1.4 million in 2007. While sport had always been a mainstay of the channel's programming, the launch of a sub-brand, SuperSport in 1988, signaled a substantial advancement in the appeal of M-Net's content (News24, 2017). A second terrestrial analogue channel, Community Services Network, was launched in the early 1990s (My Broadband, 2011). This channel offered sports programming and specialized batches of programming to 'niche minorities' such as the large community of South Africans of Indian descent, immigrants from Portugal and Italy, or Jewish, Islamic and Christian religious communities (Jacobs, 2013: 175).

The real sea change, however, arrived in 1992 with the inauguration of EuTelSat, an analogue direct-to-home (DTH) satellite transmission platform. In 1993, M-Net established a second company, MultiChoice, its name presaging the multi-channel environment that would soon be available. M-Net became a channel, later a series of channels, responsible for local production and curation of content (MultiChoiceAfrica, n.d.). The first inkling of a fully fledged digital satellite transmission was heralded by PanAmSat's PAS-4 Ku-band with a steerable transmission that enabled it to direct its signal to a specific 'spot', including services to sub-Saharan Africa and the Indian Ocean Isles. Two DTH bouquets were launched on this satellite, allowing MultiChoice to offer digital satellite service in South Africa, and to launch joint ventures in Kenya and Zambia, with a Zimbabwe franchise. This was also the start of a Greek satellite platform, MultiChoice Hellas.

The initial licensing of M-Net as an encrypted analogue channel was decided on neither technological nor public policy grounds, but largely for political purposes, with the objective of sustaining the print press, specifically the Afrikaans press (Naspers), in order to provide ideological support for the apartheid government, then under severe political pressure. The move from terrestrial analogue to satellite analogue (EuTelSat) was effected in the absence of any broadcast regulation concerning satellite transmission, as was the later move to digital satellite transmission (PanAmSat). This is an example of where 'no policy' acts precisely as a policy of enablement, but only for a niche sector of the industry, a disguised favoritism that allowed MultiChoice and later, DStv, a 'first mover advantage' that made them practically immune from later competition. This is evident in their dominant subscriber base; their ability to secure premium content with long contract times; and most importantly, their ability to practically monopolize the acquisition of high-price, high-value sports franchises.

## IBA Triple Enquiry Report, Introduction of e-TV and the Relaunch of the SABC Terrestrial Channels

The first fully democratic elections of 1994 heralded a new political dispensation that officially ended apartheid and ushered in a non-racial South Africa. The Independent Broadcasting Authority (IBA) was established ahead of the first election in order to ensure a free and fair election (RSA, 1993). Under its auspices all broadcast licensing issues, including consideration of new channels, ownership and control structures of broadcasting companies, and local content quotas, were publicly discussed, regulated and monitored. The SABC too was overhauled, with the appointment of a new board of governors following public nominations and hearings (Duncan and Glenn, 2010; Horwitz, 2001; Teer-Tomaselli, 2004).

One of the IBA's first tasks was to hold a series of public hearings to determine a new regulatory framework. The 'Triple Enquiry' covered the viability of the public broadcaster, local content and cross-media ownership and control (IBA, 1995). The outcomes significantly changed the landscape of both radio and television ownership. The broadcasting sector was divided into three tiers: public, commercial and community.

The SABC 'relaunch' of 1996 realigned its television channels, partly in response to the Triple Enquiry Report, and partly in response to the competition from e-TV (see below) and M-Net, both of which ate into the SABC's advertising revenue. The reconfiguration of these channels was arranged on two intersecting logics: first, on the axis of 'public' versus 'commercial' logic, and thereafter on language. SABC1 broadcast in mainly Nguni languages (isiZulu, isiXhosa, and Ndebele); SABC2 in the seSotho and seTswana family of languages, while also carrying Afrikaans. Afrikaans was reduced from 50% to 15% of channel coverage. These two were 'public' broadcasters, while SABC3 is a commercially driven, all English-language channel. The original idea was that the profits realized on this channel would cross-subsidize the other two language-driven channels; however, the language-driven, public service channels proved to be the most profitable and these remain the three terrestrial channels at the time of writing (2018). Organizationally, the SABC underwent several realignments (RSA, 2009); yet it has been beset by crises and scandals relating to poor governance and internal malfunctions (Duncan and Glenn, 2010; Skinner, 2011).

## Establishment of a Privately Owned Free-to-air Broadcaster: e-TV

A primary purpose of the Triple Enquiry was to break the monopoly of the SABC and open up the airwaves to a multiplicity of voices and economic opportunities. To strengthen the private commercial sector, tenders were invited for a public bidding process for one terrestrial free-to-air service (IBA, 1997). Three serious contenders were made up of local investor and an international partner (confined by law to no more than 25% shareholding), and drew on media management expertise, usually in the form of ex-SABC employees (radio and television).

Midi-Television won the bid in March 1998 and began broadcasting as e-TV in October 1998. Midi-TV was constituted from trade union investment vehicles (Hosken Consolidated Investments (HCI), 2001), small business trusts and local media production companies, an ownership structure that gave the company 'black empowerment' status. The international partner, Warner Brothers, a subsidiary of what was, at the time, Time-Warner, held a 25% stake, the maximum allowed under the IBA regulations (IBA, 1996: 13–14). Warner had hoped to buy the entire channel in time, however, their share was diluted to 20% when HCI injected more equity into the company and, thwarted by regulation, Warner Brothers sold their shares in 2001.

Technically, the station was very advanced for its time, being the first fully digitally produced and scheduled television in the country, despite its terrestrial transmission. However, the first few years of the station's existence were perilous, with a slow build-up of audiences, uneven advertising revenue and numerous staffing problems (Jacobs, 2013: 176; Teer-Tomaselli, 2011). Beginning with a broadcast schedule of only six hours a day, the channel rapidly expanded, and by 2001 had overtaken SABC3 as the most watched English-language entertainment channel (Teer-Tomaselli, 2011).

The introduction of a new channel was not made on technological grounds, but rather social engineering, driven by political and 'civil society' wishes, and orchestrated through a tough regulatory process of invited bidding, public hearings and long consultations within the IBA. The idea behind the introduction was to increase the choice available to the viewing public through breaking the monopoly of the state-sanctioned public broadcaster, the SABC. In this respect, the introduction of e-TV was not entirely dissimilar to the introduction of ITV in the United Kingdom, though there was none of the regional emphasis that played into the formation of ITV.

## The Digital Domain

New technologies and the effects of convergence begin from 2000, the date when the International Telecommunications Union (ITU) introduced Mobile Telecommunications-2000 (IMT-2000). This was based on the third generation of wireless telephony (3G), extending mobile capacity from voice and text messaging to include video calls. It also allowed for mobile access to the internet through both fixed and mobile wireless (wifi) internet access. Thus, for the first time, mobile television was a technological possibility on a widespread public scale (Jamalipour, 2003). These developments changed the way consumers accessed audio-visual contents (OECD, 2013: 8). From a business planning point of view, this produced uncertainty in the ability to predict future demand, and in turn, created real challenges in formulating and applying regulation and competition law. Within South Africa, the issue of convergence has been as thorny an issue as it has in many other countries.

In order to come to terms with the growing technological convergence between various data transmission systems that resulted in the rapid convergence of broadcasting and telecommunications services, the IBA and the South African Telecommunications Regulatory Authority, were merged into a single regulator, the Independent Communications Authority of South Africa (ICASA) in 2000 (RSA, 2000).

Convergence also meant divergence. With the now commonly accepted practice of streaming audio-visual materials in the form of entertainment and news directly via the internet and delivered on a variety of platforms – personal computers, laptop, tablets and mobile telephony – the distinction between 'broadcasting' and 'streaming' became a legislative and regulatory quagmire, while of course there are very specific differences in application, pace and consumer uptake in South Africa. The rest of this chapter will attempt to untangle some of these developments and will be divided into three sections: the expansion of satellite-transmitted programming; the delayed process of introducing digital terrestrial television (DTT); and the emerging phenomenon of streaming audio-visual material.

## The Expansion of Satellite Television

The subscription television market grew steadily from the inception of M-Net. M-Net began broadcasting via analogue satellite in 1992, and in anticipation of the launch of the PANAMSAT 4 (PAS4) satellite in 1996, the company split into two separate entities. MultiChoice was launched in 1993 as Naspers' mother company for its growing and diversifying satellite television business, and in 1995 was listed on the Johannesburg Stock Exchange as a separate company, though remaining wholly owned by Naspers. A subsidiary company, Direct Satellite Television (DStv), was established as a vehicle to manage the subscription function of the business (itWeb, 2014). PAS4 allowed for the migration of the subscription service onto a digital platform, initially comprising 16 channels, a huge advance at the time. M-Net was redefined as a content brand. The PAS4 enabled a full-scale penetration of the continental African market, the Middle East, and even Greece via satellite, with varying degrees of success. The transnational nature of MultiChoice's expansion throughout Africa resulted in the company's establishment as the pre-eminent content provider and carrier across the continent.

The subscription audience was relatively small when compared to that of the free-to-air audience. In 2012, for instance, the Department of Communication estimated that 72% of South African households relied solely on terrestrial broadcasting (OECD, 2013: 275). The fees for a full bouquet were (and remain) high; consequently, DStv experienced a ceiling to the number of viewers it could command. DStv saw technology as one way to boost subscriptions. The launch of 'interactive television' was the first step in 2002. These 'iTV' services utilized the feedback path enabled from the set-top decoders that connected households to MultiChoice's back-end service. The technology was still an analogue system, relying on the terrestrial (landline) telephone grid. A groundbreaking development was the claimed world's first launch of the dual view decoder in 2003 (upgraded in 2008), allowing the simultaneous viewing of two different channels in different rooms on the same subscription. Another innovation was the Personal Video Recorder (PVR) in 2005 (itWeb, 2014), which saved onto an attached hard drive, which could be rewound, fast forwarded and paused. Unlike the older analogue VHS system, the programmes were not transferable and could only be retrieved on the system that recorded them, thus ensuring the safety (and monopoly) of the subscription arrangement.

## DStv Compact

Still facing a subscriber-ceiling with the near-saturation of the middle class, DStv turned in 2015 to the 'emerging market' of the lower middle and upper working classes for future growth prospects, introducing the 'compact' bouquet. Confined to 25 rather than 50 channels, the bouquet offered proprietary M-Net and MultiChoice channels, some of the less expensive international channels, and an increasing amount of local programming in African languages.

To feed the appetite for local programming, DStv invested in training and supporting technical and artistic talent, and in the process has built the local television-production infrastructure. So successful were the local programmes that in 2010 DStv launched a separate channel – 'Mzanzi Magic'. 'Mzanzi' is an affectionate slang for 'South Africa', while the 'magic' played in the marketing branding of DStv. While it began life as a single channel, Mzanzi gradually built up to be the most watched sub-portion of DStv.

## Expansion of the Subscription Television Market

The IBA's Triple Enquiry considered free-to-air analogue broadcasting only, and so by 2005 there was a clear gap in the regulatory environment. As a consequence of further convergence issues, the Broadcasting Act (RSA, 1999a, 1999b) was augmented in 2005 with the Electronic Communications Act (RSA, 2005) and the accompanying Independent Communications Authority of South Africa Amendment Act. Among other issues, the legislation opened up the subscription television space, and the possibility of the introduction of competition to DStv. ICASA initiated a pay-television review and licensing exercise in 2007 (ICASA, 2007). Applications were invited from consortia wishing to launch subscription television in competition to DStv. Eighteen applications were received, of which five were approved: MultiChoice, ESAT (an offshoot of e-TV); On Digital Media; Telkom Media; Walking on Water. Telkom failed to establish a viable business; despite headhunting staff from established broadcasters and investing considerably into infrastructure and satellite rights, after attempts at securing a Chinese investment partner came to nothing, and the venture was closed in 2009 (Mail and Guardian, 2009).

ESAT is a wholly owned subsidiary of e-TV. Initial attempts at launching a stand-alone satellite service proved abortive, and they opted to trade as a channel content supplier on the established MultiChoice DStv platform. Beginning June 1, 2008, the channel expanded into a 24-hour rolling news format, rebranding itself eNews Channel Africa (ENCA), preparing the way for a launch in the UK in 2012. In 2013, the channel abandoned its hopes of establishing a stand-alone service and moved into the DStv bouquet under the brand of eNCA.com.

On Digital Media launched its satellite channel as 'Top TV'. The company punted itself as embodying the values of broad-based economic empowerment, with a predominantly black ownership (of approximately 68%). The channel was aimed at the lower end of the socio-economic commercial satellite subscription market, the same strata that DStv had identified as having growth potential. Late in 2011, it expanded its channel offerings, but still struggled to gain traction, with a daily total of 134,000 viewing households, compared to DStv's 3.49 million during the same period (IOL, 2011a). With very substantial debts to creditors, including the Disney Corporation, Warner, and Fox International, in October 2012 TopTV went into receivership and sold its assets (and license) to StarTimes, based in China. StarTimes had operations in 17 African countries, and approximately 7 million pay-television subscribers in China (IOL, 2013). Several counter-offers were made, including a Kenya-led African consortium in collusion with DStv (McLeod, 2015). Faced with this information, ICASA awarded the transfer of the license to StarTimes at the end of 2015, which rebranded as StarSat (McLeod, 2015).

## Digital Terrestrial Migration

In 2006, the ITU formalized an agenda for countries in Europe, the Middle East and Africa to migrate from analogue to digital broadcast transmission by mid-2015. The purpose of digital migration is to allow for greater signal compression, thus 'freeing' valuable electromagnetic spectrum for data distribution. The process is key for opening up more frequencies and faster mobile broadband services. South Africa committed to complete the process by 2011 to convert all free-to-air broadcasts to DTT. Pilot tests of the set-top boxes (STBs) were scheduled for 2010 ahead of the FIFA Soccer World Cup. However, the digital migration was stalled by disagreements among broadcasters, allegations of corruption, in-fighting between stakeholders, drawn-out court cases, constant switching on broadcast standards and fighting over conditional access and encryption system issues (Fin24, 2018). To date (2018), the migration has not been completed. The cost of the STBs is a matter of contention. The government has only committed to hand out 5.2 million free STBs to the neediest, but that will leave several million more precarious households to fend for themselves. If they are unable to purchase the STB, they will simply disappear off the grid as viewers.

## Audio-visual Streaming

ICASA first signaled their interest in creating a regulatory framework for Internet Protocol television (IPTV) in 2010 (IT News Africa, 2010). At the time, broadband access within South Africa was significantly less advanced than elsewhere, and very expensive. Unlike broadcasting, broadband provision is not regulated, making intervention in pricing difficult. ICASA issued its Final Internet Protocol Report in 2013. In terms of the Electronic Communications Act, IPTV is regarded as 'broadcasting' (since it is an extension of the broadcasting license and closed to all but those clients who subscribe to the channel); while Video on Demand (VOD), streamed from a publicly accessible internet service, is regarded as broadcasting, but not covered by the Act, nor any of ICASA's regulations. In response to the regulatory enablement, DStv inaugurated Digital Video Broadcasting over Internet in 2010. Initially confined to 'gated communities' using an Internet Protocol (IP) capable decoder, the service was later expanded to the newer PVRs. Soon after ICASA published the Mobile Television Regulations in 2010 (ICASA, 2010), DStv introduced 'Mobile' and 'iDrifta', mobile television decoder devices that allowed subscribers to access content directly onto mobile devices. The service was discontinued without fanfare in 2017 on the grounds that the technology had moved on and widely available broadband made the application obsolete. This is an excellent example of the way in which some technologies, so promising when they are first introduced, prove to have an unexpectedly quick obsolescence. More widely used was 'CatchUp', also launched in 2011 as a time-shifting service in which subscribers were able to access specific programming automatically stored on their PVR hard drives. This is an offline retrieval service, and therefore falls under the auspices of broadcasting.

'Box Office', again an extension of DStv, regarded as an IPTV, was introduced in 2013. Each film was paid for separately but delivered to the subscriber's hard drive (IOL, 2011b). This was supplanted by ShowMax in 2015 (Lautenbach, 2015), a streaming package accessed directly from the internet, a true VOD service. The offerings were fairly standard for the kind of service: box-set season runs of series, children's shows, documentaries and full-length films. Initially, it was available only as an add-on service, with a small monthly subscription; but as the competition in this area increased, it was offered as a free value-added service to the premium customers in 2017.

Streaming VOD services are not regulated, nor do they require a license to operate. All these 'over the top' (OTT) programming providers require are simple decoder boxes connected directly to the broadband fiber feed coming into the home. The ease of market entry saw the arrival in 2016 of Netflix as the star turn of Kwese, a platform that also includes YouTube, NBA TV and TedTalk. Kwese is owned by EcoNet Media, an African telecoms giant operating through southern and east Africa. It is unclear, however, how long this sector of the market will remain unregulated, since in 2017 the minister of communications signaled that the department was planning to present a draft white paper on the Audio-Visual and Digital Content Policy to cabinet for approval before publishing it for public comment, specifically to bring streaming services into line with the Broadcast Act (Tech Financials, 2017).

## Conclusion

South Africa is the largest television market in Africa, valued in 2016 at ZAR 40.9 billion (USD 3.27 at the current exchange rate). This figure, which has risen at the time of writing (2018) to ZAR 51.2 billion (USD 4.1 billion), includes revenue from pay-television subscriptions, physical home video/television, internet video, public license fees and advertising (PWC, 2017: 2). A little more than half of this amount is accounted for by 'end user' spend, while the remainder comes from advertising.

South Africa in 2016 had 16.9 million 'households' (StatsSA, 2016). Of these, 14.09 million owned a television set, of whom nearly 8 million are reliant on free-to-air viewing, and just over 6 million subscribe to DStv. Other options (alternative satellite options, internet viewing) account for 811,000 (this does not include viewers who supplement their primary viewing with internet streaming) (BRC, 2018: 29). Viewership of free-to-air channels is higher than indicated by these figures, since all free-to-air channels are also flighted on DStv.

The middle class is growing, with a resultant expansion of disposable income, allowing for growth in the media sector, particularly the pay-television and internet-based OTT streaming options. The delay in the introduction of DTT has acted as a brake on the growth of the digital market, halting not only the introduction of DTT channels, but also stymieing the long hoped for 'digital dividend', which would in turn allow for further development in the entire broadcast and streamed audio-visual mediascape.

The market is characterized by a strongly delineated duopoly in which the free-to-air space is dominated by the state-owned broadcaster, the SABC, and eTV. The latter is the only viable commercial free-to-air alternative to the state-owned broad broadcaster. While e-TV would like to transition into the arena, the delay of DTT (see above) has slowed these ambitions.

MultiChoice's DStv dominates the pay-television sector. New competitors in the pay-television market, including internet streaming services (notably NetFlix and Amazon) that are not legally regarded as 'broadcasters' but who function to provide audio-visual content to the viewing public, are making inroads into the television ecology, but at the time of writing remain secondary to the main players, without the 'killer content' of sports and current news broadcasts. Between them, Google TV, Apple TV, Amazon and Netflix commanded 36,500 viewers, or less than 1% of the population in 2017 (BRC, 2018: 9).

While technology does not make change inevitable, and nor does it necessarily shape the kind of social outcomes a society adopts, it is clear that technology is a significant driver in the process of change. The case study of South African television evolution over the past quarter century illustrates that well. Starting with a simple three-channel analogue, free-to-air service (SABC) and one analogue encrypted service (M-Net), some 25 years later the broadcast ecology is made up of a mix of four analogue free-to-air channels, some experimental digital terrestrial broadcast; encrypted satellite television carrying a mix of imported and locally produced programming, including wholly owned television channels belonging to other South African broadcasters; and a flurry of VOD and IPTV options.

Despite this rich range of options, 56% of South African households still depend on analogue free-to-air channels (SABC and e-TV only); 41% have a subscription to one of the DStv bouquets (many of whom use the satellite services of the FTA that are transmitted on DStv); and 4% of households rely on OTT video streaming.[1]

Yet this landscape is far from driven only by the affordances of technology. Its stop and start character has been shaped by at least four considerations. The legislative environment and regulatory oversight have played an important social engineering role which has sought to balance commercial advancement and sustainability with public interest issues, local content production and distribution, language rights, foreign ownership thresholds and cross-ownership concerns. Cost, too, has been a major factor in shaping the broadcasting landscape. The expansion of subscription television has been hemmed in by the huge disparity in income across the country. The World Bank country overview of 2018 cites a Gini coefficient of 0.63 in 2015, one of the highest in the world, where the poorest 20% of the population consume less than 3%, while the wealthiest 20% consume 65% of the country's total expenditure. Despite economic growth, the rapid population growth and very youthful profile of the population has meant that the average GDP per capita has been stagnant or even declining. The 'triple challenge' of poverty, unemployment and inequality, makes it inevitable that for a significant proportion of the population chasing after the latest technological gadgets and topping into the just-released international series is not an immediate priority. With the delayed introduction to DTT, access to television in the short term may actually fall, as the 'missing middle', those households not poor enough to qualify for a state-subsidized STB and not wealthy enough to spend money on one, find themselves excluded from free-to-air transmissions. Language and programming selection remain the biggest driver of consumer choice in television provision. Free-to-air broadcasting is dominated by 'black' viewers (86%), with isiZulu and isiXhosa speakers making up nearly half the entire viewership, while the other African languages together with English and Afrikaans, make up the rest. While subscription television has increasingly added African-language programming to its schedule, and while all the SABC channels are included on the DStv bouquet, market logic dictates that the programming with the largest number of African languages will have a head-start, despite the technology through which they are offered. Programming and content are key factors in the consumer choice for television provision. The ability to secure the rights to premium content in the fields of entertainment, news and sport, cannot be exaggerated. Here, the ready availability of money is essential, as is the 'first mover' advantage of signing long contracts, particularly in the area of sports. DStv have always used sport as a drawcard.

The broad outlines of the South African televisionscape will remain the same, with the majority watching the public broadcasters, mainly in their home language, albeit through a variety of transmission options and platforms. The long-awaited introduction of DTT will not change this much. The expansion of subscription television and OTT services will continue gradually to eat into the audiences and revenues of the public broadcaster, with the proviso that these paid-for services provide programming in indigenous languages, as well as the sought-after genres of entertainment, news and sport.

## Note

1 These figures are based on the PWC 2018 report. Free to air (FTA) 7,928,834; DSTV 6,156,745; Other video households 811,000. Ref: www.pwc.co.za/en/assets/pdf/enm-20120-chapter3.pdf. (The figures quoted from Price Waterhouse Cooper were polled in 2016, which goes some way to explaining the difference in calculations. Based on the 2016 ITU numbers, there is a television penetration of 81% of households, whereas the 2018 figure would be elevated to 87%.)

## References


Broadcasting Research Council, 2018. (BRC 2018:29). BRC TAMS Update, April, 2018. http://brcsa.org.za/brc-tams-update-april-2018/

Duncan, J. and Glenn, I. 2010. 'Turning points in South Africa's television policy and practice since 1990'. In *Media Policy in a Changing Southern African: Critical Reflections on Media Reform in a Global Age*, Eds. D. Moyo and W. Chuma. Pretoria: UNISA Press, pp. 39–72.

Fin24, 2018. 'Mokoyane appoints advisory council to speed up digital migration'. (10.07.2918) www.fin24.com/Companies/ICT/mokonyane-appoints-advisory-council-to-speed-up-digital-migration-20180710-2.

Hayman, G. and Tomaselli, R. 1989. 'Ideology and technology in the growth of South African broadcasting, 1924–1971'. In *Current of Power: State Broadcasting in South Africa*, Eds. R. Tomaselli, K. Tomaselli, and J. Muller. Denver: International Academic Publishers, pp. 23–75.

HCI, 2001. 'Sabio to amplify e-TV Stake'. www.hci.co.za/2001/12/05/sabido-to-amplify-e-tv-stake/. Accessed 29.06.2018.

Horwitz, R. 2001. *Communication and Democratic Reform in South Africa.* Cambridge: Cambridge University Press.

ICASA, 2007. *Position Paper on Subscription Broadcasting Services.* www.icasa.org.za/uploads/files/Position-Paper-Subscription-Broadcasting-Services.pdf, accessed 18.07.2018.

ICASA, 2010. *Mobile Television Regulations.* www.icasa.org.za/legislation-and-regulations/mobile-television-regulation-2010

Independent Broadcasting Authority (IBA), 1995. *Triple Inquiry Report.* Johannesburg: IBA. (not online).

Independent Broadcasting Authority (IBA), 1996. *Independent Broadcasting Authority Triple Inquiry Report.* Johannesburg: IBA. (not online).

Independent Broadcasting Authority (IBA), 1997. *Position Paper on the Introduction of the First Free-to-Air Private Television Service in South Africa.* Johannesburg: IBA. Unpublished paper. (not online).

Independent Broadcasting Authority (IBA), 1999. *Position Paper on a Definition of Advertising, the Regulation of Infomercials and the Regulation of Programme Sponsorship.* Johannesburg: IBA. Unpublished paper. (not online).

IOL, 2011a. 'Top TV into selling its airtime'. (26.10.2011) www.iol.co.za/business-report/companies/toptv-moves-into-selling-its-airtime-1165363.

IOL, 2011b. 'DStv launches video on demand services'. (21.07.2011) www.iol.co.za/business-report/companies/dstv-launches-video-on-demand-service-1103606.

IOL, 2013. 'China's StarTimes wins TopTV bid' (02.05.2013) www.iol.co.za/business-report/companies/chinas-startimes-wins-toptv-bid-1509038

IT News Africa, 2010. 'ICASA speaks up on Internet-TV' (16.03.2010). www.itnewsafrica.com/2010/03/icasa-speaks-up-on-internet-tv/.

itWeb, 2014. 'Timeline: DStv through the years'. www.itweb.co.za/content/JKjlyr7w4OzMk6am. Accessed 18.07.2018.

Jacobs, S. 2013. 'Media, citizenship and social justice'. In *State of the Nation, 2012–2013*, Ed. HSRC. 333-354. Pretoria: HSCR Press.

Jamalipour, A. 2003. *The Wireless Mobile Internet.* Hoboken, NJ: Wiley and Son.

Lautenbach, K. 2015. Showmax officially launched in South Africa' (21.08.2015) http://za.ign.com/scitech/93548/news/showmax-officially-launched-in-south-africa-here-are-the-det

Mail and Guardian, 2009. 'Telkom's broadcast bungle'. https://mg.co.za/article/2009-03-28-telkoms-broadcast-bungle. Accessed 20.07.2018.

McLeod, D. 2015. 'StarSat secures big victory at ICASA' (02.11.105) https://techcentral.co.za/StarSat-secures-big-victory-at-icasa/61148/

MultiChoiceAfrica, n.d. http://multichoiceafrica.com/about-us/history/

My Broadband, 2011. 'M-Net Community Service' https://mybroadband.co.za/vb/showthread.php/380269-Mnet

NB: All dates are cited as day.month.year. Therefore 09.11.2011 reads 9 November 2011.

News24, 2007. 'Goodbye to M-Net Open Time'. www.news24.com/Entertainment/SouthAfrica/Goodbye-to-MNet-open-time-20070325

OECD, 2013. 'Policy Roundtable: Competition issues in television and broadcasting'. www.oecd.org/daf/competition/TV-and-broadcasting2013.pdf. Accessed 16.06.2018.

Price Waterhouse Cooper (PWC), 2017. Entertainment and Media outlook, 2017–2021. www.pwc.co.za/en/assets/pdf/entertainment-and-media-outlook-2017.pdf

RSA, 1993. *Independent Broadcasting Authority (IBA) Act.* 153/1993. GG 15999 30.09.1993. www.icasa.org.za/legislation-and-regulations/acts.

RSA, 1999a Broadcasting Act (73/1976). Not online.

RSA, 1999b Broadcasting Act (4/1999) GG 20263 30.06.1999. www.gov.za/sites/www.gov.za/files/a4-99.pdf

RSA, 2000. Republic of South Africa, ICASA Amendment Act, (13/2000) GG 21154 11.05.2000. www.2.saflii.org/za/legis/hist_reg/icaosaa13o2000rangnr852975/icaosaa13o2000rangnr852a21s20151106.html

RSA Republic of South Africa, 2005. Electronic Communications Act (36/2005). Government Gazette (GG) 28743, 18.04.2006. www.saflii.org/za/legis/num_act/fsbra1996420/.

RSA Republic of South Africa, 2009. *Public Service Broadcasting Bill*, 2009. Government Gazette No. 32663 of 28 October 2009. www.icasa.org.za/legislation-and-regulations/acts

SABC, 1980. *Annual Report.* Auckland Park: SABC. (not online).

SABC, 1982. *Annual Report.* Auckland Park: SABC. (not online).

Skinner, K. 2011. 'The South African Broadcasting Corporation – the creation and loss of a citizenship vision and the possibilities for building a new one'. Unpublished Seminar, UJ Sociology, Anthropology & Development Studies. www.uj.ac.za/faculties/humanities/sociology/Seminars/2011/Skinner%202011%20Broadcasting.pdf. Accessed 21.06.2018.

Statistics South Africa (StatsSA), 2016. General Household Survey, 2016. www.statssa.gov.za/?p=9922

Tech Financials, 2017. 'South Africa plans to regulate streaming services'. https://techfinancials.co.za/2017/10/06/sa-plans-to-regulate-streaming-services-netflix-showmax/

Teer-Tomaselli, R. 2011. 'Transforming State Owned Enterprises in the global age: Lessons from broadcasting and telecommunications in South Africa'. In *Political Economy of Media Transformation in Southern Africa*, Eds. K. Tomaselli and H. Dunn. Cresswill, NJ: Hampton Press, pp. 133–166.

Teer-Tomaselli, R. 2004. 'Transforming state owned enterprises in the Global Age: Lessons from broadcasting and telecommunications in South Africa', *Critical Arts*, 18:1, 7–41. doi:10.1080/02560240485310031

Teer-Tomaselli, R. 2015. 'In the beginning: Politics and programming in the first years of South African television'. In *Tuning In: Perspectives on Television in South Africa*, Eds. M. Tager and C. Chasi. Cape Town: Pearson, pp. 1–17.

# 39
# PIRATE UTOPIA REVISITED

*Martin Fredriksson*

In the opening scene of the 2006 movie *Steal this Film* we see a huge ocean wave ushering in over Manhattan, slowly sweeping away the megalopolis. This illustrates a message superimposed over the catastrophic scene in big block letters:

> Peer to peer networks have unleashed a massive wave of change on the world

*Steal this Film* is a crowd-funded, activist documentary in two parts that discusses how media piracy, filesharing, and Peer to Peer (P2P) networks challenge the movie and media industry. The film builds its narrative around the authorities' sanctions against the filesharing hub, The Pirate Bay, which stood accused of promoting piracy by providing links to a large number of BitTorrent sites. The prosecution against The Pirate Bay was initiated in May 2006 when Swedish police confiscated The Pirate Bay's servers in a raid that was partly instigated by pressure from the American media industry. In *Steal this Film*, The Pirate Bay case serves as a starting point to discuss a changing media landscape where proprietary business models are giving way to new forms of distribution.

Today, more than ten years after its release, the movie comes across as an interesting historical document from a time when both the media industry and the pirates still thought that piracy could kill Hollywood. It is an equally utopian and dystopian narrative: while representatives of the movie industry feared that piracy would undermine the creation of art and entertainment, the pirate technologists and ideologists believed that piracy could set culture free and democratize media by providing unlimited access to art and information and make every passive consumer a potential producer. The main obstacle to this brave new world was the backward striving copyright industry's attempts to outlaw filesharing.

The people behind *Steal this Film* were not alone in voicing critique against the copyright industry. The prosecution of file sharers and the passing of stricter copyright laws caused strong resistance in the years following 2006, and the reactions intensified further in March 2009 when the owners of The Pirate Bay were sentenced to one year in prison and fined 30 million SEK (US$3.8 million) in a trial clouded in allegations of biased judges (The Economist 2009; Rydell & Sundberg 2010; Burkart 2013). Over time these debates over intellectual property rights versus public access to knowledge gave rise to a growing popular resistance against what was perceived as copyright expansionism: a development toward more extensive copyright laws, resulting in the enclosure and privatization of culture and information. This pirate movement stretched from hacktivists such as Anonymous to national political parties – so-called Pirate Parties – that seek parliamentary representation in order to protect freedom of speech, access to information and rights to privacy in a digital world (Spender 2009; Burkart 2013; Fredriksson 2015a, 2015b).

For a while The Pirate Bay became a poster boy for that movement. Although their servers were confiscated by Swedish police in 2006 and the founders were prosecuted and imprisoned three years later, the site itself never went offline. By tweaking their technological solutions and moving between servers in different parts of the world, The Pirate Bay has remained online and is still one of the largest pirate sites. The endurance of The Pirate Bay seemed to speak of a persistence of new technology over old monopolies, confirming the image of the BitTorrent technology as an unstoppable wave, washing away old media monopolies. This idea of technological persistence came to underline much of the pirate ideology.

To some extent, this belief in the predominance of technology grew from a recent tradition of digital technotopianism where people like John Perry Barlow declared the 'Independence of Cyberspace' against the 'weary giants of flesh and steel' who ruled the material world through government and corporations (Barlow 1996). In a longer perspective, the filesharing revolution also corresponded to earlier stages in the history of piracy. Copyright historian Adrian Johns, for instance, draws parallels to how extensive pirating of sheet music threatened to undermine the music publishing industry in Victorian England. During the last decades of the 19th century a combination of new technology and changing musical trends had opened up the market for music scores to a wave of unauthorized reprints. The demand for printed piano music had grown considerably in the late 19th century, but the high prices charged by the music publishers left the consumers looking for cheaper alternatives. The newly developed technology of photolithography made it significantly easier and cheaper to produce more-or-less exact copies of printed material, and particularly of loose sheets such as music scores. The only obstacle to this new printing revolution was copyright, which gave a few established publishers exclusive rights to the most popular works (Johns 2009, 327).

More recent examples of how new technologies are seen as threats to old businesses are the cassette tape and the VCR. In the 1970s and the 1980s, music and video cassettes were believed to undermine the music and the movie industries by providing consumers with the means to easily reproduce and share content with their friends. The reaction to the VCR is a frequently recalled example in the pirate debates. When Hollywood raged against The Pirate Bay many recalled how the movie studios in the 1980s tried to ban the VCR as a means of piracy and Jack Valenti, Chairman of the Motion Picture Association of America, famously argued that the 'VCR is to the American film producer and the American public as the Boston strangler is to the woman home alone'. In retrospect, people have pointed out that the VCR did not kill the movie industry but rather reinvigorated it by providing a new and highly profitable distribution channel.

Common to many narratives on contemporary and historical piracy is that they approach piracy as a technologically driven phenomenon and frame it as a conflict between grassroots users of new technology and the hegemony of old media monopolies. An underlying assumption in many of those narratives has also been that new technology is destined to prevail over the 'weary giants of flesh and steel'. This text will revisit the pirate utopia and relate it to the deregulation of the Swedish TV-market in the late 1980s, in order to initiate a critical discussion about the filesharing revolution's persisting consequences for the media landscape of today.

## The Pirate Capitalist and the Media Monopoly

But before I look closer at the filesharing revolution I would like to go further back in time and briefly discuss another (pseudo)piratical intervention in media history, namely how commercial broadcasters forced their way into the Swedish TV-market. This development was spearheaded by the television station TV3, owned by Kinnevik; a large investment corporation with roots in the Swedish forest and paper industry. In the 1980s Kinnevik gradually moved into the media and telecom industry and by the early 1990s they had become a major player in television and newspaper publishing. As early as 1984, Kinnevik invested in SES/Astra1A, Europe's first private satellite system, which, three years later, they used to launch TV3 as the first commercial TV channel aimed directly at a Swedish audience.

Since the 1920s, public radio broadcasting in Sweden had been controlled by a public service monopoly akin to that of the British Broadcasting Corporation (BBC). When television was introduced in 1956, a similar monopoly applied to that medium, and this persisted until the late 1980s. While public service TV was transmitted through air waves, a fiberoptic cable TV distribution network was established in Sweden in the late 1970s. In the 1980s both local, community-run, TV broadcasters as well as global satellite channels such as the Murdoch-owned Sky Channel, used this cable network to broadcast to Swedish viewers. As part-owners of the Luxemburg-based corporation *Société Européenne des Satellites* (SES), Kinnevik could use their satellite Astra1A to broadcast from London into the Swedish cable network. By running the station from London, TV3 could evade the still quite restrictive Swedish media laws and broadcast commercial, ad-funded content tailored for a Swedish audience. This blatant circumvention of Swedish laws was, quite accurately, perceived as an open attack on Swedish media politics, but it turned out to be impossible to stop. By the early 1990s the public service monopoly was disintegrating, the media laws rewritten, and the television market deregulated to allow for private, commercial broadcasters and ad-funded content.

A key actor in this development was Jan Stenbeck, CEO of Kinnevik, who emerged as something of a capitalist revolutionary in the Swedish media landscape of the late 1980s. Stenbeck took to the barricades with the slogan 'technology trumps politics', arguing that it is technology and not politics that drives the development: once the technology exists, politicians and legislators will merely have to adapt to the new realities (Ewertsson 2002). This pseudomaterialist, technologically deterministic perspective came to form the narrative about the fall of the Swedish public service monopoly and the rise of Kinnevik's media empire.

The way TV3 disrupted the public service television monopoly had similarities to the pirate radio stations that challenged the BBC in the 1960s and the Scandinavian public service broadcasters in the 1970s and 80s (Ewertsson 2002). Pirate radio expanded simultaneously with the state-controlled public service broadcasting system across Europe. As soon as the BBC was formed in the early 1920s, it met competition from pirate radio stations, and by 1925 the UK was estimated to have around 2.5 million pirate listeners (Johns 2010, 21). Pirate radio persisted in different forms throughout the 20th century and peaked again in the mid-1960s when a new generation of pirate radio stations, often broadcasting from ships in the North Sea, emerged. In the Scandinavian countries, similar kinds of offshore pirate radio stations challenged the state-owned public service monopolies in the 1960 and 70s, eventually provoking changes to the regulation of radio broadcasting (Ewertsson 2002; Burkhart & Fredriksson forthcoming).

Both the pirate radio stations and the commercial TV stations claimed to provide alternative content that the public service broadcasters rejected. While the British pirate DJs of the 1960s played new pop music that was rarely heard on BBC, TV3 offered a kind of light, commercial entertainment that was shunned by public service television. Both cases were entrepreneurial interventions, carving out alternatives to existing media distribution that challenged the state's attempts to control the mediascape, although from quite different positions. Both the radio pirates and TV3 relied on a combination of technology and territoriality to evade media regulations by operating across national borders. And, like Stenbeck, the radio pirates eventually provoked legislative change when the liberalization of media laws allowed for small-scale, amateur initiatives to promote a legal alternative to piracy (Burkart & Fredriksson forthcoming; Johns 2010).

Neither TV3 nor the pirate radio stations were innovative in the sense that they developed new revolutionary technology: rather, they utilized already existing technology in new ways. Both the European satellite system and the Swedish cable network had been developed by other actors, often with public funding, and the challenge facing TV3 was to get access to those socio-technological systems. This was achieved not primarily through technological innovation but by buying resources and building a wide network of alliances in the corporate world as well as in politics – and, not least, within the ruling social democratic party. Likewise, the radio pirates, particularly in the UK, grew out of an older movement of radio amateurs that had been experimenting with the new medium since the early 1900s. Pirate radio programmers thus reproduced technological innovations that had been developed by unauthorized radio enthusiasts, but they often combined this with an individualist and libertarian rhetoric.

The fall of the Swedish public service monopoly is not merely a piece of Swedish political history but connects to a restructuring of the international media industry. Jan Stenbeck had been working in the US in the 1970s, learning about the American TV-market, but he was also keeping an eye on the changes in Western Europe (Ewertsson 2002). Sweden was not the only country where new media oligopolies were born out of the fall of the public TV monopoly. Similar tendencies had been seen in other European countries. This probably inspired the Kinnevik corporation whose annual report of 1987 suggested focusing on areas where 'the situation of competition resembles oligopolies. Such possibilities can often be found in the vicinities of the public monopolies'[1] (Ewertsson 2002, 24).

The fact that the corporation had its origins in a sector alien to the media industry – the forest industry – was also significant for the new media oligarchs who often came from entirely different businesses. This external colonization of the media had begun as early as in the 1960s when corporations such as IBM and Xerox began to buy businesses in the publishing industry, but accelerated in the 1990s, when global producers of home electronics made strategic acquisitions of content producers – primarily in the film and the music industry – in order to gather hardware and software in the same corporation (Bagdikian 2000; Hemmungs Wirtén 2004). In this period, the content and media industry was thus increasingly incorporated into huge multinational conglomerations, creating a concentration of power and resources in the commercial media industry that far surpassed the size and value of public media monopolies on the fringes of Europe.

## Memories of a Free Network

Just like Jan Stenbeck and TV3, the Swedish filesharing pioneers of the early 2000s also partly relied on an existing technological infrastructure. The BitTorrent protocol, first developed by Bram Cohen, in 2001, was indeed a new and innovative way to distribute files in a decentralized way that was fast, reliable, and hard to shut down. The rapid expansion of filesharing in Sweden was, however, partly enhanced by two political initiatives in the 1990s. In the first half of the 1990s the conservative government set out to develop an internet infrastructure that would give Sweden a head start in the digital economy, and by the mid-1990s they had laid the foundations for an extensive and well-functioning broadband network that continued to grow throughout the century. In 1997 the succeeding social democratic government continued the mission by launching the so-called Home-PC reform, where people could lease computers through their employees under beneficial tax conditions. The campaign was primarily directed toward blue-collar employees who did not use computers in their everyday work and the goal was to create an information society for everyone by bringing computers into the homes of working-class families. The campaign was a great success, but as one might expect the computers were mostly put to use by the kids rather than by their parents. This provided a generation of Swedish youths with the infrastructure, the hardware, and the computer literacy to develop their piratical capabilities in the new millennium (Rydell & Sundberg 2010).

This new media revolution, however, had an entirely different ideological approach from that of the early 1990s. If the disruption of the public service monopoly was an outright capitalist intervention, then the pirate revolution was often shrouded in an anticapitalistic rhetoric, or at least in a strong criticism against corporate media. The pirate movement, thus, rallied against the commercial media monopoly (or oligopoly) that TV3 was part of establishing 20 years earlier (Bagdikian 2000). The pirate ethos of the early 21st century was largely anticommercial and the P2P revolution was seen as a way to democratize art and culture by undermining the commodification of the artwork and challenging the hegemony of the cultural industry.

The free distribution of media files was ascribed two emancipatory effects. First it was expected to break the media monopoly's control over selection and distribution of content. When P2P networks made every digitized movie or piece of music potentially available to every consumer, the viewers or listeners were no longer subject to the narrow selection of programs presented to them by TV or radio stations. Or, as they put it in the manifesto of the filesharing site Kazaa:

> 30 years of watching the movies they want you to see.
> 30 years of paying the prices they demand.
> 30 years of swallowing what they're shovelling.
> [...] Over. With one single click.
>
> *(Meese 2014, 31–32)*

Chris Andersson, chief editor of *Wired*, (2006) famously talked about the long tail of media content and argued that digitization and cheaper distribution costs would make it commercially viable to distribute a wider selection of culture and entertainment to niche audiences. Renegade distributors such as P2P networks and shadow libraries took it upon themselves to make that vision reality regardless of restrictions posed by the media monopoly or copyright owners. In one sense, the P2P pirates shared the ambitions of the radio pirates or commercial TV stations in Sweden in the 1980s to provide content that was excluded from the existing media establishment, the difference being that TV3 and other commercial networks' mainstreaming of media content now represented the establishment that the pirates revolted against.

Second, there were high hopes among the filesharing movement that digital technology would empower the users and promote a remix culture where people were no longer merely viewers or listeners who passively consumed media, but became prosumers who could create their own art and share it freely with each other. In his 2008 book, *Remix*, Lawrence Lessig argued that analogue cultural distribution had supported a Read Only culture while digital technology enables a Read/Write culture where consumers are increasingly taking part in remixing and recreating the artworks (Lessig 2008). *Steal this Film* is, in itself, an example of such a vision of decentralized DIY-creativity, as it was produced by a collective calling themselves The League of Noble Peers, funded through crowdsourcing and widely distributed by different filesharing networks (Özdemirci 2014).

The question is to what extent this ideology of piracy was actualized in piratical practices. Although the pirate ideologist tended to see filesharing as radical or countercultural, most users probably approached it as a more mundane everyday practice. Many researchers of piracy have pointed out that there is nothing intrinsically radical or countercultural in the act of filesharing. In fact, it is in most cases merely an extension of an ordinary consumer behavior – responding to the impulse to consume what you want as quickly and as cheaply as possible (Andersson Swartz 2014; Da Rimi & Marshall 2014). In the end we can see that the most popular content on major filesharing networks tends to be big-budget Hollywood productions. As Ramon Lobato points out:

> This suggests that pirate activity tends, by and large, to follow the precedents set by established forms of film marketing, underscoring the earlier point about the co-constitutive nature of piracy and film marketing. In other words, pirates like blockbusters as much as the rest of us.
>
> *(Lobato 2014, 14)*

Consequently, P2P filesharing came to be not so much about sharing rare and obscure material – although that certainly happens (Bodó 2014) – but about making mainstream entertainment more accessible. In that vein, piracy has also often been a way for audiences in marginalised national markets to get access to the latest episodes of popular TV series at the same time as they are aired in the USA or the UK. This phenomenon should be less prominent as mainstream series are increasingly distributed not through TV broadcasting, where programs are aired at fixed times, but as streamed content that can be released simultaneously anywhere in the world. In that case, however, geo-blocking poses another barrier for viewers where content is still not equally accessible across the world although it might be distributed through the same services (Lobato and Meese 2016). This opens up a market for piracy as a global equalizer, providing the same access to content in different parts of the world that are sectorized by the global media industry. In addition to this, piracy can also provide access to prohibited content in parts of the works where the media market is subjected to state censorship.

There are, however, also indications that commercial media corporations might be benefiting from filesharing networks for promotional reasons (Jenkins 2006; Lobato 2014). The American TV series *Prison Break*, for instance, was hugely popular in mainland China in spite of the fact that it was never distributed through any legal channels. The revenues on sales of *Prison Break* merchandize in China were thus largely created by piracy and You Jie argues that digital movie piracy has often served as a 'pioneer for Hollywood' in mainland China (Jie 2014, 204). It is thus unclear whether the filesharing networks actually promoted a greater variety in media consumption habits or whether they merely contributed to the global dominance of a mainstream media industry by providing alternative distribution channels in a part of the world that the capitalist distribution channels could not access.

The rise of the digital prosumer has, indeed, had significant impact on the mediascape, but the question is whether the promise of a remix culture that empowers the users has been fulfilled. In 2017 YouTube was ranked as the world's second most visited website, arguably making it the largest international content provider (Alexa 2017). Most of its popular content is, however, still produced by commercial media corporations and, while the user-generated content is certainly significant, it has been commercialized by Google to the extent that the term YouTuber is now more or less synonymous with prosumer.

## How the Media Monopoly Reclaimed the Control it Never Lost

In *Steal this Film*, Sebastian Lütgert, reflects on the P2P revolution of the early 2000s:

> People have lamented much the 'death of the author.' What we are witnessing now is far beyond. It's the becoming producer of former consumers. And that suggests a new economic model for society
>
> (Steal this Film II, *01:07:00)*

Ten years later, history has to some extent proven him right, but maybe not in the way that he imagined. It is doubtful whether the filesharing networks created a more emancipatory and egalitarian media economy, but they certainly changed the way movies and music have been commercially distributed. While Hollywood was busy suing file sharers, Silicon Valley saw the business opportunities that the established media houses missed. In 2006 the old media monopoly made the Swedish authorities take down The Pirate Bay's servers, thereby putting filesharing on the political agenda and contributing to the radicalization of a generation of file sharers. That same year, Google bought YouTube for US$1.6 billion, and the following year Apple launched the first iPhone.

The iPhone was the first in a new generation of mobile media devices that gradually came to support a new form of media consumption, based on access to cloud-stored content through streaming technologies that were, in most cases, either free or sold as subscriptions. Although piracy persists and The Pirate Bay remains a major torrent tracker, the importance of pirate sites as media distribution channels has waned since 2009. Today, Spotify promotes itself as a legal alternative to piracy (Savvides 2014), and in 2017, Netflix was listed at place 31 on Alexa's directory of the world's most visited websites, while The Pirate Bay came in at number 87.

When file sharers rallied to the support of The Pirate Bay in 2009, the mobilization was largely fueled by a sense of being deprived of a source of culture and information that was hitherto unseen and unparalleled (Fredriksson 2015a, 2015b). Today, much of that content can be accessed through commercial, legal sites for streamed media. Although none of them can even remotely match the vast selection of content found in a pirate library such as The Pirate Bay, they apparently serve the needs of most viewers. Few consumers seem to lament the lack of the long tail.

On the one hand, this new media economy appears to be more user friendly and better adapted to the needs and preferences of the individual user. But this came at the cost of filter bubbles and excessive data mining. Rasmus Fleisher argues that:

> Personalization and surveillance are two sides of the same coin. The more our online behavior can be mapped in every detail, the better are the possibilities of serving us with a selection of content that is automatically adapted to what we as individual consumers are expected to demand. Apple is obviously the best in the world when it comes to replacing the norm of liberty with the norm of convenience.
>
> *(Fleischer 2013, 22)*[2]

Ideologically motivated file sharers reject streamed media services for designing away the empowering potentials of the P2P network. While P2P services distribute files that users can download and remix, the streaming services simply provide access to entertainment in locked formats which users can only consume passively. Streamed media thus enforces a technological platform that effectively prohibits remixes and pushes culture back to a Read Only mode. For Lessig, the main obstacle to the remix culture was the threat of being sued for copyright infringement. The new streaming platforms largely make such repercussions superfluous as the technology itself designs away the potential transgressions.

The pirate ideology argued that art is not a commodity but something that should be shared freely and openly and that copyright law is a barrier to that. This insight was not unique to the pirates but was shared by a wide range of groups, from librarians to academics (Fredriksson 2015b: Fleischer and Snickars 2018). Many of us who have conducted research on copyright and piracy have been inspired by the empowering and liberating ideals of the pirate movement (Fredriksson and Arvanitakis 2014; Arvanitakis, Fredriksson & Shillings 2017), while the pirate ideology largely departed from an analysis that had been developed by an earlier generation of critical copyright scholars in the 1990s and early 2000s. Academics such as Lawrence Lessig (2001, 2008), Kembrew McLeod (2001), James Boyle (1997, 2003), and Rosemary Coombes (1998) argued that the expansion of copyright limited the public access to culture and information in ways that interfered both with public education and free speech, which have been the historical rationale behind copyright law since the 18th century. These copyright critics thus formulated the problem of regulating data flows in the information society as one of enclosure and access, where the privatization of immaterial resources, through intellectual property, creates an artificial scarcity and limits public access to culture and information.

This analysis is becoming insufficient as a basis for a critique of the media economy when the streaming services have developed new means to commodify the act of sharing instead of merely the object that is shared. In this regard Sebastian Lütgert is correct that piracy and new technology provoked a new mode of propertization and commodification of culture. Spotify is a good example of this. Not only does it cater to a consumption pattern that was bred by filesharing sites such as Kazaa, where every song ever made is expected to be accessible with one single click. It also has a more immediate pirate heritage as it was largely constructed by programmers who had previously been working with BitTorrent clients, and the content on the beta version admittedly consisted of the founders' own pirated music catalogues (Fleischer & Snickars 2018). The piratical impact on television is not as immediate, but the distribution of TV content has however changed radically since 2010, and it is obvious that companies such as Netflix have learned to capitalize on the consumption patterns fostered by filesharing. In that regard, filesharing can be seen as an avantgarde medium that served as a testing ground for new models for distribution and prepared the market and the audience for legal, commercial applications of those distribution forms, developed by a new media monopoly where Silicon Valley is also claiming a seat at the table.

These examples show that although new technology may have ways to persist and force its way upon the world, the media industry, in its turn, has its ways of domesticating and profiting from these technological changes. In the case of the streaming revolution, the tech industry saw the commercial potential of the consumer behavior fostered by the pirate networks and set out to satisfy and capitalize on those new habits. This required a change of technology, and what we witness now is an ongoing migration of users from one platform – the P2P networks – to another – streamed media – that provides better means to monitor and control the users. While copyright critics and pirate ideologists were busy fighting Hollywood, Silicon Valley was figuring out new ways to cash in and, ten years later, the wave of change unleashed by the P2P networks trickles down to a steady stream of revenues for the new media monopoly.

## Notes

1 "med en oligopolartad konkurrenssituation. Sådana möjligheter finns ofta i närheten av de offentliga monopolen"

2 "Personalisering och övervakning är alltså två sidor av samma mynt. Ju mer våra nätbeteenden kan kartläggas i detalj, desto större möjligheter ges att servera oss ett automatiskt urval som anpassats för vad vi som individuella konsumenter förväntas efterfråga. … Apple är uppenbarligen världsbäst när det kommer till att ersätta frihetsnormen med en bekvämlighetsnorm." – my translation.

## References

Alexa (2017). www.alexa.com/topsites (retrieved 2017-09-19).

Anderson, C. (2006). *The Long Tail: Why the Future of Business Is Selling Less of More*, New York: Hyperion.

Andersson Swartz, J. (2014). *Online File Sharing: Innovations in Media Consumption*, New York: Routledge.

Arvanitakis, J., Fredriksson, M. and Schillings, S. (2017). "Bellamy's Rage and Beer's Conscience: Pirate Methodologies and the Contemporary University", *Culture Unbound*, Vol. 9, Issue 3, 260–276. Available from www.cultureunbound.ep.liu.se/v9/a18/cu17v9a18.pdf [accessed 13 August 2019].

Bagdikian, B. H. (2000). *The New Media Monopoly*, Boston: Beacon Press.

Barlow, J. P. (1996). "A declaration of the Independence of Cyberspace": www.eff.org/cyberspace-independence (retrieved 2017-09-18).

Bodó, B. (2014). "Set the fox to watch the geese: Voluntary IP regimes in piratical file-sharing communities", in M Fredriksson and J Arvanitakis (Eds.) *Piracy: Leakages from Modernity*, Sacramento: Litwin Books, pp.177–194.

Boyle, J. (1997). *Shamans, Software, and Spleens: Law and the Construction of the Information Society*, Cambridge and London: Harvard University Press.

Boyle, J. (2003). "The Second Enclosure Movement and the Construction of the Public Domain", *Law and Contemporary Problems*, 66(33): 33–74.

Burkart, P. (2013). *Pirate Politics*, Cambridge, MA: MIT Press.

Burkart, P. and Fredriksson, M. (forthcoming) "Pirate Protest Music", in: E. Drott and M. Nopriko (Eds.) *Oxford handbook of Protest Music*, New York: Oxford University Press.

Coombes, R. (1998). *The Cultural Life of Intellectual Properties: Authorship, Appropriation, and the Law*, Durham: Duke University Press.

Da Rimi, F. and Marshall J. (2014). "Piracy is Normal, Piracy is Boring: Systemic disruption as Everyday Life", in: Martin Fredriksson and James Arvanitakis *Piracy: Leakages from Modernity*, Sacramento: Litwin Books 323–344.

*The Economist* (2009, September 5). Keeping Pirates at Bay. Retrieved on November 21, 2010 from www.economist.com/node/14299558.

Ewertsson, L. (2002). "TV3 och myten om teknikdriven utveckling", *Tvärsnitt*, 2002: 2.

Fleischer, R. (2013). "Nätets kontrarevolution", In: Fleisher, R. *Tapirskrift*, Stockholm: Axl Books, pp. 15–38.

Fleischer, R. and Snickars, P. (2018). *Den Svenska enhörningein: Storyn om Spotify*, Stockholm: Mondial.

Fredriksson, M. (2015a). "The Pirate Party and the Politics of Communication", *International Journal of Communication*, 9: 909–924. Available from: http://ijoc.org/index.php/ijoc/article/view/3742/1339 [accessed 29 April 2016].

Fredriksson, M. (2015b). "Pirates, Librarian and Open Source Capitalists: New Alliances in the Copyright Wars", in H. Porsdam (Ed.) *Copyrighting Creativity: Creative Values, Cultural Heritage Institutions and Systems of Intellectual Property*, Farnham: Ashgate 153–167.

Fredriksson, M. and Arvanitakis, J. (Eds.) (2014). *Piracy: Leakages from Modernity*, Sacramento: Litwin Books.

Hemmungs Wirtén, E. (2004). *No Trespassing: Authorship, Intellectual Property Rights and the Boundaries of Globalization*, Toronto: University of Toronto Press.

Jenkins, H. (2006). *Convergence Culture: Where Old and New Media Collide*, New York: NYU Press.

Jie, Y. (2014). "Cultural Resistance or Corporate Assistance: Disenchanting the Anti-Capitalist Myth of Digital Piracy", in: M Fredriksson and J. Arvanitakis (Eds.) *Piracy: Leakages from Modernity*, Sacramento: Litwin Books 195-214.

Johns, A. (2009). *Piracy: The Intellectual Property Wars from Gutenberg to Gates*, Chicago & London: University of Chicago Press.

Johns, A. (2010). *Death of a Pirate: British Radio and the Making of the Information Age*, New York & London: WW Norton & Company.

Lessig, L. (2001). *The Future of Ideas: The Fate of the Commons in a Connected World*, New York: Random House.

Lessig, L. (2008). *Remix: Making Art and Commerce Thrive in the Hybrid Economy*, New York: Penguin Press.

Lobato, R. (2014). "Piracy and Promotion: The Double-Edged Power of Crowds", in: Nolwenn Mingant, Cecilia Tirtaine, and Joel Augros (Eds.) *Film Marketing into the Twenty First Century*, London: British Film Institute/Palgrave. http://dx.doi.org/10.2139/ssrn.2568469.

Lobato, R. and Meese, J. (Eds.) (2016) *Geoblocking and Global Video Culture*. Amsterdam: The Institute for Networked Cultures: http://networkcultures.org/wp-content/uploads/2016/01/TOD18-Geoblocking-Final1.pdf.

McLeod, K. (2001). *Owning Culture: Authorship, Ownership and Intellectual Property*, New York: Peter Lang.

Meese, J. (2014). "The Pirate Imaginary and the Potential of the Authorial Pirate", in: Martin Fredriksson and James Arvanitakis (Eds.) *Piracy: Leakages from Modernity*, Sacrament, CA: Litwin Books, pp. 19–37.

Özdemirci, E. G. (2014). "BitTorrent: Stealing or Sharing Culture? A Discussion of the Pirate Bay Case and the Documentaries 'Steal This Film I & II'", in M. Fredriksson and J Arvanitakis., *Piracy: Leakages from Modernity* Sacramento: Litwin Books, pp. 157–176.

Rydell, A. and Sundberg, S. (2010). *Piraterna: Historien om the Pirate Bay, Piratpartiet och Piratbyrån*. Stockholm: Ordfront.

Savvides, L. (2014). "Spotify Australia: Our Major Competitor is Piracy", *Cnet*, June 26. www.cnet.com/au/news/spotify-australia-our-major-competitor-is-piracy/

Spender, L. (2009). *Digital culture, copyright maximalism and the challenge to copyright law*, Sydney, Australia: University of Western Sydney.

*Steal this Film I* (2006).

*Steal this Film II* (2007).

# 40

# EVOLVING PRACTICES OF INFORMAL DISTRIBUTION IN INTERNET TELEVISION

*Ramon Lobato*

The last decade has been an especially volatile period in the history of television, with far-reaching changes in technology, business models, and viewing practices. Among the most significant of these changes is the rise of internet distribution (streaming), a development that has transformed both formal and informal television markets. This chapter focuses on one of the side-effects of internet distribution – namely, streaming television piracy via unauthorized set-top boxes and apps. My focus here is on how the technology and business of streaming piracy have changed in recent years, and what this means for the international distribution of television content.

To illustrate the extent of these changes, consider how online television piracy has evolved over the course of a decade. In the mid-to-late 2000s, television buffs flocked to torrent tracker sites such as The Pirate Bay, BTJunkie, and Isohunt to find the latest episodes of their favorite television series. They used torrent clients to download AVI files and discussed their activities in online forums. This was a peer-to-peer (P2P) piracy culture characterized by user-to-user sharing, with its own social norms, archives, and quality control systems (Burkart, 2014; Crisp, 2015; De Kosnik, 2016). Pirate streaming and cloud storage services were still in their infancy, as were social media platforms and apps. File sizes, connection speeds, and encoding standards meant that it was impractical to stream video in real time, so downloads were the preferred option.

Fast forward to the present, and the pirate ecosystem has changed profoundly. Mobile-first youth now prefer illegal streaming sites, apps, and Kodi add-ons for their television fix. Industry anti-piracy campaigns target domain seizures and "fully loaded" streamer boxes – the kind sold online or in grey-market stores, which offer access to hundreds of live international channels, on-demand movies, games, and karaoke apps for a one-off price. The Motion Picture Association now speaks of "Piracy 3.0" and the "challenge of illegal streaming devices" as the next frontier of intellectual property regulation (Ernesto, 2017).

There is, in short, a general trajectory in informal online distribution away from P2P toward streaming. This is not a uniform trend, nor is it irrevocable. Nonetheless, scholars of global television should pay close attention to this structural change, because it presents several implications for how we understand global television and the role of informal distribution within it. This chapter will provide an overview of the recent changes in streaming piracy, including the rise of

Kodi and IPTV piracy; situate these changes with reference to the scholarly research on television distribution; and explore related conceptual problems for global television debates.

## Piracy, P2P, and Television Studies

Within television studies and media studies, a small but growing body of research has addressed informal television distribution. Television audience scholars have documented diverse practices of pirating, sharing, and streaming television content (Jenkins, 2006; Leaver, 2008; Gray, 2011; Newman, 2011). Media industry scholars have probed the relationship between informal and formal distribution (Holt and Sanson, 2013; Lobato and Thomas, 2015; Smith and Telang, 2016). Scholars of Asian and diasporic television have documented pirate VCD and VHS networks (Cunningham and Sinclair, 1999; Hu, 2004), cable piracy (Athique, 2008, 2014; Sundaram, 2009), and online piracy (Zhao, 2017; Tse, 2016). Looking back further, there is also a rich vein of work on the VCR as a technology of both formal and informal distribution (Ganley and Ganley, 1985; O'Regan, 1991). These and other studies provide important precedents for understanding the relationship between television distribution, informality, and global media flows.

Scholarship in this area has often tended to focus on P2P sharing of episodes of series television. The idea of a digital TV commons has been debated and theorized (Newman, 2011; Strangelove, 2015). Scholars have examined the proposition that television is becoming shareable, social, and "spreadable" (Jenkins, Ford, and Green, 2012). Piracy in these contexts can be seen as communal activity – a form of engaged sharing conducted by fans for fans. This way of thinking about piracy was reasonable when BitTorrent was the dominant distribution system, roughly between 2005 and 2015.

We must remember, however, that BitTorrent sharing is only one, historically specific form of television piracy. BitTorrent emerged from a particular technocultural context and that has specific affordances and limitations. For example, BitTorrent is a search-based system designed for users who already know what they want to watch: there are no recommendation/browsing capabilities, aside from a "most downloaded" list on some trackers. It is not suited to watching live broadcasts or time-sensitive content such as sports and news.[1]

BitTorrent has always existed alongside and interacted with other informal distribution practices, including optical disc piracy, streaming and direct-download piracy, while remaining the central "engine" in the wider system. However, in recent years a number of factors have combined to reshape how this system works, and to gradually reduce BitTorrent's dominance within it. These factors include:

- a shift from desktop computers to mobile phones as the primary device for accessing the internet, and the corresponding importance of tap-and-swipe interfaces (e.g. embedded media players in websites and apps rather than torrent client interfaces designed for desktop computers);
- increased enforcement of P2P downloads, including high-profile "copyright troll" litigations;[2]
- the appearance of cheap Android streamer boxes and open-source media players (especially Kodi/XBMC), along with popular connected-TV devices such as Amazon Fire TV Sticks and Apple TV that can be "jailbroken", or informally reprogrammed, to enable streaming piracy; and
- the emergence of sophisticated, consumer-ready pirate IPTV subscription services offering live pay-TV channels at relatively low cost.

These factors have collectively worked to swing the pendulum of pirate distribution away from downloads and toward streaming-based viewing, thus mirroring the shift in formal TV consumption toward Netflix and other SVOD services. Indeed, Sandvine estimates that between 2011 to

2015, BitTorrent decreased as a proportion of overall traffic from 20% to 5% in Europe, and from 10% to 2% in the United States.[3] The aforementioned shift in focus of anti-piracy enforcement toward streaming, streamer boxes, and apps is further evidence of changing user behaviors and industry priorities. I will now explain how this shift from P2P to streaming requires the rethinking of widely held assumptions about the nature of online TV piracy and its implications for debates about global television.

## The Maturation of Streaming Piracy and Other Structural Changes in the TV Ecology

Table 40.1 lists the key technologies of television piracy and their affordances, as of 2017. Reading down the list, we see a rough trajectory from "residual" through "dominant" to "emergent" phenomena (Williams, 1977; c.f. Lotz, 2017). BitTorrent, now approaching its teenage years, is the most established form of online TV piracy but is no longer on a growth curve; streaming services of various kinds can presently be described as dominant; and Kodi and pirate IPTV subscription services, in particular, are emergent. The power balance among these systems will of course vary geographically according to local custom and infrastructure, so only general observations can be made. Nonetheless, the overall trend here is toward further *fragmentation* of the piracy ecosystem, as more and more technologies become available, combined with deepening *interdependence* between these various technologies. For example, P2P transfer protocols are now subtly integrated into some of the streaming-like services, notably Popcorn Time.

Looking closely at the streaming/hybrid services, we can see pirate streaming now comes in numerous forms. The simplest option are pirate streaming websites such as 123Movie, SolarMovie, Project Free-TV and their various clones. These websites can be viewed in any browser and feature embedded media players that are easy to use. This is an ephemeral scene, characterized by "whack-a-mole" enforcement – domains are seized by the authorities only for the service to reappear with a slightly different URL. Live streaming of channel feeds via Periscope, Facebook Live, and YouTube Live is also emerging as a popular practice for watching major TV events such as pay-per-view boxing matches, sports, and flagship premieres such as *Game of Thrones* (Meese and Podkalicka, 2016; Rowe and Hutchins, 2017).[4] There is also the possibility of watching full TV episodes on YouTube and other UGC sites, though the increasing sophistication of YouTube's content filters has stemmed the flow of unauthorized uploads.

A different kind of distribution model can be seen in Popcorn Time, an app that offers a Netflix-like interface for watching pirated movies and TV episodes.[5] Popcorn Time users can select from hundreds of titles, with many subtitle options. Although its interface looks like a streaming website, Popcorn Time interestingly uses a P2P client to download the content. This activates as soon as the user clicks on the desired title: users download content packets in sequential order, resulting in a near-on-demand experience. Popcorn Time has been extremely popular since its release in 2014 and promotes itself as "an application for those without access to a real Streaming platform and a real catalog, for free, without ads" (Popcorn Time, 2016).

A newer player is Kodi, the popular open-source media player that was originally designed for the Xbox (its original name was Xbox Media Center, or XBMC). Over the years Kodi has evolved into a multi-purpose media center that can be used via remote control and can be installed on almost any device. Importantly, Kodi can be customized as a powerful hub for TV piracy. This is achieved by installing unofficial add-ons such as Exodus and TVAddOns that expand Kodi's capabilities and allow it to access a wide range of TV episodes and movies from cloud storage sites (cyberlockers), and even live channel feeds. Use of these add-ons is discouraged by the Kodi development team but the open-source nature of the platform means such use cannot easily be controlled. These add-ons are very popular among the young gamer

*Table 40.1* The ecology of online TV piracy circa 2017

| *Service* | *Type* | *Live TV feeds?* | *Stream, P2P or direct download* | *Enforcement risk for viewer* | *Typical device* | *Best suited to …* | *Historical predecessors* |
|---|---|---|---|---|---|---|---|
| **Downloading** | | | | | | | |
| *Usenet/ newsgroups* | Bulletin board | No | Direct download | Low | Computer | Any content | BBS |
| *BitTorrent* | P2P file-sharing | No | P2P | High | Computer | Any content | Napster, Limewire |
| *Cyberlockers* | Cloud storage | No | Direct download | Low | Any device | Any content | BBS |
| **Streaming/hybrid services** | | | | | | | |
| *Video-hosting sites* (YouTube, Dailymotion, etc.) | Unauthorized uploads to a video-hosting site | No | Stream | Negligible | Any device | Content that can get through platform filters | Video stores, community TV, home video |
| *Pirate streaming websites* (123Movie, SolarMovie, etc.) | Pirate website | No | Stream | Low | Computer, phone, tablet | Any content | Video stores |
| *Live streaming platforms* (YouTube, Periscope, Facebook Live) | Unauthorized live streaming through a free platform | Yes | Stream | Low | Any device | Sports, major TV events | Community TV, home video |
| *Popcorn Time* | BitTorrent client with integrated media player | No | P2P (but with streaming-like interface) | High | Computer | Movies, TV | Video stores |
| *Kodi* | Open-source media player (when used with pirate add-ons) | Yes | Stream, P2P, and direct download | Variable | Computer, streamer box, phone, tablet | Movies, TV, porn, live TV | PC software (VLC etc.) |
| *Pirate IPTV services* | Illegal online redistribution of live TV feeds | Yes | Stream (sometimes combined with P2P and direct download) | Medium | Computer or streamer box | Linear TV; also: sports, diasporic media | Pay-TV piracy |

demographic that is Kodi's main constituency. In the United States, a recent technical study by Sandvine (2017) found that 8.8% of households have a Kodi device installed. The majority of these devices (6%) are configured with pirate add-ons.

Kodi has spawned a fast-changing commercial ecology that includes hardware and software vendors, subscription services, and technical support services. Streamer boxes and other internet-connected TV devices that come "fully loaded" with Kodi add-ons can be purchased online and occasionally in stores, as in Figure 40.1 below (a store in Middlesbrough, United Kingdom). Ambiguity about the legal status of these boxes was clarified somewhat by a 2017 EU Court of Justice ruling that banned the sale of fully loaded boxes. Facebook has also introduced its own ban (Spangler, 2017), and eBay and Amazon conduct periodic surveillance to shut down listings. Nonetheless, enterprising vendors are able to work around these by changing the euphemisms they use to promote their goods (for example, from "fully loaded" to "gift").

Finally, there is also a separate category of geoblocking circumvention devices, such as VPNs (virtual private networks) and DNS (domain name system) proxies, which are used to access out-of-region streaming content. For example, a New Zealander who wishes to watch UK catch-up TV services can, in theory, use a VPN or proxy to get back-door access, by "spoofing" their IP address. (I have written about these technologies in more detail elsewhere [Lobato and Meese, 2016].) Since major streaming services such as Netflix and BBC iPlayer have started introducing VPN detection policies and/or personal login credentials, these tools are now less effective than they used to be. It is also worth emphasizing that these tools enable out-of-region access – a practice akin to parallel importation – rather than piracy per se. For this reason, they are not a key focus of my discussion. Nonetheless, VPNs remain popular as an identity-masking device for BitTorrent users, among other licit and illicit uses. This reflects the growing interdependency of the various technologies in the pirate TV ecology.

*Figure 40.1* A store in Middlesbrough, United Kingdom, selling fully loaded Kodi boxes
Source: Screen capture by author

## The Rise of IPTV Piracy

Looking ahead, one piracy technology stands out as particularly interesting for scholars of global television studies. IPTV (Internet Protocol Television) piracy involves *live streams/feeds of international TV channels*. The term "IPTV" conventionally refers to pay-TV delivered over managed networks, but since around 2015 IPTV has also become a euphemism for the rebroadcast and viewing of unauthorized live TV feeds. There is now a thriving online scene dedicated to the sale, use, and DIY setup of IPTV services, typically via Android streaming boxes. As a live television rebroadcast service, rather than an on-demand service, IPTV piracy is categorically distinct from other forms of piracy such as BitTorrent; it is closer in nature to satellite TV piracy, but with a much larger range of channels. While the technology of IPTV piracy is consistent, pirate IPTV markets are highly fragmented with quite distinctive dynamics. At the time of writing, there are two main sub-markets associated with this form of streaming piracy: diasporic communities seeking cheap pay-TV from home, and sports fans seeking cheap or free access to pay-TV sport broadcasts.

The experience of using pirate IPTV varies from service to service but typically features an EPG (electronic programme guide) that is navigable through a remote control or channel list, and/or a series of apps. Users can access pirate IPTV services in a few different ways. One option involves purchasing a streamer box (colloquially called a TV box) which can be found in certain electronics shops, grocery stores, street markets, and via specialty online retailers (often advertised through social media). These boxes, which typically use a modified Android operating system, are preconfigured to access a certain provider's pirate IPTV service and have a custom EPG or app system built in. Most offer more than 100 channels and attract a one-off price tag of US-$100–300. The range of channels included in these IPTV boxes varies considerably depending on which wholesaler is used, but as an example, one service targeting South Asians offers "over 100 Indian, Pakistan and Nepal TV channels" (Figure 40.2), while another box targeting Chinese users offers Korean, Japanese, Thai and Chinese channels alongside US, UK, and Canadian channels, plus dedicated apps for karaoke, games, and adult entertainment.

As an alternative to purchasing a fully loaded IPTV box, it is also possible to bring your own device and only pay for the subscription component. The service provider will then provide you with either a custom TV player app and login details or an M3U URL which can be loaded into Kodi or a VLC player, generating a channel list. These basic elements of the IPTV service can be mixed and matched by different vendors, depending on the needs of their customers. For

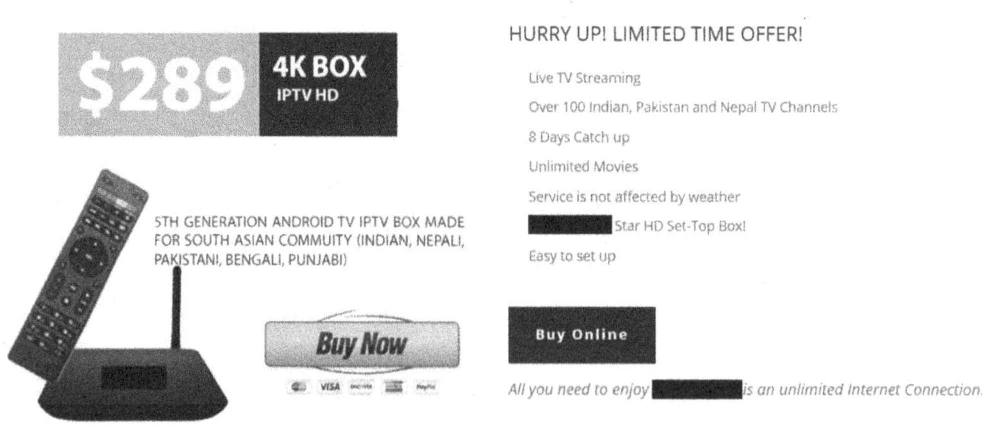


*Figure 40.2* IPTV box offering live channels for the South Asian community – name removed
Source: Screen capture by author

example, some vendors offer a fixed-price "tech support" service and will happily update your existing Kodi setup box to feature the latest add-ons ("send me your machine and I'll send it back").

Like Kodi, IPTV piracy has a complex ecology of its own that includes channel aggregators charging US$10–20 per month for 100+ channel packages; white-label resellers of these same packages; aggregator stores such as IPTVstore.com; and local intermediaries/dealers who supply boxes and/or packages and/or customer service. These intermediaries typically deal in one or two IPTV services that are most relevant to their customers, who are often from diasporic communities. For example, the Modbox IPTV system is used by Israelis; Indians, Bangladeshis, Nepalis, and Sri Lankans prefer RealTV, Jadoo and Maxx; and TVpad and EVpad are popular in the Chinese diaspora.

Chinese entrepreneurs have played an important role in the development of these markets, due to the size of the Chinese diaspora and the historically weak copyright enforcement for PRC-based TV channels. The longest-running services appear to be Chinese in origin, and there is a history of rights-holder enforcement actions against Chinese-language IPTV services stretching back to at least 2012 (Barlass, 2012). Keane (2016) notes the significance of IPTV boxes for the Chinese diaspora, remarking that "the set-top box is now the default technology for accessing Chinese programming overseas":

> Whereas a decade ago people outside China visited video shops in Chinatown areas, brought back DVDs from China to share with friends, or erected satellite dishes on roofs, an array of technological interfaces such as satellite dishes, set-top boxes, and virtual private networks (VPNs) now allow Chinese people around the world to watch and interact with Chinese-produced television shows in real time … For overseas audiences, accessing a diverse buffet of content becomes as simple as purchasing a digital set-top box on Alipay, Alibaba's international e-commerce platform.
>
> *(Keane, 2016, 5428)*

Elaine Zhao has similarly described how "jailbroken" (user-modified) Mi Box streamer boxes have become popular hardware for TV piracy in China, noting that "Many purchase the box simply to root it and install third-party apps on it" (2017, 35). In IPTV piracy, as in live streaming, Chinese markets feature an unusually dense and complex ecology of services. As these observations suggest, IPTV piracy has a cultural history partly embedded in diasporic media. It offers a new kind of long-distance cultural connectivity, mediated through the pivotal figure of the ethnic IPTV entrepreneur who re-connects the community to the homeland TV culture.

## Conceptual Implications of the Streaming Turn and IPTV Piracy

As IPTV piracy has developed over the years, its user base has grown to include other kinds of communities aside from diasporic markets. The thriving IPTV community scene concentrated around Reddit (see reddit.com/r/IPTV) includes many sports fans (who want cheap access to UK Premier League matches or Canadian ice hockey games), news junkies (who want international news channels not available in their local cable/satellite systems), and Western expats as well as diasporic viewers. Overall, IPTV piracy seems to be becoming less an "ethnic" media practice and more a "geek" hobby. Kodi has a similar trajectory: once used exclusively by gamers, it has now evolved into a multi-purpose media player used by a wider range of consumers.

Looking ahead, scholars in the field of global television studies need to ask two questions of these emergent practices. What does the growing significance of streaming piracy in general mean for the distribution of international television channels? And what might it mean for how we study and theorize global television?

As I have argued, the new IPTV piracy is a different proposition from BitTorrent file-sharing. IPTV involves a set-top box pulling content from faraway servers via IP protocol and displaying it in a channel list navigable with a remote control. IPTV has a different structure, form, and set of affordances. It is primarily concerned with *live* television feeds, rather than Netflix-like libraries of on-demand content. It is about *channels*, and *linear* television genres, especially sports and news. It is suited to remote controls and armchair viewing.

Another notable feature of IPTV piracy which distinguishes it from other popular forms of internet-distributed television – at least for sports and diasporic markets – is that IPTV piracy seems bound to more *social* rituals of viewing (gathering around the big screen TV for the Premier League game), rather than the more intimate settings envisaged by personalized streaming libraries ("Netflix and chill"). One can similarly argue that IPTV piracy may be a return to a linear model of TV viewing, in contrast to the nonlinearity of on-demand services and the more durable, digital libraries that can be amassed through P2P downloading. To be more precise, it may be a way of remaking linearity and "liveness" (van Es, 2016) for a cord-cutting age.

This fundamentally linear experience seems out of step with the on-demand viewing culture associated with Netflix and SVOD services. But it helps to remind us that television's digital transformations are going to involve the co-existence and interaction of different technologies rather than their sequential replacement. Linear TV is not a phase to be superseded by on-demand viewing but rather a highly durable televisual pleasure that complements on-demand, in both formal and informal distribution settings. Paradoxically, this means that the latest innovations can sometimes work to breathe new life into older modes of television viewing. IPTV piracy ushers in a "global" TV culture that has more in common with satellite TV of the 1990s than the SVOD services of the 2010s. In this sense, it is a compelling example of both change and continuity in global television distribution.

## Notes

1 Popular Torrent clients have recently introduced features to let users download packets in sequential order, meaning a kind of near-instantaneous playback experience can be possible if bandwidth conditions and the number of seeders permit.

2 Legal precedent in most nations has treated downloading as copyright violation and unauthorised redistribution (because of the nature of peer-to-peer), whereas streaming has a different technical basis because a user is not redistributing content at the same time. This appears to be slowly changing, however, as more jurisdictions start to deem pirate streaming as infringing (European Audiovisual Observatory 2015).

3 These figures were compiled from the quarterly Sandvine Internet Phenomena reports, 2011 to 2016. These reports are available at www.internetphenomena.com.

4 There is also a history of semi-legal startups operating in this area, notably Barry Diller's Aereo and Alki David's FilmOn (Lobato and Thomas, 2015).

5 Popcorn Time has had a complicated development process, with multiple development teams, forks, and legal challenges along the way. It has proven surprisingly resilient to enforcement efforts, despite its unambiguously illegal nature.

## References

Athique, A. (2008) "The Global Dynamics of Indian Media Piracy: Export Markets, Playback Media and the Informal Economy", *Media, Culture and Society* 30 (5): 699–717.

Athique, A. (2014) "Piracy at the Frontier: Uneven Development and the Public Sphere", *Media International Australia* 152 (1): 87–97.

Barlass, T. (2012) "Dramatic Twist: Police to Raid Suspected Pay TV Pirates", *The Age* 1 April. www.theage.com.au/it-pro/security-it/dramatic-twist-police-to-raid-suspected-pay-tv-pirates-20120331-1w542.html.

Burkart, P. (2014) *Pirate Politics: The New Information Policy Contests*, Cambridge: The MIT Press.

Crisp, V. (2015) *Film Distribution in the Digital Age: Pirates and Professionals*, Basingstoke: BFI Palgrave Macmillan.

Cunningham, S. and Sinclair, J. (eds) (1999) *Floating Lives: The Media and Asian Diasporas*, St Lucia: University of Queensland Press.

De Kosnik, A. (2016) *Rogue Archives: Digital Cultural Memory and Media Fandom*, Cambridge: The MIT Press.

Ernesto (2017) "Hollywood Sees Illegal Streaming Devices as 'Piracy 3.0'." *TorrentFreak*, 2 June, https://torrentfreak.com/hollywood-sees-illegal-streaming-devices-as-piracy-3-0-170502/

Ganley, G. D. and Ganley, O. H. (1985) *Global Political Fallout: The VCR's First Decade*, Cambridge: Harvard University Center for Information Policy Research.

Gray, J. (2011) "Mobility Through Piracy, or How Steven Seagal Got to Malawi", *Popular Communication* 9 (2): 99–113.

Holt, J. and Sanson, K. (eds) (2013) *Connected Viewing: Selling, Streaming, & Sharing Media in the Digital Age*, London: Routledge.

Hu, K. (2004) "Chinese Re-Makings of Pirated Japanese VCD Dramas", In: K. Iwabuchi (ed.), *Feeling Asian Modernities: Transnational Consumption of Japanese TV Dramas*, Hong Kong: Hong Kong University Press, 205–226.

Jenkins, H. (2006) *Convergence Culture: Where Old and New Media Collide*, New York: New York University Press.

Jenkins, H., Ford, S., and Green, J. (2012) *Spreadable Media: Creating Value and Meaning in a Networked Culture*, New York: NYU Press.

Keane, M. (2016) *The Chinese Television Industry*, London: British Film Institute.

Leaver, T. (2008) "Watching *Battlestar Galactica* in Australia and the Tyranny of Digital Distance", *Media International Australia* 126 (1): 145–154.

Lobato, R. and Meese, J. (eds) (2016) *Geoblocking and Global Video Culture*, Amsterdam: Institute of Network Cultures.

Lobato, R. and Thomas, J. (2015) *The Informal Media Economy*, Cambridge: Polity.

Lotz, A. (2017) *Portals: A Treatise on Internet-Distributed Television*, Ann Arbor: University of Michigan.

Newman, M. Z. (2011) "Free TV: File-Sharing and the Value of Television", *Television & New Media* 13 (6): 463–479.

O'Regan, T. (1991) "From Piracy to Sovereignty: International VCR Trends", *Continuum: The Australian Journal of Media and Culture* 4 (2): 112–135.

Rowe, D. and Hutchins, B. (2017) "Prize Fight Over Live-Streamed Sport Will Go On Long After the Final Bell Sounds," *The Conversation*, 8 February https://theconversation.com/prize-fight-over-live-streamed-sport-will-go-on-long-after-the-final-bell-sounds-72494

Sandvine (2017) "Spotlight: The 'Fully Loaded' Kodi Ecosystem (2017 Global Internet Phenomena report)," www.sandvine.com/resources/global-internet-phenomena/spotlight/the-fully-loaded-kodi-ecosystem.html

Smith, M. D. and Telang, R. (2016) *Streaming, Sharing, Stealing: Big Data and the Future of Entertainment*, Cambridge: MIT Press.

Spangler, T. (2017) "Facebook Quietly Adds Piracy Devices to List of Banned Sale Items," *Variety* 26 May http://variety.com/2017/digital/news/facebook-bans-kodi-piracy-devices-1202445930/

Strangelove, M. (2015) *Post-TV: Piracy, Cord-Cutting, and the Future of Television*, Toronto: University of Toronto Press.

Sundaram, R. (2009) *Pirate Modernity: Delhi's Media Urbanism*, London: Routledge.

Popcorn Time (2016) "We Are Officially Back," *Popcorn Time blog*, 24 February. https://blog.popcorntime.sh/official-statement/

Tse, Y.-T. (2016) "Watching Foreign TV in an Age of Online Sharing: The Cultural Implications of Cross-Border Television Experience," PhD Thesis, Goldsmiths, University of London http://research.gold.ac.uk/17441/

van Es, K. (2016) *The Future of Live*, Cambridge: Polity.

Williams, R. (1977) *Marxism and Literature*, Oxford: Oxford University Press.

Zhao, E. J. (2017) "The Bumpy Road Towards Network Convergence in China: The Case of Over-the-Top Streaming Services", *Global Media and China* 2 (1): 28–42.

# 41
# OFF THE LINE
## Expanding Creativity in the Production and Distribution of Web Series

*Aymar Jean Christian*

"Amazed, and in shock," Daryn Strauss wrote on her website after her nomination for a 2012 Writers Guild of America East Award in original storytelling in new media for *Downsized*.[1] *Downsized* was led by a woman and featured a racially diverse cast, all of whom felt the economic downturn in different ways. Started in 2010, the new media category was relatively new and low-profile for the WGA, an outgrowth of the 2007–2008 strike when temporarily unemployed Hollywood writers created dozens of web programs. That zeal waned, and indeed by 2018, the WGA let the category go unfilled, with zero nominees.[2] Yet for Strauss, recognition from a powerful guild meant something. Strauss was a writer and actress working as a freelance producer when she made *Downsized*, an online drama about the recession, which earned her membership to the WGA and recognition on a host of sites from the *Los Angeles Times* to web TV industry publication *Tubefilter*. Working in "Hollywood" but without creative ownership and control, Strauss went outside traditional structures and into a new medium and form: web series. By 2012 her series shared WGA nominations with indie darlings like Michael Cyril Creighton's *Jack in a Box* but also higher-budget projects like AOL's *Aim High* (from studio director McG) and NBC's *30 Rock*.

From production to story and release, Strauss' projects expanded the scope of the media industry. For her series she worked alongside underemployed actor-friends to develop storylines about the consequences of economic decline:

> It was a very collaborative process … We shoot in blocks. So we shot a couple of episodes, and then I would watch the footage and see what the footage is like and then I would write more based on what they brought.

Here, Strauss identifies a flexible mode of production driven by actors and crew, rather than clear deliverables, outside of clearly defined roles. Not content with producing, Strauss also started her own network, Digital Chick TV, to showcase and distribute web shows by and about women. Aware that dramas about women and economic collapse were hard to sell, Strauss created a distribution platform to correct YouTube's discovery problem, because of the high volume of user-generated uploads, and legacy television's high barriers to entry: "I knew offhand that there were series for women and by women, and women didn't know where to find them." From 2010 to 2015, Digital Chick TV streamlined distribution for independent series much in the way cable operators and traditional networks curate corporate content.

In their bid to break into Hollywood, independent television creators replicate but also break away from industry practices, ethics and logics, forging distinct models and narratives of success. I call this activity "off the line" for the ways in which creative workers blur the boundaries between creative and technical labor, ownership and execution, tradition and innovation or improvisation. A legacy of industrial film and television production, the "line" divides creation and craft. "Above-the-line" labor defines writers, directors, producers and stars considered most critical to bringing stories to life, whereas "below-the-line" laborers are craftspersons who technically execute the creative vision of those above – camera and boom operators, production assistants, visual effects artists, stylists, etc. "This line indexed the scarcity or surplus of so-called creativity and professionalism, two competing resources for labor value in industrial capitalism since the late 1800s" (Mayer, 2011, 4). Creativity has won out as a more valuable resource. In terms of pay, above-the-line workers are more likely to receive the highest rates of pay, a share of the project's intellectual property value, and percentages of sales, the payouts from which can easily reach into the millions for studio projects with national and global media distribution systems. This has caused tremendous inequalities in media. Vicki Mayer has further critiqued this divide by arguing there are workers who are not "on set" in Hollywood and so even further below the line: reality TV casting agents contracted by studios in New Orleans, factory workers who assemble TV sets in South America, amateur porn producers making direct-to-video specials, etc. These workers have even less job security and remuneration than the more often unionized Hollywood below-the-line professionals, and they are more likely to be women, minorities or perform work that clearly exploits them. Thus, Mayer argues, the line created a class of workers who are perceived as creative; anyone without those credits lacks claims to intellectual property, which is of the highest value in a creative industry.

The Internet emerged as an alternative distribution system in the 1990s, and from the beginning independent producers utilized it to expand, critique, and remix legacy practices to expand access to systems of value creation. They created new intellectual properties – TV shows – in new ways, redefining the very nature of production, distribution, marketing and thus TV as intellectual property. Independent television creators – and their peers in other industries, such as bloggers in journalism, mixtape artists in music and podcasters in radio – expose the chaotic and confusing tensions between new and legacy media, amateur freedom and corporate power enabled by an open distribution system without the legacy of the line and the inequality it engenders. What holds those tensions together is the chance for creators to tell a wider range of stories in a system with fewer barriers to reaching fans and profiting from their attention. Markets like these expose blind spots in scholarly and popular works about media, particularly the Internet, which tend to focus either on high-value producers making big-budget programs or under-resourced amateurs with little knowledge of distribution. Analyzing what I call "off-the-line" spaces – spaces in which industry workers try to forge what Matthew Hindman calls "the missing middle," a sustainable market for independent workers and their fans (Hindman, 2008) – contributes substantially to debates over the possibility of a creative economy focused on maximizing value for all workers and consumers. This missing middle remained elusive for the first decade or two of web original programming, and most productions were under-financed and featured the kind of precarious labor workers were trying to escape; creative freedom, however, made digital work attractive compared with legacy TV.

This chapter uses the case study of Scott Zakarin to explore how an independent production *and* open distribution promoted creativity and innovation off the line (Christian, 2018). Known to some as an original innovator of the "web series" for the early sale of *The Spot* to NBC in the mid-1990s, Zakarin's story reflects broad changes in television production online and on-air. Open distribution gave him license to experiment with how to tell stories in collaboration with fans and producers. His creative freedom fostered innovation in the production and release of serial stories but also raised questions about its effect on workers and the ethics of developing flexible storytelling practices for profit. Unlike Daryn Strauss, who used web distribution to correct labor inequalities for women and racial minorities, Zakarin's work replicated Hollywood's male gaze, suggesting that new technologies for series distribution and organizational structures for series production cannot alone correct inequalities born of the line.

## Innovation in Storytelling Production and Release

In conversation Scott Zakarin spoke excitedly about how web distribution supports creative storytelling, a combination of fan involvement, collaborative production, cultural immediacy and transmedia publishing. Zakarin positioned himself as a tinkerer, as a crusader for some masculine, unrealized vision of serialized storytelling, a journey he described with some grandiloquence: "I've spent my life chasing a Moby Dick … In chasing this, I'm truly Ahab," he told me in an interview. Starting from 1995 and continuing throughout the 2000s, Zakarin's career trajectory maps the early history of web television, from accounts of Internet culture's democratizing power to the dot-com bust, from the reinvigoration of utopianism through Web 2.0 to the subsequent desire from off-the-line producers to incorporate this activity into a structured market. If anyone can lay claim to inaugurating excitement over web series, it is Zakarin. His work also intersects with reality television, branded entertainment and advertising, social networking, text-based storytelling and video streaming, driven by fans and transmedia distribution. His projects are variously profitable and unsuccessful, transformational and unnoticed. He has acted as a producer, director, writer, executive and marketer, from creating his own series, films and television shows to starting his own independent distribution networks online.

Zakarin's focus on creativity and experimentation is crucial for understanding why many producers of independent television work outside the legacy development process. Scholars of organization communication have a number of theories for how creativity occurs within legacy structures: through broken routines and organizational conflict; under the auspices of a struggling network, a producer with past success, a new or hungry network executive; or through the constant uncertainty endemic to the production of culture (Ettema, 1982; Gitlin, 1983; Turow, 1982). Yet in legacy television, creativity always happens under institutional constraints, particularly the need to routinize labor: "a kind of balancing act between the needs by creators to search for novelty and their need to limit that search in the interests of predictability and efficiency" (Turow, 1985, 222). A networked environment with undefined rules, routines and norms allowed Zakarin to experiment with various storytelling and distribution strategies.

Zakarin's history in the web series market is distinct – really, without peer – and so warrants full accounting. He ranks among few producers who have consistently produced and innovated within serial stories for web distribution, making him an ideal site for investigating possibilities and challenges of making television outside the dominant network system. Zakarin has spent his career combining various media – personal blogs, books, photography, television, web video – and disrupting norms of serial television. His story as one the Internet's longest-working producers began in 1995 with *The Spot,* his first major endeavor since his first feature film went to Cannes in the late 1980s. A significant shift away from his roots in cinema toward a kind of television, *The Spot* was a text-based series about the lives of a group of twentysomethings living in a Los Angeles beach house, inspired by *The Real World* or *Melrose Place.* Yet Zakarin wanted to increase the audience's sense of immediacy, so, predating the rise of blogs, the site featured individual web pages – diaries – of each of the houses' roommates. In the very beginning Zakarin wrote everything, creating "layers of history" for each character. Each page included photos and, at times, videos.[3] *The Spot*, produced in-house at the ad agency Fattal & Collins, quickly became popular, logging hundreds of thousands of visitors a day. It ran for two years, spawning a companion book and eventually providing the catalyst for the short-lived online network American Cybercast. *The Spot*'s institutionalization in an advertising agency, as opposed to a studio affiliated with a television network, allowed Zakarin to take risks and play with storytelling, but he only had free rein after office hours until the project started to bring in sponsors.

For Zakarin, *The Spot* showed the potential for the web to deviate from television's static relationship with the audience, to incorporate interactivity and greater character depth and flexibility, all of which he positioned as the result of his and his collaborators' creative energy and the web's ability to integrate different media formats. He tried to advance these ideas in his next project, *Grape Jam*, which folded after one year. It had combined elements of a radio show, with chat rooms, avatars, social games and improv comedy for users, all performed by "good-looking" actor/performers who contributed heavily to the programming. "It was all sorts of fun for something that didn't make any money," Zakarin said. He then spearheaded the development of an entertainment portal for AOL called Entertainment Asylum, designed to compete with Microsoft's MSN as a hub for entertainment news, fan engagement and celebrity interviews. An expensive venture in "interactive webcasting," incorporating live video and chat with movie premieres, interviews and oddball programming, Asylum did not last long. Its end signaled a sobering period for "Siliwood,"[4] a handful of partnerships made between Silicon Valley and Hollywood (Geirland and Sonesh-Kedar, 1999). Freed from some traditional production constraints, Zakarin found monetizing something so unruly a challenge. Soon after, Zakarin moved into mainstream television and documentary production and distribution and worked there for several years.

Yet the excitement around Web 2.0 and social networking brought Zakarin back to the web in 2004, starting with a site called Fishbowl. Where once he experimented with scripted entertainment, Zakarin now jumped into reality TV. Fishbowl was a site about reality television offering original programming – blogs, video and audio podcasts, all of which featured interviews with and testimonials from past and present reality TV contestants. Former stars participated in the social network to talk about their experiences within and after their work on TV; fans of the shows joined forums to talk among themselves and with the talent. Fishbowl reflected Zakarin's desire to utilize labor undervalued by Hollywood studios, former reality stars from hits such as *The Real World* who were not unionized actors: "How'd I get the biggest names? Because nobody else wanted them." Premiering before YouTube and when both MySpace and Facebook were in their infancy, however, Fishbowl lacked the infrastructure to reach its full potential. It lasted only for a year. The following year, Zakarin made one more heavy push into reality television by producing a series for the U.S. cable channel E! called *Kill Reality* about a group of former reality stars tasked with making a feature film, eventually released on DVD as *The Scorned*.

In 2006, *lonelygirl15* alerted the mainstream media to the possibility of serialized video online and reminded producers of its ability to break the codes of storytelling through interactivity – its protagonist, Bree, responded directly to fans in the comments, in character. Seeing a connection between *lonelygirl*'s blend of reality and fiction (Christian, 2009) and his text-based series *The Spot*, Zakarin re-entered scripted entertainment. As online video reached maturity Zakarin filmed his first full-fledged video web series, *Soup of the Day*. *Soup of the Day* was an interactive scripted series about a young man dating three girls at the same time, one each for three days of the week: Monday, Wednesday and Friday. For several weeks, Zakarin and his production team released an episode on each of those days, both on YouTube, MySpace and a semi-popular video site called LiveVideo. Fans gave suggestions on how the protagonist should handle his predicament and eventually decided who he should date permanently. The series was a success, with 14 million views in total, half of which were on YouTube, and was eventually released on DVD.

*Soup of the Day* marked the beginning of a clear Zakarin web video aesthetic, mixing lo-fi realism and audience participation, sutured with a binary gendered address to clarify the gaze to prospective fans. "I need to let the audience be part of it. I need truth," he said. In his next series, Zakarin honed this aesthetic into something he would eventually repurpose and deploy with new actors, settings and contexts. Following the success of *Soup of the Day*, LiveVideo commissioned another set of series, *NoHo Girls*, about attractive young women living in Los Angeles, a series which created a number of spin-offs and permutations including *FlashBash*, *HiHo Girls*, *WeHo Girls* and *Van Nuys Guys*, all running until 2007 and some having sub-series and shows based on specific characters. The series were incredibly popular, garnering millions of views between YouTube and LiveVideo. The combination of attractive young people, a casual reality TV aesthetic and the tropes of online video – a gritty guerilla filmmaking approach mixed with vlogging – made them infectious for users looking for something more accessible than the increasingly scripted and thriller-driven *lonelygirl* franchise. The *NoHo Girls* series and its permutations featured actors who spoke to and interacted with audiences through the series, ad infinitum (the shows are not bound by seasons), a process about which Zakarin waxes idealistically as the zenith of new media creativity.

Seeing this success, MySpace, looking to compete with YouTube and the insurgent Facebook, contracted Zakarin to film a branded web series called *Roommates*, which premiered in 2007. With more money, Zakarin's aesthetic improved in technical polish but retained improvisation, self-reflexivity, the reality TV aesthetic, and vlogging. *Roommates* prefigured shows like *The Hills* on MTV, with its blend of situation comedy, mundane events and emphasis on personality over plot. MySpace and its brand partners were pleased with the series and ordered a second season. After *Roommates*, Zakarin continued to produce shows guided by the aesthetic that made him famous, including *Model Ball*, which ran for one year, and *Upstairs Girls* and *Downstairs Guys*, all of which were published on YouTube and resemble his other *Girls* and *Guys* series. Later shifting to distribution like Strauss before him, Zakarin started developing a home site, a kind of independent network, for his production company, Iron Sink Entertainment, to house a library of his recent web series alongside new episodes, forums and other ways for users to interact with the characters and content.

## Producing Independent Television: Collaboration, Flexibility and Immediacy

While Zakarin justified his projects as a break from television, his production strategies revealed more negotiation between old and new. Zakarin emphasized flattened hierarchies and shared responsibilities between writers and actors, the possibility for greater immediacy and audience interaction and the scrappy source material of some projects. To integrate flexible production strategies with fan engagement, relied on a clear gendered address, as we can see by even the titles of his later works, signaling to audiences a heteronormative gaze – *Upstairs Girls*, *Downstairs Guys*, *NoHo Girls*, *HiHo Girls*, *WeHo Girls* and *Van Nuys Guys*. Many of these elements, including binary gendered production and marketing, had been present in television production for decades, while many new media strategies for production and fan engagement were simultaneously, if slowly, being incorporated by mainstream film and television. What indie production and web distribution offered was the ability to rapidly and improvisationally experiment with who shapes the story, how, and when – to experiment with creativity itself.

*The Spot* was the genesis of and template for most of Zakarin's subsequent explorations in filmed series and interactive websites. It is the foundation of his argument for reinventing television reception and production. *The Spot* arose after Zakarin started participating in chatrooms, first on IRC (Internet Relay Chat) networks and then on AOL. On AOL Zakarin chatted with random strangers and eventually started experimenting with creating characters: "I became very popular as a 23-year old blonde-haired blue-eyed film student. I was also amazed at how stupid guys are." This combination of blending reality and fiction, the creation of personae – notably, young attractive women – and his effort to foster dialogue between producer and user/consumer became a critical part of his practice. The use of diaries in *The Spot*, the focus on the inner feelings and personal dramas of a discrete set of individuals, seemed to work online. Using text, photos and a couple of videos each month, Zakarin created not only a world – a mysterious beach house – but also a set of people prepped for interaction with fans.

Having characters interact with the audience via digital platforms was quickly taken up by a number of legacy networks in the late 1990s looking to keep fans engaged in a competitive media environment (Christian, 2018). Marketers emphasized greater consumer interaction to obtain higher rates from advertisers, with mixed success. Independents employed it to get users more invested in the characters as web projects proliferated and competed for attention. As William Boddy shows, changes in media distribution compel companies to adapt their business models and re-shape them to survive; gender routinely shapes this process, as seen in early TV when serial radio and television became gendered feminine through passive consumption and in digital innovations gendered masculine because consumption was active (Boddy, 2004). We can see in Zakarin's call for active audiences a masculine perspective on media consumption of women's bodies. For other series – and many of the most popular on YouTube – pretty women and a relentless focus on personal drama worked quite well. From *WeHo Girls* to *Downstairs Guys*, Zakarin indulged the tendency toward narcissism and petty sexual drama among a certain segment of the youth market, all while dressing it with a sexually attractive veneer. Writing about the politics of spectatorship and the gaze with women and webcams, Michele White noted how online video offers opportunities to see "real" bodies and intimate spectatorship, since viewers were presumed to be closer to the screen than for TV, and yet continually forecloses the possibility of real, material connection (White, 2003, 2006). This interplay between voyeurism, gender and intimacy lends to Zakarin's videos an intoxicating appeal, encouraging users to interact.

The immersive, immediate quality of web video lends itself to fan participation and spreadability, where users share cultural products to reflect on themselves and spark conversations in their communities (Jenkins, Green and Ford, 2013). "The ways in which people can connect with each other is revolutionary," says Marshall Herskovitz, co-creator of *thirysomething*, *My So-Called Life* and the short-lived web series, *Quarterlife*, which premiered online to much fanfare only to be cancelled once NBC aired it to low ratings. "It's just not a business yet," he told me in 2010. In a post-YouTube marketplace, scripted series producers used direct address and social media platforms and functionality to engage fans and sell them to sponsors. Zakarin's work presaged the rise of social media platforms and scripted/vlog hybrid series. As web video entered its third phase after YouTube, having attractive women engage users directly – direct address – became a trend as producers pursued engagement: most popularly with *lonelygirl15*, then with Marshall Herskovitz and Ed Zwick's *Quarterlife*, NBC's *Gemini Division*, MTV's *Valemont*, Day's *The Guild* and Pemberly Digital's *The Lizzie Bennett Diaries.*[5] Many of the serialized streaming video hits were extensions of Zakarin's gendered address, appealing to men who desire women or women who see themselves in community with the leads. As with Zakarin's, few have become years-long franchises, save *The Guild.*

To produce immersive, engaging worlds, series do best with writers who can respond to fans immediately, shaping or deepening the story without excessive intervention. Zakarin needed a team of writers and actors, each with some agency over a character's story. Zakarin wrote the basics – the history of the house, the relationships between the characters, and the general plot and character outlines – but the writers acted, or developed, the characters through text. Contrasting the writing structure with legacy television, he said they "weren't being led by committee. Scott was very hands-on about the big picture story, day, week, month and season. How it played out on journal entry level, he would be very vague." Zakarin acted as an editor for writing and plot consistency but left most of the details to the staff. In perhaps the most marked shift from television, in the beginning Zakarin acted as the network executive giving "notes" while also a showrunner running a writers' room whose writers nevertheless had more creative freedom.

Networked distribution, then, made explicit the collaborative nature of television writing, but it allowed for more producers – like actors – to participate in the process. For a number of his subsequent projects, actors were given a great deal of creative freedom. "A lot of these guys, they write their own dialogue on their feet," Zakarin told me. The demands of new media production – constantly updated content for a fickle audience – limits how much work one person can handle and benefits those with creative ideas. Zakarin privileged improvisational actors for his comedy site, Grape Jam, in part so they could shoulder some of the labor. In his two subsequent projects, Entertainment Asylum and Fishbowl, Zakarin used celebrities-as-producers. Celebrities such as Alyson Hannigan and Denise Richards, Gavin Rossdale and Dustin Hoffman were tasked with playing interactive games with audiences, answering questions while chatting with fans. Fishbowl took the notion of celebrity-as-producer to another extreme, literally hiring reality stars to host shows, blog about their lives, and participate in fan discussions. Years later Instagram and Twitter would provide current and would-be reality TV stars platforms for audience engagement, serving ancillary narratives in between legacy TV seasons.

Easy to execute in lower-cost text-based episodes Zakarin continued producer–fan interactivity in his filmed series, in spite of the high financial stakes of video production. "He was always willing to listen to ideas and really cultivate the talent," actor Tarah DeSpain told me about her time working with Zakarin as an actor on *WeHo Girls*, *Roommates* and *Model Ball*. As an improvisational actress, she appreciated the room to "breathe creatively." For a series not controlled by media outlets or large brands, actors had more control over how to interpret their characters and advance the plot, DeSpain said, comparing *Roommates* (distributed by MySpace and sponsored by brands like Ford) to the independent *WeHo* and *Model Ball*. Much of *WeHo* was improvised, and Zakarin's series included separate spaces for characters to vlog, confessional-style. So DeSpain directed herself numerous times. Character vlogs for many years became the standard response to demand for transmedia content: cheap and simple to make, corporations (legacy TV networks) and independent web series distributors followed the lead of *lonelygirl15* in breaking the "fourth wall" and having the characters address the camera in the tradition of reality TV (Miller, 2009).

Zakarin's productions reveal how collaboration, flexibility and immediacy can greatly shape how series develop in an open television market. Independent production renders unambiguous the vital role actors play in the production and circulation of media, along with the importance of writers and producers in determining the engagement factor of the final text. Independent productions focused on creating engaging shows online benefit from being flexible enough to disperse agency over the narrative to its producers and fans. Yet these changes are not necessarily progressive. One writer-producer, Rob Cesternino, noted how for *NoHo Girls* Zakarin and the crew were shooting 15 episodes (at around three to five minutes in length) each day in order to keep up with the publishing schedule. Without legacy television's multimillion-dollar licensing and production fees, budgets for these kinds of series tended to be small, as sponsors were wary of promises of engagement, even as they feared losing legacy television audiences to new, more interactive media forms.

## *Releasing Independent Television: Expanding Audience Participation*

The expansion of creative ownership and control in Scott Zakarin's series extended beyond writers, directors, and individuals and corporations sponsoring the series: it included the audience as marketers and participants. Much has been written on the participatory nature of new media, the communities of bloggers, fans, YouTubers, etc. who collectively create and remix mainstream media content (Jenkins, 1992; Lessig, 2008; Penley, 1997). Still, we have few examples of series and films actively guided by both the original creators and the audience it targets. Television history has some antecedents, particularly in unscripted shows: talent shows on television from *Star Search* to *American Idol* (Jenkins, 2006); radio, both on talk shows and call-in advertisements; and in film through the circulation of cult objects, whose fans "authenticate" which versions are most true (Jancovich, 2002).

The immediacy with which new media production – both in text and film – can respond to audience demands rivals earlier film or television products. Once again, starting at *The Spot* is instructive. In an amusing anecdote, a particularly irascible viewer sent an email diatribe to "Michelle," a character written by writer Troy Bolotnick. The reader was irked by *The Spot*'s coy real-versus-fake posture and accused the production of being "Hollywood." To prove the fakery (and corporate slant) of the entire series, the viewer dared Michelle to take a picture of herself in a bikini in front of a refrigerator holding strawberries, writer Rich Tackenberg recalled. Within several hours, the series had the actress playing Michelle take the photo and posted it online with a "fuck you ... this is the one time I'm going to do this" message. This is a salacious example of how the producers regularly incorporated user response and utilized cheaper production contexts into how they ran the series. They rapidly reacted to current events – the O.J. Simpson verdict or the death of Jerry Garcia. They monitored which characters were popular and which were not, using emails and hits as guides. One character, for instance, was almost universally detested but received the most (hate) mail and twice as much traffic; he stayed on the show. Throughout the series, fans sent emails suggesting what individual characters should do or say, and while the vast majority were "bad" ideas, a few made it into the narrative: "It felt very different from a focus group on a television show," said Tackenberg, citing a legacy TV development strategy. "It was something very organic. That was a very powerful tool that I think kept the show successful for a very long time." Here co-creative labor serves not only to "give the people what they want" and bridge the gap between creator and consumer but also to hone the marketing of a series lacking the support of legacy distribution.

The immediacy of producer response to fan demands supported their ideals of new media storytelling as collectively imagined by a community of fans, a concept well-researched in studies of new media (Askwith, 2007; Banks and Deuze, 2009; Benkler, 2007; Bruns, 2008; Jenkins, 1992; Shirky, 2008; Szulborski, 2005). Zakarin's body of work reveals a sincere respect for the audience's ability to imagine a world. In *The Spot*, for example, Zakarin often preserved discrepancies in different characters' recollections of events, allowing fans to make sense of the truth and presaging the rise of complex, ambiguous storytelling seen in "quality TV" such as *Lost*, *The Sopranos* and *Mad Men* (Mittell, 2014). "In no other medium can audiences have such a pivotal role in developing characters and story lines. We have found that audiences are eager, indeed resolute, in having their input utilized in the evolution of a show," Zakarin once said of his storytelling ambitions (Business Wire, 1996). In his first full-fledged entrée into online video, with *Soup of the Day*, Zakarin jumped headfirst into "interactivity." Each episode the lead character spoke in character with the fans on MySpace, trying to solicit feedback on which girl he should choose to date. In the end viewers chose one of the three girls to whom he should dedicate his time, but audience input occurred throughout. "We were shooting and writing with not a lot of lead time. When we saw the audience responding, we would change the story up," said producer Rob Cesternino, who said they shot each episode two weeks before publishing it on YouTube and other sites.[6]

By delegitimizing the author's hand and giving the series a "casual" tone Zakarin did away with the need for ambiguity over the story's veracity. His newer shows, including *Model Ball* and *Upstairs Girls/Downstairs Guys*, focused instead on making the characters personable, vlogging back and responding to users, all within an informal, handheld visual style. Actors spent time making their own vlogs – Zakarin says there was "no storytelling to it," meaning its function was emotive not narrative – and engaging one-to-one with users. Once again, these practices were not new. Fan participation with stars and the role of stars as authenticators of media discourse were indelible facets of film and television aesthetics and storytelling (Dyer, 1998; Schulze, White, and Brown, 1993). Zakarin's focus on developing intimate, immediate connections between producers and fans allowed him to experiment with narrative by lessening the importance of plot in scripted entertainment.

Further enriching his version of new television was Zakarin's "networking" of his shows. Since *NoHo Girls*, Zakarin created complex universes in which different series intersect with one another and expand based on how producers and fans shape the narrative. Characters from one series show up in others. Two series can run side-by-side with characters engaging in social and sexual intrigue, as in the companion series *Upstairs Girls* and *Downstairs Guys*, two separate shows that took place in the same house. Series like *Model Ball* had "second tier" shows – *Ashley's Hugtime, Sandy's Channel, Stupid Jim* – in which individual characters perform various hijinks, executing situational gags or revealing new emotional layers on a regular basis.[7] These practices of networked storytelling put the viewer in the role of assembling the plot, deciding what is important and what is not, a project too complex and nonlinear for most on-air television at the time.

While these methods of television storytelling shimmered with the gloss of the new, they carried a unique set of production issues, most notably the amount of labor needed to sustain the interactions. Toward the end of *The Spot*, when the series was regularly amassing over 100,000 visitors each show, the possibilities for one-to-one communication between producer and fan dissipated. With the newer shows, constant communication with fans took an emotional toll on the actors and some balked at the added labor. "The name of the game was to be interactive. You'd definitely respond to comments, we would make videos commenting on other comments," DeSpain said of her work on YouTube. "But it crossed a line for me as an artist. You want your work to be your work." DeSpain, who, like many of Zakarin's actresses, was valuable both for her comedic talent and her looks, found the constant engagement with drooling male fans "gross." Participatory culture was not exempt from the male gaze and gendered relations (Mulvey, 2000), and in fact, such dynamics are integral to how videos circulated at the time (Shifman, 2012).

## Promise and Peril in Off-the-Line Markets

Zakarin's series broadened the means of production and invited users into worlds, but producers still had to contend with the limitations of media production, the politics of labor and gender. Undergirding this chapter's focus on production, distribution and storytelling has been a narrative about the perils of the open market for online video. Web television was a vast sea of programming, where no one could control distribution. Unlike legacy television, sufficiently scaled to shut out most aspirants and well-endowed enough to enrich a small number of producers, online video lacked structure. In this liminal space, amidst hustling for views and producing new videos on which to serve advertisements, would-be auteurs like Scott Zakarin became entrepreneurs. They lacked the luxury of separating art from marketing, authenticity from aesthetics, the audience from the producer and creativity from labor. They had to engage fully with everything, and in their practices, they shone a light on the limits of legacy institutions while offering new possibilities for storytelling production and release.

One conversation I had with Zakarin illuminated this dynamic. Discussing his plans for the Iron Sink Network, a site intended to house a large part of his video library and to create new opportunities for engagement, Zakarin proposed an alternative business venture. Zakarin contemplated creating a network of text-based "content sites" based on particular topics – dogs, cars, etc. – all aimed at amassing a large mass of cheap advertisements from the powerful ad networks (which currently control the bulk of ads served on media websites). The strategy was not an artistic one, but a pure business plan: sponsors value large audiences and targeted content, not "quality." Now shuttered, the mere existence of the plan was revealing: online producers lack the relatively clear structures of legacy media. To keep producing, they have to do both what they creatively desire but also what the market demands. Their aesthetic and production strategies are tools for creative survival and worthy artistic contributions to television production practice, but workers must be paid, and during the time Zakarin produced web series, sponsors still spent most of their money in the old system.

Zakarin's work emerged before the rise of big data and the need for web platforms to increase scale in order to remain competitive. Today these kinds of flexible, producer- and fan-driven video series are under threat as distributors such as Netflix and Amazon increase production budgets and work with pre-existing intellectual property where characters and their worlds are already established. Zakarin's career and the many indie TV producers working at the same time might have spanned a period of experimentation before larger corporations moved in and re-standardized production for more predictable audience sizes and profits. Nevertheless, this period instructs us of the innovation possible, as well as the dangers implicit, with open access to distribution for agents outside legacy systems and structures.

## Notes

* Scott Zakarin quoted with permission – he can be contacted at scott@zoitmedia.com.

1 www.darynstrauss.com, retrieved July 17, 2012.

2 The WGA was among the first guilds or academies to give awards to new media storytelling, but eventually found actively curating online series a challenge. For more on web series awards, see the introduction and chapter four of Christian (2018).

3 Obviously, given its time period, videos were short, usually less than a minute, and took quite a bit of time to load.

4 The term describes Hollywood's interest in new/web-based technologies as an alternative distribution tool and Silicon Valley-based tech companies' investments in creative productions as they worked to create audiences for their new tech, services and platforms.

5 This was the first of five vlog adaptations of late-19th-century novels with women as protagonists, perhaps related to the serial, epistolary form and the dominance of women as both fans and actors online.

6 As before, the series also benefitted from an ambiguity over whether the characters on the show were real or fictional; yet since *Soup of the Day* intersected with *lonelygirl15*, by the end of the show skeptical viewers were hip to the game.

7 Such practices have since become commonplace on primetime television series; prominent examples include *Glee* on Fox, *Heroes* on NBC, *Ugly Betty* on ABC, and *Gossip Girl* on CW, all of which have given beloved side characters their own web series.

## References

Askwith, I. (2007). *Television 2.0: reconceptualizing TV as an engagement medium*, Master's Thesis. Cambridge, MA: Massachusetts Institute of Technology.

Banks, J. and Deuze, M. (2009). "Co-creative labour". *International Journal of Cultural Studies*, vol. 12, no. 5, pp. 419–431.

Benkler, Y. (2006). *The Wealth of Networks: How Social Production Transforms Markets*. New Haven, CT: Yale University Press.

Boddy, W. (2004). *New media and popular imagination: Launching radio, television, and digital media in the United States*. Oxford: Oxford University Press.

Bruns, A. (2008). *Blogs, Wikipedia, Second Life, and Beyond. From Production to Produsage*. New York: Peter Lang.

Business Wire (1996). The creator of "The Spot" forms LightSpeed Media Inc. *Business Wire*, May 1.

Christian, A. (2009). "Real Vlogs: The rules and meanings of personal online videos". *First Monday*, vol. 14, no. 11.
Christian, A. (2018). *Open TV: Innovation Beyond Hollywood and the Rise of Web Television*. New York: New York University Press.
Dyer, R. (1998). *Stars*. London: BFI Publishing.
Ettema, J. (1982). "The organizational context of creativity: A case study from public television", in J. S. Ettema and D. C. Whitney (eds.) *Individuals In mass media organizations: Creativity and constraint*. Beverly Hills, CA: Sage Publications, 91–106.
Geirland, J. and Sonesh-Kedar, E. (1999). *Digital Babylon: How the Geeks, the Suits, and the Ponytails Fought to Bring Hollywood to the Internet*. New York: Arcade Publishing.
Gitlin, T. (1983). *Inside Prime Time*. New York: Pantheon Books.
Hindman, M. (2008).*The myth of digital democracy*. Princeton, NJ: Princeton University Press.
Jancovich, M. (ed.) (2002). *Horror, the Film Reader*. New York: Routledge.
Jenkins, H. (1992). *Textual Poachers: Television Fans & Participatory Culture*. New York: Routledge.
Jenkins, H. (2006). *Convergence Culture: Where Old and New Media Collide*. New York: New York University Press.
Jenkins, H., Green, J., and Ford, S. (2013). *Spreadable Media: Creating Value and Meaning in a Networked Culture*. New York: New York University Press.
Lessig, L. (2008). *Remix: Making Art and Commerce Thrive in the Hybrid Economy*. New York: The Penguin Press.
Mayer, V. (2011). *Below the Line: Producers and Production Studies in the New Television Economy*. Durham, NC: Duke University Press.
Miller, L. (2009). "LG15's latest drama: Losing its new pilot". NewTeeWee, February 6. http://newteevee.com/2009/02/06/lg15s-latest-drama-losing-its-new-pilot/
Mittell, J. (2015). *Complex TV: the Poetics of Contemporary Television Storytelling*. New York, NY: NYU Press.
Mulvey, L. (2000). "The Index and the Uncanny", in C.B. Gill (eds) *Time and the Image*. Manchester: Manchester University Press.
Penley, C. (1997). *NASA/Trek: Popular Science and Sex in America*. New York: Verso.
Schulze, L., White, A., and Brown, J. (1993). ""A Sacred Monster in Her Prime": Audience Construction of Madonna as Low-Other", in C. Schwichtenberg *The Madonna Connection: Representational Politics, Subcultural Identities and Cultural Theory*. Boulder, CO: Westview Press, pp. 15–38.
Shifman, L. (2012). "An Anatomy of a YouTube meme". *New Media & Society*, vol. 14, no. 2, pp. 187–203.
Shirky, C. (2008). *Here Comes Everybody: The Power of Organizing Without Organizations*. New York: The Penguin Press.
Szulborski, D. (2005). *This Is Not a Game: A Guide to Alternate Reality Gaming*. New-Fiction Publishing.
Turow, J. (1982). "Unconventional Programs on Commercial Television", J.S. Ettema and D.C. Whitney. *Individuals in Mass Media Organizations*. Beverly Hills, CA: Sage, pp. 107–130.
Turow, J. (1985). "Learning to Portray Institutional Power: The Socialization of Creators", in Mass Media Organizations," P. Tompkins, and R. McPhee (eds.) *Organizational Communication: Traditional Themes and New Directions*. Beverly Hills, CA: Sage Publications, 211–234.
White, M. (2003). "Too close to see: men, women, and webcams". *New Media & Society*, vol. 5, no. 1, pp. 7–28.
White, M. (2006). *The Body and the Screen: Theories of Internet Spectatorship*. Cambridge, MA: MIT Press.

# INDEX

Locators in *italics* refer to figures and those in **bold** to tables.
Whilst acronyms are generally spelt out, main heading for broadcasters are entered under the acronym if universally known by it, e.g. BBC, HBO.